AF719127

WITHDRAWN

ENVIRONMENTAL IMPACT REVIEW AND HOUSING

ENVIRONMENTAL IMPACT REVIEW AND HOUSING

Process Lessons from the California Experience

Fred E. Case
Jeffrey Gale

PRAEGER

PRAEGER SPECIAL STUDIES • PRAEGER SCIENTIFIC

Library of Congress Cataloging in Publication Data

Case, Frederick E.
Environmental impact review and housing.

Bibliography: p.
Includes index.
1. Housing policy—California.
2. Environmental impact analysis—California.
I. Gale, Jeffrey. II. Title.
HD7303.C2C29 1982 363.5′8′09794 82-11272
ISBN 0-03-061744-8

HD
7303
.C2
C29
1982

Published in 1982 by Praeger Publishers
CBS Educational and Professional Publishing
a Division of CBS Inc.
521 Fifth Avenue, New York, New York 10175 U.S.A.

© 1982 by Praeger Publishers

All rights reserved

23456789 052 987654321

Printed in the United States of America

This book is dedicated to two persons who have been stalwarts in supporting real estate education and research at the University of California, Henry Block and Senator Arthur Breed. Senator Breed is responsible for the existence of the program and through the years has given advice and support that have helped shape the program to its current levels of excellence.

Henry Block served for years in the California Real Estate Commissioner's office, listening as we academics offered research proposals, asking pertinent questions that always helped improve our proposals, and then waiting patiently as we exceeded deadlines in submitting our final reports. He was particularly helpful in our search for funds for this report, which appeared as a working report in April 1980.

To both of these very fine friends and true gentlemen, thank you. We hope this will in some small measure help you understand how important you have been to California real estate education and research.

PREFACE

This study had its roots in the frustrations of the authors with the nature of the research being produced that is related to environmental reviews and housing. From our ongoing contacts with the development community, we have heard the stories of needless intrusion many times. From those economists in the academic world, we have heard that controls are inefficient and must be removed. Formal research efforts, often funded by development and building industry sources and carried out by research institutions and academics, mirror those viewpoints.

It is our view that these research efforts are somewhat misdirected. In those states that have comprehensive environmental impact review requirements, even the conservative political trend of 1980-81 has failed to significantly change the laws that require environmental review in the housing construction process. Such reviews seem popular, and while some in the society might object, most heartily support the approach, where it is in place. We believe that these requirements are going to be with us on a permanent basis, and making them work is a priority.

This study proceeds from that premise. We hope that we have been able to show something of the cost relationships within the generally adopted form of the environmental review process and to suggest some changes, based upon our inquiry, that we think would improve upon implementation in California and serve as a guide for other jurisdictions. Our work is directed to planners, economists, lawyers, builders, and developers—all who would have the process, which is so firmly in place in many parts of the country and which is often considered in others, function more efficiently yet retain its important positive effects. We hope that what we have learned from our examination of the California experience may help with implementation of environmental review processes impacting on housing wherever they may be tried.

We would like to thank a number of people for their help with the study. The California Department of Real Estate, through a grant to the University of California at Los Angeles Housing, Real Estate, and Urban Land Studies Program, funded an earlier version of this research. A large number of planners, builders, and developers participated in interviews and surveys, carried out with the assistance of Ron Rios, as a part of the research. Rose Altman and Nancy Kawata provided assistance with preparation of the manuscript.

CONTENTS

LIST OF TABLES

LIST OF FIGURES

ENVIRONMENTAL IMPACT REVIEW AND HOUSING

1

ENVIRONMENTAL IMPACT REVIEW PROCESSES AND HOUSING

The 1969 National Environmental Policy Act (NEPA) and subsequent state environmental policy acts (SEPAs), particularly in the two most populous states—New York and California—affected residential construction in ways that continue to evoke heated controversy. Detractors believe that SEPAs have created processes that unnecessarily complicate and increase the costs of residential construction. Supporters believe, just as firmly, that SEPAs have protected fragile environments and improved the quality of family life. Nowhere has the controversy been more heated than in California; however, when the California Environmental Quality Act (CEQA) was reconsidered in 1980, it was continued with only minimal discussion and opposition.

The California experiences with CEQA processes—consistently studied, sometimes emulated—have influenced Environmental Impact Report (EIR) regulations and procedures in many state and local jurisdictions. The emphasis in all of these laws is on the creation, analysis, review, and publication of the EIR. Evolution of an acceptable process under CEQA has been difficult, but the lessons learned from an analysis of experiences provide important insights about the effectiveness of the environmental impact review process (EIRP), which are presented in the remaining chapters of this book. Although many have studied environmental acts and the results of their implementation, none have scrutinized the processes associated with these acts. By focusing on the processes, we intend to explore the weak-

---

Throughout this book EIR refers only to the Environmental Impact Report. The term environmental impact review process (EIRP) refers to the various procedures connected with the environmental impact review, which may require preparation of an EIR.

nesses and the strengths, the benefits and the costs of EIRPs as they relate to residential construction. Chapter 2 focuses on the perceived impact of EIRPs on California housing construction and costs. Chapters 3 and 4 examine the California Environmental Quality Act and the processes and practices that have evolved from implementation of CEQA. Chapter 5 summarizes important theoretical foundations relating to estimating the benefits and costs of CEQA and then presents a summary of the actual costs that appear to derive from CEQA processes. Chapter 6 presents builders' and planners' perceptions of the problems and costs associated with CEQA-required processes. Chapter 7 continues the analysis of costs by using quantitative evidence collected from a variety of primary and secondary sources. In Chapter 8 we bring all of our findings together and suggest ways in which EIRPs might be used by builders and planners to reduce housing costs, review-processing times and many of the process uncertainties that tend to impact the creation of affordable housing.

ENVIRONMENTAL IMPACT REVIEW PROCESSES

The National Environment Policy Act and CEQA were enacted at a time when the limitations of the physical environment were beginning to affect the residential construction processes. At that time the inability of free market processes to protect diminishing scarce resources was becoming increasingly apparent, and ways were being explored for protecting critical natural resources.

Depending upon whose views are being expressed, an EIR can be an expensive document contributing to a complex, expensive, often frustrating review process, or it can provide useful construction and market information, which builders, buyers, lenders, and others can use to evaluate the need for housing and the kinds of housing that should be built. For planners and implementing agencies, EIRPs can add an undesirable burden to other construction review processes, or EIRs can provide useful information, which helps them to identify possible desirable and undesirable impacts on local environments and to institute necessary mitigation measures.

USES OF ENVIRONMENTAL IMPACT REVIEW PROCESSES

Although varieties of EIRPs are used throughout the United States, formal reviews are required in only a relatively few areas. However, it is suspected that the use of EIRPs has been increasing and is being meshed more and more with the usual planning, zoning, and related review processes. In California all local juris-

TABLE 1.1

Finished Lot Costs of Single-Family, Detached Homes in the United States, Selected Years 1950-76

Item	1950	1960	1964	1969	1976
Average finished lot size (in square feet)	7,558	8,932	10,312	12,839	14,707
Average price (in dollars)	1,485.00	2,808.00	4,567.00	6,183.00	10,533.00
Cost per square foot (in dollars)	0.20	0.31	0.44	0.48	0.72
	Percentage Change				
	1950-76	1960-76	1964-76	1969-76	Average Annual
Average finished lot	94.6	64.7	42.6	14.5	2.1
Average price	609.2	275.1	130.6	70.3	10.0
Cost per square foot	264.5	127.8	61.7	48.7	7.0

Source: Michael Sumichrist, Gopal Ahulwalia, and Robert J. Sheehan, Profile of the Builder (Washington, D.C.: National Association of Homebuilders, 1979), pp. 31, 37.

TABLE 1.2

Government Requirements for Physical Change by Census Region, 1976
(percent of respondents)

Type of Requirement	Total	New England	Middle Atlantic	East North Central	West North Central	South Atlantic	East South Central	West South Central	Moun-tain	Pacific
Density reduction	46.8	40.0	58.1	58.3	48.5	54.2	40.0	26.1	21.4	50.0
Development restructuring	47.0	36.4	53.3	48.3	37.9	50.0	29.2	46.4	61.3	47.0
Additional services	44.7	45.5	44.8	36.0	28.6	48.8	50.0	40.0	50.0	58.1
Mandatory dedication, community facilities	30.6	12.5	14.3	47.8	4.5	44.4	—	11.1	42.9	40.7
Mandatory dedication, parks and recreation	59.8	62.5	63.0	66.7	56.3	70.4	50.0	22.7	55.6	64.7
Streets and access roads	79.2	73.3	87.5	72.2	62.5	82.1	66.7	64.3	90.6	91.5
Other	61.8	200.0	50.0	50.0	—	75.0	—	—	100.0	75.0

Source: Michael Sumichrist, Gopal Ahulwalia, and Robert J. Sheehan, Profile of the Builder (Washington, D.C.: National Association of Homebuilders, 1979), table 96.

TABLE 1.3

Housing Costs Components as a Percentage of Total Construction Costs in the United States, Selected Years 1949-77

Component	1949	1969	1974	1977	1977 as Percent of 1949
Construction	69.0	54.6	48.4	46.7	67.7
Land development	11.0	21.4	24.6	25.0	227.7
Overhead and financing	15.0	17.0	17.0	17.5	116.7
Financing	5.0	7.0	10.0	10.8	216.0
Total	100.0	100.0	100.0	100.0	

Source: Michael Sumichrist, Gopal Ahulwalia, and Robert J. Sheehan, Profile of the Builder (Washington, D.C.: National Association of Homebuilders, 1979), Exhibit 4-1. Professional Builder's Magazine, March 1976.

dictions are required by CEQA to use some form of an impact assessment process, although some have less severe requirements.

Given the relatively small percentage of projects that were subject to full EIRPs, some questions might be raised about how burdensome EIRPs really are. Insofar as could be determined, the greatest dissatisfaction arises from the uncertainties and delays and the mitigation measures associated with what seems to be an endless review process. Developers believe that the outcome of these actions falls primarily on finished lot costs; they offer as evidence the increases in finished lot prices that occurred between 1950 and 1976 (see Table 1.1). Even then conclusive evidence of the role of EIRPs in such increases is not available.

The manner in which mitigation measures can affect home construction prices is suggested in Table 1.2. Builders point out that meeting these requirements creates "front-end" expenditures, which add to holding costs and ultimate sales prices for land or buildings placed on the lots. They also point out that, with the amount of land dedicated for public facilities and the streets and access roads built, the density reductions may be more than what the market can absorb. Apparently builders in the Pacific area (California, Washington, and Oregon) are much more likely to encounter mitigation measures involving physical changes. These mitigation measures result from impact assessments that consider natural environment; demand on public services; transportation; housing; and fiscal, social, and economic elements.

In terms of housing component costs, land development has been increasing and construction has been decreasing as a percentage of total construction costs (see Table 1.3). Overhead and financing percentages have increased by about 50 percent between 1949 and 1977, while financing has increased by more than 100 percent. Overall land development increases were greatest for the same period, closely followed by financing. These trends seem to lend some credence to builders' complaints that extensive land development requirements prior to the construction of homes, when coupled with rising financing costs, are the primary reasons for overall high housing construction costs and sales prices.

The next chapter pursues these complaints about rising prices and their relationships to EIRPs by examining California housing markets in greater detail.

NOTE

1. Michael Sumichrist, Gopal Ahulwalia, and Robert J. Sheehan, Profile of the Builder (Washington, D.C.: National Association of Homebuilders, 1979), pp. 31, 37.

2

HOME PRICES AND ENVIRONMENTAL REVIEW PROCESSES IN CALIFORNIA

The evidence is overwhelming that environmental controls have increased home prices, but the arguments continue over whether such controls are necessary and the extent to which they increase prices unnecessarily. A resolution of these arguments does not seem possible in the foreseeable future because of the poorly defined qualitative and shakily developed quantitative conclusions used by both sides in the debates. This study attempts to define the issues more clearly and substantially by bringing together the most convincing evidence and arguments available on the possible costs and benefits of the environmental impact review processes (EIRPs), particularly as they affect home prices. The need for this study became clear when EIRPs were reviewed but not modified substantially between 1979 and 1981 (when most federal, state, and local EIRPs were evaluated).

Environmental impact review processes were the focus of attention of homebuyers, builders, lenders, regulators, and others when home prices began to accelerate sharply after 1970 (see Table 2.1 and Figure 2.1). Those who opposed the EIRPs argued that they had played a substantial role in causing these price increases. Their concerns seemed real when price increases in 1980 were sufficient to make homeownership or affordable rentals available to an increasingly smaller percentage of U.S. families. On the other hand, those who favored EIRPs saw them as important means of achieving clean air and water, preserving open space, protecting rapidly disappearing critical natural resources, and increasing the quality of family living.

While the arguments continue over the costs and benefits of EIRPs, the demand for affordable family housing continues. Home prices and the costs of owning or renting a home show steady and

TABLE 2.1

Changes in Basic Housing Supply and Demand Elements in the United States, 1900–80

Year	Median Sales Price (in dollars)	Annual Family Income (in dollars)	Housing Units (in thousands)	Population (census years)	U.S. Housing Starts	Home Price Index (1967 = 100)
1900	4,881	490	15,964	75,994,575	189,000	22
1905	4,311	554	n.a.	n.a.	507,000	n.a.
1910	5,377	630	20,256	91,972,266	387,000	25
1915	5,159	687	n.a.	n.a.	433,000	n.a.
1920	6,296	1,489	24,352	105,710,620	247,000	29
1925	7,809	1,434	n.a.	n.a.	937,000	n.a.
1930	7,146	1,360	29,905	122,775,046	330,000	33
1935	6,296	1,137	n.a.	n.a.	216,000	n.a.
1940	6,558	1,300	34,855	131,669,275	530,000	30
1945	7,476	2,189	n.a.	n.a.	325,000	n.a.
1950	9,446	3,319	42,826	150,697,361	1,908,000	43
1955	13,386	4,418	n.a.	n.a.	1,627,000	n.a.
1960	16,652	5,620	53,024	180,697,596	1,296,000	76
1965	20,000	6,957	n.a.	n.a.	1,509,700	n.a.
1967	21,900	8,412	n.a.	n.a.	1,321,900	100
1970	23,400	9,867	63,450	204,765,770	1,468,000	107
1975	39,300	14,510	n.a.	n.a.	1,171,300	n.a.
1980	81,000	21,000	n.a.	n.a.	1,303,000	370

n.a. = not available or not calculated

Sources: National Association of Homebuilders, Economics Department; U.S., Department of Commerce, Construction Reports; National Association of Realtors; U.S., Bureau of the Census; and University of California-Los Angeles Forecasting Project.

FIGURE 2.1

Comparison of Rates of Change among Critical Housing Supply and Demand Elements in the United States, 1900-80

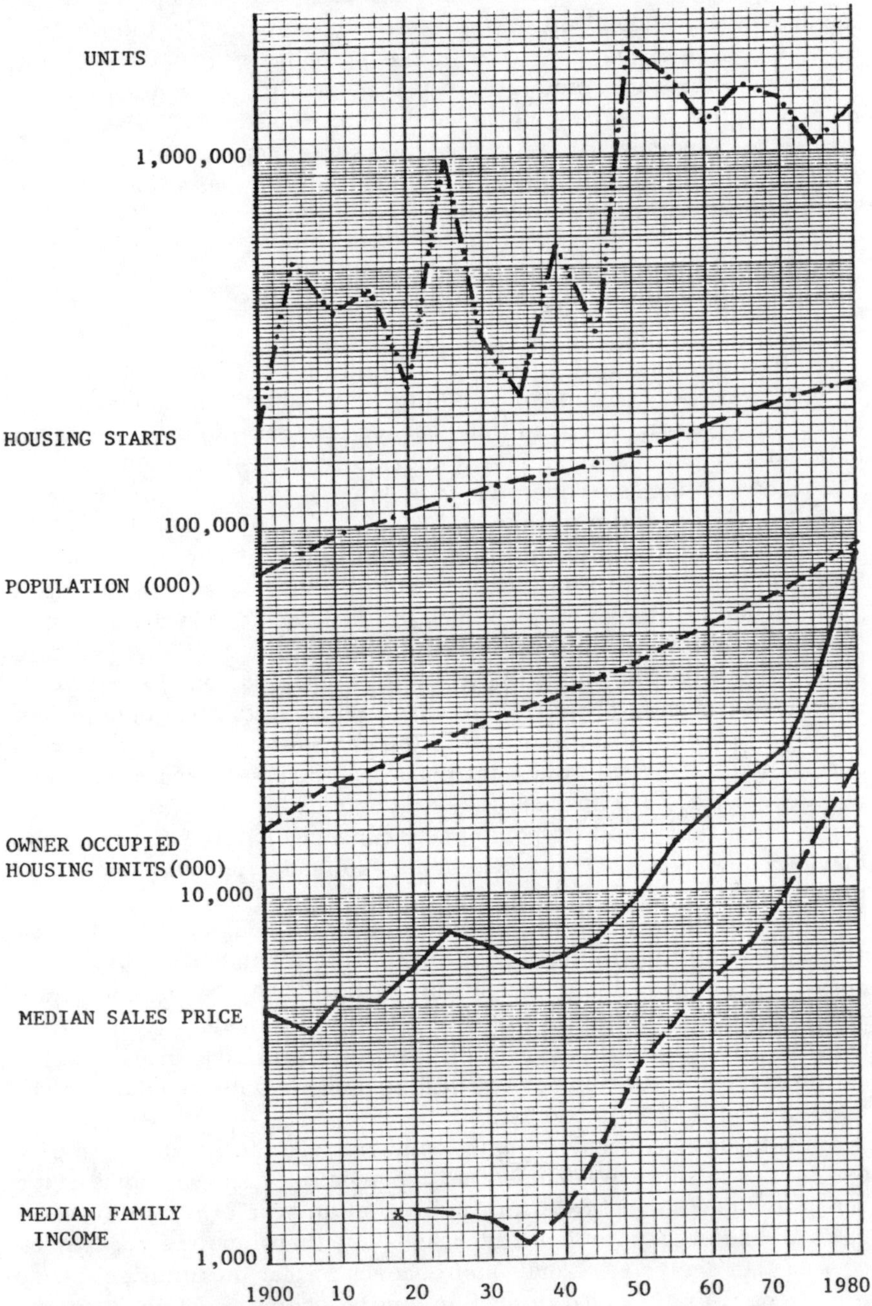

Note: 1900-20 below $1,000.

Source: Table 2.1.

important annual increases, which have now reduced the number of U. S. families able to afford a new home to less than 10 percent. With the apparent abandonment or substantial modification by the Reagan administration of the 1949 federal housing policy of providing a safe, sanitary, affordable home for every U. S. family, the role of environmental review in causing higher home prices needs analysis.

The remainder of this chapter presents macroeconomic evidence as a background against which more detailed analyses of the EIRPs and home prices will be presented in later chapters.

HOME PRICES, COSTS, AND FAMILY INCOME

Median home sales prices ranged from a high of 9.8 to a low of 5.5 times family income from 1890 to 1940, when federal programs and changing economic conditions caused this ratio to drop to ranges between 2.3 and 2.9 times family income from 1940 to 1970.[1] After 1970 the ratio began to creep up, reaching an estimated high of at least 3.5 times family income in 1980.

If all U. S. families should have an opportunity to buy homes they can afford, then concern about rising prices is legitimate, and any changes leading to lower prices are important. On the other hand, if the very low family income-home price relationships that prevailed during the 1940s and early 1950s seem an unwarranted aberration from long trends in these ratios, concerns about current high prices seem unwarranted. However, it is the changing relationships between current and historical family income, housing price, and ownership cost levels that have sparked the current debates about housing and argue for placing the EIR-home price concerns within an appropriate market context, before any judgments are made about the impacts of the EIRPs.

Since 1900 the number of owner-occupied housing units has increased slightly faster than population, but the number of persons per household has been declining. From 1970 to 1980, for example, persons per household declined from 2.9 to 2.2.[2] The supply of new housing units has moved erratically upward, with high annual levels until 1981, when the number of building permits authorized dropped to levels found in 1949 and 1925.

Although the supply of existing homes and the construction of new homes seemed to keep pace with population changes, home prices began to move up rapidly after 1940 and at an accelerating rate in 1970. The reasons for these increases are many and are not the concern of this analysis. What is of concern is that the families' capacities to pay, as reflected in median family income, kept pace with rising home sales prices until 1970, at which time prices began increasing much more rapidly than family income. It was also the year in which environmental protection was adopted as federal law and policy and imposed on state and local governments. Whether these relationships were accidental or causally related, they raised the levels of arguments over EIRPs throughout the 1970s.

California Home Price Trends

California took the lead in implementing environmental control legislation, and since new home construction in California typically equaled 10 to 15 percent of total national construction, attention was focused on the California experience. In California in 1970 the median home price was $24,200, compared with $23,200 nationally. After 1970 the California price accelerated sharply, so that by 1980 it was over $104,000, compared with a national average of only $81,000 (see Table 2.2). Some analysts felt that much of the difference in the price changes was owing to increasingly strict land-use growth and environmental controls throughout the state but particularly in a few California counties.

Analysis of housing data for California indicated that housing market trends tended to be set by only about 14 of the 58 California counties, typically those along the coast with urban population concentrations.[3] One-half of all new California homes are usually built in these counties; equally important, these 14 counties represented an acceptable cross section of approaches to land controls and the uses of EIRs.

If land-use growth or environmental controls were having any impact, then the number of housing units included in building permits would show a decline when compared with rates of construction for the state overall or for counties not using controls or using controls minimally.

In 1970 all of the northern counties selected for analysis recorded levels of new housing unit construction at rates above those for the state of California. San Mateo County, with an index in 1970 of 208, was showing the most rapid rate of growth. In 1975 San Mateo showed a declining growth rate that was the greatest among the counties, with an index of 72. Although the permits in California had increased 31 percent over 1967, none of the northern counties recorded an equivalent rate of growth, and all showed declining rates of increase after 1975.

In central California Sacramento County reported rates of increase that accelerated until 1973, declined through 1975, and accelerated again after 1976, until the universal slump in housing began in 1980.

TABLE 2.2

Indexes of Home Sales Prices for Selected Areas in Northern, Central, and Southern California, 1967-80
(1967 = 100)

	1980	1979	1978	1977	1976	1975	1974	1973	1972	1971	1970	1969	1968	1967	Price (in dollars)
Northern[a]															
Alameda (Berkeley)	440	320	276	221	198	171	150	140	136	129	122	116	106	100	33,000
Contra Costa (Western)	369	310	270	197	178	155	143	131	126	122	113	110	103	100	9,250
Marin (Southern)	522	376	322	260	210	178	151	136	127	123	113	106	101	100	40,000
San Mateo (Northern)	446	357	326	267	201	171	146	131	122	116	110	107	102	100	13,250
Santa Clara (Palo Alto)	547	431	360	303	245	211	168	153	137	131	128	120	108	100	25,000
Central[a]															
Sacramento	381	319	245	207	167	146	131	117	109	104	102	103	101	100	26,750
Southern[a]															
Los Angeles	452	348	316	262	199	160	137	122	115	112	108	105	102	100	32,000
Orange	453	395	341	268	207	176	145	135	120	117	113	108	104	100	25,800
Riverside-San Bernardino	351	298	285	244	189	159	142	130	122	120	113	109	105	100	19,800
San Diego	523	418	336	260	225	200	174	174	147	133	123	112	103	100	20,870
Santa Barbara-Ventura	442	367	323	243	194	162	142	138	122	114	110	106	103	100	21,680
State median sales price[b]	477	386	323	288	224	187	158	139	129	120	110	109	104	100	22,000

[a]In April.
[b]At year's end.

Sources: Real Estate Research Council of Northern California, Northern California Real Estate Reports; Real Estate Research Council of Southern California, Real Estate and Construction Report; University of California-Los Angeles Forecasting Project, Graduate School of Management, University of California, Los Angeles; and California Association of Realtors.

TABLE 2.3

Indexes of Housing Units in Permits, California and Selected Counties, 1967-80
(1967 = 100)

	1980	1979	1978	1977	1976	1975	1974	1973	1972	1971	1970	1969	1968	1967	Total Permits
California	161	186	207	256	205	131	112	150	181	166	104	117	127	100	111,999
Northern counties															
Alameda	85	113	116	84	69	62	53	126	209	237	183	152	140	100	5,618
Contra Costa	112	131	132	215	147	93	83	140	193	237	134	131	110	100	4,849
Marin	84	63	177	132	113	104	128	315	329	264	118	148	136	100	1,102
San Francisco	99	152	169	127	133	95	115	344	285	299	147	63	104	100	1,207
San Mateo	87	96	100	144	108	72	81	231	318	261	208	116	126	100	2,665
Santa Clara	78	76	98	126	125	78	88	126	145	189	173	182	171	100	10,241
Central county															
Sacramento	161	382	282	366	286	188	157	203	304	263	215	157	117	100	4,164
Southern counties															
Los Angeles	118	153	172	167	124	75	86	183	226	192	198	175	144	100	23,539
Orange	86	143	160	223	248	125	146	229	290	254	190	266	193	100	12,307
Riverside	333	558	599	698	396	203	170	363	431	323	206	138	130	100	2,363
San Bernardino	247	441	569	540	315	171	123	212	298	238	150	109	130	100	3,331
San Diego	90	150	222	288	231	116	127	189	309	296	181	197	170	100	12,656
Santa Barbara	93	143	129	154	120	79	101	247	212	135	123	150	122	100	1,604
Ventura	98	210	283	293	298	160	93	193	257	281	176	207	165	100	2,899

Sources: Real Estate Research Council of Northern California, Northern California Real Estate Reports; and Real Estate Research Council of Southern California, Real Estate and Construction Report.

In southern California in 1970 indexes were all above the state index and registered more rapid rates of growth. Two counties, Los Angeles and Ventura, increased housing construction rapidly from 1970 through 1973 and then decreased rapidly to lows in 1975 (86 in Los Angeles and 93 in Ventura). Orange followed the same pattern but ended 1975 with an index of 125. Only Riverside and San Bernardino counties showed some increase in 1975.

The year 1975 represented a time of depressed real estate market conditions, so that declines in all counties from 1970 were to be expected. After 1975 residential construction picked up throughout California, peaking at an index of 256 in 1977 and declining at an increasing rate after that, reflecting the new, generally depressed conditions of real estate markets. In northern California some counties—Alameda, Marin, San Francisco, San Mateo, and Santa Clara—reported increasing rates of construction from 1975 through 1978 and then declines, which by 1980 placed housing construction at 1975 levels. None of the counties had construction rates in 1980 equal to the state rate, although the Contra Costa County rate was above its 1975 rate. After 1975 Sacramento housing construction accelerated to historic high rates, declining only in 1980.

In southern California 1980 rates below those of 1975 were reported in Orange, San Diego, Santa Barbara, and Ventura counties. In both Los Angeles and Riverside counties rates were well above the 1975 rates.

In summary, counties reporting rates of housing construction that seemed to run counter to the overall state trends were Alameda, Marin, San Francisco, San Mateo, and Santa Clara in the north and Orange, San Diego, and Santa Barbara in the south. Coincidentally, these were also the counties in which concerns were being expressed about the deadening effect of increasing types of building controls.

One of the counties using strong growth and environmental controls was Santa Clara County; its index of home price changes is compared with those of California and the United States to dramatize the apparent impact on home prices of control measures (see Figure 2.2). The index of California home prices reflects to some degree the extent to which California prices exceeded national prices. High prices in California, in other words, are not just the result of growth controls but many other factors. The extent to which Santa Clara prices exceed California prices includes some impact from the EIRP, but that impact cannot accurately be isolated.

Not shown in Figure 2.1 but presented in Table 2.2 is the combined index for two counties that had minimal or no growth controls—Riverside and San Bernardino. In 1980 this index was only 10 points above the state index and 196 points below the Santa Clara index.

There are, of course, many dangers in relying too heavily on these data to determine the cost or price impacts of EIRPs, so that

FIGURE 2.2

Indexes of Home Sales Prices in the United States, California, and Santa Clara County, 1967-80
(1967 = 100)

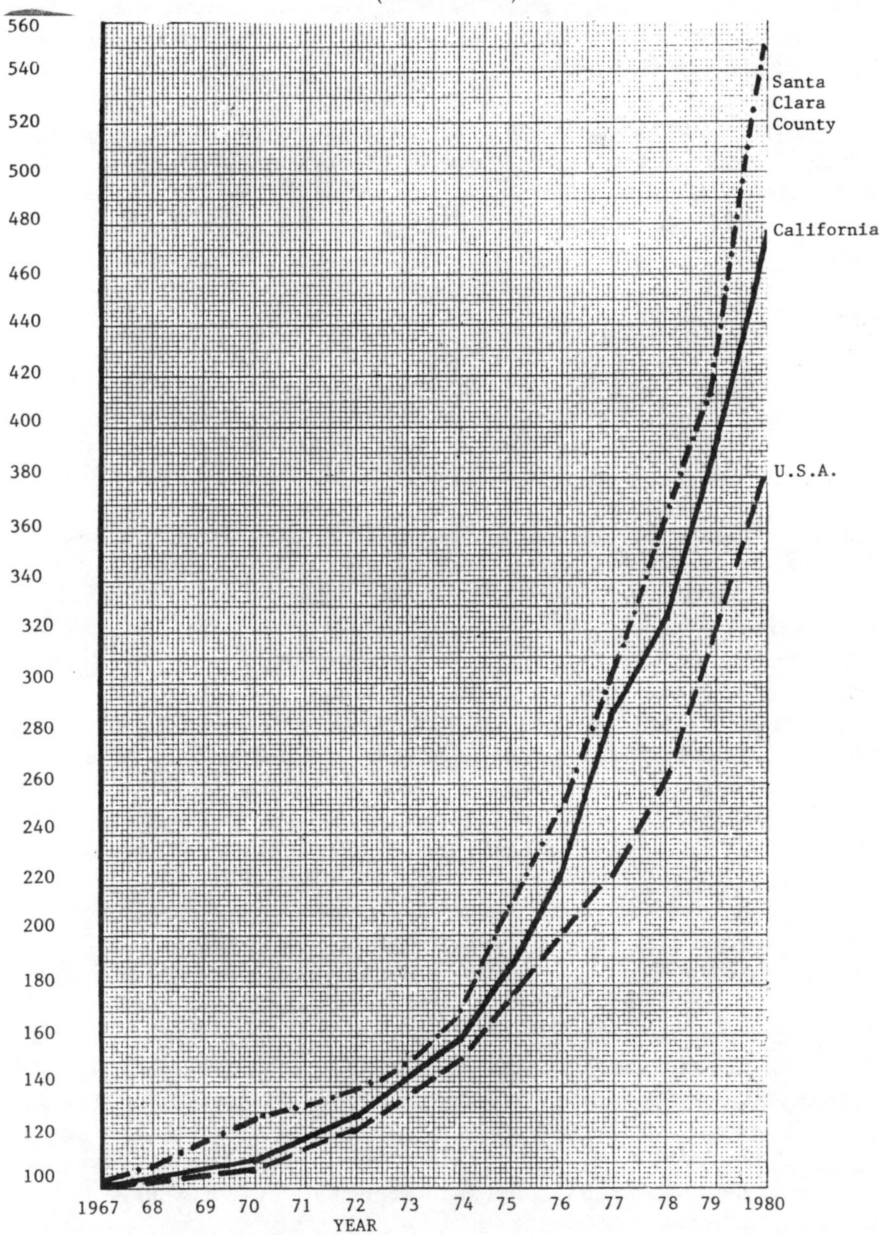

Sources: Tables 2.1 and 2.2.

the conclusions offered so far should be treated only as introductory. The regional data, for example, are from a consistent sampling of the same homes by professional appraisers. As an area grows, more homes are added to the sample. The state sample is from all homes sold. Santa Clara is a county that is growing because of the lack of space for new homes in the San Francisco area. Riverside and San Bernardino are affected only minimally by spillage from the Los Angeles or San Diego growth trends. Furthermore, San Bernardino has the largest area of any county in California.

At this point an analysis of the reasons for the differing price changes in counties will be postponed. All that is concluded is that there are differences in price changes among the counties to be studied, and higher rates of price changes appear to be related to those counties that are known to have more complete and stricter growth controls than is true statewide.

Home Prices and Construction Costs

Higher rates of price changes may occur in some counties simply because of higher construction costs. Since the major cost impact of EIRPs is on finished land, isolating housing construction costs may provide some insight into price changes. The growing markets in California since 1970, even considering the market dips in 1975 and 1980, would presumably permit builders to pass on any increased construction costs.

Indexes of home construction costs for northern and southern California are presented in Table 2.4. Northern California is defined as the urbanized area around San Francisco Bay. Southern California is the urbanized area in Los Angeles and Orange counties. The house is carefully defined in terms of plans and specifications, and costs are estimated by the same firm, Marshall and Swift. The same index year of 1967 is used, so that price and cost changes could be compared.

In northern California the typical house was redefined in 1975 from 1,395 to 1,570 square feet. The plans and specifications for northern and southern California houses are very similar for the entire period from 1967 through 1980.

Construction costs increased much more in northern California than in southern California. In part the difference is owing to the increase in the size of the sample home in northern California, since the index of per square foot costs increased less than did the index for southern California. In both areas the cost indexes increased less than the price indexes, but the differences in increases between northern and southern California are substantial.

TABLE 2.4

Indexes of Construction Costs of a Single-Family, Detached, Average Quality House, Northern and Southern California, 1967-80 (1967 = 100)

Year	Size (in square feet)	Costs (in dollars) Per Square Foot	Index	Total	Index
Northern California					
1980	1,570	34.64	273	54,373	308
1979	1,570	34.26	270	53,780	304
1978	1,570	33.96	268	53,325	302
1977*	1,570	29.75	245	46,715	276
1976	1,570	27.23	215	42,750	242
1975	1,570	24.24	191	38,056	215
1974	1,395	25.04	198	34,931	198
1973	1,395	22.14	175	30,882	175
1972	1,395	19.66	155	27,426	155
1971	1,395	18.29	144	25,520	144
1970	1,395	17.25	136	24,063	136
1969	1,395	15.42	122	21,509	122
1968	1,395	14.35	113	20,023	113
1967	1,395	12.12	100	16,911	100
Southern California					
1980	1,570	32.16	282	50,497	232
1979	1,570	28.29	248	44,418	248
1978	1,570	25.66	225	40,280	225
1977	1,570	22.27	195	34,960	195
1976	1,570	20.55	180	32,258	180
1975	1,570	18.66	164	29,302	164
1974	1,570	17.16	150	26,938	150
1973	1,570	16.14	142	25,333	142
1972	1,570	14.60	128	22,927	128
1971	1,570	13.47	120	21,513	120
1970	1,570	12.59	110	19,760	110
1969	1,570	12.13	106	19,042	106
1968	1,570	11.85	104	18,600	104
1967	1,570	11.40	100	17,900	100

*As of July 1, all others are of October 1.

Sources: Marshall and Swift Publication Company; Real Estate Research Council of Northern California, Northern California Real Estate Report; Real Estate Research Council of Southern California, Real Estate and Construction Report.

Between 1967 and 1970 cost increases were modest in both geographic areas, equaling 36 percent in northern California and 10 percent in southern California. Between 1971 and 1975 costs increased 49 percent in northern California and 36 percent in southern California. The rates of increase declined in both areas between 1976 and 1980, with costs increasing by only 27 percent in northern California and 29 percent in southern California.

The increase in northern California home construction costs was well below home price increases after 1970. Between 1967 and 1970 the average increase among northern California counties ranged between 10 percent and 28 percent, compared with cost increases in the same period of 29 percent.

Percentage increases between 1971 and 1975 in both northern and southern California were only slightly higher than the previous five-year period. In northern California they ranged between 31 percent and 61 percent; statewide they were 43 percent; in central California they were 43 percent; and in southern California they ranged between 33 percent and 42 percent. Statewide the price increases were 56 percent.

Percentage increases in home prices between 1976 and 1980 were dramatic, particularly when compared with the increases in construction costs. In northern California price increases ranged between 107 percent and 172 percent, with the high in Alameda (Berkeley) County and the low in Contra Costa County. Prices in central California increased by only 112 percent. In southern California the range increases were between 127 percent and 188 percent, with the low in the Santa Barbara-Ventura County area and the high in Orange County.

Apparently, house construction costs were not a strong force in the price increases. Again, and perhaps only coincidentally, the highest price and cost increases occurred in counties with the strongest controls—Alameda, Marin, and Santa Clara in the north; Orange and San Diego in the south. In central California, Sacramento County price and cost increases were only moderate.

If construction cost changes cannot account for any substantial portion of home price changes, perhaps land costs are responsible.

Land Costs and Home Prices

Land costs on a comparable basis are impossible to obtain. A substitute, though not completely satisfactory, can be found in the statements of land value as a percent of assessed value (see Table 2.5). Estimates of land value are derived by taking the total assessed value, which is computed based on market comparison approach, and

TABLE 2.5

Land as a Percentage of Total Assessed Value, Northern California, 1967-80

	Assessment Year (ending July 1)													
Location	1967	1968	1969	1970	1971	1972	1973	1974	1975	1976	1977	1978	1979	1980
California	40	41	45	43	43	43	42	42	41	40	39	37	37	38
Northern California														
Alameda	29	34	46	44	44	41	42	43	41	42	44	43	42	43
Contra Costa	28	32	33	31	31	32	33	35	34	33	31	31	31	32
Marin	46	46	46	44	43	42	41	39	38	37	36	35	34	34
Monterey	38	38	50	42	42	43	42	43	42	41	40	38	37	39
Napa	39	51	46	44	45	44	42	43	41	38	37	35	35	35
San Francisco	39	37	38	36	36	36	36	35	36	34	34	32	32	32
San Mateo	41	41	43	42	41	39	37	35	32	33	28	28	30	31
Santa Clara	34	35	39	40	40	39	38	37	34	33	32	31	31	30
Solano	43	42	43	40	39	38	38	36	35	33	31	29	29	34
Sonoma	37	39	43	43	43	44	42	41	40	39	37	36	36	37
	Indexes of Assessed Land Values (1967 = 100)													
California	100	108	118	130	137	146	152	166	179	199	223	228	267	321
Northern California														
Alameda ($519,164)	100	128	209	215	230	232	255	275	286	325	395	395	402	460
Contra Costa ($369,393)	100	113	123	134	144	155	180	211	230	254	285	312	334	391
Marin ($235,806)	100	105	106	113	125	130	140	157	167	192	222	203	233	263
Monterey ($188,505)	100	106	114	137	147	156	177	206	231	252	293	305	295	372
Napa ($51,727)	100	108	135	147	178	194	211	243	261	280	325	314	365	418
San Francisco ($756,032)	100	99	98	101	103	107	111	112	140	141	154	165	176	190
San Mateo ($544,740)	100	107	120	125	130	137	140	146	157	166	174	182	214	251
Santa Clara ($680,232)	100	111	137	163	173	182	189	200	215	234	262	265	294	334
Solano ($109,447)	100	111	126	128	131	135	143	151	165	179	203	208	248	366
Sonoma ($139,881)	100	112	123	147	155	176	192	220	242	265	306	356	404	495

Note: Based on assessed valuation of state-assessed tangible property subject to general property taxes.

Source: Annual Reports, California State Board of Equalization, Sacramento, California.

TABLE 2.6

Land as a Percentage of Total Assessed Value, Central and Southern California, 1967-80

Location	Assessment Year (ending July 1)													
	1967	1968	1969	1970	1971	1972	1973	1974	1975	1976	1977	1978	1979	1980
Central California														
Fresno	44	43	49	43	41	40	39	42	40	38	37	34	34	37
Sacramento	40	37	38	36	35	33	32	31	30	27	27	26	25	26
Southern California														
Imperial	66	63	66	62	61	59	58	57	58	56	56	50	50	50
Los Angeles	41	43	45	45	44	44	43	43	42	41	40	39	39	40
Orange	38	39	41	42	41	43	41	39	41	42	43	39	42	41
Riverside	44	46	48	47	47	46	44	43	41	40	38	36	35	36
San Bernardino	31	32	44	34	36	40	36	35	33	31	28	25	25	26
San Diego	43	45	47	46	46	45	44	43	43	43	43	39	39	40
Santa Barbara	45	46	47	45	45	45	43	44	42	40	38	38	38	41
Ventura	52	52	56	52	50	48	45	44	41	37	33	33	35	36
	Indexes of Assessed Land Values (1967 = 100)													
Central California														
Fresno ($367,111)	100	100	98	102	110	113	119	146	152	165	173	206	238	318
Sacramento ($382,268)	100	101	101	103	109	113	117	121	127	134	155	181	211	257
Southern California														
Imperial ($106,838)	100	114	103	104	109	109	111	116	136	144	175	159	172	189
Los Angeles ($6,137,230)	100	108	114	125	129	134	134	142	149	166	171	182	218	259
Orange ($997,171)	100	110	122	154	160	190	195	218	254	315	396	357	438	518
Riverside ($422,054)	100	108	115	125	135	143	150	158	164	173	198	228	276	350
San Bernardino ($387,714)	100	110	116	130	147	161	172	178	176	181	195	202	242	296
San Diego ($882,084)	100	111	124	143	166	180	198	226	256	286	334	321	395	471
Santa Barbara ($255,112)	100	106	109	113	118	123	126	142	144	162	197	219	249	315
Ventura ($431,389)	100	102	117	126	130	133	129	139	141	151	174	171	208	269

Note: Based on assessed valuation of state-assessed tangible property subject to general property taxes.

Source: See Table 2.5.

subtracting the costs of the homes (based on the Marshall and Swift cost handbook also used for the data in Table 2.4). The State Board of Equalization is required to keep assessed valuations in all counties at roughly the same percent of market value, usually 25 percent.

The assessed value evidence does not permit conclusive findings but does provide additional insights into home prices and land values. In northern California land value did not change appreciably as a percent of total assessed value (see Table 2.5). This might reflect the degree to which land values tended to keep pace with and approximate home prices. However, by 1980 counties with reputed restrictive growth practices (northern bay and peninsula areas) reported land values equal to 37 percent to 38 percent of total assessed values, as compared with 30 percent to 31 percent for the other northern California areas. Sacramento, with few restrictive practices, maintained a consistent ratio of land value equal to 24 percent to 28 percent of assessed value.

Selected evidence for southern California showed somewhat the same trends as northern California for the period 1967-76 (see Table 2.6). Particularly interesting is the extent to which land values in San Diego and Orange counties increased much more rapidly than did overall housing prices. The land values as a percent of assessed values also reflected some increases for the period, but this was true of all of the counties analyzed. Land value increases were much sharper than construction cost increases in Orange, Riverside, San Diego, and San Bernardino counties.

HOME PRICES AND GROWTH CONTROLS

Economic theory indicates that if population and personal income are increasing and other things are held equal, the demand for homes will increase. If demand increases relative to supply, then home prices will increase. If demand pressures increase, but the supplies of new homes do not, then prices will accelerate more than the rate at which supplies increase.

In Table 2.7 population changes are used as a surrogate for demand changes and construction of new homes as a surrogate for supplies. The home price index is used as a surrogate for price schedules. Using these assumptions, if housing supplies respond to population changes at less than clearance rates, then prices will increase, and the index will be high. If supplies respond at faster rates than population changes, then prices will decline. If both housing supplies and population increases are being controlled at rates below market requirements, then lesser rates of housing construction will produce higher prices. Such assumptions are tested in Table 2.7, using state changes as a measure of free market, competitive responses.

TABLE 2.7

Home Price Changes Compared with Changes in Population and New Housing Unit Construction, 1967–79

Location	Average Home Price Index[a]	Relationship between Prices and New Housing Construction Permits[b]	Annual Increase[c]		Average Annual Increase (in percent)	
			Population	New Housing Units	Population	New Housing Units
California	183	+	—	—	1.02	2.8
San Diego	214	+	+	+	3.37	5.4
Santa Clara	204	-	±	±	1.47	3.3
San Mateo	190	-	-	-	0.00	1.5
Ventura	188	+	±	+	3.56	5.1
Orange	185	-	+	+	2.04	4.3
Santa Barbara	180	-	-	±	1.07	1.9
Marin	178	-	±	-	0.00	0.0
San Bernardino	178	+	+	+	5.60	4.9
Alameda	176	-	-	-	0.80	0.9
Los Angeles	171	-	-	±	0.23	1.0
Riverside	164	+	+	+	5.79	5.2
Contra Costa	161	+	±	+	1.60	3.4
Sacramento	157	+	+	+	3.44	5.3

[a] 1975–79 (1967 = 100).

[b] Correlation between housing prices and permit changes: + = positive and - = negative.

[c] Annual rate of increase compared with state average: + = above, ± = equal, - = below.

San Diego, Orange, San Bernardino, Riverside, and Sacramento counties recorded population and housing supply changes above the state average, and all reported prices rising as supplies increased. Other market evidence indicated that these were all major growth areas in California, so that apparently housing supplies were not keeping pace with demand requirements.

San Mateo and Alameda counties reported both population and housing changes below state averages; however, as the number of permits dropped, housing prices continued upward. These are counties reputed to engage in strong growth control measures.

Santa Clara County reported population and housing supply changes at rates close to those for the state, but housing prices continued upward as permits dropped. This is a county that has previously reflected housing market trends that would coincide with an area facing strong growth pressures but seeking to resist them.

Ventura and Contra Costa counties showed housing supply rates above average compared with the state and population changes about equal to those in the state. Nevertheless, both counties reflected prices that were increasing, even though new housing construction was also increasing. Apparently the housing market pressures were continuing and not being relieved. Ventura has modest and Contra Costa somewhat stronger growth control programs.

Marin County population changes matched those of the state, but new housing construction was below state averages, so that prices were increasing, even though new construction was continuing. Marin is a county with reputedly very strong growth controls. Santa Barbara County has the same reputation as Marin with respect to growth controls, and it also reported an inverse relationship between housing price changes and permits, even though population was increasing at below and housing at about the same as state rates. Contra Costa, with controls that might be rated as lighter than those in Ventura County, had the same change patterns as Ventura.

These very crude analyses do lend additional support to the assumption that any area that invokes growth controls that subvert market changes will affect home prices adversely. In an effort to test this hypothesis more rigorously, home price changes were regressed on building permit changes for the 14 counties being analyzed for the period from 1960 to 1979. The results are reported in Table 2.8. Counties believed to have relatively strong growth controls—Alameda, Marin, San Mateo, and Santa Clara—showed relatively strong negative correlations between housing prices and permit changes. Counties with relatively modest growth controls—Sacramento, Riverside, and San Bernardino—showed relatively strong positive relationships between housing prices and new construction changes.

Another measure of these same trends is found when comparing the mean indexes for housing prices and building permits. In Contra

TABLE 2.8

Relationships between Housing Price Changes and Building Permit Changes, Selected Counties in California, 1960–79

Location	Housing Prices Change as Dependent on Number of Building Permit Changes			Mean of Index for County: Housing Prices		Mean of Index for County: Building Permits	
	r[a]	r^2	F[b]	Mean	Standard Deviation	Mean	Standard Deviation
California	.3425	.11731	1.26103	183.0	94.3	179.5	49.3
Alameda	-.3207	.10285	1.26103	176.3	84.5	126.5	56.8
Contra Costa	.0811	.00657	0.07279	160.6	68.0	142.6	46.4
Marin	-.3436	.11806	1.47247	178.0	92.1	163.6	84.7
Sacramento	.6709	.45007	9.00268	156.9	70.9	228.6	84.0
San Mateo	-.4531	.20528	2.84131	189.8	102.6	151.4	78.7
Santa Clara	-.6655	.44287	8.74399	200.8	95.9	129.4	39.9
Los Angeles	-.0273	.00074	0.00820	171.4	94.2	153.5	45.8
Orange	-.2438	.05943		185.2	105.1	197.5	59.5
Riverside	.8176	.66847		163.7	72.0	351.9	196.3
San Diego	.0661	.00436		213.8	131.1	198.2	68.6
San Bernardino	.8395	.70476		177.9	76.6	261.2	162.8
Santa Barbara	-.0177	.00031		180.2	95.5	140.1	45.7
Ventura	.4248	.18045		188.1	103.0	214.5	61.4

[a]-1.0 or 1.0 indicates strong relationships.
[b]Significance at .01 = 3.96, at .05 = 2.60.

Costa and Marin counties the rates of change in housing prices were much greater than the rates of increase in building permits. In San Bernardino and Ventura counties the rates of increase in building permits were much greater than the increases in housing prices.

Every county with a mean price index above the state average—San Mateo, Santa Clara, Orange, San Diego, Santa Barbara, and Ventura—is also a county in which growth controls range from at least moderate to severe. However, such counties as Marin, with very severe growth controls, showed a below-average price index. For these reasons any efforts to relate housing prices to growth controls must be done carefully and within the context of the market environments prevailing over time. Another crude but theoretically correct way of stating the conclusions would be that the relationships between demand-supply elements in California housing markets in 1970 produced a median sales price of $24,200. Increasing demand pressures, as compared with market supply responses, pushed the median prices to $41,100. If these same demand-supply relationships, which occurred between 1970 and 1975, had continued to 1979, the median price would have been $84,920. However, in counties where growth controls were invoked but where market pressures continue, market prices would have been approximately $110,000. In counties where market pressures continued and growth was accommodated with relatively modest growth controls, the prices would have been approximately $60,000.

Some support for these estimates can be found in the median prices that prevailed in growth controls counties.[4] In Marin, for example, median prices in October 1979 averaged between $90,000 and $200,000. In Santa Clara (including Palo Alto), an area of very strict controls, prices averaged between $105,000 and $225,000. In San Mateo, with modest controls, prices averaged between $80,000 and $90,000. In Sacramento prices tended to average between $40,000 and $90,000.

HOME PRICES AND ENVIRONMENTAL IMPACT REVIEW PROCESSES

California home prices have increased sharply since 1970 and at rates that have been accelerating much faster than national average price rates. Furthermore, the acceleration in California prices coincided with the introduction in 1969 of the National Environmental Protection Act (NEPA) and in 1970 of the California Environmental Quality Act (CEQA). However, during this same period many California counties introduced an increasingly broad array of growth control measures. In many cases these measures were linked with environmental impact review processes, with the EIRPs receiving the chief blame for any extraordinary costs.

The macromarket data presented in this chapter support the view that growth controls do increase prices when markets and economic conditions are improving. The sharpest price increases tended to be in the counties reputed to have the most stringent growth controls. Since the demand for housing continues throughout California and is particularly strong in the more desirable coastal locations, attractive areas in major urbanizing areas that use growth controls have high and sustained housing prices. Unfortunately, the data do not permit isolating the impacts of EIRPs on housing prices. In the next two chapters, the details of the EIRPs in California will be examined and evaluated. In Chapter 5, a model, which could be used in evaluating the cost (price) impacts of EIRPs, is presented and associated with some cost estimates. Finally, the chapter includes a review of other work on the costs of environmental controls and presents cost-benefit and cost-effectiveness models, which might be used in evaluating EIRPs and their impacts on housing costs and prices.

The findings in this chapter about the relationships between housing prices and EIRPs are reinforced by other studies.[5] When housing markets are active and demand strong, controls may push up prices, but builders can pass these costs on to the buyers. When markets are weak or demand slight, the extra costs of EIRPs cannot be passed on, so most builders will refuse to provide housing in those markets. In markets where demand pressures and EIRPs are strong, some builders will be willing to learn how to adjust to such markets. The few builders who do this operate under semimonopoly conditions, which allow them to pass through the extra EIRP costs. Unfortunately, demand pressures in markets with limited entry force prices to levels that are beyond low and moderate income families. Perhaps the protection afforded the environment by the EIRPs is worth this trade-off, but this possibility has yet to be tested.

NOTES

1. Economics Department, National Association of Homebuilders, U.S. Department of Commerce, National Association of Realtors.

2. U.S., Department of Commerce, Bureau of the Census, Census of Population and Housing, 1980.

3. The 14 counties are northern California—Alameda, Contra Costa, Marin, San Francisco, San Mateo, Santa Clara; central—Sacramento; southern—Los Angeles, Orange, Riverside, San Diego, San Bernardino, Santa Barbara-Ventura (combined). They usually provide at least 50 percent of all new California residential housing annually.

4. Real Estate Research Council of Northern California, Northern California Real Estate Report.

5. Urban Land Institute and Gruen and Gruen Associates, Effects of Regulations on Housing Costs: Two Case Studies, Research Report no. 27 (Washington, D.C.: Urban Land Institute, 1977); and Stephen R. Seidel, Housing Costs and Government Regulations: Confronting the Regulatory Maze (New Brunswick, N.J.: Rutgers University, Center for Urban Policy Research, 1978).

3

THE EVOLUTION OF ENVIRONMENTAL REVIEW UNDER CALIFORNIA LAW

As a result of significantly increased environmental concern by a broad range of citizens, Congress passed the National Environmental Policy Act (NEPA)[1] in 1969 to take effect on January 1, 1970. The scope of NEPA was, however, limited to federal government actions.

In a number of states, including California, pressures built for a more localized response to environmental degradation as well. Over half of the states passed state environmental policy acts (SEPAs) based upon the federal model. The range of these SEPAs varied, but the environmental reviews mandated by these state acts are the source of most environmental review related to housing construction, which is still a private sector function.

This chapter examines the evolution of one of those SEPAS—the California Environmental Quality Act (CEQA). After a brief look at NEPA and other SEPAs, the development of CEQA from its simple beginnings through its stormy expansion in scope to the present is traced. Although detailed discussion of the CEQA-mandated review process is left until Chapter 4, this chapter looks at both the structure of the act and the changes in it related to housing. Finally, the act is related to other major controls on land use and housing.

THE NATIONAL ENVIRONMENTAL POLICY ACT

As a congressional response to the environmental concern of its constituents, NEPA proceeded in three important ways. It stated a national policy of environmental protection to be reflected in governmental activities. It sought to assure administrative response through a series of "action-forcing" mechanisms. Finally, it set up a Council on Environmental Quality (CEQ) to act as a watchdog over the environmental review process.

The act opens with a section spelling out both purpose and basic policy. The stated purpose is to declare "a national policy which will encourage productive and enjoyable harmony between man and his environment." The national environmental policy is spelled out in six specific mandates to the federal government in Section 101(b). These give some substantive content to the opening broad statements of policy.

It is, however, the action-forcing Section 102 that is the heart of NEPA. That section requires that to the fullest extent possible all federal policies, regulations, and laws will be interpreted in accordance with NEPA policies. Further, all agencies of the federal government are to use a systematic interdisciplinary approach to decision making and are to develop ways of appropriately considering environmental amenities in those decisions. Section 102(c) is the basic foundation of the review process and is worth repeating in full.

> [Federal agencies shall] include in every recommendation or report on proposals for legislation and other major Federal actions significantly affecting the quality of the human environment a detailed statement by the responsible official on:
>
> (i) The environmental impact of the proposed action,
> (ii) Any adverse environmental effects which cannot be avoided should the proposal be implemented,
> (iii) Alternatives to the proposed action,
> (iv) The relationship between local short-term uses of man's environment and the maintenance and enhancement of long-term productivity, and
> (v) Any irreversible and irretrievable commitments of resources which would be involved in the proposed action should it be implemented.

A duty to consult with and obtain comments from other federal agencies with special competence is imposed on those responsible for the statement. Those comments are to accompany the statement through the review process. Agencies are to implement procedures, based largely on Council on Environmental Quality (CEQ) guidelines. The council was created by Title II of the act.

The scope of NEPA's effect was, at first, unclear, but several developments provided it with the teeth that were in doubt at the beginning. The CEQ published its guidelines rather quickly, and they detailed a strong process. But perhaps the most significant was the

Calvert Cliffs Coordinating Committe, Inc. v. Atomic Energy Commission[2], which strongly stated the significance of NEPA and clarified what was required for compliance and the consequences of an agency not complying—the invalidation of the action.

Given this federal antecedent to the state acts and their adoption of the environmental impact review process (EIRP) of NEPA as their approach as well, it will be useful to briefly review how that particular process was incorporated into NEPA in the first place. Both Anderson[3] and Liroff[4] provide a legislative history narrative filled with compromise. According to Liroff, the environmental impact statement (EIS) requirement was the result of a compromise between Senators Edmund Muskie of Maine and Henry Jackson of Washington. The theory behind the bill, as it emerged from Jackson's Interior Committee, was that the addition of the environmental analysis requirement of agency decision making would lead to some organizational learning on the parts of the agency and, thus, to less adverse environmental consequences. The Office of Management and the Budget (OMB) was to be the oversight agency. Senator Muskie, on the other hand, believed that agencies could not be trusted with environmental responsibilities and that a detailed impact statement should be required rather than Jackson's proposed "finding" of environmental impact. The Muskie approach held that agency habits were so firmly in place that far more external oversight would be required; thus, the impact statement was developed and circulated to other agencies, which would necessarily include those with an environmental expertise and constituency. Thus, as described by Liroff, two theories of administrative change were embodied in the compromise NEPA, one emphasizing external oversight and one internal reform.

As the result of some seven years of experience with the impact statement requirements and their review, the CEQ revised the impact statement requirements in 1977 to make the EIS more efficient. Among the changes made were limiting the statements to 150 pages (or 300 if an unusually complex project), with a concentration on alternatives; the use of "scoping"—a process to early on determine how long an impact statement should be as well as what its central issues should be and the division of labor among the agencies preparing it; and a tightening of the preparation schedules of EISs.[5]

This all too brief review of a significant federal statute, which has spawned over 600 court cases and reams of commentary, sets the stage for turning to the state acts substantially patterned on NEPA. For, while NEPA only related to federal actions, it is the states and their local government branches that are the dominant forces in the control of land and, thus, in housing development.

STATE ENVIRONMENTAL POLICY ACTS

NEPA applied only to actions of the federal government. This meant that, while a model for more localized activity was provided, the legal framework for environmental consideration at the state level did not yet exist. The time period of the early 1970s was marked by adoption at the state level of acts based upon NEPA in over 30 states.[6] Indeed, as one commentator has remarked, this was a significant departure from traditional notions of federalism: in the instance of NEPA, "the federal government has served as an experimental laboratory for the states."[7]

Eighteen of these states and Puerto Rico adopted similar governmental machinery from NEPA on a comprehensive basis, requiring environmental impact statements or reports (EIRs) on projects that may significantly affect the environment. Those implementing these requirements by statute were California, Connecticut, Hawaii, Indiana, Maryland, Massachusetts, Minnesota, Montana, New York, North Carolina, South Dakota, Virginia, Washington, and Wisconsin. Michigan, New Jersey, and Utah have accomplished the equivalent through use of executive orders, while Texas has instituted it through administrative action.[8]

Other states have confined their acts to more limited applicability, generally to a defined geographic area. These include Alabama, Delaware, Mississippi, New Hampshire, Pennsylvania, and Rhode Island.[9]

Finally, a number of cities and counties in various states have instituted their own versions of the NEPA process. These include Transylvania County, North Carolina; Blaine County, Idaho; Bowie, Maryland; Lakewood, Colorado; and Glastonbury, Connecticut.[10]

Concentrating on the 18 states with general-applicability SEPAs and drawing upon in-depth studies by Pearlman[11] and Hart and Enk[12], some generalizations about their ties to NEPA may be made. Virtually all of the statutes in those states with statutory authorization are drawn very closely from the federal statute. Those based on executive order also parallel the federal statute. There are broad declarations of environmental policy. There are requirements that the policy be implemented. Finally, there are the action-forcing requirements for the preparation of impact statements. Indeed, the five provisions recited in NEPA for impact statement contents are generally repeated, though there is some variation on this point. Just as NEPA failed to provide an agency with strong supervisory powers, but provided instead the CEQ, which was to draft guidelines and be a "watchdog," so too did the SEPAs fail to create a strong agency.[13] Most placed coordination in either a specially created environmental council, board, or committee; a department of the executive branch of the state government, such as a natural resources department; or in the governor's office.[14] Virtually all prepared guidelines.

There are, however, some major differences among the states in how their environmental reviews impact land use and housing. This is owing to the rather sticky problem of the applicability of the acts to political divisions of the state as well as the state government itself and the application of the acts to government permitting as well as governmental projects. The former do not exist under NEPA. The latter, in the federal context, have been resolved in favor of the applicability to permitting. Our concerns as to the impacts on housing of the state acts then revolve around four questions:

Are local governments (cities and counties) included under the act?
Is permitting activity included under the act as well as governmental projects?
Are growth-inducing impacts of actions or projects to be considered?
Are mitigation measures to be addressed and considered in EIRs?

Table 3.1 indicates those dimensions for the 18 states with general application SEPAs. The most crucial is the inclusion of local government permitting activity within the act, as most housing and land-use decisions are made at that level.[15] As may be seen, only three states have applied the SEPA to all such activities—California, New York, and Washington.[16]

Four other states have applied the EIR requirements to local government in a more limited manner. In Hawaii, the requirement applies to actions involving state or county land or funds and to private actions within several designated areas of the state land-use plan—this being among the few states with such a plan. The Massachusetts act is applicable to activities or projects with state agency financing, initiation, or permitting or to actions of housing authorities, redevelopment authorities, and development commissions that involve state funds or state permits. North Carolina requires Environmental Impact Reports (EIRs) for actions that use public funds and authorizes local requirements for EIRs (but does not mandate them) for major private development projects. Finally, Minnesota requires an EIR on any "major governmental action or for any major private action of more than local significance."[17]

The remainder of these states do not apply the SEPA to local government activities. This does not quite end housing impacts, however. A number of these states, as well as some of the others that have adopted more limited versions of environmental reviews, apply the EIRP to critical areas of environmental sensitivity. Often these are coastal areas. Consequently, some further applicability of the process to housing circumstances through special state permitting

TABLE 3.1

State Environmental Policy Acts Coverage in Selected States

State	SEPA Adoption		Local Government Approval of Private-Project Permits	EIR Coverage	
	Year	Manner		Mitigation	Required Growth
California	1970	Statute	Yes	Yes	Yes
Connecticut	1973	Statute	Unclear	Yes	No
Hawaii	1974	Statute	Limited	No	No
Indiana	1972	Statute	No	Yes	No
Maryland	1973	Statute	No	No	No
Massachusetts	1972	Statute	Limited	Yes	No
Michigan	1971	Executive order	State level only	Yes	No
Minnesota	1973	Statute	Limited	No	No
Montana	1971	Statute	State level only	Yes	Yes
New Jersey	1973	Executive order	State level only	Yes	No
New York	1976	Statute	Yes	Yes	Yes
North Carolina	1971	Statute	Limited	Yes	No
South Dakota	1974	Statute	State level only	Yes	No
Texas	1975	Administrative action	State level only	Yes	No
Utah	1975	Executive order	State level only	Yes	No
Virginia	1972	Statute	No	Yes	No
Washington	1971	Statute	Yes	Yes	No
Wisconsin	1971	Statute	State level only	Yes	No

Source: Stuart L. Hart and Gordon A. Enk, *Green Goals and Greenbacks: State-Level Environmental Review Programs and Their Associated Costs* (Boulder, Colo.: Westview Press, 1980).

for areas of special concern is also present in such states as Delaware and Alabama.[18]

The consideration of growth-inducing impacts and mitigation measures is also significant. The former brings the issue of housing to the forefront in virtually all projects that generate substantial employment. The latter requires that measures to ease environmental effects (often less density) be considered in the review.[19]

While this discussion has been much too brief to adequately cover the various NEPA progeny that have sprung up in the states, it has served to indicate the range of applicability of the EIR requirement to housing development in a substantial number of jurisdictions. With that broader applicability in mind, we turn to a more detailed examination of the focus of the study—the California Environmental Quality Act—the most litigated and most visible of the SEPAs.

THE CALIFORNIA ENVIRONMENTAL QUALITY ACT

California became the first state to act on a state environmental policy act; it did so only three months after NEPA was passed.[20] This was a part of a flurry of environmental law activity in the 1970 California legislature. It included enactment of a statute directing preparation of a comprehensive state environmental goals and policy report, with priority given to the development of a statewide land-use policy.[21] An act that encouraged development of educational programs for the training of teachers and protection of the quality of the environment was also passed.[22]

The California Environmental Quality Act[23] was to become the most significant environmental legislation of that session. As enacted on September 17, 1970, CEQA was quite simple in structure and approach. Following a declaration of policy and findings, the act contained two operative chapters.

The first chapter applied to all state agencies, boards, and commissions. It required a report on any project they proposed that would have a significant effect on the environment.[24] The report was to include a statement similar in coverage to that required under NEPA.[25] No state division could fund or request funds for a project unless the statement was provided.[26] An exception was provided for projects involving only planning. The State of California Office of Planning and Research (OPR) was to work with state, regional, and local agencies to develop provisions for implementation.[27] The body preparing the statement, known in California as the Environmental Impact Report (EIR) rather than the EIS, which is the NEPA designation, was to consult with other knowledgeable state agencies before issuing the statement.[28] The EIR and comments from these agencies were to be used in project decision making and be available to the legislature and the general public. Divisions of the state were to include in their budget requests funds to protect the environment from their activities and to let the governor and legislature know of deficiencies in their present authority that would hinder CEQA compliance.[29]

The other operative chapter dealt largely with the act's application to local agencies. State divisions allocating state or federal funds to local agencies on a project-by-project basis were to require, unless exempted by the implementing guidelines, an EIR from the local agency before disbursing funds for other than planning purposes.[30] Also, legislatures of all cities and counties officially adopting a conservation element of a general plan were to make a finding that any project they intended to carry out having a significant environmental effect was in accordance with that element. The act continued with the following:

> All other local governmental agencies shall make an environmental impact report on any project they intend to carry out which may have a significant effect on the environment and shall submit it to the appropriate local planning agency as a part of the report required by . . . the Government Code.[31]

A noted commentator has suggested that this second chapter, which is replete with ambiguity as to what local governments and what actions are covered under the act, delayed local attention to the act. In the same legislative session a state deadline imposed on cities and counties requiring them to adopt conservation elements was, in his opinion, more threatening.[32]

Friends of Mammoth

In its early days, CEQA did not have much impact on the land development process. The act was interpreted by public agencies to apply only to projects constructed or carried out by government agencies.[33] During the last half of 1971, the California Resources Agency promulgated some interim guidelines, which would have limited CEQA applications to local decisions.[34] This was due to the definition of project within the proposed guidelines: "'Project' includes any major work segment involving siting, land purchases, design, or construction activities, utilizing state or federal funds carried out by any or all levels of government which could have a significant effect on the environment of the state."[35] The California attorney general was so incensed that he and his staff filed a formal petition challenging this and other provisions of the proposed guidelines as contrary to the letter and spirit of CEQA.[36]

It was the courts, however, that ultimately clarified the applicability of CEQA at the local level. In June 1971, the Board of Supervisors of Mono County affirmed a decision of the county planning commission to issue a conditional use permit for construction of a 184-unit condominium, and a building permit was subsequently issued. A local group decided to challenge the actions and asked the courts to issue a stay, since all permits had been granted. While the lower courts refused, the California Supreme Court did issue a stay taking jurisdiction in January 1972.

On September 21, 1972, the court announced its decision in Friends of Mammoth v. Board of Supervisors[37], providing CEQA with applicability beyond that of the interim guidelines. The primary question decided in Friends of Mammoth was whether CEQA applied to private activities for which permits are required. As Mono County did not have a conservation element, the key issue was whether the act's term project, which was undefined in the act, included that private activity with governmental permitting. Faced with a split of opinion, including coauthors of CEQA on each side of the issue, the court emphasized the intent of the statute as well as the CEQA guidelines adopted for NEPA at the federal level in coming to the following conclusion:

> The granting or denying of a permit is an act which a governmental authority "carries out." Accordingly, we construe the phrase following "project" to mean only that before an environmental impact report becomes required the government must have some minimal link with the activity, either by direct proprietary interest or by permitting, regulating, or funding private activity.[38]

As a result of the Friends of Mammoth case, it now became clear that CEQA would apply to local government activities directly impacting on the land development process. Any changes that were to occur would now have to emanate from the legislature.

Legislative Response to Friends of Mammoth

In the wake of Friends of Mammoth, the legislature moved quickly in its efforts to amend CEQA; on December 1, 1972, Governor Reagan signed a heavily debated and amended bill introduced by Assemblyman John Knox, one of the authors of the original CEQA. The amendments to the bill considerably complicated the original form.[39]

Under the amendments, the coverage of the act includes "activities involving the issuance to a person of a lease, permit, license, certificate or other entitlement for use."[40] The act is to apply to "discretionary projects" but not to "ministerial projects." Discretionary projects explicitly include "enactment and amendment of zoning ordinances, the issuance of zoning variances, the issuance of conditional use permits and the approval of tentative subdivision maps."[41] All public agencies are to adopt guidelines consistent with those authorized to be adopted by the Secretary of the Resources Agency, including criteria to determine if projects will have a significant effect on the environment. Significant effect projects are defined as the following:

> (a) A proposed project has the potential to degrade the quality of the environment, curtail the range of the environment or to achieve short-term, to the disadvantage of long-term environmental goals;
> (b) The possible effects of a project are individually limited but cumulatively considerable;
> (c) The environmental effects of a project will cause substantial adverse effects on human beings, either directly or indirectly.[42]

Environment was narrowly defined as only physical environment.[43]

Other items relevant for this examination include the addition of the coverage of "the growth inducing impact of the proposed action," to the list of items covered in an EIR.[44] Public agencies are allowed to charge fees for the preparation of EIRs.[45] Local agencies are to comply in the same manner as state agencies. The former exemption of those agencies with conservation elements in their general plans is eliminated.[46] Finally, local agencies can require the proposer of the private project under consideration to submit information necessary for adequate review.[47]

A number of other provisions were introduced in this law, including a clearinghouse at the state level for EIRs as well as publication of the 102 Monitor (subsequently titled California EIR Monitor) to provide a communications medium for guidelines and announcements of completed EIRs.[48] Time limits were set at 180 days for a court challenge to a project for failure to determine the significant effect and 30 days if the determination is alleged to be erroneous.[49]

This brief summary does not capture nearly all of the changes brought by Assembly Bill (AB) 889 or all of the bitterness engendered. As a part of the compromise in its passage, a moratorium on its application was included, which allowed permits issued before the effective date of the amendments as well as publicly permitted private projects for a period up to 121 days after the effective date. The complexity introduced in such a large number of amended sections being developed so quickly led to a myriad of ambiguities and inconsistencies, which would be clarified over much of the rest of the decade. On February 3, 1973, some 28 months after the passage of CEQA, the guidelines for its implementation became final.[50]

Continued Development

In the time since 1973 there have been one general and several minor amendments to CEQA in each legislative session. These major amendments have dealt largely with measures to limit and change the EIRP in the light of experience. This chapter will quickly review some major changes of interest to complete the account of CEQA's evolution.

In an effort to prevent shielding from the public of an exemption decision, AB 2338 was signed into law in 1974. It allows a public notice to be filed if no EIR is required because a project is ministerial (categorically exempted under the guidelines of emergency or disaster repairs). Either the agency or the person receiving the approval may file. If the notice is filed, any challenge must come within 35 days; if no notice is filed the time is 180 days.[51]

In the closing days of the 1976 legislative session, AB 2679 was passed. The bill, sponsored by Senator Knox, amended a number of CEQA sections. It declared a policy of mitigation measures and alternatives. It encouraged use of incorporation of other document contents by reference and added a definition to CEQA of <u>Negative Declaration</u>, which is prepared when an agency finds no significant effect and makes provision for an agency charging fees for preparation of Negative Declarations. It provided that EIRs must be prepared directly by or under contract to the public agency. Its most important sections, however, were provisions for public notice by one of three means—publication, posting of notices, or direct mail to adjacent property owners—and the requirement of local agencies for setting time limits of no more than one year for the completion of all environmental reviews on private projects.[52]

Passed in 1977, AB 884 continued the emphasis upon streamlining the process. Lead agencies (defined under the guidelines) are limited to 45 days from acceptance of the application as complete to deciding whether an EIR or Negative Declaration would be issued. Increased and earlier consultation with other responsible agencies is required. The one-year requirement for EIRs is extended to all agencies, and the Negative Declaration time limit is changed to 105 days for all agencies.[53]

Three other amendments dealt directly with housing issues. In 1979 Senate Bill (SB) 606 added to the policy sections of CEQA a state policy of providing a decent home and satisfying living environment for every Californian.[54] Passed in 1981, AB 1185 was designed to prevent duplication in EIRs by providing that when an EIR is prepared for a zoning action or community plan, given certain specified circumstances, then a later EIR for a subdivision map or residential project that is consistent with the zoning or community plan need only address effects particular to the parcel or project that were not addressed in the earlier EIR as being significant.[55]

An effort to encourage planning related to housing was embodied in SB 003, passed in 1978. Coming out of suggestions from the Office of Planning and Research, this amendment exempts from EIR or Negative Declaration requirements housing construction in an area for which a specific plan has been adopted that has an adequately detailed EIR. The specific plan, as authorized in California Government Code, Sections 65450-65453, contains detailed standards for development of a portion of the area covered by the required general plan under California Government Code, Section 65302. In short, the legislature built in an exemption from CEQA for housing projects if the area is covered by a plan addressing the concerns in detail and limiting development strictly to the planned standards.[56]

In the years since CEQA's birth in the early 1970s, it has grown to be a complex, often confusing, and pervasive influence on the development process in California. We have tried to give some flavor of the evolution of the law in this section. This chapter closes with a brief discussion meshing CEQA with the more traditional land-use controls it overlays.

CEQA and Private Housing Construction

The construction of residential dwelling units is still a private decision. It is, however, a process heavily regulated through local government specifications and approvals. In California CEQA applies to local government decisions on private projects. It provides a procedural overlay—through the environmental review process—and a substantive policy consideration to these decisions. This concluding section contains details of the applications of CEQA to the government approvals required for private housing construction.[57] We argue that the application of environmental review to these decisions is the point of impact of environmental laws on housing costs.

Table 3.2 contains a summary of the applicability of CEQA to various decisions in the process. If housing is divided into three classes—single family, multifamily, and manufactured—it may be seen that the application of CEQA impacts on decisions for all three types.

TABLE 3.2

CEQA Application to the Private Housing Construction Process

Government Decision/Permit	Subject to CEQA[a]	Authority
Subdivision		
Tentative tract map	Yes	Cal. Pub. Res. Code § 21080(a)
Final map	No	14 Cal. Admin. Code § 15073
Condominium conversion	No	14 Cal. Admin. Code § 15101
Zoning action	Yes	Cal. Pub. Res. Code § 21080(a)
Map amendment		
Variance[b]		
Conditional-use permit		
Permits		
Demolition	No	14 Cal. Admin. Code § 15101
Grading[c]	Yes	<u>Day</u> v. <u>City of Glendale</u> 51 Cal. App. 3d 817 (1975)
Building	No	14 Cal. Admin. Code § 15073
Coastal development	Yes	Cal. Pub. Res. Code §§ 21065, 21080(a)

[a]Under 14 Cal. Admin. Code § 15103, new construction of one single-family residence or one multifamily structure of four units or less (six units or less in urbanized areas) are exempt.

[b]Minor lot variances are exempt under 14 Cal. Admin. Code § 15105.

[c]Minor grading on lots with a slope of less than 10 percent are exempt under 14 Cal. Admin. Code § 15104.

The determination of applicability of CEQA comes from three sources. First, the act itself enumerates some projects that are to be considered under the review and some that are not. Second, the guidelines published by the Resource Agency carrying out the act's directive include certain classes of projects considered categorically exempt from the application of the act, meaning that they have been determined to have no significant effect on the environment.[58] Finally, the act does not cover ministerial projects, which under the guidelines are defined as including "those activities . . . which are undertaken or approved by a governmental decision which a public officer or public agency makes upon a given state of facts in a prescribed manner in obedience to the mandate of legal authority"—in other words, without personal judgment.[59]

Several categorical exemptions are important for housing construction. The most important is the exemption for "New Construction of Small Structures." Included under this exemption is the construction of individual single-family residences. Also included is the construction of multifamily residential structures of four units or less (six or less in urban areas), if only one is being built. Small accessory structures are also exempt.[60] Other exemptions include minor grading of land,[61] minor lot adjustments (variances),[62] demolition of buildings with some exception,[63] and conversion from existing multifamily rental to condominium.[64] These categorical exemptions are virtually all related to a minor scale of construction.

The subdivision process is the first step toward a larger scale of development and involves a substantial amount of discretion. Thus, it is explicitly under the review requirements of CEQA.[65] The approval of the final map, however, is considered ministerial under the guidelines and, thus, does not require another review.[66] Condominiums, it should be noted, are considered subdivisions under California law.[67]

Zoning actions, including map amendments, variances, and conditional use permits, are explicitly subject to the requirements for environmental review.[68] This is quite significant for the construction of trailer parks, which almost always require a conditional use permit; planned unit developments, which generally involve some form of rezone or use permit; and large-scale multifamily developments, which generally involve some variance measures.

Building permits, by virtue of the guidelines, are generally considered to be ministerial.[69] Grading permits, on the other hand, are subject to review unless categorically exempt.[70]

Several other qualifying points should be remembered. For example, the major categorical exemptions discussed are to be considered in light of the environmental sensitivity of the location of the project, as the project may be deemed not exempt under certain conditions.[71] Also, all exemptions are inapplicable if the cumulative impact of successive projects of the same type in the same place over time is significant.[72] A project may still be exempt without falling into any of the classifications we have discussed if it does not have a significant effect on the environment.[73]

One final control must be mentioned. The California Coastal Act of 1976,[74] which superseded the Coastal Zone Conservation Act of 1972 (passed by voter initiative), provides for a planning and control structure for a coastal zone generally extending 1,000 yards inland from mean high tide. With some very minor exceptions, all building within this zone requires a special coastal development permit granted either by the local government (if an approved local coastal program is in place) or by the appropriate regional or state commission (if not). Issuance of these permits is subject to CEQA requirements as well.[75]

We have traced the development of environmental review in California and elaborated upon its application to the housing construction process. The California legal framework shares much with that of other states and localities, which have taken their environmental review from the NEPA model. The next chapter contains the development of a model and the review of literature on the economics of EIR and housing costs before directing its attention to the California implementation experience and its broader lessons.

NOTES

1. 42 U.S.C. §§ 4321-4347 (1970).

2. 449 F.2d 1108 (D.C. Cir., 1971).

3. Frederick R. Anderson, NEPA in the Courts (Washington, D.C.: Resources for the Future, 1973).

4. Richard A. Liroff, "NEPA—Where Have We Been and Where Are We Going?" Journal of the American Planning Association, April 1980, pp. 154-61.

5. Council on Environmental Quality, "Regulations for Implementing the Procedural Provisions of the National Environmental Policy Act," 43 Federal Register 55978-56007 (November 29, 1978), 40 C.F.R. 1500-08.

6. Robert W. Burchell and David Listokin, The Environmental Impact Handbook (New Brunswick, N.J.: Rutgers University, 1975), pp. 7-36.

7. Nicholas C. Yost, "NEPA's Progeny: State Environmental Policy Acts," Environmental Law Reporter 3 (1974): 50090.

8. California—Cal. Pub. Res. Code §§ 21000-21193; Connecticut—Conn. Gen. Stat. Ann Ch. 439 §§ 22a-1 to 22a-27; Hawaii—Hawaii Rev. Stat. Ch. 344; Indiana—Ind. Stat. Ann tit. 13, art. 1 chap. 10; Maryland—41 Md. Ann Code §§ 447-51; Massachusetts—Mass. Ann Laws ch. 30 §§ 61-62; Minnesota—Minn. Stat. Ann ch. 116D; Montana—Mont. Code Ann tit. 75, ch. 1; New York—N.Y. ECL 8-0101 to 8-0117; North Carolina—N.C. Gen. Stat. ch. 113A; South Dakota—S.D. Comp. Laws 1967 ch. 11-1A; Virginia—Va. Code Ann tits. 10-17.107 to 10-17.112 and §§ 10-177 to 10-185; Washington—

Wash. Rev. Code ch. 43.21C; Wisconsin—Wisc. Stat. Ann ch. 1 §§ 1.11-1.12. Executive Order states are Michigan—Michigan Exec. Order 1974-4; New Jersey—New Jersey Exec. Order No. 53; and Utah—State of Utah Exec. Order, August 27, 1975. Texas administrative adoption—"Policy for One Environment"—was adopted by the Interagency Council on Natural Resources and Environment on March 7, 1972. In addition, New Mexico adopted a SEPA but repealed it.

9. A listing of states and localities that have adopted SEPA-like statutes for regional and general use may be found in Burchell and Listokin, Environmental Impact Handbook.

10. Ibid.

11. Kenneth Pearlman, "State Environmental Policy Acts: Local Decision-making and Land Use Planning," Journal of the American Institute of Planners, January 1977, pp. 42-53.

12. Stuart L. Hart and Gordon A. Enk, Green Goals and Greenbacks: State-Level Environmental Review Programs and Their Associated Costs (Boulder, Colo.: Westview Press, 1980).

13. Ibid., p. 22.

14. Ibid., p. 24.

15. Pearlman, "State Environmental Policy Acts," p. 47.

16. Hart and Enk, Green Goals and Greenbacks, pp. 18-21.

17. Ibid.

18. Burchell and Listokin, Environmental Impact Handbook, pp. 14-36.

19. Hart and Enk, Green Goals and Greenbacks, pp. 13-16.

20. The following discussion of CEQA draws heavily from Donald G. Hagman, "NEPA's Progeny Inhabit the States—Were the Genes Defective?" Urban Law Annual 7 (1974): 3-56.

21. Cal. Gov. Code §§ 65041-65049.

22. Cal. Ed. Code §§ 565-570.4.

23. Cal. Pub. Res. Code §§ 21000-21174 (reprinted in Appendix C).

24. Cal. Pub. Res. Code § 21100.

25. The NEPA § 102 requirements are discussed in the first pages of this chapter.

26. Cal. Pub. Res. Code § 21101.

27. Ibid., § 21103.

28. Ibid., § 21105.

29. Ibid., § 21107 (this section was later repealed).

30. Ibid., § 21150.

31. Ibid., § 21152.

32. Hagman, "NEPA's Progeny," p. 7.

33. James Longtin, California Land Use Regulations (Westlake Village, Calif.: Local Government Publications, 1977), p. 357.

34. Proposed Guidelines for the Preparation and Evaluation of Environmental Impact Statements under the California Environmental Quality Act of 1970 (Sacramento: Resources Agency, 1971).

35. As repeated in Attorney General of the State of California, Petition: "In re Proposed Guidelines for the Preparation and Evaluation of Environmental Impact Statements under the California Environmental Quality Act of 1970," September 3, 1971, p. 9.

36. Ibid.

37. 8 Cal. 3d 247, 502 P.2d 1089, 104 Cal. Rptr. 761 (1972).

38. Ibid., 262, 502 P.2d 1059, 104 Cal. Rptr. 771.

39. AB 889, 1972.

40. Cal. Pub. Res. Code § 21065(c).

41. Ibid., § 21080.

42. Ibid., § 21083.

43. Ibid., § 21060.5.

44. Ibid., § 21100(g).

45. Ibid., § 21089.

46. Cal. Pub. Res. Code §§ 21000-21174 (reprinted in Appendix C).

47. Cal. Pub. Res. Code § 21160.

48. Ibid., § 21088.

49. Ibid., § 21167.

50. Resources Agency of California, Guidelines for Implementation of the California Environmental Quality Act of 1970, February 5, 1973. Cal. Admin. Code §§ 15000-15203 (1973).

51. "New Law on Notices for Project Approval," California EIR Monitor 1 (1974): 1-2.

52. "Legislature Passes AB 2679," California EIR Monitor 3 (1975): 1-2; and Barbara Sahm, "Project Approval under the California Environmental Quality Act: It Always Takes Longer Than You Think," Santa Clara Law Review 19 (1979): 581-83.

53. "Summary of AB 884," California EIR Monitor 4 (1977): 1-3.

54. "Enacted 1979 Legislation Dealing with CEQA," California EIR Monitor 6 (1979): 1-3.

55. "1981 Amendments to CEQA," California EIR Monitor 8 (1981): 1-5.

56. Barry Steiner and Peter M. Detwiler, "SB2003—Common Sense or Housing Developer's Dream," California EIR Monitor 6 (1979): 1-5.

57. We will discuss the details of the review process itself in Chapter 5. This section can only give a broad outline. For a discussion of the nuances of the characterizations presented here, see Longtin, California Land Use Regulations, ch. 4.

58. Cal. Pub. Res. Code §§ 21084-21085.

59. 14 Cal. Admin. Code § 15032.

60. Ibid., § 15103.
61. Ibid., § 15104.
62. Ibid., § 15105.
63. Ibid., § 15101(l).
64. Ibid., § 15101(k).
65. Cal. Pub. Res. Code § 21080(a).
66. 14 Cal. Admin. Code § 15073.
67. Cal. Gov. Code § 66424.
68. Cal. Pub. Res. Code § 21080.
69. 14 Cal. Admin. Code § 15073 (there is a difference of opinion among experts as to how definitive this may be).
70. Day v. City of Glendale 51 Cal. App. 3d 817 (1975).
71. 14 Cal. Admin. Code § 15100.2(a).
72. Ibid., § 15100.2(b).
73. Cal. Pub. Res. Code §§ 21100, 21151.
74. Cal. Pub. Res. Code §§ 30000-30900.
75. Ibid., §§ 21065, 21080(a); for a further description of the development and application of the Coastal Act, see Fred E. Case and Jeffrey Gale, "The Impact on Housing Costs of the California Coastal Zone Conservation Act," American Real Estate and Urban Land Economics Journal, Winter 1981, pp. 345-66.

4

FORMS OF THE REVIEW PROCESS

The development of the California Environmental Quality Act (CEQA), as traced in Chapter 3, included a framework for impact review that set basic ground rules for localities in developing an environmental impact review process (EIRP) for their own use. There is some amount of local discretion that remains, and it is useful to examine the differences among the forms chosen and the results of the different implementation forms.

This chapter contains a review of the basic framework for environmental review under CEQA as set by the State Resources Agency guidelines. All agencies of the state were required to adopt procedures consistent with these guidelines.[1] The chapter then examines several research studies of CEQA initiated by the state and used to revise the basic process. The results are reported of survey work designed to explore the differences in the process adopted by a number of California cities and counties. Finally, more detailed case studies are provided of several of these jurisdictions and how they have approached the problem of environmental review.

THE ENVIRONMENTAL IMPACT REVIEW PROCESS UNDER THE GUIDELINES

The process outlined in the CEQA guidelines proceeds in four general steps. The determination must first be made as to whether CEQA applies to the project in question. Then it must be determined whether an Environmental Impact Report (EIR) is necessary or whether a Negative Declaration will suffice. If an EIR is required, a draft EIR must be prepared and circulated. Finally, the final EIR must be adopted under the law. The guidelines include a schematic diagram

FIGURE 4.1

Environmental Impact Report Process Flow Chart

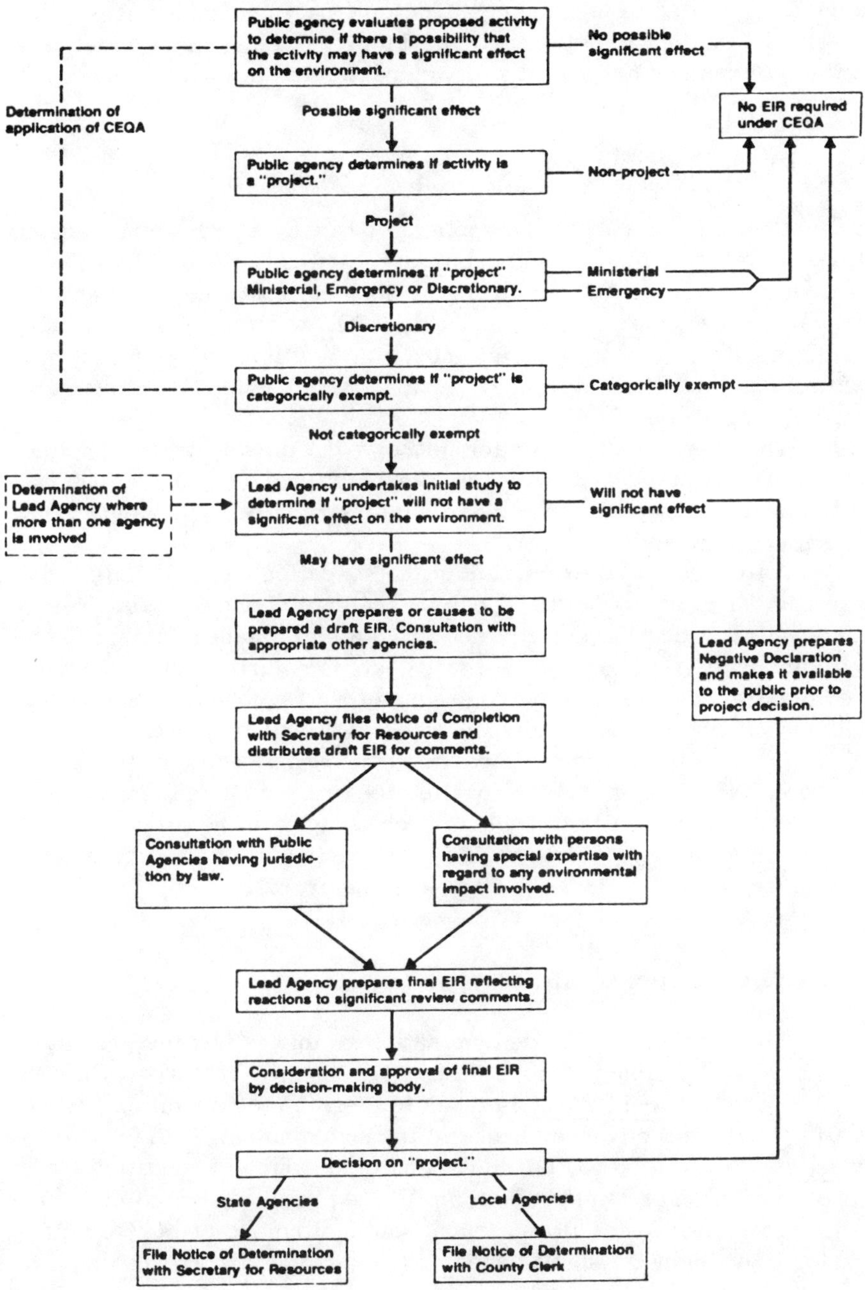

of the process (see Figure 4.1). The first step in the process is, however, not included in the diagram. Under CEQA, there must first be a determination as to the "lead agency"—the agency with principal responsibility for carrying out or approving the project. For most decisions related to housing construction, only one public agency is involved, and this is not a major problem. For example, most permit decisions normally rest with one particular individual or body.[2]

Does CEQA Apply?

Chapter 3 reviewed the application of CEQA to housing construction. The terms project and public agency were shown to be broad, including licensing and permitting by all local governments as well as the state government. As discussed, CEQA applies to discretionary decisions but not to ministerial ones.[3] In addition to the ministerial exemption, specific emergency repairs[4] and feasibility and planning studies are excluded.[5]

The categorical exemption concept was introduced in Chapter 3 as well. It includes 15 classes of exempt projects, including, as discussed earlier, construction of small structures and minor variances and grading.

A final possible exemption is the so-called General Rule, which exempts "a project wherein it can be seen with certainty that there is no possibility that the activity may have a significant effect on the environment."[6] One commentator suggests that jurisdictions have been reluctant to use this approach because of its vagueness and the "calculated risk" that noncompliance with CEQA will be found.[7]

If an exemption is claimed, the agency may file a Notice of Exemption. If the agency does not file, the private project sponsor may file. Such filing limits the potential challenges to the propriety of exclusion to 30 days. If no notice is filed, the statute of limitations is 180 days after project approval or disapproval.[8]

Is There a Significant Effect?

The next step is the determination as to whether the project will have a significant effect on the environment. In making the determination, significant effect is taken to mean a substantial or potentially substantial adverse change in the environment.[9] The term environment is limited to "physical conditions which exist within the area which will be affected by the proposed project, including land, air, water, minerals, flora, fauna, ambient noise, and objects of historic or aesthetic significance."[10] A finding of significance is mandatory if the project has the potential to degrade the environment, has the potential to achieve short-term environmental goals to the disadvantage of long-term environmental goals, has possible environmental effects that are individually limited but cumulatively considerable, or will cause substantial adverse effects on human beings.[11]

The determination of significant effect is to be made based upon an Initial Study. If it is clear that the project does have a significant effect, an EIR must be prepared and the Initial Study may be forgone.[12] The Initial Study is a preliminary analysis prepared by the lead agency to determine whether an EIR or Negative Declaration must be prepared. The study's purposes and contents, which are generally specified in the guidelines, are designed to provide information for the determination of significance.[13] The guidelines (see Appendix D) include an environmental information form for the applicant of a private project to complete and an environmental checklist which will, when completed by the agency, be adequate as the written Initial Study.[14] Both forms are merely suggested.

The review at this point may result in the issuance of a Negative Declaration or in continuation of the EIRP. A Negative Declaration is appropriate when the decision is made as to no significant effect. It is first issued by the agency in draft form, with a notice by publication, posting, or direct mail to contiguous property owners,[15] so that the required public review and comment are possible.[16] Circulation to other interested agencies for comment is also required.[17] The Initial Study is included as a part of the Negative Declaration as well as a finding of no significant effect. A Mitigated Negative Declaration may also be issued if there is an enforceable commitment from the private sponsor to include mitigation measures and if it is found that the project, including those measures, will have no significant effect.[18]

Time limits have been set by the amendment of the original CEQA. The determination of whether an EIR or Negative Declaration will be required must be made within 45 days of the project application being accepted as complete.[19] The adoption of a Negative Declaration must be completed within 105 days.[20] It is required that a notice of determination be filed with either the state secretary for resources or the county clerk.[21] This filing starts a 30-day statute-of-limitation period for challenging the determination.[22]

It is clear that a substantial amount of discretion is involved in the determination of significant effect. Indeed, the guidelines recognize that it is a highly subjective decision. However, the bias toward a finding of significant effect and the subsequent preparation of an EIR is contained in both the guidelines and in the result—voiding of the project decision—if the determination is challenged and found in error.[23]

Preparation of the Environmental Impact Report

Once it has been determined that the project will have a significant effect, the agency must prepare or cause to be prepared an EIR. The EIR is a detailed statement setting out environmental effects and other considerations required under CEQA. It is an information document to be used in informing the public of effects in a timely manner and in the evaluation of projects by public officials. The draft EIR may be prepared either by the public agency itself or under contract to the agency by consultants.[24] The agency may charge a private sponsor a reasonable fee for this preparation, and it may charge for the Negative Declaration as well.[25] It is the agency, though, that is responsible for the contents.[26] If the information supplied by a private sponsor is in the form of a draft EIR, it may not be used as the agency draft without independent analysis.[27]

CEQA sets forward the required contents of an EIR; the guidelines expand upon them.[28] These guidelines consist of the following:

1. A description of the project;
2. A description and discussion of the environmental setting;
3. A direct description and discussion and indirect significant effects of the project on the environment, giving consideration to both short-term and long-term effects;
4. A description of significant environmental effects that cannot be avoided if the project is implemented;
5. A description and discussion of mitigation measures proposed to minimize adverse impacts;
6. A description and discussion of all reasonable alternatives to the project and reasons for their rejection;
7. A description and discussion of the cumulative and long-term effects of the project that adversely affect the environment;
8. A description and discussion of significant irreversible environmental changes involved in the project should it be implemented;
9. A discussion of the growth-inducing impacts of the proposed action; and
10. A description and discussion of reasons why various possible significant effects were found not to be significant.

The guidelines encourage technical and scientific data inclusion in the appendix rather than the body. Incorporation of other documents on the public record by reference is permitted.[29] EIRs are to focus upon important issues, not including all possible topics available; emphasis is to be upon alternatives and mitigation.[30]

The issue of the timing of consultations with other agencies and interested parties has been stormy. The lead agency is to consult with other responsible agencies before the draft EIR is issued.[31] However, consultation at this stage with other persons or organizations likely to conflict, which was made mandatory by a court decision in 1980,[32] has now been made permissive by law in 1981.[33]

When the draft EIR is completed and acceptable to the lead agency, it is published and distributed for public and agency review. Notice of completion must be given to both the state and the public through at least one newspaper publication, posting, or direct mail to owners of contiguous property.[34] The draft is made available to the public with the opportunity for comments. Comments are to be solicited from public agencies and experts.[35] The review period must be at least 30 days.[36] No public hearing is required under state law.[37]

Finalizing the EIR and Project Approval

After this review, the EIR must be made final. A final EIR includes the revision of the draft EIR; comments and recommendations made on the draft, either in whole or summary; a list of those commenting; and the responses of the public agency to the comments.[38] The decision-making body of the lead agency must certify that after evaluation the final EIR is in compliance with CEQA and the guidelines.[39] A notice of determination must be filed.

With the final EIR certified, the agency may make the decision on the project. Realizing that environmental concerns are not always the sole consideration, the guidelines recognize a balancing test.[40] Findings must be made, and if a project is allowed that permits significant environmental effects, it must be supported in writing, including reasons for the approval.[41]

The time limit for completion and certification of an EIR is one year from the date on which the project application is accepted as complete.[42] Any challenge of the EIR's compliance with CEQA must be filed within 30 days after the notice of determination.[43]

PROCESS STUDIES

The evolution of CEQA has not occurred in a vacuum. In addition to the typical legislative feedback mechanism of citizen communication, CEQA has been the object of three formal studies at the state government level, which have led to changes in the law and its accompanying guidelines. Two of these have been carried on by a San Diego-based consultant for State Assembly committees. The third was the product of the Office of Planning and Research.

Assembly Committee on Local Government Study

This study[44] was commissioned in late 1974 by the committee chaired by CEQA coauthor John T. Knox. Undertaken by Bob Small and associates of a San Diego consulting firm, the study was to make an evaluation of the effectiveness of CEQA, emphasizing local government, before it had become so institutionalized as to be rigidly unchangeable; the study was also to make recommendations as to appropriate changes. The primary source of data for the study was a survey of the 90 percent of CEQA-related activity conducted by local governments and reported in the California EIR Monitor for 1974, as required under the law. Other sources were in-depth reviews of 185 EIRs prepared by 23 cities and counties as well as interviews.

The findings of the report illuminated a number of different aspects of the CEQA experience to that point. Of the approximately 3,800 EIRs prepared in 1974, 90 percent were prepared by local governments and roughly two-thirds were for private projects (see the distribution in Table 4.1). Of the impacts analyzed in the 185 EIRs studied in depth, over 1,100 were examined; but the average per

TABLE 4.1

Approximate Number of EIRs Prepared Annually for Private Projects, by Type of Discretionary Decision Involved

Type	Number	Percent of Total
Tentative tract map	650	26
Tentative parcel map	100	4
Use permit	625	25
Zone change	475	19
Variance	75	3
Grading/site development	175	7
Specific plan approval	250	10
Planned development (planned residential development, planned unit development)	75	3
Miscellaneous	75	3
Total	2,500	100

Source: Environmental Analysis Systems, The California Environmental Quality Act (Sacramento: California Assembly Committee on Local Government, 1975), vol. 1, p. 256.

TABLE 4.2

Costs of Environmental Review

	Private Projects		
Category of Cost	Cost to Private Applicant or Consumer*	Cost to Public	Public Projects, Cost to Public
Document preparation, administration, and review	Results from supplying information, document preparation, and/or fees	Document preparation, administration, and review	Document preparation, administration, and review
Delay			
Inflation effect	Affects price, true effect on cost must take into account increased ability to pay	None	Affects price, true effect on cost must take into account increased ability to pay
Carrying or forgone opportunity	Carrying cost during period of delay and forgone opportunity on benefits of use	None	Forgone net benefits from project and carrying cost in certain cases
Uncertainty	Amortization of sunk costs in denied or abandoned projects; forgone benefits on projects not undertaken	None	Amortization of sunk costs in denied or abandoned projects; forgone benefits on projects not undertaken
Mitigation	Cost of design or operating changes to mitigate impacts	Cost to mitigate impacts	Costs of design or operating changes to mitigate impacts

*Costs in this column will be absorbed by the developer in the short run but may be assumed to be passed on to the consumer in the long run.

Source: Environmental Analysis Systems, The California Environmental Quality Act (Sacramento: California Assembly Committee on Local Government, 1975), vol. 1, p. 39A.

TABLE 4.3

Estimated Approximate Incremental Costs of Implementing the CEQA

Category of Cost	Private Projects: Cost to Private Applicant or Consumer	Private Projects: Cost to Public	Public Projects, Cost to Public
Document preparation, administration, and review	$10 million	$9 million	$15 million
Delay			
Inflation effect	$30 million impact on price	None	Less than $11 million impact on price
Carrying or forgone opportunity	$21 million	None	Less than $11 million
Uncertainty	Perhaps $2 million owing to project abandonment, remainder unknown	None	Unknown, but believed to be low
Mitigation	Less than $2.5 million	Unknown, but believed to be low	Unknown, but may be substantial

Source: Environmental Analysis Systems, The California Environmental Quality Act (Sacramento: California Assembly Committee on Local Government, 1975), vol. 1, p. 39B.

EIR was six, and about 30 general types of impacts accounted for 70 percent of those identified, led by traffic, air pollution, aesthetic considerations, general habitat degradation, and water runoff and pollution. Small found some evidence of mitigation: about 31 percent of the projects examined included at least one instance of developer-initiated mitigation and 30 percent of the projects were approved under conditions of some mitigation.

Small found several areas where the impact on practice was less than desired. He found few examples of increased environmental awareness being translated into content of plans or policies. He also found few examples of the negative impacts of plans or policies being identified or mitigated as a result of the EIRP. He attributed much of the latter to the project-by-project bias of the techniques developed.

Small found three areas of difficulty with CEQA analysis. One was the feasibility of mitigation because of the regional nature of some impacts and because of the effectiveness of actions that local governments could insist on. A second was the utility of the analysis and the amount of marginally relevant data present in EIRs. The third was the area of costs, for which a framework and some estimates (reproduced here as Tables 4.2 and 4.3) were assembled, the result of a long series of estimates made in one volume of the study. The estimate by Small of the effect of CEQA on the cost of an "average dwelling unit" was about $150.[45]

The report also indicated a series of findings on some other aspects of CEQA operations. It found public hearings with substantial input on roughly 50 percent of the EIRs. It suggested that in roughly 25 percent of those cases studied, the negative impacts were balanced by the mention of positive ones as well.

Generally, then, the 1975 report found CEQA to be affecting decision making through systematic identification of impacts, while the costs of preparation were moderate. A series of recommendations were made suggesting differentiating regulatory aspects of the guidelines from advisory provisions, developing minimum standards for impacts to be used by local governments, the removal of the analysis of short-term versus long-term trade-offs from the analysis of private projects, and the setting of time standards for processing of the environmental review. This last recommendation formed the foundation for the series of time limits imposed on the process by 1975 amendments.

The Office of Planning and Research Study

When the Brown administration took office in 1975, one of the first issues it addressed was CEQA. The Office of Planning and Research (OPR) initiated a study of the CEQA experience, conducted primarily internally, which was published, along with a set of proposed guideline amendments, in March of 1976.[46] The study was based upon OPR surveys of state and local government CEQA experiences, a series of four workshops held around the state, and a commissioned field study of CEQA operations in 15 local agencies conducted by the Public Policy Research Organization at the University of California at Irvine.[47] In addition, the study made use of the Small study described above.

The study generally found that governmental decisions had improved in environmental sensitivity as the result of CEQA and that the flexibility in implementation was a positive aspect of the law. A series of deficiencies were also identified. EIRs were found to be overly technical and lengthy. Analysis requirements were the same regardless of scale and type of activity. The analysis often came late in the project design process. Some uncertainty was present regarding legal adequacy of review. Intergovernmental review and comment often was not helpful in the analysis. Environmental review was criticized as not integrated with land-use planning and regulation.

A series of proposals were made in the report, several of which were adopted in guidelines. The report recommended the development of master environmental assessments at both state and local levels to provide a basis for EIR preparation and for incorporation by reference. It also recommended an increased state effort in EIR monitoring, review, and comment. Finally, in a series of proposals adopted to a large extent in the guidelines, the report recommended the use of Initial Studies to support the significant effect decision, the limitation of EIRs to potentially significant effects, joint agency preparation of EIRs where jointly responsible, and the provision for a phased EIR for large projects, with a general project EIR and supplementary detailed EIRs for phases.

Assembly Committee on Rules Study

In 1979, the Assembly Committee on Rules commissioned a second study to be headed by Bob Small, which was to be a replication of the earlier study after five years.[48] That study, published in June 1980, provided a longitudinal review of the changes in CEQA over the period of its study and made a series of recommendations for changes as well. The study was based on a survey of the same 23 cities and counties of the earlier study and on a review of EIRs from 1979, with 116 studied in depth.

In a series of findings on the nature of the reviews, the study reported that while there was a 14 percent reduction in public project

EIRs, the number of private ones remained the same. The number of Negative Declarations on public projects decreased slightly, while the number on private projects increased dramatically, owing, it is suggested, to the use of conditional Negative Declarations rather than full EIRs. The number of impacts being analyzed increased. More than half the jurisdictions were found to be complying with the time limits set under the 1976 amendments.

Some resistance to change was found in the process. While the report suggested that EIRs had improved in quality, there was little improved data support. Responsible agencies continued to contribute little through review and comment; little state comment occurred either. Small and associates indicated that the lack of ability to adopt policies and criteria for determining significance created much of the attendant uncertainty.

As in the earlier study, the cost framework developed was applied. The summary table from the report comparing the 1974 and 1979 experiences is reproduced here as Table 4.4. Costs were found to have shifted to private parties from local governments, but net cost per EIR had not changed significantly. Delay costs had increased substantially with Negative Declaration average time reduced to about one month from two, but EIR's delay increased from four months to seven. The largest increase was in housing and land development activities. The costs of mitigation were found to be small and relatively unchanged. Rate of abandonment had not changed significantly. No method for calculating forgone project cost was attempted.

The report concluded that local agency review tended to impose special conditions on a project-by-project basis, and that no translation into policies or ordinances of the increased environmental awareness was occurring. This leads to similar projects receiving vastly different treatment from agency to agency as well as significantly different reviews of different classes of projects in the same agency. Regional or cumulative impacts continued to be poorly treated. The state agencies' comments provided little assistance. The report paints the local government review as essentially autonomous, with full responsibility for weighing and treating impacts on the local level, a reduced sense of responsibility on the parts of local agencies as the result of the shift of costs of EIRs to the private proposer, and with a view of the time limits as minimum rather than maximum standards.

The study identified three options the legislature might use to respond to the findings: continuation of local autonomy, increased state agency guidance on regional environmental impacts, and increased local responsibility through use of the General Plan Amendment process. This third option was the report's favored response. A proposed approach recommended development of a complex series of standards and criteria for evaluating impacts of types of projects,

TABLE 4.4

Findings Relative to Volumes and Costs of Document Preparation and Processing, from a Sample of California Cities and Counties

Factor	1974[a]	1979	Findings
Average number of private project EIRs per agency per year	24.2	23.2	No significant difference overall but substantial changes in certain agencies
Average number of public project EIRs per agency per year	9.1	5.5	Significantly lower in 1979
Average number of EIRs (both public and and private) per agency per year	33.3	28.8	Significantly lower owing to public projects
Average number of private project NDs per agency per year	141.9	222.6	Significantly greater
Average number of public project NDs per agency per year	61.0	28.0	Significantly fewer
Average number of NDs (both public and private) per agency per year	202.9	250.6	Significantly higher owing to increase in private project NDs
Average net costs to agencies for private projects	$64,045 ($96,387)	$10,957	1979 costs significantly lower
Average net costs to agencies for public projects	$71,943 ($102,274)	$60,090	1979 average is lower but not statistically significant
Average net costs to agencies for both private and public projects	$135,988 ($204,661)	$71,048	1979 costs are significantly lower

Total document preparation and processing costs to private applicants	$90,825 ($136,691)	$218,754	1979 costs are significantly higher
Total document processing for private projects	$154,870 ($233,078)	$229,711	No significant difference
Overall average cost per EIR for private projects	$6,655 ($10,015)	$12,420	1979 average is higher but not statistically significant
Overall average cost per EIR for public projects	$6,054 ($9,111)	$14,188	1979 average is significantly higher
Overall average cost per EIR for all types of projects	$6,568 ($9,888)	$12,504	1979 average is higher but not statistically significant
Weighted cost per EIR for private projects	$6,339 ($9,540)	$9,880	Significantly higher[b]
Weighted cost per EIR for public projects	$7,906 ($11,899)	$10,925	1979 significantly higher[b]
Weighted cost per EIR for all types of projects	$6,803 ($10,238)	$10,077	1979 significantly higher[b]

[a]1974 costs expressed in 1979 dollars are shown in parentheses in the 1974 column. The "California CPI," as prepared by the State Department of Finance, was used for this adjustment, providing an index of 142.6 in mid-1974 and an index of 214.6 in mid-1979. Accordingly, 1974 costs in 1979 dollars = 1974 costs × 1.505. Two types of unit costs were calculated. Overall average unit costs are obtained by taking the simple arithmetic average of the average unit costs calculated for each agency in the sample. Weighted average unit costs are obtained by weighting each agency's average unit cost by the ratio of the EIRs prepared by that agency to the total EIRs for all agencies in the sample.

[b]The authors disagree with this report finding.

Source: Small and Knust, Inc., California Local Government and CEQA, 1979 (Sacramento: California Assembly Committee on Rules, June 1980), pp. 815-16.

including standard mitigation measures. The 1981 legislature did little to implement any of the suggestions presented though the amendment of CEQA by Assembly Bill (AB) 1185 for residential projects and community plans appears to be partially inspired by the report.

The three reports prepared for the state give a view of CEQA as a living, evolving body of law. The most recent report would, however, suggest that local government autonomy over the process has some negative effects in its current form. It is, of course, this local government autonomy that is of most interest to this study on housing costs. Consequently, the chapter now turns to a survey of the range of this local activity in a number of areas CEQA leaves as discretionary.

A SURVEY OF CITIES AND COUNTIES

Given the relative local government autonomy identified in the last Small report and the locus of housing-related approval at the local level, additional research was conducted into the nature of the implementation of the process at this local level. The only detailed work in the area was conducted by researchers at the University of California at Irvine early in the history of CEQA.[49]

In addition to a series of more in-depth case examinations of selected jurisdictions, a telephone survey of planning officials responsible for environmental review in 13 counties and 19 cities, containing over 70 percent of California's population, was conducted; it concerned itself with the jurisdiction's approach to a number of discretionary choices within California's EIR process. The jurisdictions are listed in Table 4.5. A series of questions were posed regarding the processing of a hypothetical 50-lot subdivision; such a decision, at the preliminary approval stage, is the one that impacts the greatest number of single-family housing units directly and is the largest single class of private project approvals falling under CEQA, according to the Small study.

Concentrating upon the review process, the survey was constructed to ask about jurisdiction activities at major discretionary steps. As may be recalled, the CEQA guidelines specify a set of steps and a framework for review but leave individual agencies free to make choices of form within that framework. The 1979 Small report to the Assembly Committee on Rules cited this as a major concern while not reporting the variations.

Findings

Results of the survey reveal that on many of the discretionary steps there is a substantial commonality of approach among the local

TABLE 4.5

Jurisdictions Surveyed by Environmental Review Procedures under CEQA

Counties	Cities	
Alameda	Anaheim	Pomona
Contra Costa	Antioch	San Bernardino
Los Angeles	Camarillo	San Diego
Marin	Carlsbad	San Jose
Napa	Emeryville	Santa Rosa
Orange	Escondido	Ventura
San Bernardino	Irvine	
San Diego	Long Beach	
San Francisco*	Los Angeles	
San Mateo	Oakland	
Santa Clara	Oceanside	
Sonoma	Oxnard	
Ventura	Palo Alto	

*San Francisco City and County are one government.

Source: Survey by the authors, 1982.

governments. They also indicate some choices in which a wide disparity of experience is present.

One way in which a residential project is not required to proceed with the CEQA review is for it to fall under the specific-plan approach. Under the 1978 amendments to CEQA, no EIR is necessary in a residential construction project in an urban area if a specific plan (as defined under the state government code adopted within the last five years) accompanied by an EIR is in place covering the parcel.[50] The idea is that the specific plan includes a detailed setting of land use by parcel in the covered area, and detailed regulations and policies were developed to cover the uses. Of the jurisdictions surveyed, over half (18) indicated that the specific-plan approach was in use. Of those not currently using it, another six indicated plans to utilize this means of avoiding the project-by-project approach.

The Initial Study is the first document prepared in the process and is utilized to determine if any significant effect will occur. The guidelines provide a set of suggested forms for that study (reproduced in Appendix D). Of those jurisdictions surveyed, 10 utilize the guideline forms, while 22 have produced their own version largely based on the forms. The Initial Study draft is prepared by 23 of the jurisdictions, while developers prepare the draft in seven (two have joint mechanisms). There is an additional flow of information from developer and/or agency in virtually all jurisdictions surveyed.

The openness of the process and use of the Initial Study can determine both the public response later and the coverage of reviews subsequently carried out. In 20 of the 32 jurisdictions surveyed, public input is solicited in some form at this stage. Rarely, however, is this input in a public hearing. While some requirement for such input did exist for a while after the Woodland Hills case, it is no longer a necessity under the law. Virtually all jurisdictions examined utilize the Initial Study to focus the subsequent EIR and attendant review on particularly relevant impacts. Almost all make suggestions as to changes in projects at this early stage before the intense formal review begins. An attempt was made to discover what percentage of these changes, in the respondents' opinions, would reduce density of the proposed hypothetical subdivision. Of the 17 jurisdictions answering, eight indicated density reduction in 20 percent or more of the changes, while nine indicated density reduction in 5 percent or less. However, five of six northern California jurisdictions answering were in the 20 percent group, while eight of eleven southern California replies were in the 5 percent or less category. This would indicate, in an informal way, that density would be a larger concern in northern California decisions at this stage. Finally, 22 of the jurisdictions have adopted a formal internal comment process for other agency input at this stage.

The Negative Declaration decision—that a project has no significant effect on the environment—is a major exit from the process. Indeed, the Small report indicated that many jurisdictions were utilizing this decision to avoid the review required.[51] Our survey indicates that this important decision is made by the planning staff in 23 of our subject jurisdictions, by planning commissions in two, and by environmental review committees in three others. Our survey indicated that in roughly three-fourths of the projects reviewed, Negative Declarations were issued. The breakdown of averages by city/county and by part of the state is indicated in Table 4.6. Northern California counties appear to utilize this Negative Declaration least, thus yielding full EIRs in the greatest percentage of cases. Finally, the use of the Mitigated Negative Declaration—in which the government basically said to a developer that "if these changes are made, there will be no significant effect"—was formally included in the guidelines in January 1982, after informal use for some time. Of the jurisdictions surveyed, 27 of 32 utilize the Mitigated Negative Declaration, with only two indicating no future plans for its use and many indicating planned increased usage.

TABLE 4.6

Environmental Review Process Survey, Selected City/County and Regional Average Comparisons

	Negative Declarations (in percent)	Time to Negative Declaration (in days)	Time to EIR (in months)
Northern California			
Counties	66 (N = 7)	73 (N = 7)	6¼ (N = 7)
Cities*	81 (N = 7)	45 (N = 7)	9½ (N = 5)
Southern California			
Counties	80 (N = 5)	37 (N = 5)	7 (N = 4)
Cities	74 (N = 13)	44 (N = 13)	6½ (N = 10)
Sample averages	75 (N = 32)	49 (N = 32)	7 (N = 26)

*San Francisco, owing to its built-up structure and lack of room to expand, is considered a city in this analysis despite its legal status as both city and county.

Source: Survey by the authors, 1982.

The EIR itself, if required, may be approached in a number of ways. Its draft preparer is subject to the choice of the agency. In our survey, 29 of 32 jurisdictions utilize consultants to prepare the EIR draft. In only three is the draft report prepared internally. The Small report of 1980 criticized the manner in which private developers are charged for the costs of this essentially public document. In our survey, only one jurisdiction did not charge a fee for the preparation. Flat fees and schedules of fees based on development size were charged by eight of the respondents. Some 21 of the jurisdictions charge developers for the EIR on a cost or cost-plus-markup or cost-plus-hourly-fee-for-staff-time basis. The costs of the preparation are, thus, being passed through in most cases.

The contents of the EIR itself in terms of emphasis and approach, while generally described in the law and guidelines, are left to the local agency. While our survey indicated few special foci for EIRs in subject jurisdictions, those that did respond with special concerns (seven) all included traffic congestion as the one most troublesome, followed by erosion and runoff. No particular methodology was favored in the reports. As required by CEQA, 28 of 31 indicated a major emphasis upon mitigation in the reports. In 20 of 30 responding jurisdictions density was considered a major factor, with northern California jurisdictions (12 of 14) more likely to consider it major than southern California jurisdictions (8 of 16). Incorporation of other documents, especially of other EIRs, by reference in EIRs is a way in which the bulk of the documents may be reduced. While 24 of 31 responding agencies indicate attempts to encourage this, it appears to be infrequent given the tone of the answers.

Environmental Impact Report review and comment procedures are the heart of the provision for public input on environmental evaluation of a project. The review of the draft EIR for adequacy occurs in-house at the staff level in all jurisdictions. Formal policies for circulation of the drafts to other interested agencies exist in 22 of them. Public access to the documents may be provided under the legal framework through at least one of three means: posting on the property, mail notification to adjacent landowners, or newspaper publication of announcements. Our survey indicates, however, that jurisdictions have chosen to provide wider circulation of the draft EIR than required. In the survey, 27 of the agencies provide for at least two means of public access. Five of these provide for three, and three provide for four. The most popular means for public access are placement of drafts in public libraries and newspaper ads. Only two jurisdictions utilize posting as permitted by the law. All of the jurisdictions solicit written comments, and 28 of 32 hold hearings on the draft EIR even though that is not required under CEQA.

Finally, the total time to process the Negative Declaration and EIR certification was examined. The city/county breakdown and regional breakdowns are reported in Table 4.6. The average time for a Negative Declaration is 45 days in our sample; the average time for a certified EIR is seven months. This is roughly in keeping with the times reported in both anecdotal evidence gathered for this examination and that reported in other research. Northern California counties appear to be the slowest in granting Negative Declarations, while northern California cities appear slowest in producing completed EIRs.

Discussion

Though our findings are somewhat sketchy and the data are insufficient for detailed statistical analysis owing to both sample size and data quality, we can make some observations regarding the implementation of CEQA by those jurisdictions surveyed. While not providing all necessary data, the survey provides a profile of the approach taken by those communities to the discretionary steps in the process.

It would appear that there is a substantial intention on the part of those surveyed to utilize the specific-plan approach. This requires a significant effort on the part of the local governments to be carried out. Specific plans, if they conform to the legal framework, provide a detailed bridge between the required general plan and the more detailed regulation. However, the specific plan, treating a number of properties in a detailed manner and carrying some regulatory weight itself, will be subject to close scrutiny by all involved; its implementation and successful defense in court, especially in areas of environmental concern, could well be more difficult than the current project-by-project approach to environmental review. And, of course, it is inapplicable except in urban areas. Nevertheless, this expansion could lead to a decrease in required EIRs.

The Initial Study has developed much as it would appear its designers would have wanted. Guideline forms have been adopted or formed a basis for the study. Some public input is solicited, even though none is legally required currently at this stage; perhaps the period when it was required after Woodland Hills has left a residue. The Initial Study is used to focus the later inquiry and to suggest changes in the project before proceeding to the main stages of review, hopefully cutting down on the expense and frustration.

Density considerations are an important focus for the Initial Study and for the review process as a whole. This is because decreases in density have been found to be a contributing factor to land-use impacts on housing costs. In our survey we asked for indications of density concerns at both the Initial Study stage and in the EIR. Our results indicate that in the northern California portion of our sample, which falls mostly in the environmentally conscious and more densely populated San Francisco Bay area, jurisdictions appear to be more concerned with the density impacts of projects both at the Initial Study stage and on EIR review. This density focus may be attributable to local political needs, geographic considerations in the bay area, ecological sensitivity of the reviewers, or a combination of the above. We would suggest that the openness of the development review process, as encouraged by the environmental review, would be an important contextual factor contributing to this emphasis on density.

It is at the Negative Declaration that the bulk of those projects that fall under CEQA exit the review process. It is, in most jurisdictions, a staff decision that occurs in roughly 45 days for our hypothetical subdivision. Since most subdivisions occur at the urban fringe in cities or at that fringe in unincorporated areas of counties, the county data here are significant. It would appear that the northern California counties are most reserved in their use of the Negative Declaration and most deliberate in their reviews. This is indicated by an average percentage of Negative Declarations lower than that of other parts of the sample and a mean processing time that is somewhat higher. The new Mitigated Negative Declaration, which allows an exit from the EIR in teturn for a set of changes in the project, promises to increase the use of the Negative Declaration in general while somewhat extending time, as a negotiation process between developer and jurisdiction is certain to develop here.

The EIR itself, for those projects that fall under the requirement, is a major delay. Our data show great disparity between localities in expected times for an EIR certification for the hypothetical project used. They run from two months to almost a year. The process in virtually all communities is very open, with a greater circulation of reports than required and public hearings. There is little incentive for the localities to tightly control preparation, as costs are generally passed directly back to the developer. As required by the law, mitigation measures are emphasized, and density is, as we have indicated, often a major concern. However, the efforts to encourage incorporation of information in other documents, especially EIRs, is not proceeding in the manner that framers of the provision of the act desire.

In an institutional sense, the data would seem to indicate that a very important aspect of dealing with the review process for both developer and local government is the close, ongoing contact of developer and planning staff. Indeed, the rapport between the two may be crucial to the flow of information needed to negotiate the process successfully. Changes in the project at the initial study phase, granting of a Negative Declaration by the staff, and content of the EIR are all greatly dependent upon this contact. One can expect, with the probable increased use of Mitigated Negative Declarations, which generally bypass public hearings and broad distribution and are destined to be largely mitigated in accordance with planning staff concerns, that contact with the staff will be even more significant.

One perhaps unintentional side effect of this importance of the staff-developer interface is the creation of a set of barriers for entry into the market. The experienced, well-connected developer has an advantage over the out of town or new developer. This may well return extra profits to certain individuals in some areas. As a counterpoint to this, a new profession is growing up in Los Angeles and some other jurisdictions studied. An individual, well connected in necessary planning circles in the locality, sells services as an "environmental facilitator" and provides the communications linkage and knowledge base necessary to negotiate the process successfully.

Though not a major force at this time, such individuals may become essential to successful development in the future.

This chapter has reviewed the process mandated by CEQA, those studies of the process that have been carried out for the state, and our supplementary survey, which provides some increased information as to the choices made by selected localities in implementing the discretionary steps of the process. We turn from here to a theoretical consideration of the process costs before evaluating the California process for potential changes.

NOTES

1. Cal. Pub. Res. Code § 21082.
2. Ibid., §§ 21067, 21165.
3. Ibid., §§ 21085, 21060.3, 21172.
4. 14 Cal. Admin. Code § 15071.
5. Ibid., § 15072.
6. Ibid., § 15060.
7. Selma Bendix, "A Short Introduction to the California Environmental Quality Act," Santa Clara Law Review 19 (1979): 521.
8. Cal. Pub. Res. Code § 21167.
9. Ibid., § 21068 and 14 Cal. Admin. Code § 15040.
10. Cal. Pub. Res. Code § 21060.5 and 14 Cal. Admin. Code §§ 15026, 15040.
11. Cal. Pub. Res. Code § 21083 and 14 Cal. Admin. Code § 15082.
12. 14 Cal. Admin. Code § 15080(a).
13. Ibid., § 15080.
14. Ibid., § 15080(f).
15. Ibid., § 15083(d)(1).
16. Ibid., §§ 15083(d)-15083(e).
17. Ibid., § 15066.
18. Ibid., § 15080(d)(2).
19. Cal. Pub. Res. Code § 21080.2.
20. Ibid., § 21151.5.
21. Ibid., § 21152.
22. 14 Cal. Admin. Code § 15083(f).
23. Ibid., § 15081(a).
24. Cal. Pub. Res. Code § 21082.1.
25. 14 Cal. Admin. Code § 15053.
26. Ibid., § 15050.
27. Ibid., § 15061(b).
28. Cal. Pub. Res. Code § 21100 and 14 Cal. Admin. Code §§ 15140-15146.

29. 14 Cal. Admin. Code § 15149.

30. Ibid., § 15011.6(d)-15011.6(e).

31. Cal. Pub. Res. Code § 21080.3.

32. Woodland Hills Residents Association v. City Council 26 C.3d 938 (1980).

33. AB 710 (1981), codified as Cal. Pub. Res. Code §§ 21083, 21087.

34. 14 Cal. Admin. Code § 15085(d)(2).

35. Ibid., § 15085.

36. Ibid., § 15060.

37. Ibid., § 15165.

38. Ibid., § 15146.

39. Ibid., § 15085(g).

40. Ibid., § 15089(a).

41. Cal. Pub. Res. Code § 21081.

42. Ibid., § 21151.5.

43. Ibid., § 21167(b).

44. Environmental Analysis Systems, The California Environmental Quality Act, 3 vols. (Sacramento: California Assembly Committee on Local Government, 1975).

45. Ibid., vol. 1, p. 78.

46. California Office of Planning and Research, The California Environmental Quality Act: A Review (Sacramento: State of California, Office of Planning and Research, 1976).

47. For other parts of the Public Policy Research Organization research, see Ralph Catalano and Richard Reich, "County Processing of EIR's," California EIR Monitor 1 (1974): 4-20; and Ralph Catalano and Richard Reich, "Local Government and the Environmental Impact Report: The California Experience," Urban Lawyer 8 (1976): 367-74.

48. Small and Knust, Inc., California Local Government and CEQA, 1979 (Sacramento: California Assembly Committee on Rules, June 1980).

49. See notes 46 and 47.

50. Cal. Pub. Res. Code § 21080.7 and Cal. Gov't Code §§ 65450-65453.

51. Small and Knust, Inc., Local Government and CEQA.

5

INFORMATION AND ENVIRONMENTAL IMPACT REPORT CONTENTS

The literature on the relationships between environmental impact review processes (EIRPs) and housing prices is polarized between that which is strongly anecdotal and empirical and that which is highly theoretical—with most of the theoretical materials derived from some form of welfare economics reasoning. However, an important conclusion can be derived from the theoretical literature: if consumers were better informed about the costs and benefits associated with externalities, they would probably accept higher home prices in return for some environmental controls. When home buyers and homeowners can see that relatively modest price increases (and costs of owning) produce clearly apparent environmental improvements, their willingness to pay (that is, marginal willingness) will increase.

The anecdotal evidence is a rich mine of information about what costs builders will accept and what prices buyers will pay for given levels of environmental attractiveness under varying market conditions. We also observed that builders, even land-use planners and regulators, were basically unaware of this willingness of home buyers to pay more. We also believe that buyers and sellers thought home prices were unnecessarily high because of their ignorance about environmental conditions, mitigation measures, and the possible trade-offs between Environmental Impact Report (EIR)-associated costs and improved environmental benefits.

This chapter presents an amalgam of theoretical statements that attempt to explain why buyers might continue to demand housing, even when prices are continuing upward, and how better information might produce better informed, more rational actions by buyers, builders, and planners. Then the environmental impact and review processes could be used more efficiently and have less impact on housing prices. In this chapter we make use of some earlier anecdotal materials as well as some that will be used again in later chapters. A basic assumption in the analysis is that even though the Environmental Impact Review (EIR) is a single purpose document, it can be a valuable source of all kinds of information, which would permit buyers and builders to understand and perhaps accept higher home prices in return for a much improved environment. A well-organized EIR will present a systematic appraisal of all benefits and costs of contemplated courses of action. In other words, an EIR should be "nothing more than a description of a rational behavior. . . . Benefits are the value of a good or service to a consumer [or society] and . . . costs are benefits foregone [private or public]. . . . Most environmental problems arise because modern production and consumption techniques of firms and households impose waste disposal costs on society. . . . In the broadest sense benefit cost means a systematic evaluation of all the advantages and disadvantages of any actual or hypothetical change in society's production and consumption arrangements."[1]

MARKETPLACE AND WELFARE ECONOMICS

EIRs have developed primarily because marketplace activities were not dealing adequately with the problems of using and allocating scarce natural resources.[2] Population and growth pressures were seen to be using scarce, limited natural resources at rates that threatened extinction or the serious impairment of vital natural resources. "Despite considerable rhetoric describing the benefits of an unencumbered market, past experiences, buttressed by considerable economic theory, suggest that the private market will not adequately preserve our environmental resources. . . . Only those goods which can be owned and used exclusively by the purchaser can be allocated by the marketplace."[3] A primary reason for this failure of the marketplace can be traced to the ability of private developers to shift to others the external costs created by their actions, thus realizing profits from adding costs to others.

ECONOMICS AND INFORMATION

Economists assume that a free, perfect market works only because buyers and sellers are fully informed, which means that EIRs have an important theoretical function—providing information. Perfect market economics also assumes some substitutability among goods if one becomes too high priced. In the case of environmental goods there are no really acceptable substitutes and, if the current, readily available environmental goods are destroyed, they may not be replaceable.[4] Equally important is the danger that if market solutions are relied upon too heavily, they might, in the short run, destroy natural resources that were created on a time scale, which for human purposes must be considered nonrenewable.[5] An EIR should be able to present the trade-offs between an irreplaceable, or replaceable only at high cost, environmental feature and the costs of building and using housing.

In Figure 5.1 we develop a theoretical explanation of the relationships among construction costs, environmental review processes, and information. After that we summarize some findings about costs in order to indicate that the kinds of information assumed necessary in the theoretical model are available and could be used. In the model we proceed from a market in which builders are able to avoid all costs of environmental mitigation and to pass on all of their costs to the buyers (Situation A) to a market in which an effective EIR process provides the information that permits informed actions by buyers, builders, and regulators so that housing prices can be kept down and maximum environmental protection is also provided.

Situation A is like those found in many California communities, where the EIR process is used in a minimal way. New home prices are based on the builder's actual project costs. Later homeowners will have to pay for the public services and facilities and the environmental protections and improvements they want. In such a market, the initial purchase price (P_1) is relatively low or, if high, not as high as it might have been had more externalities been identified and charged to the project. However, as the external costs begin to emerge and become embedded in taxes and other fees and costs, the long-term price of homeownership (PO_1) begins to rise.

In Situation B, markets remain about the same, but the builders must now accept some costs of mitigating some of their project impacts; they are, however, able to pass on these added costs because demand pressures are strong. We found this happening in counties with relatively strong environmental attractions and some modest EIR-related costs.

In Situation C, market demands begin to weaken somewhat; public or external costs have to be absorbed by the buyer somewhat more than was true in Situation B. Home ownership costs (PO_3) continue to rise slightly as owners are forced to pay for more environmental improvement and protection costs. In this market the EIR process becomes known, and several builders leave the market. The builders who remain have a semimonopoly position, because they know how to meet local requirements. Although demand is slightly weaker, new buyers recognize that the environment is improving and are willing to pay the higher costs just to live in that environment.

In Situation D, a full EIRP is used, and information is provided on all of the costs and benefits associated with a given residential

FIGURE 5.1

Housing Market Responses and Environmental Reviews

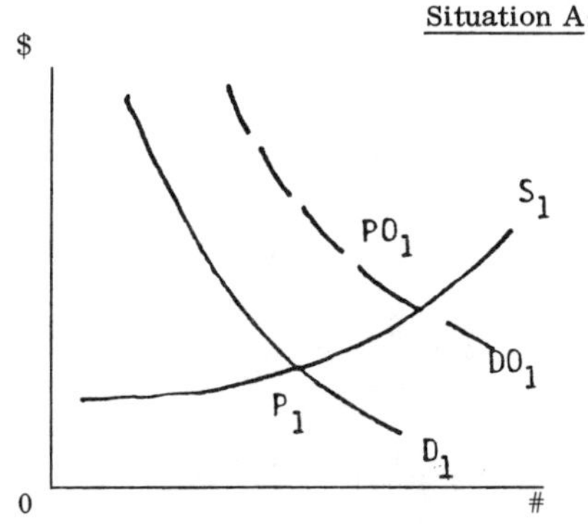

Situation A: Market demand is strong, supply is constrained, and no environmental review processes are used. Builder can and does shift all external costs to present and future buyers.

Results: Builder passes all costs that are project related to initial buyers, including profit that builder wants to make. Eventually, external costs are added to price of owning through taxes, fees, and the like, or levels of public services and facilities decay because neither the builder nor the homeowners want to pay for them.

D_1 = Initial demand
S_1 = Builder supply schedule, only project costs are included
P_1 = Price buyers will pay; price includes only builder costs
DO_1 = Buyer responses once externalities become part of home-owning costs
PO_1 = Eventual price reflecting inclusion of ownership costs with externalities as part of the costs

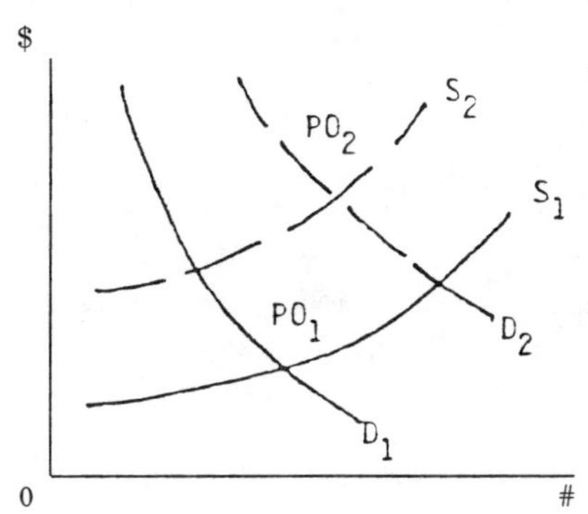

Situation B: Market demand is strong, supply is constrained, and minimal environmental review processes cause some external costs to be transferred to present and future buyers.

Results: Builder can pass on some of the added costs from externalities because the area is still environmentally attractive and demand pressures are strong. Buyers do not resist higher prices too much, because they see environmental controls actually creating a more favorable living environment. At the same time more and more buyers are being excluded by the higher prices.

S_2 = New builder supply schedule reflects added costs because external costs are now added partially to sales price
D_2 = New demand responses reflecting reactions to the slightly higher initial and future ownership costs
PO_2 = New price schedule is slightly higher than in Situation A*

Situation C: Market demand weakens, supply is still relatively constrained, and market conditions require builder to absorb some public-related costs. Environment quality is much like in Situation B.

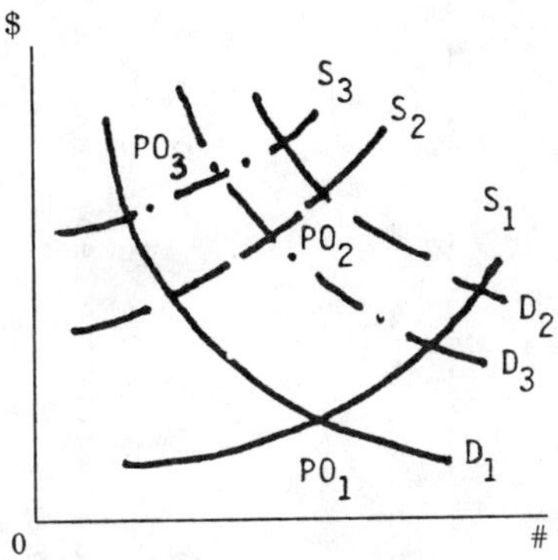

Results: Prices continue up because of continued demand pressures, but rate of increase slows. Buyers receive more value per dollar of price, because improved environment (that is, marginal price increases) match marginal willingness to pay more for environment. Prices are higher than in Situation B.

S_3 = New builder supply schedule
PO_3 = New ownership costs
D_3 = New demand responses reflecting new PO_3

Situation D: Market demand continues to weaken, but supply is still somewhat constrained, and full EIRPs are invoked. Unlike Situation C, the EIRPs provide an equitable distribution of external costs between current and future and other community home buyers.

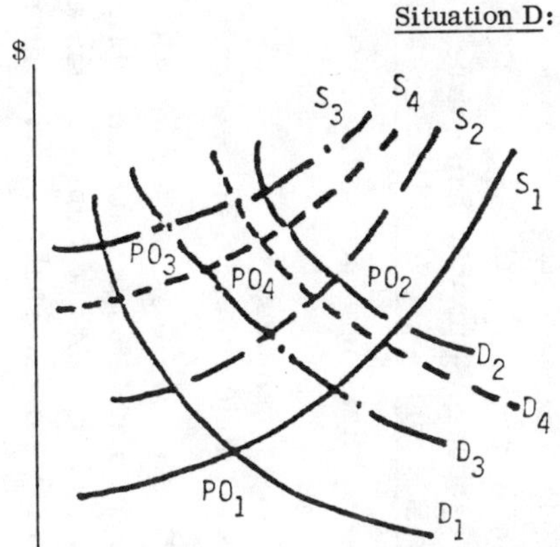

Results: Prices are higher than in Situations A and B but less than in Situation C. Improved public facilities and services maintain neighborhood quality and home prices. More buyers can enter market than in Situation C.

S_4 = New builder supply schedule
PO_4 = New ownership costs
D_4 = New demand responses reflecting new PO_4

*Presumably higher prices should shift the demand schedule to the left from its position in Situation A; however, the comparative environmental attractions of some areas will keep demand strong even though prices increase quite sharply. This was quite evident in those California counties in which we identified situations analagous to the one described as Situation B.

TABLE 5.1

Value of Environmental Impact Review Processes to the Community

Situation A:	$V_{p1} = R_p - C_1(p + e)$	Public revenues are increased somewhat by new construction but are offset by increased public costs because of project externalities.
Situation B:	$V_{p2} = R_{p2} - C_2$(less than C_1); V_{p2} exceeds V_{p1}	Net public revenues increase over Situation A, but externalities from project still keep down public environment and values.
Situation C:	V_{p3} where R_{p3} exceeds R_{p2} C_3 is less than C_2 V_{p3} exceeds V_{p2}	Environment continues to improve, but future owners may not carry all costs that they should. Heaviest burdens for paying for externalities fall on current buyers.
Situation D*:	$V_{p4} = R_{p4}$ which exceeds R_{p3}; C_4 is less than C_3; V_{p4} exceeds V_3 private homeownership costs, which are less than C.	

V_p = net value of residential construction for a given development in terms of assessed values and revenues to the local government.

R_p = revenues to local government
C = costs—private and public (internal and external)

*Result of Situation D: home prices and -ownership costs are higher than in situations A and B but less than in C. Improved environment maintains private and public land values, and costs are more equitably, efficiently distributed between public and current and future homeowners. A balance is reached between the marginal willingness of home buyers to pay for external costs and improved environment.

project. Since costs are now estimated more carefully, pricing of environmental impact mitigation measures becomes more efficient, and public costs are distributed more equitably among buyers, future homeowners, and local government than was true in Situation C. With a more efficient and accurate assignment of costs and better recognition of the full range of benefits associated with the EIRP, home prices tend to decline below the levels of Situation C but remain higher than in Situations A and B.[6]

Conventional economic wisdom might conclude the analysis at this point; however, since one of our objectives is to examine the costs and benefits of EIRPs, Table 5.1 indicates benefits that might accrue to the community. The analysis suggests that as the EIRP improves the overall environment through better analysis of a given project's impacts, public lands and facilities will increase in value, and the property tax base will have a greater value. Also, the EIRPs have provided mitigation measures, which reduce problems and costs of maintaining public services and facilities. The general overall improved public services and facilities help maintain the values of all properties, thus creating a willingness by homeowners to pay some share of the external costs they create.

The point of this theoretical discussion is that public and private decisions about the treatment of natural resources must be based on the best possible information about costs and outcomes. Armed with this knowledge, decision makers can make choices about what they are willing to pay to accomplish given outcomes. The role of the EIR becomes that of providing the information on which such decisions can be made. The remainder of this chapter is directed toward identifying costs associated with private developments so that their impact on housing prices can be anticipated or identified by decision makers. After identifying some of the major types of costs that should be included as information in an EIR, we bring our findings together in the last chapter by adding some ideas about the benefits information that should also be included. Our goal in that last chapter, as it is in the entire book, is to suggest how to formulate informed behavioral rules for organizations to follow to achieve some acceptable level of social efficiency. An example would be the rules that an environmental agency should follow in various situations."[7]

IDENTIFYING COSTS IN THE ENVIRONMENTAL IMPACT REVIEW PROCESS

Required information needed to guide decision makers, who are interested in the role of EIRPs in levels of housing prices and costs, includes an understanding of the process steps. Steps involved in such an analysis are the following:

1. An inventory at the project site must be made of the physical, biological, and cultural environments. The physical environment inventory must consider geology, topography, subsurface and surface water resources, water quality, air quality, and climatology, among other elements. The biological environmental inventory must consider flora and fauna, including tree species, grasses, fish, herpetofauna, birds, and mammals. Reference must be made to any rare or endangered plant or animal species. The cultural environmental inventory must report on human population trends and distribution, historic and archeological sites, and economic welfare indicators.

2. An environmental assessment must be made, which evaluates the consequences of the proposed project on each of the environmental elements described in the environmental inventory. The assessment requires technical input from many disciplines—engineering, biological, geological, geographical, and so forth—on the possible impact of the project on the items listed in the inventories (for example, possible impacts on air or water quality, sewage disposal and treatment, noise levels, educational facilities, and road systems).

3. A written report must be prepared, which enumerates the actions to be taken based on the information provided in the environmental inventory and assessment analyses. This statement is first prepared in draft form for comment by various affected public agencies and the residents in the communities likely to be impacted by the proposed development.

The actual content of an EIR may vary, depending upon local needs, the specific project, and the legal requirements. After public hearings, a final report is prepared, which includes comments of those appearing at the public hearings and a discussion of the concerns raised by those who have reviewed the report.

Although the requirements for environmental review are set by federal and/or state regulations, implementation of the requirements is carried out by local agencies, usually the Planning Department and Commission. Because specific standards cannot be enumerated in connection with EIRPs and reports, reviews may become judicial in nature. Since comments on the process and reports may be made at several points in the processes by any affected public agency, private citizen, organization, or business, the processes can be costly and time-consuming. The processes often become quite meticulous and slow because of the numerous places where intervention can occur and because of the potential that any case being considered might be taken through court proceedings to the U.S. Supreme Court.

EIRP COSTS AND OTHER CONSTRUCTION REVIEW PROCESSES

Environmental review processes are interwoven with more conventional planning and building construction review processes; an often overlooked cost of the EIRP is that conventional local processes have become more complex and costly to review, implement, and comply with. Typically then, a construction project requiring an environmental review must also include the following.

1. Conformity with a general and specific plan and requests for variance when project plans do not conform to planning requirements,
2. Conformity with local zoning and land-use requirements or request for variance from these,
3. Grading and building permits,
4. Periodic inspection and compliance requirements,
5. Certificates of occupancy before any sales can be made,
6. Approval by the California Department of Real Estate for compliance with the California Subdivided Land Acts if a residential development includes five or more lots (in California).

The net impact of these total requirements is that approvals and reviews, which once could be completed in days or weeks, sometimes stretch into months and years. The contents for a typical EIR are shown in Table 5.2. The actual contents are usually specified by an agency designated as the "lead agency." The specified format must be followed strictly, but additional subject matter can be added under the required headings. In some cases the applicant must supply all of the information; in other cases some information may be provided by the local jurisdiction. Preparation can require weeks to months and can cost a few hundred to several thousand dollars. A major project involving hundreds of housing units or large-scale industrial or commercial developments can literally require EIRs of several volumes, which extend for two or three or more feet on a bookshelf, cost thousands of dollars, and take at least a year to prepare.

The extent to which the EIRP can be embedded in the overall construction review processes is suggested in Table 5.3. Processes cited here are those used in Orange County, California, and reflect those which might be found in any county with a reasonably large population, housing inventory, and growth potential. A builder may face these requirements in any of at least eight construction situations: seeking an amendment to the general plan; requesting a zone change; planning a subdivision; when an EIR has been found necessary; re-

TABLE 5.2

Typical Table of Contents for an Environmental Impact Report

Summary

Section I. Description of the project

- A. Statement of objectives
- B. Location and boundaries
- C. Description of project characteristics

Section II. General description of environmental setting

- A. Overview of the environmental setting
- B. Related projects

Section III. Environmental impact analysis

- A. Physical features
- B. Land use and zoning
- C. Circulation
- D. Air
- E. Noise
- F. Service systems
- G. Public facilities
- H. Paleontological, archeological, and historical
- I. Socioeconomic

Section IV. Long-term, irreversible, and growth-inducing impacts

- A. Short-term uses versus long-term productivity
- B. Irreversible environmental changes
- C. Growth-inducing impacts
- D. Cumulative impacts

Section V. Alternatives to the proposed action

- A. Alternatives
    1. No project
    2. Changes of intensity
    3. Changes of land use
    4. Alternative sites
- B. Project justification

Section VI. Organizations and persons consulted

TABLE 5.3

Local Permit and Review Processes for Residential Construction
(Orange County, California)

I. General Plan Amendment

(CP)	A. Amendment to general plan is initiated by the developer
(CP)	B. Applicant prepares petition to amend general plan and draft initial study
#(CP)	C. Applicant reviews EIR requirement with environmental services division
(CP)	D. Environmental Management Agency (EMA) prepares review schedule in January for ensuing 18 months
(CP)	E. Schedule is submitted to Planning Commission for review and comment
(CP)	F. Schedule is submitted to Board of Supervisors, and schedule is published
#(EMA)	G. EMA staff processes the land-use element
#(EMA)	1. Other elements of the general plan are processed using the same procedures. The EIR may be combined in the procedures
(SR)#(EMA)	H. Simultaneous processes
#(EMA)	1. Staff review of amendment proposals
(SR)	a. Alternatives to proposed amendment developed
#(EMA)	b. Staff report prepared on amendment proposals, including alternatives
(SR)	2. Initial Study prepared
(SR)	a. Initial Study prepared
#(SR)	b. Determination of need for an EIR (Negative Declaration will stop EIR review processes)
#(SR)	c. Draft EIR prepared and circulated
#(SR)	d. Final EIR prepared
(EMA)	I. Planning commission meetings for amendments, sometimes termed a study session. A public meeting with informal public discussion
	1. Planning commission may have a final hearing
(SR)	a. Planning commission makes recommendation on EIR and amendments
(SR)	b. Board of Supervisors has hearing on amendments
(SR)	c. Board of Supervisors takes final action
#(EMA)	2. After hearing, Planning Commission may choose to reschedule for further consideration, with no action at this time
(SR)	3. Board of Supervisors approves the EIR
(SR)	4. Effective immediately, two-week delay until zone change may be initiated

II. Zone change

(CP)	Developer meets with advance planning section, environmental services division to determine if EIR is required
(CP)	A. Determination that EIR is required
(CP)	1. Draft EIR must be prepared, or
(CP)	2. Categorical exemption provided, or
(CP)	3. Negative Declaration is made
#(EMA)	B. Developer prepares and submits application for zone change
#(EMA)	C. Formal application for zone change may not be filed until draft EIR has been submitted for review (if required)
#(EMA)	D. Application reviewed by current regulation division of EMA
(CP)	E. Staff determines whether the application is complete, consistent with general plan, and completes the necessary EIR procedures

---

(continued)

#(EMA)	F.	Request or application is issued a case number
#(EMA)		1. Referral letters requesting comments are sent to various affected agencies and organizations
		2. Scheduled for public hearing
#(CP)		a. Hearing date will be affected by a 35-day review period of a draft EIR (if required)
#(CP)		b. Notices are sent to the developer as well as property owners within 300 feet of proposed project
#(EMA)		3. Current planning staff prepares written analysis of zone change request
#(EMA)		4. Final staff report
(CP)		5. Responses from agencies or organizations
#(CP)	G.	Draft EIR (if required)
	H.	All returned responses, final staff report, and EIR submitted to the Planning Commission
#(CP)	I.	Planning Commission hearing
		1. Case may be continued for additional information, or
		2. Denial of zone change (case considered closed unless appealed in 14 days), or
		3. Commission recommends approval
(CP)	J.	Planning Commission makes recommendations
(CP)	K.	Distribution of required hearing notices
(CP)	L.	Case is transmitted to Board of Supervisors
(CP)		1. Denied zoning appeal may be appealed directly to Board of Supervisors
(CP)	M.	Board of Supervisors hearing
		1. Case continued, or
		2. Case denied (this decision is final or, if plan is modified, it must be referred back to the Planning Commission), or
	N.	Board of Supervisors adopts Planning Commission report
(CP)	O.	The ordinance establishing the new zoning district will become effective 30 days after its approval

III. Subdivision flow chart

Requirement before filing		
#(CP)	A.	Environmental determination (EIR)
(CP)	B.	Map needs to be consistent with general plan
(CP)	C.	Map needs to be consistent with zoning regulations
(SR)	D.	Letter from water district confirming service
(SR)	E.	Letter from sewer district confirming service
(SR)	A.	File screen check map components from various involved county or private organizations and agencies
(SR)	B.	If five initial requirements have been completed, the map may be filed
(SR)	C.	Subdivision process is costed at $740 (for each lot, $1.20 extra)
(CP)	D.	Once map is filed, action must be taken within 50 days. If problems arise, the time may be extended. Thirty copies of map are submitted
(SR)		1. Staff report is prepared, which includes tentative condition of approval
(SR)		2. Copies of the map are sent to various landowners or agencies required by law for recommendations

(SR) a. School districts, building and safety, traffic, public utilities, health, road, flood, fire, and others

#(EMA) b. Replies are sent to subdivision committee

(CP) 3. 20-day period until a hearing is scheduled

(SR) E. Subdivision committee meeting

1. Approves and accepts the proposal with conditions, or
2. Disapproves with conditions or changes to proposal

(SR) F. Subdivision committee makes their recommendations to Planning Commission

(CP) G. Planning Commission hearing

1. Approves—tentative tract map approved, developer has 24 months in which to file final tract.
2. Disapproves—appeal to Board of Supervisors

a. Board disapproves
b. Request ends or project is revised

#(EMA) H. Rough grading is permitted after tract map approved

(CP) I. Final map approval

#(EMA) 1. Clearance letters obtained from Subdivision Committee members (letters returned to Board of Supervisors)

#(EMA) 2. Developer makes correction to final map and submits original to title company

#(EMA) 3. Title company obtains and/or verifies signatures, obtains tax letter, tax collector signature, and tax bond and submits original for final check

#(EMA) 4. Title letter and estimates from other agencies received

(CP) 5. Developer obtains bonds and signs agreements

(SR) J. Improvement plan approved

1. Required studies

#(EMA) 2. Hydrology—general study of drainage and water service

#(EMA) 3. Improvement plans must pass an average one to three checks, depending upon the plan

a. First check—20 working days
b. Second check—10 working days
c. Third check and each additional check—5 working days

#(EMA) K. Building permit process may begin after grading approved

#(EMA) L. County counsel clearance of bonds and agreements obtained

#(EMA) M. Board of Supervisors agenda package prepared when all maps, bonds, agreements, letters, and so on, obtained

(CP) N. Board of Supervisors approves final map and bonds and signs agreements

(CP) O. Recordation of map

(CP) P. Lots may be sold legally

IV. EIR procedure

(CP) A. Developer prepares and submits environmental information form

(CP) B. Director evaluates data submitted for adequacy

(CP) 1. Additional data supplied until considered adequate

(CP) C. Information considered adequate

(SR) D. Director conducts Initial Study and consults with all affected parties

(SR) E. Director determines if there are significant effects

(continued)

TABLE 5.3 (continued)

(SR)		1. Director files Negative Declaration—prepared, posted, and approved
		a. Negative Declaration appealed to Planning Commission
		b. If Negative Declaration is approved, it becomes official
		c. Decision maker considers Negative Declaration content and makes appropriate findings
(SR)	F.	Feasible mitigation measures available
(SR)	G.	Developer revises project to avoid or mitigate all significant effects
(SR)	H.	Director requires EIR
(CP)		1. Determination appealed
		a. Appealed directly to Planning Commission, which either approves a Negative Declaration or disapproves and requires an EIR
(EMA)		b. Notice of preparation is filed with responsible agencies
(CP)	I.	Prepreparation meeting
(CP)	J.	Screen check EIR prepared and submitted by developer, county staff, or county consultant as determined by the environmental management director
(CP)		1. Major revisions if necessary
(SR)	K.	Director consults with all responsible agencies and reviews screen check EIR for adequacy
		1. Minor revisions if necessary
(CP)	L.	Screen check acceptable as draft EIR
(SR)	M.	Notice of completion filed with state
(SR)	N.	Draft EIR distributed by director to all affected parties
(SR)	O.	Review and comment received from all interested parties
(SR)	P.	Director responds to comments received, finalizes EIR project report, and forwards it to planning commission
		1. Necessary revisions are made
		2. No revisions necessary
(CP)	Q.	Planning Commission approves the EIR
		1. Yes, for projects where Planning Commission is final decision maker (tentative tract map, variances, and so on)
		a. Planning Commission certifies EIR and takes action on project, or
(CP)		2. Yes, for projects where decision maker is other than Planning Commission (zone change, general plan)
		a. Board of Supervisors certifies EIR and takes action on project
(SR)	R.	Notice of determination filed with clerk of board and Secretary of Resources as required
#(EMA)	S.	Project implemented if approved by final decision maker
V.		Variance, use permit, conditional permit processing
#(EMA)	A.	Developer obtains filing instructions and discusses them with staff
#(EMA)	B.	Screen check—purpose is to assume filing documents are properly prepared
(SR)	C.	Required filing materials
		1. Plot plan
		2. Letter of justification
		3. Legal description
		4. Letter of authorization
		5. Draft Initial Study

(CP)	D.	EIR clearance obtained
		1. If not obtained developer seeks environmental clearance, Negative Declaration granted or draft EIR must be reviewable
(CP)	E.	Project found consistent with general plan
		1. May be appealed
(CP)	F.	Application checked for acceptability for filing
#(EMA)		1. If not acceptable, then developer reverses
#(EMA)	G.	Application is accepted for filing and fees paid
#(EMA)	H.	Application reviewed and evaluated by the staff
#(EMA)	I.	Project is ready for hearing—or any problem is solved before hearing
(CP)	J.	Item set for hearing
(CP)	K.	Public notification requirements met
(CP)	L.	Planning Commission or zoning administrator action—or new hearing is set and project continued
(CP)	M.	Action by Planning Commission is final unless appealed by either party
(CP)	N.	Project appealed after approval
(CP)	O.	Board of Supervisors make final decision
(SR)	P.	If supervisors approve, permit is issued after 14-day appeal period; upon fifteenth day, decision is final

VI. Grading permits

(SR)	A.	Applicant applies for grading permit and submits required documents
#(CP)	B.	Environmental services reviews Initial Study and determines need for EIR
(CP)		1. Draft EIR prepared and circulated for review
(CP)		2. Final EIR prepared and transmitted to Planning Commission
(CP)		a. EIR revised as required
(CP)		3. EIR considered adequate
		a. Mitigation measures recommended
(SR)	C.	Land use verifies legal status of site, establishes intended use, provides project additions
(CP)		1. Land-use clearances include Coastal Commission
#(EMA)		2. Contact city, special district, homeowner association clearances, and other clearances as required
		3. Clearances obtained
(CP)	D.	Land-use clearance obtained
(SR)	E.	Applicant submits plans and pays map check fees
(SR)	F.	Site preinspection made to verify conditions compatible with plans submitted
	G.	Plans acceptable and revisions made by applicant
	H.	Grading plan reviewed for compliance with
(SR)		1. Good engineering practice
(SR)		2. County excavation/grading code
(SR)		3. Planning Commission or subdivision committee condition of approval
#(CP)		4. EIR mitigation measures
(SR)	I.	Upon plan approval, grading permit fees paid, grading bond received, and permit issued
(EMA)	J.	Grading takes place
(SR)	K.	Grading completed and final inspection made
(SR)	L.	Upon final grading inspection clearance, grading permit finalized and bonds released

(continued)

TABLE 5.3 (continued)

(EMA)	M.	If building starts, file for precise grading permit indicating exact location of buildings after approval building permits may be pulled
VII. Building permits		
(SR)	A.	Developer applies for building permit and submits required plans/ documents
(SR)	B.	Application and other documents acceptable, or applicant revises documents/plans until acceptable
(CP)		1. Land use reviews application and documents
(SR)		2. Fees established
(EMA)		3. Applicant informed of other permits, including electrical, plumbing, heating/mechanical, local park fees, plastering permit, grading permit; all processed concurrently
(SR)	C.	Land use verifies legal status of building site, establishes uses intended, provides legal address
	D.	Developer informed of special requirements if applicable
(CP)		1. Coastal Commission
#(EMA)		2. Homeowners associations
#(EMA)		3. Miscellaneous districts
#(EMA)		4. Contract cities
(SR)	E.	Applicant pays applicable fees and plans accepted for checking
#(EMA)	F.	Plans checked as adequate, or applicant revises plans as required
#(EMA)		1. Adequacy of plans for good engineering practice
(SR)		2. Meeting uniform building codes
(CP)		3. Meeting state sound attenuation laws
(CP)		4. Meeting state energy requirements
(CP)		5. Meeting state handicapped provisions
(SR)		6. Meeting uniform plumbing code
(SR)		7. Meeting mechanical code
(SR)		8. Meeting national electrical code
(EMA)		9. Verification of adequate water and sewage disposal facilities
	G.	Applicant picks up plans
	H.	Other requirements as applicable
(SR)		1. Grading permit required—satisfied as necessary
(CP)	I.	Variance appeal time applicable—time delay satisfied
(SR)	J.	Plumbing clearance as required—satisfied as necessary
(CP)	K.	Coastal zone clearance required—satisfied as necessary
(CP)	L.	Zoning approval required—satisfied as necessary
#(EMA)	M.	All other clearance requirements met—satisfied as necessary
(SR)	N.	Remaining permit fees paid
(SR)	O.	Building permit issued
VIII. Building inspections		
(SR)	A.	Building inspections commence after rough grading
(SR)	B.	Concrete may not be poured until following applicable items have been signed

(SR)	C. Plumbing (drain/waste) pipe, electrical ground electrode, setback forms trenches
#(EMA)	D. Do not install subfloors or pour floor slab until the following applicable items have been signed
(SR)	1. Electrical (groundwork)
(SR)	2. Plumbing (drain/waste pipe)
(SR)	3. Plumbing (water pipe)
(SR)	4. Mechanical (groundwork)
(SR)	5. Preslab (compaction/membrane/reinforced)
(SR)	6. Underpinning (floor joist/girders)
#(EMA)	E. Do not cover or conceal the following items until items have been signed
(SR)	1. Fireplace
(SR)	2. Decorator appliance (vent/chimney)
(SR)	3. Roof (sheathing diaphragm) frame
(SR)	4. Masonry (reinforced/bond beam)
(SR)	5. Rough plumbing
(SR)	6. Rough heating, venting, refrigeration piping
(SR)	7. Rough electrical
(SR)	8. Frame
(CP)(SR)	9. Insulation (sound/energy)
(EMA)	F. Do not tape or plaster until the following items have been signed
(SR)	1. Electrical ceiling heat rough
(SR)	2. Dry wall
(SR)	3. Interior lath
(SR)	4. Exterior lath
	G. General items
(SR)	1. Exterior plaster (brown coat)
(SR)	2. Water service
(SR)	3. Gas test (building stubout or valve)
(SR)	4. Seepage pits, leaching field
(SR)	5. Septic tank
(SR)	6. Sewer
(SR)	7. Sign (interior-structural) electricity
(SR)	8. Permanent electrical pole
(EMA)	H. All applicable items must be signed before utilities will be released
(SR)	1. Land use final
(SR)	2. Grading final
(SR)	3. Structural final
(SR)	4. Plumbing final
(SR)	5. Heating/ventilating final
(SR)	6. Electrical final
(SR)	I. Certificate of Occupancy is issued

IX. Residence may be occupied or sold

Note: (SR) = State requirement; (CP) = Adopted county procedure; (EMA) = County Environmental Management Agency policy; # = EIR-related processes.

Source: Californians for an Environment of Excellence, Full Employment, and a Strong Economy through Planned Development (CEEED), Newport Beach, California, 1979.

TABLE 5.4

Selected Money and Time Cost Estimates for Other Than a Minor Subdivision, Sonoma County, California, 1978

	Estimated Costs
Type of planning department action	
Tentative subdivision map	
Filing fee	$ 350 + $2/lot + $47 + $2/lot (health)
Extension of time, approved map	100
Map revision (subdivider request)	100
Final subdivision map	
Plan check and survey fee	100 + $25/lot
Improvement inspection fee	100 minimum (1 percent/ thousand dollars)
Variances	175
Appeals	75
General plan administrative appeal	90
General plan amendment	10 per acre
EIR preparation	100 minimum (8 percent)
Using only these costs and assuming 40 acres gross, 4 housing units per acre, 100 new housing units to be sold for $90,000, the following costs are incurred:	
Tentative subdivision map	797
Final subdivision map	3,500
Variance	175
Appeals (at least one)	75
General plan appeal (at least one)	90
General plan amendment	400
EIR preparation (900 × 100 units)	90,000
Total	95,037 ($950 per home)

questing variances, use permits, or conditional use permit processing; filing for grading permits; requesting building permits (in specified circumstances); or asking for building inspections prior to sale or occupancy of the finished properties.

The intertwining of EIR and other construction inspection requirements can produce delays and extra costs, which are not often included as the costs of the EIRPs. The extent to which they could add to costs is suggested by the listings in Table 5.4.[8]

COSTS ASSOCIATED WITH THE ENVIRONMENTAL IMPACT REVIEW PROCESSES

In order to identify the kinds of costs associated with the EIRPs previously outlined, varieties of anecdotal evidence were assembled in this and the next chapter and organized to reflect the most common categories of costs. This anecdotal evidence is then used to provide an overall general statement of the categories of costs that have to be considered.

The examples presented are ones that were identified rather consistently throughout a large collection of anecdotal materials. These range from the front-end costs of Sonoma County through duplicative review costs to imposition of social costs. Examples are from many states, although the focus is on California. The findings are then summarized at the end of the recitation.

Front-End Costs

Sonoma County is located on the San Francisco Bay and is included in the land-use planning activities of the Association of Bay Area Governments (ABAG). Sonoma County supervisors were not pleased with ABAG plans for making the county an access corridor, so they instituted various kinds of plans for reserving open space and reducing the quantity and impact of density and industrial and commercial planning.

In order to assist orderly land-use planning and development, Sonoma County officials created an extensive data bank. The data bank includes, among other items, information on soil conditions, open space characteristics, land-use studies related to roads, land uses, open space, public services and facilities, and density transfer potentials. The Planning Department also may contract with a quasi-public EIR consultant group to write preliminary final draft reports. The consultants may use the data bank for a small fee; the fee is used to support the EIR review budget and to encourage development of a uniform EIR format.

Sonoma County planning officials put emphasis on design standards. Both residential and nonresidential developments are evaluated by design review boards, which can specify such things as materials to be used, color, height, and blending of the buildings with the surrounding area. Applications that do not require environmental review are usually approved within two weeks, but the others require four or five weeks review by agencies other than the Planning Department. At least 15 agencies are included in the design review processes.

Fees for design review are $40 for filing, plus $100 if environmental considerations are involved. An additional $24 fee is required if Health Department clearances are needed. Other costs relate to needs for one full size and seven reduced copies of the site plan, photos of the site from different locations, and appropriate portions of a United States Geological Survey (USGS) topographical map.

All projects are evaluated in terms of consistency with the general plan, using a check list containing 21 major categories and 84 subcategories. Given the broad nature of the goals statement in the general plan, almost all projects would require an appeal, which would cost $45 initially and $45 for any subsequent appeal. Any project involving a need for amending the general plan costs $10 per acre, with a minimum fee of $150 and a maximum of $1,000. Applications requiring variance and use permits cost $199, and use permits cost $199. If the project is not categorically exempt, the environmental review costs an additional $100.

Projects exempt from environmental review are subject to a 10-day posting period in which others can make comments on the project, after which the Board of Zoning Adjustment holds public hearings. Uses cannot begin until at least 12 days after a hearing or decision. If the project is subject to environmental review, the developer must hire a private consultant under contract to the county but paid for by the applicant. Processing time can take "a few to several weeks."

In an effort to reduce some of the economic hardships of their planning and environmental processes, planners have considered density transfers. The purpose of density transfers would be to serve open space more effectively, compensate landholders for lost development opportunities, and accomplish important planning goals without significant public costs.

The front-end costs for a major subdivision in Sonoma County in 1978 are presented in Table 5.3. The cost of $950 per home seems reasonable in comparison to the averages reported in the more general California study.

Duplicative Environmental Reviews

Many states have duplicative environmental review processes, which can add to costs. Important among those states are California,

Florida, and Vermont, with at least 30 coastal states having some form of mixed state and local EIR requirements.[9] Estimates of the cost of environmental review processes range from $17 to $25 per residential unit in California to $30 per unit in Florida.[10]

An example from the California Coastal Commission files exemplifies the problems builders face in working under multiple environmental jurisdiction rules. The commission is expected to assure "access to publicly owned or used beaches, recreation areas, and natural resources" (Proposition 20). In most coastal communities these areas are also under local (county or city) governmental controls.

A vacant 8.5-acre parcel in southern California was purchased in 1973 for $385,000. Streets and utilities were in place, and no grading of the site was needed. Zoning permitted the construction of 104 condominium units (12 units per acre). After public hearings and review by the Coastal Zone Commission, only 69 units (7.8 per acre) were approved. The firm spent 1974 making corrections and revisions required by local reviewing authorities. Throughout 1975 the firm's officers held conferences with various review authorities seeking information on the kind of project that would be approved. In December 1975 a conditional-use permit was approved by local authorities for the construction of 58 units; however, during the next year the builder was required to produce a totally new project, since approvals given in 1975 were never implemented.

Some of the costs incurred by the builder-developer included a filing fee of $250, EIR preparation of $4,000, and consultant fees of $1,550. In the meantime the original investment was earning no interest.

After the imposition of Coastal Zone Commission regulations, a comparison of home prices for homes within the commission's purview and homes outside its jurisdiction indicated prices averaging 47 percent more for coastal than noncoastal homes.[11] Such a difference must be treated with care, however, since coastal properties have always commanded premium prices, and such land is in short supply. Even so, duplicative reviews by city, county, and state agencies undoubtedly increase prices.

Impact of Extended Time

Coastal growth control zones are located all along the eastern coast of the United States, with particularly strong controls in New Jersey. An examination of coastal protection costs was provided in some detail through a study of 982 dwellings on 719.7 acres.[12] Costs of the subdivision process under the Coastal Area Facility Review

Act (CAFRA) were $4,720 and required 24 months for approval. A subdivision of equivalent type and not under the CAFRA required 22 months at a cost of $4,585. A site review plan under the CAFRA required 23 months for approval and cost $2,310, compared with 19 months and $2,195 for a similar project not under the CAFRA. Overall coastal review processes added approximately 10 percent more time and 3 percent to 5 percent more in direct costs.

"Extra" Requirements

Requests for exceptions to existing general plans or zoning often produce trade-offs between the builder and the regulators, always creating "extra" requirements. When the EIRP is added in, the trade-offs become almost routine. A review of typical agenda items from the city of Los Angeles involving the EIRP included the following (material in parentheses refers to cases filed with the Los Angeles City Planning Commission [CPC]):

1. Setback requirements beyond those called for in zoning regulations (Tract 25210).
2. Construction of block wall fences to protect views or landscaping to hide developments or to reduce sound or both (Tract Map 30729; CPC 27610).
3. Specification of particular types of plantings, usually to be installed by a licensed landscape architect. Such plantings could often be purchased by the builder through a wholesale nursery at one-fourth the cost of the required planting (CPC 28055).
4. Installation of sidewalks, street widening, flood channels of a size larger than required to service the development but sufficient for future development needs in other projects, and street lighting (CPC 27901).
5. Construction of public alleys and reservation of easements for alleys or future streets, even when existing conditions indicate they are not needed at the time nor in the foreseeable future (CPC 28051).
6. Construction of public improvements or bonding for libraries, schools, and parks, even when the affected city agencies or departments see no need for them or lack budgets to service the facilities or to build new ones (EIR 254-77-Sub).

An ad hoc committee appointed by the mayor of Los Angeles found that review times in the city took two to three times longer than in adjacent jurisdictions.[13] Among the items found adding to process times were: re-review of already approved items; inconsistent interpretations of applicable codes or regulations by the same inspector or by different inspectors; changes required by field inspectors that were not required in the approved plans; duplicate processing by various departments of the same required reports; significant differences in estimations of plan check; inability of concerned departments to provide key information needed for compliances; temporary interpretations of requirements, which became permanent without adequate review or approval; and lengthy and costly new materials checking, even though other responsible agencies had reviewed and approved the materials.

Public Participation

Citizen participation is provided throughout the EIRPs in most jurisdictions, and scare stories abound about this requirement. One in particular highlights the more spectacular type of problem.

Alvin J. Hall, a southern California builder, spent $600,000 for plans and reviews for 165 homes. While the project was being reviewed by the City Council, an almost casual comment (source unknown) focused attention on a large rock formation in the middle of the planned subdivision. Although citizens' groups had paid no particular attention to the rock, the mention of it created a movement to save the rock and oppose the subdivision plans. Ultimately the builder received the permit needed for the subdivision, but he estimates that the delay cost him an additional $150,000, all of which was passed on to the home buyers.

Market Skewing

In 1968 Mountain Properties purchased 3,500 acres for $17.5 million, expecting to develop a subdivision of approximately 1,200 homes similar to those already completed in other projects. Before approvals were received, the company was required to build a 1.9-mile road into the project ($5,600,000). After selling and dedicating 2,100 acres for park purposes, Mountain Properties was given permission to plan residential developments on the remaining 1,400 acres. The first phase of development called for 397 detached single-family homes and 559 town homes on 250 acres. These homes had been planned to sell for $65,000 to $85,000 if completed in 1969-70. When the homes were finally placed on the market in 1978, prices started at over $100,000 and by 1980 were at over $500,000.

The interrelationships between market conditions and EIRPs are highlighted in two studies completed for the Urban Land Institute in San Jose, California, and Jacksonville, Florida.[14]

In San Jose housing demand was strong, and costs were passed on to home buyers. Local growth management policies were identified as having caused 20 percent to 40 percent of the price increases. For one builder home prices, because of cost increases, rose by 80.2 percent and for another builder 121.3 percent. In Jacksonville controls were estimated to have added $1,000 to $1,500 to house prices. Market demand in Jacksonville was weak. The San Jose builders were able to pass on all increased costs, the Jacksonville builders only a small portion.

Social Issues

Some EIR review bodies use the process to resolve social issues having only a peripheral relationship to environmental problems. In Los Angeles a large bank planned to build a new service center by assembling several blocks of decaying residential-commercial properties near the central business district. The area selected was populated by Spanish-speaking families occupying very old, obviously dilapidated, single-family and apartment units. The area was characterized by small lots and blocks and numerous short, often deadend, streets. Land assemblage was considered to be reasonably easy, since 60 percent of the properties were owned by one family.

The building to be constructed was considered to be an important addition to the local economy and particularly important for minority communities, since 2,000 of the 6,000 persons employed in the building would be members of minority groups. The construction required the removal of 250 families.

At public hearings considerable evidence was provided that showed that the land assemblage would produce a reasonable profit for the assemblers; the requested zone change needed to permit construction of the building would provide decided economic advantages to the bank. Since street closings and major construction were involved, the EIR was rather lengthy.

Through a long and complicated bargaining process, including several public hearings, the proposed construction was approved, but the bank had to place $250,000 in trust for disbursement to families who had to move. An additional $5 million was pledged for loans to be used in the redevelopment of the adjacent area, according to plans to be developed by the local residents.

Summary and Some Conclusions about EIR Costs and Processes

These are only a few of the numerous anecdotes cited in other studies of EIR processes, costs, and home prices. Fortunately, these reports, although from various parts of the United States, permit some aggregation into groupings from which some general observations can be made about EIR impacts on housing prices. All, for example, suggest that EIR processes add to costs, but the extent to which they add to prices depends upon market conditions at the times the homes are sold. In almost all instances generally improving economic conditions appeared to keep consumers' abilities to pay equal to the increased housing prices until late 1980 and through 1981. Almost all of the reports focus on gross costs and make no effort to subtract private or public benefits. In part the reluctance to consider benefits lies in the difficulties of identifying and measuring the benefits. The purpose of this chapter is to suggest a model that could be used in identifying all costs and benefits, private and public, quantifiable and nonquantifiable.

As a first step in creating a cost-benefit model, findings from the surveys of both anecdotal and quantitative information permit the identification of the following critical elements in any such analysis:

1. Fees, permits, and other costs must be paid immediately, without guarantee the project will be approved either as requested or in modified form; however, evidence to this point does not suggest that these costs are particularly onerous;
2. Duplicated reviews, since EIRPs are embedded in total building processes; the net result is delay and uncertainty;
3. Delays occasioned by uncertainty on the part of regulators about matters in the EIR—how to interpret and enforce new kinds of conditions; regulators become judges as well as administrators and must use personal judgement in lieu of rules or regulations;
4. Extra costs because of added requirements arising from trade-offs between what the builder wants and what the regulators believe is needed;
5. Public participation, which takes time and often creates added requirements and unexpected delays;
6. Intrusion of social issues, which adds requirements that produce higher costs but not necessarily higher prices;
7. Dedications and other land and financial contributions as a part of the approval process;
8. Bureaucratic conflicts arising from the need for numerous comments and reviews by governmental departments concerned with residential construction;
9. Dedication of land for public purposes without compensation;
10. Reduction in the number of units from the number originally planned or preliminarily approved;
11. Down-zoning or annexation proceedings.

Preparation Costs

Once all of the data are assembled, they must be organized into an EIR. Report preparation can vary from a few hundred to thousands of dollars, depending upon contents, lead agency requirements, experts used, and the problems to be dealt with on the project. Since a lead agency determines the kind of report, its needs become paramount. If the local agency places emphasis on protecting the environment, an initial report may be very extensive, even if the potential impacts of the project are expected to be insignificant.

The first element in preparing an EIR is the creation of a data base. Sometimes the applicant must do this; sometimes all or a major portion of the data are provided by the lead agency. For these reasons this process can take days to months. Once the data are prepared, an expert must evaluate the data to determine possible impacts. If more than one expert is used (and they do not agree), this, too, can delay EIRPs by days, weeks, or perhaps months. Once the EIR is ready, subsequent reviews by public agencies and in public hearings can take additional time, so that one, two, or more years' preparation time for major projects are understandable, even if not desirable. An expert has said: "A good EIR is not just a good report. It should represent an environmental analysis conducted prior to project definition, if possible, so that the environmental setting can influence the project design."[15]

Under such conditions the EIR becomes a valuable guide for the developer and the architects in designing a project that incorporates mitigation measures and can be processed quickly by the lead agency. In developed areas, such as Los Angeles County, one of the greatest costs associated with environmental impact reviews can be mitigation measures, because most developable land has problems relating to the fragility or difficult contouring of the terrain. An EIR that incorporates measures for dealing with these problems rather than just information that presents the problems would reduce construction costs.

The EIR costs may be increased because the developer and the lead agency may both be required to prepare EIRs. In such cases the lead agency may let the developer prepare the first EIR and then build its report around the developer report. In other cases an approved organization may be asked to prepare an EIR, or the developer may be required to hire one of the approved agencies. Some local governments are experimenting with providing the basic data from which an EIR can be prepared or incorporated into private EIRs by reference. There is no evidence at present as to which method produces the most efficient, effective, or least costly EIR. Whatever the method of preparation, the costs of an EIR preparation include the following:

1. Actual preparation costs—data collection, processing, expert review, reproduction, and distribution;

2. Use of consultants to prepare the report, recommend mitigation measures, and give testimony at public hearings;

3. Revisions and additional preparations as the review processes continue.

What determines the level of costs of an EIR?

The fragility of the environment, which may require complex and expert analyses
Reviewing agency requirements
Availability of needed data
Project size

What can create larger costs for a developer? Waiting is a primary cost, although mitigation measures can add even more cost. The extent to which the builder presents a project more suitable to its environmental setting, identifies possible development problems, and is sensitive to demands the project may place on public infrastructure, the lesser the costs. For the public the major costs of the EIR preparation processes are written appeals, reviews by staff, site inspections, and public hearings.

The emphasis in the preparation phase is on costs. Benefits are not considered or evaluated, so that an eventual mismatch between costs and benefits can evolve from the review processes. An expensive mitigation measure that produces only a minor environmental improvement or protection at a much higher housing construction cost may go unchallenged. Costs to local governments compared with potential benefits from a project may be overlooked as the search for mitigation measures includes density reductions, which produce lower levels of assessable properties than are necessary to produce tax revenues to support the new construction.

Several attempts have been made to measure the actual dollar costs of EIR preparation. One survey estimated that these costs in California in 1974 varied between $4,000 and $10,000 or approximately $17 to $25 per unit. In Florida the costs were estimated to average $30 per unit or from $50,000 to $200,000 per project.[16] Studies for the California legislature, based on sampling several hundred projects, indicated that per unit costs surveyed ranged between $962 to $20,100 for all projects, $565 to $24,900 for public projects, and $1,078 to $16,900 for private projects.[17] The average unit cost was estimated at $6,800 for all projects in all districts.

In addition to the actual costs mentioned, additional costs were estimated for the delays, uncertainties, and mitigation measures associated with the preparation processes. Included were assumptions about the impact of inflation on costs. It was estimated that the overall costs of implementing EIRPs (CEQA) were between $50 million and $75 million per year. This equaled less than 0.8 percent of total construction costs and approximately 0.5 percent of total project costs. For the state this equaled $2 to $3 per person. In terms of residential construction an estimated $20 million in costs were incurred (for the 125,000 units constructed in that year, a cost of $150 per unit). What was not estimated was the actual cost added for the projects for which CEQA reports had to be prepared. Equally important, no evidence was developed as to the extent to which developers passed these costs on to the home buyers; however, the analysts did point out that inflation, which increased costs, also increased the abilities of home buyers to pay them.

A discussion of issues relating to environmental benefits is beyond the scope of this book.

Delays, Uncertainty, and Mitigation Costs

In addition to the actual costs of EIR preparation, other costs arise because the project proceeds more slowly than the developer believes it should or could; costs paid but not recovered because projects were denied or abandoned; and design and operating changes required to mitigate an undesirable and potentially adverse impact.

Delay costs are important only when they are rising faster than the ability of home buyers to pay, as evidenced by their changes in personal income. As uncertainty increases, builders increase costs by requiring a higher rate of return on their investments. Mitigation measures are the most easily measured, because they are actual outlays designated to reduce the impact of an unfavorable aspect of the proposed project.

The great difficulties in measuring these additional costs have made an accurate estimate of their total financial impact impossible, even though these costs may well be the most important of all. However, anecdotal evidence does provide some insights into their nature and extent.

EIRP IMPACTS ON HOUSING COSTS

The costs that must be considered can be placed into five major classes: preacquisition, land acquisition, land preparation, construction, and marketing and sales (see Table 5.5). EIR-associated costs begin with land acquisition and the decision as to whether an EIR will be required. At this point the only costs are associated with permits.

TABLE 5.5

Processes and Costs Associated with Residential Development and Environmental Impact Review Processes

Processes	Costs, Fees
Phase I: Preacquisition	
Feasibility and market studies	Options, holding costs, taxes, appraisals, consulting
Phase II: Land acquisition	Land costs, escrow fees and costs, commissions, title policies, legal, land taxes
Design	Engineering studies
Predevelopment	Architect, soil and geological specialists, land planner, special consultants
EIRPs	Maps, plans, analytical reports, surveys, fees, "expediters"
Negative Declaration	
Draft, final reports	
Public reviews	
Appeals	
Certification	
Zoning requests	Filing fees, master plan, subdivision and grading plans, special analyses
Standard permits	
Subdivision maps	Architects, engineers, and permit charges; building, plumbing, and electrical fees; assessments; hookup fees
Grading plans	
Landscaping plans	
Design (architectural review)	
Building permits	
Possible redesign	Architects, engineers, legal
Phase III: Land preparation and improvement	
Preparation and dedication	
Grading	Labor, materials, loans, bonds, deposits, land dedication, construction loans, access roads, grading, drainage, flood control, utilities, landscaping, irrigation, traffic signals, street lights, park preparations, telephone
Utilities	
Public improvements (streets, lights, sewers, walks, landscaping)	
Improvement	
Off-site	
On-site	
Phase IV: Construction	
Foundations	Construction loans, labor, materials, property taxes; community amenities—schools, parks, open space, sidewalks, inclusionary housing, recreational facilities, open space
Framing	
Enclosing	
Final landscaping	
Finished models	
Phase V: Marketing and sales	
Marketing studies	Takeout financing, labor, materials, commissions, fees, charges, loan fees and charges, transfer costs
Advertising, selling	

Source: Compiled by the authors.

TABLE 5.6

Types and Sources of Environmental Impacts and Possible Mitigation Measures

Environmental Element	Associated Impact (social costs)	Mitigation Measures (private costs)*
Physical		
Air	Destruction, deterioration	Use reduction
Water	Destruction, deterioration	Use reduction, processing
Sound	Increased volume	Source elimination, reduction
Flora/fauna	Destruction, elimination	No development, move from area
Topography	Contour changes, grading	Reduce, eliminate grading
Public systems, service		
Services (fire, police, cleaning)	Quality deterioration, quantity reduction	Reduce construction, add fees for services
Traffic flows	Increased air pollution, noise, congestion	New routes, increase capacity of existing facilities, reduce construction
Public facilities (sewers, utilities, parks, open space)	Deterioration of quality and quantity of services	Furnish facilities, fees, bonds, dedication of land
Schools and cultural facilities	Crowding, deterioration of quality and quantity	Land dedication, new facilities by builder, fees and bonds, reduce need
Public land and buildings	Overuse, crowding, rapid deterioration	Provide new facilities, fees, bonds, reduce need
Private property		
Homes, employment and commercial centers, properties	Higher property prices; higher property taxes; less demand for old, underutilized facilities in older areas	Reduce quantity, relax application of codes and planning requirements, inclusionary zoning, fees, bonds, stricter codes, planning standards
Existing land uses	Increased traffic, congestion, pollution, noise, delays	Reduce construction, require public improvements, provide "protective" barriers (open space, parks, and so on)
Social		
Population density Employment access Shopping, recreation, churches, clubs	Life-style changes, higher-living costs, lower-living quality	Density reductions, builder provides new facilities, land dedication, add private services, facilities

*Included are delay and public hearing costs.

Source: Compiled by the authors.

After that, a project plan must be prepared and submitted for determination as to whether the project is exempt from further environmental impact reviews. Included in the processes are the preparation of drafts and final reports; presentations at public hearings and reviews; appeals from the findings, which are published after the public review processes; and final certification.

After EIR certification has been received, the final EIR serves as a blueprint, which guides all future land preparation and housing construction. It is even possible that additional changes may be required in the EIR-related processes, if construction reveals undesirable environmental impacts, particularly if these impacts are observed and appealed by the general public.

It is the intertwining of EIR requirements with other building and construction requirements that often makes EIR-associated costs appear larger than what they really are. On the other hand, as has been suggested earlier, it is that intertwining that has created the costs and perhaps needs consideration when evaluating EIR benefit costs.

The costs listed in Table 5.5 are those that can be identified and measured rather easily. Costs not included are those related to a reduction in project sizes, delays, duplicative reviews, and other items mentioned earlier. The sources of these costs lie in the impacts, which are presumed to be created for a given project. These costs are difficult to identify and quantify; considerable time can be spent on those tasks. From these considerations of social costs flow mitigation requirements, which affect the ultimate costs of the housing. The impact on price of these costs is also difficult to measure. Depending upon anticipated market conditions, they may be avoided by abandonment of the project, or they may be passed on in part or totally to future home buyers. Once their scope is defined, they can be costed and finally can add considerable amounts to home prices; they are essentially front-end costs, which over time—given the time value of money—can inflate considerably. All of these considerations are summarized in Table 5.5 and, with Table 5.6, provide the base on which the cost portion of a cost-benefit model can be developed.

In the next chapter we examine the varieties of review processes in an effort to identify a common model into which the costs of this chapter can be fit.

NOTES

1. Joseph J. Seneca and Michael K. Taussig, Environmental Economics (Englewood Cliffs, N.J.: Prentice-Hall, 1974), p. 66.

2. Paul Samuelson, Economics, 9th ed. (New York: McGraw-Hill, 1973), p. 633.

3. Stephen R. Seidel, Housing Costs and Government Regulations: Confronting the Regulatory Maze (New Brunswick, N.J.: Center for Urban Policy Research, 1978), pp. 236-37.

4. Samuelson, Economics, p. 778.

5. Jeremy Rifkin, Entropy: A New World View (New York: Bantam Books, 1980), pp. 35-37, and Bernard J. Frieden, "Environmental Politics," Urban Land, March 1977, pp. 3-8.

6. For a more complete description of this theoretical approach, see Krister Hjalte, Karl Lidgren, and Ingema Stahl, Environmental Policy and Welfare Economics (Cambridge: At the University Press, 1977).

7. Hjalte, Lidgren, and Stahl, Environmental Policy, p. 12.

8. Edward M. Bergman, "Development Controls and Housing Costs: A Policy Guide to Research," in Management and Control of Growth, ed. Randall W. Scott, vol. 3 (Washington, D.C.: Urban Land Institute, 1975), pp. 527-36.

9. American Institute of Planners, Survey of State Land Use Planning Activity (Washington, D.C.: Department of Housing Development, 1976).

10. Seidel, Housing Costs, p. 248.

11. Security Pacific Bank, Coastal Zone Housing Prices (Los Angeles: Security Pacific Bank, Research Department, 1980).

12. Seidel, Housing Costs.

13. Ad Hoc Committee on Construction Processes, Report (Los Angeles, California: Offices of the Mayor, June 1976).

14. Urban Land Institute and Gruen and Gruen Associates, Effects of Regulations on Housing Costs: Two Case Studies, Report 27 (Washington, D.C.: Urban Land Institute, 1977).

15. Don M. Demars, "Environmental Impact Report" (Speech delivered at the Society of Real Estate Appraisers Seminar, Newport Beach, Calif., March 7, 1965).

16. Seidel, Housing Costs, p. 248.

17. A report prepared for the Assembly Committee on Local Government, California State Legislature, The California Environmental Quality Act (CEQA), vol. 3, Discussion of Costs and Delays Related to Environmental Review (Sacramento, November 1975).

6

PROCESSES AND PARTICIPANT PERCEPTIONS

Since the quantitative evidence relating to environmental impact review processes (EIRPs) is minimal, anecdotal evidence provides an excellent means of evaluating some of the findings in the earlier chapters, particularly since the anecdotes are from primary sources, are well documented, and are numerous. In this chapter we have selected anecdotes that are most representative of the entire group. We believe we have been able to collect and evaluate most the available anecdotal evidence. We also feel reassured, because the anecdotes provide the color and background that illuminate much of the quantitative data already presented. We present them in an order and grouping that conform to the major findings of the quantitative analyses.

LOS ANGELES COUNTY

Los Angeles County is one of the most populous counties in California. The city of Los Angeles is the largest city in the county and has been a leader in land-use planning and use of the EIRP. Some of the largest construction projects in the state subject to Environmental Impact Reports (EIRs) have been planned and built in and near Los Angeles.

The Processes and Costs of Delays

Mountain Properties purchased 3,500 acres for $17.5 million in January 1968. In the late 1970s the company had sold 2,100 acres to the State of California, including some land that was dedicated to the city of Los Angeles. The project was planned in phases, with a zone change planned and requested (with a fee of $5,100). A six-month delay ensued at the request of city council members, but engineering studies were completed at a cost of $190,073. The first phase went through various hearings and reviews. It began with a plan for over 800 units—397 single-family homes to be priced between $85,000 and $165,000 and 559 townhouses priced between $65,000 and $105,000. After numerous public hearings and some court suits, 160 acres were dedicated for public open space; the units were completed and sold in 1978 for prices 25 percent higher than planned.

Phase II was then begun, calling for approximately 1,200 new housing units. This request met with numerous planning and environmental impact reviews, including dealing with various property owner groups; finally 800 units were approved. However, throughout 1980 and 1981 the project was still under public protest.

The corporation estimates that all land dedication, public improvements, open-space dedication, permits, reports, and fees totaled $19,480,000. In late 1981 the earlier units were being resold at prices beginning at $400,000.

Interagency Conflicts

The project was planned in 1972 to be an 18-story, 252-unit apartment project for elderly citizens. By 1976 the project had been changed to a three-story, 99-unit project for the elderly. Conflicts soon developed between the city agencies and the Department of Housing and Urban Development (HUD), which was providing financing underwriting. Ultimately, the project was approved after the developer agreed to relocate families being displaced at a total cost of $26,000. After final agreements among all agencies, at least 10 percent to 15 percent were estimated as added construction costs owing to delays and agencies' requirements.

Dedication and Special Requirements

An analysis of 25 projects in Los Angeles revealed a pattern of requirements that tended to follow EIR reviews. The requirements would be arrived at by a combination of trade-offs between the applicant builder and the planning department reviewer or added requirements arising from a chain of reviews between the planning director, planning commission, public hearings, city council, planning committees, mayor's office reviews, and other city departments. Typical of some of the special requirements from particular cases filed with the Los Angeles City Planning Commission (CPC) are the following:

1. Setback requirements beyond those called for in zoning regulations (Tract 25210)
2. Construction of block wall (decorative required), fences, and landscaping to hide proposed developments, reduce sound from the development, or both (Tract Map 30729; CPC 27610)
3. Specification of particular types of plantings (CPC 28055)
4. Installation of sidewalks, street widening, flood channels (of a size larger than required to service the development), street lighting, and curbs (CPC 27901)
5. Construction of public alleys, reservations of easement for alleys or future streets, with no apparent need (CPC 28051)
6. Construction of public improvements or bonding for libraries, schools, and parks (sometimes the related agencies lacked budgets to service the facilities) (EIR 254-77-SUI).

SOUTHERN CALIFORNIA

Surveys among builders in southern California provided six illustrations of the problems and costs associated with EIRPs. Some of the largest builders indicated that they did not build in counties in which EIRPs were too strict or too onerous. Among some of the comments are the following:

> As of 1980 this project is still going through approval processes although construction was slated in 1978. If we are fortunate it may commence construction in December of 1980. This letter is not simply a "developer whines about the process" story. It is, rather, an objective reporting of some of the hidden factors in current costs of housing.
>
> The project in question was originally proposed to the city of Oxnard in March of 1976 and was to include 884 condominium units on a parcel containing 115 acres of land and a density of 7.7 units per acre. In addition, a school site, a park site and a public beach were to be dedicated to the public. The city of Oxnard engaged an independent consultant in July 1976 for the purpose of preparing an EIR. Preparation of the EIR, under the direction of the city of Oxnard, took until September 1977, a period of 14 months, and cost the applicant in excess of $25,000.

The EIR stating that all adverse impacts could be avoided or mitigated was certified by the city council in November of 1977, 20 months after the project was first proposed to the City. Upon certification of the EIR, and in response to its findings, the applicant submitted a revised plan reducing the density plan from 884 units to 514 units, and increasing public areas from 34.5 acres to 44.5 acres.

The City then undertook a "study" of the Tentative Map which resulted in the City Council in October 1978, denying the map in order to study possible acquisition of the site for a park. The City hired a different independent consultant in June of 1979 to study the site. The finding of this consultant was that the appropriate use of the site was 500 dwelling units on 45 acres of land, a hotel on 8 acres, and the remaining 61 acres of land to be devoted to public uses.

Thus after spending 43 months and hundreds of thousands of dollars for studies, the city, in October of 1979, approved a plan which the EIR, two years earlier, had said was right. We now face the Coastal Commission. In preliminary discussions with the Commission staff, they state the EIR may be sufficient for CEQA purposes, but not for their purpose and that additional studies may be required. In addition, we now stand between the City's demand for all high-priced housing and the Coastal Commission's demand for 25 percent low and moderate-income housing.

What this saga demonstrates, is not so much the actual cost of the EIR, but the costs that can result through the use of the EIR as a delaying tool. In the case of the subject project, the units were originally proposed to sell at a minimum price of $35,000 and the average price of $50,000. Now they will have to sell for a minimum price of $95,000 and an average price of $175,000. The developer's return on investment will be just about the same. It will be the home buyers who suffer.

Another developer related the following:

The following are general observations based on our experiences as active real estate developers:

1. Cost: An EIR will cost somewhere between $10,000 to $50,000 per project in actual fees to consultants. In addition, the time delay in preparing the EIR

> and having it reviewed by the municipality plus numerous hearings will probably add between 5 percent and 10 percent in cost to the initial purchase price of the land.
>
> 2. Problems: a major problem is that after we have spent all this money doing an EIR it is often not even read but merely placed in a file. The EIR generally discloses problems which will be caused by the project and alternatives such as not building the project. They fail to discuss the benefits of the project, such as providing badly needed shelter which normally overrides the environmental concerns.
>
> We are presently doing an EIR which will be completed next month (February-March 1980) that, so far, has cost us in excess of $100,000.

Yet another developer made the following statement:

> The EIR process was used in this project to further the political aims of those on the Planning Commission and the city council. There was no expert testimony included aside from project proposals given by the developer; most of the structuring came from the EIR firm, who obviously, wants to sell more EIR services to the city.

The briefest reply came from a large builder whose stationery has the logo Leaders in Housing since 1892 and the comment "We do not build in areas which require an EIR." Statements from other builders included the following:

> Agency guidelines have to be specific and consolidated considering <u>all</u> various items to be discussed in the EIR so that supplemental additions are minimum. Agency should expedite the reviewing process.
>
> Eliminate the EIR Process in established areas when no change in environment will take place.
>
> I think the idea of a "focused EIR" covering specific problems as relates to a specific site can be a valuable planning tool both to the builder and the public agency. The idea of creating a 10 lb. report with 9 1/2 lbs. of filler to fulfill the letter of the law makes no sense and is a waste of everyone's time and money.

EIRPs AND LIFE QUALITY

San Mateo has an environmental review process typical of counties in which reasonably strict environmental review standards are followed to protect the quality of life. The process for a single project can take several weeks to a year and involves numerous agencies' reviews and approvals. The county has been seeking ways to simplify and shorten EIRP-related processes. In 1979, costs of environmental impact reviews, charged on the basis of direct labor costs plus expenses, could have equaled the following:

Minimum fee	$750
Subdivision applications	450-500 + $50 per lot over three
Variance applications	300
Use permit application	450
Architectural review	200

An analysis of the development review process, completed in 1978, proposed a "comprehensive program to increase control and accountability over the development review process with joint goals of reducing processing times and increasing applicant convenience."

ORANGE COUNTY AND EMBATTLED DEVELOPERS

Perhaps the richest supply of anecdotal evidence on the extra costs of review processes, including EIRs, are found in reports developed by Orange County developers.[1]

Dedication and Lower Densities

A firm planned to develop homes on land claimed by the city of Yorba Linda; an old irrigation canal had been built on the land 80 to 90 years earlier. The city wanted the land for a bicycle trail. The builder submitted a plan and a proposed EIR; however, the city protested the plan and indicated that, through law suits and other processes, it was going to stop the project.

The firm feels that it was coerced into allowing the canal site to be annexed by the city. In order to complete the project the builder was required to reduce the dwelling density from 280 to 240 units, build part of an equestrian trail and landscape it at a cost of $50,000, and dedicate park sites. The process took one and one-half years.

Public Intervention

Broadmore Homes experienced a three-year delay in developing a major housing project primarily because of public hearings and related interventions. An EIR took 18 months to prepare, providing additional delays. A homeowner group intervened, and the project density was reduced to one home per 20,000 square feet. A conflict developed and continued between the densities the protesters wanted and those in the general plan. The firm accumulated losses connected with delays and changes of an estimated $600,000.

EIRPs AND CONTROLLED GROWTH

A number of planning jurisdictions have combined EIRPs with their overall general plans as a means of controlling growth. Napa and Marin counties and the city of Palo Alto are examples.

Napa County

Napa County has a well-articulated, controlled-growth policy. The plan, effective until the year 2000, calls for keeping agricultural and related activities as the primary land uses, with urban land uses concentrated in existing cities and urban areas. Future population size and its distribution is to be controlled, so that the quality of life can be maintained. Conflicts between human uses and natural resources are to be minimized. Work is to continue on all community elements to assure achievement of the overall growth policy. The goals and the policies to be followed in achieving them are shown in Table 6.1 and Figure 6.1.

Housing quality is to be maintained and improved through the use of minimum standards and enforcement, using the following as a guide: the provision of low-cost and moderate-cost housing in proportion to the number of low-income and moderate-income householders; population distribution and residential growth concentration preferably in existing incorporated and urban areas; an average annual population growth of approximately 1 percent but with the rate also related to national, state, and regional growth rates; flexibility, economy, and variety in housing, without destroying environmental quality; and urban expansion limited to the Residential Urban Limit lines in the 1975 Napa City General Plan.

TABLE 6.1

Napa County Planning Goals to the Year 2000

Goal	
1	To plan for agriculture and related activities as the primary land uses in Napa County and concentrate urban uses in the county's existing cities and urban areas.
2	To develop and implement a set of planning policies which combine to define a population size, rate of population growth and the geographic distribution of that population in such a manner that the desired quality of life is achieved.
3	To determine what the land is best suited for; to match man's activities to the land's natural suitability; to take advantage of natural capabilities and minimize conflict with the natural environment.
4	To work with cities, other governmental units, citizens and the private sector to plan for services, facilities and accommodations, including housing, transportation, economic development, parks and recreation, open space and other total county needs.

Source: Napa County Planning Department.

Napa County Fees

In connection with residential subdivision and EIRPs, the major fees include the following (effective January 12, 1979, representing substantial increases):

Parcel map	\$2,000 + \$20 per lot
Tentative subdivision map	400 + \$10 per lot
Final subdivision map	100
Time extensions	One-half the filing fee
Preliminary environmental review (categorical exemption or general rule)	15
Preliminary environmental review (Initial Study)	125
EIR contract negotiations	650
Renegotiation	250
Administration and review	1,500
Contract sum deposits	Equal to contract sum

FIGURE 6.1

Napa County Planning Goals–Policies Relationship to the Year 2000

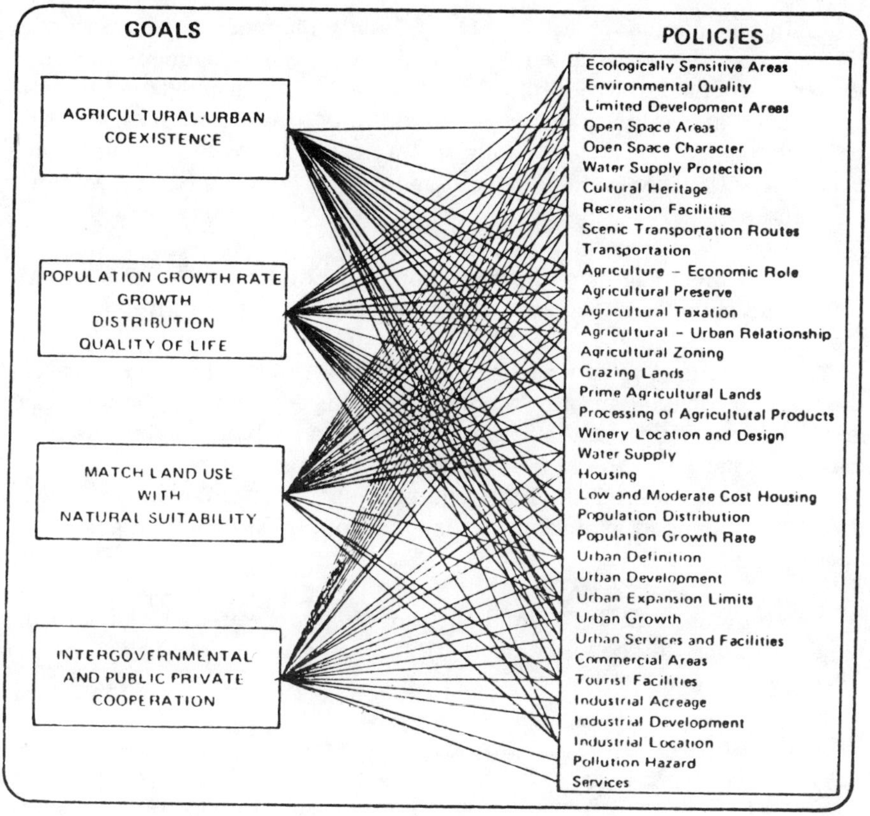

Source: Napa County Planning Department.

Marin County

In 1973 Marin County adopted a general plan, which has guided development since then. The plan indicates that in 1970 the county's population was an estimated 209,574, and that, if growth were not controlled, the population could reach 260,000 by 1980 and 300,000 by 1990. After several public meetings, the desired rate of population increase was set at 1.9 percent per year, with a population potential of 260,000 by 1990. In January 1980 the California Department of Finance estimated that the population would be 224,300 by July 1, 1980, and would reach only 233,300 by July 1, 1985, well below the limits set in the plan. The achievement of levels below those planned reflects the ability of the Marin County Planning Department to limit growth. Another measure of the effectiveness of the controls is the 81 percent decline in new housing permits between August 1978 and August 1979.

In the plan adopted, the Marin County planners recommended that actions should be taken by "city and county governments acting together to change direction in Marin—by initiating programs to acquire open space, to discourage precipitous growth, to diversify the housing supply, to increase jobs in Marin and to provide transportation that will support these developments." The plan has three general goals.

1. Discourage rapid or disruptive population growth, while encouraging social and economic diversity within communities and in the county as a whole.

2. Achieve greater economic balance for Marin, by increasing the number of jobs and the supply of housing for people who hold them and becoming more self-sufficient economically by reducing the present heavy reliance on the commute to San Francisco.

3. Achieve high quality in natural and built environments, through a balanced system of transportation, land use, and open space.

The plan is strictly enforced, making very effective uses of the usual planning devices and the EIRP.

City of Palo Alto

In 1976 Palo Alto adopted a general plan, which was intended to maintain the general low densities in its residential areas; slow employment growth; maintain existing housing to provide some new housing for low-, moderate-, and middle-income households; reduce automobile traffic growth; and change the appearance and function of El Camino Real. Since then the plan has been amended several times, but the original goals remain the same. In 1979 an important amendment reviewed the plans for better natural resource management. Among the goals of the plan, which generally was to maintain the supply of natural resources from which were derived community utilities at the lowest costs and in the most environmentally sound manner, were the following:

1. Maximize conservation and efficiency in the use of utilities and resources;

2. Encourage the appropriate use and implementation of alternative energy technologies; and

3. Manage solid waste disposal to reduce the amount of solid waste generated.

In the comprehensive plan, heavy emphasis is placed on the conservation of natural resources and the overall improvement of the visual appearance of the city, particularly the downtown business district. Environmental concerns are focused on natural water runoff, use of waste water, floods, the negative impact of human activities on the environment, the control of access to environmentally sensitive areas, the use of pesticides, and the protection of prime agricultural land.

The approval and review processes are intended to conform closely with the rather strict conditions associated with the comprehensive plan. Encouragement to follow the plan is provided through density bonuses, which are key to the providing of below-market rate housing.

Palo Alto Fees

Fees (estimated for 1979) can be increased by hundreds to thousands of dollars when modifications are requested or required for construction changes, grading, and other items beyond the usual requirements in the review processes. Fees in Palo Alto are the following:

EIR assessment	$ 75
EIR	1,000
Parcel maps, preliminary	120-$250 (relates to size)
Parcel map, final	50-$90
Appeals	90
Minor map changes	100
Zone changes	50

A typical five-acre residential development with four homes per acre could produce the following fees:

Zone change	$ 500
Parcel maps	170
EIRs (exclusive of reports for developer use)	1,075
Architectural review board	40-45

MITIGATION MEASURES AND NEGATIVE DECLARATIONS

Located on the San Francisco Bay just north of Alameda County, Contra Costa County currently serves as a "bedroom" community, since it lacks major sources of industrial employment (except for oil refineries in Martinez) and a transportation system sufficient to serve a large population. In many of the rural areas, topography often isolates sections of the land, which are also without adequate public facilities and services. Development is controlled through strict land-use regulations contained in an open-space plan, specific plans, and grading and ridgetop development regulations.

The environmental review process figures prominently in all residential development. Originally, the county planned to prepare all EIRs through its own resources or by contract with developers, but Proposition 13 cutbacks forced it to change the process in some cases. The county may require that a private person or organization furnish the data and other information from which the county can prepare an Initial Study. Among the fees relating to a residential subdivision, the following are the key ones:

Preliminary and tentative tract maps	\$2,000 + \$300 for each lot shown
Final map	500
Improvement plan	6 percent cash deposit, based on estimated improvement costs
Geological and soils report review	
For the EIR	\$ 500
For an environmental assessment	100

Typically, the county requires minimum deposits for services in excess of the anticipated total fees.

Contra Costa developers may sometimes shortcut the EIR preparation processes by securing a Negative Declaration (ND) with mitigation measures. Typical of this process is the report of a development by Edward F. Biggs. Biggs was seeking approval for the construction of 18 townhouse units, changed to 17. In return for approval to construct the 17 units, Biggs agreed to the following:

Reduce the units from the requested 18 to 17
Rotate units 90° from the direction planned
Buffer the site with trees and landscaping on all sides, together with a three-foot-wide planter strip along the western property line

Keep the development as a multi-family rental development and not convert it to a condominium
Provide a protective drainage ditch, maintainable from public property
Dedicate a 15-foot right of way along a major street
Construct a 6' 6" sidewalk, 18 feet of pavement widening, curbs, and drainage along another major street
Construct a downstream drainage facility, if the county decided it was needed
Place all utility distribution underground

There were other requirements, but these were the main ones, growing out of efforts to protect the environment from the adverse effects of construction of the development.

In another case a single-family home was to be changed to office uses. The developer was required to do the following:

Provide 22 parking spaces
Provide a ramp for the handicapped
Save existing trees and add more landscaping, as approved by the county
Use earth-tone paints
Comply with irrigation and landscape requirements stipulated by appropriate agencies
Install underground service facilities
Construct a 6' 6" sidewalk with curb along a major street
Provide storm drainage
Install street lights along the major street

Work on the project was begun in mid-1978, and final approvals were received in October 1979.

MITIGATION MEASURES AND DELAYS

Large portions of central Oakland are in clear need of renewal or redevelopment, and federal aid is being relied on heavily for financial assistance in doing some redeveloping. For this reason growth in Oakland is focused on redeveloping the central business areas. Planned unit developments and density bonuses are being used to attract new kinds of residential developments.

One major Planned Unit Development (PUD) was attempted in Oakland but failed after three versions were planned; the EIRP played an important role in what happened. The project was to be located in the hilly portion of Oakland, which is topographically separated from central Oakland.

In the first development, higher-priced homes were to be placed around an artificial lake. The total development was to include 560 acres. The EIRP uncovered potential seismic activity, which would threaten the lake, so the project was abandoned as planned.

In the second plan for the development, the city asked for a donation of 150 acres for a city park. In return the developer was given density bonuses for residential properties in the remainder of the 560 acres. Approval delays forced the developer to abandon this project.

In the final plan very strict mitigation measures provided for cluster development and a small golf course. By this time inflation, rising construction costs, and other forces required that the projects be planned for upper income families, which in Oakland, with a large minority population in the central area, meant essentially white families.

Fees charged by Oakland city planners, most of them directly related to residential construction and EIRPs, included the following:

Subdivision fees (in connection with an approved PUD or use permit)	$ 250
Parcel map	250
Design review	100
EIR	350
Negative Declaration	50
All appeals from permit decisions	125
Total	$1,325

EIRPS AS CRITICAL ELEMENTS IN PLANNING

The EIRP in Santa Clara, as in all counties, plays a most important role in residential subdivisions of five or more parcels. First a tentative and then a final map is prepared, public hearings are conducted, and a preliminary report or finding is made. This process may take as much as 50 days, after which the approved map, together with all conditions of approval, must be implemented within 18 months.

The cost of a typical 100-lot subdivision under these processes could be as much as the following:

Subdivision application fees ($1,000 + $50/lot)	$6,000
Subdivision map checking fees ($500 + $10/lot)	1,500
Map recording fee ($5 first sheet, $2 each additional)	205
Environmental assessment (if required)	100
Total	7,805

If the staff determines that a project may have a significant impact on the environment, an EIR is required. When an EIR is required, the applicant pays an additional $2,000 and meets with the environmental planning specialist to prepare a preliminary EIR. The staff reviews the preliminary EIR for completeness and then instructs the applicant in the preparation of a draft EIR, which is distributed and reviewed at a public hearing. After any reactions to the EIR are brought out at the hearing, the needed action must be undertaken before the completed EIR can be approved and the project begun.

As is true in other jurisdictions, additional fees can be incurred for use permits when the use may not be permitted under existing planning regulations. Other fees can be incurred for the following:

Initiation of zone changes	$ 600 filing fee 400 environmental assessment
Zoning variance	50
Appeal, when rehearing is requested on previously approved proposal zone, architectural site, historical or archeological requirement	50
Encroachment permit (involving public land)	75 minimum
Grading permit	250
Environmental assessment ($100 + EIR $1,500)	1,600
Geological hazard review	variable

OTHER KINDS OF IMPACT ANALYSIS

As environmental impact reviews became more intertwined with other land-use-related decisions and analyses, more and more planning jurisdictions experimented with extending environmental impact analysis to fiscal, social, economic, public service and facility, housing, and transportation impacts. Limited budgets, Proposition 13, and changing economic conditions put strains on local governmental budgets, making these added impact analyses more and more attractive.

In an effort to determine the extent to which these extended analyses were included as a part of the EIRPs, we contacted several major builders, asking them what types of impact analyses had to be considered in connection with projects on which they were working. The

TABLE 6.2

Seven Builders' Reports on Environmental Impact Review Processes, 1979

Item	Builder/County						
	1 San Mateo	2 Orange	3 Los Angeles	4 Los Angeles	5 Orange	6 Orange	7[a]
Number of units							
Permitted under zoning	150	25	17	272	548	23	
Unit type	Condominium	Single-family detached	Single-family detached	Single-family detached	Condominium	Single-family detached	
Number applied for	150	25	17	136	548	23	
Number built	68	25	17	Not start	548	23	400
Construction time							
Months		10	10		24	8	
EIRPs							
Approval time (in months)	33	Negative Declaration		20	19	Negative Declaration	
Weeks to prepare EIR	20			8	20		
Pages in EIR	95			75	228		
Cost to prepare	$15			$9	$10		
Types of impact assessments required[b]							
Natural/environmental	A			A	C		
Fiscal	A			A	C		
Social	A			A	C		
Economic	A			B	C		
Public service/facility	A			A	C		
Housing	B			B	C		
Transportation (traffic)	A			A	C		
Physical changes required[c]							
Street construction	X			$500			
Open space dedication	X			75 acres	$185		
Reduced units	X			50 percent			
Landscaping	$11		$15	$50	$493		
Schools payment/dedication				Pending			
Other (noise)					$274		

[a]Does not build where EIRs are required

[b]A = adverse, B = positive, C = both.

[c]X means builder responded affirmatively but no indication of cost was included.

Note: Blank spaces indicate that no response was received.

Source: Builder reports.

experiences of seven are presented in Table 6.2. One of the seven avoids jurisdictions where extensive EIRPs exist. In fact, at the time he was contacted, he mentioned that he was considering leaving California and conducting business in adjacent states without EIRPs.

Three builders indicated they were required only to present the usual environmentally related impact analyses. Three reported encountering additional analyses. One, for example, was required to analyze both the positive and adverse impacts related to his project. Two builders were essentially required to consider the adverse impacts of their projects. However, both had to be concerned only about the positive impacts on housing.

SUMMARY OF BUILDERS' PERCEPTIONS OF EIRPs

Generally builders reflect a feeling that EIRPs are not well-organized or consistent. A single or limited objective in requiring EIRPs is rarely provided as a guide to help them determine what information and analyses they should provide in the EIRPs, even though California law seems to have some rather well-defined goals and processes. Local jurisdictions do appear to have relatively wide latitude in setting requirements for EIRPs. However, a careful analysis of the examples provided in this chapter (along with others considered but not discussed), reveals a commonality of activities, which permit the classification of major costs in EIRPs into the following categories:

1. Direct fees, preparation, and presentation costs for reports, forms, and so on;
2. Related to other goals, but not necessarily connected with EIR considerations; these include requirements for meeting general and specific plans, controlled growth goals, and protection of a perceived existing quality of living;
3. Interagency conflicts, which require duplication of reviews, reports, and hearings;
4. Delays related to interagency reviews and requirements, public intervention, and uncertainties in review goals or purposes;
5. Mitigation measures arising from identification of adverse or cumulative impacts. Usually mitigation measures require reduction of housing densities, dedication of space, provision of or funding for public services and facilities.

Many such costs are related to the added kinds of impact analysis being required in more and more jurisdictions. In the next chapter we use additional organized statistical evidence to estimate what the costs may be for EIRPs.

NOTE

1. William J. Tater, An Appraisal of Environmental Protection Legislation and Administration upon the Economy in General and upon Real Estate Development in Orange County in Particular (Fullerton: California State University, School of Business Administration and Economics, December 1975); and The Housing Crisis (Orange County, Calif.: Californians for an Environment of Excellence, Full Employment, and a Strong Economy through Planned Development, 1980).

7

ENVIRONMENTAL IMPACT REVIEW PROCESSES—WHERE THE COSTS ARE

The earlier chapters have analyzed California housing price and cost trends, traced the development of the California Environmental Quality Act (CEQA), reviewed the structure of the environmental impact review processes (EIRPs) mandated by CEQA, discussed the theoretical aspects of the EIRP costs and benefits, and presented anecdotal evidence on the builders' and planners' views of the EIRPs. This chapter presents a more explicit statistical and anecdotal review of CEQA processes and costs. The goal is to identify places in the process where it might be changed, modified, supported, and improved. We accept the existence of EIRPs and the preparation of Environmental Impact Reports (EIRs); we believe that they are here permanently, having something to offer those seeking better quality in residential living at more moderate costs, and we will offer means of improving the processes.

The analyses in the previous chapter indicated that CEQA-related costs can be divided into four specific classes, each of which are reviewed in this chapter.

1. Direct charges and fees: costs charged to the builder/developer for the preparation and processing of documents, particularly the EIR;
2. Process-directed mitigation costs: the costs created when impacts have been identified and something must be done about them;
3. Indirect process effects, delays, and abandonments: these are not direct nor easily calculated costs, but they reflect various elements of multiple reviews, opportunities delayed or forgone, and complexities of the processes;
4. Market effects: EIRPs tend to change market processes, sometimes speeding them up, sometimes slowing them down, sometimes diverting them, sometimes affecting housing markets beyond the regulated area.

After reviewing the evidence on each class of cost, the chapter will profile the review processes under CEQA, estimating the probabilities of having to incur or avoid additional costs at each step and the possible costs that might be incurred. Using this profile, we then recommend places where management of the processes might be improved. As always, our concern is for a better functioning, more effective EIRP.

ESTIMATED INCIDENCE OF EIRPs

CEQA conditions must be met for almost every kind of residential process, but in 44 percent of the cases no review is required (see Table 7.1). For the projects that must be reviewed one-half will be subject to required reviews and one-half to discretionary processes. For private projects (the focus of this study) only 15 percent will be subject to full-scale environmental impact review, 20 percent to focused EIRs, and the remainder to some form of Negative Declaration. We estimate that less than 4 percent of all new California residential construction is subjected to EIRPs annually.

There are ten major places in the EIRP where private project proposals can encounter costs (see Table 7.2). The first occurs when a developer is proposing a residential project and confers with the appropriate review authorities about it. Consistently about 70 percent of all residential projects will require some form of EIR, with most only requiring one submittal of it. This step, as will be explained later, is probably the least costly step, since only a proposal, not a full EIR, is required at this point.

The proposal is evaluated to determine whether CEQA regulations will apply; almost one-half (45 percent) will be exempt from further review. Of the projects subject to further review, only about one-third (35 percent) will be subject to the preparation of an EIR. The remainder will be given a Negative Declaration (ND), with almost all of the NDs imposing no further requirements.

An EIR can be costly to prepare; fortunately only one-half will have to be revised and resubmitted. The usual causes for additional submissions relate to inadequate or missing specialists' reports relating to possible impacts, the scope of the impacts, and possible mitigation measures. When impacts are identified, almost two-thirds will relate to local or neighborhood conditions. The most frequent impact, which often requires expensive additional studies, relates to traffic congestion or degradation of local native flora and fauna.

Under CEQA, project applications are to be accepted within 30 days and EIRs fully reviewed and certified within one year of acceptance of the application. Unfortunately, only one-half of all applica-

TABLE 7.1

Estimated Incidence of Environmental Impact Review Processes in California, 1979

Process	Incidence (in percent)
Private residential projects subject to review	56
Required reviews	28
Discretionary*	28
Discretionary projects subject to full review	38
Required	28
Actually reviewed	10
Private projects actually reviewed	
Subject to full-scale EIR and review	15
Focused EIRs	20
Conditional Negative Declaration	7
Nonconditional Negative Declaration	58
Total	100

*Discretionary here means a project not requiring an EIR as a matter of course in the jurisdiction.

Sources: Small and Knust, Inc., California Local Government and CEQA, 1979, a report submitted to the Assembly Committee on Rules, California Legislature (San Diego, Calif.: Small and Knust, Inc., 1979); Professional Builder Association, Economics Research Department, Washington, D.C.; Environmental Analysis Systems, The California Environmental Quality Act (CEQA): An Evaluation Emphasizing Its Impact upon California Cities and Counties with Recommendations for Improving Its Effectiveness (San Diego, Calif.: Environmental Analysis Systems, 1975).

tions are accepted, and only 80 percent of EIRs are certified in the required time. In many cases the extended time results from agreements between the submitter and the reviewers that more time is required because of special problems associated with a particular project. Sometimes extended reviews rather than required legal time reviews will be accepted by developers rather than risk complete rejection of a project.

Only one-third of the applicants (35 percent) will be responsible for preparing the EIR, and even then two-thirds will be required to use consultants from a list supplied by the review agency. Very useful information and clues as to the treatment of various EIR elements can be found either in earlier EIRs or from agency-maintained files; however, such data sources are available only about 50 percent of the time.

Jurisdictions tend to have widely varying attitudes about public participation. Jurisdictions with fragile environments and active property owners and environmental groups are likely to have extensive public reviews. Whatever the attitudes of the agencies, those who use public notification will receive public comments in only about 30 percent of the cases.

Once reviewed, 77 percent of the projects will be approved, 4 percent will be withdrawn (usually to be submitted in a new format at a later date), 7 percent will be denied, and 13 percent receive other unknown or not reported fates.

BUILDER REPORTS ON COSTS AND PROFITS

Accurate information about home builders' costs and profits is difficult to obtain, primarily because each residential project is so unique that comparison among projects would be difficult. Some clues to cost trends can be found in the standard cost estimating handbooks.[1] Examination of the records of individual builders reveals little consistent information because of a lack of standardized cost allocations. The best overall information is provided in the surveys reported by Professional Builder (see Table 7.3).

Major builder cost elements are identified as land development, construction overhead, profit, and financing. Among these costs the most notable increases were reported in financing and land development (the most likely place for added EIR costs to surface). Construction costs declined as a percentage of total costs between 1977 and 1979. Overhead and profit remained at approximately 15 percent to 17 percent of total costs.

The increase in land development costs from 11 percent to 25 percent of total costs could reflect the increased costs associated with environmental impact review processes; unfortunately, the survey provides no information on this. To some extent EIR-associated costs could be reflected in the doubling of finance costs, since these financing costs usually begin with land acquisition. In builder language, a continuing problem is that the financial clock is "ticking" throughout the project planning and development stages.

Other views of builder costs and profits are provided in Table 7.4. The information was collected from several hundred Orange County builders, reflecting their reports on the sources of costs in

TABLE 7.2

Frequency of Occurrence of Selected EIRP Elements in California, 1979
(in percent)

Step/Process	Frequency	Probability
1. Initial proposal		
Percentage of residential projects requiring proposals		70
Number of submittals to determine need for an EIR		
One	70	
More than one submittal required	30	
Subtotal	100	
2. Review for CEQA applicability		
Exempt from further review (categorical or other)		45
Subject to further review		55
Total		100
3. For projects subject to further review		
Subject to an EIR		35
Full-scale EIR	43	
Focused EIR	57	
Subtotal	100	
Received Negative Declaration (no significant effect)		65
Conditional[a]	11	
Nonconditional	89	
Subtotal	100	
Total		100
4. Frequency of proposed draft EIR submittals[b]		
Once		50
More than once, but not continuous		20
Continuous consultation and review[c]		30
5. Occurrence of impacts[d]		
Local or neighborhood		64
Regional		36
Occurrence of inadequate analysis of submittals		
Traffic congestion	30	
Degradation of native habitat	20	
Noises from project	18	
Archeological	12	
Energy consumption	10	
Adequacy of public services	10	
Subtotal	100	
Total		100
6. Review agencies' response times		
Met deadline (30 days) for acceptance of application		50+
Met deadline (one year) for certifying EIR		80
7. Preparation of EIR documents		
Applicant responsibility		35
Select from agency list and contact consultant	14	
Select consultant independently, direct contact	28	
Select consultant from agency list, agency contacts	58	
Subtotal	100	

(continued)

TABLE 7.2 (continued)

Step/Process	Frequency	Probability
Agency responsibility		65
Staff prepares EIR or uses consultant specialists	23	
Agency contacts consultants, contracts with consultants	69	
Other (or not known)	8	
Subtotal	100	
Total		100
8. Data sources		
Earlier EIRs or reference data file		50+
9. Public notification[e]		
Percent of comments on impacts of all types		
By public	31	
Prior to public hearings	9	
By public on unique issues	5	
By responsible public agencies	5	
By public agencies on overall issues		
Local agencies	12	
State agencies	4	
10. At the EIR state, findings related to mitigation measures and disposition of impacts		
Mitigation proposed		63
No mitigation		22
Impact considered favorable		2
Regional impact and not mitigable		13
Total		100
11. Disposition of project applications		
Approved		77
Standard conditions	29	
Special conditions	17	
No conditions	44	
Overriding considerations	10	
Subtotal	100	
Withdrawn (project)		4
Environmental reasons	25	
Other reasons	75	
Subtotal	100	
Denied		7
Environmental reasons	86	
Other	14	
Subtotal	100	
Other disposition (or not known)		13
Total		100

[a]Revised process permits Mitigated Negative Declaration, see Chapter 4.

[b]Number of submissions of proposed draft EIRs before agency adoption of the draft EIR.

[c]Causes for continuous consultation: inadequate analysis of traffic congestion, soils, geology, seismic hazards, archeology, wildlife habitat, public service elements.

[d]All impacts classified under these headings.

[e]Methods in order of use: special-interests mailing list, newspaper publication, direct mail to contiguous owners, posting on site, and posting off site.

Note: See flow chart in Figure 4.1.

Sources: Small and Knust, Inc., California Local Government and CEQA, 1979, a report submitted to the Assembly Committee on Rules, California Legislature (San Diego, Calif.: Small and Knust, Inc., 1979); Professional Builder Association, Economics Research Department, Washington, D.C.; Environmental Analysis Systems, The California Environmental Quality Act (CEQA): An Evaluation Emphasizing Its Impact upon California Cities and Counties with Recommendations for Improving Its Effectiveness (San Diego, Calif.: Environmental Analysis Systems, 1975).

TABLE 7.3

Distribution of Builder Costs for Selected Years, 1949-79

Type of Cost	Percent of Total Costs 1949	1969	1974	1977
Construction	69.0	54.6	48.4	46.7
Land development	11.0	21.4	24.6	25.0
Financing	5.0	7.0	10.0	10.8
Overhead and profit	15.0	17.0	17.0	17.5
Total	100.0	100.0	100.0	100.0

Source: Professional Builder, 1978.

residential construction in that county. These analyses were originated by the builders to help explain why average home prices in Orange County had risen from less than $30,000 in 1965 to almost $150,000 in 1980.[2] Their conclusions were that overregulation, including EIRPs, were an important source of increased costs and land shortages. For example, their major conclusions were that this overregulation restricted the amount of developable land, created an "expensive, tedious and time consuming permits process," provided minor monopolies for some builders, and transferred too many community development and service costs to new homes.

Using the same costs categories as Professional Builder, the analysts reported that between 1972 and 1978 finished lot costs rose from 28 percent to 35.3 percent, structural costs dropped from 47 percent to 34.1 percent, other costs rose from 9 percent to 18.8 percent (fees 3 percent to 4.1 percent, financing from 3 percent to 4.7 percent, and marketing remained steady at 3 percent), and profit and overhead rose from 16 percent to 18.8 percent (despite development profits declining from 8.3 percent to 7.5 percent). Inflation was estimated to have caused 13 percent of the price increase for that six-year period. Judging from their analyses, EIRPs probably had the greatest impact on raw land cost (14 percent of the total 1978 price) and the necessity of providing "public and community improvements that historically have been paid for by the community at large through bond issues and taxes." Included in community improvement costs were parks, open space, greenbelts, and bicycle trails.

The place where EIRPs fit within typical residential development processes and the place of the EIR within these processes are suggested in Table 5.5. This table reinforces the conclusion that the major im-

TABLE 7.4

Distribution of Housing Costs by Type

I. Building Site

  A. Land Acquisition

    1. Market study
    2. Preliminary engineering studies
    3. Appraisal
    4. Escrow fees
    5. Brokerage commission
    6. Title policy
    7. Legal
    8. Miscellaneous

  B. Predevelopment Costs

    1. Design costs

      a. Land planner
      b. Civil engineer
      c. Soils and geology
      d. Architect
      e. Landscape architect
      f. Special consultants
      g. Miscellaneous

    2. Governmental processing fees and assessments

      a. Master plan
      b. EIR
      c. Zoning
      d. Subdivision plan (tract map)
      e. Grading permit
      f. Special (that is, Coastal Commission)
      g. Building permit
      h. HVAC (heating, ventilation, and air conditioning) permit
      i. Plumbing permit
      j. Electrical permit
      k. Assessments for water, sewer, flood control, fire protection, schools, parks, open space, and special
      l. Other charges, fees, and assessments
      m. Legal

    3. Builder's overhead and staff time during the predevelopment period for coordination and processing of the above activities

C. Off-site Improvements

1. Access roads
2. Grading
3. Drainage and flood control
4. Utility extensions
5. Landscaping and irrigation
6. Traffic signals and signs

D. On-site Improvements

1. Grading
2. Streets
3. Sidewalks
4. Street lights
5. Underground electrical distribution
6. Underground telephone distribution
7. Water distribution
8. Gas distribution
9. Sewer
10. Landscaping and irrigation
11. Underground CATV distribution

E. Community Amenities

1. Dedications for schools, parks, open space, utilities, installations, street widening, and so on
2. Construction of parks, greenbelts, open space, and recreational elements

II. Housing Unit and Appurtenant Structures

A. Basic Structure: Labor and Materials

1. Excavation
2. Foundations and slab floor
3. Fireplace and chimney
4. Rough carpentry
5. Roofing
6. Stucco, exterior walls, and soffits
7. Gypsum board, interior walls, and ceilings
8. Windows
9. Doors and trim
10. Wardrobe doors
11. Cabinets
12. Painting

(continued)

TABLE 7.4 (continued)

13. Finish floors
14. Finish carpentry
15. Plumbing (including fixtures)
16. Electrical (including fixtures)
17. Heating
18. Insulation
19. Rough hardware
20. Finish hardware
21. Driveway and walks
22. Miscellaneous

B. Amenities: Labor and Materials

1. Range and oven
2. Dishwasher
3. Disposer
4. Carpets
5. Air conditioner
6. Special detail and finish, interior/exterior
7. Landscaping, walks, patio, irrigation

III. Financing

A. Cost of Equity Funds
B. Cost of Construction Loan(s)
C. Cost of Permanent Loan(s)
D. Legal Fees
E. Insurance and Bonds
F. Title Fees
G. Escrow Costs
H. Recording Costs

IV. Marketing

A. Model Complex
B. Brochures, Models, and Graphics
C. Advertising
D. Promotion
E. Sales Commissions
F. Closing Costs

V. Project Administration and Builder's Overhead

VI. Profit and Contingency

Source: Orange County Cost of Housing Committee, The Cost of Housing in Orange County (Orange County, Calif.: Orange County Board of Supervisors, February 20, 1975).

pact of EIR-associated processes is on land preparation costs; when EIR costs can be estimated, the direct costs are a very insignificant part of finished lot costs, but indirect, delay, and mitigation costs can be a significant portion of finished lot costs.

The questions of how significant some of these EIR-related costs might be and how they impact costs are presented in the remainder of this chapter. The observations about costs tend to add to and shed more light on the cost and price data presented in Chapter 2.

DIRECT CHARGES AND FEES

The most obvious costs imposed on development by CEQA are those that must be paid out of the pockets of the developer and local government for the EIRP. These include the costs paid for preparation of the needed documents and the costs of processing those documents. Different jurisdictions have chosen to distribute those costs between the public and private sectors in different ways.

The cost of a Negative Declaration, prepared when a project is subject to the CEQA but will have no significant impact, is paid largely by the local government. Little research as to the makeup of the effort that goes into the preparation of a Negative Declaration has been carried out, but an early study done for the city of Los Angeles estimates the hours and personnel required as being those displayed in Table 7.5. Our interviews indicate that this effort remains substantially the same in most cases, but that Mitigated Negative Declarations or unusually complex projects may well require more. If the costs estimated in this early study are adjusted for inflation, an average cost to the local government of about \$140 (in 1981 dollars) is reached. Our survey indicates a fee of about \$50 is normally charged developers for preparation and processing of the Negative Declaration.

Should an EIR be required, the costs in most cases must be borne directly by the developer, as indicated in our survey reported in Chapter 4. Of course, the costs of preparation will vary greatly according to the environmental sensitivity of the site, the scale of the project, and the competence of the preparer. A number of cost estimates for EIRs may be reported from past research. James and Muller of the Urban Institute reported in 1977 that the costs of EIRs for the projects of San Diego area builders they had interviewed was between \$4,000 and \$10,000.[3] Our survey of builders done for this report indicated a range between \$9,000 and \$15,000 for such EIRs. Finally, in the reports to the state legislature done by Small and his colleagues, an average cost, in real dollars, of a private EIR was reported as \$12,420 for 1979, which was an increase from \$6,655 in 1974.[4] Thus, an estimate of \$14,000 to \$15,000 as an average would

TABLE 7.5

Los Angeles City Planning Department's Preparation of Negative Declarations, 1973
(in 1973 dollars)

Description of Steps	Level of Personnel	Time (in hours)	Hourly Rate	Cost
1. Pick up application and environmental assessment form	Planning assistant	0.5	7.10	3.55
Check application form for completeness				
Assign file number and describe in log the type of project, location, and so on				
Code according to type of application				
Record in map books				
2. Review file for completeness	City planner	0.25	9.88	2.47
3. Make up file jacket and organize data within file	Senior clerk stenographer	0.25	4.46	1.11
Type file card				
4. Perform appraisal of project	City planning associate	1.0–3.0	8.38	8.38–25.14
5. Final review of project file prior to presentation to ERC	City planner	0.25	9.88	2.47
6. Consideration of project by the Environmental Review Commission	City planner	0.25	9.88	2.47
	City planning associate	0.25	8.38	2.09
	Principal city planner II (2)	0.25	14.12	7.06
	Chief zoning administrator	0.25	14.89	3.72
7. Type Negative Declaration	Senior clerk stenographer	0.50	4.46	2.23
Prepare copies for distribution and mail				
Separate application from file and return to counter				
Update file card and place file in filing system				

Note: Direct labor cost = \$35.55–\$52.31
Indirect cost (at 47.2 percent) = \$16.78–\$24.70
Total cost = \$52.33–\$77.01
Average cost = \$64.67

Source: Ultrasystems, Analysis of Environmental Procedures within the City of Los Angeles Planning Department (Newport Beach, Calif.: Ultrasystems, 1973), p. 53.

not be out of line for a project. This is a project estimate and must be taken in that light when considered.

The final class of costs for the EIRP is the cost of processing the EIR through to its conclusion. These costs vary greatly by project, and we have no useful overall estimate of the average per project costs involved. However, we may make some observations based on data accumulated. The 1979 Small report made an estimate of the local government yearly costs of about $11,000 for private project review. This compared with a yearly net cost for the review of such projects in 1974 of $64,000 ($96,000 in equivalent 1979 dollars). This decrease was because of the greater imposition of processing fees on developers in that time period, the total cost remained unchanged at about $230,000 per year, including both public and developer costs.[5] James and Muller estimated the costs of environmental impact review in San Diego at about $1,500 to $3,000.[6]

Different communities have imposed different processing fees upon the developer for EIR processing and review. The range may be seen in Table 7.6, which reports the EIR preparation and review

TABLE 7.6

County and Private Project, Environmental Impact Report Charges and Processing Fees

Jurisdiction	Fees
Northern California counties	
Alameda	Cost
Contra Costa	Cost + 25 percent
Marin	Cost + 25 percent
Napa	Cost + $1,600
San Mateo	Cost + $750 + $20/hour staff time
Santa Clara	Cost + $1,500
Sonoma	Cost + 10 percent
Southern California counties	
Ventura	Cost + $43/hour staff time
Orange	Cost + fee schedule ($850 maximum)
San Diego	Cost
Los Angeles	Fee schedule ($1,264-$6,502)
San Bernardino	Cost + $600

Source: Survey by the authors.

TABLE 7.7

Environmental Impact Report Fees and Total Development Fees for a 50-Unit, Single-Family, Detached Residential Development, Orange County Development Fee Survey, 1980

	Negative Declaration	EIR Review	Total Fees
Minimum county fees	$1.00	$34.00	$4,384.90
Average minimum city fees	1.65	9.23	3,449.93

Source: Alfred Gobar and Associates, Housing Costs Analysis —Orange County (Brea, Calif.: Alfred Gobar Associates, 1980).

costs for counties from our survey. Percentage processing fees of up to 25 percent can add as much as $4,000 to the cost of the project for the developer. However, a charge of $2,000 would be closer to the norm based on our data.

The direct costs of the EIR and the environmental assessment process may be accurately described as fairly small. Though varying widely from project to project, the relative amounts are modest. Most projects are subject to Negative Declarations, which we estimate cost the local government about $140 and for which the developer is charged $50. Should an EIR be required, it will cost upward of $15,000 in all likelihood and will be paid for by the developer. A processing fee will also be charged, and that will generally be in the range of $2,000. Little of the cost of review will be borne by the local government. James and Muller estimated a "typical" California development as having 100 to 300 units and calculated average costs of EIR preparation at about $20 per housing unit and review costs at about $6 per unit, assuming 300 units in a project.[7] This data was from the 1974 time period. Making similar assumptions, we may estimate current per unit costs at $50 for EIR preparation and roughly $6.50 for reviews charged to the developer. In order to put this magnitude in perspective, Table 7.7 indicates EIR fees and total development fees from a survey of fees in Orange County.

In summation, the direct costs of the EIRP are small in the context of overall development fees and in terms of average housing costs exceeding $100,000. Most projects are subject to a Negative Declaration, which imposes minimal direct fees. This may increase owing to increased use of Mitigated Negative Declarations. This change, which will likely reduce overall costs of the process because of transfers of projects from full EIR preparation, will only make the minimal costs even smaller.

MITIGATION COSTS

Mitigation costs are created when an adverse impact has been identified and something must be done about it, except when the impact is considered regional in scope and so not mitigable by anything that the developer could do on a specific project. Extensive studies of mitigation costs between 1974 and 1979 for California projects caused the consultants to conclude that "the actual costs of mitigation of the typical private project is low and . . . has not changed substantially from 1974 . . . [particularly] in relation to site selection and project design, actions taken prior to submittal to reduce potential environmental impacts."[8] Frequently the design changes required because of adverse impact reports would have been made without the review processes.

Usually review agencies will not reject a proposed project outright but will suggest ways in which the adverse impacts might be reduced or eliminated. Such recommendations usually include density reductions, open-space dedications—perhaps for parks, trails, or the construction (or payment through a bond) of streets, traffic lights, sidewalks, schools, and sewers. Best estimates of these costs in California in 1977 were that they equaled $165 per housing unit or 0.6 percent of the final sales price.[9] A telephone survey in 1976 by the Rutgers University Center on Urban Policy Research of 133 builders indicated that at least three-fourths had accepted design changes or made them to ensure quicker reviews and approvals.

Small and Knust also reported that more than one-half of all projects reviewed in San Diego in 1976 encountered some type of adverse impact report requiring mitigation measures. Among these, in order of occurrence, were traffic congestion, air pollution, aesthetic degradation, wildlife habitat degradation, water runoff or pollution, noise, and growth inducement.

Perhaps the most controversial and frequent mitigation measure is that requiring density reductions. Carefully structured research reports contained evidence that density reductions did increase home prices and reduce limited supplies of developable land.[10] New house price differences between communities were found to be $1,300 more for equivalent units in communities with density limitations than in those without limitations. Even one acre of developable land removed by limitations increased housing prices. However, they also pointed out that the precise impacts of growth controls and density restrictions were difficult to measure because of other market influences, particularly inflation.

In evaluating all forms of land-use policies in San Francisco counties, Dowall and Landis concluded that density restrictions tended to exert the most "substantial inflationary effect" and that the results of the San Francisco study supported the theoretical finding that controlled land uses do increase housing prices.

DELAY, UNCERTAINTY, AND PROJECTS FORGONE

Delays create a variety of measured costs, including inflation, carrying and finance costs, inability to proceed on other projects, or actual abandonment of a project. The previous analyses suggest that, insofar as can be determined, none of these costs is very large. Equally important consultants found that in 1979 many projects, which had been approved by CEQA processes, were not built or started within permitted time limits.[11] They also emphasized that large projects, which were to be built over periods of years, suffered no significant cost through review delays. In many cases the analysts also found that approved projects were delayed in starting because of a lack of funding.

In contrast to this view Gobar found that in Orange County, where inflation and financing costs were increasing very rapidly, a delay of even months could increase land development costs significantly.[12] Gobar also pointed out that higher land costs induced builders to construct more expensive homes in order to keep the land-house ratio in balance and estimated that this caused home prices to increase by 16.3 percent between 1972 and 1978, if inflation costs were also considered. Rising land costs also encourage builders to buy and hold vacant land, further reducing the supply of available land. Similar findings were also reported by the Construction Industry Research Board for Los Angeles in 1975.[13]

In many projects, delays in one part of the review process will be compensated for by more attention to other parts of the project; the net result is that there is not total delay time. The extent of the delay is also influenced by the point in the project where the delay(s) occur. If delays occur after financing has been secured, plans completed, and contracts made with subcontractors, delay costs can add up rapidly.

An early review of California project delays suggests that the average delay in discretionary decisions amounts to 3.87 months for EIRs and 1.79 months for Negative Declarations.[14] Inflation was estimated to add between 6.5 percent and 13.0 percent to project sales costs; total of holding, net inflation, and overhead costs added 13.7 percent to 21.2 percent annually to project sales costs.

A frequently mentioned delay cause is the increased intervention in EIRPs because of the requirements for public inputs or hearings.

The overall data suggest that this is still not an important cost, but that public participation and comment are increasing. The most frequent comments tend to relate to traffic congestion, noise, and densities. Public inputs that seem to stem from EIR analyses are frequently more properly related to plan or zoning changes or variance requests.

A lack of data or incomplete data reports can also contribute to delays. Delays related to data needs can run from a few days to several months, particularly when the data to be collected are complex and difficult to interpret. Unfortunately much of the data collected for EIRs is rarely examined or relied upon for decisions about projects. Surveys among builders revealed that many felt the bulk of data included in an EIR is not needed or ever referred to.

A 1979 report suggested that delays were becoming increasingly long and beginning to contribute significantly to overall costs.[15] The overall increase over 1976 was estimated at 200 percent, representing delay increases from four months (in 1976) to seven months (in 1979) associated with the preparation and processing of Environmental Impact Reports, while delays in receiving Negative Declarations had been reduced from two months to one month.

MARKET EFFECTS

The final class of costs for the CEQA is the most difficult to evaluate. The impacts of the EIRP fall not only on the individual project but also on all other prices within the relevant housing market. These effects stem from two sources—the added costs of building and the decrease, if any, in the supply of land and housing units.

The EIRP is somewhat different to evaluate than other so-called environmental controls. Most studies of the costs of environmental regulations focus upon coastal zone regulations and growth management plans (as indicated in our review in Appendix E). These are control forms usually rooted in a master plan concept; generally, they involve predictable outcomes in the case of individual decisions. The EIRP is different. It is designed to provide added information upon which to make those discretionary decisions to which it applies and to make certain that environmental concerns are adequately considered. It mandates no particular outcome. Indeed, those mitigation measures, which sometimes follow from the environmental review, might well have been imposed prior to the CEQA.

A second factor that makes it difficult to evaluate the effects of the CEQA is the multitude of other forces that have played a role in the massive increase in housing prices since the early 1970s.[16] In addition to changes in demographics, various regions have been subject to differing economic conditions and are at different points in the shift fron rural to urban. Also, regulatory efforts such as growth controls[17] and the coastal controls imposed by voter initiative in 1972[18] are concurrent. While statistical analysis, through regression methods with hedonic pricing, may yield some estimate of the magnitude of the effect, large numbers of simplifying assumptions about market structures, economic trends, and process forms would be required. However, the most important barrier to such analysis is the lack of data readily available.[19]

A final difficulty with estimating price impacts in a housing market is the differing impact of controls in differing market conditions. Reasonably clear evidence of the extent of this is provided in two studies completed for the Urban Land Institute in San Jose, California, and Jacksonville, Florida.[20] Though these were growth management studies, the same principle applies. In the case of San Jose, restricted housing supplies produced by the controls in the face of strong demand pressures permitted builders to pass on extra costs and to protect profit margins. However, in Jacksonville, where the market demand for housing was not as strong, builders could not pass along many of the increased costs to lot or home buyers. The study emphasizes that generalizations could not be made from the two case studies because of that difference.

We have attempted to compute statistical correlations between measures of environmental review intensity derived from survey data jurisdiction responses to process questions and from expected process time calculated from times and likelihoods of EIR and Negative Declaration requirements. These were correlated with housing price changes and land value changes calculated from assessment records for ten of the counties. Housing cost and assessment data are primarily available on a countywide basis, and these were used on the basis that countywide areas are reasonable, if not entirely accurate, boundaries of housing markets. County jurisdictions were examined, as the greatest EIR application is to subdivisions, which are most likely to be under county jurisdiction. Taking weak nonparametric correlations (gamma, Spearman's rho), no statistically significant relationships could be found. This is not surprising in light of the small cost impacts of the EIRP on those housing units being built each year and the small percentage of new units to total units.[21]

We can make no statements about the general market impacts of EIRPs from any data we have been able to find or develop. The lack of data, coupled with the nonregulatory nature of the process, proves a difficult, if not intractable, conceptual research problem. Our lack of even a weak relationship in those test runs indicates that the effects of review intensity, as measured by expected processing time (chosen because delay is the largest component of process cost), might well be considered weak in comparison with other forces operating in the housing markets. Even were there to be a relationship, it would be taken circumspectly owing to differences in the housing markets considered.

We would, however, like to make several points regarding market effects. As we observed, EIRP effects on the housing market may be divided into two classes. The increased costs directly borne by the developer, which may be passed on in whole or in part to the buyer (depending upon market demand conditions), are not estimated to be of a substantial magnitude for the new homes constructed and are even less significant in their effect on general market conditions owing to the small part of the available housing supply they affect in any one year.

As James and Muller point out, it is the density changes that may arise from the environmental impact process that are more significant. They believe that "it is very likely that the costs of EIR measured in terms of increased housing costs or diminished housing availability are likely to bear down most heavily on moderate income families."[22] They cite a Florida case in which an environmental impact review and subsequent density reduction led to an 11 percent increase in the price of the homes finally built—necessary, one would assume, to cover profit margins. In addition to the decrease in the number of units for the developer to spread the costs over, there may also be a decrease in overall housing supply. As previously cited, a recent econometric study of the bay area suggests that land-use controls increase the cost of housing about $1,300 for each one-unit-per-acre reduction in density.[23] We must once again stress that such reductions are not necessarily an outcome of the EIRP nor would they not have occurred in the absence of the process.

CONCLUSIONS ABOUT EIRP-RELATED COSTS

The EIRP-related costs appear to vary widely, becoming significant in only a minor number of projects. Costs can vary primarily because of the project size, requirements of local review agencies, and sensitivity to the area's physical environment. Our own surveys of counties with reputations for the most extensive EIRPs, delays, and mitigation measures centered around the smaller counties in the San Francisco area. For example, San Bernardino County is larger in area than all of the smaller counties combined. These counties also contain environmentally sensitive areas. However, many developers find that key environmental issues may not have been identified, so that the EIRPs may be used to uncover them through difficult and slow review processes.

TABLE 7.8

Environmental Impact Review Processes Related to Process Costs

Process Step	Costs		
	Direct	Delay, Uncertainty	Mitigation
1. Initial proposal	Applications	30 days maximum	None
2. CEQA applicability determination	Fees, licenses	Revisions, new added reports	Project rejected or changed
3. Further review			
Negative Declaration conditional	Filing fees, forms	1-2 months delay	Project redesign
	Added studies	Revised or added reports, 3-6 months	
EIR prepared	Consultant fees		
Full-scale	$50 to $50,000 or more	24-36 months	Added consultant reports
Focused	Less than $50,000	1-12 months: Projects abandoned, 2.0 percent; Projects withdrawn, below 2.0 percent	
4. Draft submittals			
Once	30 days review, report preparation	Incomplete, inadequate, inaccurate reports revised	Special consultants studies, non-EIR-related studies, project redesign
More than once or continuous	Report preparation, distribution		
5. Impact identification			
Local		Additional consultant reports, public hearings, other departmental inputs requiring attention	50 percent of properties affected, average project cost in California in 1979 is $2,000
Regional			Usually ignored
Traffic		Traffic counts	New streets, sidewalks, signals, lighting, street widening
Natural habitat		Additional site surveys	Open-space dedication
Noise		Additional noise measures	Density reductions, landscaping, street rerouting, open space

Archeological		Six months or more for more site analysis by experts	Space reservations, open space
Energy		Energy surveys	Insulation, solar orientation
Public services, facilities		Additional other-department comments, reviews, requirements	Performance bonds, other bonds, space dedication, construction
6. Other agencies' reviews		30 days minimum per agency, may need more time or studies	See items under no. 5
7. EIR documents		Final price impacts of delays before acceptance: Holding, 1.2 percent-2.2 percent Net inflation, 6.5 percent-13.0 percent Overhead, 6.0 percent-6.5 percent Total effect, 13.7 percent-21.2 percent; For land development only: Holding, 21.6 percent Inflation, 6.0 percent Forgone opportunities, 1.4 percent Projects abandoned, 2.0 percent	Land banking by developers, with minimal expected mitigation requirements Density trade-offs
8. Data sources	Collection, analysis, presentation at $150 average per unit in project	Inaccurate, incomplete, 3 days to months where exotic data are needed	New or increased measurements or impacts not found important
9. Notification	Mailing, announcements	30 days minimum for replies, one-third of public notified will respond	More reports or see no. 5
10. Findings			
11. Disposition		Delays in reports from agencies related to reviews once EIR leaves lead agency	Reports returned for additional work

Source: Compiled by the authors.

An authority very familiar with EIRs and EIRPs suggested that "a good EIR is not just a good report. It should represent an environmental analysis conducted prior to project definition, if possible, so that the entire environmental setting can influence the project design. The results can be a tool for the developer and his architects and a report that incorporates mitigation measures sufficient to reduce the adverse impacts to insignificant levels. The approach represents good environmental planning and would be more in accordance with the CEQA requirements than the generation of 'after the fact reports.'"[24] These very early comments were repeated in other reports, which contained analyses of EIRPs. These later reports stressed the importance of more clearly defined environmental goals and standards and better consideration of the usefulness of the required information. Of concern to builders was the need for preparing duplicative environmental analyses, which they had to submit, since the agencies tended not to accept earlier information or reports on those conditions.

Included in the surveys for this report were observations about the uses of the EIRs. Invariably bulky, they often had to be read and digested in a relatively short time by agency reviewers, so that much of the report was simply ignored. As local governmental staffs in California have decreased because of Proposition 13 and the growing costs of government, the reviewers have been more and more casual in their studies of EIRs, often relying on public hearings to bring out the points to which they should direct their attention.

An effort to relate costs (Table 7.4) to processes (Table 5.5) is presented in Table 7.8. Ideally, these costs would be relatively precise dollar amounts; this was not possible, for reasons already enumerated. However, the extent to which these averages will be realized on a given project depend on such factors as the attitudes of local governments on growth, the fragility of the natural environment, the project size, the care with which the project is planned, and the familiarity of the developer with local requirements.

Another way of approaching the costs associated with EIRPs is suggested in Figure 7.1. The data provided earlier in the chapter have been translated into the probability of a cost being incurred at a particular point in the processes. For example, there is a 70 percent probability that any given residential project will require some form of environmental submittal. These are, of course, very rough approximations and should be subjected to further analysis. They are an improvement over the present approaches used by builders and planners to "shepherd" a project through the EIRP maze. Equally important, they provide some perspective on processes and costs. For example, there is a 45 percent probability that a project will not be subject to further review after the initial submittal.

This analysis also pinpoints the places where improvements could be made in the processes. The more common mitigation mea-

FIGURE 7.1

Probability of Incurring a Cost at a Given Environmental Impact Review Process Step, Private Residential Projects

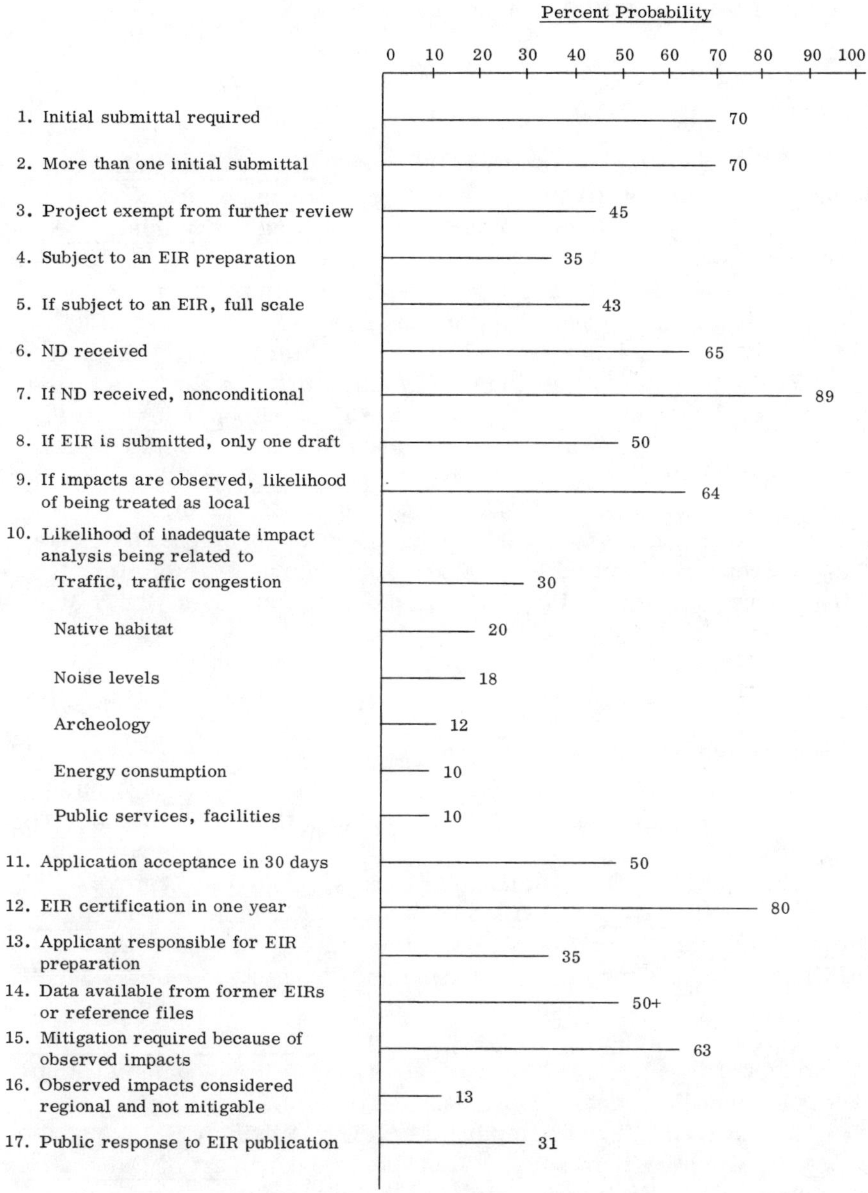

Source: Compiled by the authors.

sures—traffic congestion, native habitat degradation, and noise levels—suggest that if developers spent more time measuring the extent of such impacts and planning ahead of time for their mitigation, the remainder of the EIRP might be much shorter and less expensive. On the other hand, some of the review agencies might consider why they are not meeting legally required review times.

NOTES

1. For example, see Marshall and Swift, Residential Cost Handbook, yearbook (Los Angeles: Marshall and Swift).

2. Alfred Gobar and Associates, The Housing Crisis: The Cause and What We Can Do about It (Newport Beach, Calif.: Californians for an Environment of Excellence, Full Employment and a Strong Economy through Planned Development, 1980).

3. Franklin J. James and Thomas Muller, "Environmental Impact Evaluation, Land Use Planning and the Housing Consumer," Journal of the American Real Estate and Urban Economics Association, Fall 1977, pp. 279-301. That study is based on data reported in Thomas Muller, State-Mandated Impact Evaluation: A Preliminary Assessment (Washington, D.C.: Urban Institute, 1976).

4. Small and Knust, California Local Government and CEQA, 1979, a report submitted to the Assembly Committee on Rules, California Legislature (San Diego, Calif.: Small and Knust, Inc., June 1980), app. B.

5. Ibid., p. B-3.

6. James and Muller, "Environmental Impact Evaluation," pp. 284-85.

7. Ibid.

8. Small and Knust, Local Government and CEQA, p. 30.

9. James and Muller, "Environmental Impact Evaluation," pp. 279-301.

10. David E. Dowall and John Landis, Land-Use Controls and Housing Costs: An Examination of San Francisco Bay Area Communities, Working Paper 81-24, Berkeley and Los Angeles: University of California, Center for Real Estate and Urban Economics, March 1981, pp. 16, 17.

11. Environmental Analysis Systems, The California Environmental Quality Act, vol. 3 (San Diego, Calif.: Environmental Analysis Systems, November 1975), pp. 18-20.

12. Alfred Gobar and Associates, The Housing Crisis, pp. 7-8.

13. Construction Industry Research Board, Cost of Delay Prior to Construction, mimeographed (Los Angeles: CIRB, March 11, 1975).

14. Environmental Analysis Systems, California Environmental Quality Act, vol. 3.

15. Don M. Demars, "Environmental Impact Report" (Speech delivered at the Society of Real Estate Appraisers Seminar, Newport Beach, Calif., March 7, 1965).

16. See Chapter 2. Also see Leo Grebler and Frank G. Mittelbach, The Inflation of House Prices (Lexington, Mass.: Lexington Books, 1979).

17. Randall W. Scott, ed., Management and Control of Growth, 3 vols. (Washington, D.C.: Urban Land Institute, 1975).

18. Fred E. Case and Jeffrey Gale, "The Impact on Housing Costs of the California Coastal Zone Conservation Act," Journal of the American Real Estate and Urban Land Economics Journal, Winter 1981, pp. 345-66.

19. The lack of data on mitigation measures is particularly troublesome, as all reviews would have to be examined to determine actions taken.

20. Urban Land Institute and Gruen and Gruen Associates, The ULI Study: Effects of Regulation on Housing Costs: Two Case Studies (Washington, D.C.: Urban Land Institute, 1977).

21. An additional problem for this effort was the focus upon the process itself and thus a data sample somewhat by number of jurisdictions studied.

22. James and Muller, "Environmental Impact Evaluation," pp. 298-99.

23. Dowall and Landis, Land Use Controls.

24. Demars, Environmental Impact Report.

8

TOWARD MORE EFFECTIVE ENVIRONMENTAL IMPACT REVIEW PROCESSES

The previous chapters contain analyses that suggest that the overall private and public costs of current environmental impact review processes (EIRPs) may be relatively small compared with the total housing costs. When the total costs of EIRPs are considered, they come to rather large amounts. This emphasis on costs does not, however, indicate the extent to which benefits are created and possibly offset some or all of the costs. Some benefits are obviously incalculable, as are some costs; however, we will try to provide some balance as a means of identifying ways of improving the EIRP.

BALANCING COSTS AND BENEFITS

All controls over housing construction processes are imposed for the purpose of promoting some combination of public welfare, health, and safety. These considerations, however, are not mentioned in statements of the EIRP goals, nor do any of the analyses of EIRPs indicate the extent to which these standards are or are not achieved. Environmental Impact Reports (EIRs) have been shown to be useful in protecting natural environments and public interests by mitigation measures and requiring builders to absorb some of the external costs they create. The EIR has also proved to be a useful document by providing builders with information on which projects can be marketed and lead agencies with a basis for estimating accumulating impacts on projects they are reviewing.[1]

In general, costs are related to direct fees, charges, costs, and report preparation; delays and uncertainties; and mitigation measures (see Table 8.1). These costs are paid for by private persons, society, or governmental agencies. Equally important, each of these

TABLE 8.1

Major Cost-Benefit Elements and Factors as They Relate to the Cost of Housing

Cost	Benefit
Private	**Private**
Processing time	Market identification
Special fees and reports	Monopolistic market opportunities
Report preparation	Personal enjoyment of scarce resources
Loss of "traditional freedoms"	Personal safety, health protection
Lack of economic scale	Better quality products
Higher costs/prices/rents	Helping others get "fair shares"
Inefficiencies in production, sales	
Social	**Social**
Citizen participation costs	Many scarce resources preserved for later generations
Exclusionary for some groups	Those who benefit pay
Limits on innovation, newer products, construction methods	Better neighborhoods
Abandonment	Creation of public good expenditures (good can be consumed by one person without detracting from consumption by another)
Crowding, wide-open space	Reduction in ownership externalities
Market monopolies	Better housing for larger share of population (less advantaged groups get their "fair share")
Segregation, polarization	Better quality housing, neighborhoods, cities
Job cuts, racial shifts	Less pollution (air, water, sight, sound)
Government expenditures crowd out private expenditures	Conservation of scarce, irreplaceable natural resources
Limitations on right to travel, social mobility	
Public	**Public**
Development of entrenched bureaucracy	New revenue sources/cost sharing/high property taxes
Public promotion costs to gain acceptance of program	Better neighborhoods
Coordination of mandated activities	Lower public service/capital improvements costs
Implementing organizations	Restraints on some kinds of governmental expenditures
Intergovernmental coordination (super agency, special agency, coordinating agency)	Reduced costs of police, fire, health, and so on)
Code, law, regulation preparation	Preserves homogeneous neighborhoods
Supplementary regulations (subdivision laws, dedication, in-lieu fees, bonuses, lot consolidation, transfer of development rights)	Protects and improves the property tax base
	Resource oriented management of public assets versus demand orientation
	Growth relates to servicing capabilities of local governments

Source: Compiled by the authors.

costs can be offset by various kinds of benefits (also listed in the table), which are rarely mentioned in an evaluation of EIRPs. In some cases the benefits might be quantified; in most cases quantification would not be possible or, if attempted, would produce only the most general answers.

EIRPS, HOME PRICES, AND EXTERNALITIES

In the San Francisco Bay Area, the EIRPs were often tied into growth control measures, so that a project had to be evaluated in terms of its growth impacts on the local general welfare. Livermore, California, enacted such an ordinance and was sued by a builder.[2] The court upheld the ordinance but pointed out that the ordinance had to be judged from the viewpoint of all those who might be affected, including not just the residents of Livermore but also the adjacent nonresidents, emphasizing that any program that affected the supply and demand in a region was to be evaluated in terms of its impact on the entire region. The case was returned to the lower court, because insufficient evidence was provided about the impact and duration of the growth restriction. The need to protect the environment and the supply of public facilities was to be balanced against the need to provide housing and places to live for the regional population. Overall, the thrust of the court decision seemed to be that growth may be limited through environmental review actions but that the limitations must be based on well-established evidence of the costs and benefits for public welfare.

Niskanen and Hanke make the additional point that focusing on house price changes to measure the impacts of environmental controls overlooks the values that may be added to public properties.[3] They conclude that "because the public value of land is not incorporated into existing models and empirical studies, benefits from projects and programs that increase the productivity of land have been consistently and substantially underestimated." Since more productive land will have higher prices, and higher prices create higher assessed values and larger tax bills, an additional cost from EIRPs is imposed on home prices.

An example can show the importance of this concept. An individual homeowner can buy enough land to provide some open park-like space; or the owner may join with neighbors in providing a common park area; or, an EIRP can require a developer to provide park space for not only the newly planned subdivision but for the community generally. In the first two cases the home buyers voluntarily pay more to have open space and probably enjoy higher home values as a result. In the last case, the home buyers probably pay more for the open space, but the open space also influences home values of those who did not pay for the space, and the city has a valuable open park, which adds to its assets. Isolating the impact of the EIRPs under such conditions would be difficult, but clearly there is an overall public benefit, which may be overlooked in concentrating only on the higher home prices in the new development. Clearly, any cost-benefit model of the EIRP must consider this potential.

IMPROVING THE ENVIRONMENTAL IMPACT REVIEW PROCESS

A major thrust of this study has been the analysis of the EIRP with an eye toward steps that may be taken to improve it. As we have suggested, this process, in its general form, is applicable in the legal framework developed in a number of jurisdictions around the country. Indeed, a recent comprehensive study of state environmental policy acts recommends that they be made applicable to private projects.[4]

We believe that, in general, the process developed under the CEQA is a reasonable starting point for implementation of an EIRP in a local government and that the overall costs of such an effort, even if carried out as currently handled in California local governments studied, would be tolerable. Though the costs fall unevenly, with significant expenditures on some developers, it is nevertheless a basically sound process.

This does not mean, however, that the process as instituted in California is not capable of improvement. Gains in both effectiveness and efficiency may be made through more clearly defining the goals and subject matter, better consideration of regional concerns, and improving support and review process structures. We divide our recommendations into five classes: clarification of the nature of the process, improved planning-EIRP coordination, increased support, a revised review process, and a revised assessment procedure.

Clarification of the Nature of the Process

A major difficulty in dealing with the EIRP is the lack of direction in the CEQA framework. An early attempt at enunciating a developer perspective stressed this in terms of whether the purpose of the process is informational or substantive.[5] We believe that this concern has been largely answered in California and that the process, as it has developed, is both. We would suggest that the informational aspect is dominant but that the inclusion of environmental concerns and their weighting in decision making is an important aspect as well. Our recommendation on this matter is that, while both informational and substantive purposes are appropriate for the process, substantive applications need adequate specification of use to be effective. For example, there needs to be some indication to decision makers as to how important nonmitigable environmental effects are when weighed in project approvals. A lack of such clarity leads to uncertainty in the process itself and thus to longer reviews, less useful documents, and consequent cost increases.

Improved Planning-EIRP Coordination

The EIRP, as currently implemented, suffers from much the same problem as that from which zoning has long been accused—the project-by-project, case-by-case evaluation. Even when competently and efficiently carried out, the process, when emphasizing a project focus, often does not adequately bring into the analysis regional considerations, cumulative impacts, and the comprehensive totality of impacts. It would be unfair to expect that the review of an individual project could accomplish these tasks.

Other studies of the process have emphasized the need for coordination.[6] The Small study of 1979 suggests that the major emphasis of CEQA reviews at the local government level be shifted to the General Plan Amendment process, requiring local governments to take the following actions:

1. To identify, by type of project or type of discretionary decision, those environmental impacts likely to be caused by such project or decision types;
2. To establish standards and criteria which specify the degree to which identified environmental impacts are to be treated in environmental documents that may be prepared in conjunction with the review of various types of projects or discretionary decisions;
3. To identify the types of mitigating measures that may be required as a condition of approval of various types of projects and discretionary decisions;
4. To make a finding whether public projects which cause impacts similar to those caused by private projects are to be treated any differently in terms of the analysis of such impacts in environmental documents and the requirement of mitigating measures to treat such impacts.[7]

This approach would go a step beyond the specific-plan approach. It would not exempt projects from review but rather provide more certainty for review by specifying standards and criteria as a part of the general plan, which is required in California. Regional input would be required. We believe that this concept is both laudable and ambitious. Conceptually, such an approach is a positive step in that it both handles regional issues and increases specificity of EIRP requirements, cutting uncertainty. However, the problems of full-scale implementation of this effort in a planning process, which is generally underdeveloped in most jurisdictions and which will, we believe, lead to battles over the environmental results of such plans, makes it, in its totality, not practicable.

However, we believe that partial adoption of such an approach would be extremely useful. Our studies indicate that the incidence of the EIRP burden falls on particular types of projects and particular environmentally sensitive geographic areas. The plan-based standards approach described above would, we believe, be a useful exercise in those environmentally sensitive geographic areas, which tend to require the full EIRP for development. It is here that the intense effort involved would pay off most fully in affecting heavy-cost EIR preparation.

Our planning-related recommendations are that the use of specific plans, as currently developed in the CEQA and the guidelines, be increased and encouraged as a means of consolidating reviews of residential development. We also recommend the development of a general plan-based standards and criteria approach by local governments for those areas that are environmentally sensitive and more likely to require a full EIRP.

Increased Support

Our detailed analysis of the frequency of occurrence of certain types of actions at different process stages was reported in Chapter 7. Within these actions are a set of delay-inducing, multiple-submittal occurrences. Our survey indicates that information for these steps is largely drawn from the data supplied by local governments. Thus, one of our recommendations is for increased support of the process through better design and maintenance of environmental data bases at the local level.

This step may seem like an anomaly—saving on the process through spending funds for increased data support, but the entire process might be made more efficient through better data access and a common data source and definition. A recent *Journal of the American Planning Association* article suggests a large-scale monitoring system as a means of reducing cost effects of land-use controls.[8] It may be that the environmental data base for use in the EIRP is best developed as a part of the data base necessary for such a system.

Revising the Review Process

There are a number of steps that might be taken within the context of the current review process to improve the processing of the projects subject to it. The main thrust of our recommendations is to decrease the uncertainty of the process and to emphasize full environmental impact review only where effects are significant enough to warrant such review. Recalling that delay is the largest component of process cost, the emphasis is upon shift from full EIR preparation to other paths through the process.

Changes we have already suggested, if implemented, would serve to decrease the number of project EIRs through increased use of specific plans, more narrowly define the review process for environmentally sensitive areas (cutting the uncertainty and cost of review), and provide better information support for those persons required to traverse the EIRP. Changes suggested in this section (as summarized in Table 8.2), would ease other process-related delays and costs identified in the study.

Proceeding through the process, we would make three suggestions for changes before the process even begins. As discussed in Chapter 4, the relationship between the planning staff and the developer is crucial to the successful review. We would recommend that the planning staff seek ways to encourage and improve this preprocess consultation. This would serve to inform the developer earlier in the project of those problems likely to arise. It would also coordinate environmental review with the early phase of project review, improving on the often observed phenomenon of the review occurring after significant funds are expended. We recommend an effort to inform the private developer of what might be expected in the process. This should decrease the uncertainty of the builder to some extent. We also recommend that the local government explicitly identify the types and scales of projects likely to be deemed environmentally significant and require an EIR. Our study indicates that certain classes of projects are almost always subject to the process in its entirety, while others are rarely subject. An effort to identify these would increase the predictability of the project approval process for the private developer.

At the Initial Study stage, efforts to improve should focus upon the processing time and reporting on progress. We recommend adopting procedures for concurrent agency reviews to cut review time, establishing target dates for completion of the step (and all other steps in the review), and providing a means of tracking the progress of the entire review for the developer. These steps would aid in the management of the review process, encourage early resolution of difficulty, and provide the developer with information

TABLE 8.2

EIRP-Improvement Recommendations

Step	Recommendation
Preprocess	Encourage and improve preprocess staff-developer consultation
	Government effort to inform developer of process details
	Local government development of classification of projects likely to require full EIRP by scale and type
Initial Study	Adopt procedures for concurrent agency review
	Establish target dates for completion of all project review steps
	Provide a mechanism for developer to track progress of review
	Continue efforts to focus and scope EIRs at this stage
Negative Declaration	Encourage expanded use of Mitigated Negative Declaration
	Institute structured public review
	Identify in Negative Declaration exclusionary effects, if any, of mitigation measures
EIR	Monitor consultant performance
	Identify and analyze exclusionary aspects of mitigation measures
	Encourage incorporation by reference of other documents, especially plans and plan EIRs
	Provide multiple accessibility to draft EIRs for public
	Confine public hearing to one consolidated hearing for all interested parties and agencies
	Adopt formal procedures for concurrent agency review of draft EIRs

Source: Compiled by the authors.

about progress in the review. Our work indicates that the Initial Study is being utilized to focus subsequent EIRs, and we recommend that efforts to scope and focus the reviews be further expanded, as costs of well-defined narrower studies are significantly less than costs of full-coverage studies. With the increase in California of Negative Declarations, this will be more significant.

We believe that the Negative Declaration, and especially the Mitigated Negative Declaration, is the best means of finding a middle ground for those many projects with mitigable significant effects. Every project review that shifts from an EIR to a Mitigated Negative Declaration will save the developer and the government an average of five months of review. We recommend that the use of the Mitigated Negative Declaration be encouraged as a step that will ensure both review and mitigation. We also recommend that communities institute some form of structured public review, though not necessarily hearings, at the Negative Declaration stage to preserve the openness of the EIRP, which we believe is a positive aspect. We also believe that some identification and analysis of the exclusionary effects of mitigation measures adopted should be included in the Mitigated Negative Declaration. As the mitigation taken is often a density change, which affects supply negatively (as discussed in Chapter 7), we believe that this must be addressed—indeed it would seem mandated—by the housing goal with the CEQA for California.

The early criticisms of the EIR itself were of the bulk of the document and its attention to irrelevant issues. We believe that the efforts described in Chapter 3 have somewhat blunted that criticism as regards the CEQA. Nevertheless, a number of useful steps may be taken. Consultants, according to those studies we have found and our own, perform the great majority of the writing of EIRs, and the costs are passed to the developer generally. Almost half of all EIRs done are submitted more than once by the consultant. We recommend an effort to monitor consultant performance to improve this record. We would expect the better data base we propose to help in this performance as well as the standards and criteria for environmentally critical areas in the general plan, which we recommend; given the passing through of document costs, we believe that this is one of several ways in which rising EIR costs may be limited, since this is essentially a third-party payer situation. Development of standards and criteria for certain classes and scales of developments would also be useful. We again recommend that mitigation measures adopted be subjected to scrutiny for exclusionary effects as described above. To decrease bulk and increase comprehensiveness of review, we recommend further encouragement of incorporation of past EIRs and other relevant documents by reference in an EIR, and we especially encourage such incorporation of plans and plan EIRs.

The public review part of the EIRP has been subject to scathing criticism in our discussions with builders. However, we believe that the positive aspects of this participation are an integral part of the process. More definition of the process would, we believe, serve to lessen the antagonism. We recommend that multiple accessibility to Draft EIRs be provided and that there be one consolidated hearing for all agencies reviewing a project. It appears to us that better public document access, coupled with a limited multiagency forum, would ease some of the bitterness that currently results from frequent hearings by different agencies on the same EIR. We also recommend that formal procedures for agency review of the EIR be adopted to encourage concurrent review. As in all steps, we believe that time targets for processing and a means of monitoring progress must be established.

As a last comment on the process, we emphasize that numerous other changes to improve the effort might well be recommended both now and in the future. It is important to monitor the process itself, with an attitude that it is to be improved, when improvement can be made. The local control of the process, as permitted by the CEQA, is a significant aid in shaping the process to the needs of the community and should be retained.

A Revised Assessment Approach

Table 8.3 provides a format for analyzing costs and benefits based on the findings in this study. In effect it provides a form of summary of how the EIR relates to housing production.

The types of impact that are identified in the EIRP may be divided into physical, economic, and social, even though the CEQA does not require identification of these. Social impacts are included, because many mitigation measures are clearly efforts to correct apparent social imbalances being created by a particular project. If an impact of a housing project, for example, is to be mitigated by lowering densities, then this can have an impact on the population densities, abilities of families to pay for the housing, and the location of lower income families. The types of impacts listed for consideration are those most frequently encountered during our research.

At this point, impact analysis could be preliminary or supported by detailed consultant reports. The next step is to make a preliminary rating of the impact as favorable, which could trigger a Negative Declaration; mitigable, which could allow for a focused EIR; or irreversible, which could be an early warning to abandon the project or redesign it extensively.

The many ways in which an impact can be mitigated are often not known or considered, so the advantage of asking for consideration

TABLE 8.3

A Structured Approach to the Impact Assessment of Private Residential Construction Projects

Type of Impact[a]	Impact Rating[b]	Mitigation Proposed[c]	Mitigation Measures—Costs/Benefits[d]			
			Costs		Benefits	
			Lump Sum	Flows	Lump Sum	Flows
Physical						
Natural						
Private						
Public						
Economic						
Private						
Public						
Social						
Private						
Public						

[a]Type of impacts to be considered (one time and cumulative):

A. Physical

1. Natural environmental—air, water, sight, sound, native habitat, light and glare, geology, flora, fauna, aesthetics, potential hazards, other
2. Private—open space, facilities, densities, development rights, land uses
3. Public—service systems, police, fire, trash, traffic-transportation, recreation, educational, planning, zoning, other departments and services, facilities, utilities, buildings, streets, parks, open space

B. Economic

1. Private—housing, construction costs and housing prices, markets, employment, energy sources and uses
2. Public—assessed values, facilities, services, revenues-expense ratios, dedications, removals

C. Social

1. Private—demographic, cultural, health
2. Public—demographic (location, distribution, densities, growth)

[b]Impact ratings: A = favorable, B = mitigable, C = irreversible.

[c]Mitigations proposed: A = physical change, B = cash payments, C = bonds, D = dedications, E = construction or renewal of existing or creation of new.

[d]Costs/benefits: Best estimates of the actual costs of the mitigation measures proposed; benefits measured in terms of property value changes or rentals; discount rate to be current prime mortgage rate, assuming that mortgage money would be needed to finance construction and mitigation efforts.

of all possible means of mitigating an impact can be realized through the format proposed.

Cost-benefit analysis is often avoided by review agencies simply because the reviewers do not know how to make such an analysis. In one sense the process is relatively simple. When mitigation measures are known, costs can be estimated either as lump sums or flows (that is, a series of continuous payments). The final step of bringing all future costs or benefits back to a present value requires consideration of a discount rate. Since the processes all involve residential construction, mortgage lending rates (for construction or takeout loans) would provide an acceptable standard.

In recommending this type of cost-benefit analysis, we recognize that some kinds of environmental impacts (particularly regional) cannot be easily converted to money terms; that the choice of a discount rate is always controversial but critical to the final estimates; that indirect and induced effects cannot be measured; and that the uniqueness of each project makes project comparisons extremely difficult.[9]

Admittedly, this approach is crude and difficult to use initially; however, it does provide a structured approach to the EIRP not currently found. The data presented support the notion that EIRPs have reduced some of the more extreme forms of "unconstrained development—have prevented increased traffic congestions or the degradation of important wildlife habitats. . . . Although . . . it is difficult to compare benefits with costs since the benefits fall mainly on existing residents while the costs fall primarily on developers and new buyers. . . . Higher housing costs and diminished housing availability are most likely to fall on moderate or low income groups."[10]

At the heart of environmental concerns is the extent to which many natural resources are being destroyed, which cannot be replaced or renewed within a human time frame.[11] Such resources are irreplaceable for all practical purposes. As irreplaceable resources are destroyed, more and more resources (energy) will be needed to produce the same amount of housing, leaving fewer and fewer resources for future development.[12] There is already concern that resources constraints in the 1980s will "reshape the average new home and change the living patterns of people everywhere."[13]

ARE ENVIRONMENTAL IMPACT REVIEW PROCESSES WORTHWHILE?

Regulations controlling land uses and protecting the environment add to the costs of housing; the question is, Do the added costs provide equal or equivalent quality in home living? Many of the costs can be identified easily; others can only be estimated, so that total cost estimates related to land-use controls tend to vary widely. However, the question is not whether such controls add to costs (everyone who has studied the subject agrees that they do) but, How much do they add beyond what might be considered reasonable?

In the broad sense the control costs are reasonable, because they are intended to protect, and often do, public health, welfare, and safety. In the narrowest sense these costs are excessive because of the restraints they place on families' freedom of choice as to where they want to live and the kinds of dwellings they prefer. In other words, in a private enterprise, free market economy, land-use and environmental controls tend to substitute to an increasing extent the "collective judgments of society" for individual consumer judgments, even though these social judgments may be poorly informed and ill advised.

Those who support controls argue for them by invoking a larger array of social and individual costs for markets in which these controls are needed but do not exist. Those who oppose controls do just the opposite. In neither case have the goals associated with invoking the controls been sufficiently well articulated to provide clues for deciding how much health, welfare, and safety is being protected nor how well the added costs equal the added benefits for either individuals or society.

In other words, the current tendency to focus on costs is like using only one blade of a scissors. For every cost there is likely to be a benefit; the question is, Do the benefits associated with a particular cost equal or exceed that cost? The benefit side of the control problem has been sadly neglected, thus leaving an unparalleled opportunity for regulators to continue to increase their activities without successful challenge. Admittedly, trying to quantify the benefits involves varieties of personal value judgments, which heighten the arguments on control rather than resolving them. On the other hand, identifying costs is equally fraught with imprecision and large numbers of value judgments.

NOTES

1. Don M. DeMars, Environmental Impact Report, Report at Society of Real Estate Appraisers Seminar (Newport Beach, Calif.: Ultrasystems, March 7, 1975).

2. "California Supreme Court Upholds Livermore Growth Law," Practicing Planner, September 1977, pp. 12-13.

3. William A. Niskanen and Steven H. Hanke, "Land Prices Substantially Underestimate the Value of Environmental Quality," Review of Economics and Statistics 59 (August 1977): 375-77.

4. Stuart L. Hart and Gordon A. Enk, Green Goals and Greenbacks: State-Level Environmental Review Programs and Their Associated Costs (Boulder, Colo.: Westview Press, 1980), p. 21.

5. George A. Rice, "Reform of CEQA—The Developer's Perspective," California EIR Monitor, December 15, 1975, pp. 5-6.

6. Hart and Enk, Green Goals and Greenbacks, pp. 60-63.

7. Small and Knust, California Local Government and CEQA, 1979, A report submitted to the Assembly Committee on Rules, California Legislature (San Diego, Calif.: Small and Knust, Inc., 1980, pp. 39-40.

8. David E. Dowall, "Reducing the Cost Effects of Local Land Use Controls," Journal of the American Planning Association, April 1981, pp. 145-53.

9. Joseph E. Haring and Allen Van Deventer, "Measuring the Impact of Environmental Regulations: Indirect and Induced Benefit-Cost Analysis" (Paper presented at the Fifteenth Annual Meeting of the Western Regional Science Association, February 27-29, 1976).

10. Lawrence Katz and Kenneth T. Rosen, The Effect of Land Use Controls on Housing Prices, Working Paper 80-13 (Berkeley: University of California, Center for Real Estate and Urban Economics, September 1980), p. 36.

11. Paul A. Samuelson, Economics, 9th ed. (New York: McGraw-Hill, 1973), p. 778.

12. Jeremy Rifkin, Entropy: A New World View (New York: Bantam Books, 1980), pp. 35-37.

13. Bruce Stokes, Global Housing Prospects: The Resource Constraints (Washington, D.C.: Worldwatch Institute, September 1981), p. 5.

APPENDIX A

CALIFORNIA DATA AND FACTS

Counties included in study are underlined.

TABLE A.1

Land and Water Areas of California Counties

County	Water Area		Land Area		Total Area	
	Acres	Square Miles	Acres	Square Miles	Acres	Square Miles
Total	1,356,780	2,120.5	100,206,720	156,573.0	101,563,500	158,693.2
Alameda	59,150	92.4	469,120	733.0	528,270	825.4
Alpine	2,310	3.6	462,720	723.0	465,030	726.6
Amador	5,390	8.3	379,520	593.0	384,810	601.3
Butte	1,170	1.8	1,064,320	1,663.0	1,065,490	1,664.8
Calaveras	6,010	9.4	657,280	1,027.0	663,290	1,036.4
Colusa	1,820	2.8	737,920	1,153.0	739,740	1,155.8
Contra Costa	40,920	63.9	469,760	734.0	510,680	797.9
Del Norte	—	—	641,920	1,003.0	641,920	1,003.0
El Dorado	58,080	90.8	1,096,960	1,714.0	1,155,040	1,804.8
Fresno	21,860	34.2	3,816,960	5,964.0	3,838,820	5,998.2
Glenn	1,280	2.0	842,880	1,317.0	844,160	1,319.0
Humboldt	16,970	26.5	2,286,720	3,573.0	2,303,690	3,599.5
Imperial	200,580	313.4	2,741,760	4,284.0	2,942,340	4,597.4
Inyo	4,400	6.9	6,458,240	10,091.0	6,462,640	10,097.9
Kern	11,720	18.3	5,217,280	8,152.0	5,229,000	8,170.3
Kings	25,990	40.6	892,800	1,395.0	918,790	1,435.6
Lake	45,120	70.5	803,840	1,256.0	848,960	1,326.5
Lassen	91,700	143.3	2,910,080	4,547.0	3,001,780	4,690.3

Los Angeles	12,330	19.3	2,598,400	4,060.0	2,610,730	4,079.3
Madera	2,000	3.1	1,372,160	2,144.0	1,374,160	2,147.1
Marin	43,500	68.0	332,800	520.0	376,300	588.0
Mariposa	3,490	5.5	931,200	1,455.0	934,690	1,460.5
Mendocino	2,360	3.7	2,244,480	3,507.0	2,246,840	3,510.4
Merced	16,450	25.7	1,268,480	1,982.0	1,284,930	2,007.7
Modoc	158,990	248.4	2,618,880	4,092.0	2,777,870	4,340.4
Mono	48,030	75.0	1,937,920	3,028.0	1,985,950	3,103.0
Monterey	70	0.1	2,127,360	3,324.0	2,127,430	3,324.1
Napa	24,890	38.9	485,120	758.0	510,010	796.9
Nevada	9,090	14.2	625,920	978.0	635,010	992.2
Orange	1,960	3.1	500,480	782.0	502,440	785.1
Placer	52,780	82.5	911,360	1,424.0	964,140	1,506.5
Plumas	30,980	48.4	1,644,800	2,570.0	1,675,780	2,618.4
Riverside	42,260	66.0	4,593,280	7,177.0	4,635,540	7,243.0
Sacramento	20,660	32.3	629,120	983.0	649,780	1,015.3
San Benito	710	1.1	893,440	1,396.0	894,150	1,397.1
San Bernardino	21,120	33.0	12,883,840	20,131.0	12,904,960	20,164.0
San Diego	16,360	25.6	2,723,200	4,255.0	2,739,560	4,280.6
San Francisco	29,500	46.1	28,800	45.0	58,300	91.1
San Joaquin	17,420	27.2	901,760	1,409.0	919,180	1,436.2
San Luis Obispo	6,560	10.2	2,122,240	3,316.0	2,128,800	3,326.2
San Mateo	49,130	76.8	290,560	454.0	339,690	530.8
Santa Barbara	4,260	6.7	1,752,320	2,738.0	1,756,580	2,744.7
Santa Clara	8,880	13.9	833,280	1,302.0	842,160	1,315.9
Santa Cruz	400	0.6	280,960	439.0	281,360	439.6
Shasta	33,420	52.2	2,430,720	3,798.0	2,464,140	3,850.2

TABLE A.2

Population of California Cities, January 1979 and 1980

County and City	Total Population, 1979	Total Population, 1980	County and City	Total Population, 1979	Total Population, 1980
Alameda	1,098,800	1,098,500	Contra Costa (continued)		
Alameda	72,700	69,200	El Cerrito	22,700	22,650
Albany	15,400	15,200	Hercules	3,750	5,825
Berkeley	110,400	110,400	Lafayette	20,100	20,250
Emeryville	4,490	4,380	Martinez	22,100	22,850
Fremont	123,100	127,300	Moraga	15,300	15,550
Hayward	94,300	95,100	Pinole	15,300	15,200
Livermore	48,550	48,150	Pittsburg	29,050	31,150
Newark	30,700	31,000	Pleasant Hill	25,900	26,150
Oakland	329,700	327,300	Richmond	69,700	70,800
Piedmont	10,450	10,350	San Pablo	18,850	18,850
Pleasanton	34,500	34,750	Walnut Creek	51,600	54,600
San Leandro	67,400	67,100	Unincorporated	188,100	186,600
Union City	34,950	36,550	Del Norte	17,300	17,650
Unincorporated	122,100	121,600	Crescent City	2,710	2,710
Alpine	1,100	1,110	Unincorporated	14,550	14,950
Unincorporated	1,100	1,110			

Amador	18,050	19,050
Amador	160	180
Ione	2,350	2,480
Jackson	2,660	2,690
Plymouth	670	680
Sutter Creek	1,760	1,800
Unincorporated	10,450	11,200
Butte	135,300	141,200
Biggs	1,460	1,490
Chico	26,000	26,950
Gridley	3,890	4,000
Oroville	8,150	8,350
Paradise	—	23,000
Unincorporated	95,800	77,400
Calaveras	19,900	21,300
Angels	2,510	2,440
Unincorporated	17,400	18,850
Colusa	13,050	13,150
Colusa	4,300	4,280
Williams	1,610	1,610
Unincorporated	7,125	7,250
Contra Costa	631,800	645,300
Antioch	39,900	41,750
Brentwood	3,830	3,890
Clayton	3,710	4,030
Concord	102,000	105,100
El Dorado	79,700	87,300
Placerville	7,150	7,225
South Lake Tahoe	21,900	22,950
Unincorporated	50,700	57,100
Fresno	482,300	490,800
Clovis	30,300	32,050
Coalinga	6,375	6,400
Firebaugh	3,920	3,900
Fowler	2,190	2,430
Fresno	199,600	207,300
Huron	2,560	2,710
Kerman	3,850	3,920
Kingsburg	4,740	4,920
Mendota	4,160	4,340
Orange Cove	3,810	3,780
Parlier	2,580	2,580
Reedley	10,250	10,300
Sanger	11,300	11,750
San Joaquin	1,640	1,770
Selma	9,550	9,750
Unincorporated	185,500	183,000
Glenn	21,050	21,800
Orland	3,880	4,040
Willows	4,810	5,075
Unincorporated	12,350	12,700

(continued)

TABLE A.2 (continued)

County and City	Total Population, 1979	Total Population, 1980	County and City	Total Population, 1979	Total Population, 1980
Humboldt	107,200	107,500	Los Angeles	7,104,000	7,163,100
Arcata	12,900	12,750	Alhambra	61,400	62,500
Blue Lake	1,240	1,300	Arcadia	46,900	46,750
Eureka	24,300	24,200	Artesia	14,900	15,300
Ferndale	1,420	1,400	Avalon	1,790	1,910
Fortuna	7,900	7,825	Azusa	26,050	26,650
Rio Dell	2,870	2,830	Baldwin Park	47,150	47,950
Trinidad	400	400	Bell	22,700	23,000
Unincorporated	56,100	56,700	Bellflower	52,100	52,200
Imperial	91,900	94,500	Bell Gardens	31,700	32,050
Brawley	14,300	14,350	Beverly Hills	33,050	33,250
Calexico	13,700	14,300	Bradbury	860	870
Calipatria	2,550	2,570	Burbank	84,600	84,700
El Centro	24,650	25,600	Carson	78,700	79,100
Holtville	4,630	4,680	Cerritos	50,200	51,200
Imperial	3,400	3,460	Claremont	28,350	28,500
Westmoreland	1,580	1,630	Commerce	10,000	10,050
Unincorporated	27,100	27,900	Compton	75,100	75,700
Inyo	18,450	18,850	Covina	33,200	33,250
Bishop	3,520	3,550	Cudahy	16,900	17,300
Unincorporated	14,900	15,300	Culver City	38,200	38,000

Kern	373,300	383,300	Downey	85,900	86,700
Arvin	6,225	6,375	Duarte	15,300	15,950
Bakersfield	91,800	98,400	El Monte	67,700	68,400
California City	2,620	2,800	El Segundo	15,000	15,100
Delano	15,500	15,850	Gardena	44,550	44,650
Maricopa	760	810	Glendale	134,600	135,100
McFarland	4,710	4,620	Glendora	36,850	37,100
Ridgecrest	15,050	15,750	Hawaiian Gardens	9,600	10,200
Shafter	6,600	6,725	Hawthorne	54,900	54,500
Taft	4,870	4,980	Hermosa Beach	19,100	19,150
Tehachapi	3,970	3,990	Hidden Hills	1,720	1,730
Wasco	9,125	9,300	Huntington Park	39,350	39,800
Unincorporated	212,000	213,800	Industry	630	630
Kings	71,700	71,700	Inglewood	90,900	91,200
Avenal	—	3,560	Irwindale	770	750
Corcoran	5,850	6,050	La Canada Flintridge	19,350	19,300
Hanford	19,550	20,300	Lakewood	80,400	80,200
Lemoore	8,000	8,350	La Mirada	44,150	44,100
Unincorporated	38,300	33,500	Lancaster	46,600	48,100
Lake	32,950	35,200	La Puente	30,200	30,550
Lakeport	3,770	4,010	La Verne	21,950	23,450
Unincorporated	29,150	31,150	Lawndale	23,550	23,750
Lassen	20,600	21,500	Lomita	19,800	20,000
Susanville	6,900	7,150	Long Beach	345,400	345,300
Unincorporated	13,700	14,350	Los Angeles	2,801,900	2,817,800

(continued)

TABLE A.2 (continued)

County and City	Total Population, 1979	Total Population, 1980	County and City	Total Population, 1979	Total Population, 1980
Los Angeles (continued)			Marin (continued)		
Lynwood	40,200	40,850	Mill Valley	13,550	13,500
Manhattan Beach	31,550	31,250	Novato	39,400	42,550
Maywood	18,200	18,500	Ross	2,570	2,520
Monrovia	29,800	30,450	San Anselmo	12,650	12,450
Montebello	49,750	49,900	San Rafael	44,950	44,300
Monterey Park	51,100	52,300	Sausalito	6,550	6,850
Norwalk	84,000	83,700	Tiburon	7,125	7,000
Palmdale	11,950	12,700	Unincorporated	66,700	65,400
Palos Verdes Estates	14,750	14,850	Mariposa	10,300	11,150
Paramount	31,200	31,800	Unincorporated	10,300	11,150
Pasadena	106,800	107,600	Mendocino	64,500	66,700
Pico Rivera	50,700	50,500	Fort Bragg	5,150	5,400
Pomona	86,400	87,600	Point Arena	520	520
Rancho Palos Verdes	37,800	38,200	Ukiah	12,600	12,950
Redondo Beach	63,600	63,600	Willits	4,120	4,420
Rolling Hills	2,080	2,110	Unincorporated	42,100	43,400
Rolling Hills Estate	7,725	7,875	Merced	126,500	128,700
Rosemead	40,300	40,750	Atwater	15,750	16,700
San Dimas	21,300	24,150	Dos Palos	2,920	2,940

San Fernando	15,050	15,250	Gustine	3,000	3,030
San Gabriel	29,250	29,500	Livingston	4,370	4,630
San Marino	13,400	13,450	Los Banos	9,600	10,000
Santa Fe Springs	15,350	15,350	Merced	33,750	34,300
Santa Monica	90,300	88,600	Unincorporated	57,100	57,100
Sierra Madre	12,050	12,100	Modoc	8,550	8,650
Signal Hill	6,250	6,550	Alturas	2,960	2,980
South El Monte	15,900	15,950	Unincorporated	5,600	5,675
South Gate	60,900	61,600	Mono	8,300	9,075
South Pasadena	23,200	23,400	Unincorporated	8,300	9,075
Temple City	29,800	30,000	Monterey	279,000	281,300
Torrance	127,500	126,700	Carmel-by-the-Sea	4,790	4,760
Vernon	230	230	Del Rey Oaks	1,550	1,510
Walnut	11,950	11,950	Gonzales	2,680	2,720
West Covina	75,900	78,700	Greenfield	3,610	3,840
Whittier	69,800	69,500	King City	5,400	5,450
Unincorporated	923,800	942,000	Marina	19,500	18,950
Madera	55,900	59,500	Monterey	27,550	27,600
Chowchilla	4,810	4,960	Pacific Grove	16,450	16,400
Madera	19,450	20,750	Salinas	78,200	79,800
Unincorporated	31,650	33,800	Sand City	190	170
Marin	224,400	225,200	Seaside	34,550	35,100
Belvedere	2,530	2,470	Soledad	5,525	5,475
Corte Madera	7,975	7,775	Unincorporated	79,000	79,500
Fairfax	7,650	7,500	Napa	93,600	93,400
Larkspur	12,700	12,850	Calistoga	3,540	3,530

(continued)

TABLE A.2 (continued)

County and City	Total Population, 1979	Total Population, 1980	County and City	Total Population, 1979	Total Population, 1980
Napa (continued)			Plumas	16,200	17,350
Napa	48,700	48,800	Portola	1,650	1,840
St. Helena	4,540	4,550	Unincorporated	14,550	15,500
Yountville	3,010	2,950	Riverside	620,300	650,600
Unincorporated	33,800	33,600	Banning	13,000	13,250
Nevada	45,500	49,300	Beaumont	6,675	6,725
Grass Valley	6,025	6,000	Blythe	7,250	7,375
Nevada City	2,430	2,390	Coachella	8,700	8,750
Unincorporated	37,050	40,900	Corona	37,100	37,400
Orange	1,854,800	1,896,200	Desert Hot Springs	4,780	5,375
Anaheim	209,100	211,700	Lake Elsinore	5,650	5,875
Brea	27,050	28,650	Hemet	20,850	22,700
Buena Park	63,600	63,900	Indian Wells	1,390	1,600
Costa Mesa	80,100	81,600	Indio	19,450	20,700
Cypress	40,250	40,400	Norco	19,150	19,400
Fountain Valley	54,500	54,200	Palm Desert	12,000	14,900
Fullerton	100,800	101,900	Palm Springs	33,350	36,150
Garden Grove	118,700	120,100	Perris	6,525	6,675
Huntington Beach	167,200	172,200	Rancho Mirage	7,250	8,225
Irvine	53,100	63,500	Riverside	162,900	164,600

Laguna Beach	16,700	16,900	San Jacinto	6,850	7,175
La Habra	44,150	44,800	Unincorporated	247,400	263,800
La Palma	15,050	15,050	Sacramento	749,100	770,200
Los Alamitos	11,050	11,300	Folsom	10,450	11,350
Newport Beach	65,200	65,300	Galt	5,275	5,400
Orange	87,100	87,900	Isleton	920	930
Placentia	33,950	33,300	Sacramento	265,700	274,400
San Clemente	25,950	26,400	Unincorporated	466,700	478,100
San Juan Capistrano	17,700	18,400	San Benito	21,900	22,400
Santa Ana	187,000	189,000	Hollister	9,700	9,900
Seal Beach	26,450	26,800	San Juan Bautista	1,090	1,130
Stanton	23,450	23,300	Unincorporated	11,100	11,350
Tustin	33,700	33,950	San Bernardino	795,600	833,700
Villa Park	7,325	7,375	Adelanto	2,180	2,110
Westminster	70,000	70,800	Barstow	17,200	16,800
Yorba Linda	26,800	29,150	Chino	37,550	37,900
Unincorporated	248,700	256,500	Colton	19,650	20,750
Placer	110,600	116,100	Fontana	26,100	29,500
Auburn	7,325	7,675	Grand Terrace	8,075	7,825
Colfax	960	950	Loma Linda	9,625	9,425
Lincoln	3,960	4,430	Montclair	21,650	22,200
Rocklin	6,500	7,275	Needles	4,040	3,930
Roseville	23,650	24,250	Ontario	74,300	77,500
Unincorporated	68,200	71,500	Rancho Cucamonga	48,100	52,900

(continued)

TABLE A.2 (continued)

County and City	Total Population, 1979	Total Population, 1980	County and City	Total Population, 1979	Total Population, 1980
San Bernardino (continued)			San Mateo (continued)		
Redlands	38,050	40,250	Colma	440	450
Rialto	33,200	34,150	Daly City	73,900	74,000
San Bernardino	107,100	106,500	Foster City	23,250	23,050
Upland	43,550	45,850	Half Moon Bay	7,025	7,075
Victorville	13,550	13,750	Hillsborough	10,400	10,650
Unincorporated	291,700	312,400	Menlo Park	26,150	25,850
San Diego	1,769,000	1,808,200	Millbrae	20,200	20,100
Carlsbad	33,600	35,500	Pacifica	37,400	37,050
Chula Vista	81,400	82,200	Portola Valley	5,225	5,225
Coronado	21,300	20,750	Redwood City	55,600	56,000
Del Mar	5,225	5,225	San Bruno	37,800	37,300
El Cajon	71,100	72,100	San Carlos	27,250	27,050
Escondido	60,000	62,500	San Mateo	80,100	80,300
Imperial Beach	20,850	21,100	South San Francisco	49,550	49,650
La Mesa	50,100	50,900	Woodside	5,425	5,450
Lemon Grove	19,950	20,050	Unincorporated	65,700	65,600
National City	48,100	46,950	Santa Barbara	292,500	295,100
Oceanside	72,400	77,800	Carpinteria	10,100	10,500
San Diego	825,600	842,200	Guadalupe	3,030	3,110

San Marcos	14,600	15,400	Lompoc	26,250	26,750
Vista	33,450	35,150	Santa Barbara	73,500	73,900
Unincorporated	411,300	420,300	Santa Maria	37,050	37,350
San Francisco	647,400	642,900	Unincorporated	142,600	143,500
San Francisco	647,400	642,900	Santa Clara	1,244,400	1,265,200
San Joaquin	317,000	325,100	Campbell	25,000	25,550
Escalon	2,880	3,020	Cupertino	23,500	35,200
Lodi	33,450	34,400	Gilroy	20,100	21,550
Manteca	21,650	23,150	Los Altos	26,350	26,150
Ripon	3,190	3,420	Los Altos Hills	7,475	7,575
Stockton	131,400	137,700	Los Gatos	25,550	28,200
Tracy	16,400	17,000	Milpitas	33,950	35,700
Unincorporated	108,100	106,400	Monte Sereno	3,340	3,420
San Luis Obispo	144,400	149,600	Morgan Hill	16,150	16,600
Arroyo Grande	10,500	10,950	Mountain View	57,600	58,200
El Paso de Robles	8,375	9,325	Palo Alto	54,000	54,100
Grover City	8,275	8,400	San Jose	607,900	610,400
Morro Bay	8,775	8,775	Santa Clara	83,700	85,000
Pismo Beach	5,150	5,200	Saratoga	29,800	30,050
San Luis Obispo	34,500	35,150	Sunnyvale	107,100	107,200
Unincorporated	68,900	71,800	Unincorporated	122,900	120,300
San Mateo	590,500	589,200	Santa Cruz	175,200	178,300
Atherton	7,850	7,800	Capitola	8,825	9,300
Belmont	26,250	26,100	Santa Cruz	39,900	40,500
Brisbane	3,020	2,990	Scotts Valley	6,550	6,850
Burlingame	27,850	27,650			

(continued)

TABLE A.2 (continued)

County and City	Total Population, 1979	Total Population, 1980
Santa Cruz (continued)		
Watsonville	21,150	21,900
Unincorporated	98,800	99,800
Shasta	110,300	115,400
Anderson	7,050	7,200
Redding	42,400	44,000
Unincorporated	60,800	64,200
Sierra	3,240	3,270
Loyalton	940	940
Unincorporated	2,300	2,330
Siskiyou	38,950	40,350
Dorris	860	860
Dunsmuir	2,300	2,310
Etna	720	750
Fort Jones	520	560
Montague	1,080	1,090
Mount Shasta	2,850	3,090
Tulelake	990	960
Weed	2,900	2,920
Yreka	5,500	5,800
Unincorporated	21,250	22,000
Solano	213,600	225,500
Benicia	12,800	14,000
Stanislaus (continued)		
Waterford	2,460	2,700
Unincorporated	91,700	90,200
Sutter	49,600	50,400
Live Oak	2,880	2,970
Yuba City	17,050	17,650
Unincorporated	29,650	29,750
Tehama	36,250	38,200
Corning	4,520	4,980
Red Bluff	9,325	9,650
Tehama	420	420
Unincorporated	21,950	23,150
Trinity	11,800	12,200
Unincorporated	11,800	12,200
Tulare	228,700	234,800
Dinuba	9,775	10,100
Exeter	5,350	5,400
Farmersville	4,900	5,175
Lindsay	6,875	6,900
Porterville	18,400	19,450
Tulare	20,600	21,550
Visalia	44,750	47,750

Dixon	6,625	7,025	Woodlake	4,080	4,250
Fairfield	55,400	58,100	Unincorporated	113,900	114,300
Rio Vista	2,940	3,030	Tuolumne	34,150	35,300
Suisun City	9,450	11,050	Sonora	3,860	3,840
Vacaville	39,450	42,450	Unincorporated	30,300	31,450
Vallejo	72,100	75,400	Ventura	491,400	510,300
Unincorporated	14,750	14,500	Camarillo	33,450	37,100
Sonoma	277,700	284,400	Fillmore	8,950	8,925
Cloverdale	3,870	3,950	Ojai	6,375	6,425
Cotati	2,830	2,830	Oxnard	98,300	102,100
Healdsburg	6,925	7,025	Port Hueneme	18,500	17,950
Petaluma	32,600	32,750	San Buenaventura	70,100	70,400
Rohnert Park	20,250	22,550	Santa Paula	19,450	19,600
Santa Rosa	76,100	78,300	Simi Valley	75,700	77,400
Sebastopol	5,075	5,100	Thousand Oaks	68,300	73,800
Sonoma	5,875	5,925	Unincorporated	92,300	96,500
Unincorporated	124,200	126,000	Yolo	108,700	110,400
Stanislaus	249,900	255,100	Davis	36,950	37,800
Ceres	11,050	11,800	Winters	2,630	2,640
Hughson	2,680	2,690	Woodland	28,050	28,250
Modesto	99,500	103,400	Unincorporated	41,050	41,700
Newman	2,600	2,640	Yuba	48,500	49,450
Oakdale	8,300	8,500	Marysville	10,050	10,150
Patterson	3,960	3,880	Wheatland	1,500	1,460
Riverbank	5,025	4,920	Unincorporated	36,950	37,850
Turlock	22,650	24,350	State total	22,507,000	22,911,000

Note: Totals may not add due to independent rounding.

Source: Department of Finance, Population Research Unit.

TABLE A.3

Consumer Price Index for All Urban Consumers in the United States and Three California Areas

Component	Index Level, April 1981	Average Annual Percent Change 1977-81	1979-81	April 1980 to April 1981	Forecast April 1981 to April 1984
All items					
United States	280.7	10.2	11.7	9.6	4.7
Los Angeles	281.8	10.4	12.9	10.3	4.6
San Diego	321.3	13.4	15.1	15.1	4.3
San Francisco	295.5	11.5	14.4	16.6	3.7
Housing					
United States	304.3	11.5	13.4	11.1	3.5
Los Angeles	310.3	11.3	15.3	11.8	3.0
San Diego	394.6	16.7	18.7	20.0	3.3
San Francisco	338.4	14.0	18.6	25.5	1.3
Homeownership					
United States	367.2	13.9	15.8	11.6	1.1
Los Angeles	377.6	13.2	18.4	13.3	0.6
San Diego	556.2	20.2	22.0	24.9	-0.2
San Francisco	424.1	16.5	22.6	35.7	-2.4
Food and beverages					
United States	270.2	8.6	7.5	5.0	6.5
Los Angeles	262.4	8.7	7.3	5.4	6.7
San Diego	268.7	8.8	7.6	4.3	6.9
San Francisco	262.8	8.5	7.0	4.1	7.0
Transportation					
United States	288.7	11.3	15.2	11.6	2.6
Los Angeles	292.8	11.3	15.2	11.5	2.7
San Diego	284.1	11.4	14.9	11.9	2.6
San Francisco	287.5	11.3	14.8	9.8	3.1
Apparel and upkeep					
United States	191.1	5.1	6.2	3.7	4.4
Los Angeles	176.1	4.4	6.3	2.4	4.9
San Diego	192.0	6.3	6.7	6.1	4.7
San Francisco	196.2	5.8	6.4	4.7	5.3
Medical care					
United States	307.7	10.0	10.7	12.2	7.8
Los Angeles	332.9	11.6	13.3	16.9	7.5
San Diego	291.3	9.5	9.9	10.6	7.4
San Francisco	304.6	9.8	10.5	12.3	8.0
Miscellaneous					
United States	245.9	8.2	9.5	10.3	7.0
Los Angeles	237.6	9.0	9.5	10.3	8.2
San Diego	238.5	8.4	8.7	8.5	7.8
San Francisco	243.3	8.4	8.5	10.4	7.1
Entertainment					
United States	226.5	6.9	8.0	7.2	4.9
Los Angeles	203.3	5.8	8.2	7.2	3.9
San Diego	220.6	8.2	8.9	8.4	5.5
San Francisco	233.5	8.3	10.7	12.3	4.0

APPENDIX B

GLOSSARY OF ENVIRONMENTAL IMPACT REPORT TERMINOLOGY

TO: All Persons Interested in Environmental Impact Reports

FROM: Ron Bass

For those who may not be entirely clear about the different types of environmental documents under C.E.Q.A. and N.E.P.A., the following list should be instructive.

I. TYPES OF E.I.R.s

- Environmental Impact Report
  (State E.I.R. Guidelines Sec. 15027)

  A detailed statement setting forth the environmental effects and consideration pertaining to a particular project. Must be prepared when there may be significant environmental impacts that cannot be mitigated.

- "Focused E.I.R."
  (Term not used in guidelines)

  All E.I.R.s are supposed to be focused on the significant issues only. May even be a single issue E.I.R. (air quality for instance). It is not necessary to analyze insignificant issues so long as there is an explanation of why they are insignificant.

- Subsequent E.I.R.
  (State E.I.R. Guidelines Sec. 15067)

  A second E.I.R. on a project for which one E.I.R. has already been prepared. Required when there are changes in the project or new information requiring important revisions to the original E.I.R. due to new significant impacts.

- Supplement to an E.I.R.
  (State E.I.R. Guidelines Sec. 15067).

  A revision to a previously prepared E.I.R. Required when only minor additions or changes would be necessary to make

---

This material is reprinted from the California EIR Monitor, vol. 7, no. 16 (August 19, 1980). It contains explanations of the most common environmental documents under the California Environmental Quality Act and the National Environmental Protection Act. It is reproduced as originally published so readers are cautioned that citations to sections of statutes and codes may not be current, as in the case of the mitigated Negative Declaration.

previous E.I.R. adequate. Only the inadequate part of the previous E.I.R. need be changed and circulated for review.

- Master Environmental Assessment
(State E.I.R. Guidelines Sec. 15069.6)

A document used to identify and organize environmental information for a region or area in order to provide information which may be used or referenced in other documents.

- Program E.I.R.
(State E.I.R. Guidelines Sec. 15069.8)

An E.I.R. prepared for an integrated series of actions that are related either: a) geographically; b) as logical parts of a chain of actions; c) in connection with the issuance of rules, regulations, or plans; d) which fall under a common authorizing action.

- General Plan E.I.R.

A program E.I.R. that is prepared in connection with or as a part of a General Plan element, specific plan, area plan or amendment to any of these.

II. OTHER DOCUMENTS UNDER C.E.Q.A.

- Initial Study
(State E.I.R. Guidelines Sec. 15029.5)

A preliminary analysis prepared to determine whether an E.I.R. or negative declaration must be prepared.

- Negative Declaration
(State E.I.R. Guidelines Sec. 15033)

A written statement briefly describing the reasons that a proposed project, although not otherwise exempt, will not have a significant effect on the environment and therefore does not require an E.I.R.

- "Mitigated" Negative Declaration
(Term not used in guidelines)

A negative declaration prepared for a proposed project when significant environmental impact have been identified but have been or will be mitigated, by project redesign or conditions on approval, leaving no significant unmitigated impacts.

III. N.E.P.A. DOCUMENTS

- Environmental Impact Statement
  (40 CFR 1508.11)

  A detailed document describing the environmental impacts of any major Federal actions significantly affecting the environment.

- Finding of No Significant Impact (F.O.N.S.)
  (40 CFR 1508.13)

  The Federal equivalent to the Negative Declaration, briefly presenting the reasons why an action will not significantly affect the environment.

- Environmental Assessment
  (40 CFR 1508.9)

  The Federal equivalent to the Initial Study, permitting the federal agency to determine whether or not an E.I.S. is necessary.

IV. JOINT DOCUMENTS

- Joint E.I.R./E.I.S.
  (40 CFR 1506.2 and 15069 E.I.R. Guidelines)

  An environmental document prepared cooperatively by a state or local agency and a Federal agency serving as an E.I.R. under C.E.Q.A. and an E.I.S. under N.E.P.A.

- Joint E.I.R./F.O.N.S.
  (Term not found in either E.I.R. Guidelines or N.E.P.A. Regulations)

  An environmental document serving as an E.I.R. under C.E.Q.A. and a F.O.N.S. under N.E.P.A.

- Joint Negative Declaration/F.O.N.S.
  (Term not found in either E.I.R. Guidelines or N.E.P.A. Regulations)

  An environment document serving as a negative declaration under C.E.Q.A. and a F.O.N.S. under N.E.P.A.

- Joint Negative Declaration/E.I.S.
  (Term not found in either E.I.R. Guidelines or N.E.P.A. Regulations)

  Unlikely to ever happen.

APPENDIX C

CALIFORNIA ENVIRONMENTAL QUALITY ACT

(California Public Resources Code, Division 13-Environmental Quality; Text as amended through 1979, With 1980-81 amendment attached)

DIVISION 13. ENVIRONMENTAL QUALITY

Chapter 1. Policy

21000. The Legislature finds and declares as follows:

(a) The maintenance of a quality environment for the people of this state now and in the future is a matter of statewide concern.

(b) It is necessary to provide a high-quality environment that at all times is healthful and pleasing to the senses and intellect of man.

(c) There is a need to understand the relationship between the maintenance of high-quality ecological systems and the general welfare of the people of the state, including their enjoyment of the natural resources of the state.

(d) The capacity of the environment is limited, and it is the intent of the Legislature that the government of the state take immediate steps to identify any critical thresholds for the health and safety of the people of the state and take all coordinated actions necessary to prevent such thresholds being reached.

(e) Every citizen has a responsibility to contribute to the preservation and enhancement of the environment.

(f) The interrelationship of policies and practices in the management of natural resources and waste disposal requires systematic and concerted efforts by public and private interests to enhance environmental quality and to control environmental pollution.

(g) It is the intent of the Legislature that all agencies of the state government which regulate activities of private individuals, corporations, and public agencies which are found to affect the quality of the environment, shall regulate such activities so that major consideration is given to preventing environmental damage.

21001. The Legislature further finds and declares that it is the policy of the state to:

(a) Develop and maintain a high-quality environment now and in the future, and take all action necessary to protect, rehabilitate, and enhance the environmental quality of the state.

(b) Take all action necessary to provide the people of this state with clean air and water, enjoyment of aesthetic, natural, scenic, and historic environmental qualities, and freedom from excessive noise.

(c) Prevent the elimination of fish or wildlife species due to man's activities, insure that the fish and wildlife populations do not drop below self-perpetuating levels, and preserve for future generations representations of all plant and animal communities and examples of the major periods of California history.

(d) Ensure that the long-term protection of the environment shall be the guiding criterion in public decisions.

(e) Create and maintain conditions under which man and nature can exist in productive harmony to fulfill the social and economic requirements of present and future generations.

(f) Require governmental agencies at all levels to develop standards and procedures necessary to protect environmental quality.

(g) Require governmental agencies at all levels to consider qualitative factors as well as economic and technical factors and long-term benefits and costs, in addition to short-term benefits and costs and to consider alternatives to proposed actions affecting the environment.

21002. The Legislature finds and declares that it is the policy of the state that public agencies should not approve projects as proposed if there are feasible alternatives or feasible mitigation measures available which would substantially lessen the significant environmental effects of such projects, and that the procedures required by this division are intended to assist public agencies in systematically identifying both the significant effects of proposed projects and the feasible alternatives or feasible mitigation measures which will avoid or substantially lessen such significant effects. The Legislature further finds and declares that in the event specific economic, social, or other conditions make infeasible such project alternative or such mitigation measures, individual projects may be approved in spite of one or more significant effects thereof.

21002.1. In order to achieve the objectives set forth in Section 21002, the Legislature finds and declares that the following policy shall apply to the use of environmental impact reports prepared pursuant to the provisions of this division:

(a) The purpose of an environmental impact report is to identify the significant effects of a project on the environment, to identify alternatives to the project, and to indicate the manner in which such significant effects can be mitigated or avoided.

(b) Each public agency shall mitigate or avoid the significant effects on the environment of projects it approves or carries out whenever it is feasible to do so.

(c) In the event that economic, social, or other conditions make it infeasible to mitigate one or more significant effects of a project on the environment, such project may nonetheless be approved or carried out at the discretion of a public agency, provided that the project is otherwise permissible under applicable laws and regulations.

(d) In applying the policies of subdivisions (b) and (c) to individual projects, the responsibility of a public agency which is functioning as a lead agency shall differ from that of a public agency which is functioning as a responsible agency. A public agency functioning as a lead agency shall have responsibility for considering the effects, both individual and collective, of all activities involved in a project. A public agency functioning as a responsible agency shall have responsibility for considering only the effects of those activities involved in a project, which it is required by law to carry out or approve. The provisions of this subdivision shall apply only to decisions by a public agency to carry out or approve a project and shall not limit the scope of the comments such agency may wish to make pursuant to Section 21104 or 21153.

21003. The Legislature further finds and declares that it is the policy of the state that:

(a) Local agencies integrate the requirements of this division with planning and environmental review procedures otherwise required by law or by local practice so that all such procedures, to the maximum feasible extent, run concurrently, rather than consecutively.

(b) Documents prepared pursuant to this division be organized and written in such a manner that they will be meaningful and useful to decision makers and to the public.

(c) Environmental impact reports omit unnecessary descriptions of projects and emphasize feasible mitigation measures and alternatives to projects.

(d) Information developed in individual environmental impact reports be incorporated into a data base which can be used to reduce delay and duplication in preparation of subsequent environmental impact reports.

(e) Information developed in environmental impact reports covering large geographical areas be used to contribute to information required in specific environmental impact reports.

Chapter 2. Short Title

21050. This division shall be known and may be cited as the California Environmental Quality Act.

Chapter 2.5. Definitions

21060. Unless the context otherwise requires, the definitions in this chapter govern the construction of this division.

21060.3. "Emergency" means a sudden, unexpected occurrence, involving a clear and imminent danger, demanding immediate action to prevent or mitigate loss of, or damage to, life, health, property, or essential public services. "Emergency" includes such occurrences as fire, flood, earthquake, or other soil or geologic movements, as well as such occurrences as riot, accident, or sabotage.

21060.5. "Environment" means the physical conditions which exist within the area which will be affected by a proposed project, including land, air, water, minerals, flora, fauna, noise, and objects of historic or aesthetic significance.

21061. "Environmental impact report" means a detailed statement setting forth the matters specified in Section 21100 and 21100.1; provided that information or data which is relevant to such a statement and is a matter of public record or is generally available to the public need not be repeated in its entirety in such statement, but may be specifically cited as the source for conclusions stated therein; and provided further that such information or data shall be briefly described, that its relationship to the environmental impact report shall be indicated, and that the source thereof shall be reasonably available for inspection at a public place or public building. An environmental impact report also includes any comments which are obtained pursuant to Section 21104 or 21153, or which are required to be obtained pursuant to this division.

An environmental impact report is an informational document which, when its preparation is required by this division, shall be considered by every public agency prior to its approval or disapproval of a project. The purpose of an environmental impact report is to provide public agencies and the public in general with detailed information about the effect which a proposed project is likely to have on the environment; to list ways in which the significant effects of such a project might be minimized; and to indicate alternatives to such a project.

In order to facilitate the use of environmental impact reports, public agencies shall require that such reports contain an index or table of contents and a summary. Failure to include such index, table of contents, or summary shall not constitute a cause of action pursuant to Section 21167.

21061.1. "Feasible" means capable of being accomplished in a successful manner within a reasonable period of time, taking into account economic, environmental, social, and technological factors.

21062. "Local agency" means any public agency other than a state agency, board, or commission. For purposes of this division a redevelopment agency and a local agency, formation commission are local agencies, and neither is a state agency, board or commission.

21063. "Public agency" includes any state agency, board, or commission, any county, city and county, city, regional agency, public district, redevelopment agency, or other political subdivision.

21064. "Negative declaration" means a written statement briefly describing the reasons that a proposed project will not have a significant effect on the environment and does not require the preparation of an environmental impact report.

21065. "Project" means the following:

(a) Activities directly undertaken by any public agency.

(b) Activities undertaken by a person which are supported in whole or in part through contracts, grants, subsidies, loans, or other forms of assistance from one or more public agencies.

(c) Activities involving the issuance to a person of a lease, permit, license, certificate, or other entitlement for use by one or more public agencies.

21065.5. "Geothermal exploratory project" means a project as defined in Section 21065 composed of not more than six wells and associated drilling and testing equipment, whose chief and original purpose is to evaluate the presence and characteristics of geothermal resources prior to commencement of a geothermal field development project as defined in Section 65928.5 of the Government Code. Wells included within a geothermal exploratory project must be located at least one-half mile from geothermal development wells which are capable of producing geothermal resources in commercial quantities.

21066. "Person" includes any person, firm, association, organization, partnership, business, trust, corporation, company, district, county, city and county, city, town, the state, and any of the agencies and political subdivisions of such entities.

21067. "Lead agency" means the public agency which has the principal responsibility for carrying out or approving a project which may have a significant effect upon the environment.

21068. "Significant effect on the environment" means a substantial, or potentially substantial, adverse change in the environment.

21069. "Responsible agency" means a public agency, other than the lead agency, which has responsibility for carrying out or approving a project.

Chapter 2.6. General

21080. (a) Except as otherwise provided in this division, this division shall apply to discretionary projects proposed to be carried out or approved by public agencies, including, but not limited to, the enactment and amendment of zoning ordinances, the issuance of zoning variances, the issuance of conditional use permits and the approval of tentative subdivision maps (except where such a project is exempt from the preparation of an environmental impact report pursuant to Section 21166).

(b) This division shall not apply to the following:

(1) Ministerial projects proposed to be carried out or approved by public agencies.

(2) Emergency repairs to public service facilities necessary to maintain service.

(3) Projects undertaken, carried out, or approved by a public agency to maintain, repair, restore, demolish, or replace property or facilities damaged or destroyed as a result of a disaster in a disaster-stricken area in which a state of emergency has been proclaimed by the Governor pursuant to Chapter 7 (commencing with Section 8550) of Division I of Title 2 of the Government Code.

(4) Specific actions, necessary to prevent or mitigate an emergency.

(5) Projects which a public agency rejects or disapproves.

(6) Actions undertaken by a public agency relating to any thermal power plant site or facility, including the expenditure, obligation, or encumbrance of funds by a public agency for planning, engineering, or design purposes, or for the conditional sale or purchase of equipment, fuel, water (except ground water), steam, or power for such a thermal powerplant, if the powerplant site and related facility will be the subject of an environmental impact report or negative declaration or other document, or documents, prepared pursuant to a regulatory program certified pursuant to Section 21080.5, which will be prepared by the State Energy Resources Conservation and Development Commission, by the Public Utilities Commission, or by the city or county in which the powerplant and related facility would be located; provided that such environmental impact report, negative declaration or other document, or documents, shall include the environmental impact, if any, of the action described in this paragraph.

(7) Activities or approvals necessary to the bidding for, hosting or staging of, and funding or carrying out of, an Olympic games under the authority of the International Olympic Committee, except for the construction of facilities necessary for such Olympic games.

(8) The establishment, modification, structuring, restructuring, or approval of rates, tolls, fares or other charges by public agencies which the public agency finds are for the purpose of (1) meeting operating expenses, including employee wage rates and fringe benefits, (2) purchasing or leasing supplies, equipment or materials, (3) meeting financial reserve needs and requirements, (4) obtaining funds for capital projects, necessary to maintain service within existing service areas, or (5) obtaining funds necessary to maintain such intracity transfers as are authorized by city charter. The public agency shall incorporate written findings in the record of any proceeding in which an exemption under this paragraph is claimed setting forth with specificity the basis for the claim of exemption.

(9) Actions taken prior to January 1, 1982, by a public agency to implement the transition from the property taxation system in effect prior to June 1, 1978, to the system provided for by Article XIII A of the California Constitution. Such actions shall be limited to projects defined in subdivision (a) or (b) of Section 21065 which initiate

or increase fees, rates, or charges charged for any existing public service, program, or activity; reduce or eliminate the availability of an existing public service, program, or activity; close publicly owned or operated facilities; or reduce or eliminate the availability of an existing publicly owned transit service, program, or activity.

(c) In the event that a lead agency determines that a proposed project, not otherwise exempt from the provisions of this division, does not have a significant effect on the environment, such lead agency shall adopt a negative declaration to that effect.

21080.1. The lead agency shall have the responsibility for determining whether an environmental impact report or a negative declaration shall be required for any project subject to the provisions of this division. Such determination shall be final and conclusive on all persons, including responsible agencies, unless challenged as provided in Section 21167.

21080.2. In the case of a project described in subdivision (c) of Section 21065, the determination required by Section 21080.1 shall be made within 45 days from the date on which an application for a project has been received and accepted as complete by the lead agency.

21080.3. (a) Prior to determining whether a negative declaration or environmental impact report is required for a project, the lead agency shall consult with all responsible agencies and with any other public agency which has jurisdiction by law over natural resources affected by the project which are held in trust for the people of the State of California.

(b) In order to expedite the requirements of subdivision (a), the Office of Planning and Research, upon request of a lead agency, shall assist such lead agency in determining the various responsible agencies for a proposed project. In the case of a project described in subdivision (c) of Section 21065, such a request may also be made by the project applicant.

21080.4. (a) In the event that a lead agency determines that an environmental impact report is required for a project, such lead agency shall immediately send notice of such determination by certified mail or an equivalent procedure to each responsible agency and to those public agencies having jurisdiction by law over natural resources affected by the project which are held in trust for the people of the State of California. Upon receipt of such notice, each responsible agency shall specify to the lead agency the scope and content of the environmental information which is germane to such a responsible agency's statutory responsibilities in connection with the proposed project and which, pursuant to the requirements of this division, shall be included in the environmental impact report. Such information shall be specified in writing and shall be communicated to the lead agency by certified mail or equivalent procedure not later than 45 days after receipt of the notice of the lead agency's determination. The lead agency shall request similar guidance from appropriate federal agencies.

(b) In order to expedite the requirements of subdivision (a), the lead agency or any responsible agency or public agency having jurisdiction by law over natural resources affected by the project which are held in trust for the people of the State of California may request one or more meetings between representatives of such agencies for the purpose of assisting the lead agency to determine the scope and content of the environmental information such responsible agency may require. In the case of a project described in subdivision (c) of Section 21065, such a request may also be made by the project applicant. Such meetings shall be convened by the lead agency as soon as possible, but no later than 30 days, after they have been requested.

(c) In order to expedite the requirements of subdivision (a), the Office of Planning and Research, upon request of a lead agency, shall assist such lead agency in determining the various responsible agencies and any federal agencies which have responsibility for carrying out or approving a proposed project. In the case of a project described in subdivision (c) of Section 21065, such a request may also be made by the project applicant.

(d) In the event that a state agency is a responsible agency or a public agency having jurisdiction by law over natural resources affected by the project which are held in trust for the people of the State of California. Subject to the requirements of subdivision (a), the Office of Planning and Research shall ensure that the information required by subdivision (a) is transmitted to the lead agency within the required time period.

21080.5. (a) When the regulatory program of a state agency, board, or commission requires a plan or other written documentation, containing environmental information and complying with the requirements of paragraph (3) of subdivision (d) of this section, to be submitted in support of any of the activities listed in subdivision (b), such plan or other written documentation may be submitted in lieu of the environmental impact report required by the division; provided that the Secretary of the Resources Agency has certified the regulatory program pursuant to this section.

(b) The provisions of this section shall apply only regulatory programs or portions thereof which involve:

(1) The issuance to a person of a lease, permit, license, certificate, or other entitlement for use; or

(2) The adoption or approval of standards, rules, regulations or plans for use in the regulatory program.

(c) A regulatory program certified pursuant to this section is exempt from the provisions of Chapter 3 (commencing with Section

21100) and Chapter 4 (commencing with Section 21150) of this division and Section 21167.

(d) In order to qualify for certification pursuant to this section, a regulatory program shall require utilization of an interdisciplinary approach which will ensure the integrated use of the natural and social sciences in decision making and shall meet all of the following criteria;

(1) The enabling legislation of the regulatory program shall:

(i) Include protection of the environment among its principal purposes.

(ii) Contain authority for the administering agency to promulgate rules and regulations for the protection of the environment, guided by standards set forth in the enabling legislation.

(2) The rules and regulations adopted by the administering agency shall:

(i) Require that an activity will not be approved or adopted as proposed if there are feasible alternatives or feasible mitigation measures available which would substantially lessen any significant adverse impact which the activity may have on the environment.

(ii) Include guidelines for the orderly evaluation of proposed activities and the preparation of the plan or other written documentation in a manner consistent with the environmental protection purposes of the regulatory program.

(iii) Require the administering agency to consult with all public agencies which have jurisdiction, by law, with respect to the proposed activity.

(iv) Require that final action on the proposed activity include the written responses of the issuing authority to significant environmental points raised during the evaluation process.

(v) Require the filing of a notice of the decision by the administering agency on the proposed activity with the Secretary of the Resource Agency. Such notices shall be available for public inspection, and a list of such notices shall be posted on a weekly basis in the office of the Resources Agency. Each such list shall remain posted for a period of 30 days.

(vi) Require notice of the filing of the plan or other written documentation to be made to the public and to any person who requests, in writing, such notification. The notification shall be made in a manner that will provide the public or any such person with sufficient time to review and comment on such filing.

(3) The plan or other written documentation required by the regulatory program shall:

(i) Include a description of the proposed activity with alternatives to the activity and mitigation measures to minimize any significant adverse environmental impact.

(ii) Be available for a reasonable time for review and comment by other public agencies and the general public.

(e) The Secretary of the Resources Agency shall certify a regulatory program which the secretary determines meets all the qualifications for certification set forth in this section and withdraw certification on determination that the regulatory program has been altered so that it no longer meets such qualifications. Certification and withdrawal of certification shall occur only after compliance with Chapter 4.5 (commencing with Section 11371) of Part I of Division 3 of Title 2 of the Government Code.

In determining whether or not a regulatory program meets the qualifications for certification set forth in this section, the inquiry of the Secretary of the Resources Agency shall extend only to the question of whether the regulatory program meets the generic requirements of subdivision (d). The inquiry shall not extend to individual decisions to be reached under such regulatory program, including the nature of specific alternatives or mitigation measures which might be proposed to lessen any significant adverse environmental effects of the activity.

In the event that the Secretary of the Resources Agency determines that the regulatory program submitted for certification does not meet the qualifications for certification set forth in this section, the secretary shall adopt findings setting forth the reasons for such determination.

(f) After a regulatory program has been certified pursuant to this section, any proposed change in such program which could affect compliance with the qualifications for certification specified in subdivision (d) may be submitted to the Secretary of the Resources Agency for a review and comment. The scope of the secretary's review shall extend only to the question of whether the regulatory program meets the generic requirements of subdivision (d). The review shall not extend to individual decisions to be reached under such regulatory program, including specific alternatives or mitigation measures which might be proposed to lessen any significant adverse environmental effects of the activity. The secretary shall have 30 days after receipt of such proposed change to notify the state agency, board, or commission whether the proposed change will alter the regulatory program so that it no longer meets the qualification for certification established in this section and will result in a withdrawal of certification as provided in this section.

(g) Any action or proceeding to attack, review, set aside, void, or annul a determination or decision of a state agency, board, or commission approving or adopting a proposed activity under a regulatory program which has been certified pursuant to this section on the basis that the plan or other written documentation prepared pursuant to paragraph (3) of subdivision (d) does not comply with the provisions of this section shall be commenced no later than 30 days from the date of the filing of notice of the approval or adoption of the activity.

(h) Any action or proceeding to attack, review, set aside, void or annul a determination of the Secretary of the Resources Agency to certify a regulatory program pursuant to this section on the basis that the regulatory program does not comply with the provisions of this section shall be commenced within 30 days after certification by the secretary.

In any action brought under this subdivision, the inquiry shall extend only to whether there was a prejudicial abuse of discretion by the Secretary of the Resources Agency. Abuse of discretion is established if the secretary has not proceeded in a manner required by law or if the determination is not supported by substantial evidence.

(i) For the purposes of this section, any county agricultural commissioner shall be considered a state agency.

21080.7. (a) No environmental impact report or negative declaration shall be required for any project involving the construction of housing or neighborhood commercial facilities in an urbanized area if the lead agency does all of the following:

(1) Finds, after giving notice pursuant to subdivision (b) or (c) of Section 21092 and following the procedure prescribed by law or regulation which would be necessary to make a determination pursuant to Section 21080.1.

(i) That such project is consistent with a specific plan which has been adopted pursuant to Article 8 (commencing with Section 65450), Article 9 (commencing with Section 65500), and Article 10 (commencing with Section 65550) of Chapter 3 of Title 7 of the Government Code or, in the coastal zone, a local coastal program certified pursuant to Article 2 (commencing with Section 30510) of Chapter 6 of Division 20.

(ii) That such plan or program was adopted pursuant to the procedure established by Article 9 (commencing with Section 65500) of Chapter 3 of Title 7 of the Government Code not more than five years prior to the finding made pursuant to this section, and,

(iii) That such plan or program has been the subject of an environmental impact report, and

(iv) That the environmental impact report is sufficiently detailed so that the significant adverse effects of the project on the environment and measures necessary to mitigate or avoid any such effects can be determined, including significant physical effects on existing structures and neighborhoods of historical or aesthetic significance if any exist in the area covered by the plan or program, and measures necessary to mitigate such effects.

(2) Makes one or more of the findings as required pursuant to Section 21081.

(3) Files a notice of the decision on the proposed activity with the county clerk. Such notices shall be available for public inspection, and a list of such notices shall be posted on a weekly basis in the office of the county clerk. Each such list shall remain posted for a period of 30 days.

(b) As used in this section:

(1) "Neighborhood commercial facilities" means those commercial facilities which are an integral part of a project involving the construction of housing and which will serve the residents of such housing.

(2) "Urbanized area" means a central city or cities and surrounding closely settled territory, as defined by the United States Department of Commerce Bureau of the Census in the Federal Register, Volume 39, Number 85, for Wednesday, May 1, 1974, at pages 15202-15203 and as periodically updated.

21080.9 This division shall not apply to activities and approvals by any local government, as defined in Section 30109, necessary for the preparation and adoption of a local coastal program pursuant to Division 20 (commencing with Section 30000); provided, however, that certification of a local coastal program by the California Coastal Commission pursuant to Chapter 6 (commencing with Section 30500) of Division 20 shall be subject to the requirements of this division. For the purpose of Section 21080.5, a certified local coastal program constitutes a plan for use in the California Coastal Commission's regulatory program.

21081. Pursuant to the policy stated in Section 21002 and 21002.1, no public agency shall approve or carry out a project for which an environmental impact report has been completed which identifies one or more significant effects thereof unless such public agency makes one, or more, of the following findings:

(a) Changes or alterations have been required in, or incorporated into, such project which mitigate or avoid the significant environmental effects thereof as identified in the completed environmental impact report.

(b) Such changes or alterations are within the responsibility and jurisdiction of another public agency and such changes have been adopted by such other agency, or can and should be adopted by such other agency.

(c) Specific economic, social, or other considerations make infeasible the mitigation measures or project alternatives identified in the environmental impact report.

21082. All public agencies shall adopt by ordinance, resolution, rule, or regulation, objectives, criteria, and procedures for the evaluation of projects and the preparation of environmental impact reports and negative declarations pursuant to this division. A school district, or any other district, whose boundaries are coterminous with a city, county, or city and county, may utilize the objectives, criteria, and procedures of the city, county, or city and county, as may be applicable, in which case, the school district or other district need not adopt objectives, criteria and procedures of its own. The objectives, criteria, and procedures shall be consistent with the provisions of this division and with the guidelines adopted by the Secretary of the Resources Agency pursuant to Section 21083. Such objectives, criteria, and procedures shall be adopted by each public agency no later than 60 days after the Secretary of the Resources Agency has adopted guidelines pursuant to Section 21083.

21082.1. Any environmental impact report or negative declaration prepared pursuant to the requirement of this division shall be prepared directly by, or under contract to, a public agency.

This section is not intended to prohibit, and shall not be construed as prohibiting, any person from submitting information to the public agency responsible for preparing an environmental impact report or negative declaration. Such information may be submitted in any format and may be included in whole or in part in any such report or declaration.

21083. The Office of Planning and Research shall prepare and develop proposed guidelines for the implementation of this division by public agencies. Such guidelines shall include objectives and criteria for the orderly evaluation of projects and the preparation of environmental reports and negative declarations in manner consistent with this division.

Such guidelines shall specifically include criteria for public agencies to follow in determining whether or not a proposed project may have a "significant effect on the environment". Such criteria shall require a finding that a project may have a "significant effect on the environment" if any of the following conditions exist:

(a) A proposed project has the potential to degrade the quality of the environment, curtail the range of the environment, or to achieve short-term, to the disadvantage of long-term, environmental goals.

(b) The possible effects of a project are individually limited but cumulatively considerable. As used in this subdivision, "cumulatively considerable" means that the incremental effects of an individual project are considerable when viewed in connection with the effect of past projects, the effects of other current projects, and the effects of probable future projects.

(c) The environmental effects of a project, will cause substantial adverse effect on human beings, either directly or indirectly.

Such guidelines shall also include procedures for determining the lead agency pursuant to the provisions of Section 21165.

Such guidelines shall also include criteria for public agencies to use in determining when a proposed project is of sufficient statewide, regional, or areawide environmental significance that it should be submitted to appropriate state agencies for review and comment prior to completion of an environmental impact report or negative declaration thereon.

The Office of Planning and Research shall develop and prepare such proposed guidelines as soon as possible and shall transmit them immediately to the Secretary of the Resources Agency. No later than 60 days after the effective date of this section the Secretary of the Resources Agency shall certify and adopt such guidelines pursuant to Chapter 4.5 (commencing with Section 11371) of Part I, Division 3, Title 2, of the Government Code, which shall become effective upon the filing thereof, provided that such guidelines shall not be adopted without compliance with Sections 11423, 11424 and 11425 of the Government Code.

21083.5 The guidelines prepared and adopted pursuant to Section 21083 shall provide that when an environmental impact statement has been, or will be, prepared for the same project pursuant to the requirements of the National Environmental Policy Act of 1969 and implementing regulations thereto, all or any part of such statement may be submitted in lieu of all or any part of an environmental impact report required by this division, provided that such statement, or the part thereof so used, shall comply with the requirements of this division and the guidelines adopted pursuant thereto.

21083.6 In the event that a project requires both an environmental impact report prepared pursuant to the requirements of this division and an environmental impact statement prepared pursuant to the requirements of the National Environmental Policy Act of 1969, an applicant may request and the lead agency may waive the time limits established pursuant to Section 21100.2 or 21151.5 if it finds that additional time is required to prepare a combined environmental impact report-environmental impact statement and that the time required to prepare such a combined document would be shorter than that required to prepare each document separately.

21083.7. In the event that a project requires both an environmental impact report prepared pursuant to the requirements of this division and an environmental impact statement prepared pursuant to the requirements of the National Environmental Policy Act of 1969, the lead agency shall, whenever possible, use the environmental impact statement as such environmental impact report as provided in Section 21083.5. In order to implement the provisions of this section, each lead agency to which this section is applicable shall consult, as soon as possible, with the agency required to prepare such environmental impact statement.

21084. (a) The guidelines prepared and adopted pursuant to Section 21083 shall include a list of classes of projects which have been determined not to have a significant effect on the environment and which shall be exempt from the provisions of this division. In adopting the guidelines, the Secretary of the Resources Agency shall make a finding that the list or classification of projects referred to in this section do not have a significant effect on the environment.

(b) No project of the Department of Transportation which will result in damage to scenic resources, including, but not limited to trees, historic buildings, rock outcroppings, or similar resources, with an official state scenic highway, as designated pursuant to Section 262 of the Streets and Highways Code, shall be exempted from this division pursuant to subdivision (a).

21085. All classes of projects designated pursuant to Section 21084, shall be exempt from the provisions of this division.

21085.5. A project for the institution or increase of passenger or commuter service on rail lines already in use, including the modernization of existing stations and parking facilities, shall be exempt from this division.

21085.6. A project for the development of a regional transportation improvement program or the state transportation improvement program is exempt from this division.

21086. A public agency may at any time, request the addition or deletion of class of projects, to the list to designated pursuant to Section 21084. Such a request shall be made in writing to the Office of Planning and Research and shall include information supporting the public agency's position that such class of projects does, or does not, have a significant effect on the environment.

The Office of Planning and Research shall review each such request and, as soon as possible, shall submit its recommendation to the Secretary of the Resources Agency. Following the receipt of such recommendation, the Secretary of the Resources Agency may add or delete the class of projects to the list of classes of projects designated pursuant to Section 21084 which are exempt from the requirements of this division.

The addition or deletion of a class of projects, as provided in this section, to the list specified in Section 21084 shall constitute an amendment to the guidelines adopted pursuant to Section 21083 and shall be adopted in the manner prescribed in Section 21083, 21084, and 21087.

21087. The Office of Planning and Research shall periodically review the guidelines adopted pursuant to Section 21083 and shall recommend proposed changes or amendments to the Secretary of the Resources Agency. Changes or amendments to the guidelines shall be adopted by the Secretary of the Resources Agency in conformance with Chapter 4.5 (commencing with Section 11371) of Part 1, Division 3, Title 2 of the Government Code.

21088. The Secretary of the Resources Agency shall provide for the timely distribution to all public agencies of the guidelines and any amendments or changes thereto. In addition, the Secretary of the Resources Agency may provide for publication of a bulletin to provide public notice of the guidelines, or any amendments or changes thereto, and of the completion of environmental impact reports prepared in compliance with this division.

21089. A public agency may charge and collect a reasonable fee from any person proposing a project subject to the provisions of this division in order to recover the estimated costs incurred by the public agency in preparing a negative declaration or an environmental impact report for such project.

21090. For all purposes of this division all public and private activities or undertakings pursuant to or in furtherance of a redevelopment plan shall be deemed a single project.

21090.1. For all purposes of this division, a geothermal exploratory project shall be deemed to be separate and distinct from any subsequent geothermal field development project as defined in Section 65928.5 of the Government Code.

21092. Any public agency which is preparing an environmental impact report or a negative declaration shall provide public notice of such fact within a reasonable period of time prior to final adoption by the public agency of such environmental impact report or negative declaration. Notice shall be given to all organizations and individuals who have previously requested such notice and shall also be given by at least one of the following procedures.

(a) Publication, no fewer times than required by Section 6061 of the Government Code, by the public agency in a newspaper of general circulation in the area affected by the proposed project.

(b) Posting of notice by the public agency on- and offsite in the area where the project is to be located.

(c) Direct mailing to owners of contiguous property.

The alternatives for providing notice specified in subdivisions (a) to (c), inclusive, shall not preclude a public agency from providing additional notice by other means if such agency so desires, nor shall the requirements of this section preclude a public agency from providing the public notice required herein at the same time and in the same manner as public notice otherwise required by law for such project.

Chapter 3. State Agencies, Boards and Commissions

21100. All state agencies, boards and commissions shall prepare, or cause to be prepared by contract, and certify the completion of an environmental impact report on any project they propose to carry out or approve which may have a significant effect on the environment. Such a report shall include a detailed statement setting forth the following:

(a) The significant environmental effects of the proposed action.

(b) Any significant environmental effects which cannot be avoided if the proposal is implemented.

(c) Mitigation measures proposed to minimize the significant environmental effects including, but not limited to, measures to reduce wasteful, inefficient, and unnecessary consumption of energy.

(d) Alternatives to the proposed action.

(e) The relationship between local short-term uses of man's environment and the maintenance and enhancement of long-term productivity.

(f) Any significant irreversible environmental changes which would be involved in the proposed action should it be implemented.

(g) The growth-inducing impact of the proposed project.

The report shall also contain a statement briefly indicating the reasons for determining that various effects of a project are not significant and consequently have not been discussed in detail in the environmental impact report.

21100.1 The information described in subdivisions (e) and (f) of Section 21100 shall be required only in environmental impact reports prepared in connection with the following:

(a) The adoption, amendment, or enactment of a plan, policy, or ordinance of a public agency.

(b) The adoption by a local agency formation commission of a resolution making determinations.

(c) A project which will be subject to the requirement for preparing an environmental impact statement pursuant to the requirements of the National Environmental Policy Act of 1969.

21100.2. Each state agency shall establish, by resolution or order, time limits, not to exceed one year for completing and certifying environmental impact reports, and 105 days for negative declarations, for projects described in subdivision (c) of Section 21065. Such time limits shall apply only to those circumstances in which such state agency is the lead agency for a project. Such resolutions or orders may establish different time limits for different types or classes of projects, but all such limits shall be measured from the date on which an application requesting approval of such project is received and accepted as complete by the state agency. The resolutions or orders required by this section may provide for a reasonable extension of such time period in the event that compelling circumstances justify additional time and the project applicants consent thereto.

21101. In regard to any proposed federal project in this state which may have a significant effect on the environment and on which the state officially comments, the state officials responsible for such comments shall include in their report a detailed statement setting forth the matters specified in Section 21100 prior to transmitting the comments of the state to the federal government. No report shall be transmitted to the federal government unless it includes such a detailed statement as to the matters specified in Section 21100.

21102. No state agency, board, or commission shall request funds, nor shall any state agency, board, or commission which authorizes expenditures of funds, other than funds appropriated in the Budget Act, authorize funds for expenditure for any project, other than a project involving only feasibility or planning studies for possible future actions which the agency, board, or commission has not approved, adopted or funded, which may have a significant effect on the environment unless such request or authorization is accompanied by an environmental impact report.

Feasibility and planning studies exempted by this section from the preparation of an environmental impact report shall nevertheless include consideration of environmental factors.

21103. [Repealed]

21104. Prior to completing an environmental impact report, the state lead agency shall consult with, and obtain comments from, each responsible agency and any public agency which has jurisdiction by law with respect to the project, and may consult with any person who has special expertise with respect to any environmental impact involved.

The state lead agency shall consult with, and obtain comments from, the State Air Resources Board in preparing an environmental impact report on a highway or freeway project, as to the air pollution impact of the potential vehicular use of such highway or freeway.

21105. The responsible state lead agency shall include the environmental impact report as a part of the regular project report used in the existing review and budgetary process. It shall be available to the Legislature. It shall also be available for inspection by any member of the general public, who may secure a copy thereof by paying for the actual cost of such a copy. It shall be filed by the state lead agency with the appropriate local planning agency of any city, county, or city and county which will be affected by the project.

21106. All state agencies, boards, and commissions shall request in their budgets the funds necessary to protect the environment in relation to problems caused by their activities.

21107. [Repealed]

21108. (a) Whenever a state agency, board or commission approves or determines to carry out a project which is subject to the provisions of this division, it shall file notice of such approval or such determination with the Secretary of the Resources Agency. Such notice shall indicate the determination of the agency, board or commission whether the project will, or will not, have a significant effect on the environment and shall indicate whether an environmental impact report has been prepared pursuant to the provisions of this division.

(b) Whenever a state agency, board, or commission determines that a project is not subject to the provisions of this division pursuant to subdivision (b) of Section 21080 or pursuant to Section 21085 or 21172, and it approves or determines to carryout such project, it, or the person specified in subdivision (b) or (c) of Section 21065, may file notice of such determination with the Secretary of the Resources Agency. Any notice filed pursuant to this subdivision by a person specified in subdivision (b) or (c) of Section 21065 shall have a certificate of determination attached to it issued by the state agency, board, or commission responsible for making the determination that a project is not subject to the provisions of this division pursuant to subdivision (b) of Section 21080 or pursuant to Section 21085 or 21172. The certificate of determination may be in the form of a certified copy of an existing document or record of the state agency, board, or commission.

(c) All notices filed pursuant to this section shall be available for public inspection, and a list of such notices shall be posted on a weekly basis in the Office of the Resources Agency. Each such list shall remain posted for a period of 30 days.

Chapter 4. Local Agencies

21150. State agencies, boards, and commissions, responsible for allocating state or federal funds on a project-by-project basis to local agencies for any project which may have a significant effect on the environment, shall require from the responsible local governmental agency a detailed statement setting forth the matters specified in Section 21100 prior to the allocation of any funds other than funds solely for projects involving only feasibility or planning studies for possible future actions which the agency, board, or commission has not approved, adopted, or funded.

21151. All local agencies shall prepare, or cause to be prepared by contract and certify the completion of an environmental impact report on any project they intend to carry out or approve which may have a significant effect on the environment. When a report is required by

Section 65402 of the Government Code, the environmental impact report may be submitted as part of that report.

21151.5. Each local agency shall establish, by ordinance or resolution, time limits, not to exceed one year, for completing and certifying environmental impact reports, and 105 days for completing negative declarations for projects described in subdivision (c) of Section 21065. Such time limits shall apply only to those circumstances in which such local agency is the lead agency for a project. Such ordinances or resolutions may establish different time limits for different types or classes of projects, but all such limits shall be measured from the date on which an application requesting approval of such project is received and accepted as complete by the local agency. The ordinances or resolutions required by this section may provide for a reasonable extension of such time period in the event that compelling circumstances justify additional time and that the project applicant consents thereto.

21152. (a) Whenever a local agency approves or determines to carry out a project which is subject to the provisions of this division, it shall file notice of such approval or such determination with the county clerk of the county, or counties, in which the project will be located. Such notice shall indicate the determination of the local agency whether the project will, or will not, have a significant effect on the environment and shall indicate whether an environmental impact report has been prepared pursuant to the provisions of this division.

(b) Whenever a local agency determines that a project is not subject to the provisions of this division pursuant to subdivision (b) of Section 21080 or pursuant to Section 21085 or 21172, and it approves or determines to carry out such project, it, or the person specified in subdivision (b) or (c) of Section 21065, may file notice of such determination with the county clerk of the county, or counties, in which the project will be located. Any notice filed pursuant to this subdivision by a person specified in subdivision (b) or (c) of Section 21065 shall have a certificate of determination attached to it issued by the local agency responsible for making the determination that a project is not subject to the provisions of this division pursuant to subdivision (b) of Section 21080 or pursuant to Section 21085 or 21172. The certificate of determination may be in the form of a certified copy of an existing document or record of the local agency.

(c) All notices filed pursuant to this section shall be available for public inspection, and a list of such notices shall be posted on a weekly basis in the office of the county clerk. Each such list shall remain posted for a period of 30 days.

21153. Prior to completing an environmental impact report, every local lead agency shall consult with, and obtain comments from, each responsible agency and any public agency which has jurisdiction by law with respect to the project, and may consult with any person who has special expertise with respect to any environmental impact involved.

21154. Whenever any state agency, board, or commission issues an order which requires a local agency to carry out a project which may have a significant effect on the environment, any environmental impact report which the local agency may prepare shall be limited to consideration of those factors and alternatives which will not conflict with such order.

21155. The Secretary of the Resources Agency shall develop an economic practices manual which may be used by local agencies in assessing the economic efficiency and fiscal impact of alternative land use proposals by local agencies and as an aid in evaluating private land use proposals.

The manual shall present economic and fiscal criteria for local agencies to use in developing comprehensive and multidisciplined programs and procedures for evaluating alternative land use policies and private land use proposals. The manual shall include a step-by-step program which local agencies may follow, including, but not limited to, sources of data, methods of summarizing and using the data, formulas for evaluating the economic efficiency and fiscal impact of land use decisions, a guideline on how to prepare a simple statement of results, and a means for equating economic factors with social and environmental factors.

The Secretary of the Resources Agency may obtain private counseling in the preparation of the manual. On or before January 1, 1975, the Secretary of the Resources Agency shall submit a proposed manual to the Legislature for its review. It is the intent of the Legislature hereby to permit the Legislature to review the manual and to comment thereon prior to its final adoption by the Secretary of the Resources Agency and its distribution to local agencies.

Chapter 5. Submission of Information

21160. Whenever any person applies to any public agency for a lease, permit, license, certificate, or other entitlement for use, the public agency may require that person to submit data and information which may be necessary to enable the public agency to determine whether the proposed project may have a significant effect on the environment or to prepare an environmental impact report.

If any or all of the information so submitted is a "trade secret" as defined in Section 6254.7 of the Government Code by those submitting that information, it shall not be included in the impact report or otherwise disclosed by any public agency. This section shall not be construed to prohibit the exchange of properly designated trade secrets between public agencies who have lawful jurisdiction over the preparation of the impact report.

21161. Whenever a public agency has completed an environmental impact report, it shall cause a notice of completion of such report to be filed with the Secretary of the Resources Agency. The notice of completion shall briefly identify the project and shall indicate that an environmental impact report has been prepared. Failure to file the notice required by this section shall not affect the validity of a project.

Chapter 6. Limitations

21165. When a project is to be carried out or approved by two or more public agencies, the determination of whether the project may have a significant effect on the environment shall be made by the lead agency; and such agency shall prepare, or cause to be prepared by contract, the environmental impact report for the project, if such a report is required by this division. In the event that a dispute arises as to which is the lead agency, any public agency or in the case of a project described in subdivision (c) of Section 21065 the applicant for such project, may submit the question to the Office of Planning and Research, and the Office of Planning and Research shall designate, within 21 days of receiving such request, the lead agency, giving due consideration to the capacity of such agency to adequately fulfill the requirements of this division.

21166. When an environmental impact report has been prepared for a project pursuant to this division, no subsequent or supplemental environmental impact report shall be required by the lead agency or by any responsible agency, unless one or more of the following events occurs:

(a) Substantial changes are proposed in the project which will require major revisions of the environmental impact report.

(b) Substantial changes occur with respect to the circumstances under which the project is being undertaken which will require major revisions of the environmental impact report.

(c) New information, which was not known and could not have been known at the time the environmental impact report was certified as complete, becomes available.

21167. Any action or proceeding to attack, review, set aside, void, or annul the following acts or decisions of a public agency on the grounds of noncompliance with this division shall be commenced as follows:

(a) An action or proceeding alleging that a public agency is carrying out or has approved a project which may have a significant effect on the environment without having determined whether the project may have a significant effect on the environment shall be commenced within 180 days of the public agency's decision to carry out or approve the project, or, if the project is undertaken without a formal decision by the public agency, within 180 days after commencement of the project.

(b) Any action or proceeding alleging that a public agency has improperly determined whether a project may have a significant effect on the environment shall be commenced within 30 days after the filing of the notice required by subdivision (a) of Section 21108 or subdivision (a) of Section 21152.

(c) Any action or proceeding alleging that an environmental impact report does not comply with the provisions of this division shall be commenced within 30 days after the filing of the notice required by subdivision (a) of Section 21108 or subdivision (a) of Section 21152 by the lead agency.

(d) Any action or proceeding alleging that a public agency has improperly determined that a project is not subject to the provisions of this division pursuant to subdivision (b) of Section 20180 or pursuant to Section 21085 or 21172 shall be commenced within 35 days after the filing by the public agency, or person specified in subdivision (b) or (c) of Section 21065, of the notice authorized by subdivision (b) of Section 21108 or subdivision (b) of Section 21152. If such notice has not been filed, such action or proceeding shall be commenced within 180 days of the public agency's decision to carry out or approve the project, or, if a project is undertaken without a formal decision by the public agency, within 180 days after commencement of the project.

(c) Any action or proceeding alleging that any other act or omission of a public agency does not comply with the provisions of this division shall be commenced within 30 days after the filing of the notice required by subdivision (a) of Section 21108 or subdivision (a) of Section 21152.

21167.1. In all actions brought pursuant to Section 21167, including the hearing of any such action on appeal from the decision of a lower court, all courts wherein such actions are or may hereafter be pending, shall give such actions preference over all other civil actions therein, in the matter of setting the same for hearing or trial, and in hearing the same, to the end that all such actions shall be quickly heard and determined.

21167.2. If no action or proceeding alleging that an environmental impact report does not comply with the provisions of this division is commenced during the period prescribed in subdivision (c) of Section 21167, the environmental impact report shall be conclusively presumed to comply with the provisions of this division for purposes of its use by responsible agencies, unless the provisions of Section 21166 are applicable.

21167.3. If an action or proceeding alleging that an environmental impact report does not comply with the provisions of this division is commenced during the period described in subdivision (c) of Section 21167, responsible agencies shall assume that the environmental impact report for a project does comply with the provisions of this division and shall issue a conditional approval or disapproval of such project according to the timetable for agency action in Article 5 (commencing with Section 65950) of Chapter 4.5 of Division I of Title 7 of the Government Code. A conditional approval shall constitute permission to proceed with a project when and only when such action or proceeding results in a final determination that the environmental impact report does comply with the provisions of this division.

21167.5 Proof of prior service by mail upon the public agency carrying out or approving the project of a written notice of the commencement of any action or proceeding described in Section 21167 identifying the project shall be filed concurrently with the initial pleading in such action or proceeding.

21167.7. Every person who brings an action pursuant to Section 21167 shall comply with requirements of Section 389.6 of the Code of Civil Procedure. Every such person shall also furnish pursuant to Section 389.6 of the Code of Civil Procedure a copy of any amended or supplemental pleading filed by such person in such action to the Attorney General. No relief, temporary or permanent, shall be granted until a copy of the pleading has been furnished to the Attorney General in accordance with such requirements.

21168. Any action or proceeding to attack, review, set aside, void or annul a determination, finding, or decision of a public agency, made as a result of a proceeding in which by law a hearing is required to be given, evidence is required to be taken and discretion in the determination of facts is vested in a public agency, on the grounds of noncompliance with provisions of this division shall be in accordance with the provisions of Section 1094.5 of the Code of Civil Procedure.

In any such action, the court shall not exercise its independent judgment on the evidence but shall only determine whether the act or decision is supported by substantial evidence in the light of the whole record.

21168.3. In all actions brought pursuant to Section 21167, including the hearing of any such action on appeal from the decision of a lower court, all courts wherein such actions are or may hereafter be pending, shall give such actions preference over all other civil actions therein, in the matter of setting the same for hearing or trial, and in hearing the same, to the end that all such actions shall be quickly heard and determined.

21168.5. In any action or proceeding, other than an action or proceeding under Section 21168, to attack, review, set aside, void or annul a determination, finding, or decision of a public agency on the grounds of noncompliance with this division, the inquiry shall extend only to whether there was a prejudicial abuse of discretion. Abuse of discretion is established if the agency had not proceeded in a manner required by law or if the determination or decision is not supported by substantial evidence.

21168.6. In any action or proceeding under Sections 21168 or 21168.5 against the Public Utilities Commission the writ of mandate shall lie only from the Supreme Court to such commission.

21168.7. Sections 21168 and 21168.5 are declaratory of existing law with respect to the judicial review of determinations or decisions of public agencies made pursuant to this division.

21169. Any project defined in subdivision (c) of Section 21065 undertaken, carried out or approved on or before the effective date of this section and the issuance by any public agency of any lease, permit, license, certificate or other entitlement for use executed or issued on or before the effective date of this section notwithstanding a failure to comply with this division, if otherwise legal and valid is hereby confirmed, validated and declared legally effective. Any project undertaken by a person which was supported in whole or part through contracts with one or more public agencies on or before the effective date of this section, notwithstanding a failure to comply with this division, if otherwise legal and valid, is hereby confirmed, validated and declared legally effective.

21170. (a) Section 21169 shall not operate to confirm, validate or give legal effect to any project the legality of which was being contested in a judicial proceeding in which proceeding the pleadings prior to the effective date of this section, alleged facts constituting a cause of action for, or raised the issue of, a violation of this division and which was pending and undetermined on the effective date of this section; provided, however, that Section 21169 shall operate to confirm, validate or give legal effect to any project to which this subdivision applies if, prior to the commencement of judicial proceedings and in good faith and in reliance upon the issuance by a public agency of any lease, permit, license, certificate or other entitlement for use, substantial construction has been performed and substantial liabilities for construction and necessary materials have been incurred.

(b) Section 21169 shall not operate to confirm, validate or give legal effect to any project which has been determined in any judicial proceeding on or before the effective date of this section to be illegal, void or ineffective because of noncompliance with this division.

21171. This division, except for Section 21169, shall not apply to the issuance of any lease, permit, license, certificate or other entitlement for use for any project defined in subdivision (c) of Section 21065 or to any project undertaken by a person which is supported in whole or in part through contracts with one or more public agencies until the 121st day after the effective date of this section. This section shall not apply to any project to which Section 21170 is applicable or to any successor project which is the same as, or substantially identical to, such a project.

This section shall not prohibit or prevent a public agency, prior to the 121st day after the effective date of this section, from considering environmental factors in connection with the approval or disapproval of a project and from imposing reasonable fees in connection therewith.

21172. This division shall not apply to any project undertaken, carried out, or approved by a public agency to maintain, repair, restore, demolish or replace property or facilities damaged or destroyed as a result of a disaster in a disaster stricken area in which a state of emergency has been proclaimed by the Governor pursuant to Chapter 7 (commencing with Section 8550) of Division 1, Title 2 of the Government Code.

21172.5. Until the 121st day after the effective date of this section, any objectives, criteria and procedures adopted by public agencies in compliance with this division shall govern the evaluation of projects defined in subdivisions (a) and (b) of Section 21065 and the preparation of environmental impact reports of such projects when required by this division.

Any environmental impact report which has been completed or on which substantial work has been performed on or before the 121st day after the effective date of this section, if otherwise legally sufficient, shall, when completed, be deemed to be in compliance with this division and no further environmental impact report shall be required except as provided in Section 21166.

21173. If any provision of this division or the application thereof to any person or circumstances is held invalid, such invalidity shall not affect other provisions or applications of this division which can be given effect without the invalid provision or application thereof, and to this end the provisions of the division are severable.

21174. No provision of this division is a limitation of restriction on the power or authority of any public agency in the enforcement or administration of any provision of law which it is specifically permitted or required to enforce or administer, including, but not limited to, the powers and authority granted to the California Coastal Commission or any regional coastal commission pursuant to Division 20 (commencing with Section 30000) of the Public Resources Code. To the extent of any inconsistency or conflict between the provisions of the California Coastal Act of 1976, Division 20 (commencing with Section 30000) of the Public Resources Code, and the provisions of this division, the provisions of Division 20 (commencing with Section 30000) of the Public Resources Code shall control.

21175. In the event that a local agency formation commission, acting pursuant to the provisions of Chapter 6.6 (commencing with Section 54773) of Part I of Division 2 of Title 5 of, or pursuant to Division 1 (commencing with Section 56000) of Title 6 of the Government Code, has approved a project without complying with this division, such approval is hereby confirmed, validated, and declared legally effective notwithstanding the failure to comply with this division; provided, that such approval shall have occurred prior to February 7, 1975.

21176. (a) Section 21175 shall not operate to confirm, validate, or give legal effect to any project, the legality of which was being contested in a judicial proceeding in which proceeding the pleadings, prior to February 7, 1975, alleged facts constituting a cause of action for, or raised the issue of, a violation of this division, and which was pending and undetermined on February 7, 1975.

(b) Section 21175 shall not operate to confirm, validate, or give legal effect to any project which had been determined in any judicial proceeding, on or before the effective date of this section, to be illegal, void, or ineffective because of noncompliance with this division.

DIVISION 13.5. CALIFORNIA ENVIRONMENTAL PROTECTION PROGRAM

21190. There is in this state the California Environmental Protection Program, which shall be concerned with the preservation and protection of California's environment. Funds expended pursuant to this division shall be used to support identifiable projects and programs of state agencies, cities, counties, and districts, the University of California, and private research organizations which have a clearly defined benefit to the people of the State of California and which have one or more of the following purposes:

(a) The control and abatement of air pollution, including all phases of research into the sources, dynamics, and effects of environmental pollutants.

(b) The acquisition, preservation, restoration, or any combination thereof, of natural areas or ecological reserves.

(c) Purchase of real property for park purposes on an opportunity basis or the acquisition of public accessways to coastal areas.

(d) Environmental education, including formal school programs and informal public education programs.

(e) Enhancement of renewable and nonrenewable resources.

(f) Protection of nongame species and rare and endangered plants and animals.

(g) Protection of wildlife habitat, including review of the potential impact of development projects and land use changes on such habitat.

21191. (a) The California Environmental License Plate Fund which supersedes the California Environmental Protection Program Fund is continued in existence in the State Treasury, consisting of the moneys deposited in the fund pursuant to any provisions of law, including all revenue derived from fees provided for in Sections 5106 and 5108 of the Vehicle Code.

(b) Amounts equal to the actual costs incurred by the Department of Motor Vehicles in performing its duties pursuant to Article 8.5 (commencing with Section 5100) of Chapter 1 of Division 3 of the Vehicle Code are available to the department, when appropriated by the Legislature, from the moneys deposited in the California Environmental License Plate Fund, and shall be transferred to, and in augmentation of the appropriation from which such costs were paid.

(c) The balance of the moneys in the California Environmental License Plate Fund shall be available for expenditure for the purposes and uses of the California Environmental Protection Program upon appropriation by the Legislature.

(d) All proposed appropriations for the program shall be summarized in a section in the Governor's Budget for each fiscal year and shall bear the caption "California Environmental Protection Program." The section shall contain a separate description of each project for which an appropriation is made. All such appropriations shall be made to the department performing the project and accounted for separately.

21192. The funds provided for in subdivision (c) of Section 21191 may be used in a manner which will allow the state to qualify for any funds which may be available from any source for the purpose of carrying out the provisions of this division.

21193. The program established by this division shall be administered by the Secretary of the Resources Agency who shall:

(a) On or before November 1 of each year, forward those projects and programs recommended for funding to the Governor for inclusion in the Budget Bill, together with a statement of the purpose of each such project and program, the benefits to be realized, and the secretary's comments thereon.

(b) Periodically review projects subsequent to their funding under this division and report thereon to the Governor and the Legislature.

1980-81 AMENDMENTS

1. Section 21000(g) is amended (79-947) to read:

(g) It is the intent of the Legislature that all agencies of the state government which regulate activities of private individuals, corporations, and public agencies which are found to affect the quality of the environment, shall regulate such activities so that major consideration is given to preventing environmental damage, while providing a decent home and satisfying living environment for every Californian.

2. Section 21001(d) is amended (79-947) to read:

(d) Ensure that the long-term protection of the environment, consistent with the provision of a decent home and suitable living environment for every Californian, shall be the guiding criterion in public decisions.

3. Section 21002 is amended (80-676) so that the last sentence reads:

The Legislature further finds and declares that in the event specific economic, social, or other conditions make infeasible such project alternatives or such mitigation measures, individual projects may be approved in spite of one or more significant effects thereof.

4. Section 21080(b) is amended (80-526) by adding the following:

(10) All classes of projects designated pursuant to Section 21084.

(11) A project for the institution or increase of passenger or commuter service on rail lines already in use, including the modernization of existing stations and parking facilities.

(12) A project for the institution or increase of passenger or commuter service on high-occupancy vehicle lanes already in use, including the modernization of existing stations and parking facilities.

(13) Facility extensions not to exceed four miles in length which are required for transfer of passengers from or to exclusive public mass transit guideway or busway public transit services.

(14) A project for the development of a regional transportation improvement program or the state transportation improvement program.

(15) Any project or portion thereof located in another state which will be subject to environmental impact review pursuant to the National Environmental Policy Act of 1969 or similar state laws of that state. Any emissions or discharges that would have a significant effect on the environment in the State of California are subject to this division.

and the following (81-402):

(16) Projects undertaken by a local agency to implement a rule or regulation imposed by a state agency, board, or commission under a certified regulatory program pursuant to Section 21080.5. Any site-specific effect of the project which was not analyzed as a significant effect in the plan or other written documentation required by Section 21080.5 is subject to this division.

5. Section 21080.5(j) is added (81-480) to read:

(j) For purposes of this section, any air quality management district shall be considered a state agency, except that any activities of an air quality management district undertaken in order to comply with the provisions of Chapter 2.5 (commencing with Section 41650) of Part 4 of Division 26 of the Health and Safety Code shall not be subject to the provisions of this section.

6. Section 21080.9 is amended (79-961) to read as follows:

21080.9. This division shall not apply to activities and approvals by any local government, as defined in Section 30109, or any state university or college, as defined in Section 30119, as necessary for the preparation and adoption of a local coastal program or long-range land use development plan pursuant to Division 20 (commencing with Section 30000); provided, however, that certification of a local coastal program or long-range land use development plan by the California Coastal Commission pursuant to Chapter 6 (commencing with Section 30500) of Division 20 shall be subject to the requirements of this division. For the purpose of Section 21080.5, a certified local coastal program or long-range land use development plan constitutes a plan for use in the California Coastal Commission's regulatory program.

7. Section 21080.10 is added (80-1154) as follows:

21080.10. This division shall not apply to the following:

(a) An extension of time, granted pursuant to Section 65302.6 of the Government Code for the preparation and adoption of one or more elements of a city or county general plan.

(b) Actions taken by the Department of Housing and Community Development to provide financial assistance for the development and construction of residential housing for persons and families of low or moderate income, as defined in Section 50093 of the Health and Safety Code; provided, that the project which is the subject of the application for financial assistance will be reviewed pursuant to the provisions of this division by another agency.

8. Section 21082.1 is amended (81-480) to read as follows:

21082.1 Any environmental impact report or negative declaration prepared pursuant to the requirement of this division shall be prepared directly by, or under contract to, a public agency.

This section is not intended to prohibit, and shall not be construed as prohibiting, any person from submitting information or other comments to the public agency responsible for preparing an environmental impact report or negative declaration. The information or other comments may be submitted in any format, shall be considered by the public agency, and may be included, in whole or in part, in any report or declaration.

9. Section 21083.3 is added (81-480) as follows:

21083.3. (a) If a parcel has been zoned to accommodate a particular density of residential development or has been designated in a community plan to accommodate a particular density of residential development and an environmental impact report was certified for that zoning or planning action, the application of this division to the approval of any subdivision map or other project that is consistent with the zoning or community plan shall be limited to effects upon the environment which are peculiar to the parcel or to the project and which were not addressed as significant effects in the prior environmental impact report. However, all public agencies with authority to mitigate the significant effects shall undertake or require the undertaking of any feasible mitigation measures specified in the prior environmental impact report relevant to a significant effect which the project will have on the environment or, if not, then the provisions of this section shall have no application to that effect. The lead agency shall make a finding, at a public hearing, as to whether those mitigation measures will be undertaken. An effect of a project upon the environment shall not be considered peculiar to the parcel or to the project, for purposes of this subdivision, if uniformly applied development policies or standards have been previously adopted by the city or county, with a finding based upon substantial evidence, which need not include an environmental impact report, that the development policies or standards will substantially mitigate that environmental effect when applied to future projects.

Where a community plan is the basis for application of this section, any rezoning action consistent with the community plan shall be a project subject to exemption from this division in accordance with the provisions of this subdivision.

(b) As used in subdivision (a), "community plan" means a part of the general plan of a city or county which (1) applies to a defined geographic portion of the total area included in the general plan, (2) complies with Article 5 (commencing with Section 65300) of Chapter 3 of Division 1 of Title 7 of the Government Code by including or referencing each of the mandatory elements specified in Section 65302 of the Government Code, and (3) contains specific development policies adopted for the area included in the community plan and identifies measures to implement those policies, so that the policies which will apply to each parcel can be determined.

(c) No person shall have standing to bring an action or proceeding to attack, review, set aside, void, or annul a finding of a public agency made at a public hearing pursuant to subdivision (a) with respect to the conformity of the project prepared for the zoning or planning action, unless he or she has participated in that public hearing. However, this subdivision shall not be applicable if the local agency failed to give public notice of the hearing as required by law. For purposes of this subdivision, a person has participated in the public hearing if he or she has either submitted oral or written testimony regarding the proposed determination, finding, or decision prior to the close of the hearing.

(d) Any plan adopted prior to January 1, 1982, which does not comply with the definitional criteria specified in subdivision (b) may be amended to comply with that criteria, in which case, the plan shall be deemed a "community plan" within the meaning of subdivision (b); provided, that, (1) an environmental impact report was certified for adoption of the plan and (2) at the time of the conforming amendment, the environmental impact report has not been held inadequate by a court of this state, and the environmental impact report is not the subject of pending litigation challenging its adequacy.

10. Sections 21085, 21085.5, and 21085.6 are repealed (79-697).

11. Section 21092(c) is amended (80-131) as follows:

(c) Direct mailing to the owners of contiguous property as such owners are shown on the latest equalized assessment roll.

12. Section 21100 is amended (81-803) by adding the following:

For purposes of this section, any significant effect on the environment shall be limited to substantial, or potentially substantial, adverse changes in physical conditions which exist within the area as defined in Section 21060.5.

13. Section 21151 is amended (81-803) by adding the following:

For purposes of this section, any significant effect on the environment shall be limited to substantial, or potentially substantial, adverse changes in physical conditions which exist within the area as defined in Section 21060.5.

14. Section 21167.3 is amended (80-131) to read:

21167.3. (a) If an action or proceeding alleging that an environmental impact report or a negative declaration does not comply with the provisions of this division is commenced during the period described in subdivision (b) or (c) of Section 21167, and if an injunction or stay is issued prohibiting the project from being carried out or approved pending final determination of the issue of such compliance, responsible agencies shall assume that the environmental impact report or the negative declaration for the project does comply with the provisions of this division and shall issue a conditional approval or disapproval of such project according to the timetable for agency action in Article 5 commencing with Section 65950) of Chapter 4.5 of Division 1 of Title 7 of the Government Code. A conditional approval shall constitute permission to proceed with a project when and only when such action or proceeding results in a final determination that the environmental impact report or negative declaration does comply with the provisions of this division.

(b) In the event that an action or proceeding is commenced as described in subdivision (a) but no injunction or similar relief is sought and granted, responsible agencies shall assume that the environmental impact report or negative declaration for the project does comply with the provisions of this division and shall approve or disapprove the project according to the timetable for agency action in Article 5 (commencing with Section 65950) of Chapter 4.5 of Division 1 of Title 7 of the Government Code. Such approval shall constitute permission to proceed with the project at the applicant's risk pending final determination of such action or proceeding.

15. Section 21167.4 is added (80-131):

21167.4. In a writ of mandate proceeding alleging noncompliance with this division, the petitioner shall request a hearing within 90 days of filing the petition or otherwise be subject to dismissal on the court's own motion or on the motion of any party interested therein.

16. Section 21174 is amended (80-676) to read as follows:

21174. No provision of this division is a limitation or restriction . . .

APPENDIX D

ENVIRONMENTAL INFORMATION FORM AND ENVIRONMENTAL CHECKLIST FOR USE IN THE INITIAL STUDY

ENVIRONMENTAL INFORMATION FORM
(to be completed by applicant)

Date filed ______________________________

General Information

1. Name and address of developer or project sponsor: ____________
   ____
2. Address of project: ______________________________________
   Assessor's block and lot numbers: __________________________
3. Name, address, and telephone number of person to be contacted concerning this project: ________________________________
   ____
4. Indicate number of the permit application for the project to which this form pertains: _____________________________________
5. List and describe any other related permits and other public approvals required for this project, including those required by city, regional, state and federal agencies: _________________
   ____
6. Existing zoning districts: _______________________________
7. Proposed use of site (project for which this form is filed):
   ____

Project Description

8. Site size.
9. Square footage.
10. Number of floors of construction.
11. Amount of off-street parking provided.
12. Attach plans.
13. Proposed scheduling.
14. Associated project.
15. Anticipated incremental development.

This form is taken from Title 14, California Administrative Code, secs. 15001-15203, app. H.

16. If residential, include the number of units, schedule of unit sizes, range of sale prices or rents, and type of household size expected.

17. If commercial, indicate the type: whether neighborhood, city, or regionally oriented; square footage of sales area; and loading facilities.

18. If industrial, indicate type, estimated employment per shift, and loading facilities.

19. If institutional, indicate the major function, estimated employment per shift, estimated occupancy, loading facilities, and community benefits to be derived from the project.

20. If the project involves a variance, conditional use, or rezoning application, state this and indicate clearly why the application is required.

Are the following items applicable to the project or its effects? Discuss below all items checked yes (attach additional sheets as necessary

Yes	No	
_____	_____	21. Change in existing features of any bays, tidelands, beaches, lakes or hills, or substantial alteration of ground contours.
_____	_____	22. Change in scenic views or vistas from existing residential areas or public lands or roads.
_____	_____	23. Change in pattern, scale, or character of general area of project.
_____	_____	24. Significant amounts of solid waste or litter.
_____	_____	25. Change in dust, ash, smoke, fumes, or odors in vicinity.
_____	_____	26. Change in ocean, bay, lake, stream, or groundwater quality or quantity, or alteration of existing drainage patterns.
_____	_____	27. Substantial change in existing noise or vibration levels in the vicinity.
_____	_____	28. Site on filled land or on slope of 10 percent or more.
_____	_____	29. Use of disposal of potentially hazardous materials, such as toxic substances, flammables, or explosives.
_____	_____	30. Substantial change in demand for municipal services (police, fire, water, sewage, etc.).
_____	_____	31. Substantially increased fossil fuel consumption (electricity, oil, natural gas, etc.).
_____	_____	32. Relationship to a larger project or series of projects.

Environmental Setting

33. Describe the project site as it exists before the project, including information on topography, soil stability, plants, and animals, and any cultural, historical, or scenic aspects. Describe any existing structures on the site and the use of the structures. Attach photographs of the site. Snapshots or polaroid photos will be accepted.
34. Describe the surrounding properties, including information on plants and animals and any cultural, historical, or scenic aspects. Indicate the type of land use (residential, commercial, etc.), intensity of land use (one-family, apartment houses, shops, department stores, etc.), and scale of development (height, frontage, setback, rear yard, etc.). Attach photographs of the vicinity. Snapshots or polaroid photos will be accepted.

Certification

I hereby certify that the statements furnished above and in the attached exhibits present the data and information required for this initial evaluation to the best of my ability, and that the facts, statements, and information presented are true and correct to the best of my knowledge and belief.

________________________ ________________________

Date Signature

For ________________________

(Note: This is only a suggested form. Public agencies are free to devise their own format for initial studies.)

ENVIRONMENTAL CHECKLIST FORM
(to be completed by lead agency)

I. Background

1. Name of proponent______________________________
2. Address and phone number of proponent________________

____

____

3. Date of checklist submission________________________
4. Agency requiring checklist__________________________
5. Name of proposal, if applicable______________________

II. Environmental Impacts

(Explanations of all "yes" and "maybe" answers are required on attached sheets.)

	Yes	Maybe	No
1. Earth. Will the proposal result in:			
a. Unstable earth conditions or in changes in geologic substructures?	___	___	___
b. Disruptions, displacements, compaction, or overcovering of the soil?	___	___	___
c. Change in topography or ground surface relief features?	___	___	___
d. The destruction, covering, or modification of any unique geologic or physical feature?	___	___	___
e. Any increase in wind or water erosion of soils, either on or off the site?	___	___	___
f. Changes in deposition or erosion of beach sands, or changes in siltation, deposition, or erosion that may modify the channel of a river or stream or the bed of the ocean or any bay, inlet, or lake?	___	___	___
g. Exposure of people or property to geological hazards such as earthquakes, landslides, mudslides, ground failure, or similar hazards?	___	___	___

This form is taken from Title 14, California Administrative Code, secs. 15001-15203, app. I.

	Yes	Maybe	No
2. Air. Will the proposal result in:			
a. Substantial air emissions or deterioration of ambient air quality?	___	___	___
b. The creation of objectionable odors?	___	___	___
c. Alteration of air movement, moisture or temperature, or any change in climate, either locally or regionally?	___	___	___
3. Water. Will the proposal result in:			
a. Changes in currents, or the course or direction of water movements, in either marine or fresh waters?	___	___	___
b. Changes in absorption rates, drainage patterns, or the rate and amount of surface runoff?	___	___	___
c. Alterations to the course of flow of flood waters?	___	___	___
d. Change in the amount of surface water in any water body?	___	___	___
e. Discharge into surface waters or in any alteration of surface water quality, including but not limited to temperature, dissolved oxygen, or turbidity?	___	___	___
f. Alteration of the direction or rate of flow of groundwaters?	___	___	___
g. Change in the quantity of groundwaters, either through direct additions, or withdrawals, or through interception of an aquifer by cuts or excavations?	___	___	___
h. Substantial reduction in the amount of water otherwise available for public water supplies?	___	___	___
i. Exposure of people or property to water-related hazards such as flooding or tidal waves?	___	___	___
j. Significant changes in the temperature, flow, or chemical content of surface thermal springs?	___	___	___
4. Plant life. Will the proposal result in:			
a. Change in the diversity of species or number of any species of plants (including trees, shrubs, grass, crops, and aquatic plants)?	___	___	___

	Yes	Maybe	No
B. Reduction of the numbers of any unique, rare, or endangered species of plants?	____	____	____
C. Introduction of new species of plants into an area, or in a barrier to the normal replenishment of existing species?	____	____	____
D. Reduction in acreage of any agricultural crop?	____	____	____
5. <u>Animal life</u>. Will the proposal result in:			
A. Change in the diversity of species, or numbers of any species of animals (birds, land animals, including reptiles, fish and shellfish, benthic organisms, or insects)?	____	____	____
B. Reduction of the numbers of any unique, rare, or endangered species of animals?	____	____	____
C. Introduction of new species of animals into an area, or result in a barrier to the migration or movement of animals?	____	____	____
D. Deterioration to existing fish or wildlife habitat?	____	____	____
6. <u>Noise</u>. Will the proposal result in:			
A. Increases in existing noise levels?	____	____	____
B. Exposure of people to severe noise levels?	____	____	____
7. <u>Light and glare</u>. Will the proposal produce new light or glare?	____	____	____
8. <u>Land use</u>. Will the proposal result in a substantial alteration of the present or planned land use of an area?	____	____	____
9. <u>Natural resources</u>. Will the proposal result in:			
A. Increase in the rate of use of any natural reresource?	____	____	____
B. Substantial depletion of any nonrenewable natural resource?	____	____	____
10. <u>Risk of upset</u>. Will the proposal involve:			
A. A risk of an explosion or the release of hazardous substances (including, but not limited to, oil, pesticides, chemicals, or radiation) in the event of an accident or upset conditions?	____	____	____

	Yes	Maybe	No
b. Possible interference with an emergency response plan or an emergency evacuation plan?	___	___	___
11. <u>Population</u>. Will the proposal alter the location, distribution, density, or growth rate of the human population of an area?	___	___	___
12. <u>Housing</u>. Will the proposal affect existing housing or create a demand for additional housing?	___	___	___
13. <u>Transportation/circulation</u>. Will the proposal result in:			
a. Generation of substantial additional vehicular movement?	___	___	___
b. Effects on existing parking facilities or demand for new parking?	___	___	___
c. Substantial impact upon existing transportation systems?	___	___	___
d. Alterations to present patterns of circulation or movement of people and/or goods?	___	___	___
e. Alterations to waterborne, rail, or air traffic?	___	___	___
f. Increase in traffic hazardous to motor vehicles, bicyclists, or pedestrians?	___	___	___
14. <u>Public services</u>. Will the proposal have an effect upon, or result in, a need for new or altered governmental services in any of the following areas:			
a. Fire protection?	___	___	___
b. Police protection?	___	___	___
c. Schools?	___	___	___
d. Parks or other recreational facilities?	___	___	___
e. Maintenance of public facilities, including roads?	___	___	___
f. Other governmental services?	___	___	___
15. <u>Energy</u>. Will the proposal result in:			
a. Use of substantial amounts of fuel or energy?	___	___	___
b. Substantial increase in demand upon existing sources of energy or require the development of new sources of energy?	___	___	___

	Yes	Maybe	No
16. Utilities. Will the proposal result in a need for new systems or substantial alterations to the following utilities:			
A. Power or natural gas?	___	___	___
B. Communications systems?	___	___	___
C. Water?	___	___	___
D. Sewer or septic tanks?	___	___	___
E. Storm water drainage?	___	___	___
F. Solid waste and disposal?	___	___	___
17. Human health. Will the proposal result in:			
A. Creation of any health hazard or potential health hazard (excluding mental health)?	___	___	___
B. Exposure of people to potential health hazards?	___	___	___
18. Aesthetics. Will the proposal result in the obstruction of any scenic vista or view open to the public, or will the proposal result in the creation of an aesthetically offensive site open to public view?	___	___	___
19. Recreation. Will the proposal result in an impact upon the quality or quantity of existing recreational opportunities?	___	___	___
20. Cultural resources.			
A. Will the proposal result in the alteration of or the destruction of a prehistoric or historic archaelogical site?	___	___	___
B. Will the proposal result in adverse physical or aesthetic effects to a prehistoric or historic building, structure, or object?	___	___	___
C. Does the proposal have the potential to cause a physical change that would affect unique ethnic cultural values?	___	___	___
D. Will the proposal restrict existing religious or sacred uses within the potential impact area?	___	___	___
21. Mandatory findings of significance.			
A. Does the project have the potential to degrade the quality of the environment, substantially reduce the habitat of a fish or wildlife			

	Yes	Maybe	No
species, cause a fish or wildlife population to drop below self-sustaining levels, threaten to eliminate a plant or animal community, reduce the number or restrict the range of a rare or endangered plant or animal, or eliminate important examples of major periods of California history or prehistory?	___	___	___
B. Does the project have the potential to achieve short-term, to the disadvantage of long-term, environmental goals? (A short-term impact on the environment is one that occurs in a relatively brief definitive period of time, while long-term impacts will endure well into the future.)	___	___	___
C. Does the project have impacts that are individually limited, but cumulatively considerable? (A project may impact on two or more separate resources where the impact on each resource is relatively small, but where the effect of the total of those impacts on the environment is significant.)	___	___	___
D. Does the project have environmental effects that will cause substantial adverse effects on human beings, either directly or indirectly?	___	___	___

III. Discussion of Environmental Evaluation

IV. Determination
(to be completed by the lead agency)

On the basis of this initial evaluation:

_____ I find that the proposed project could not have a significant effect on the environment, and a Negative Declaration will be prepared.

_____ I find that although the proposed project could have a significant effect on the environment, there will not be a significant effect in this case because the mitigation measures described on an attached sheet have been added to the project. A Negative Declaration will be prepared.

_____ I find the proposed project may have a significant effect on the environment, and an Environmental Impact Report is required.

_______________________________ _______________________________

Date   Signature

For _______________________________

(Note: This is only a suggested form. Public agencies are free to devise their own format for initial studies.)

APPENDIX E

EMPIRICAL RESEARCH ON COST IMPACTS OF ENVIRONMENTAL CONTROLS

In the course of this study, we examined a number of empirical studies of cost impacts of environmental and other land-use controls. Of those examined, only two were at all related to impacts of the environmental impact review process (EIRP). Most of the others were directed to one of three topics—growth management, coastal area controls, or general land-use/government regulation. Studies were either case studies of limited areas; surveys of builders or planners; or regression or econometric studies, often using hedonic pricing techniques.

Because of the nature of EIRPs as supplemental to other controls, most of the results of these studies were of little relevance for this study. We have utilized those results that we found helpful in the text. Because of the unreliability of data and the weakness of statistical techniques in sorting out effects of the almost infinite number of housing-impacting variables, these uses are few.

We have listed a selected group of the studies reviewed in Table E.1. The reader is directed to the Dowall and Mingitton bibliography[1] for further sources and to the study by Seidel[2] for a comprehensive overall study.

NOTES

1. David E. Dowall and Jessie Mingitton, Effects of Environmental Regulations on Housing Costs (Chicago: Council of Planning Librarians, 1979).

2. Stephen R. Seidel, Housing Costs and Government Regulations: Confronting the Regulatory Maze (New Brunswick, N.J.: Rutgers University, Center for Urban Policy Research, 1978.

TABLE E.1

Selected Empirical Studies on Effects of Environmental Regulation on Housing Costs

Author	Year	Topic	Approach	Relevant Results
Adams	1978	Growth control and costs	Rank correlation of prices	Controls have effect on rankings
Construction Industry Research Board	1975	CEQA and costs	Builder survey	Holding costs increase 10.75 percent to 15.00 percent of raw land value Overhead to builder increased 3 percent to 9 percent Net inflationary impact 8 percent to 14 percent (single family); 8 percent to 12 percent (multifamily) One-month delay cost = $1,027
Dowall and Landis	1981	Land-use controls and housing costs (San Francisco Bay area)	Econometric model	Land availability and density controls directly and systematically affect prices of new homes but not as much as often alleged A one-unit-per-acre increase in density cuts $1,300 off new home prices A program of increasing density of one-unit per acre, reducing development fees 50 percent, and doubling supply of vacant land would lead to $6,000 price cut in average new home
Elliott	1981	Growth control and housing prices	Regression utilizing averaged data to derive hedonic prices	Isolated suburban controls have less impact than if surrounding communities also control For 1969-76, housing prices increased 35 percent more in rate-controlled areas and 20 percent more in quality-controlled areas than other areas
Frech and Lafferty	1976	Coastal zone controls and land values	Various statistical	Reduced by 15 percent values of undeveloped land in coastal versus noncoastal areas

Gleeson	1978	Growth management system and land values	Case study with regression	Two-thirds of difference in mean value between developable and nondevelopable portions can be attributed to growth management system; holds for large parcels but not small ones
ICF Associates	1976	Coastal plan impacts (California)	Case studies	Larger developments will be more costly in relation to smaller ones Some small and medium-sized developers will be politically disadvantaged Visual control policies add 5 percent to 10 percent to project cost
James and Muller	1977	State environmental protection acts and housing cost	Case studies	Costs in two classes: review and mitigation For Broward County, Florida, price increase is $194 per unit For San Diego, California, price increase is $50 per unit
Kneisel	1979	Coastal zone controls and housing cost (California)	Econometric/regression	Examines impacts on single-family costs in coastal and border areas Maximum impact would be 11.71 percent in coastal areas and 6.76 percent in border areas Effects are regressive No decline occurred in coastal construction No substitution effect in border areas
Lincoln, Coddington, and Penberthy	1976	Government intervention and housing costs (Colorado)	Comparative case studies	Added $1,500-$2,000 to new home cost during 1970-75 Half of increase in site development Average delay increased four months during 1970-75
Orange County Cost of Housing Committee	1975	Housing costs including regulation	Builder questionnaire	Government fees add 3 percent to housing price Delays add 1.1 percent per month to housing price

(continued)

TABLE E.1 (continued)

Author	Year	Topic	Approach	Relevant Results
Peterson	1974	Zoning impacts on land and housing costs	Regression	Fiscal and externality effects tend to offset land-use constraint effects
Real Estate Research Corporation	1975	Regulatory impacts on costs (St. Louis, Missouri)	Cost estimate for "typical" home	From 1970 to 1975, $1,600 to $2,500 added by regulatory requirements
Richardson	1976	Coastal area control and housing costs (New Jersey)	County case study	Added $135 direct costs and $1,100 with multiplier effect to average single-family home price Added $125 direct costs and $800 with multiplier effect to multifamily unit price
Seidel	1978	Government regulation and housing costs	Survey and literature review	Comprehensive study concludes that controls have impact (too detailed results to list)
Schwartz, Hansen, and Green	1981	Growth control and housing price	Quasi-experimental design before and after hedonic pricing	Estimate increase in Petaluma, California, as $6,000-$9,000 more than control cities or 7.8 percent of precontrol price
Sagalyn and Sternlieb	1972	Zoning and housing costs	Survey and regression	Zoning in New Jersey is exclusionary in effect (detail in study too great to list)
Stull	1975	Zoning and land values	Regression	Premium prices paid for zoning protection of single-family home character of surrounding area
Urban Land Institute	1979	Delay costs (Houston, Texas)	Survey and cost estimate	Delays of five and one-half months average due to regulation Cost of $560-$840 per lot
Urban Land Institute and Gruen and Gruen Associates	1977	Regulation and housing costs (San Jose, California, and Jacksonville, Florida)	Case studies	San Jose: 20 percent to 30 percent increase by government intervention; $4,000 to $8,000 over ten years; Jacksonville: adds $1,000-$1,500 to cost of building, but weak market prevents full pass through

BIBLIOGRAPHY

BOOKS

Abt and Associates. Integration of Environmental Considerations in the Comprehensive Planning and Management Process. National Technical Information Service, PB-300 129. Washington, D. C.: U. S. Department of Commerce, August 1977.

Ad Hoc Committee on Construction Processes. Report. Los Angeles: Office of the Mayor, June 1976.

Adams, John S. The Evaluation of Changes in House Prices within Urban Housing Submarkets: A Method of Testing for the Effects of Local Growth Control Systems. Hubert H. Humphrey Institute of Urban Affairs, NSF Grant no. AEN-7606857. Minneapolis: University of Minnesota, March 1978.

Alchian, Armen A., and William R. Allen. University Economics. Belmont, Calif.: Wadsworth, 1964.

Alfred Gobar and Associates. Housing Costs Analysis, Orange County. Brea, Calif.: Alfred Gobar Associates, 1980.

_____. The Housing Crisis: The Cause and What We Can Do about It. Newport Beach, Calif.: Californians for an Environment of Excellence, Full Employment, and a Strong Economy through Planned Development, 1980.

American Institute of Planners. Survey of State Land Use Planning Activity. Washington, D. C.: Department of Housing Development, 1976.

Anderson, Frederick R. NEPA in the Courts. Baltimore: Johns Hopkins University Press, 1973.

Andrews, Frank M., Laura Klem, Terrence N. Davidson, Patrick M. O'Malley, and Willard L. Rodgers. A Guide for Selecting Statistical Techniques for Analyzing Social Science Data. Ann Arbor: University of Michigan, Institute for Social Research, 1974.

Bergman, Edward M. A Policy Guide to Evaluations of Policy Related Research on Development Controls and Housing Costs: Final Report. Chapel Hill, N. C.: Center for Urban and Regional Studies, 1974.

Bosselman, Fred, David Callies, and John Banta. The Taking Issue. Washington, D. C.: Superintendent of Documents, 1973.

Bosselman, Fred, Duane A. Feurer, and Charles L. Siemon. The Permit Explosion: Coordination of the Proliferation. Washington, D. C.: Urban Land Institute, 1976.

Burchell, Robert W., and David Listokin. The Environmental Impact Handbook. New Brunswick, N. J.: Rutgers University, 1975.

California Administrative Code. Title XIV, Sections 15000-15203. Sacramento: State of California, 1982.

California Department of Real Estate. Real Estate Law. Sacramento: Department of Real Estate, 1977.

California Office of Planning and Administration. Local Government Planning Survey. Sacramento: State of California, 1979.

California Office of Planning and Research. The California Environmental Quality Act: A Review. Sacramento: State of California, Office of Planning and Research, 1976.

_____. Specific Plans: How Some California Communities Use Them. Sacramento: State of California, Office of Planning and Research, 1981.

California Office of the Attorney General. Land Use in California. Sacramento: Office of the Attorney General, 1975.

California Public Resources Code. Sections 21000-21176. St. Paul: West, 1982.

Californians for an Environment of Excellence, Full Employment, and a Strong Economy through Planned Development. Local Government Permit Process (Orange County). Newport Beach, Calif.: CEEED, 1979.

Canter, Larry W. Environmental Impact Assessment. New York: McGraw-Hill, 1977.

Construction Industry Research Board. "Cost of Delay Prior to Construction." Mimeographed. Los Angeles: CIRB, March 11, 1975.

Das Amiya K. Reducing Housing Costs. Tallahassee: State of Florida, Department of Community Affairs, Division of Technical Assistance, 1977.

Demars, Don M. "Environmental Impact Report." Speech delivered at the Society of Real Estate Appraisers Seminar, Newport Beach, Calif., March 7, 1965.

Dowall, David E. The Suburban Squeeze: A Study of the Effects of Land Use and Environmental Regulation on Housing Costs in the San Francisco Bay Area. Working Paper. Berkeley: University of California Institute of Urban and Regional Development, 1980.

Dowall, David E., and Jesse Mingitton. Effects of Environmental Regulations on Housing Costs. Chicago: Council of Planning Librarians, May 1979.

Dowall, David E., and John Landis. Land-Use Controls and Housing Costs: An Examination of San Francisco Bay Area Communities. Working Paper 81-24. Berkeley: University of California, Center for Real Estate and Urban Economics, March 1981.

Environmental Analysis Systems. The California Environmental Quality Act, vols. 1-3. San Diego: Environmental Analysis Systems, November 1975.

Federal Home Loan Bank of San Francisco. The Cost of Housing. Proceedings of the Third Annual Conference. San Francisco: FHLB of San Francisco, 1978.

Frieden, Bernard J. The Environmental Protection Hustle. Cambridge, Mass.: MIT Press, 1979.

Gleeson, Michael E. The Effects of an Urban Growth Management System on Land Values. Minneapolis: University of Minnesota, Hubert H. Humphrey Institute of Public Affairs, 1978.

Grebler, Leo, and Frank G. Mittelbach. The Inflation of House Prices. Lexington, Mass.: Lexington Books, 1979.

Grieson, Ronald E. Public and Urban Economics. Lexington, Mass.: D. C. Heath, 1976.

Hagman, Donald G. Urban Planning and Land Development Control Law. St. Paul: West, 1975.

Hagman, Donald G., John H. Larson, and Charles R. Martin. California Zoning Practice. Berkeley: California Continuing Education of the Bar, 1969.

Hart, Stuart L., and Gordon A. Enk. Green Goals and Greenbacks: State-Level Environmental Review Programs and Their Associated Costs. Boulder, Colo.: Westview Press, 1980.

Haskell, Elizabeth H., and Victoria S. Price. State Environmental Management: Case Studies of Nine States. New York: Praeger, 1973.

Hughes, Heiss, and Associates. Final Report on the Study of the Development Review Process, San Mateo County. Redwood City, Calif.: San Mateo County Planning Department, February 28, 1978.

Johnston, Robert A., et al. Public Mechanisms for Controlling Land Conversion in California. Report no. 32. Davis: University of California, Institute of Governmental Affairs, September 1976.

Jokela, Arthur. Self-Regulation of Environmental Quality: Impact Analysis in California Local Government. Claremont: Center for California Public Affairs, 1975.

Katz, Lawrence, and Kenneth T. Rosen. The Effect of Land Use Controls on Housing Prices. Center for Real Estate and Urban Economics, Working Paper no. 80-13. Berkeley: University of California, September 1980.

Kneese, Allen V., and Blair T. Bower. Environmental Quality Analysis. Baltimore: Johns Hopkins University Press, 1972.

Kneisel, Robert P. "The Impact of the California Coastal Zone Conservation Commission on the Local Housing Market: A Study of the South Coast Regional Commission." Ph.D. dissertation, University of California-Riverside, 1979.

Knight, Robert L., and Mark D. Menchik. Residential Environmental Attitudes and Preferences: Report of a Questionnaire Survey. Madison: University of Wisconsin, Institute of Environmental Studies, October 1974.

Kolis, Annette, ed. Thirteen Perspectives on Regulation Simplification. Urban Land Institute Research Report no. 29. Washington, D. C.: Urban Land Institute, 1979.

Krister, Hjalte, Karl Lidgren, and Ingema Stahl. Environmental Policy and Welfare Economics. Cambridge, Mass.: Harvard University Press, 1977.

Lerner, Allen W. The Politics of Decision Making: Strategy, Cooperation and Conflict. Beverly Hills, Calif.: Sage, 1976.

Lincoln, James R., Dan Coddington, and John Penberthy. An Analysis of State and Local Intervention on the Homebuilding Process in Colorado, 1970-1975. Denver: Bickert, Browne, Coddington and Associates, 1976.

Liroff, Richard A. A National Policy for the Environment: NEPA and Its Aftermath. Bloomington: Indiana University Press, 1976.

Longtin, James. California Land Use Regulations. Westlake Village, Calif.: Local Government Publications, 1977.

_____. California Land Use Regulations, 1980-81 Supplement. Westlake Village, Calif.: Local Government Publications, 1981.

Los Angeles City Planning Department. EIR Manual for Private Projects. Los Angeles: Department of City Planning, August 1975.

McGowan, Anne. Government Regulations and the Cost of Housing: A Partially Annotated Bibliography. Chicago: Council of Planning Librarians, Bibliographies, 1979.

Merewitz, Leonard, and Richard Parker. The Economics of Environmental Restrictions on the Use of Land. Berkeley: University of California, Institute of Urban and Regional Development, 1973.

Muller, Thomas. State-Mandated Impact Evaluation: A Preliminary Assessment. Washington, D. C.: Urban Institute, 1976.

National Association of Homebuilders. Progress Report on the Cost of Government Regulations in the Housing Industry. Washington, D. C.: NAHB, 1976.

Orange County Cost of Housing Committee. The Cost of Housing in Orange County. Orange County, Calif.: Orange County Board of Supervisors, February 20, 1975.

Peterson, George E. The Influence of Zoning Regulations on Land and Housing Prices. Washington, D. C.: Urban Institute, 1974.

Polinsky, A. Mitchell, and Daniel L. Rubinfeld. Property Values and the Benefits of Environmental Improvements: Theory and Measurement. Ann Arbor: University of Michigan, Institute of Public Policy Studies, 1975.

Real Estate Research Corporation. The Costs of Sprawl. Washington, D. C.: U. S. Government Printing Office, 1975.

_____. New Home Construction Cost Increase—St. Louis, Missouri. Chicago: Real Estate Research Corporation, 1975.

Richardson, Dan K. The Cost of Environmental Protection. New Brunswick, N. J.: Rutgers University, Center for Urban Policy Research, 1976.

Rifkin, Jeremy. Entropy: A New World View. New York: Bantam Books, 1981.

Sagalyn, Lynne B., and George Sternlieb. Zoning and Housing Costs: The Impact of Land Use Controls on Housing Price. New Brunswick, N. J.: Rutgers University, Center for Urban Policy Research, 1972.

Samuelson, Paul A. Economics. 9th ed. New York: McGraw-Hill, 1973.

Schaenman, Philip S., and Thomas Muller. Measuring Impacts of Land Development: An Initial Approach. Washington, D. C.: Urban Institute, 1974.

Scott, Randall W., ed. Management and Control of Growth. 3 vols. Washington, D.C.: Urban Land Institute, 1975.

Security Pacific Bank. Coastal Zone Housing Prices. Los Angeles: Security Pacific Bank, Research Department, 1980.

Seidel, Stephen R. Housing Costs and Government Regulations: Confronting the Regulatory Maze. New Brunswick, N.J.: Rugters University, Center for Urban Policy Research, 1978.

Seneca, Joseph J., and Michael K. Taussig. Environmental Economics. Englewood Cliffs, N.J.: Prentice-Hall, 1974.

Siegel, Sidney. Nonparametric Statistics for the Behavioral Sciences. New York: McGraw-Hill, 1956.

Small and Knust, Inc. California Local Government and CEQA, 1979. A report submitted to the Assembly Committee on Rules, California Legislature. San Diego, Calif.: Small and Knust, Inc., June 1980.

Solomon, Arthur P. The Effects of Land Use and Environmental Controls on Housing: A Review. Cambridge, Mass.: Joint Center for Urban Studies of MIT and Harvard, 1976.

Stokes, Bruce. Global Housing Prospects: The Resource Constraints. Washington, D.C.: Worldwatch Institute, September 1981.

Sumichrist, Michael, Gopal Ahulwalia, and Robert J. Sheehan. Profile of the Builder. Washington, D.C.: National Association of Homebuilders, 1979.

Tater, William J. An Appraisal of Environmental Protection Legislation and Administration upon the Economy in General and upon Real Estate Development in Orange County in Particular. Fullerton: California State University, School of Business Administration and Economics, December 1975.

Thompson, Fred. More and Better Analysis? The Case of Health, Safety and Environmental Regulation. Graduate School of Management Working Paper 14. Los Angeles: University of California-Los Angeles, 1980.

Trzyna, Thaddeus. Environmental Impact Requirements in the States: NEPA's Offspring. Claremont: Center for California Public Affairs, 1974.

Trzyna, Thaddeus, and Arthur W. Jokela. The California Environmental Quality Act. Claremont: Center for California Public Affairs, 1974.

Ultrasystems. Analysis of Environmental Procedures within the City of Los Angeles Planning Department. Newport Beach, Calif.: Ultrasystems, 1973.

United States Code. Title 42, Sections 4321-4347. Washington, D.C.: U.S. Government Printing Office, 1981.

Urban Land Institute. The Cost of Delay Due to Government Regulation in the Houston Housing Market. Research Report no. 28. Washington, D.C.: Urban Land Institute, 1979.

Urban Land Institute and Gruen and Gruen Associates. Effects of Regulations on Housing Costs: Two Case Studies. Report no. 27. Washington, D.C.: Urban Land Institute, 1977.

U.S., Senate, Committee on Governmental Affairs. Study on Federal Regulation. 95th Cong., 1st sess. Washington, D.C.: U.S. Superintendent of Documents, 1977.

Ventura County. Administrative Supplement to the State Resources Agency Guidelines for Implementation of California Environmental Quality Act of 1970. Ventura, Calif.: Ventura County Planning Department, February 13, 1979.

Warden, Richard E., and W. Tim Dagodag. A Guide to the Preparation and Review of Environmental Impact Reports. Los Angeles: Security World, 1976.

Wengert, Norman. The Political Allocation of Benefits and Burdens. Berkeley: University of California, Institute of Governmental Studies, 1976.

Williams, Charles J. Supplement to California Zoning Practice. Berkeley: California Continuing Education of the Bar, 1978.

ARTICLES

Andrews, Robert, Jr. "Aftermammoth: Friends of Mammoth and the Amended California Environmental Quality Act." Ecology Law Quarterly 3 (1973): 349-89.

Bendix, Selina. "A Short Introduction to the California Environmental Quality Act." Santa Clara Law Review 19 (1979): 521-39.

Bergman, Edward M. "Development Controls and Housing Costs: A Policy Guide to Research." In Management and Control of Growth, edited by Randall W. Scott, vol. 3, pp. 527-36. Washington, D. C.: Urban Land Institute, 1975.

Burchell, Robert W., and David Listokin. "Local Environmental Impact Statement: The State of the Art." Real Estate Issues, Winter 1977, pp. 61-71.

Case, Fred E. "The Impact of Land Use and Environmental Controls on Housing Costs." In The Cost of Housing, edited by Federal Home Loan Bank of San Francisco, pp. 141-76. San Francisco: Federal Home Loan Bank of San Francisco, 1978.

Case, Fred E., and Jeffrey Gale. "The Impact on Housing Costs of the California Coastal Zone Conservation Act." American Real Estate and Urban Land Economics Journal, Winter 1981, pp. 345-66.

Catalano, Ralph, and Joseph DiMento. "Local Government Response to State Environmental Impact Assessment Requirements: An Explanation and a Typology." Environmental Law 7 (1976): 25-50.

Catalano, Ralph, and Richard Reich. "County Processing of EIR's." California EIR Monitor 1 (1974): 4-20.

_____. "Local Government and the Environmental Impact Report: The California Experience." Urban Lawyer 8 (1976): 367-74.

Dowall, David E. "Reducing the Cost Effects of Local Land Use Controls." Journal of the American Planning Association, April 1981, pp. 145-53.

Elliott, Michael. "The Impact of Growth Control Regulations on Housing Prices in California." American Real Estate and Urban Economics Association Journal 9 (1981): 115-33.

"Enacted 1979 Legislation Dealing with CEQA." California EIR Monitor 6 (1979): 1-3.

Fishkind, Henry H. "The Effects of Growth Management Policies on New Home Prices: An Econometric Approach." In Urban Land Markets, Urban Land Institute Research Report no. 30. Washington, D. C.: Urban Land Institute, 1978.

Frech, H. E., and Ronald N. Lafferty. "The Economic Impact: Land Use and Land Values." In The California Coastal Plan: A Critique, pp. 69-92. San Francisco: Institute for Contemporary Studies, 1976.

Frieden, Bernard J. "Environmental Politics." Urban Land, March 1977, pp. 3-8.

Hagman, Donald G. "NEPA's Progeny Inhabit the States—Were the Genes Deficient?" Urban Law Annual 7 (1974): 3-56.

Hildreth, Richard G. "Environmental Impact Reports under the California Environmental Quality Act: The New Legal Framework." Santa Clara Law Review 17 (1977): 805-26.

ICF Associates. "Assessing the Impact of the California Coastal Plan on Commercial and Residential Construction." In Review of the California Coastal Plan, vol. 2. Sacramento: Office of the California Legislative Analyst, 1976.

James, Franklin J., and Thomas Muller. "Environmental Impact Evaluation, Land Use Planning and the Housing Consumer." Journal of the American Real Estate and Urban Economics Association, Fall 1977, pp. 279-301.

"Legislature Passes AB 2679." California EIR Monitor 3 (1975): 1-2.

Liroff, Richard A. "NEPA—Where Have We Been and Where Are We Going?" Journal of the American Planning Association, April 1980, pp. 154-61.

"New Law on Notices for Project Approval." California EIR Monitor 1 (1974): 1-2.

"1981 Amendments to CEQA." California EIR Monitor 8 (1981): 1-5.

Niskanen, William A., and Steven H. Hanke. "Land Prices Substantially Underestimate the Value of Environmental Quality." Review of Economics and Statistics 59 (August 1977): 375-77.

Ohls, James C., Richard C. Weisber, and Michelle J. White. "The Effect of Zoning on Land Values." Journal of Urban Economics 1 (1974): 428-44.

Pearlman, Kenneth. "State Environmental Policy Acts: Local Decision-Making and Land Use Planning." Journal of the American Institute of Planners, January 1977, pp. 42-53.

Rice, George A. "Reform of CEQA—The Developer's Perspective." California EIR Monitor, December 15, 1975, pp. 4-9.

Sahm, Barbara. "Project Approval under the California Environmental Quality Act: It Always Takes Longer Than You Think." Santa Clara Law Review 19 (1979): 579-617.

Schwartz, Seymour I., David E. Hansen, and Richard Green. "Suburban Growth Controls and the Price of New Housing." Journal of Environmental Economics and Management, December 1981, pp. 303-20.

"Statement by Resources Secretary Claire T. Dedrick at a Press Conference on CEQA—March 17, 1976." California EIR Monitor 3 (1976): 2-5.

Steiner, Barry, and Peter M. Detwiler. "SB2003—Common Sense or Housing Developers' Dream." California EIR Monitor 6 (1979): 1-5.

Stimson, J. Michael. "If There's a Way out of the Impasse among Housing, the Community, and the Environmentalists, the Way Is Impact Zoning." House and Home, August 1972, pp. 58-67.

Stull, William J. "Community Environment, Zoning, and the Market Value of Single Family Homes." Journal of Law and Economics, October 1975, pp. 337-47.

"Summary of AB 884." California EIR Monitor 4 (1977): 1-3.

United States Council on Environmental Quality. "Regulations for Implementing the Procedural Provisions of the National Environmental Policy Act." Federal Register 43 (November 29, 1978): 55978-56007. (Codified at 40 C.F.R. 1500-1508)

Weidenbaum, Murray L. "Government Regulations Blamed for Much of Housing Cost Increase." Mortgage Banker, March 1978, pp. 12-16.

Yost, Nicholas C. "NEPA's Progeny: State Environmental Policy Acts." Environmental Law Reporter 3 (1974): 50090.

INDEX

ABOUT THE AUTHORS

FRED E. CASE is an Urban Land Economist who received his Ph.D. at Indiana University, School of Business in 1951. He has been with the Graduate School of Management, University of California at Los Angeles since that time in a variety of teaching and administrative capacities. He is a member of the Urban Resources Study Center in the Graduate School of Management, a center devoted to analyzing and reporting on current housing, real estate, and urban land-use problems.

In the professional field outside of the university, Professor Case has served in numerous advisory capacities to national and local business and real estate trade associations and national, state, and local governments. Recently, he has completed studies for the California Legislature, the Legislative Analyst's Office, the California Real Estate Commissioner's Office, the Office of the Mayor of Los Angeles, and the Los Angeles Planning Commission.

Recent research studies have included _The California Real Estate Industry to 2000_, _California Housing Market Trends to 1980_, _California Real Estate Brokerage Operations 1950-1975_, _The Impact on Housing Costs of Coastal Environmental Legislation_, and _Simplification of Forms in Buying and Selling Homes_. Among his more recent publications are _Real Estate Investment and Appraising_, 2d ed. (with Sanders Kahn); _Real Estate Financing_ (with John Clapp); and _Using the Home for Investment Purposes_, _The Real Estate Investing Game_, and _Real Estate Economics and Real Estate Brokerage: A Management Approach_.

Presently, he is working on environmental issues as they affect the supply and sales of housing and on methods of managing public service organizations for greater benefit to the public. He was a member of the Los Angeles City Planning Commission and is on the Board of Directors of Financial Federation, a $2 billion savings and loan holding company.

JEFFREY GALE is an Assistant Professor of Business, Government, and Society at the Graduate School of Business Administration at the University of Washington. He holds a Ph.D. and a J.D. from the University of California-Los Angeles.

His teaching and research interests are in the fields of urban development, public policy analysis, political economy, and public and private planning. His current research is on environmental and land-use regulations and on self-regulation. He is a member of the California Bar.

AF719128

DO NOT REMOVE FROM
LAW LIBRARY
GOLDEN GATE UNIVERSITY

2022
The Supreme Court Review

"Judges as persons, or courts as institutions, are entitled to no greater immunity from criticism than other persons or institutions . . . [J]udges must be kept mindful of their limitations and of their ultimate public responsibility by a vigorous stream of criticism expressed with candor however blunt."

—*Felix Frankfurter*

". . . while it is proper that people should find fault when their judges fail, it is only reasonable that they should recognize the difficulties. . . . Let them be severely brought to book, when they go wrong, but by those who will take the trouble to understand them."

—*Learned Hand*

THE LAW SCHOOL
THE UNIVERSITY OF CHICAGO

Volume 2022

The Supreme Court Review

EDITED BY
DAVID A. STRAUSS
GEOFFREY R. STONE
JUSTIN DRIVER
AND WILLIAM BAUDE

DO NOT REMOVE FROM
LAW LIBRARY
AUG 07 2023
GOLDEN GATE UNIVERSITY

THE UNIVERSITY OF CHICAGO PRESS
CHICAGO AND LONDON

The Supreme Court Review, Volume 2022

Published annually by The University of Chicago Press.
www.journals.uchicago.edu/SCR/

© 2023 by The University of Chicago

All rights reserved. No part of this book may be reproduced in any form by any electronic or mechanical means (including photocopying, recording, or information storage and retrieval) without permission in writing from the publisher. Chapters may be copied or otherwise reused without permission only to the extent permitted by Sections 107 and 108 of the U.S. Copyright Law. Permission to copy articles for personal, internal, classroom, or library use may be obtained from the Copyright Clearance Center (www.copyright.com). For all other uses, such as copying for general distribution, for advertising or promotional purposes, for creating new collective works, or for resale, please contact Permissions Coordinator, Journals Division, University of Chicago Press, 1427 E. 60th Street, Chicago, IL 60637 USA. Fax: (773) 834-3489. E-mail: journalpermissions@press.uchicago.edu.

Subscriptions: For individual and institutional subscription rates, visit www.journals.uchicago.edu, email subscriptions@press.uchicago.edu, or call (877) 705-1878 (US) or (773) 753-3347 (international). Free or deeply discounted institutional access is available in most developing nations through the Chicago Emerging Nations Initiative (www.journals.uchicago.edu/inst/ceni).

Please direct subscription inquiries to Subscription Fulfillment, 1427 E. 60th Street, Chicago, IL 60637-2902. Telephone: (773) 753-3347 or toll free in the United States and Canada (877) 705-1878. Fax: (773) 753-0811 or toll-free (877) 705-1879. E-mail: subscriptions@press.uchicago.edu.

Standing orders: To place a standing order for this book series, please address your request to The University of Chicago Press, Chicago Distribution Center, Attn. Standing Orders/Customer Service, 11030 S. Langley Avenue, Chicago, IL 60628. Telephone toll free in the U.S. and Canada: 1-800-621-2736; or 1-773-702-7000. Fax toll free in the U.S. and Canada: 1-800-621-8476; or 1-773-702-7212.

Single-copy orders: In the U.S., Canada, and the rest of the world, order from your local bookseller or direct from The University of Chicago Press, Chicago Distribution Center, 11030 S. Langley Avenue, Chicago, IL 60628. Telephone toll free in the U.S. and Canada: 1-800-621-2736; or 1-773-702-7000. Fax toll free in the U.S. and Canada: 1-800-621-8476; or 1-773-702-7212. In the U.K. and Europe, order from your local bookseller or direct from The University of Chicago Press, c/o John Wiley Ltd. Distribution Center, 1 Oldlands Way, Bognor Regis, West Sussex PO22 9SA, UK. Telephone 01243 779777 or Fax 01243 820250. E-mail: cs-books@wiley.co.uk.

The University of Chicago Press offers bulk discounts on individual titles to Corporate, Premium, and Gift accounts. For information, please write to Sales Department—Special Sales, The University of Chicago Press, 1427 E. 60th Street, Chicago, IL 60637 USA or telephone 1-773-702-7723.

This book was printed and bound in the United States of America.

ISSN: 0081-9557
E-ISSN: 2158-2459
ISBN: 978-0-226-82805-3 (cloth)
E-ISBN: 978-0-226-82806-0 (ebook)

TO

OWEN FISS

A dedicated and always generous
scholar, teacher, and colleague

CONTENTS

JODY FREEMAN AND
MATTHEW C. STEPHENSON

THE ANTI-DEMOCRATIC MAJOR QUESTIONS DOCTRINE

West Virginia v. Environmental Protection Agency[1] is the Supreme Court's most important administrative law decision in decades. The opinion's significance is due principally to the Court's embrace of an aggressive version of the so-called "major questions doctrine" (MQD), which appears to require unusually explicit statutory authorization before agencies may undertake "major" regulatory actions. The *West Virginia* Court claims that this strong MQD is based on longstanding precedent, and that its use has salutary effects on the policymaking process. Neither claim is accurate. In Part I of this Article, we show that the strong version of the MQD embraced by the *West Virginia* Court is in fact relatively new; the extent of the doctrinal innovation is obscured by the fact that the MQD label has been unhelpfully attached to several related but distinct interpretive techniques, which we disentangle. In Part II, we turn to the impact of this new MQD on the policymaking process, focusing in particular on democratic accountability. While the

Jody Freeman is the Archibald Cox Professor of Law at Harvard Law School. Matthew C. Stephenson is the Henry L. Shattuck Professor of Law at Harvard Law School.

Authors' note: We are grateful to Dan Deacon, Ben Eidelson, Blake Emerson, Jon Gould, Leah Litman, David Strauss, and Cass Sunstein for helpful comments on earlier drafts, and to Nicholas Gladstone and Thomas Nielsen for excellent research assistance.

[1] 142 S. Ct. 2587 (2022).

The Supreme Court Review, 2023.
© 2023 The University of Chicago. All rights reserved. Published by The University of Chicago Press. https://doi.org/10.1086/724919

MQD's proponents claim that this doctrine protects separation-of-powers principles and the prerogatives of Congress, in fact the new MQD is more likely to weaken democratic accountability by shifting power from the elected branches to the courts, undermining transparency, and exacerbating the already excessive tendency toward minoritarian obstruction in Congress. The *West Virginia* Court's aggressive MQD would likely have other effects; perhaps most importantly, this version of the MQD makes it much more difficult for the federal government to address new problems under broadly worded statutes. Both the MQD's supporters and its detractors anticipate that the doctrine will result in less, and less aggressive, federal regulation. For purposes of this Article, though, our critique of the MQD focuses less on its impact on policy *outcomes* (though we think this is very important), and more on the impact of the MQD on the policymaking *process*, especially the extent to which the MQD makes that process less democratic.

I. The Multiple Major Questions Doctrines

In *West Virginia*, the Court held that the Environmental Protection Agency (EPA) had exceeded its authority under the Clean Air Act (CAA) when setting greenhouse gas emissions standards for power plants. The EPA, in its Clean Power Plan (CPP), had set these standards based on the emissions reductions that could be achieved by shifting electricity generation away from fossil fuels to cleaner alternatives, rather than basing the standards solely on what could be achieved by implementing pollution control measures at individual facilities. The EPA asserted that the relevant section of the CAA, which instructs the agency to set emissions standards based on the "best system of emissions reduction" that "has been adequately demonstrated," allowed the agency to base its standards on the possibility of such "generation shifting."[2] The Court disagreed. Although Chief Justice Roberts's majority opinion conceded that "[a]s a matter of definitional possibilities, generation shifting can be described as a system,"[3] the Court nevertheless held that the CAA could not be read to authorize the EPA's rule. The Court expressed "skepticism" that Congress could have intended to delegate to the EPA the expansive

---

[2] *See* Brief of Respondents at 8–9, *West Virginia*, 142 S. Ct. 2587 (2022) (No. 20-1530).

[3] *West Virginia*, 142 S. Ct. at 2614 (internal citation and quotation marks omitted).

authority to "substantially restructure the American energy market";[4] the Court further asserted that, under the MQD, the EPA could overcome this skepticism only if it could "point to clear congressional authorization to regulate in that manner."[5] This requirement of clear congressional authorization, which does not ordinarily apply when agencies exercise authority under broadly-worded statutes, was based, the *West Virginia* Court explained, on a presumption "that Congress intends to make major policy decisions itself, not leave those decisions to agencies."[6] Finding the requisite clear authorization lacking, the Court held that the EPA did not have the authority to set emissions standards based on the possibility of generation shifting.[7]

Although *West Virginia* was the first case in which a Supreme Court majority opinion used the phrase "major questions doctrine," neither the term nor the general idea was new. In 1986, then-Judge Breyer observed that "Congress is more likely to have focused upon, and answered, major questions, while leaving interstitial matters to answer themselves in the course of the statute's daily administration."[8] The "major questions" terminology gained traction in the academic literature in the 2000s in the wake of decisions like *MCI Telecommunications Corp. v. American Telephone & Telegraph Co.*,[9] *Food and Drug Administration v. Brown & Williamson Tobacco Corp.*,[10] and *Gonzales v. Oregon.*[11] In commentaries on those cases, scholars noted the possibility that the Court might be crafting some sort of "major questions exception" to the deferential standard of review announced in *Chevron U.S.A., Inc. v. Natural Resources Defense Council, Inc.*[12] In the ensuing two decades, the

---

[4] *Id.* at 2610, 2614.

[5] *Id.* (internal citation and quotation marks omitted).

[6] *Id.* at 2609 (internal citation and quotation marks omitted).

[7] *West Virginia* followed closely on the heels of two summary per curiam opinions that invoked the MQD to support emergency injunctions of Biden Administration efforts to arrest the spread of Covid-19: the Centers for Disease Control and Prevention's eviction moratorium, Ala. Ass'n of Realtors v. Dep't of Health & Hum. Servs., 141 S. Ct. 2485 (2021) (per curiam), and the Occupational Safety and Health Administration's test-or-vaccinate rule for large employers, Nat'l Fed'n of Indep. Bus. v. Occupational Safety & Health Admin., 142 S. Ct. 661 (2022) (per curiam).

[8] Stephen Breyer, *Judicial Review of Questions of Law and Policy*, 38 ADMIN. L. REV. 363, 370 (1986).

[9] 512 U.S. 218 (1994).

[10] 529 U.S. 120 (2000).

[11] 546 U.S. 243 (2006).

[12] 467 U.S. 837 (1984). *See, e.g.*, Cass R. Sunstein, Chevron *Step Zero*, 92 VA. L. REV. 187, 231–34 (2006); Jacob E. Gersen, *Overlapping and Underlapping Jurisdiction in Administrative*

"major questions doctrine" (or, occasionally, the "major rules doctrine") label was used by commentators and some lower courts to characterize the Court's apparent reluctance—in rare but notable cases like *Utility Air Regulatory Group v. Environmental Protection Agency*[13] and *King v. Burwell*[14]—to defer to agency interpretations on fundamental legal questions that Congress could not plausibly have delegated to an agency.[15] In summing up these developments, Chief Justice Roberts's majority opinion in *West Virginia* characterized the MQD "label" as "refer[ring] to an identifiable body of law that has developed over a series of significant cases all addressing . . . [the] recurring problem [of] agencies asserting highly consequential power beyond what Congress could reasonably be understood to have granted."[16]

That characterization, though not exactly wrong, is misleading. As numerous other scholars have noted, the MQD label has been applied to several different principles, with different doctrinal structures and underlying justifications.[17] Lumping these quite distinct doctrinal techniques together under the MQD label has obscured important differences among them, and facilitated a kind of doctrinal "creep," whereby an initially modest and defensible proposition about conventional inferences from statutory structure has metamorphosed into a novel clear statement rule that not only ignores but inverts longstanding administrative law principles.

Law, 2006 Sup. Ct. Rev. 201, 226, 241; Abigail R. Moncrieff, *Reincarnating the "Major Questions" Exception to* Chevron *Deference as a Doctrine of Noninterference (or Why* Massachusetts v. EPA *Got It Wrong)*, 60 Admin. L. Rev. 593 (2008).

[13] 573 U.S. 302 (2014).

[14] 576 U.S. 473 (2015).

[15] *See, e.g.*, Blake Emerson, *Administrative Answers to Major Questions: On the Democratic Legitimacy of Agency Statutory Interpretation*, 102 Minn. L. Rev. 2019 (2018); Michael Coenen & Seth Davis, *Minor Courts, Major Questions*, 70 Vand. L. Rev. 777 (2017); Jonas J. Monast, *Major Questions About the Major Questions Doctrine*, 68 Admin. L. Rev. 445 (2016); Nathan Richardson, *Keeping Big Cases from Making Bad Law: The Resurgent "Major Questions" Doctrine*, 49 Conn. L. Rev. 355 (2016); Michael Sebring, *The Major Rules Doctrine: How Justice Brett Kavanaugh's Novel Doctrine Can Bridge the Gap Between the* Chevron *and Nondelegation Doctrines*, 12 N.Y.U. J. L. & Liberty 189 (2018).

[16] *West Virginia*, 142 S. Ct. at 2609.

[17] *See, e.g.*, Cass R. Sunstein, *There Are Two "Major Questions" Doctrines*, 73 Admin. L. Rev. 475 (2021); Daniel Deacon & Leah Litman, *The New Major Questions Doctrine*, 109 Va. L. Rev. (forthcoming 2023); Kevin O. Laske, *Major Questions About the "Major Questions" Doctrine*, 5 Mich. J. Env't & Admin. L. 479 (2016); Alison Gocke, Chevron*'s Next Chapter: A Fig Leaf for the Nondelegation Doctrine?*, 55 U.C. Davis L. Rev. 955, 966–88 (2021); Nathan Richardson, *Antideference: COVID, Climate, and the Rise of the Major Questions Canon*, 108 Va. L. Rev. Online 174, 175–78 (2022); Sebring, *supra* note 15, at 192, 212.

We can discern in the case law three different versions of the MQD, which can be distinguished from one another in part by their relationship to the *Chevron* doctrine. *Chevron* established a deferential standard of review for courts to employ when reviewing a challenge to an administrative agency's interpretation of a statute the agency administers: if Congress has expressed a clear intent on the precise question at issue, then the agency is bound by Congress's instructions, but if the statute is unclear, the reviewing court should accept the agency's reasonable resolution of the ambiguity.[18] The standard justification for this deferential approach is that resolving statutory ambiguities entails policy judgments, not merely legal judgments, and courts should therefore presume—likely as a kind of legal fiction—that Congress's use of ambiguous language was intended as an implicit delegation of policymaking authority to the agency, which has greater expertise and is more politically accountable than the judiciary.[19] That said, courts defer under *Chevron* only to *reasonable* agency interpretations.[20] Additionally, the Court has made clear, most notably in *United States v. Mead Corp.*,[21] that *Chevron*'s deferential standard of review does not apply to *all* agency interpretations. Rather, a reviewing court must first consider a preliminary question—sometimes described as *Chevron* Step Zero[22]—as to whether the decision at issue is one where it makes sense to presume that Congress intended to delegate interpretive authority to the agency.[23] *Mead*, for example, held that such a presumption is usually unreasonable when the agency issues interpretations in informal guidance documents rather than in notice-and-comment rules or formal orders.[24]

One version of the MQD operates entirely *within* the traditional *Chevron* framework. Under this "within-*Chevron*" MQD, the agency's reasonable interpretations of its statutory authority are still entitled to *Chevron* deference, but the "majorness" of the agency's assertion of power weighs in the court's determination of whether the agency's

---

[18] *See* Chevron U.S.A., Inc. v. Nat. Res. Def. Council, Inc., 467 U.S. 837, 842–45 (1984).

[19] *Id.* at 843–45, 865–66.

[20] *Id.* at 842–43 & n.9.

[21] 533 U.S. 218 (2001).

[22] *See* Cass R. Sunstein, Chevron *Step Zero*, 92 VA. L. REV. 187 (2006); *see also* Thomas W. Merrill & Kristin E. Hickman, Chevron's *Domain*, 89 GEO. L.J. 833 (2001).

[23] *See Mead*, 533 U.S. at 227–29.

[24] *See id.* at 230–34.

interpretation is indeed reasonable. A second version of the MQD operates at *Chevron* Step Zero. Under this version, if the issue is sufficiently major, then the reviewing court must decide *de novo* whether the agency's action is consistent with the statute, without any presumption for or against the correctness of the agency's view. A third version of the MQD maintains that, on major questions, the court should not only abandon but *reverse* the usual *Chevron* presumption that statutory ambiguity constitutes an implicit delegation to the agency. Under this anti-delegation version of the MQD, courts should presume that Congress does not intend to delegate "major" powers to agencies and so should read ambiguous statutes to *prohibit* agencies from exercising such powers, even if the statute could be read—indeed, even if it would more naturally be read—to grant such powers.

These three versions of the MQD have a family resemblance. At a high level of generality, they are all, as the *West Virginia* Court put it, concerned with preventing agencies from "asserting highly consequential power beyond what Congress could reasonably be understood to have granted."[25] But they are not the same. Their doctrinal structures, practical implications, and normative underpinnings are importantly distinct, and appreciating these distinctions makes clear just how much the MQD, which at least arguably began as a tame and conventional form of statutory interpretation informed by structural inferences, has metastasized into the tool by which the Court seems poised to upend nearly forty years of administrative law.

A. THE "WITHIN-*CHEVRON*" MAJOR QUESTIONS DOCTRINE

As noted above, although *Chevron* instructs courts to defer to agencies' reasonable resolution of statutory ambiguities, a reviewing court is still obligated to determine whether an agency's interpretation is in fact reasonable. An agency interpretation that contravenes the clear meaning of the statute is entitled to no deference. And the determination as to whether the statute indeed speaks clearly to the precise question at issue is to be made, as *Chevron* itself notes, using the "traditional tools of statutory construction."[26] If the text of a statutory provision, read in context and in light of the overall structure of the statute, clearly indicates that the agency's powers are meant to be narrow and

[25] West Virginia v. Env't Prot. Agency, 142 S. Ct. 2487, 2609 (2022).

[26] Chevron U.S.A., Inc. v. Nat. Res. Def. Council, Inc., 467 U.S. 837, 843 n.9 (1984).

limited, then an interpretation that would give the agency broad transformational authority may well be unreasonable, even under *Chevron*'s deferential standard. As the Court sometimes puts the idea, Congress does not "hide elephants in mouseholes."[27] If the statutory provision at issue is unambiguously a mousehole, then the court would properly reject an agency's purported discovery of an elephant hiding within. On that understanding, the so-called MQD is not really a distinct doctrine so much as a label for a particular sort of structural inference, which is part of the conventional statutory interpretation exercise that the Court must conduct even under *Chevron*.[28] While one can always take issue with any particular use of this sort of inference, its underlying justification is obvious and entirely compatible with *Chevron*: Congress has the ultimate responsibility to decide how much authority to delegate to an agency, and an agency cannot operate beyond the bounds that Congress has set.

A clear example of this version of the MQD can be found in *MCI*.[29] The statutory provision at issue in that case, found in the Communications Act, gave the Federal Communications Commission (FCC) the authority to "modify" certain rules concerning long distance carriers' rate-filing obligations. The FCC argued that it could use this authority to *eliminate* the rate-filing requirement for smaller carriers. In rejecting that claim, Justice Scalia's opinion for the Court emphasized that the word "modify" implies a small or incremental change.[30] He then asserted that, because the rate-filing requirement is central to the overall structure of the Communications Act, the elimination of this requirement (even for a subset of carriers) could not reasonably be considered a "modification."[31] In making this point, Justice Scalia observed—in a passage that proved important to the future development of the MQD—that "[i]t is highly unlikely that Congress would

---

[27] Whitman v. Am. Trucking Ass'ns, 531 U.S. 457, 468 (2001).

[28] In the canonical formulation of the *Chevron* doctrine, the reviewing court's inquiry proceeds in two steps: at the first step, the court asks whether Congress has spoken to the precise question at issue, and if the answer is no, the court asks, at the second step, whether the agency's interpretation is reasonable. The within-*Chevron* structural inference version of the MQD might take place at either Step One or Step Two, and indeed we do not think that there is a meaningful difference between *Chevron*'s two steps, at least in this context. *See* Matthew C. Stephenson & Adrian Vermeule, Chevron *Has Only One Step*, 95 VA. L. REV. 597 (2013).

[29] 512 U.S. 218 (1994).

[30] *See* MCI Telecomms. Corp. v. Am. Tel. & Tel. Co., 512 U.S. 218, 225–28 (1994).

[31] *See id.* at 229–32.

leave the determination of whether an industry will be entirely, or even substantially, rate-regulated to agency discretion—and even more unlikely that it would achieve that through such a subtle device as permission to 'modify' rate-filing requirements."[32] In context, it is clear that the *MCI* Court decided the key question—"whether [the agency's elimination of the rate-filing requirements for non-dominant carriers] can be considered a mere 'modification'"[33]—entirely *within* the *Chevron* framework. Thus, the *MCI* Court's approach, though certainly contestable, is neither novel nor radical. Rather, it is a straightforward form of structural inference, one that does not really require a special name.

B. THE "*CHEVRON* STEP ZERO" MAJOR QUESTIONS DOCTRINE

A second version of the MQD, distinct from Justice Scalia's approach in *MCI*, is best understood as part of what has come to be known as *Chevron* Step Zero, articulated most clearly and influentially in *Mead*. In *Mead*, the Court held that the *Chevron* presumption—that a statutory ambiguity should be treated as an implicit delegation by Congress of authority to the agency—only applies "when it appears that Congress delegated authority to the agency generally to make rules carrying the force of law, and that the agency interpretation claiming deference was promulgated in the exercise of that authority."[34] That general holding requires courts to develop more specific criteria for when *Chevron*'s ambiguity-equals-delegation assumption should apply and when it should not. In elaborating on the relevant considerations, the *Mead* Court focused principally on the procedural formality of the agency's decision, and, secondarily, on whether the decision had issued from the agency's central leadership or from dispersed lower-level bureaucrats.[35] But *Mead* held open the possibility that other factors might bear on whether *Chevron* applies.[36] The second version of the MQD treats the "majorness" of the interpretive question as one of these other factors.

---

[32] *Id.* at 231.

[33] *Id.*

[34] United States v. Mead Corp., 533 U.S. 218, 226–27 (2001).

[35] *See id.* at 229–34.

[36] *See id.* at 230–31.

The idea here is that, just as it is implausible or inappropriate to presume that Congress intends to delegate to agencies the power to determine statutory meaning through informal statements put out by low-level bureaucrats, it is also implausible or inappropriate to presume that Congress intends to delegate to agencies the power to determine statutory meaning on questions of great importance. In such cases, the Step Zero version of the MQD instructs courts to treat statutory ambiguities on major issues as legal questions for the courts to resolve using the standard array of interpretive techniques, rather than presuming that these ambiguities are implicit delegations to the agency. Just as in *Mead*, the agency's view on the meaning of the statutory provision in question is not entitled to *Chevron* deference, but neither is the agency's interpretation treated with special skepticism. The agency, like any other litigant, may present its arguments, but it is the court's job to wrestle with the statute and pronounce what the disputed provision means.[37]

The case that most clearly illustrates this version of the MQD is *King v. Burwell*,[38] in which the Court had to decide whether to uphold the IRS's interpretation of a provision of the Affordable Care Act (ACA) concerning the availability of tax credits for health insurance purchased on federally-created exchanges. At the outset of the Court's opinion, Chief Justice Roberts considered and rejected the notion that the Court should apply *Chevron*'s deferential framework to the IRS's interpretation. The reason, the Court explained, is that the availability of tax credits on federal exchanges is "a question of deep economic and political significance that is central to this statutory scheme," and that if "Congress wished to assign that question to an agency, it surely would have done so expressly."[39] In short, the Court declared, "[t]his is not a case for the IRS," but "[i]t is instead [the Court's] task to determine the correct reading of" the relevant statutory provision.[40] Notably, the *King* Court acknowledged that the text of that provision

---

[37] The only difference suggested by the relevant major questions cases is that unlike in *Mead*, the court would resolve these issues de novo, rather than applying an intermediate standard of deference such as that embraced in *Skidmore v. Swift & Co.*, 323 U.S. 134 (1944).

[38] 576 U.S. 473 (2015).

[39] *Id.* at 486 (internal citations and quotation marks omitted).

[40] *Id.*

is ambiguous,[41] but the Court ultimately concluded, in light of the overall structure of the ACA, that the IRS's reading was correct.[42]

What is the underlying justification for a *Chevron* Step Zero version of the MQD? We can imagine two related but distinct possibilities. First, perhaps one might suppose—as an empirical matter—that if the issue is sufficiently important, Congress almost certainly *did* have a specific intent on the precise question at issue; the idea that Congress either deliberately left such an important matter to agency discretion or failed to consider the issue is, on this view, simply too implausible to entertain.[43] If that is so, the reviewing court must do its best to recover what Congress was trying to say, even if the statute appears unclear. But the empirical assertion that Congress simply must have had a specific intent on a sufficiently major question is unconvincing, at least as a general matter. It is not unheard of for Congress to overlook or avoid even quite important issues. That seems just as likely as the possibility that Congress thought about the question and had a specific intent but failed to express this intent clearly. Furthermore, the assertion that the issue is so fundamental that Congress simply must have had a specific intent as to its resolution works only for those issues that would have been considered fundamental *at the time of enactment*. Yet advocates of the Step Zero version of the MQD rarely if ever suggest such a limit on the doctrine's scope.[44] Moreover, even if we stipulate that Congress really did have a specific intent, if we also stipulate that the statute is so ambiguous that it is hard to discern what that intent actually was, then the court (or any other interpreter) may well misconstrue Congress's meaning, or allow its own policy preferences to influence its reading. If that is so, then treating the question *as if* it is one where Congress lacked a clear intent, thus leaving the matter to the more expert and accountable agency, would be better than allowing the Court to guess.

[41] *Id.* at 491–92.

[42] *Id.* at 492–98.

[43] *See* Cary Coglianese, Chevron*'s Interstitial Steps*, 85 GEO. WASH. L. REV. 1339, 1359 (2017).

[44] *Cf.* U.S. Telecom Ass'n v. Fed. Commc'n Comm'n, 855 F.3d 381, 422 (D.C. Cir. 2017) (Kavanaugh, J., dissenting from denial of rehearing en banc) (defending the MQD in part on the grounds that, as an empirical matter, legislative drafters do not intend to leave major questions unresolved, even though the substantive issue in that case—whether the FCC had the statutory authority to mandate "net neutrality"—was not an issue that the drafters of the 1996 Telecommunications Act could reasonably have anticipated).

An alternative justification for the Step Zero MQD maintains that even if one accepts the standard *Chevron* arguments for giving agencies primary interpretive authority over ordinary questions of statutory construction, it is better—on policy grounds—for courts to have exclusive authority when it comes unusually fundamental questions of statutory meaning. But it is hard to see a good reason why the policy arguments in favor of *Chevron* deference would not apply with as much or more force to major questions. The two leading arguments in favor of *Chevron*, after all, are agencies' greater expertise and greater political accountability (through the President). If a question on which Congress has not spoken clearly is especially consequential, it would seem even more important that this question be resolved by the more expert and accountable decision-maker.[45]

We will not dwell further on the question whether the justifications for the *Chevron* Step Zero version of the MQD make sense. We simply mean to bring them to the surface and highlight the ways in which they differ from each other and from the justifications for other versions of the MQD.

As should be readily apparent, the Step Zero version of the MQD is quite different, in terms of both structure and justification, from the within-*Chevron* version. Yet language in some of the Court's key cases unhelpfully conflates the two. Perhaps the best example of this is the case that did the most to catalyze the evolution of the MQD along its current trajectory: *Brown & Williamson*. In that case, the Court considered a challenge to the FDA's conclusion that the Food, Drug and Cosmetics Act (FDCA) gave the agency jurisdiction over tobacco products.[46] Justice O'Connor's majority opinion acknowledged that the FDCA definitions of "drug" and "drug delivery device" would seem to cover tobacco products, yet the opinion nonetheless concluded that, even under the deferential *Chevron* standard, the FDA's assertion of authority over tobacco was unreasonable in light of the structure of the statute and the history of tobacco regulation (and non-regulation). Near the end of the opinion, in a passage that would prove crucial to the development of the MQD, the Court further noted that the "inquiry into whether Congress has directly spoken to the precise question at issue is shaped, at least in some measure, by the nature of the

[45] *See* Coglianese, *supra* note 43, at 1357; Sunstein, *supra* note 12, at 232–33.

[46] *See* Food & Drug Admin. v. Brown & Williamson Tobacco Corp., 529 U.S. 120 (2000).

question presented," and that the question whether the FDA may regulate tobacco is extraordinarily significant.[47]

In invoking this consideration, was the *Brown & Williamson* Court relying on the within-*Chevron* version of what we now call the MQD, or the Step Zero version? It is difficult to say. On one hand, the opinion purported to analyze the case under *Chevron*.[48] Furthermore, in discussing the extraordinary nature of the FDA's asserted authority, the *Brown & Williamson* Court relied for authority principally on *MCI*, and *MCI*, as we have seen, operated entirely within the *Chevron* framework.[49] The *Brown & Williamson* Court held that, taking into account the extraordinary nature of the FDA's claimed authority and in light of statutory history and structure, it was "clear . . . that Congress has directly spoken to the question at issue."[50] That language indicates that *Brown & Williamson* was deploying what we have called the within-*Chevron* version of the MQD—a conclusion buttressed by the fact that the section of the Court's opinion that discusses the significance of the issue comes near the end, after the Court had completed its extensive analysis of the statute.[51]

On the other hand, in explaining why the significance of the question was relevant, the *Brown & Williamson* Court also declared that "[d]eference under *Chevron* to an agency's construction of a statute that it administers is premised on the theory that a statute's ambiguity constitutes an implicit delegation from Congress to the agency to fill in the statutory gaps," and that "[i]n extraordinary cases, . . . there may be reason to hesitate before concluding that Congress has intended such an implicit delegation."[52] That language implies—and has been subsequently understood to embrace—something more akin to the *Chevron* Step Zero version of the MQD: not the idea that the significance of the question can contribute to a court's determination that the agency's claim of (broad, unusual) authority cannot be squared with

---

[47] *Id.* at 159.

[48] *Id.* at 132.

[49] *Id.* at 160.

[50] *Id.* at 160–61.

[51] *See id.* at 159–61. The opinion did not indicate that the significance of the question was a decisive factor in the case, but noted only that the Court's "inquiry into whether Congress has spoken to the precise question at issue" (the *Chevron* inquiry) "is shaped, at least in some measure, by the nature of the question presented." *Id.* at 159.

[52] *Id.* at 159.

the (modest, constraining) statutory provision even under *Chevron*'s deferential standard, but rather the idea that, in light of the issue's significance, the court should decide that issue *without* adopting the usual *Chevron* presumption that Congress intended agencies to resolve statutory ambiguities. When read in context, we think *Brown & Williamson* Court intended the former meaning. But as often happens, the passage quoted above took on a life of its own.[53]

C. THE ANTI-DELEGATION MAJOR QUESTIONS DOCTRINE

A third and more aggressive version of the MQD inverts the usual *Chevron* presumption that a statutory ambiguity constitutes an implicit delegation to the responsible agency. This version proceeds from the same starting point as the Step Zero version—the idea that Congress is unlikely to delegate major issues to agencies through ambiguous or broad statutory terms—but takes it further. Under this anti-delegation version of the MQD, when a major question is at issue, not only should the court decline to apply the usual *Chevron* presumption that statutory ambiguity implies delegation to the agency, but instead the court should embrace the opposite presumption: an ambiguous or broadly worded statute, on this version of the MQD, should be interpreted as *not* delegating to the agency the power to take an "extraordinary" action, even if the statute would more naturally be read as authorizing that action. This non-delegation presumption can only be overcome by especially clear and express statutory authorization.[54]

Until the Supreme Court's 2021–22 major questions trilogy (*Alabama Association of Realtors v. Department of Health and Human Services*,[55] *National Federation of Independent Business v. Occupational Safety and Health Administration*,[56] and *West Virginia*) the two cases that came closest to suggesting something like this stronger version of the MQD were the so-called *Benzene* case[57] and *Utility Air Regulatory Group*

[53] Notably, Chief Justice Roberts's majority opinion in *King* cited this passage from *Brown & Williamson* as precedent when explaining why the *Chevron* framework was inappropriate when addressing major questions. *See* King v. Burwell, 576 U.S. 473, 485 (2015) (quoting *Brown & Williamson*, 529 U.S. at 159).

[54] West Virginia v. Env't Prot. Agency, 142 S. Ct. 2587, 2617–18 (2022) (Gorsuch, J., concurring).

[55] 141 S. Ct. 2485 (2021) (per curiam).

[56] 142 S. Ct. 661 (2022) (per curiam).

[57] Indus. Union Dep't, AFL–CIO v. Am. Petroleum Inst., 448 U.S. 607 (1980) (plurality opinion).

(UARG) v. Environmental Protection Agency.[58] On closer examination, though, neither case really embraced the anti-delegation MQD that has emerged in the more recent decisions.

In *Benzene*, decided back in 1980, Justice Stevens's plurality opinion held that the Occupational Safety and Health Administration (OSHA) could not tighten a workplace safety standard without first making a threshold showing that affected workplaces were currently "unsafe."[59] Though the plurality purported to derive this requirement from the text of the statute, that reading was hardly straightforward. In defending its strained interpretation of the statute, the plurality emphasized that OSHA's understanding of its statutory authority would give the agency "power to impose enormous costs that might produce little, if any, discernible benefit."[60] The plurality then intimated that granting such sweeping authority would raise concerns under the constitutional nondelegation doctrine, and asserted that "[a] construction of the statute that avoids this kind of open-ended grant [of authority] should certainly be favored."[61] Given this language, it is possible to argue—as Justice Gorsuch did in his *West Virginia* concurrence—that *Benzene* embraced something like the anti-delegation version of the MQD: Even though the statute would more naturally be read as supporting the agency's position, the reviewing court should read the statute more narrowly so as to avoid the conclusion that Congress delegated extraordinary powers to the agency.[62] But *Benzene* is not terribly strong authority for the anti-delegation MQD.

This is in part because *Benzene* is a pre-*Chevron* case, but more importantly because *Benzene* suggested that OSHA's reading of the statute might *actually* be unconstitutional under the nondelegation doctrine.[63] This fact suggests that the case is better understood as an application (though perhaps an unconvincing one) of the constitutional avoidance canon. If we understand the case that way, then

---

[58] 573 U.S. 302 (2014).

[59] *Indus. Union Dep't*, 448 U.S. at 642.

[60] *Id.* at 645.

[61] *Id.* at 646.

[62] *See* West Virginia v. Env't Prot. Agency, 142 S. Ct. 2587, 2619 (2022) (Gorsuch, J., concurring).

[63] *See Indus. Union Dep't*, 448 U.S. at 646 ("If the Government was correct in [its assertion of statutory authority], the statute would make such a sweeping delegation of legislative power that it might be unconstitutional." (internal quotation marks omitted) (citing Schechter Poultry Corp. v. United States, 295 U.S. 495 (1935); and Panama Refining Co. v. Ryan, 293 U.S. 388 (1935)).

unless we also think that any delegation of a "major question" is potentially unconstitutional, the modern cases embracing the anti-delegation MQD must be doing something different from what Justice Stevens's *Benzene* plurality meant to do.[64]

UARG might also be read as suggesting the anti-delegation version of the MQD. In *UARG*, Justice Scalia's opinion for the Court held that even under *Chevron*'s deferential standard, the EPA could not permissibly interpret one of the CAA's permitting programs to be triggered solely by greenhouse gas emissions. Such an interpretation, the Court explained, would be "inconsistent with—in fact, would overthrow—the Act's structure and design."[65] That reasoning sounds like conventional structural inference, fitting comfortably within *Chevron*'s deferential framework, and along the lines of Justice Scalia's opinion for the Court in *MCI*. Indeed, the *UARG* opinion purported to operate entirely within the *Chevron* framework.[66] But in a widely-cited passage, the *UARG* Court—citing *MCI*, *Brown & Williamson*, and *Benzene*—stated that "EPA's interpretation is also unreasonable because it would bring about an enormous and transformative expansion in EPA's regulatory authority without clear congressional operation."[67] Elaborating on this point, the Court declared that "[w]hen an agency claims to discover in a long-extant statute an unheralded power to regulate a significant portion of the American economy, we typically greet its announcement with a measure of skepticism. We expect Congress to speak clearly if it wishes to assign to an agency decisions of vast economic and political significance."[68] That language has been read as endorsing the stronger, anti-delegation version of the MQD. Read in context, though, it does

---

[64] It is also worth noting that *Benzene* did not deprive the agency of the power to issue important regulations on major questions; rather, it required the agency to first find that such regulations would address a sufficiently serious risk to merit the imposition of large regulatory costs. Holding that Congress is unlikely to have delegated to an agency the authority to impose extremely burdensome regulations without first establishing that those regulations address a significant problem is quite different from holding that Congress is unlikely to have delegated to an agency the authority to adopt major regulations *at all*.

[65] Util. Air Regul. Grp. v. Env't Prot. Agency, 573 U.S. 302, 321 (2014).

[66] *See id.* at 315 ("We review EPA's interpretations of the Clean Air Act using the standard set forth in *Chevron*. . . ."); *see also id.* at 321 ("Even under *Chevron*'s deferential framework, agencies must operate within the bounds of reasonable interpretation," which "must account for both the specific context in which language is used and the broader context of the statute as a whole." (internal citations and quotation marks omitted)).

[67] *Id.* at 324.

[68] *Id.*

not seem that the *UARG* Court meant to go that far. Indeed, as noted above, the *UARG* majority expressly embraced and applied the *Chevron* framework. The Court declined to defer to the EPA's reading of the statute because the Court concluded, in light of context and statutory structure, that the agency's interpretation was "patently unreasonable."[69] The Court did not say, or even suggest, that if a *reasonable* reading of the statutory language granted the agency sweeping authority, the Court should nonetheless reject that interpretation if the statute lacked a sufficiently clear statement. Nevertheless, the case has been cited for that latter proposition.[70] Thus, as with *Brown & Williamson*, we see in *UARG* some of the imprecise language that contributes to doctrinal creep: On one hand, *UARG* could be read as of a piece with *MCI*, finding that the breadth and impact of the authority claimed by the agency is so out of harmony with the text and structure of the statute that the interpretation is unreasonable even under *Chevron*'s deferential standard. On the other hand, *UARG* could also be read to suggest that Congress must speak especially clearly to authorize agencies to take actions that have vast economic and political significance.

Putting aside for the moment the question of how well the antidelegation MQD is grounded in pre-2021 precedent, what is the underlying justification for this version of the doctrine? As with the Step Zero version, there is both an empirical and a policy-based rationale for refusing to read even statutory language that appears to delegate authority over major questions to in fact do so, unless Congress has made the delegation unusually clear.

The empirical argument would be that while Congress may be happy to delegate to agencies the power to address and resolve a host of ordinary matters, Congress would be unlikely to want to give agencies broad power to address unusually significant issues. But if that claim as understood as a genuine empirical claim—a claim about the result that Members of Congress intended to bring about—then the claim is questionable at best. For much of the last century, especially during the New Deal, Congress seemed quite happy to hand off important issues

---

[69] *Id.*

[70] *See, e.g.*, West Virginia v. Env't Prot. Agency, 142 S. Ct. 2587, 2608–09, 2614, 2621, 2623 (2022) (Gorsuch, J., concurring); Ala. Ass'n of Realtors v. Dep't of Health & Hum. Servs., 141 S. Ct. 2485, 2489 (2021) (per curiam).

to agencies.[71] The legislative history of the APA, moreover, suggests that the Members of Congress who enacted that statute fully expected that agencies could and would make decisions on important matters.[72] Furthermore, to insist, based on conjectures about congressional policy preferences, that the courts should require a clear statement to delegate even when the *better* reading of the (ambiguous) statutory text would favor such a delegation would seem to require a great deal of confidence in one's ability to know what Congress really wanted.

Another way to justify the anti-delegation MQD relies not on assertions about what Congress (probably) wanted, but on what would better promote good governance. (On that understanding, the rhetoric that grounds the MQD in "congressional intent" should be understood not as an empirical claim but as a legal fiction, similar to the legal fiction used to justify *Chevron*.) This is the position that Justice Gorsuch staked out in his *West Virginia* concurrence, as well as his dissent in *Gundy v. United States*.[73] As he put the point in *Gundy*, "[a]lthough [the MQD] is nominally a canon of statutory construction, we apply [it] in service of the constitutional rule that Congress may not divest itself of its legislative power by transferring that power to an executive agency."[74] For Justice Gorsuch, this is not because Congress would not *want* to transfer its power in this way; rather, Justice Gorsuch believes that the MQD is important precisely because Congress is all too enthusiastic about delegating policy decisions to the executive, thus avoiding political accountability.[75] Overly broad delegations of power to resolve major questions, he emphasized, threaten to undermine separation-of-powers principles and democratic accountability.[76]

In Part II, we take up the normative question whether this anti-delegation version of the MQD in fact promotes good and politically accountable governance. For now, though, it is important to observe that while the justifications for the Step Zero and anti-delegation versions of the MQD might appear similar, they are meaningfully

---

[71] *See* JOHN F. MANNING & MATTHEW C. STEPHENSON, LEGISLATION AND REGULATION: CASES AND MATERIALS 552–56 (4th ed. 2020).

[72] *See* Blake Emerson, *"Policy" in the Administrative Procedure Act: Implications for Delegation, Deference, and Democracy*, 97 CHI.-KENT L. REV. 113, 133–35 (2022).

[73] 139 S. Ct. 2116, 2141–42 (2019) (Gorsuch, J., dissenting).

[74] *Id.* at 2142.

[75] *Id.* at 2134–35.

[76] *West Virginia*, 142 S. Ct. at 2617–18 (Gorsuch, J., concurring).

different. While the Step Zero version requires the reviewing court to determine the statute's meaning, the court may ultimately accept the agency's interpretation of an ambiguous provision; the anti-delegation version of the MQD forecloses this possibility. The empirical justification for the Step Zero version assumes that Congress almost certainly *had a specific intent*, which the reviewing court must try to ascertain, even if the statute appears ambiguous. The empirical justification for the anti-delegation MQD is that on a major question, Congress almost certainly *did not intend to delegate to the agency*. These justifications sometimes overlap, but not always: In some cases, the best reconstruction of Congress's (unclearly expressed) intent might be to allow or even to require the agency to exercise broad powers. As for the policy-based justifications, the argument for the Step Zero MQD is that courts are better suited than agencies to resolving interpretive questions of extraordinary significance, while the argument for the anti-delegation MQD is that delegations (at least of major questions) are so normatively undesirable that courts should avoid finding that they have occurred unless Congress has compelled that conclusion. That is, the normative argument for the Step Zero MQD is grounded in a claim about *comparative institutional competence*, while the normative argument for the anti-delegation MQD is grounded in opposition to (or at least deep skepticism of) delegation.

D. WHICH MAJOR QUESTIONS DOCTRINE DID THE *WEST VIRGINIA* COURT ENDORSE?

As we have seen, the MQD is not a single doctrine, but a label that has been attached to at least three thematically related but analytically distinct ways in which the "majorness" of an agency's action may bear on whether that action is authorized by statute. Which of these three versions of the MQD did the *West Virginia* Court embrace? It's hard to say. The concurring and dissenting opinions are easier to categorize: Justice Kagan's dissent clearly endorsed the within-*Chevron* version, asserting that the cases the Court has cited for the so-called MQD in fact employed a particular form of structural inference—the stuff of "normal statutory interpretation."[77] Justice Gorsuch's concurrence just as clearly endorsed the anti-delegation version, insisting that the

[77] *Id.* at 2633 (Kagan, J., dissenting).

MQD requires a clear statement of congressional intent before a court may read a statute to confer sweeping powers on the agency.[78]

The majority opinion in *West Virginia* is more difficult to categorize. Part of the challenge is that although we have discussed each of the three versions of the MQD with reference to *Chevron* (a framing that we think is both doctrinally appropriate and analytically helpful), the *West Virginia* Court did not mention *Chevron*, even to reject its applicability.[79] But the larger difficulty in figuring out what the Court is doing is that the majority opinion both contains language suggesting that it is using the more modest within-*Chevron* MQD,[80] and also contains broader language that appears to endorse the more aggressive anti-delegation MQD.[81] That latter language has led most commentators to conclude, quite reasonably, that *West Virginia* has embraced the anti-delegation MQD.[82] We are also inclined to this view, though

---

[78] *Id.* at 2616 (Gorsuch, J., concurring).

[79] This may be further evidence that *Chevron*'s status at the Supreme Court is tenuous at best, even though the case continues to supply the governing standard in circuit and district courts. *See, e.g.*, Kristen E. Hickman & Aaron Nielson, *The Future of* Chevron *Deference*, 70 DUKE L.J. 1015, 1017 (2021); Nathan Richardson, *Deference Is Dead (Long Live* Chevron*)*, 73 RUTGERS U. L. REV. 441 (2021).

[80] For example, in portions of the opinion the majority emphasizes that there is such a mismatch between the authority the EPA asserted and the CAA's text, structure, and overall design that the EPA's interpretation is simply unreasonable. *See, e.g.*, *West Virginia*, 142 S. Ct. at 2609 (emphasizing that "extraordinary grants of regulatory authority are rarely accomplished through modest words, vague terms, or subtle devices" (internal citations and quotation marks omitted)); *id.* at 2610 (emphasizing that the CAA provision upon which the EPA relied is "an ancillary provision . . . that was designed to function as a gap filler" (internal citations and quotation marks omitted)); *id.* at 2611–12 (arguing that the "consistent understanding" of the relevant statutory language has been that it endorsed a "technology-based approach . . . that focuses on improving the emissions performance of individual sources," and that EPA's decision in the CPP to read this standard as allowing consideration of generation-shifting was "unprecedented" and "effected a fundamental revision of the statute" (internal citations and quotation marks omitted)).

[81] *See, e.g.*, *id.* at 2609 (asserting that the Court "presume[s] that Congress intends to make major policy decisions itself, not leave those decisions to agencies" (internal citations and quotation marks omitted)); *id.* (elaborating that to persuade the Court that Congress has delegated an agency extraordinary powers, "something more than a merely plausible textual basis for the agency action is necessary," in the form of "clear congressional authorization for the power [the agency] claims" (internal citation and quotation marks omitted)); *id.* (explaining that "the major questions doctrine is distinct" from "routine statutory interpretation" due to the "requirement of clear congressional authorization") (internal citation and quotation marks omitted)); *id.* at 2614 (restating the claim that in order "[t]o overcome [the Court's skepticism that Congress intended to delegate such consequential power to the EPA], the Government must—under the major questions doctrine—point to clear congressional authorization to regulate in that manner" (internal citations and quotation marks omitted)).

[82] *See, e.g.*, Deacon & Litman, *supra* note 17, at 60; Jonathan H. Adler, West Virginia v. EPA*: Some Answers About Major Questions*, 2022 CATO SUP. CT. REV. 37, 53–54. Further

we think it would be possible for the Court to walk this position back while claiming fidelity to *West Virginia*. (As we have seen in cases like *Brown & Williamson* and *UARG*, the fact that the MQD comes in multiple forms, and that the similar language can be used with respect to these different versions, not only makes it harder to figure out exactly what a given opinion means but also makes it easier to recharacterize that opinion later.) Assuming for now that *West Virginia* does indeed embrace an aggressive, anti-delegation version of the MQD, what are the likely consequences? We turn to that question next.

II. The Anti-Democratic Effects of the New Major Questions Doctrine

Having shown that the most aggressive version of the MQD is in fact a novel judicial innovation—the product of doctrine creep enabled by the Court's conflation of three distinct ideas under a single label—we turn now to its normative implications. A full normative evaluation of the MQD, in any of its versions, would have to consider a wide range of factors well beyond what we can address in this short essay. We therefore focus here on one specific aspect of this more general normative evaluation: the likely implications of the anti-delegation MQD for democratically accountable policymaking.

How to define democratic accountability is the subject to considerable debate, one that has generated a vast literature.[83] For our purposes, we posit a few simple claims, which we think are fairly uncontroversial: that policymaking is more democratic when policy decisions are made by elected officials or by those who are supervised by elected officials; when citizens and interest groups can more easily attribute credit or blame for government action (or inaction) to particular

---

support for this conclusion comes from the two Covid-related cases immediately preceding *West Virginia*. *See* Ala. Ass'n of Realtors v. Dep't of Health & Hum. Servs., 141 S. Ct. 2485, 2489 (2021) (per curiam) ("Even if the test were ambiguous, the sheer scope of [the agency's] claimed authority under [the statute] would counsel against the Government's interpretation. We expect the government to speak clearly when authorizing an agency to exercise powers of vast economic and political significance.") (internal citations and quotation marks omitted); *see also* Nat'l Fed'n of Indep. Bus. v. Occupational Safety & Health Admin., 142 S. Ct. 661, 665 (2022) (per curiam).

[83] *See generally*, *e.g.*, Democracy, Accountability, and Representation (Adam Przeworski et al. eds., 1999); Robert D. Behn, Rethinking Democratic Accountability (2001); Craig T. Borowiak, Accountability and Democracy: The Pitfalls and Promise of Popular Control (2011).

elected officials or political parties; and when the outcomes of government decision-making processes are, at least on average and over time, more likely to reflect the preferences and interests of political majorities.[84] Democratic accountability is one value among many, and we do not claim that a constitutional system, or an individual legal principle like the MQD, should be evaluated *solely* with respect to whether it maximizes democratically accountable policymaking. But we expect most readers will agree that this is an important value to consider.

The anti-delegation MQD undermines democratic accountability in at least three respects. First, although the doctrine is portrayed by its proponents as protecting the prerogatives of Congress from encroachment by unelected bureaucrats,[85] in fact the MQD shifts substantial policy discretion to unelected federal judges. Second, because the MQD makes it legally risky for the executive branch to tout ambitious regulatory initiatives as a central part of the President's strategy for addressing salient public issues, the doctrine is likely to encourage executive branch actors to downplay significant rules, make them seem as technocratic as possible, and obscure or understate the connection between individual regulations and the President's larger agenda. This result is a loss for democratic accountability, which is typically enhanced when presidents take explicit ownership of significant administrative actions and when agencies are fully transparent about the goals they are seeking to accomplish.[86]

Third, a robust MQD exacerbates one of the most significant pathologies of the U.S. federal lawmaking system as it exists today: the minoritarian obstructionism that goes well beyond anything the Constitution's Framers could have intended or anticipated. The ability of the executive branch to act unilaterally, relying on open-ended but non-specific terms in old statutes to address pressing national problems, provides a kind of safety valve that offsets, albeit imperfectly, this

---

[84] Some of these factors might be considered part of the *definition* of democratically accountable policymaking, while others might be more accurately characterized as contributors to or expected consequences of democratically accountable policymaking. While important in other contexts, distinguishing among causes, consequences, and constitutive elements is not essential, as we simply want to identify characteristics that we think are properly associated with democratic policymaking, in order to have a rough benchmark against which to assess how a legal doctrine—in this case the MQD—might affect democratic accountability.

[85] *See, e.g.*, Sebring, *supra* note 15, at 234–41; Riley T. Svikhart, Note, *"Major Questions" as Major Opportunities*, 92 NOTRE DAME L. REV. 1873, 1892–1903 (2017).

[86] *See* Elena Kagan, *Presidential Administration*, 114 HARV. L. REV. 2245, 2331–34 (2001).

undemocratic feature of the legislative process.[87] Additionally, the *possibility* of unilateral executive action may sometimes be the catalyst for genuine legislative bargaining and compromise that might not otherwise take place. By closing this safety valve, the anti-delegation MQD will exacerbate the minoritarian obstructionism that is one of our current constitutional system's least democratic features.

A. JUDICIAL AGGRANDIZEMENT

The most obvious and important way in which the new MQD undermines democratic accountability is the extent to which it empowers unelected judges to decide significant public policy issues, likely in ways that are influenced, consciously or subconsciously, by the judges' policy preferences. This problem has already received substantial attention,[88] and we generally agree with the thrust of the prior critical commentary. But the concern is important enough that we will elaborate here the reasons the new MQD is so problematic in this respect.

Proponents of the MQD would bristle at the suggestion that the doctrine authorizes judges to decide cases based on the judges' own policy preferences. They would insist, to the contrary, that the MQD ensures fidelity to *Congress's* choices by returning the most important policy questions to Congress for a clear expression of intent.[89] That benign portrayal is misleading, however, when one considers how the anti-delegation MQD operates. Consider a statute that is written in open-ended terms, or that uses sufficiently ambiguous or capacious language that the statute *could* be read to authorize the agency's interpretation but does not do so explicitly. Under *Chevron*, since Congress has not resolved the precise question at issue, the agency's reading would be upheld. But under the new, anti-delegation MQD, if the court decides the rule is "major," then the agency almost certainly loses, because the statute does not *clearly* authorize the agency's rule. Determining that an issue is "major," then, is like flipping a switch to

---

[87] *See, e.g.*, *id.* at 2248, 2345; Lauren Gilbert, *Obama's Ruby Slippers: Enforcement Discretion in the Absence of Immigration Reform*, 116 W. VA. L. REV. 255, 261 (2013); David E. Pozen, *Self-Help and the Separation of Powers*, 124 YALE L.J. 2, 7–8 (2014).

[88] *See, e.g.*, Deacon & Litman, *supra* note 17; Marla D. Tortorice, *Nondelegation and the Major Questions Doctrine: Displacing Interpretive Power*, 67 BUFF. L. REV. 1075, 1119–26 (2019); Richardson, *supra* note 17, at 195–206.

[89] *See, e.g.*, West Virginia v. Env't Prot. Agency, 142 S. Ct. 2587, 2617 (2022) (Gorsuch, J., concurring); Nat'l Fed'n of Indep. Bus. v. Occupational Safety & Health Admin., 142 S. Ct. 661, 668 (2022) (Gorsuch, J., concurring).

kill what would otherwise be a valid rule, without any serious and sustained effort to wrestle with the text and structure of the statute. This is bad for democratic accountability for two related reasons.

First, one of the most important justifications for the *Chevron* doctrine is the recognition that where Congress has not expressed itself clearly, resolution of an interpretive question inevitably involves a discretionary policy judgment, and it is more democratic for such decisions to be made by agencies—which are part of an executive branch headed by an elected President, and subject to ongoing oversight by Congress—rather than by unelected judges.[90] Thus, part of the democratic accountability objection to the MQD is simply an extension of the democratic accountability argument in favor of *Chevron*.

But there is a bigger problem with the MQD than with other limits or exceptions to *Chevron* deference. That problem derives from the Court's inability or unwillingness to articulate an objective test, or even consistently applied guidelines, for determining when a rule is sufficiently major to trigger the MQD, and when a statute contains a sufficiently clear delegation of authority to decide major questions. This gives judges an extraordinary degree of flexibility. Without a clearer demarcation of the MQD's domain, it is perfectly reasonable to worry that judges will be influenced (perhaps subconsciously) by their hostility to a particular regulation—or to a particular agency or administration—when determining that a rule is extraordinary enough to require extra-clear congressional authorization.

This problem is far greater with the anti-delegation version of the MQD than with the within-*Chevron* version. To be sure, even under the latter version the reviewing court must determine whether the asserted agency authority is too "big" to fit within the statutory scheme. But at least the court would need to do the analytical work of explaining *why* the regulatory elephant doesn't fit into the statutory mousehole, and (assuming that *Chevron* is still the governing standard) the court would do so with a thumb on the scale in favor of the agency's view. Even under the Step Zero version of the MQD, the court would still have to offer a reasoned explanation—under a *de novo* standard of review, without any presumption in favor or against the agency—as to why the statute does not authorize the agency rule, leaving open the possibility that the agency's interpretation would prevail. But under the

[90] *See* Chevron U.S.A., Inc. v. Nat. Res. Def. Council, Inc., 467 U.S. 837, 865–66 (1984); *see also* MANNING & STEPHENSON, *supra* note 71, at 1114–15.

anti-delegation MQD, the reviewing court needs to conduct only the most perfunctory statutory analysis. Once the court declares the agency's action "major," the game is all but over: the agency loses unless the statutory authorization is "clear"—itself a highly subjective determination.

West Virginia illustrates the problem. The dispute there concerned whether the term "best system of emissions reduction" (BSER) could incorporate emissions reductions made through shifting electricity generation from fossil fuels to cleaner sources. This is a hard question. The government, and Justice Kagan's dissent, presented a series of plausible arguments as to why this reading of "system" was not only permissible but correct.[91] There are also reasonable arguments against that conclusion[92]—but Chief Justice Roberts's majority opinion did not really make them. Though the majority gestured toward some of these arguments in responding to the claim that the CAA clearly authorized the EPA's rule, the Court seemed content to rely principally on the government's failure to demonstrate enough statutory clarity to overcome the high bar set by the MQD.[93] Might the Court have reached the same conclusion had it reviewed the EPA's assertion of authority either deferentially or *de novo*? Possibly. But at least the Court would have had to do the hard work of explaining why the EPA's interpretation was *unreasonable* (in the former case) or at least *incorrect* (in the latter). And the need to do that analysis might provide some constraint—if not in this case, then at least in other cases—on the influence that the Justices' policy preferences may have on their statutory analysis. The anti-delegation MQD removes that check.

This concern would be mitigated if the Court could articulate some reasonably objective, judicially manageable standard for determining when an issue is sufficiently "major" to require especially clear congressional authorization. But the Court has not done so, either in *West Virginia* or in earlier cases. Even the doctrine's proponents have acknowledged that the "majorness" inquiry is basically an "I-know-it-when-I-see-it" test.[94] The *West Virginia* Court did purport to offer some guidelines, but they are insufficient.

---

[91] *West Virginia*, 142 S. Ct. at 2628–33 (Kagan, J., dissenting).

[92] *See, e.g.*, Nathan Richardson, *Trading Unmoored: The Uncertain Legal Foundation for Emissions Trading Under § 111 of the Clean Air Act*, 120 Penn St. L. Rev. 181 (2015).

[93] *West Virginia*, 142 S. Ct. at 2614–16.

[94] *See* U.S. Telecom Ass'n v. Fed. Commc'ns Comm'n, 855 F.3d 381, 423 (D.C. Cir. 2017) (Kavanaugh, J., dissenting from denial of rehearing en banc).

For starters, both Chief Justice Roberts's majority opinion and Justice Gorsuch's concurrence suggest the commonsense notion that a rule's economic impact is relevant to whether the rule counts as major.[95] Fair enough, but the Court does not seem to apply this idea in a consistent way. For example, while the EPA estimated that the CPP would impose significant compliance costs and have far-reaching impacts—a point that the *West Virginia* Court emphasized[96]—several prior agency rules that did *not* trigger the MQD had higher costs and more sweeping impacts—a fact that the Justices in the *West Virginia* majority simply ignored, even though it had been pointed out to them.[97]

A second consideration that both the *West Virginia* Court and Justice Gorsuch's concurrence suggested as relevant to the "majorness" inquiry is whether Congress had considered but failed to pass legislation similar to the challenged rule.[98] This factor is puzzling on its face—Congress considers and rejects legislation on many topics for a variety of reasons, some of which have nothing to do with the substance of the policy under consideration, making legislative rejection a weak proxy for "majorness." Furthermore, agencies often regulate in areas where Congress has considered but not taken legislative action, but courts rarely point to this fact as a reason to require some special clear statement from Congress before agencies may regulate.[99] Indeed, as we discuss further below, we suspect that despite the Court's framing, the fact that a rule seeks to advance a policy that had stalled in Congress has little to do with whether the rule is "major," but is instead motivated by a distinct (and, our view, misplaced) set of concerns about circumventing the legislative process.[100]

---

[95] *See West Virginia*, 142 S. Ct. at 2612; *id.* at 2621 (Gorsuch, J., concurring).

[96] *See id.* at 2604 (majority opinion).

[97] *See* Brief for Richard L. Revesz as Amicus Curiae Supporting Respondents at 8–9, West Virginia v. Env't Prot. Agency, 142 S. Ct. 2587 (2022) (No. 20-1530) (observing that, even if one uses the high end of the EPA's cost estimates, "[t]he estimated costs of the [CPP] were . . . unexceptional for EPA pollution-control rules," and substantially lower than the costs of several other rules recently adopted by other agencies).

[98] *See West Virginia*, 142 S. Ct. at 2610, 2614; *id.* at 2620–21 (Gorsuch, J., concurring).

[99] Despite the way the Court presents this issue, we suspect that the Court's concern with the fact that the agency rule advances a policy resembling legislative proposals that Congress had failed to enact has little to do with the question's "majorness," but instead signals the Court's concern with unilateral executive action. *See infra* Section II.C.

[100] *See id.*

Yet another factor that the *West Virginia* Court suggests might indicate that a rule is "major" is whether the rule concerns a matter outside of the agency's traditional domain. That consideration also seems to have little to do with "majorness" as such: the question whether a rule is important is distinct from the question whether the agency that promulgated that rule possesses relevant expertise or has strayed from the areas the agency traditionally regulates. But even putting that aside, the question whether an agency is acting within its "traditional" domain is quite subjective, and depends on the level of generality at which one characterizes that domain.[101] Furthermore, many agencies implement their broad statutory mandates in ways that overlap with the domains of other agencies,[102] and this is not typically viewed as a reason to deny agencies the authority those statutes appear to confer. Here again, the Court stresses a factor that *might* trigger the MQD, but often (indeed, usually) does not.

An additional factor that might bear on the "majorness" inquiry—suggested more clearly in Justice Gorsuch's concurrence than in Chief Justice Roberts's majority opinion—is whether the agency's rule concerns a matter of great *political* significance.[103] But presidential administrations take actions on politically controversial issues all the time—indeed, it is hard to succeed in being elected President without promising to adopt policies that some percentage of the public would view as of "great political significance." It seems strange to require unusually clear statutory authorization before a President can do important things, at least when the relevant statutes seem to confer quite broad authority. Furthermore, even if the level of political controversy it generates might be relevant to a rule's "majorness," how controversial is controversial enough? Suppose that several members of Congress, twenty state Attorneys General, and one powerful trade association oppose a rule—does that mean *ipso facto* that the rule is so politically controversial that it requires unusually explicit congressional

---

[101] Indeed, in *West Virginia*, the Court characterized the EPA's use of generation shifting to set standards as involving the EPA in a question of *national energy policy* beyond the agency's expertise, whereas the EPA saw itself as taking action that falls squarely within the agency's *pollution control* mandate. *Compare West Virginia*, 142 S. Ct. at 2612–13, *with id.* at 2636–38 (Kagan, J., dissenting); *see also id.* at 2623 n.5 (Gorsuch, J., concurring).

[102] *See generally* Jody Freeman & Jim Rossi, *Agency Coordination in Shared Regulatory Space*, 125 Harv. L. Rev. 1131, 1155–81 (2012) (providing examples of authority shared between or among agencies).

[103] *See West Virginia*, 142 S. Ct. at 2620–21 (Gorsuch, J., concurring).

authorization? Political significance is highly subjective, and it is doubtful that courts would or could consistently apply it.

Moreover, incorporating the political controversy over a rule into the inquiry as to whether the rule triggers the MQD is especially problematic from a democratic accountability perspective for two reasons. First, a court's evaluation of a rule's "political salience" is especially likely to be unduly influenced by judicial policy preferences: the *extent* of political controversy over an agency rule is easy to conflate with the *perceived reasonableness* of that controversy.[104] Second, the "political controversy" criterion invites organized minorities to thwart the will of political majorities by ginning up controversy about a rule that would otherwise seem to fall within the scope of the agency's statutory authority. This may give rise to a kind of heckler's veto, in which minorities can block agency action without any of the procedural safeguards associated with either the rulemaking process or the legislative process.[105]

In short, the Court has failed to provide anything resembling reasonably definite criteria for distinguishing "major" questions from ordinary questions, and the factors it has offered are highly subjective and inconsistently applied. Moreover, the Court has failed to provide guidance on whether one of these factors alone, if large enough, is sufficient to trigger the MQD, or whether several are required to do so. This leaves open innumerable MQD-triggering combinations (high political controversy + beyond the agency's domain, even if moderate economic impact; large economic impact + novel exercise of authority, even if low political salience; congressional consideration and inaction + political controversy, even if economically unremarkable), making it virtually impossible to know when the Court might invoke the doctrine.

On top of all that, the MQD incorporates another highly subjective inquiry. If a reviewing court does decide, based on some or all of the preceding factors, that the agency action in question triggers the MQD, the court must then decide whether the statute authorizes that action with sufficient clarity. What sort of statutory language satisfies this "clear statement" requirement? Neither the *West Virginia* majority nor the concurrence seems to require that the statute explicitly authorize the specific agency action at issue. The Court, quite sensibly,

---

[104] *See* Deacon & Litman, *supra* note 71, at 62.

[105] *See id.*

appears to acknowledge that general statutory language can sometimes delegate sufficiently broad discretion for an agency to take major actions. But when? We don't know.

Importantly, the Justices in the *West Virginia* majority (including the two concurring Justices) have written or joined numerous opinions that have found consequential and controversial exercises of executive power lawful on the ground that the relevant statutes delegate broad authority to the executive branch. Although these cases did not involve an express MQD challenge, it is nonetheless notable that these Justices seemed untroubled by *those* sweeping delegations of power to the executive. Indeed, in voting to uphold those exercises of executive authority, these Justices have emphasized the broad, open-ended terms of the statutory delegations without so much as hesitating.

Consider, for example, *Department of Commerce v. New York*,[106] in which the Court invalidated the Commerce Department's decision to add a citizenship question to the census. Chief Justice Roberts's controlling opinion based this holding on the conclusion that the Department had acted arbitrarily and capriciously because its stated reasons for adding the question were pretextual.[107] For present purposes, though, the more interesting portion of the opinion addresses whether the Department has the statutory authority to add the citizenship question. The Court said that it did. In so holding, Chief Justice Roberts's majority opinion—joined on this point by Justices Thomas, Alito, Gorsuch, and Kavanaugh—emphasized that Congress has "virtually unlimited discretion" in conducing the census, and that Congress, by statute, had "delegated [that] broad authority over the census to the Secretary [of Commerce]."[108] This conclusion is striking considering the later MQD cases. As Justice Alito's *Department of Commerce* concurrence acknowledged (albeit critically), the question whether to include a citizenship question on the census had "become a subject of bitter political controversy."[109] Yet Justice Alito—who joined Justice Gorsuch's *West Virginia* concurrence—did not

[106] 139 S. Ct. 2551 (2019).

[107] *Id.* at 2575–76.

[108] *Id.* at 2566 (quoting Wisconsin v. City of New York, 517 U.S. 1, 19 (1996)); *see also id.* at 2598–99 (Alito, J., concurring) (asserting that this statutory grant of discretion to the Secretary is so broad as to render the Secretary's decisions regarding the content of the census to be judicially unreviewable).

[109] *Id.* at 2596.

treat this as a reason to require some more specific statutory authorization for the inclusion of the citizenship question.

And *Department of Commerce* is not unique: In several other cases, the Court, in opinions joined by the Justices in the *West Virginia* majority, has emphasized the (permissible) breadth of the statutory delegation as the reason to uphold some controversial executive branch action.[110] To be clear, no MQD objection was explicitly raised in these cases. Yet these decisions—and the fact that these Justices seem so untroubled by asserting that statutes grant sweeping discretion to the executive branch to issue decisions on socially and politically controversial issues—naturally invites the question whether a judge will be more inclined to find a broad grant of statutory power insufficiently specific to authorize a "major" action when the judge opposes that action, and more likely to find that a statute delegates the authority to decide a major issue when the judge thinks that sort of broad delegation is appropriate.

In short, the MQD criteria suggested by the *West Virginia* Court are too subjective and manipulable to yield anything like a manageable judicial standard. Many average or unremarkable rules could be characterized as "major" or "extraordinary" by skillfully framing their economic impacts, political significance, novelty, or ambition.

[110] Deacon & Litman, *supra* note 17, provides several additional illustrations: First, in *Little Sisters of the Poor v. Pennsylvania*, 140 S. Ct. 2367 (2020), the Court upheld the Trump Administration's statutory authority, under the Affordable Care Act (ACA), to create religious and moral exemptions from rules issued pursuant to the ACA that required certain employers provide contraceptive health coverage to their employees. In so holding, the majority opinion—joined by four of the five Justices in the *West Virginia* majority—emphasized that the statute gave the agency "broad discretion" and that it would be improper "to impos[e] limits on an agency's discretion that are not supported by the text." *Id.* at 2381. As Deacon and Litman point out, the question whether to create religious and moral exemptions to the contraception mandate was both "politically controversial" and "economically significant." Deacon & Litman, *supra* note 17, at 44. (In fairness, the same could be said of the decision to create the contraception mandate in the first place—but the Court never suggested that doing so was a "major" decision that required special authorization.) Second, in *Hamdi v. Rumsfeld*, 542 U.S. 507 (2004), the Court concluded that the broadly worded Authorization of the Use of Military Force (AUMF) gave the Bush Administration the authority to detain American citizens—a subject of extraordinary political controversy that the AUMF did not explicitly discuss. Third, in *Trump v. Hawaii*, 138 S. Ct. 2392 (2018), the Court—in another majority opinion joined by all four Justices in the *West Virginia* majority who were on the Court at the time— upheld the Trump Administration's so-called "Muslim ban," which had been refashioned as a facially neutral set of entry restrictions on foreign nationals from eight foreign countries (six Muslim-majority countries, plus North Korea and Venezuela), ostensibly due to deficiencies in the vetting processes used in those countries. In upholding the policy, the Court emphasized that the relevant statutory provision "exudes deference to the President in every clause." 138 S. Ct. at 2408. That may be, but this policy was at least as high profile and politically controversial as the CPP—indeed, far more so. Yet there was no suggestion that the statute would need to speak with unusual clarity to authorize the President to adopt this policy.

The same statute could easily be characterized as authorizing or not authorizing a major agency action. Thus, the strong version of the MQD embraced in *West Virginia* enhances judicial discretion to arbitrarily decide which questions are beyond agency authority without workable guidelines or limiting criteria. And because the anti-delegation MQD operates not as tool of interpretation but as a reversal of the usual presumption in favor of the agency, one that has such a subjective trigger, it relieves courts of the need to wrestle in any serious way with whether the *better* reading of the statute in fact authorizes the agency's action—what Justice Kagan referred to as a "get-out-of-text free" card.[111] The potential for judicial usurpation of the policymaking powers of both Congress and the President seems obvious.

A defender of the MQD might respond that even if we are correct about the largely unconstrained discretion the MQD gives to judges, we are exaggerating the degree to which this effects a judicial usurpation of policymaking power, because Congress can always enact legislation expressly authorizing the agency to take the major action at issue (or major actions generally). Indeed, the response continues, the whole point of the MQD is to ensure that *Congress*—not the agency and not the judiciary—decide whether the agency should have the power to take major action on a major question. That response, however, is problematic. For one thing, as the Court has emphasized in other contexts, the views and decisions of the *current* Congress are not always informative as to the meaning of a statute enacted by a previous Congress.

In many cases where the anti-delegation MQD would apply, at least arguably including *West Virginia*, that previous statute is best read as effecting a broad delegation to an agency, thereby affording the agency the flexibility it needs to address new and unforeseen problems as they arise. The fact that the current or future Congress may not be able to enact legislation reaffirming the earlier Congress's preference for broad agency authority should not have any bearing on the better reading of that previously enacted statute. But we can put that objection to the side. The far more important problem with the claim that the anti-delegation MQD merely shifts the issue to Congress is that this facile assertion overlooks the difficulty—indeed, in many cases the practical impossibility—of Congress overriding the Court's decision,

[111] West Virginia v. Env't Prot. Agency, 142 S. Ct. 2587, 2641 (2022) (Kagan, J., dissenting).

even if majorities in both chambers favor delegating the agency the power at issue.[112] And of course the Court is well aware of this.

Furthermore, as a formal matter, there is no need for the Court to deploy the MQD to "return" the issue to Congress, because Congress already has the power to react to agency overreaching by enacting legislation that overrides the agency's rule (perhaps using the fast-track, filibuster-free procedures authorized by the Congressional Review Act[113]) or narrows the scope of the agency's delegated authority. The use of the anti-delegation of the MQD does not give Congress authority or opportunities that Congress does not already have. In addition, Congress can also use other tools, such as oversight hearings and the appropriations power, to chasten an agency that oversteps. Unlike a reviewing Court, Congress can tailor its response to its level of dissatisfaction with the agency's action. By invoking the anti-delegation MQD to deprive the agency of authority that appears to be granted under the best reading of the relevant statutory provisions, the Court short-circuits that more nuanced back-and-forth between the legislative and executive branches, which is itself an important part of the democratic process. In sum, when the Court invokes the anti-delegation MQD, it is not simply and modestly insisting that Congress decide whether the agency should have major authority; the Court is deciding for itself—on the basis of vague, easily manipulable factors—that notwithstanding the best or most natural reading of the statutory text, the agency shall not have the power to adopt the regulation in question.

B. LOSS OF TRANSPARENCY

In addition to transferring substantial policymaking authority from Congress and the executive branch to the federal judiciary, the new MQD may have a more subtle and indirect effect on political accountability through the impact the doctrine is likely to have on how the White House and the agencies design and describe their regulatory initiatives.

Consider the incentives that the anti-delegation MQD creates. Even if the executive branch believes that it has statutory authority for some policy or program, it can never be sure whether a court

---

[112] *See infra* Section II.C.

[113] *See* 5 U.S.C. §§ 801–08.

might decide, retrospectively, that this policy is so "major" that clearer congressional authorization is required. A rule that attempts to address some problem in a more comprehensive fashion—and that therefore may cost more, produce greater benefits, affect more industry participants, drive greater change, and impact more people—is more likely to seem major. Furthermore, because the Court, in *West Virginia* and elsewhere, has suggested that a rule is more like to be considered major if it is *politically* salient, a rule that is explicitly framed as part of an effort to advance the President's policy agenda on a high-profile topic may be more vulnerable to invalidation. Rhetoric from presidents and senior administration officials that has become commonplace—about how the administration would use statutes "creatively," or about how they "don't need Congress" to accomplish some ambitious policy change[114]—now seems much more legally risky, potentially giving opponents material for a major questions challenge.[115]

In light of this, executive branch lawyers might well start advising agencies to make their rules seem smaller in scope, more incremental, and more technocratic, and suggesting that the President, agency officials, and other supporters avoid talking about how these rules contribute to some larger policy agenda or help address some big national problem. Naturally, this advice goes against the grain, since most political appointees are eager to tout their regulatory accomplishments, and presidents want to demonstrate strong leadership by

[114] *See, e.g., Press Briefing by Principal Deputy Press Secretary Karine Jean-Pierre*, WHITE HOUSE (Oct. 21, 2021, 2:43 PM), https://www.whitehouse.gov/briefing-room/press-briefings/2021/10/21/press-briefing-by-principal-deputy-press-secretary-karine-jean-pierre-2 ("The President will advance his climate agenda using every tool at his disposal and can make significant progress in curbing emissions, growing our economy, and good-paying union jobs. And so he could do that without Congress. . . . what we're saying is we don't need the—we don't need Congress. We can do it without Congress.").

[115] While it may be *politically* advantageous to describe executive branch initiatives as creative or transformational, or as necessary to substitute for congressional inaction, after *West Virginia*, such claims could increase legal risk. One need only look at the oral arguments from the *West Virginia* litigation in the D.C. Circuit to see why: during oral argument, Judge Brown noted that the "Administrator [Gina McCarthy] . . . refer[ring] to [the CPP] as transformative" was evidence that the CPP was "a major question." Transcript of Oral Argument at 51, West Virginia v. Env't Prot. Agency (D.C. Cir. 2016) (No. 15-1363); *cf.* Evan Gerstman, *How a White House Tweet May Doom Biden's Workplace Vaccine Mandate*, FORBES (Jan. 8, 2022), https://www.forbes.com/sites/evangerstmann/2022/01/08/how-a-white-house-tweet-may-doom-bidens-workplace-vaccine-mandate (noting that in the Supreme Court oral argument on the challenge to OSHA's test-or-vaccinate rule, the Justices noted that President Biden's Chief of Staff, Ron Klain, had retweeted a tweet describing this rule as a "work-around," presumably of congressional refusal to adopt such a mandate legislatively).

publicizing and claiming credit for what their agencies are doing; for example, by emphasizing how their administration is pursuing a "whole of government" approach to pressing issues through a series of interrelated agency actions.[116] But in light of the MQD, it may be more legally prudent for agency officials to adjust their messaging—and perhaps the design of the rules themselves—so that these rules can be portrayed as narrow technocratic interventions. It may be similarly prudent for presidents not to talk much about agency rulemaking activities, and to avoid in particular discussing individual rulemakings as part of a more comprehensive strategy to address high-profile agenda items, such as climate change, public health, or immigration.[117]

To be sure, the Court's embrace of a more aggressive MQD may not produce a radical change in agency or presidential behavior overnight. Given both legal constraints and efficiency considerations, agencies are likely to continue to promulgate broad, comprehensive rules. And there are strong political reasons for presidents and agency officials to continue to tie agency rulemaking activities to larger policy goals. But in the wake of *West Virginia*, executive branch officials may become convinced that such talk creates legal risk, and may try to dissuade the President from playing up how individual rulemakings connect to the White House's larger agenda. Indeed, there are already anecdotal accounts of agencies reconsidering some of their proposed rules, scrubbing them for potential MQD vulnerability and contemplating ways in which to scale them back or to frame them to

---

[116] *See, e.g.*, Janet Yellen, Secretary, U.S. Dep't Treasury, Remarks by Secretary of the Treasury Janet L. Yellen Addressing the Threat of Climate Change to the Coalition of Finance Minsters for Climate Action (Apr. 6, 2021), https://home.treasury.gov/news/press-releases/jy0104 (describing the Biden Administration's "whole-of-government approach to tackle the climate challenge"); Exec. Order No. 14,067, 87 Fed. Reg. 14,143 (Mar. 9, 2022) (ordering interagency consultation to further a "whole-of-government" effort to ensure the responsible development of digital assets). *See generally* Kagan, *supra* note 86, at 2302.

[117] Blake Emerson has made a somewhat different but related argument that, to the extent that the MQD induced agencies to characterize certain actions as technocratic rather than "political," the doctrine may make agencies less likely to engage in deliberative-democratic processes, both internally and with the public. *See* BLAKE EMERSON, THE PUBLIC'S LAW: ORIGINS AND ARCHITECTURE OF PROGRESSIVE DEMOCRACY 18 (2019) ("If courts may deprive agencies of their deliberative discretion whenever they think the issue is a significant one, agencies will have strong incentives to treat every regulatory matter as clerical and non-controversial. Agencies will then withdraw further into a technocratic mindset rather than growing into the role of public engagement. Important questions of political value will either be pushed beneath the surface or passed into the hands of judges.").

make them at least *seem* less "major," while trying to maintain as much of the substance as possible.[118]

These sorts of changes might not seem like that much of a problem, at least if the substance of the agencies' activities does not change too much.[119] But we think these changes in framing, messaging, and rule design would be bad for democratic accountability. If agencies downplay the real reasons for their rules, segment actions to make them seem more innocuous, or try to portray significant policy changes as technocratic, it becomes more difficult for the public to know what the executive branch is doing (and why), and therefore harder to hold the President to account for it.[120] Likewise, democratic accountability is undermined when presidents avoid talking about their comprehensive vision for addressing important problems, especially if agencies are (now secretly) working on policies to achieve it.

From an accountability perspective it is far better for the public to be aware of the President's big-picture plans on, say, immigration, climate change, public health, or trade policy than it is for citizens to be left to guess, or to piece together for themselves what that larger vision might be by tracking what various agencies are doing. That kind of sleuthing is beyond most people. Indeed, the potential loss of transparency we describe is likely to be asymmetric in its impact:

---

[118] *See* Kevin Bogardus, *EPA's Scramble Post-Supreme Court Climate Ruling*, E&E News: Greenwire (Aug. 30, 2022, 1:57 PM), https://www.eenews.net/articles/epas-scramble-post-supreme-court-climate-ruling; Kevin Bogardus & Sean Reilly, *Meet EPA's Team Behind the Next Power Plant Rule*, E&E News: Greenwire (July 13, 2022, 1:34 PM), https://www.eenews.net/articles/meet-epas-team-behind-the-next-power-plant-rule (describing EPA struggle to craft new rules "under the yoke of . . . *West Virginia*").

[119] Of course, as many other commentators have argued, the new and more aggressive MQD is very likely to have a substantial chilling effect on agency action, a result that we think is normatively problematic as well. *See, e.g.*, Deacon & Litman, *supra* note 17, at 55–58; Richardson, *supra* note 17, at 202–05. Our argument here is that changes in *messaging and presentation* are normatively problematic even independent of the effect on regulatory substance.

[120] *Cf.* Wendy Wagner, *The Science Charade in Toxic Risk Regulation*, 95 Colum. L. Rev. 1613, 1617 (1995) (arguing that the tendency of agencies to "camouflage[e] controversial policy decisions as science" is a way for agencies to "avoid accountability for the underlying policy decisions"); K. Sabeel Rahman, Note, *Envisioning the Regulatory State: Technocracy, Democracy, and Institutional Experimentation in the 2010 Financial Reform and Oil Spill Disputes*, 48 Harv. J. on Legis. 555, 571 (2011) (arguing that a "technocratic model of regulation . . . obscures the reality of the normative disputes underlying seemingly technical policy questions," and that the "avoidance of normative debate" tends to obscure "the underlying sources of political debate and disagreement. . . , leaving many affected social groups either unaware of the key implications of regulatory debates or unable to engage in the debate itself, which has retreated behind a smokescreen of technocratic discourse").

Sophisticated, well-organized interest groups are more likely than ordinary citizens to understand how a collection of seemingly technocratic, incremental regulations contribute to some larger policy agenda. In sum, to the extent the MQD affects the incentives of agencies and presidents in the way we describe above, we think the doctrine will lead to less transparency and hence be democracy-eroding.

C. EXACERBATING MINORITY OBSTRUCTIONISM

While every modern President has sometimes used regulation unilaterally to advance domestic policy priorities, presidents have done so more often since the mid-1990s as it has become substantially harder to enact significant legislation in Congress.[121] This increased difficulty of passing legislation is attributable to a variety of factors, including increasing political polarization,[122] legislative reforms that make it more difficult to "buy" compromise through earmarks or other targeted provisions,[123] changes in the media environment[124] and electoral incentives for legislators,[125] and perhaps most importantly, the transformation of the Senate filibuster from an occasionally

---

[121] *See* Kagan, *supra* note 86, at 2248, 2311; Michael Gentithes, *Gobbledygook: Political Questions, Manageability, & Partisan Gerrymanders*, 105 IOWA L. REV. 1081, 1118–19 (2020); Edward G. Carmines & Matthew Fowler, *The Temptation of Executive Authority: How Increased Polarization and the Decline in Legislative Capacity Have Contributed to the Expansion of Presidential Power*, 24 IND. J. GLOB. LEGAL STUD. 369, 379–84 (2017). To be clear, it has become *harder* for Congress to enact significant legislation, not impossible. The 117th Congress, for example, was quite productive. But this should not obscure the well-documented point that enacting legislation in the U.S. Congress, which has always been difficult, has become substantially more difficult over the last generation. And much of the major legislation enacted by recent Congresses, including the 117th Congress, has been enacted through the budget reconciliation process, which is limited to the raising and spending of revenue. It remains quite rare for Congress to substantively update old regulatory laws, which have been left to languish because of the factors we discuss in the text, especially hyper-polarization and the filibuster.

[122] *See* Carmines & Fowler, *supra* note 121, at 376–78.

[123] Congress imposed a moratorium on earmarks in 2011, which made passing major legislation harder. *See* ZACHARY COURSER & KEVIN R. KOSAR, AM. ENTER. INST., RESTORING THE POWER OF THE PURSE: EARMARKS AND RE-EMPOWERING LEGISLATORS TO DELIVER LOCAL BENEFITS 2, 11–12 (2021). Note, however, that Congress recently lifted the earmark ban. *See* Paul Kane et al., *Congress Revives Earmarks in Hopes of Bipartisan Deals on Infrastructure, Budgets*, WASH. POST (Mar. 21, 2021, 6:06 PM), https://www.washingtonpost.com/powerpost/earmarks-biden-infrastructure/2021/03/21/5a9d15ca-88c6-11eb-bfdf-4d36dab83a6d_story.html.

[124] *See* Matthew S. Levendusky, *Why Do Partisan Media Polarize Viewers?*, 67 AM. J. POL. SCI. 611, 611–12 (2013).

[125] *See* Nolan McCarty, *Polarization, Congressional Dysfunction, and Constitutional Change*, 50 IND. L. REV. 223, 232, 235 (2016).

employed device into a de facto sixty-vote requirement for almost all ordinary legislation.[126] Presidents and their allies would typically prefer to secure policy reforms through legislation, which is more durable and typically more far-reaching. But when Congress proves unwilling or unable to act, the White House will often encourage or direct agencies to find ways to use existing statutory provisions to promulgate rules that achieve at least some of the objectives of the stalled or rejected legislation.[127]

In *West Virginia*, the Court implied that this tactic is illegitimate, or at least suspect. More specifically, both the majority and concurring opinions suggested that adopting a regulation similar to legislation that Congress debated but failed to enact is one of the factors indicating that an agency's regulation implicates a "major" question, and therefore requires especially clear congressional authorization.[128] These statements seem to suggest that prior or ongoing congressional debates over bills that resemble the policy in the agency's rule are a kind of proxy for the rule's economic and political significance. But that possibility seems an unlikely explanation for why the Court is making so much of the fact that the agency's rule resembles (or can be characterized as resembling) legislative proposals that Congress failed to enact. As noted above, although there may be a correlation between the substantive importance of an issue and its discussion in Congress, that correlation is likely to be a weak one.[129] Congress debates and rejects all sorts of legislative proposals of modest importance, and Congress has also been known to ignore pressing matters altogether.

---

[126] *See, e.g.*, Adam Jentleson, Kill Switch: The Rise of the Modern Senate and the Decline of American Democracy (2021).

[127] *See* Carmines & Fowler, *supra* note 121, at 384–95.

[128] West Virginia v. Env't Prot. Agency, 142 S. Ct. 2587, 2610 (2022) (asserting that the CPP presents a major question in part because the EPA's "discovery" of the power to compel generation shifting "allowed it to adopt a regulatory program that Congress had conspicuously and repeatedly declined to enact itself"); *see also id.* at 2614 ("[W]e cannot ignore that the regulatory writ EPA newly uncovered conveniently enabled it to enact a program that, long after the dangers posed by greenhouse gas emissions had become well known, Congress had considered and rejected multiple times." (internal quotation marks and citations omitted)); *id.* at 2621 (Gorsuch, J., concurring) (stating that one of the main factors courts should look to in determining whether an agency's rule implicates a "major question" is whether "Congress has considered and rejected bills authorizing something akin to the agency's proposed course of action," on the grounds that this "may be a sign that an agency is attempting to work around the legislative process to resolve for itself a question of great political significance" (internal citations and quotation marks omitted)).

[129] *See supra* Section II.A.

Given that the Court can assess a rule's economic and social significance more directly, it is hard to believe that the Court thinks past congressional debate of an issue is relevant mainly as an indirect indicator of "majorness."

There are other reasons, however, why the Court might be skeptical about an agency's claimed authority to regulate when Congress has considered the issue but not acted. The Court may think, for example, that congressional inaction is a signal that existing statutes do not authorize the agency to adopt a similar policy, or that unilateral agency action would take the issue out of Congress's hands prematurely, ending a debate that ought to play out through the legislative process.[130] But both of these propositions are problematic. The former claim contravenes the Court's oft-repeated assertion that congressional failure to act is not usually probative as to the meaning of statutes enacted by a previous Congress.[131] As for the latter suggestion, agency action on an issue that Congress has considered, or is considering, does not take the issue away from Congress. It simply changes the status quo against which Congress may legislate. If sufficient congressional majorities disapprove of the agency's rule, they can override it legislatively.[132] Congress may also enact legislation that differs both from the old status quo and from whatever new policy the agency has adopted. Congressional debate and political contestation can go on, just as before.[133]

---

[130] Neither the majority nor the concurrence says this explicitly, though the concurrence hints at this idea. *See West Virginia*, 142 S. Ct. at 2620 (Gorsuch, J., concurring) (suggesting that the MQD applies when "an agency claims the power to . . . end an earnest and profound debate," and offering, as an illustration, OSHA's test-or-vaccinate rule, which purportedly displaced ongoing legislative debates over Covid-19 vaccine mandates). Some scholars have suggested that this might be the most plausible normative justification for the MQD. *See* Moncrieff, *supra* note 12.

[131] Indeed, in response to Justice Kagan's assertion that the Court is using congressional inaction in this improper fashion, *West Virginia*, 142 S. Ct. at 2641 (Kagan, J., dissenting), Justice Gorsuch insisted that the Court has "not pointed to failed legislation to resolve what a duly enacted statutory text means, only to help resolve the antecedent question whether the agency's challenged action implicates a major question." *Id.* at 2623 n.5 (Gorsuch, J., concurring).

[132] *See supra* text accompanying note 113.

[133] That said, some cases that invoke the MQD involve federal regulations that preempt state law, and such federal regulations, if sustained, would indeed cut off deliberation in state legislatures. *See* Gonzales v. Oregon, 546 U.S. 243 (2006). But the possibility that federal regulation may resolve issues being debated in the states is an established feature of our constitutional system. Furthermore, other canons of interpretation—which Justice Gorsuch's concurrence confusingly conflated with the MQD, *see West Virginia*, 142 S. Ct. at 2621, instruct courts to avoid reading federal statutes to displace traditional state authority, and so it is not at all clear why one would need the MQD in addition to these other canons. And of

Because these explanations for the Court's concern with agencies promulgating rules that address issues that Congress has failed to address seem so weak on their face, we think the *West Virginia* majority and concurrence are really getting at another worry: that agencies (and the White House) are improperly circumventing the legislative process. Though that concern would not be logically limited to major questions, we suspect that the Court believes that the *consequences* of such workarounds are greater—and worse—when the issue in question is of great significance, perhaps due to the belief that on major issues, working through the grueling gauntlet that is the federal legislative process is especially important. Justice Gorsuch's concurring opinion in *West Virginia* and his dissenting opinion in *Gundy* put this concern at the forefront,[134] and we think it is the best way to understand what the *West Virginia* majority is anxious about as well. The thinking here, we submit, is not really that congressional failure to enact a similar policy *demonstrates* that an issue is "major"; rather, the idea seems to be that *when* an issue is major (as determined by other criteria), it is *especially inappropriate* for agencies to work around the legislative process by purporting to find authority in older statutes. The MQD, on this view, helps to ensure that the U.S. lawmaking process is working as it should, and as the Framers intended.

That position is coherent but, we believe, incorrect. Our disagreement with the Court turns on our different sense of the more serious problem besetting the U.S. lawmaking system today. The U.S. constitutional system, like that of every other constitutional democracy of which we are aware, seeks to balance a general commitment to majoritarian democracy against the need to safeguard the interests of political minorities. The goal is to avoid a tyranny of the majority without enabling excessive minoritarian obstruction. Relatedly, the design of lawmaking procedures inevitably entails an attempt to balance the interest in preventing the enactment of bad laws—an interest furthered, for example, through mechanisms that add procedural hurdles, disperse authority, and impose implicit or explicit

course even if one is open to the (contestable) view that it might be a problem, from a democratic accountability perspective, for a federal agency to interpret its existing statutory authority in a way that cuts off debate on some issue in state legislatures, this has nothing to do with the impact of a federal regulation on the prerogatives of *Congress*, which remains free to override any federal agency action if it so chooses.

[134] *See West Virginia*, 142 S. Ct. at 2617-18, 2624–25 (Gorsuch, J., concurring); Gundy v. United States, 139 S. Ct. 2116, 2134–35, 2141–42 (2019) (Gorsuch, J., dissenting).

super-majority requirements—against the interest in facilitating the passage of good laws.[135] The *West Virginia* Court—and even more clearly Justice Gorsuch's concurrence—seem to believe that the more significant dangers facing the United States are the tyranny of the majority and the prospect that Congress will enact an excess of bad laws, thereby threatening "liberty."[136]

In our view, the far larger problem today is minoritarian obstructionism run amok, which—especially when coupled with extreme political polarization—has produced congressional paralysis and dysfunction.[137] Stated that way, it becomes clear that different assessments of the MQD may turn on questions of political philosophy. To those who subscribe to a particular form of constitutional libertarianism—in which federal government action should be extremely rare and occur only when supported by supermajorities—the anti-delegation MQD may be attractive. But it is important to underscore that such a justification for the MQD would not be grounded in *democratic* arguments, but rather in libertarian arguments for making government action—even action supported by majorities of citizens and representatives—more difficult.

Now, defenders of the anti-delegation MQD might respond (correctly) that the United States is not a pure majoritarian democracy, and then assert that the MQD preserves and supports the checks and balances that the Constitution's Framers intended. But that latter claim is not correct. Indeed, because it is so common for those defending doctrines like the anti-delegation MQD to invoke the Framers' interest in making federal lawmaking difficult by building in sufficient checks and balances to constrain pure majoritarianism, it is worth emphasizing that the type and extent of minority obstructionism that currently plagues federal lawmaking is not at all what the

---

[135] *Cf.* THE FEDERALIST No. 73 (Alexander Hamilton) (observing that "the power of preventing bad laws includes that of preventing good ones").

[136] *See West Virginia*, 142 S. Ct. at 2616; *id.* at 2618 (Gorsuch, J., concurring); *cf. also* THE FEDERALIST No. 73 (Alexander Hamilton) (stating, in the context of defending the President's veto power, that "[t]he injury which may possibly be done by defeating a few good laws, will be amply compensated by the advantage of preventing a number of bad ones").

[137] *See, e.g.*, JENTLESON, *supra* note 126; STEVEN S. SMITH, THE SENATE SYNDROME: THE EVOLUTION OF PROCEDURAL WARFARE IN THE MODERN U.S. SENATE 2–3 (2014); Sarah Binder, *The Dysfunctional Congress*, 18 ANN. REV. POL. SCI. 85, 93, 96–98 (2015) (providing evidence of congressional dysfunction and its roots in minority obstructionism and partisan polarization); *see also* Jane Mansbridge, *On the Importance of Getting Things Done*, 45 P.S.: POL. SCI. & POL. 1 (2012).

Framers envisioned. The Senate filibuster, which is probably the institutional mechanism most responsible for empowering minorities to obstruct broadly popular legislation, was created largely by accident and clearly contravenes the Framers' intent that both chambers of Congress would operate under the principle of majority rule.[138] The Framers also did not anticipate the emergence of organized political parties,[139] much less the extreme partisan polarization that currently exists, nor did they anticipate the extent and sophistication of modern partisan gerrymandering, which can prevent even large swings in underlying voter choices from being reflected in similar shifts in partisan control of House seats.[140]

Furthermore, even if we put aside the specifics of the institutional and political features of modern U.S. politics that the Framers did not foresee, we think that, notwithstanding cherry-picked quotations from *The Federalist Papers*,[141] the degree of minoritarian obstruction and legislative paralysis that pervades the current federal policymaking process is not consistent with the balance that the Framers sought to strike, or with a rational and effective system. In light of the overwhelming evidence about how our political system currently operates, it is hard to treat as credible the position that the modern Congress passes too many laws on too many important subjects, and that political minorities have too few tools at their disposal to prevent legislation that would threaten their interests. To the contrary, under current conditions, even when substantial majorities of voters and legislators favor a given policy change, and do so consistently over multiple election cycles, Congress often fails to act.[142] That is a problem

---

[138] *See* Emmet J. Bondurant, *The Senate Filibuster: The Politics of Obstruction*, 48 HARV. J. ON LEGIS. 467, 470–73 (2011); Catherine Fisk & Erwin Chemerinsky, *The Filibuster*, 49 STAN. L. REV. 181, 190–94 (1997).

[139] *See* Daryl J. Levinson & Richard H. Pildes, *Separation of Parties, Not Powers*, 119 HARV. L. REV. 2311, 2316–25 (2006).

[140] *See* Nicholas O. Stephanopoulos, *The Causes and Consequences of Gerrymandering*, 59 WM. & MARY L. REV. 2115, 2147 (2018); Christopher T. Kenny et al., *Widespread Partisan Gerrymandering Mostly Cancels Nationally, but Reduces Electoral Competition* (2022) (unpublished manuscript).

[141] *See supra* note 136.

[142] For example, large majorities of Americans support creating a pathway to lawful status for undocumented immigrants brought to the United States as children, *see, e.g.*, Jens Manuel Krogstad, *Americans Broadly Support Legal Status for Immigrants Brought to the U.S. Illegally as Children*, PEW RSCH. CTR. (June 17, 2020), https://www.pewresearch.org/fact-tank/2020/06/17/americans-broadly-support-legal-status-for-immigrants-brought-to-the-u-s-illegally-as-children, authorizing Medicare to negotiate prescription drug prices, *see, e.g.*, Carl Bialik &

for American democracy, even if one believes—as we do—that some limits on pure majoritarianism are necessary and appropriate.

We acknowledge that unilateral executive action also has its problems. Thanks to the electoral college, a candidate can win the presidency without winning the popular vote, and presidential policy preferences may be more extreme than those of the median voter, or even than the President's median supporter.[143] And even if current congressional procedures offer too many opportunities for opponents to delay or block actions that threaten their interests, executive branch rulemaking procedures may offer too few such opportunities. Yet on balance, we think that depriving the White House of the ability to proceed through agency rulemaking to address major issues that Congress has failed to address—in cases where there is plausible statutory authority for the agency's action—would be, on balance, bad for democratic governance, for three reasons.

The first and most straightforward reason builds on the observations above about excessive minoritarian obstructionism in Congress. Notwithstanding the problems with treating the President as

---

Taylor Orth, YouGovAmerica (July 14, 2022, 3:28 PM), https://today.yougov.com/topics/health/articles-reports/2022/07/14/americans-support-gov-negotiation-drug-prices, and requiring background checks for gun buyers, *see, e.g.*, Monique Beals, *88 Percent in New Poll Support Background Checks on All Gun Sales*, Hill (May 26, 2022, 7:44 AM), https://thehill.com/homenews/state-watch/3502285-88-percent-in-new-poll-support-background-checks-on-all-gun-sales. These majorities have existed for at least a decade. *See, e.g.*, *Section 3: Opinions About Immigration*, Pew Rsch. Ctr. (June 24, 2010), https://www.pewresearch.org/politics/2010/06/24/section-3-opinions-about-immigration; Jonathan Voss, *Medicare and Social Security in the 2012 Elections: Findings from Survey of 1200 Registered Likely 2012 Voters*, Nat'l Comm. to Pres. Soc. Sec. & Medicare 17 (Nov. 15, 2012), https://www.ncpssm.org/wp-content/uploads/2017/07/post-election-polls.pdf; Steve Gorman, *Gun Owners Support Wider Background Checks: Poll*, Reuters (Jan. 18, 2011, 11:11 PM), https://www.reuters.com/article/us-usa-shooting-survey/gun-owners-support-wider-background-checks-poll-idUSTRE70I0QO20110119. Yet Congress has not turned any of these preferences into law. *See also* Jonathan S. Gould et al., *Democratizing the Senate from Within*, 13 J. Legal Analysis 502, 517–18 (2021) (providing examples of broadly popular legislation with majority support in Congress that failed to pass due to the filibuster).

[143] *See* Matthew C. Stephenson, *Optimal Political Control of the Bureaucracy*, 107 Mich. L. Rev. 53, 55 (2006); John O. McGinnis & Michael B. Rappaport, *Presidential Polarization*, 83 Ohio St. L. J. 5, 19–20 (2022); Michael A. Bailey, *Comparable Preference Estimates Across Time and Institutions for the Court, Congress, and Presidency*, 51 Am. J. Pol. Sci. 433, 444 fig.7 (2007) (finding that in every year in dataset spanning 1951–2002, the President has been more ideologically extreme than the median senator, median member of the House, and median Supreme Court Justice); Larry M. Bartels, *Failure to Converge: Presidential Candidates, Core Partisans, and the Missing Middle in American Electoral Politics*, 667 Ann. Am. Acad. Pol & Soc. Sci. 143, 155–58, 160–61 (2016) (finding that presidential candidates are typically as or even more extreme than their base).

an ideal avatar of the "will of the people,"[144] it remains that case that the President's electoral accountability to a national constituency—and the desire of the President's party to retain control of the White House even when the sitting President is in her second term—gives presidentially-directed administrative initiatives a degree of democratic legitimacy.[145] Moreover, because voters, rightly or wrongly, tend to attribute the federal government's action (and inaction) to the sitting President,[146] it is sensible to give the executive branch a fair degree of latitude in setting policy through administrative action.

To be clear, we are not arguing that the President should have unfettered control over the administrative state. Indeed, we think that some constraints—including procedural safeguards, oversight by Congress and the courts, and an "internal separation of powers"[147]—all perform a salutary function.[148] Nor do we mean to suggest that agencies ought to be able to act in excess of clearly delineated statutory authority. But the anti-delegation version of the MQD applies not only in cases where existing statutory authority *cannot* be read to authorize the agency's action, or only in cases where existing statutory authority is *more plausibly* read as not authorizing the agency's action; this version of the MQD prohibits the agency from addressing a major question even if existing statutory authority is more plausibly read to *authorize* the agency's action, if the statute does not do so with sufficient clarity.

The issue, then, is not whether an agency may act without a plausible claim to statutory authority. It may not. The issue is whether agency action on a major question requires an unusually clear statement from Congress—that is, whether the courts, through a canon of

---

[144] *See* Stephenson, *supra* note 143; McGinnis & Rappaport, *supra* note 143.

[145] *See* Kagan, *supra* note 86, at 2331–39.

[146] *See id.* at 2310; Brad T. Gomez & J. Matthew Wilson, *Political Sophistication and Attributions of Blame in the Wake of Hurricane Katrina*, 38 PUBLIUS 633, 637 (2008).

[147] Neal K. Katyal, *Internal Separation of Powers: Checking Today's Most Dangerous Branch From Within*, 115 YALE L.J. 2314 (2006); *see also* Matthew C. Stephenson, *The Qualities of Public Servants Determine the Quality of Public Service*, 2019 MICH. ST. L. REV. 1177, 1184–89.

[148] *See* Jody Freeman & Sharon Jacobs, *Structural Deregulation*, 135 HARV. L. REV 585 (2021) (describing ways in which presidents can use unilateral executive power to undermine democratic accountability, and advocating various constraints on presidential power); Stephenson, *supra* note 143 (arguing that a degree of bureaucratic insulation from presidential control can improve the degree to which agency policy reflects majority preferences); Stephenson, *supra* note 147, at 1184–88 (suggesting that the semi-autonomous agency officials and bureaucrats can provide a salutary check on presidential control of the administrative state).

statutory construction, ought to put a heavy thumb on the scale *against* finding that agencies may address issues of significant public concern, even if a broad statute would seem to authorize this. If one concludes, as we do, that on balance minoritarian obstructionism within Congress poses a greater threat to effective democratic governance than do executive branch efforts to achieve policy reforms through regulation, then creating such a special canon is unwise and undemocratic.

A second reason to object to the Court's suggestion that the MQD is more likely to apply when Congress has considered but failed to enact legislative proposals is that this approach will create a disincentive for presidents to seek legislative solutions, even when they think legislation would be preferable to agency action. Let us stipulate that, all else equal, it would be better—both as a matter of democratic legitimacy and as a matter of public policy—if major public problems were addressed through legislation than through unilateral agency action. All else equal, the President would usually (though perhaps not always) prefer legislation as well. The President may be able to accomplish more of her policy agenda with a statute, and a statute is more likely to remain in place if partisan control of the White House flips to the other party.

But if, as *West Virginia* suggests, a failed legislative effort increases the chances that a court will treat a subsequent agency rule as implicating the MQD, then the President may think twice before pursuing the legislative route first. After all, if the Court really meant what it said—that the MQD is more likely to apply if Congress had considered but rejected a legislative solution—then the administration may be on a stronger legal footing in defending an ambitious new regulation if Congress has never taken up the issue. The President and her advisors might conclude that the administration would be better off proceeding immediately to unilateral executive branch rulemaking. After all, the odds of successfully securing new legislation are low even in the best of circumstances, which means that the administration might rationally decide that it is not prudent to sacrifice even a marginally helpful litigating position in the longshot hope of securing congressional legislation.[149]

---

[149] To be sure, there are political reasons that the President might be inclined to seek a legislative solution first, even if congressional action is unlikely and even if doing so might make subsequent unilateral action more legally vulnerable. But at the margin, the emergence of the more aggressive MQD might shift the White House's calculations.

Third, the odds of Congress negotiating a legislative solution may sometimes be higher if the executive branch retains the option of proceeding unilaterally if legislative negotiations fail. This possibility may initially seem counterintuitive. Indeed, some commentators have argued that presidents have less incentive to seek legislative solutions if they know the executive branch can act unilaterally.[150] But as noted earlier, the President often has strong incentives to pursue her policy agenda through legislation even when it is possible to act unilaterally, both because the President may get more of what she wants from a new statute than from existing statutes (even if she has to compromise to some degree on the content of the new statute), and because statutes are more durable. Often, we suspect, the main obstacle to a negotiated legislative solution is not that the President has too little incentive to bargain in good faith, but rather that congressional opponents, even if in the minority, have too little incentive to bargain in good faith. When that is the case, the possibility (and implicit threat) of unilateral presidential action may increase the chances that Congress will negotiate and pass a legislative solution.

To see the logic of the argument, consider a situation in which the President and her congressional allies would prefer to negotiate a legislative solution to a high-priority agenda item, even if doing so involves making some policy concessions, but the White House would prefer unilateral executive action to no action at all. (So, the preferences of this bloc are *Legislative Compromise* > *Unilateral Executive Action* > *No Action*.) Assume further that a critical mass of Members of Congress—a minority, but enough to block new legislation—would like to retain the status quo if possible, but would prefer negotiating a compromise legislative solution, in which they might be able to extract some concessions, to unilateral administrative action. (So, the preferences of this bloc are *No Action* > *Legislative Compromise* > *Unilateral Executive Action*.) In this scenario, the *possibility* of unilateral executive action may result in a legislative compromise that would not otherwise occur.

The bloc in Congress that prefers inaction has no incentive to negotiate with the President and her congressional allies if the only way for the administration to address the problem at issue is through legislation. The minority can simply obstruct. But if members of this bloc are aware that intransigence will eventually provoke the administration

---

[150] *See., e.g.*, William P. Marshall, *Actually We Should Wait: Evaluating the Obama Administration's Commitment to Unilateral Executive-Branch Action*, 2014 Utah L. Rev. 773, 789.

to act unilaterally, then they have a reason to come to the table and try to hammer out a deal. If one believes, as we do, that the ability of minority factions to obstruct popular legislative initiatives is excessive, and that it would often be better if opponents worked with proponents through the legislative process to craft negotiated solutions, then holding open the possibility of unilateral executive action is a good thing. Taking that possibility away, for example through an aggressive version of the MQD, would undermine the legislative process that the MQD's proponents purport to extol.

To be clear, we recognize that the circumstances we sketch above will not always obtain. After all, the background threat of unilateral action has often failed to produce legislative compromises, including with respect to greenhouse gas regulations of the sort at issue in *West Virginia*. We are certainly not arguing that the possibility of unilateral executive action will usually induce the parties to come to the table and work out a bargain that can secure a House majority and sixty votes in the Senate. Our point is that taking away the possibility of unilateral executive action may well make negotiated solutions even *less* likely.

Indeed, the *West Virginia* opinion may have sent exactly the wrong signal to legislators who at some future date may find themselves considering whether to try to work with the President and their colleagues across the aisle on a bipartisan bill to address some issue on the President's agenda. If *West Virginia* had come out the other way—upholding rather than rejecting the CPP—it would have served as a cautionary tale: legislators who refused to come to agreement on a climate bill during the Obama Administration would be seen, in retrospect, to have made a strategic mistake. The lesson of this alternative history would have been that coming to the table and wringing some policy concessions from the majority in a compromise bill is often wiser than remaining intransigent and eventually provoking the executive to take unilateral action that disregards your interests. The lesson from the actual *West Virginia* case is the opposite: obstruction pays off—and negotiation is for suckers—because at the end of the day there is nothing that the administration can do, even if it has a majority of the people and a majority of Members of Congress on its side.

Furthermore, as we noted earlier, unilateral executive action does not take an issue away from Congress; it simply alters the regulatory status quo against which Congress operates.[151] The administration's

---

[151] *See supra* text accompanying notes 113, 132–133.

ability to alter that regulatory status quo unilaterally can sometimes spur legislative action that might not otherwise take place, by shifting the burden of inertia and inaction. So even if the *threat* of unilateral executive action does not prompt political minorities from refusing to bargain in good faith over a comprehensive legislative solution, the *reality* of unilateral executive action may do so. We emphasize again that we are must assuredly *not* arguing for unconstrained presidential authority to do whatever the administration likes. Checks and balances are essential, and any valid agency rulemaking activity must be authorized by statute. The question at issue is whether an aggressive, anti-delegation version of the MQD improves the legislative process—by ensuring that all "major" agency actions are grounded in unusually clear statutory language—or worsens that process by strengthening the hand of already-too-strong minority blocs that oppose any effort to address pressing issues through new legislation. We submit that the latter is more plausible.

For the reasons sketched above, the aggressive anti-delegation version of the MQD that *West Virginia* appears to have adopted will likely exacerbate the worst tendencies of the federal lawmaking process. Not only does it choke off an important safety valve that reduces the undemocratic consequences of minoritarian obstructionism, but it may also make negotiated legislative solutions—which MQD proponents claim to favor—less likely. This is hardly a win for democracy.

III. Conclusion

The *West Virginia* Court appears to have embraced a strong, anti-delegation version of the MQD. Although the Court asserted that this MQD is grounded in longstanding precedent, that is not the case. This anti-delegation MQD is a recent invention, the product of a kind of "doctrine creep," whereby the same label (originally developed by scholars to describe an inchoate or unspoken theme in certain judicial opinions, and then picked up by the courts) is used to describe related but distinct ideas. On closer examination, the underlying justifications for moderate versions of the MQD do not simply carry over into this novel anti-delegation MQD.

Of course, the fact that the new MQD is the product of doctrine creep is not in and of itself a reason to oppose it. Doctrines evolve in unexpected directions all the time, often for the better. But in this case, the aggressive anti-delegation MQD that the *West Virginia* concurrence advocated, and that the *West Virginia* majority appeared to embrace, is likely to have bad consequences. As other critics have pointed out, this anti-delegation MQD, if deployed broadly, would prevent agencies from addressing unforeseen and important problems, even when existing statutes appear designed to give agencies the authority to do precisely that. Additionally, as we have emphasized in this Article, the anti-delegation MQD is far more likely to weaken democratic accountability than strengthen it, by shifting considerable discretionary power to the judiciary, undermining executive branch transparency, and exacerbating minority obstructionism in Congress.

The current Court may dismiss these concerns as unconvincing and unimportant. Nevertheless, we suspect that the Court may come to have second thoughts about the most expansive understanding of the new MQD. The subjectivity and manipulability of this doctrine is likely to create considerable legal uncertainty and confusion in the lower courts. Litigants opposed to agency action are likely to invoke the MQD profligately, against even routine agency rules, orders, and enforcement actions, in the hope that the reviewing court will be persuaded to deploy it. Indeed, this is already starting to occur.[152] And given the subjectivity of the doctrine and the potential for forum shopping by litigants, we suspect that the MQD will lead to what at least some of the Justices in the *West Virginia* majority will view as overreaching by lower courts. We therefore expect that the Supreme Court may need to intervene again, probably sooner rather than later, to clarify the scope and rationale of the MQD.

If the Court does so, the issue that we presented as a problem in Part I of this Article—that the Court tends to conflate distinct versions of the MQD, and often uses language that would be compatible

---

[152] *See e.g.*, Louisiana v. Becerra, No. 3:21-CV-04370, 2022 WL 4370448, at *11 (W.D. La. Sept. 21, 2022) (holding that the under the MQD the Department of Health and Human Services lacked statutory authority to impose masking rules for Head Start program); Brief for Petitioners at 3–4, 50–59, Concerned Household Elec. Consumers v. Env't Prot. Agency, Nos. 22-1139, 22-1140, 2022 WL 9333252 (D.C. Cir. Oct. 14, 2022) (challenging EPA's denial of motion to reconsider EPA's earlier finding, upheld in *Massachusetts v. EPA*, 549 U.S. 497 (2007), that greenhouse gases endanger human health and the environment within the meaning of the CAA, on the grounds that under the MQD as explained in *West Virginia*, *Massachusetts* would have come out differently); Motion to Dismiss Counts One and Three, United States v. DiMezzo, No. 1:21-cr-41-JL-01, 2022 WL 3147642 (D.N.H. July 29, 2022) (defendant charged with violating Treasury Department rule requiring registration of cryptocurrency business moves to dismiss on the ground that whether to regulate cryptocurrency is a major question); Complaint, Nebraska v. Biden, No. 22-cv-01040, 2022 WL 4594457 (E.D. Mo. Sept. 29, 2022) (challenging Biden Administration student loan relief plan on MQD grounds).

with more than one version—could prove to be an advantage. While we agree with those commentators who have suggested that the *West Virginia* Court is best understood as having embraced the anti-delegation version of the MQD, the majority opinion is sufficiently ambiguous that it would be possible for the Court to walk back that position while claiming fidelity to the precedent. The Court should seize that opportunity when it arises—for the sake of legal stability, but more importantly for the sake of democracy.

DARRELL A. H. MILLER AND
JOSEPH BLOCHER

MANUFACTURING OUTLIERS

Last term, the Supreme Court issued its first major Second Amendment decision in more than a decade, *New York State Rifle and Pistol Association, Inc. v. Bruen*.[1] The case concerned a challenge to New York's century-old "may-issue" regulation, which required applicants to show "proper cause" to receive a license to carry a concealed handgun in public. Petitioners described New York's may-issue law as an outlier compared to forty-three other states with more relaxed "shall-issue" or permitless carry laws. At oral argument, Paul Clement, representing the petitioners, framed the case as a simple request: "[W]e'd like what they're having."[2]

In a 6-3 majority opinion striking down New York's law, Justice Clarence Thomas embraced petitioners' characterization of the regulation as a contemporary outlier—and went further, casting it as a historical outlier as well. New York and supporting amici had amassed a

Darrell A. H. Miller is the Melvin G. Shimm Professor of Law at Duke University Law School. Joseph Blocher is the Lanty L. Smith '67 Professor of Law at Duke University Law School.

Authors' note: Thanks to Jake Charles, Eric Ruben, Larry Solum, and Andrew Willinger for comments, and to Noah Jacobs, Luisa Gomez, and Benjamin Rossi for their research assistance. Thanks also to the participants at the University of Colorado faculty workshop, the University of Chicago Law School public law workshop, the University of Illinois faculty workshop, and the University of Houston Law Center colloquium for their helpful questions and contributions.

[1] 142 S. Ct. 2111 (2022).

[2] Transcript of Oral Argument at 50, N.Y. State Rifle & Pistol Ass'n, Inc. v. Bruen, 142 S. Ct. 2111 (2021) (No. 20-843) (comments of Paul Clement).

The Supreme Court Review, 2023.
© 2023 The University of Chicago. All rights reserved. Published by The University of Chicago Press. https://doi.org/10.1086/725159

substantial record of Anglo-American regulations from the medieval period to the early twentieth century to show New York's law was part of a deep and long tradition of public carry regulation. One by one, the majority characterized each of these historical regulations as outliers. Some were too new; some were too old. Some were outliers because they were passed by territorial governments; some were outliers because they were passed by Reconstruction governments. Some were outliers because they weren't adequately enforced; some because they weren't enforced criminally; some were outliers because they governed a population too small, or too regional. None of these regulations, according to the majority, were probative of a longstanding historical tradition of regulating public carry into which the New York law fit.

Bruen's description of New York's law as a contemporary and historical outlier raises fundamental questions about how the Justices define that term and the jurisprudential significance of that characterization. After all, there was nothing atypical about New York's licensing law even thirty years ago, when most states had may-issue permitting, or prohibited concealed carry altogether. Instead, the law became a minority position because of an aggressive and successful campaign to change state law and public norms regarding public carry, and to change the perception of American history pertaining to that practice.[3] Further, to suggest that New York's permitting law is an outlier compared to a longstanding tradition of public carry regulation depends entirely on what counts as constitutive of that tradition.

Of course, enforcing constitutional rights against outlier jurisdictions is not unusual. In contexts as varied as school segregation, contraception, the death penalty, and LGBTQ+ equality, the Justices have effectively nationalized a single standard as constitutional law when state laws and norms come to reflect a putative consensus, curbing jurisdictions that appear to break out in an atypical or retrogressive fashion.[4] Scholars across the ideological spectrum, including Justin Driver,[5]

---

[3] *See infra* notes 100–116 and accompanying text.

[4] Justin Driver, *Constitutional Outliers*, 81 U. Chi. L. Rev. 929 (2014).

[5] *Id.* Professor Driver's taxonomy of "constitutional outliers" is an illuminating contribution and does an excellent job of categorizing the kind of outlier regulations the Court tends to throttle. Our focus here is on the antecedent question of how a law becomes coded as an outlier in the first place, the inherent normativity and judicial discretion involved in that process, and the potential hazards that accompany the "outlier" framing as a reason for decision.

Keith Whittington,[6] Michael Klarman,[7] Cass Sunstein,[8] Brannon Denning, and Glenn Reynolds,[9] have all written on the tendency of the Court to use constitutional rights to prune outlier regulations.

The Court's invocation of outliers in *Bruen* warrants renewed attention to the topic for several reasons. First, *Bruen* highlights the current Court's inclination to cast its work in exclusively investigative, empirical terms—as a matter of finding something, rather than fashioning it. This maneuver masks assumptions that are essential to the decision and allows the Court to characterize its holding as the dispassionate evaluation of objective facts, uncontaminated by normative choice. Second, given the present majority's methodological tendencies, overt or covert conceptions of "outliers" are likely to appear in other constitutional domains. Many forms of originalism treat history as a matter of fact which judges can discover.[10] This historical-empiricism is frequently touted to distinguish originalist jurisprudence from other approaches that are purportedly more judge-empowering. As Justice Antonin Scalia put it, "[t]exts and traditions are facts to study, not convictions to demonstrate about."[11] "Outlier" arguments are exemplary of this framing, and of its weaknesses. *Bruen*

---

[6] Keith E. Whittington, Political Foundations of Judicial Supremacy 105 (2007).

[7] Michael J. Klarman, From Jim Crow to Civil Rights 453 (2004) ("More constitutional law than is commonly supposed reflects this tendency to constitutionalize consensus and suppress outliers.").

[8] Cass R. Sunstein, *Second Amendment Minimalism:* Heller *as* Griswold, 122 Harv. L. Rev. 246, 260–64 (2008).

[9] Brannon P. Denning & Glenn H. Reynolds, *Five Takes on* McDonald v. Chicago, 26 J.L. & Pol. 273, 284 (2011) (citing McDonald v. City of Chicago, 561 U.S. 742 (2010), as "another data point in favor of the theory that the Court is a lagging, rather than a leading indicator of popular preference, often constitutionalizing a national consensus and enforcing that consensus against outliers").

[10] Richard H. Fallon, Jr., *Are Originalist Constitutional Theories Principled, or Are They Rationalizations for Conservatism?*, 34 Harv. J.L. & Pub. Pol'y 5, 8 (2011) ("Proponents of originalism agree that historical facts at the time of a constitutional provision's adoption normally determine its meaning."); Mark A. Graber, *Original Expectations*, 52 Conn. L. Rev. 1573, 1579 (2021) ("Both original public meaning and original intentions/expectations purport to be facts about constitutional politics at the time the constitution was ratified."); Lawrence B. Solum, *The Fixation Thesis: The Role of Historical Fact in Original Meaning*, 91 Notre Dame L. Rev. 1 (2015). Will Baude and Steve Sachs maintain that the investigation is not into historical fact generally, but into that part of historical fact that counts as "original law" which judges then are bound to enforce as law in the present until the original law is lawfully changed. William Baude & Stephen E. Sachs, *Originalism and the Law of the Past*, 37 L. & Hist. Rev. 809, 809 (2019).

[11] Planned Parenthood of Se. Pa. v. Casey, 505 U.S. 833, 1000 (1992) (Scalia, J., concurring in the judgment in part and dissenting in part).

reveals that such arguments are replete with discretion, whether in selecting the relevant jurisdictions, grouping social phenomena (including law), setting a temporal duration, or elevating or minimizing a set of historical data points. *Bruen* demonstrates the Court's tendency to curate a historical record and then to treat it as an objective basis for decision.

Part I of this Article analyzes *Bruen* with particular focus on how the majority describes contemporary or historical regulations as "outliers." The majority asserts that its test, which supposedly considers only text, history and tradition, is more "administrable" than the tiers-of-scrutiny test it replaced.[12] But *Bruen* both engages in and encourages forms of judicial intuitionism and discretion that, if left unguided, are likely to disrupt Second Amendment law significantly.

Sections II.A and B then identify and elaborate two principles—neither of them evident in *Bruen*—that might help bring more discipline to outlier arguments. The first is transparency: Unlike the *Bruen* majority, courts should be clear about their assumptions and how they shape the baselines against which purported outliers will be measured. The second is rigor: Although law is not statistics and may not lend itself to the same kinds of coding and robustness checks one sees in other disciplines,[13] characterizing something as an outlier, as *Bruen* does, makes an empirical claim that must be subject to some kind of verification.[14] We do not advance a trans-substantive legal rule by which one can identify outliers, but we are confident that *Bruen*'s approach is plainly inadequate, and that, at the very least, some general guiding principles are warranted if outlier analysis is to be anything beyond a rhetorical device. Section II.C briefly explains why an outlier analysis modeled on *Bruen* poses a threat to history and tradition as a modality of constitutional argument. In Part III we conclude by recognizing that our critique of *Bruen*'s

---

[12] N.Y. State Rifle & Pistol Ass'n, Inc. v. Bruen, 142 S. Ct. 2111, 2130 (2022).

[13] *Cf.* Alexander M. Bickel, The Supreme Court and the Idea of Progress 97 (1978) ("Even when the law pretends to be a science, it is not, after all, mathematics."); Steven J. Burton, *Judge Posner's Jurisprudence of Skepticism*, 87 Mich. L. Rev. 710, 713 (1988) ("Law is not a science to be understood or criticized in scientific terms.").

[14] We agree that quantitative and qualitative analysis can both be empirical and subject to a set of best practices. *See* Lee Epstein & Andrew D. Martin, An Introduction to Empirical Legal Research (2014) (arguing that "data can be words, not numbers" that can "be organized into categories" from which to "identify patterns").

outlier argument is not confined to Second Amendment adjudication but is applicable to other areas of constitutional doctrine.

I. BRUEN

Gun rights advocates placed all their bets on *Bruen* and hit the jackpot. They wanted a declaration that the right to "bear arms" meant a right to carry guns in public. They got it. They wanted a declaration that New York's "may issue" discretionary licensing law was unconstitutional. They got it. And the big prize: they wanted a declaration that judges must abandon conventional tiers of scrutiny and analyze Second Amendment cases solely through text, history, and tradition. It appears they got that, too.

A. THE DECISION

Bruen was a case nearly fifteen years in the making. In 2008, the Court decided *District of Columbia v. Heller*,[15] the first Supreme Court case to hold that the Second Amendment guarantees a right to keep and bear arms for personal purposes like self-defense in the home. It was shortly followed by *McDonald v. City of Chicago*,[16] which, by incorporating the Second Amendment through the Due Process Clause, vastly expanded the reach of the Second Amendment to cover every state and local jurisdiction in the nation. Although *Heller* and *McDonald* were undeniably watersheds in Second Amendment doctrine, they were Delphic in their guidance to lower courts.

Lower court judges scrambled to put together a doctrine that could implement this new right.[17] They quickly converged on a two-step framework.[18] At the first step, the lower courts used an approach—strongly influenced by history and *Heller*'s categorical distinctions[19]—to decide whether the Second Amendment covered the challenger's activity. If it did not, then that was the end of the inquiry. Felons caught

[15] 554 U.S. 570 (2008).

[16] 561 U.S. 742 (2010).

[17] *See* RICHARD H. FALLON, JR., IMPLEMENTING THE CONSTITUTION (2001).

[18] *See* JOSEPH BLOCHER & DARRELL A.H. MILLER, THE POSITIVE SECOND AMENDMENT: RIGHTS, REGULATION, AND THE FUTURE OF *HELLER* 100–18 (2018) (describing structure and development of the framework).

[19] Joseph Blocher, *Categoricalism and Balancing in First and Second Amendment Analysis*, 84 N.Y.U. L. REV. 375, 405 (2009) (describing *Heller*'s categories).

with firearms,[20] or challenges to guns in "sensitive places" like commercial airliners,[21] usually failed at this step.

If the history was unclear, or the category underspecified, or the judge was cautious, courts would address step two, and apply a conventional means-end analysis calibrated by how close the regulation came to the "core" of the right. In doing so, courts would evaluate the stated government interest (typically described as public safety) and examine how closely the regulation fit with this stated goal compared to the expected burden on otherwise core Second Amendment conduct. Often, but not always, this analysis took the form of intermediate scrutiny. For example, the Third Circuit found firearms with obliterated serial numbers potentially within the coverage of the Second Amendment, because *Heller* did not categorically exclude them and because, unsurprisingly, there was no clear historical record of prohibiting ownership of unserialized firearms. Still, the prohibition survived, because preventing use of untraceable firearms in crime outweighs the minimal impact on a law-abiding person's ability to use other weapons—in particular firearms with serial numbers—for the core purpose of self-defense in the home.[22]

By the time the Court granted certiorari in *Bruen*, every federal circuit to directly address the issue had adopted this two-part framework.[23] But there were critics, including then-Judge Brett Kavanaugh of the U.S. Court of Appeals for the D.C. Circuit. In *Heller v. District of Columbia* (*Heller II*), Dick Heller—the successful plaintiff in the 2008 Supreme Court case—challenged the District of Columbia's regulation of assault rifles and large capacity magazines as a violation of the Second Amendment.[24] The majority used the two-part framework and concluded that the District's regulation was constitutional. In dissent, then-Judge Kavanaugh read *Heller* as forbidding any type of means-end

[20] United States v. Vongxay, 594 F.3d 1111, 1115 (9th Cir. 2010) ("[F]elons are categorically different from the individuals who have a fundamental right to bear arms. . . .").

[21] *See* United States v. Davis, 304 F. App'x 473, 474 (9th Cir. 2008) (finding 49 U.S.C. § 46505 constitutional).

[22] United States v. Marzzarella, 614 F.3d 85 (3d Cir. 2010).

[23] Along with Eric Ruben of SMU Dedman School of Law, we filed an amicus brief in support of neither side urging the Court to adopt the two-part framework adopted throughout the federal courts of appeals, rather than this historical approach. *See* Brief of Second Amendment Law Professors as Amici Curiae in Support of Neither Party, N.Y. State Rifle & Pistol Ass'n, Inc. v. Bruen, 142 S. Ct. 2111 (2021) (No. 20-843), 2021 WL 3144391.

[24] Heller v. Dist. of Columbia (*Heller II*), 670 F.3d 1244, 1271, 1282 (D.C. Cir. 2011) (Kavanaugh, J., dissenting).

scrutiny and argued instead that Second Amendment cases should be resolved solely by reference to "text, history, and tradition."[25] A few other federal judges echoed his skepticism of the second part of the two-part framework,[26] and this methodological issue became a centerpiece of the briefing in *Bruen*.

Bruen dealt with a challenge to New York's "may issue" permitting law. To obtain an unrestricted license to carry a handgun in New York, a person had to demonstrate "proper cause," meaning some kind of need different than the self-defense interests of the general population. Two plaintiffs from upstate New York, Brandon Koch and Robert Nash, challenged the regulation as a violation of the Second Amendment. They were joined in the action by the National Rifle Association's New York affiliate, the New York Rifle and Pistol Association, Inc. The tone of the Justices' questions at oral argument made it apparent that New York's regulation was going to fail; the only question was how broad the ruling would be, and on what grounds.[27] The options before the Court ranged from the factual and procedural—a remand for discovery about how licensing authorities were routinely using their discretion, for example—to the broad and disruptive—a complete revision of how Second Amendment cases should be decided.

The Court chose the latter course, overturning not only the Second Circuit's ruling, but also a decade and a half of Second Amendment doctrine. Justice Thomas began his opinion for the Court by casting New York's law as a departure from a national consensus: "In 43 States, the government issues licenses to carry based on objective criteria. But in six States, including New York, the government further conditions issuance of a license to carry on a citizen's showing of some additional special need."[28] The federal courts of appeals, with

---

[25] *Id.* at 1271 (Kavanaugh, J., dissenting); *see also* Darrell A.H. Miller, *Text, History, and Tradition: What the Seventh Amendment Can Teach Us About the Second*, 122 YALE L.J. 852 (2013) [hereinafter Miller, *Text, History, and Tradition*] (exploring, *inter alia*, the risks and rewards of a historical approach).

[26] *See, e.g.*, United States v. McGinnis, 956 F.3d 747, 761 (5th Cir. 2020) (Duncan, J., concurring) ("I write separately to reiterate the view that we should retire this [two-part] framework in favor of an approach focused on the Second Amendment's text and history."); Tyler v. Hillsdale Cnty. Sheriff's Dep't, 837 F.3d 678, 703 (6th Cir. 2016) (Batchelder, J., concurring in part).

[27] *See generally* Transcript of Oral Argument, *supra* note 2.

[28] N.Y. State Rifle & Pistol Ass'n, Inc. v. Bruen, 142 S. Ct. 2111, 2122 (2022).

the exception of the D.C. Circuit, had routinely upheld these "proper cause" laws using the conventional two-part framework.[29] But the majority of courts were wrong, according to the majority in *Bruen*.

"Despite the popularity of [the] two-step approach, it is one step too many,"[30] wrote Justice Thomas. The first part of the test is "broadly consistent with *Heller*" because it is rooted in "text, as informed by history."[31] But the second part of the two-part framework was foreclosed by *Heller*, which "did not invoke any means-end test such as strict or intermediate scrutiny."[32] Means-end scrutiny is balancing, and balancing, according to the Court, is judge-empowering.[33] The proper balance of gun rights and regulation was set "by the traditions of the American people,"[34] and any contemporary assessment of costs and benefits is impermissible policy-making by the Third Branch.[35]

Having rejected the circuits' consensus approach, the Court announced a new test: "When the Second Amendment's plain text covers an individual's conduct, the Constitution presumptively protects that conduct. The government must then justify its regulation by demonstrating that it is consistent with the Nation's historical tradition of firearm regulation."[36] Under this new test, it isn't enough for the government to simply "posit that the regulation promotes an important interest";[37] it must "affirmatively prove that its firearm regulation is part of the historical tradition that delimits the outer bounds of the right to keep and bear arms."[38]

Of course, not every modern regulation has a historical duplicate. If it doesn't, a court is to examine whether the modern regulation is

---

[29] *See id.* at 2124 (citing Gould v. Morgan, 907 F.3d 659, 677 (1st Cir. 2018); Kachalsky v. Cnty. of Westchester, 701 F.3d 81, 101 (2d Cir. 2012); Drake v. Filko, 724 F.3d 426, 440 (3d Cir. 2013); United States v. Masciandaro, 638 F.3d 458, 460 (4th Cir. 2011); Young v. Hawaii, 992 F.3d 765, 773 (9th Cir. 2021) (en banc)). *But see* Wrenn v. Dist. of Columbia, 864 F.3d 650, 668 (D.C. Cir. 2017).

[30] *Bruen*, 142 S. Ct. at 2127.

[31] *Id.*

[32] *Id.* at 2129.

[33] *Id.*

[34] *Id.* at 2131.

[35] *See id.*

[36] *Id.* at 2129–30.

[37] *Id.* at 2126.

[38] *Id.* at 2127.

"relevantly similar" to a historical one.[39] The Court declined to offer an "exhaustive survey" of what makes one regulation "relevantly similar" to another, but it did offer "two metrics: how and why the regulations burden a law-abiding citizen's right to armed self-defense."[40] The "central" considerations in this analogical process will be "whether modern and historical regulations impose a comparable burden on the right of armed self-defense and whether that burden is comparably justified."[41] The court emphasized that there need not be a historical "twin" to the modern regulation, but that any analogue must be "well-established and representative."[42] Upholding any remotely comparable historical regulation "'risk[s] endorsing outliers that our ancestors would never have accepted.'"[43] This history-focused test, the Court reasoned, would rescue the Second Amendment from second-class status and restore it to equal dignity among the Bill of Rights.[44]

Having articulated the new test, the Court applied it, concluding that the plain text of "bear arms" in the Second Amendment, according to original public understanding, means to carry weapons for self-defense.[45] The Second Amendment thus presumptively covered the plaintiffs' conduct.

The burden then shifted to New York to show that its permitting law operated within a national tradition of regulation. New York and amici had amassed more than 700 years of Anglo-American regulation regarding public carry of firearms in an effort to show that New York's law was no historical anomaly. Indeed, the New York law itself

---

[39] *See id.* at 2132 (quoting Cass Sunstein, *On Analogical Reasoning*, 106 HARV. L. REV. 741, 773 (1993) (internal quotation marks omitted)).

[40] *Id.* at 2133.

[41] *Id.*

[42] *Id.*

[43] *Id.* (quoting Drummond v. Robinson, 9 F.4th 217, 226 (3d Cir. 2021)).

[44] *Id.* at 2130, 2156; *see* Joseph Blocher & Eric Ruben, *"Second-Class" Rhetoric, Ideology, and Doctrinal Change*, 110 GEO. L.J. 613 (2021).

[45] Though the parties did not really contest this conclusion, there is good reason to doubt it as a linguistic matter. Experts using eighteenth century documents and big-data techniques unavailable to the Court when it decided *Heller* have concluded that "bear arms" was overwhelmingly used at the time in a military or collective sense, not to connote the carrying of weapons for private purposes. *See* Dennis Baron, *Corpus Evidence Illuminates the Meaning of* Bear Arms, 46 HASTINGS CONST. L.Q. 509, 510 (2019); Alison L. LaCroix, *Historical Semantics and the Meaning of the Second Amendment*, PANORAMA (Aug. 3, 2008), http://thepanorama.shear.org/2018/08/03/historical-semantics-and-the-meaning-of-the-second-amendment.

had been in place for more than a century, since at least 1911.[46] Nevertheless, the majority combed through all these regulations and explained away each one as an outlier with respect to a tradition of regulating public carry.

In doing so, the Court was bracingly candid: "not all history is created equal."[47] It chopped the historical evidence into five categories: medieval to early modern England, colonial and early American Republic, antebellum, Reconstruction, and late nineteenth to early twentieth century. Having broken the historical evidence into pieces, the Court rejected it case by case, statute by statute.

The Court began with medieval and early-Modern English law. Although the Second Amendment reflects a "pre-existing" right inherited from our English ancestors, regulations that "long predate []" the Second Amendment may not be indicative of a tradition "if linguistic or legal conventions changed in the intervening years."[48] Therefore, the copious regulation of public arms from the thirteenth century to early modern England failed to establish a relevant tradition: they were too old, they had fallen into desuetude,[49] they were limited to armor and medieval weapons,[50] they were the product of partisan or sectarian hostility[51] or domestic upheaval,[52] or they required a *mens rea* of carrying weapons "to the terror of the people."[53]

Colonial regulations contemporary to 1791—the date of the Second Amendment's ratification—did not fare any better. In the Court's estimation, New York had produced only three regulations on public carry, and two of them were identical.[54] These regulations only proscribed "dangerous and unusual" weapons, and twenty- first century handguns are no longer "dangerous and unusual."[55] The

[46] *Bruen*, 142 S. Ct. at 2122.

[47] *Id.* at 2136; *cf.* Peruta v. Cnty. of San Diego, 742 F.3d 1144, 1155 (9th Cir. 2014) ("All [Second Amendment precedents] are, in a broad sense, equally relevant, for every historical gloss on the phrase 'bear arms' furnishes a clue of that phrase's original or customary meaning. Still, some cases are more equal than others."), *rev'd on reh'g en banc*, 824 F.3d 919 (9th Cir. 2016).

[48] *Bruen*, 142 S. Ct. at 2136.

[49] *Id.* at 2140–41 & n.10.

[50] *Id.* at 2140.

[51] *Id.* at 2141.

[52] *Id.* at 2139.

[53] *Id.* at 2141.

[54] *Id.* at 2142.

[55] *Id.* at 2143.

historical regulations only touched on concealed carry, and only of weapons with three to four inch barrel lengths;[56] they didn't prohibit the carrying of long guns.[57] They also required a mens rea of "causing terror."[58]

Although New York had produced substantial evidence of common law and statutory regulation of public carry, this too was insufficient to form a tradition supporting New York's law. The Court found that the antebellum cases all required a showing of terror, and there was "no evidence indicating that these common-law limitations impaired the right of the general population to peaceable public carry."[59] Further, some regulations permitted open carry.[60] All were too new, as they were enacted seven decades after ratification of the Second Amendment.[61] Some were enacted by territorial jurisdictions that were not yet states.[62] The ten jurisdictions which passed surety statutes—statutes that required a bond to carry weapons in public—only addressed persons "threatening to do harm"[63] and required a citizen complaint (except for two antebellum laws in Virginia and West Virginia, which were, according to the Court, outliers[64]). Further, these surety laws were not clearly shown to be enforced, or were enforced pretextually, or were enforced with insufficient criminal sanction.[65] All of this made them "poor analogues to New York's proper-cause standard."[66]

Reconstruction regulations also failed to establish a tradition. A general order by a Union commander in the Carolinas to prevent the carrying of deadly weapons, except by those in active military service, was a wartime exigency subject to military jurisdiction, and therefore inapt.[67]

---

[56] *Id.*

[57] *Id.* at 2144.

[58] *Id.* at 2145.

[59] *Id.*

[60] *Id.* at 2142, 2144.

[61] *Id.* at 2147 n.22.

[62] *Id.* at 2147 n.22.

[63] *Id.* at 2148

[64] *Id.* at 2148 n.24.

[65] *Id.* at 2149.

[66] *Id* at 2149 n.25.

[67] *Id.* at 2152 n.26. Notably, the Court found an order by this same Union general, allowing the carrying of weapons openly (not concealed) except among vagrants, the disorderly, or disturbers of the peace, as prima facie evidence of a right to bear arms for personal self-defense. *Id.* at 2152.

Some regulations were applied in a racially partial manner, and (presumably) suspect for that reason.[68] The Court "acknowledge[d]" that two late 1800s Texas cases and a West Virginia Supreme Court case all upheld statutes requiring "reasonable grounds" to carry weapons, which were analogous to "New York's proper-cause requirement," but these were dismissed as outliers because they were only a "single state statute and a pair of state-court decisions."[69]

Late nineteenth-century regulations, like prohibitions on firearms in towns, cities and villages, suffered from several defects. First, they were too remote from the Founding; second, "the vast majority" of the evidence "came from the Western Territories,"[70] even then, the regulations were too "localized," and governed "less than 1% of the American population" at the time.[71] Further, the regulations were not often litigated, obscuring their "perceived legality"; they excepted certain types of weapons, and those that were upheld conflicted with the decision one hundred years later in *Heller*.[72] These territorial restrictions were thus "passing regulatory efforts by not-yet-mature jurisdictions on the way to statehood, rather than part of an enduring American tradition of state regulation."[73] One state regulation potentially analogous to New York's was a 1881 Kansas statute obliging cities with more than 15,000 residents to regulate arms. But, according to the Court, that only covered Kansas City, Topeka, and Wichita, which "[a]ccounted for only 6.5% of Kansas's total population."[74]

In sum, according to the Court, the seven hundred years of history that New York and amici had produced could not demonstrate a long American tradition of regulating arms with a "proper cause" standard—at least not once the outliers were removed. Hence, the law violated the Second Amendment.

B. Analysis and Open Questions

Whatever one's view of New York's regulation, *Bruen*'s new methodology represents a problematic shift in Second Amendment doctrine,

[68] *Id.* at 2152 n.27.

[69] *Id.* at 2153.

[70] *Id.* at 2154.

[71] *Id.* at 2154–55.

[72] *Id.* at 2155.

[73] *Id.* at 2155.

[74] *Id.* at 2156.

and perhaps constitutional law more generally. Just how big of a problem remains unclear, as it is not apparent how broadly *Bruen* sweeps. Three Justices in the majority wrote or joined concurrences that appear to trim the ambition of Justice Thomas's opinion—either as to the topics it covers or the methods it employs or both.

Justice Samuel Alito suggested that the historical analysis only goes to the unconstitutionality of a permitting law like New York's: "That is all we decide."[75] According to Alito, the ruling says nothing about who may buy firearms, or what they must do to buy them; it doesn't decide anything about the kinds of arms that are constitutionally protected, "[n]or have we disturbed anything that we said in [*Heller* or *McDonald*] about restrictions that may be imposed on the possession or carrying of guns."[76]

Justice Brett Kavanaugh, joined by Chief Justice John Roberts, wrote a separate concurrence that also seemed to cool *Bruen*'s revolutionary fire. On his account, there are "limits" to the Court's opinion. The decision "does not prohibit States from imposing licensing requirements for carrying a handgun in self-defense" along the lines of the forty-three other states with "shall-issue" laws that use more objective metrics. Further, "[p]roperly interpreted, the Second Amendment allows a 'variety' of gun regulations."[77] To emphasize the point, Justice Kavanaugh reproduced assurances from *Heller* and *McDonald* that the Court was not scrapping all gun regulations. "Like most rights, the right secured by the Second Amendment is not unlimited," he reiterated:

> From Blackstone through the 19th-century cases, commentators and courts routinely explained that the right was not a right to keep and carry any weapon whatsoever in any manner whatsoever and for whatever purpose. . . . [N]othing in our opinion should be taken to cast doubt on longstanding prohibitions on the possession of firearms by felons and the mentally ill, or laws forbidding the carrying of firearms in sensitive places such as schools and government buildings, or laws imposing conditions and qualifications on the commercial sale of arms. [Footnote 26: We identify these presumptively lawful regulatory measures only as examples; our list does not purport to be exhaustive.][78]

---

[75] *Id.* at 2157 (Alito, J., concurring).

[76] *Id.*

[77] *Id.* at 2162 (Kavanaugh, J., concurring).

[78] *Id.* (quoting and citing Dist. of Columbia v. Heller, 554 U.S. 570, 626–27 & n.26 (2008) (citations and quotation marks omitted); and McDonald v. City of Chicago, 561 U.S. 742, 786 (2010) (plurality opinion)).

Nor does the *Bruen* opinion, according to Kavanaugh, implicate prohibitions on weapons that are "not in common use at the time," a limitation supported by the prohibition on "carrying of dangerous and unusual weapons."[79] These precatory statements and assurances by three of the six Justices who formed the majority raise nettlesome questions as to what the case actually decides.

Then there's the uncertainty within the *Bruen* framework itself. Justice Thomas stated that so long as an individual's conduct falls within the "plain text" of the Second Amendment, it is presumptively protected and the burden shifts to the party defending the regulation to show that it is part of a longstanding American tradition. But that's almost certainly not right. A seventeen year old bringing a live hand grenade into a high school cafeteria fits within the plain text of "people" and "arms" and "bear."[80] It cannot be that such behavior raises a *prima facie* Second Amendment issue such that the school district must prove a longstanding tradition of keeping minor children from bringing explosives to school. Indeed, the majority itself suggests elsewhere that the first step is not just "plain text" but "text, as *informed by history*."[81] If that's the case, then it appears incumbent upon the challenger of a regulation to show more than that his conduct falls within the dictionary meaning of the Second Amendment's words; he must show how his conduct falls within a textually specified historical practice that was widely understood to be a matter of right. Only then would the burden shift to the defenders to establish a tradition of regulation.

Perhaps most perplexing, *Bruen* relied heavily on analogy but never specified criteria for determining when a present-day regulation is relevantly, as opposed to trivially, analogous to one in the past.[82] The majority assures us that its demand for analogues is neither a "regulatory straightjacket" nor a "blank check," but provides slender guidance on how to pilot between these two hazards. Instead, we get a set of abstractions: an analogue must be "well established and

---

[79] *Id.*

[80] *Heller*, 554 U.S. at 582, 584 (2008) ("bear" simply means "carry"; "keep arms" means simply "have weapons"). Again, this interpretation of "bear arms" as a matter of original public meaning is almost certainly wrong according to linguists and historians. *See supra* note 45.

[81] *Bruen*, 142 S. Ct. at 2118–19 (emphasis added).

[82] Joseph Blocher & Eric Ruben, *Originalism-by-Analogy and Second Amendment Adjudication*, 133 YALE L.J. (forthcoming 2023) (on file with authors).

representative" instead of an "outlier."[83] For a "general societal problem that has persisted since the 18th century," the analogue must operate in "materially" similar ways and not have been "rejected on constitutional grounds" by some undetermined number of jurisdictions, or else it's a historical anomaly that cannot support the contemporary regulation.[84]

So how is a court supposed to analyze firearms on subways or airplanes? How should one describe the general societal problem in that context, and with what evidence? Is it gun violence generally? Is it terrorism? What does a "representative historical analogue" look like before the advent of railroads and airplanes? A boat? A buggy?[85] Are regulations forbidding the firing of guns from the decks of riverboats[86] "materially different" from regulations forbidding the possession of loaded guns in carry-on luggage? Or should the analogue be weapon regulation in areas of intense congestion, like fairs and markets? The Court says that judges are to look to "how and why" the analogous regulation burdens a right to armed self-defense, but that, as Justice Breyer notes in dissent, is exactly what means-end scrutiny is supposed to do.[87]

Working with analogues may be possible when there's a historical equivalent. But how courts will find, much less assess, "representative historical analogue[s]" to modern prohibitions on guns on subway cars and jet airplanes, or possession of untraceable firearms,[88] or possession by domestic abusers, remains distressingly indefinite.[89]

---

[83] *Id.* at 2133.

[84] *Id.* at 2131.

[85] *Cf.* United States v. Jones, 565 U.S. 400, 420 (2012) (Alito, J., concurring) (noting, in a case about whether the government's installation of a GPS device in a car constituted a Fourth Amendment search, "it is almost impossible to think of late–18th-century situations that are analogous to what took place in this case. (Is it possible to imagine a case in which a constable secreted himself somewhere in a coach and remained there for a period of time in order to monitor the movements of the coach's owner?)").

[86] An Ordinance to Prevent Accidents from the Firing of Cannon or Other Guns on Boats, in Front of the City of Cincinnati § 1 (1828) ("[I]t shall not be lawful for any person or persons having charge or being on board of any boat upon the Ohio river, when passing by, stopping at, or leaving the city of Cincinnati, to cause any cannon, gun or other fire-arms, to be so fired as to discharge its contents towards the city. . . .").

[87] *Bruen*, 142 S. Ct. at 2179 (Breyer, J., dissenting).

[88] *See id.* at 2181 (Breyer, J., dissenting).

[89] Justice Breyer feared the Court had engineered a Second Amendment "one-way ratchet" where courts would examine prima facie claims at a very high level of generality (a "person," "keep[ing]" or "bear[ing]" an "arm"), but require any "representative historical analogue" the government supplies to be nearly identical to the modern regulation. *Id.* at 2180 (Breyer, J., dissenting). Initial post-*Bruen* cases show this fear is warranted, and that lower courts are not

II. Manufacturing Outliers

It will take decades for courts and scholars to work through *Bruen*'s problems. In the remainder of this Article, we focus on one—its recurrent reliance on the idea of an "outlier." As shown above, *Bruen* places substantial analytical weight on its characterization of laws as outliers. New York's permitting law is an outlier. Colonial regulations are outliers. Territorial laws are outliers. Reconstruction-era laws are outliers. In a broader context, the Court suggests the Second Amendment is itself being treated as an outlier among the guarantees in the Bill of Rights. *Bruen* uses the term outlier as a statement of fact and as a justification for action. But what does the Court mean by the term, and what considerations should guide its use?

A. Transparency

A central defect of *Bruen* is the suggestion that its "outliers" were simply found. They weren't. They were created. *Bruen*'s outliers are the product of decisions both inside and outside the Court, motivated by express and assumed judgments about how to count, and what counts. But the Court does not explain, much less justify, how it is constructing its baseline or how evidence deviates meaningfully from that baseline. The result is an impressionistic and seemingly arbitrary set of conclusions that at best mask the values animating the analysis, and at worse appear to be post-hoc reasoning for a pre-determined outcome.

There are, in fact, varying accounts of why the Supreme Court might nationalize legal standards against outliers.[90] The Court might trim outliers because state laws are a rough proxy for political consensus.[91] The Court might trim outliers because it's attempting to ascertain

---

yet adopting equivalent levels of generality on both sides of the rights/regulation equation. *See* Andrew Willinger, Stickley v. Winchester, *State Analogues, and the Folly of Narrow Historical Focus*, Second Thoughts Blog (Oct. 7, 2022), https://firearmslaw.duke.edu/2022/10/stickley-v-winchester-state-analogues-and-the-folly-of-narrow-historical-focus.

For exploration of historical tests that would not fall into this trap, see Miller, *Text, History, and Tradition*, *supra* note 25 (arguing that if courts adopt analogues, the level of generality must be the same for both the rights and regulations); and Darrell A.H. Miller, *Second Amendment Equilibria*, 116 Nw. U. L. Rev. 239 (2021) [hereinafter Miller, *Second Amendment Equilibria*] (same).

[90] We offer three, although there may be other reasons why the Court trims outliers. *See* Driver, *supra* note 4, at 951–63.

[91] *Cf.* Jenna Bednar & William N. Eskridge, Jr., *Steadying the Court's "Unsteady Path": A Theory of Judicial Enforcement of Federalism*, 68 S. Cal. L. Rev. 1447, 1478 (1995) ("As an

the original public understanding of a constitutional provision. Or the Court might trim outliers because it becomes satisfied from sub-constitutional consensus that nationalization will be net positive and will not generate catastrophic outcomes.[92]

Whatever the reason, identifying outliers requires choosing a reference set, and that choice is freighted with value judgments.[93] For example, using the States as the unit of measure for political consensus assumes the states are a superior proxy for the political preferences of their residents than other proxies. Discerning original public meaning from litigated cases assumes litigated cases are a reliable metric for public understanding. All these assumptions are unstated, and all do a lot of work in *Bruen*'s definition of outliers.

Take the Court's insistence that New York's "may issue" law was anomalous compared to that of forty-three other states. The first, unstated assumption is that states are the right reference set, even if some of these states are thinly populated or lack urban centers. If the Court had used population as the relevant denominator, then New York's law is not an outlier—good cause permitting covers fully a quarter of the American population, including some of the highest population densities in the country.[94] In fact, elsewhere in the opinion the Court *does* invoke population as a relevant metric, dismissing various gun laws in the territories as outliers because of "the miniscule territorial populations who would have lived under them."[95] It is not clear from the opinion why population is relevant in one context but not the other.

If at-large population is the wrong denominator, why should states be the right one? New York's law would have looked far more typical if the denominator had been large municipal governments.[96] If you

institution of national power, the Supreme Court might be expected to enforce national consensus politics against outlier or shirking states.").

[92] Roderick M. Hills, Jr., *Counting States*, 32 HARV. J.L. & PUB. POL'Y 17, 25 (2009) (state experimentation "provides information to national decision makers about how the 'maverick' norms will operate on the ground, allowing them to decide whether to nationalize the norms after they have proven themselves to be sound policies").

[93] *See* Ernest A. Young, *Foreign Law and the Denominator Problem*, 119 HARV. L. REV. 148, 148 (2005) (setting a denominator "requires choices about which . . . jurisdictions are relevant").

[94] N.Y. State Rifle & Pistol Ass'n, Inc. v. Bruen, 142 S. Ct. 2111, 2172–73 (2022) (Breyer, J., dissenting).

[95] *Id.* at 2154 (majority opinion).

[96] *See* Brief of Former Major City Police Chiefs as *Amici Curiae* in Support of Respondents, *Bruen*, 142 S. Ct. 2111 (No. 20-843) (brief filed by leaders of police departments of Boston,

combine the populations of Maine, Rhode Island, Montana, Delaware, South Dakota, North Dakota, Alaska, Vermont, and Wyoming it still doesn't equal the population of the City of New York. And even *that* approach might under-represent political preferences, considering that nearly all the forty-three states with "shall issue" or permitless carry also preempt local governments from enacting their own gun regulations.[97] That means that major urban centers like Houston and Phoenix may have their preferences unaccounted for when the unit of measure is legal regulation, because they're prohibited from registering their preferences as law. Further, some of this legislation (or absence of it) is the product of either a heavily gerrymandered state legislative map[98] or an equally skewed direct democracy mechanism.[99] The result is a data set that provides more voice to rural and less diverse populations compared to urban and multi-racial populations.

Finally, as Justice Breyer remarked in dissent, New York is only an outlier given a "snapshot" of the law in 2022.[100] Look back just forty years and the ratio is completely reversed: "shall issue" was limited to just five states[101] and permitless carry was the law of just

---

Buffalo, Chicago, Los Angeles, Milwaukee, Newark, New York City, Philadelphia, Seattle, and Washington, D.C.); Brief of City of Chicago and Eleven Other Cities as *Amicus Curiae* in Support of Respondents, *Bruen*, 142 S. Ct. 2111 (No. 20-843) (amici on behalf of Chicago, Baltimore, Cincinnati, Columbus, Dayton, Ohio, Los Angeles, Philadelphia, Portland, San Diego, San Francisco, Seattle, and St. Paul).

[97] Joseph Blocher, *Cities, Preemption, and the Statutory Second Amendment*, 89 U. Chi. L. Rev. 557 (2021).

[98] Paul A. Diller, *Toward Fairer Representation in State Legislatures*, 33 Stan. L. & Pol'y Rev. 135, 136 (2022) ("Gerrymandering and the uneven translation of votes into seats has entrenched political parties in control of the legislature in many states across election cycles despite often lacking a majority of statewide support.").

[99] Some states require the petitions for direct democracy initiatives to meet a geographic distributional requirement within the state, often giving rural communities a veto over direct democracy efforts supported by large urban centers. *See* Bernbeck v. Gale, 829 F.3d 643, 650 (8th Cir. 2016) (Kelly, J., dissenting) (plaintiff bringing Equal Protection claim on basis that "because Nebraska's counties vary widely in population . . . the signature-distribution requirement [to place an initiative on the ballot] gives disproportionate influence to voters in sparsely-populated counties"); *see also Initiative and Referendum Processes*, Nat'l Conf. of State Legislatures (Jan. 4, 2022), available at https://www.ncsl.org/research/elections-and-campaigns/initiative-and-referendum-processes.aspx ("To make it more difficult to place initiatives on the ballot and to ensure initiatives do not represent just the interests of heavily populated areas, some states have created a requirement that signatures be gathered from across the state.").

[100] *Bruen*, 142 S. Ct. at 2172 (Breyer, J., dissenting).

[101] In 1983, Indiana, Maine, New Hampshire, Rhode Island, and Washington all had shall issue licensing. Rhode Island's shall issue was limited to a reciprocity provision.

one.[102] New York went from being fairly representative to an outlier thanks to lobbying by gun-rights advocates, legislation by pro-gun state assemblies, and naked appeals to popular notions of the Constitution, history, and tradition.[103]

As Professor Jake Charles has documented, as recently as 1980 fully one-quarter of states outlawed concealed carry altogether, with most of the other states operating a "proper cause" or similar "may issue" licensing regime.[104] But "[t]his paradigm began to erode dramatically in the 1980s."[105] Florida kicked off a wave of state law changes when it adopted a "shall issue" law that retained certain requirements, but removed the "special need" or "proper cause" element.[106] By the time *Bruen* was decided, there was not a single jurisdiction in the United States that did not allow some form of concealed carry.[107]

A second wave of gun-rights legislation began to build shortly before the *Heller* decision in 2008, and soon overtook established institutions of gun-rights advocacy like the NRA. The permitless or "constitutional carry" movement was initiated in an Arizona diner around 2005,[108] and by 2020 had spread to sixteen states. With the repeal of permitting in Alabama and Florida in 2023, over half of the states are now permitless.

These legislative innovations often speak in a specific constitutional register, as the "constitutional carry" terminology attests. Washington's permitting law, for instance, has a preface that "[t]he applicant's constitutional right to bear arms shall not be denied, unless . . ."[109] The Governor of Georgia, signing a permitless carry act in 2022 announced, "The Constitution of the United States gives

---

[102] In 1983, Vermont was the only state with permitless carry.

[103] In this way, it was simply a specific iteration with public carry of a larger phenomenon of "originalism as popular constitutionalism" that Reva Siegel aptly identified in *Heller*. Reva B. Siegel, *Dead or Alive: Originalism as Popular Constitutionalism in* Heller, 122 HARV. L. REV. 191, 192 (2008).

[104] Jacob D. Charles, *Securing Gun Rights by Statute: The Right to Keep and Bear Arms Outside the Constitution*, 120 MICH. L. REV. 581, 596 (2022).

[105] *Id.*

[106] *Id.*

[107] Adam Weinstein, *Meet the Gun Rights Absolutists Bringing 'Constitutional Carry' to a State Near You*, TRACE (Feb. 8, 2017), https://www.thetrace.org/2017/02/constitutional-carry-gun-rights-absolutists.

[108] *Id.*

[109] WASH. REV. CODE ANN. § 9.41.070 (West 2022).

us that right [to carry without a permit] – not the government."[110] For proponents of "constitutional carry" in particular, the idea that the only permit necessary to carry a firearm was signed in 1791 is not a slogan, it's a legal and historical reality. Although the historical record supporting a tradition of permitless concealed carry is thin to non-existent—*Heller* itself recognized as much[111]—that does not stop adherents from arguing that American tradition and constitutional law is on their side. The majority leader of the New Hampshire Senate, for example, explained his support this way: "We have historically allowed people to openly carry a pistol. I don't see why you have to get a second permit if you're a law-abiding citizen and legally entitled to own a gun."[112] A sponsor of a permitless carry bill in Utah said: "I have that right to protect myself, the Constitution says we have the right. Why are we putting a barrier for law-abiding citizens?"[113] The sponsor of the Ohio permitless carry bill made similar remarks: "Responsible gun owners should not be punished for lawfully practicing their constitutional rights."[114] The attitude can be summarized by one journalist who interviewed "constitutional carry" advocates: "The concept, rooted in constitutional originalism, assumes that the authors of the Second Amendment envisioned an unfettered right to wield a gun for personal defense."[115] To these adherents, "*any* limitation on an individual's right to carry guns, however small, is unjust. Full stop. As such, passing constitutional-carry legislation is seen by proponents as a restoration, not an expansion, of gun freedoms."[116]

---

[110] Press Release, Off. of the Governor, Gov. Kemp Signs Georgia Constitutional Carry Act into Law (Apr. 13, 2022), https://gov.georgia.gov/press-releases/2022-04-13/gov-kemp-signs-georgia-constitutional-carry-act-law.

[111] Dist. of Columbia v. Heller, 554 U.S. 570, 626 (2008) ("[T]he majority of the 19th-century courts to consider the question held that prohibitions on carrying concealed weapons were lawful under the Second Amendment or state analogues.").

[112] Katie Zezima, *More States Are Allowing People to Carry Concealed Handguns Without a Permit*, WASH. POST (Feb. 24, 2017, 9:15 AM), https://www.washingtonpost.com/news/post-nation/wp/2017/02/24/more-states-are-allowing-people-to-carry-concealed-handguns-without-a-permit.

[113] Ray Nothstine, *How Far Will North Carolina Fall Behind on Gun Rights?*, CAROLINA J. (Jan. 25, 2021), https://www.carolinajournal.com/opinion/how-far-will-north-carolina-fall-behind-on-gun-rights.

[114] Press Release, The Ohio Senate, Senate Concurs with Changes to Johnson Permitless Carry Bill (Mar. 2, 2022), https://ohiosenate.gov/senators/johnson/news/senate-concurs-with-changes-to-johnson-permitless-carry-bill.

[115] Weinstein, *supra* note 107.

[116] *Id.*

The normativity of those arguments is evident—though unrecognized—in the Court's own treatment of the historical record.

Bruen says the historical laws supporting New York's regulation are outliers compared to the tradition of public carry regulation. But *Bruen creates* that tradition through a series of choices about the level of generality at which to identify a practice, the level of generality at which to identify a regulation, and the relevant denominator of practices, laws, groups and political units at which to define the tradition.[117]

Consider how *Bruen* dismisses the stringent regulations on public carry in the Western territories. Justice Thomas describes these regulations as outliers because they did not survive the transition into statehood.[118] That might be defensible if states were good proxies for consensus on the original understanding of the Second Amendment. But it is indefensible if the best measure of original understanding are those jurisdictions subject to the Second Amendment before the advent of incorporation. In those circumstances, the most probative reference set would be the territories, because they—*unlike* states—were subject to the Bill of Rights. In that sense, as Andrew Willinger notes, "the five territorial public-carry laws the Court declined to credit in *Bruen* may not be outliers at all, but in fact may hold crucial clues to the meaning of the *federal* Second Amendment in the 19th century."[119]

Similarly, the Court treats historical laws banning the carrying of concealable weapons as outliers when compared to four states that struck down similar regulations in the South.[120] But why are litigated cases the relevant metric for accurately assessing the legal tradition? Why wouldn't the relevant metric include all those jurisdictions where regulations went unchallenged (or unenacted) because the practice was socially frowned upon—and therefore rare[121]—or unchallenged

---

[117] *See* N.Y. State Rifle & Pistol Ass'n, Inc. v. Bruen, 142 S. Ct. 2111, 2184, 2186–87 (Breyer, J., dissenting).

[118] *Id.* at 2155 (the territorial laws "appear more as passing regulatory efforts by not-yet-mature jurisdictions on the way to statehood, rather than part of an enduring American tradition of state regulation").

[119] Andrew Willinger, *Territorial Gun Regulation and the "Lost" History of the Federal Second Amendment*, Second Thoughts Blog (Aug. 8, 2022), https://firearmslaw.duke.edu/2022/08/territorial-gun-regulation-and-the-lost-history-of-the-federal-second-amendment; *see also* Andrew Willinger, *The Territories Under Text, History and Tradition*, 101 Wash. U. L. Rev. (forthcoming 2024) (on file with authors).

[120] *See Bruen*, 142 S. Ct. at 2146–47.

[121] The Court's citation of *State v. Huntly*, 25 N.C. 418 (1843), is important in this regard. *Huntley* does say that the state cannot criminalize *all* forms of public carry (which not even

because no one thought the Second Amendment had anything to say about the regulation?[122] In fact, the Court takes precisely this kind of approach when reiterating the constitutionality of restrictions on guns in "sensitive places" like polling places and courthouses, saying "we are also aware of no disputes regarding the lawfulness of such prohibitions. We therefore can assume it settled that these locations were 'sensitive places' where arms carrying could be prohibited consistent with the Second Amendment."[123] If the relevant metric for tradition is not "what was litigated" but "what did people do and believe" then all kinds of historical sources, including newspaper articles, opinion pieces, letters and other materials should be evidence of the relevant "tradition,"[124] not just litigated cases available on Westlaw.

The difficulty here goes beyond the challenge of excavating historical sources, dealing with historical silences, and managing the disjunction between the craft of historians and that of judges. The Court has failed to acknowledge the inevitable role of "creativity and interpretive choice" that shapes its treatment of the historical record.[125] By dismissing so much of the historical evidence, and then regarding the remainder as a self-evident, pre-existing, and neutral baseline, the Court confuses an exercise in discretion for a process of discovery.

B. Rigor

If *Bruen*'s first shortcoming is a lack of candor about its own role in manufacturing outliers, its second shortcoming is a lack of rigor in

---

New York's law did). But it also says that public carry is an unusual practice to be discouraged. *Id.* at 422 ("A gun is an 'unusual weapon,' wherewith to be armed and clad. No man amongst us carries it about with him, as one of his every day accoutrements—as a part of his dress—and never we trust will the day come when any deadly weapon will be worn or wielded in our peace loving and law-abiding State, as an appendage of manly equipment.").

[122] *Bruen*, 142 S. Ct. at 2189 (Breyer, J., dissenting).

[123] *Id.* at 2133 (majority opinion).

[124] *See* Miller, *Second Amendment Equilibria*, *supra* note 89, at 247 (arguing that a more accurate measure of traditions could be adduced from "a wide range of sources, including folkways, state and local customs, law enforcement records, newspaper reports, private correspondence, and policy statements, as well as statutes, regulations, and state and federal court decisions").

[125] H. Jefferson Powell, *Rules for Originalists*, 73 Va. L. Rev. 659, 660–61 (1987) ("My specific concern is to argue that the turn to history does not obviate the personal responsibility of the originalist interpreter for the positions he takes, because historical research itself, when undertaken responsibly, requires of the interpreter the constant exercise of judgment. Historical judgments, while by no means exercises in unconstrained or subjective creativity, necessarily involve elements of creativity and interpretative choice.").

defining them.[126] Calling something an outlier suggests the categorization is based on some kind of empiricism, something objective, falsifiable and roughly quantitative, based on a method that submits to a disciplinary standard. This is consistent with the general claim that originalism, in particular, can supply a kind of empirical determinacy with history and tradition that other methods of constitutional interpretation cannot.[127] An "I know it when I see it" standard for outliers is unsuitable to a claim that purports to an empirical statement of fact.

Where might one begin to articulate the minimum rigor for outlier analysis in constitutional law?[128] Given that the term itself seems borrowed from empirically-oriented disciplines—and is being used to similar ends—one might take some guidance from outside the law. The National Institute of Standards and Technologies defines outlier this way:

> An *outlier* is an observation that lies an abnormal distance from other values in a random sample from a population. In a sense, this definition leaves it up to the analyst (or a consensus process) to decide what will be considered abnormal. Before abnormal observations can be singled out, it is necessary to characterize normal observations.[129]

Three implications flow from this definition. First, there needs to be a sufficient sample to determine whether there's a distribution. A single point can't be an outlier. Second, the sample has to fall within an

---

[126] For more on the need for more empirical rigor in statements about the state of the law among lawyers, judges and legal academics, see William Baude, Adam S. Chilton & Anup Malani, *Making Doctrinal Work More Rigorous: Lessons from Systematic Reviews*, 84 U. CHI. L. REV. 37 (2017).

[127] We say "history" here, but most varieties of originalism purport to be constrained by some objective criteria, whether it is linguistics, methods, or intentions. *See supra* note 10 and sources cited therein. *But cf.* Calvin Massey, *Elites, Identity Politics, Guns, and the Manufacture of Legal Rights*, 73 FORDHAM L. REV. 573, 581 (2004) (discussing the possibility that "text and tradition are simply rhetorical tropes" that enable judicial officers "to respond to the identity politics of the political right without overtly appearing to do so").

[128] Driver, *supra* note 4, at 930 n.2 (doing some of this definitional work); *see also* Baude, Chilton & Malani, *supra* note 126, at 37 (looking to "systematic reviews" in other disciplines with similar problems of transparency and falsifiability as guides to make doctrinal analysis more rigorous).

[129] HANDBOOK OF STATISTICAL METHODS, NAT'L INST. OF STANDARDS & TECH., https://www.itl.nist.gov/div898/handbook/prc/section1/prc16.htm; *see also* CHARLES YOE, PRINCIPLES OF RISK ANALYSIS: DECISION MAKING UNDER UNCERTAINTY 489 (2019) ("An outlier is an observation that 'appears' to be inconsistent with other observations in the data set. It is usually left to the analysts to define what is unusual."); Osage Tribe of Indians of Okla. v. United States, 96 Fed. Cl. 390, 418 (2010) ("In its most general sense, an outlier is defined as '[a]n observation that appears to deviate markedly from the other members of the sample in which it occurs.'" (quoting B.S. EVERITT, THE CAMBRIDGE DICTIONARY OF STATISTICS 274 (2d ed. 2002)).

identifiable grouping—there are no outliers when every point is an outlier. And, although researchers exercise some discretion in designating a data point as "abnormal" relative to the distribution, the discretion is not unbounded. There may be a "consensus process" to determine whether to recognize the data point as an outlier; there may be norms of the profession to justify whether something lies outside the distribution. Finally, outliers must be the result of observations. At the very least, an outlier designation can be challenged by other researchers based on a set of pre-existing criteria. Moreover, in any empirical project it's highly unorthodox—and usually forbidden—to cherry pick the data to fit a distribution you seek, rather than have the data determine the distribution.

Observations also have to be grouped according to some set of criteria or coding.[130] The norms of empirical disciplines vary, but the process tends to be designed to generate reliable data that minimizes bias or error,[131] is transparent and, if possible, is replicable.[132] Second, the more potential for bias to creep into the exercise, the more it needs to be distanced from the researcher.[133] In some empirical disciplines, for example, it's preferable to have blind coding—where the coders are given a set of parameters, but not given any information about the theory for which they are coding.[134] In other disciplines, the coding is done by two different individuals, with a third resolving disputes between the codes based on a set of ex-ante metrics.[135]

Describing the outlier analysis as a search for "tradition" does not eliminate the need for rigor.[136] Invocation of a "tradition" also presumes

---

[130] Epstein & Martin, *supra* note 14, at 10 ("Once researchers decide how and how much data to collect, they must code the data—that is, translate the information into a usable form for analysis.").

[131] *Id.* at 87–94 (discussing methods of avoiding bias in selecting data).

[132] *Id.* at 59 ("[A]ll empirical studies should adhere to the *replicability standard*: anyone should be able to understand, evaluate, build on, or reproduce the research without any additional information from the author.").

[133] *Id.* at 95, 96, 106 (discussing general principles for coding both quantitative and qualitative data to ensure the coding is reliable and replicable).

[134] Kimberly A. Neuendorf, The Content Analysis Guidebook 158 (2d ed. 2017); Gene Williams, Applied Qualitative Research 184 (2019).

[135] Lucienne T.M. Blessing & Amaresh Chakrabarti, DRM, a Design Research Methodology 120 (2009).

[136] *See* Epstein & Martin, *supra* note 14, at 9 (discussing how part of empirical research is reducing abstract and conceptual concepts like "judicial independence" or "economic freedom" to the concrete, observable and testable).

some kind of empiricism. Sociologists define tradition as "[a] set of social practices which seek to celebrate and inculcate certain behavioural norms and values, implying continuity with a real or imagined past, and usually associated with widely accepted rituals or other forms of symbolic behaviour."[137] Although the data sets and coding may be different than more quantitative projects, identifying a "tradition" still requires designation of a relevant community and some set of phenomena, practices, or regulations sufficient in number and similarity to form a distribution, and some duration of time.[138] Practices that deviate meaningfully from this grouping are outliers.

This is an incomplete survey, and we do not think that courts must necessarily adopt the definitions and norms of these disciplines.[139] But the alternative for law cannot be the kind of erratic, impressionistic approach on display in *Bruen*. Empirical standards exist among these other disciplines and share common features, and law should expect empirical conclusions about outliers to rely on processes of comparable rigor.[140]

Indeed, law already aspires to this kind of rigor in many different ways, employing techniques that effectively treat claims about outliers as matters of empirical fact. The content of customary international law, for instance, has long been defined as "a general and consistent practice of states followed by them from a sense of legal obligation."[141] At least one component of this doctrine looks to a

---

[137] John Scott & Gordon Marshall, A Dictionary of Sociology (3d ed. 2009); *cf.* David Jary & Julia Jary, Collins Dictionary of Sociology (4th ed. 2006) (defining custom as "any established pattern(s) of behaviour within a community or society").

[138] Richard A. Epstein, *Rediscovering the Classical Liberal Constitution: A Reply to Professor Hovenkamp*, 101 Iowa L. Rev. 55, 57 (2015) ("[N]o tradition is defined as a single point in time. . . ."); *see also* Epstein & Martin, *supra* note 14, at 9.

[139] Although, to the extent that courts are instructed by the Supreme Court analyze constitutional matters historically, it may be that courts will of necessity have to rely on the norms and consensus of historians instead of their own historical intuitions. United States v. Bullock, No. 3:18-CR-165-CWR-FKB (S. D. Miss. Aug. 21, 2018) ("Not wanting to itself cherry-pick the history, the Court now asks the parties whether it should appoint a historian to serve as a consulting expert in this matter. (citing Fed. R. Evid. 706)), https://s3.documentcloud.org/documents/23255743/us-v-bullock-historian-order.pdf.

[140] Baude, Chilton & Malani, *supra* note 126, at 50 (noting that disciplinary differences "justify caution when translating elements of systematic reviews to doctrinal work, but do not necessarily justify ignoring entirely the lessons of the methodology"); *see also* Epstein & Martin, *supra* note 14, at 96 (articulating a set of "best practices" for coding, while recognizing that the difficulty or ease of coding may vary across data sets).

[141] Doe v. Exxon Mobil Corp., 391 F. Supp. 3d 76, 89 (D.D.C. 2019) (quoting Restatement (Third) of Foreign Relations Law § 102(2) (Am. L. Inst. 1987)).

generalizable practice among nations,[142] followed over some duration of time—a determination that is supposed to be an empirical exercise.[143] Tort law routinely pegs a duty of care for professional service providers—like physicians—to a standard of care common to comparable experts.[144] And there are theories of the "reasonable person" that depend on observed behavior of a representative sample of people, rather than on a normative fiction of what a person should do.[145] Outside the Second Amendment, in other areas of constitutional law, judgments rest on empirical assertions of what is "cruel and *unusual* punishment,"[146] or what offends "contemporary *community* standards."[147]

Although the sophistication of empirical analysis varies across these domains, we would expect that, to the extent any of them make factual assertions about the world, they would engage in a methodologically rigorous process that produces, within some set of professionally

---

[142] J. Patrick Kelly, *The Twilight of Customary International Law*, 40 VA. J. INT'L L. 449, 500 (2000) ("In order to determine customary norms, one needs to know what counts as state practice and how to interpret that evidence.").

[143] Michael D. Ramsey, *The Empirical Dilemma of International Law*, 41 SAN DIEGO L. REV. 1243, 1247 (2004) (determining international customary law "is fundamentally a factual inquiry, seeking an objective answer about the state of the real world"); Tara Helfman, *The Dread Pirate* Who*? Challenges in Interpreting Treaties and Customary International Law in the United States*, 90 TUL. L. REV. 805, 818 (2016) ("Chief Justice John Marshall described the process of ascertaining a rule of customary international law as an empirical enterprise that requires 'consulting the works of jurists, writing professedly on public law; or by the general usage and practice of nations; or by judicial decisions recognising and enforcing that law.'" (quoting United States v. Smith, 8 U.S. (5 Wheat.) 153, 160–61 (1820)); *see also* Gregory Shaffer & Tom Ginsburg, *The Empirical Turn in International Legal Scholarship*, 106 AM. J. INT'L L. 1, 11 (2012).

[144] Mark A. Hall et al., *Measuring Medical Practice Patterns: Sources of Evidence from Health Services Research*, 37 WAKE FOREST L. REV. 779, 779–80 (2002) ("From a scientific perspective, determining a custom-based professional standard of care is partially or entirely an empirical question about actual physician behavior.").

[145] Alan D. Miller & Ronen Perry, *The Reasonable Person*, 87 N.Y.U. L. REV. 323, 371 (2012) (specifying the empirical essentials for a positive theory of a reasonable person as (1) "observations of the society, and not from the views of the researcher"; (2) more data means more accuracy; (3) "researchers may not make ex ante distinctions among individuals in the society," and (4) "it must be possible, at least on some occasion, to find that someone has behaved unreasonably").

[146] Young, *supra* note 93, at 148 ("American states split thirty to twenty on the legitimacy of the juvenile death penalty. On the international plane, however, the United States stood alone in condoning the practice.").

[147] Daniel L. Chen & Susan Yeh, *Distinguishing Between Custom and Law: Empirical Examples of Endogeneity in Property and First Amendment Precedents*, 21 WM. & MARY BILL RTS. J. 1081, 1104 (2013) (concluding that "judges are applying the *Miller* community standards test and relying on local sexual norms" to decide obscenity cases); *see also* Young, *supra* note 93, at 160 (noting that "in order to employ any community standard, one must first consider whether a coherent consensus can be identified within the proposed frame of reference").

agreed-upon parameters, an empirically falsifiable project. Like the more-transparently empirical disciplines discussed above, they begin with an identifiable grouping of sufficient size (nations, professionals, states) compare them according to the variable of interest (practice, standard of care, punishments), using standards not selected by the legal decisionmaker itself.

Bruen does none of this.[148] At least, it doesn't do so in any way that's obvious or conforms to the best practices common to empirical work. None of the quality control techniques one would expect of other disciplines appear to have been employed in *Bruen*. To the extent that customs, traditions, and baselines have been established at all, they seem to have been done through amicus briefing on appeal—not, for example, through dispassionate accumulation and characterization of the evidence, or for that matter even through adversarial testing of claims (whether historical or otherwise) at trial.

To the contrary, the Court makes confident generalizations based on phenomenon that might *themselves* be outliers. The two petitioners were denied licenses, but their Second Amendment challenge was decided on the pleadings, meaning that there is no record evidence to support the majority's conclusions that there is "limited" judicial review of license denials, that New York's law is too "demanding" and that this makes New York an outlier compared to the "vast majority of states."[149] It is entirely possible that the petitioners' experience was completely unrepresentative.

Further, the Court treats historical cases that are arguably outliers as indicative of broad historical trends. For example, the majority calls *Nunn v. State* "particularly instructive."[150] In that case, the Georgia Supreme Court applied the Second Amendment to state law at a time when Georgia had no similar provision in its state constitution, and just over a decade after the Court held that the Bill of Rights was not applicable to the states.[151] By nearly any measure,

---

[148] To be fair, the lack of rigor in specifying outliers is not unique to the Second Amendment. *See* Hills, *supra* note 92, at 21 (speaking of the Eighth Amendment prohibition on executing persons with mental incapacity: "It defies common sense to believe that the legal norms followed in 60% of the states representing roughly half the nation's population are somehow 'objective evidence' that the norms followed by the rest of the country (that is, in 40% of the states representing the other half of the nation's population) violate 'national standards.' ").

[149] N.Y. State Pistol & Rifle Ass'n, Inc. v. Bruen, 142 S. Ct. 2111, 2123 (2022).

[150] *Id.* at 2147 (citing Nunn v. State, 1 Ga. 243 (1846)).

[151] Barron v. Baltimore, 32 U.S. (7 Pet.) 243 (1833).

Nunn's understanding of the Second Amendment was an outlier in antebellum America. And yet the majority reads it to "indicate[] that it was considered beyond the constitutional pale in antebellum America to altogether prohibit public carry."[152] If a single state case can stand for the scope of a constitutional right "in antebellum America" then how can the Court dismiss, for example, Texas's Reconstruction-era regulations as outliers? Or, for that matter, why should *Nunn* count so much more than the "handful" of jurisdictions with laws that, even under the Court's own telling, supported New York?[153]

Consider, too, Justice Alito's assertion that the two-part doctrinal framework adopted throughout the federal courts of appeal "places no firm limits on the ability of judges to sustain any law restricting the possession or use of a gun. Two examples illustrate the point."[154] Considering that more than 1,000 Second Amendment challenges were resolved in the eight years after *Heller*—and probably more than 1,500 by the time *Bruen* was decided—one might expect more than two examples to sustain such a broad generalization, much less justify the doctrinal chaos *Bruen* threatens to unleash.

C. MANUFACTURING OUTLIERS AND THE INTEGRITY OF HISTORICAL REASONING

The risk of non-transparent and improvised outlier analysis is not simply that it may result in unjustifiable or erratic case outcomes. It threatens the integrity of history as a modality of constitutional interpretation.[155] A chief selling point of originalism has always been that history is an objective and falsifiable metric that can constrain judicial decision-making. But you need not buy all the judicial-constraint optimism of Ed Meese-era originalism to believe that statements of historical fact pronounced by the Court and statement of historical fact rendered by professional historians need share the same epistemic universe. A Court that repeatedly elevates one set of historical premises and denigrates the other, and then suffocates contrary historical evidence as "irrelevant" based solely upon its prior

[152] *Bruen*, 142 S. Ct. at 2147.

[153] *Id.* at 2138.

[154] *Id.* at 2160 (Alito, J., concurring).

[155] Philip Bobbitt, Constitutional Fate: Theory of the Constitution (1982).

decrees about history, risks an ever-increasing separation of law-office history from actual history.[156]

For example, in *Heller*, the Court was faced with a record containing many historical examples of strict and even prohibitionist gun regulation. Some courts applying *Heller* treated these historical laws as tainted, since some of them were based on the traditional view—rejected in *Heller*—that the right to keep and bear arms extended only to the organized militia.[157] *Bruen* engages in this same foreshortened historical investigation as well, discounting nineteenth century historical evidence when it is inconsistent with what *Heller* said over a century later.[158]

Every time the Court cuts off one of these lines of tradition as incompatible with its own historical construction, it shifts the baseline and the center of gravity for judicial history drifts further from the corpus of actual history.[159] Iterative pronouncements based on manufactured outliers are likely to continue in this context and in others, as the Court treats its holdings as precedential not only as to matters of law, but also matters of historical fact.[160]

This ratcheting effect may lead constitutional law further and further from *both* a historical baseline and contemporary best practices. As described above, "may issue" is an outlier as a result of a careful program of legislation and cultivation of constitutional memory.[161] But once you lop off "may issue," then "shall issue" becomes exposed. Employing the very same mechanisms that led to *Bruen* could lead another court to describe "shall issue" as the outlier and argue that permitless carry is what's required by the history and tradition of the

---

[156] Reva B. Siegel, *The Politics of Constitutional Memory*, 20 Geo. J.L. & Pub. Pol'y 19, 24 (2022) ("[C]onstitutional memory plays a special role in legitimating the exercise of authority when constitutional memory systematically diverges from constitutional history.").

[157] *See* Peruta v. Cnty. of San Diego, 742 F.3d 1144, 1155 (9th Cir. 2014), *rev'd on reh'g en banc*, 824 F.3d 919 (9th Cir. 2016).

[158] *Bruen*, 142 S. Ct. at 2155.

[159] *See* Stephen M. Griffin, *Optimistic Originalism and the Reconstruction Amendments*, 95 Tul. L. Rev. 281, 321 (2021) ("Public meaning originalism appears to produce outlier views from a historical perspective, views that can be understood as projections of an alternate reality had things been different.").

[160] On the topic of facts in constitutional adjudication, see Joseph Blocher & Brandon Garrett, *Originalism and Historical Fact-Finding*, 112 Geo. L.J. (forthcoming 2023) (on file with authors).

[161] On the difference between constitutional history and constitutional memory, see Siegel, *supra* note 156, at 24. For more on traditionalism's ratchet problem, see Sherif Girgis, *Living Traditionalism*, 98 N.Y.U. L. Rev. (forthcoming 2023) (on file with authors).

Second Amendment. Nor is that kind of ratcheting confined to the issue of public carry. We can see the appeal of this tool to gun rights supporters who want to target other regulations—whether those are prohibitions on large capacity magazines and "M-16 rifles and the like,"[162] age restrictions on who may possess firearms; extreme risk protection orders; training requirements to carry; and so forth. All you have to do is pull the legislative levers with receptive legislatures, support the position with a particular form of constitutional memory, present the target jurisdiction as a contemporary and historical outlier, and what had—to invert Jack Balkin's framework—formerly been "on the wall" is rendered "off the wall."[163] The result may be an American firearm policy that ironically becomes less and less the work of legislatures and representative branches of government and more and more the exclusive task of judicial officers using crude adjudicative tools and an invented historical tradition.[164]

Conclusion

We come to neither praise nor bury judgments that trim outliers; there is an appropriate role for this form of analysis in constitutional law. At one time, suppressing outliers was seen as part of the Warren Court's project to "modernize" the law for twentieth century.[165] Most recently, on the left, the push for marriage equality adopted many of the same dynamics as the campaign against "may issue" in *Bruen*.[166]

---

[162] Dist. of Columbia v. Heller, 554 U.S. 570, 627 (2008).

[163] Jack M. Balkin, *From Off the Wall to On the Wall: How the Mandate Challenge Went Mainstream*, Atlantic (June 4, 2012), www.theatlantic.com/national/archive/2012/06/from-off-the-wall-to-on-the-wall-how-the-mandate-challenge-went-mainstream/258040.

[164] Genevieve Lakier, *The Invention of Low-Value Speech*, 128 Harv. L. Rev. 2166, 2168 (2015) ("The term 'invented tradition' refers to novel social practices that are justified on the basis of an alleged, but ultimately fictitious, continuity with the past." (citing Eric Hobsbawm, *Introduction: Inventing Traditions*, *in* The Invention of Tradition 1, 1 (Eric Hobsbawm & Terence Ranger eds., 1992))).

[165] For an articulation and critique of this approach, see David A. Strauss, *The Modernizing Mission of Judicial Review*, 76 U. Chi. L. Rev. 859, 862 (2009).

[166] *See* Neil S. Siegel, *Federalism as a Way Station:* Windsor *as Exemplar of Doctrine in Motion*, 6 J. Legal Analysis 87, 133–34 (2014); Neil S. Siegel, *Reciprocal Legitimation in the Federal Courts System*, 70 Vand. L. Rev. 1183, 1197 (2017) (discussing the process whereby Court decisions affect "judicial precedent, legislation trends, and public opinion" which are then invoked for further change).

Nor do we think that *Bruen* is the only example of problematic outlier analysis. *Bruen* provides a useful illustration because of its salience, its insistence on historical objectivity, and its aggressive certitude. But the same lack of transparency and rigor so clearly on display in *Bruen* can found in decisions addressing other constitutional matters.[167]

Our more limited point is to caution that outlier analysis—to the extent it seeks to become relevant to, or even dispositive of[168] matters of constitutional law—is going to have to mature beyond the "open ended, rough-and-tumble"[169] fashion in which it's currently employed and become something far less opaque and far more exacting.

[167] *Cf.* Kennedy v. Louisiana, 554 U.S. 407, 448 (2008) (Alito, J., dissenting) ("In assessing current norms, the Court relies primarily on the fact that only 6 of the 50 States now have statutes that permit the death penalty for this offense. But this statistic is a highly unreliable indicator of the views of state lawmakers and their constituents."); *see also* Hills, *supra* note 92, at 21 (criticizing the Court's reasoning in an Eighth Amendment case).

[168] The dispositive nature of outliers is what's contestable. Shoddy empirics can show a theory false; but excellent empirics can't always prove a theory. Lopping off historical outliers can narrow the range of uncertainty or disagreement in much the same way that constitutional text can form a focal point, see David A. Strauss, *Common Law Constitutional Interpretation*, 63 U. Chi. L. Rev. 877, 911 (1996); but resolution of the residual disagreement will have to be supplied by other kinds of commitments, whether Thayerian deference or other forms of judicial minimalism, Ely's representational reinforcement, Dworkinian moral readings of the Constitution, or some other theory.

[169] *See* Ysursa v. Pocatello Educ. Ass'n, 555 U.S. 353, 360 n.2 (2009) (Roberts, C.J.) (internal quotation marks and citations omitted).

DAVID A. STRAUSS

RIGHTS, REMEDIES, AND TEXAS'S S.B. 8

It is not every day that a state enacts a law that is designed to prevent people from exercising a constitutional right. The Texas statute known as S.B. 8 is such a law. It prohibited pre-viability abortions at a time when that prohibition was unquestionably inconsistent with the Supreme Court's decisions. S.B. 8 then accompanied the prohibition with procedural rules that served no discernible purpose except to make it very difficult for anyone to challenge the law. The question that a law like that raises—or ought to raise—is whether our system of constitutional remedies has the resources to prevent a state from doing what Texas did.

In *Whole Woman's Health v. Jackson*,[1] the case in which the Supreme Court upheld S.B. 8, the Court did not ask that question. As a result, the Court gave the wrong answers to the questions it did ask. The answers it gave may also suggest that the Court is prepared to limit federal courts' power to enforce the Constitution in ways that go beyond the specific circumstances of S.B. 8. Those limits, if the Court were to follow through on the suggestions in *Whole Woman's Health*, would be inconsistent with established principles that are central both to enforcing constitutional rights and to the supremacy of federal law.

David A. Strauss is the Gerald Ratner Distinguished Service Professor of Law at the University of Chicago Law School.

Author's note: I thank Will Baude, Richard Fallon, and James Pfander for very helpful comments and the Burton and Adrienne Glazov Faculty Fund for financial support.

[1] 142 S. Ct. 522 (2021).

The Supreme Court Review, 2023.
© 2023 The University of Chicago. All rights reserved. Published by The University of Chicago Press. https://doi.org/10.1086/725213

The law governing federal remedies for constitutional violations is an amalgam of doctrines that are drawn from various sources and that developed at different times in history. Some of these doctrines have plausible policy justifications. Others, such as sovereign immunity, are with us today primarily because of their historical pedigree. Some combine those features, as principles like sovereign immunity have adjusted, but often only partially, to allow remedies for constitutional violations. In addition, these doctrines have developed through evolving lines of precedent, with the uncertainties that often accompany such development. As a result, "it is hard to provide a coherent account of the seemingly disparate parts of this aspect of our constitutional tradition."[2]

One appealing candidate for a unifying principle—stated by Blackstone and quoted in *Marbury v. Madison*—is that there must be a remedy for every legal wrong.[3] That is too unequivocal, as *Marbury* itself recognized; sovereign and official immunities, the lack of a cause of action, limitations on the form of relief, as well as familiar features of the law like statutes of limitations and preclusion rules, can cause violations to go unremedied. To put the point more generally, the government, to further its legitimate interests, may limit or shape the remedies that are available for alleged violations, with the result that there may be no remedy in some categories of cases.[4] But Blackstone's maxim does at least state an other-things-equal aspiration that, in turn, suggests a way to approach questions about constitutional remedies.[5] If a violation of a constitutional right is to go without a remedy, there should

---

[2] Richard H. Fallon, Jr. & Daniel J. Meltzer, *New Law, Non-Retroactivity, and Constitutional Remedies*, 104 Harv. L. Rev. 1731, 1786 (1991).

[3] *See* Marbury v. Madison, 5 U.S. (1 Cranch) 137, 163 (1803), quoting 3 William Blackstone, Commentaries *23:

> The very essence of civil liberty certainly consists in the right of every individual to claim the protection of the laws, whenever he receives an injury . . . "[I]t is a general and indisputable rule, that where there is a legal right, there is also a legal remedy by suit or action at law, whenever that right is invaded." . . .
>
> The government of the United States has been emphatically termed a government of laws, and not of men. It will certainly cease to deserve this high appellation, if the laws furnish no remedy for the violation of a vested legal right.

[4] *See* Henry M. Hart, Jr., *The Power of Congress to Limit the Jurisdiction of Federal Courts: An Exercise In Dialectic*, 66 Harv. L. Rev. 1362, 1366 (1953) (referring to "the great variety of possible remedies and the even greater variety of reasons why in different situations a legislature can fairly prefer one to another").

[5] *See* Richard H. Fallon, Jr., *Bidding Farewell to Constitutional Torts*, 107 Calif. L. Rev. 933, 936 (2019).

be a reason for allowing that to happen. That approach is reflected in a long tradition of courts designing remedies to provide for the more effective protection of individuals' rights when there is no good reason to deny protection. When a state enacts a law, like S.B. 8, that has "the clear purpose and actual effect" of preventing people from exercising a federal constitutional right,[6] courts should, at least, resolve the inevitable uncertainties in our system of constitutional remedies in favor of allowing a remedy.

As it happens, identifying a remedy in *Whole Woman's Health* was, in the end, not that difficult, and the Court's holding to the contrary was mistaken. It is unclear how significant the specific holding of *Whole Woman's Health* will be: the strategy that Texas used will not always be effective, and states may not want to use it even when it is. Among other things, states may not want to prevent individuals from exercising their constitutional rights. But beyond the specific holding, aspects of the Court's opinion suggest an approach to constitutional remedies that threatens the supremacy of the Constitution and the principle reflected in Blackstone's maxim—that courts should, to the extent that other values permit, shape constitutional remedies to protect constitutional rights.[7]

I. S.B. 8

S.B. 8 provides that "a physician may not knowingly perform or induce an abortion" after cardiac activity is detected in the embryo.[8] Cardiac activity can be detected at approximately six weeks of pregnancy, which is about four months before the fetus is viable. S.B. 8 contains an exception for "medical emergenc[ies]" that "prevented compliance with" the statute,[9] but otherwise has no exception for pregnancies that affected the health of the woman, and no exception for pregnancies resulting from rape or incest.

---

[6] *Whole Woman's Health*, 142 S. Ct. at 545 (Roberts, C.J., concurring in the judgment in part and dissenting in part).

[7] *See, e.g.*, James E. Pfander & Jacob P. Wentzel, *The Common Law Origins of* Ex parte Young, 72 Stan. L. Rev. 1269, 1359 (2020) (referring to "a venerable tradition ... [which] holds that remedies serve to some extent as substitutes for one another, filling gaps in the law that might arise either from misadventures or misguided legislative interventions").

[8] Tex. Health & Safety Code Ann. §§ 171.204(a), 171.205(a) (West Cum. Supp. 2021).

[9] *Id.* § 171.205(a).

Under the law at the time, the restriction that S.B. 8 imposed was unconstitutional. *Roe v. Wade*[10] and *Planned Parenthood v. Casey*[11] held that abortions cannot be entirely prohibited before fetal viability. Of course the Supreme Court later overruled those cases, in *Dobbs v. Jackson Women's Health Organization*.[12] But when S.B. 8 was enacted, and when *Whole Woman's Health* was decided, *Roe* and *Casey* were still good law, and the prohibition imposed by S.B. 8 cannot be reconciled with those cases.[13]

There are, ordinarily, two ways to challenge an allegedly unconstitutional state measure in federal court. Before it is enforced, an individual against whom it might be enforced can seek an injunction or a declaratory judgment against its enforcement. Several other states enacted prohibitions on pre-viability abortions like that of S.B. 8, and, before *Dobbs*, they were uniformly enjoined by lower federal courts. Alternatively, an individual might violate the law and raise the unconstitutionality of the law as a defense to an enforcement action. But that is more risky; if the law is upheld, the individual will be liable. The advantage of the pre-enforcement challenge is that the individuals who believe their rights are being violated do not have to gamble on making a correct prediction about how courts will rule on the constitutional issue. For that reason, a pre-enforcement challenge is, in some circumstances, the only way for an individual to assert a constitutional right: people cannot be expected to exercise a putative constitutional right if they have to risk punishment in order to do so.

S.B. 8 was designed to be immune from both kinds of challenges—by deterring anyone from violating the statute in order to test its constitutionality and by making federal pre-enforcement challenges effectively impossible as a legal matter. S.B. 8 specifies that its prohibition "shall be enforced exclusively through . . . private civil

---

[10] 410 U.S. 113 (1973).

[11] 505 U.S. 833 (1992).

[12] 142 S. Ct. 2228 (2022).

[13] S.B. 8 provided that a defendant in a suit brought under the statute could assert, as an affirmative defense, that "the relief sought by the claimant will impose an undue burden on that woman or that group of women seeking an abortion." TEX. HEALTH & SAFETY CODE ANN. § 171.209(b)(2). But the defense that S.B. 8 allowed was significantly narrower than that required by *Casey*, *see Whole Woman's Health*, 142 S. Ct. at 543 n.1 (Roberts, C.J., concurring in the judgment in part and dissenting in part), and the statute provided that the defense would not be available if *Roe* or *Casey* were overruled. *See also infra* text accompanying notes 59–64, on the chilling effect of uncertainty about when a defendant would be held liable.

actions."[14] Those civil actions can be brought by any person other than a state or local official.[15] The private plaintiff who brings such an action does not have to have any other connection to the case.[16] Enforcement actions under S.B. 8 can be brought not just against someone who performs a prohibited abortion but against anyone who "aids or abets" the performance of an abortion—apparently including nurses, receptionists, people providing transportation, and anyone who paid for the abortion.[17] S.B. 8 explicitly suspends Texas's normal venue rules and provides that those enforcement actions can be brought in the county where the plaintiff resided, even if that county has no connection to the defendant or to where the abortion was performed, and that the case cannot be transferred without the plaintiff's consent.[18] If the suit is successful, the court is required to award the plaintiff a bounty of "not less than" $10,000 for each abortion, plus costs and attorney's fees.[19]

A defendant who prevails in an S.B. 8 action, unlike a prevailing plaintiff, does not recover costs or attorney's fees.[20] S.B. 8 also provides that a defendant cannot take advantage of non-mutual claim or issue preclusion—so an abortion provider who prevailed could be sued again by a different plaintiff, even on a claim arising out of the same abortion.[21] The statute, evidently anticipating that *Roe* and *Casey* might be overruled, made it "not a defense" that the defendant relied on a "court decision that has been overruled," even if the decision "had not been overruled when the defendant engaged in the conduct that" violated the statute.[22]

The prohibition against pre-viability abortions in S.B. 8 applied only to abortion providers, not to women seeking abortions. So a woman who wanted an abortion could not test the constitutionality of

---

[14] TEX. HEALTH & SAFETY CODE ANN. § 171.207(a).

[15] *Id.* § 171.208(a). A "person who impregnated the abortion patient through an act of rape" or incest also may not bring such an action. *Id.* § 208(j).

[16] *See id.* § 171.208(a).

[17] *Id.* § 171.208(a)(1)-(2).

[18] *Id.* § 171.210.

[19] *Id.* § 171.208(b)(2)-(3).

[20] *Id.* § 171.208(i).

[21] *See id.* § 171.208(e)(5); *id.* § 171.208(c) (providing that the result of a prior suit precludes relief only if "the defendant previously paid the full amount of statutory damages" for the "particular abortion" at issue).

[22] *Id.* § 171.208(e)(3).

the statute by violating it even if she were willing to take a chance that it would not be upheld; she would have to find a provider (and people aiding the provider) willing to take that chance. The Supreme Court had repeatedly ruled that abortion providers can assert the rights of women who seek abortions[23]—S.B. 8 sought to limit their ability to do so, apparently again anticipating a change in the law[24]—but it is easy to see why an abortion provider would be reluctant to violate S.B. 8 in order to challenge its constitutionality. The provider could be sued anywhere in the state. Even a provider who prevailed would not recover the costs of litigating and could be sued again by any other plaintiff. And given the possibility that *Roe* and *Casey* would be overruled, the risk of losing and incurring crushing liability—at least $10,000 for each abortion, even those performed while *Roe* and *Casey* were good law—was substantial.[25] A pre-enforcement challenge was the only realistic option for providers and, therefore, for women in Texas who might need an abortion.

S.B. 8 was designed to prevent both state and federal pre-enforcement challenges. To limit state litigation, S.B. 8 provided that the state, its subdivisions, and "each officer and employee of this state or a political subdivision" were immune from "any action, claim, or counterclaim or any type of legal or equitable action that challenges the validity of any provision or application of this chapter, on constitutional grounds or otherwise."[26]

A federal court pre-enforcement challenge to the constitutionality of a state law would be brought under 42 U.S.C. § 1983, which provides that "[e]very person who, under color of [state law]" deprives an individual of rights secured by the Constitution "shall be liable to the party injured in an action at law, suit in equity, or other proper proceeding for redress."[27] An abortion provider who wanted to bring a pre-enforcement challenge to enjoin the enforcement of S.B. 8 would have to identify a

---

[23] *See, e.g.*, June Med. Servs., LLC v. Russo, 140 S. Ct. 2103, 2119–20 (2020) (citing cases).

[24] *See* Tex. Health & Safety Code Ann. § 179.209(1)-(2).

[25] Justice Kavanaugh, concurring in *Dobbs*, said that the Due Process Clause or the Ex Post Facto Clause would provide a defense for an individual who performed an abortion while the right to an abortion was still constitutionally protected. *See* Dobbs v. Jackson Women's Health Org., 142 U.S. 2228, 2309 (2022) (Kavanaugh., J., concurring). But that is not certain, so an abortion provider would still be at risk of substantial liability.

[26] Tex. Health & Safety Code Ann. § 179.211(b).

[27] 42 U.S.C. § 1983.

defendant who acted under color of state law. The state itself cannot be sued under Section 1983: that is because of the doctrine of state sovereign immunity[28] and because a state is not a "person" within the meaning of Section 1983, an interpretation of the statute influenced by sovereign immunity.[29] So ordinarily Section 1983 actions that seek relief from unconstitutional state action are brought against state officials.

A suit to enjoin a state official from acting on behalf of the state seems functionally indistinguishable from a suit against the state itself, but in 1908, in *Ex parte Young*,[30] the Supreme Court held that a state Attorney General could be enjoined from enforcing an unconstitutional law, notwithstanding the state's sovereign immunity. The Court's explanation in *Ex parte Young*, often described as a fiction, was that because a state's power is limited by the Constitution, an official who is shown to have violated the Constitution is "stripped of his official or representative character" and can be enjoined because "[t]he State has no power to impart to him any immunity from responsibility to the supreme authority of the United States."[31]

S.B. 8, though, specifically provided that only private parties, not public officials, can enforce the prohibition on abortions. That limit—obviously drafted with Section 1983 and *Ex parte Young* in mind—created the obstacles to a pre-enforcement challenge. The state cannot be enjoined, because of sovereign immunity. There is no clear basis for enjoining a prosecutor or other state official, because only private parties have the power to enforce the statute. An injunction could not name every private party who could potentially enforce the statute.[32] And, in any event, the plaintiff would have to show that the private enforcement of the statute somehow involved action under color of state law.

There was no apparent justification for Texas's decision to adopt the S.B. 8 enforcement regime other than to prevent challenges to the

[28] On state sovereign immunity, see, for example, Richard H. Fallon, Jr., John F. Manning, Daniel J. Meltzer & David L. Shapiro, Hart & Wechsler's The Federal Courts and the Federal System 905–06 (7th ed. 2015) [hereinafter Hart & Wechsler]; *infra* note 51.

[29] Will v. Mich. Dep't of State Police, 491 U.S. 58, 63–70 (1989); Quern v. Jordan, 440 U.S 332 (1979).

[30] 209 U.S. 123 (1908).

[31] *Ex parte* Young, 209 U.S. 123, 160 (1908).

[32] *See, e.g.*, Vuitton et Fils S. A. v. Carousel Handbags, 592 F.2d 126, 129 (2d Cir. 1979); Alemite Mfg. Corp. v. Staff, 42 F.2d 832, 832 (2d Cir. 1930) (L. Hand, J.) ("[A] court . . . cannot lawfully enjoin the world at large").

prohibition on pre-viability abortion. There are, of course, legitimate reasons for a state to prefer private enforcement to civil or criminal enforcement by the government. But when S.B. 8 was enacted, Texas had on its books a so-called "trigger" law: a near-total criminal prohibition on abortion that was to take effect if the Supreme Court overruled *Roe* or otherwise authorized states to forbid abortion.[33] So Texas apparently did not have any reservations about the criminal enforcement of abortion laws.

Instead, in S.B. 8 Texas abjured government enforcement but then established a system of private enforcement that had few of the advantages of private enforcement but that mimicked the public enforcement of a criminal law as closely as possible. S.B. 8 did not contain a standing requirement that limited enforcement to private parties who were injured in some specified way, as private remedies characteristically do but criminal laws do not. The damages remedy was effectively a fine; it was not designed to compensate the plaintiff or to force a potential defendant to internalize the costs of its actions—again, reasons for a private remedy that were lacking from S.B. 8. Instead, S.B. 8 gave any private party an incentive to enforce the law against any abortion provider. Other features of S.B. 8 were also designed to be a complete deterrent to abortions, as are criminal laws, or civil penalties enforced by the government. There is no reason for a state to do what Texas did with S.B. 8, except to prevent abortions and insulate the prohibition from a constitutional challenge.

II. Whole Woman's Health v. Jackson

Whole Woman's Health was a pre-enforcement challenge to the constitutionality of S.B. 8, brought under Section 1983 by a group of abortion providers. The plaintiffs' argument was that as long as they were threatened by the severe sanctions and burdensome litigation procedures imposed by S.B. 8, they would be deterred from providing abortions. That effectively deprived women of their constitutional right to a pre-viability abortion. The question was whether the obstacles that S.B. 8 had created were sufficient to deny a pre-enforcement remedy that Section 1983 would otherwise allow.

---

[33] *See* Human Life Protection Act of 2021, H.B. 1280, 87th Leg. § 3 (Tex. 2021), https://capitol.texas.gov/tlodocs/87R/billtext/pdf/HB01280F.pdf.

To overcome those obstacles, the complaint named, among other defendants, a clerk of one of the state courts. The providers said that they intended to seek certification of a defendant class of state court clerks. The clerks would be responsible for docketing any action filed under S.B. 8 and issuing a summons to the defendant,[34] so an injunction against the clerks could remove the threat of enforcement. The clerks certainly acted under color of state law. And, the providers argued, the immunity-stripping logic of *Ex parte Young* applied to the clerks, so they could not assert sovereign immunity.

The Court ruled that the plaintiffs could not sue the clerks. The Court's unsystematic discussion offered three reasons for reaching that conclusion. One reason was that, in a variety of ways, the clerks were, according to the Court, not suitable defendants in an action for an injunction. The Court said, for example, that clerks "serve to file cases as they arrive, not to participate as adversaries in those disputes";[35] that if a court entered an injunction against docketing S.B. 8 complaints, clerks might be unable to identify the complaints they were not supposed to docket ("would clerks have to assemble a blacklist of banned claims subject to immediate dismissal?"[36]); that states "might have to hire independent legal counsel for their clerks";[37] and that a clerk whose decision was challenged could not "defend himself consistent with his ethical obligation of neutrality."[38]

A second reason was related to the first: it was part of the Court's response to the apparent novelty of allowing a suit against clerks. If the clerks could be sued in *Whole Woman's Health*, the Court suggested, state court clerks could routinely be sued whenever a party wanted to forestall private litigation. The *Whole Woman's Health* plaintiffs, the Court said, "have not offered any meaningful limiting principles."[39] If the plaintiffs prevailed, nothing "would stop federal judges from prohibiting state courts and clerks from hearing and docketing disputes" involving other statutes or common law claims.[40]

[34] *See* Tex. R. Civ. P. 22, 99.

[35] Whole Woman's Health v. Jackson, 142 S. Ct. 522, 532 (2021).

[36] *Id.* at 533.

[37] *Id.*

[38] *Id.*

[39] *Id.* at 532.

[40] *Id.*

The Court's third reason has the most far-reaching implications. The Court ruled that the plaintiffs' claim was barred by sovereign immunity, despite *Ex parte Young*. The Court asserted that *Ex parte Young* established only "a narrow exception" to sovereign immunity, one that was "grounded in traditional equity practice" and that only "allows certain private parties to seek judicial orders in federal court preventing state executive officials from enforcing state laws that are contrary to federal law."[41] "[S]tate-court judges or clerks," the Court said "do not enforce state laws as executive officials might; instead, they work to resolve disputes between parties."[42] Therefore, the Court concluded, the *Ex parte Young* principle did not extend to the plaintiffs' claims.

The Court gave the plaintiffs an ostensible partial victory, holding that some Texas state officials who were among the defendants had the authority, under Texas law, to license and discipline physicians; those officials, the Court said, could be enjoined from revoking the plaintiffs' licenses for violations of S.B. 8.[43] That holding apparently enabled the providers to get pre-enforcement review of the constitutionality of S.B. 8, but in a way that, even in principle, would have failed to secure the rights of many women seeking abortions.

An injunction against the licensing officials would not have bound private parties, either in terms or by its preclusive effect; private parties could still sue a provider who performed an abortion. A federal district court's (or federal court of appeals') holding that S.B. 8 was unconstitutional would not be binding precedent in Texas state courts. In some circumstances, state courts might follow a decision of a lower federal court even though they are not bound, so as to avoid creating an intra-jurisdiction conflict. Or a state might even amend its law once a federal court, especially a court of appeals, ruled that it was unconstitutional. But in the atmosphere of hostility to abortion rights that produced S.B. 8, that was unlikely to happen, especially because there was a widespread expectation that the Supreme Court would limit or overturn *Roe* and *Casey*.

So even if the abortion providers successfully sued the licensing officials, they would very likely not be protected from civil liability in state court, and women would, as a practical matter, continue to be

---

[41] *Id.*

[42] *Id.*

[43] *Id.* at 535–37.

denied their constitutional rights. A pre-enforcement suit by the providers against the licensing officials could, ultimately, reach the Supreme Court, and that would accomplish one important purpose: it would prevent S.B. 8 from thwarting judicial review entirely so that the state could simply disregard the limits imposed by the Court's decisions.[44] But the providers would not be protected from civil liability until, and unless, the Court ruled in their favor. For someone who needs an abortion, delay amounts to denial, so even an ultimate victory by the providers might not be enough. Litigation delays are unavoidable, but under the law at the time, S.B. 8 unquestionably violated women's constitutional rights. The decision to allow suits against licensing officials did not prevent that violation from continuing while the litigation ran its course.

The holding that the licensing officials could be sued also did nothing to address the issue that potentially makes *Whole Woman's Health* significant even after *Dobbs*: whether a state can prevent pre-enforcement challenges to an arguably unconstitutional law by specifying that the law is to be enforced only by private parties, and—crucially—by any private party.[45] If the licensing officials could be sued, that was only because the architects of S.B. 8 inadvertently allowed a vestige of government enforcement to remain. A state that wanted to use Texas's strategy in the future could just avoid that mistake.

As it happened, there was, apparently, no mistake. Even though the Supreme Court seemed to hold that the officials had the authority under state law to revoke the plaintiffs' licenses for violating S.B. 8,[46] the Fifth Circuit, on remand, certified the state law question to the Texas Supreme Court, which ruled that the officials lacked that

---

[44] *See* Richard H. Fallon, Jr., *Constitutional Remedies: In One Era and Out the Other*, 136 Harv. L. Rev. 1300, 1310 & nn. 57, 58 (2023) ("[T]he Constitution requires an adequate scheme of rights to sue and judicial remedies to ensure that federal and state officials will stay generally within constitutional bounds." (citing Fallon & Meltzer, *supra* note 2, at 1789)).

[45] See *infra* text accompanying notes 75–76, on how the absence of a standing requirement in S.B. 8 made a declaratory judgment remedy unavailable in *Whole Woman's Health*.

[46] Four Justices from the majority said, on this point (joined as to the conclusion by the Justices who dissented from the principal holdings), "we hold that sovereign immunity does not bar the petitioners' suit against these named defendants at the motion to dismiss stage," *Whole Woman's Health*, 142 S. Ct. at 535–36, and specifically took issue with Justice Thomas's assertion, in his opinion dissenting on this point, that the officials lacked enforcement authority under state law, *see id.* at 539–41 (Thomas, J., concurring in part and dissenting in part).

authority.[47] The U.S. Supreme Court, over three dissents, refused to take any further action.[48] So S.B. 8 remained in effect, and effectively immune from a pre-enforcement challenge.

III. S.B. 8 and Pre-enforcement Remedies

Were it not for sovereign immunity, the logical defendant in a suit to enjoin the enforcement of S.B. 8's ban on pre-viability abortions would have been the state of Texas. A state can violate the Constitution by enforcing a judgment in a private civil action; *Shelley v. Kraemer*[49] and *New York Times v. Sullivan*[50] establish that. But an individual whose rights were violated by S.B. 8 could not sue Texas, because of state sovereign immunity.[51]

This situation—the government has violated an individual's right, but the government itself cannot be sued because of sovereign immunity—is not new. Courts have confronted its equivalent for centuries. Their response—apparently since the time of Edward I of England, who reigned from 1272 to 1307—has been to allow an action in damages or a suit for specific relief not against the government itself (or, in England, the sovereign) but against the official whose acts violated the individual's rights.[52] Dicey considered officials' susceptibility to such

[47] *See* Whole Woman's Health v. Jackson, 23 F.4th 380 (5th Cir. 2022) (certifying question); Whole Woman's Health v. Jackson, 642 S.W.3d 569 (Tex. 2022) (answering certified question); Whole Woman's Health v. Jackson, 31 F.4th 1004 (5th Cir. 2022) (remanding case with instructions to dismiss challenges to private enforcement provisions).

[48] *In re* Whole Woman's Health, 142 S. Ct. 701 (2022) (denying petition for a writ of mandamus).

[49] 334 U.S. 1 (1948).

[50] 376 U.S. 254 (1964).

[51] State sovereign immunity was at one time identified with the Eleventh Amendment (as it was in *Ex parte Young*), but more recently the Court has said that "the sovereign immunity of the States neither derives from, nor is limited by, the terms of the Eleventh Amendment." Alden v. Maine, 527 U.S. 706, 713 (1999). In fact, there does not seem to be any specific textual basis for state sovereign immunity; according to the Court, it is "a fundamental aspect of the sovereignty with the States enjoyed before the ratification of the Constitution, and which they retain today . . . except as altered by the plan of the Convention or certain constitutional Amendments." *Id.* On the various bases for state sovereign immunity, and the relationship between the Eleventh Amendment and state sovereign immunity, see William Baude & Stephen E. Sachs, *The Misunderstood Eleventh Amendment*, 169 U. Pa. L. Rev. 609 (2021).

[52] *See* Louis L. Jaffe, *Suits Against Governments and Officers: Sovereign Immunity*, 77 Harv. L. Rev. 1, 3 (1963) (suggesting that such actions have been available "[f]rom the reign of Edward I on").

suits to be integral to the rule of law.[53] That is the kind of suit that *Ex parte Young* allowed. When *Ex parte Young* was decided, so-called officer suits of this kind were familiar, which may explain why the Court adopted what now seems like a fiction.[54]

This approach to remedies—to allow a suit against an official when an individual has been wronged and the government itself cannot be sued—is in keeping with Blackstone's maxim. "The availability of suit against an officer . . . was the result of a deliberate effort to protect the citizen from governmental misuse of authority."[55] S.B. 8 was a governmental misuse of authority: it imposed a plainly unconstitutional restriction on the abortion providers. The role of the government was obscured, because S.B. 8 specified that only private parties could sue the providers; that is why the providers' version of the officer suit was a suit against the clerks. But like the plaintiffs whose claims gave rise to officer suits, the providers were subjected to an unlawful action by the government, and sovereign immunity prevented them from suing the government.

In these circumstances, the approach of the *Whole Woman's Health* Court was not like that of the courts that enabled officer suits; it was more nearly the opposite. Rather than identifying ways in which remedial principles might protect individuals from an unconstitutional law, the *Whole Woman's Health* Court overstated the novelty of the

---

[53] *See* A.V. DICEY, THE LAW OF THE CONSTITUTION 189 (7th ed. 1908):

> We mean . . . when we speak of the "rule of law" as a characteristic of our country not only that no man is above the law, but (what is a different thing) that here every man, whatever be his rank or condition, is subject to the ordinary law of the realm and amenable to the jurisdiction of the ordinary tribunals. . . . The Reports abound with cases in which officials have been brought before the Courts, and made, in their personal capacity, liable to punishment, or to the payment of damages, for acts done in their official character but in excess of their lawful authority.

[54] *See, e.g.*, HART & WECHSLER, *supra* note 28, at 927 ("At the time of the Court's decision in Ex parte Young, the tradition of 'officer suits' as a means of redress for official misconduct was already well established."); Anne Woolhandler, *The Common Law Origins of Constitutionally Compelled Remedies*, 107 YALE L.J. 77, 99, 108, 121 (1997).

There was a comparable background of providing specific relief through the common law writs of certiorari, mandamus, and prohibition. *See* Pfander & Wentzel, *supra* note 7, at 1335 ("[C]ommon law courts granted specific relief against Crown officials from the early seventeenth century onward, both to prevent and to compel official action, without worrying about the English tradition of sovereign immunity."). For an argument that the writ of prohibition, where available, could provide relief against a law like S.B. 8, see James E. Pfander, *Judicial Review of Unconventional Enforcement Regimes*, 102 TEX. L. REV. (forthcoming 2023).

[55] LOUIS L. JAFFE, JUDICIAL CONTROL OF ADMINISTRATIVE ACTION 237 (1965).

relief that the plaintiffs sought and, at times, seemed to go out of its way to find reasons to deny a remedy. A suit against the clerks of the court was not different in principle from suits that the Court has allowed; the Court's explanations for why the clerks were not suitable defendants were unpersuasive, sometimes to the point of seeming contrived. The apparent (but not real) absence of a limiting principle was the result not of the providers' claim but of the provision of S.B. 8 that allowed any private party to sue, without a standing requirement—one of the provisions designed to prevent pre-enforcement review. And the Court put forward a version of *Ex parte Young* that is at odds with the way the principle of that case has been understood for decades.

The Court's discussion of sovereign immunity and *Ex parte Young* is especially problematic. The Court's treatment of the other two issues—whether the clerks were appropriate defendants whether there is a limiting principle—will have implications if other states attempt to evade federal constitutional requirements in the way that Texas did. The principle of *Ex parte Young* reaches further. Sovereign immunity is, to say the least, a difficult doctrine to justify.[56] An argument can be made that it should, nonetheless, not be uprooted, because it is an established principle and other remedial principles have grown up around it and adapted to it. But the principle of *Ex parte Young* is one of those remedial adaptations that have limited sovereign immunity in a way that makes it more acceptable. For the Court to impose new limits on *Ex parte Young*, and in that way expand sovereign immunity—without explaining why that should be done—would be a troubling development, one that is unsupported by precedent and that threatens to undermine principles that are central to our constitutional system.

A. PROPER DEFENDANTS

It is unusual to name the clerk of the court as a defendant in a suit to enforce constitutional rights. As the dissenters in *Whole Woman's Health* said, though, that was a necessary response to the unusual

[56] *See, e.g.*, Erwin Chemerinsky, *Against Sovereign Immunity*, 53 Stan. L. Rev. 1201 (2001); *id.* at 1201 n.3 ('"[T]he principle has never been discussed or the reasons for it given, but it has always been treated as an established doctrine."' (quoting United States v. Lee, 106 U.S. 196, 207 (1882)) (citations omitted)); David P. Currie, *Sovereign Immunity and Suits Against Government Officers*, 1984 Sup. Ct. Rev. 149, 168 ("Sovereign immunity in an unattractive doctrine that does not belong in an enlightened constitution. Unfortunately, however, it is a part of ours.").

character of S.B. 8.[57] Other defendants that the providers named—state judges (against whom they also intended to pursue a class action) and, possibly, the state Attorney General—might also have been appropriate defendants.[58] But the clerks present the strongest case.

The providers' claim was that because of S.B. 8, there was an ongoing violation of women's constitutional rights. There is no question that a threat of liability, by itself, can deter individuals from exercising a constitutional right; when it does, pre-enforcement relief is an appropriate remedy.[59] Violating S.B. 8 and defending an enforcement action was not a "'meaningful' avenue of relief" because, as the Court has said in a related context, it would have required the providers "to 'bet the farm'"[60] by risking liability if the enforcement court did not agree with their position. In all these respects, the providers' suit (assuming, as the Court did, that they could assert the rights of women seeking abortions) resembled the claim in *Ex parte Young*.[61]

In fact, the state statute in *Ex parte Young*—a regulation of railroad rates—was the S.B. 8 of its time. It imposed severe penalties for a violation specifically in order to deter the railroads from challenging the constitutionality of the rates.[62] The proponents of the scheme in

---

[57] *See, e.g.*, 142 S. Ct. at 545 (Roberts, C.J., concurring in the judgment in part and dissenting in part).

[58] S.B. 8 put private parties in the position of enforcing the law in a way that closely resembled the enforcement of the criminal law. *See supra* text accompanying note 33. That might have been a basis for making the state's chief law enforcement officer responsible for the private parties' actions.

[59] The Court has said this many times. *See, e.g.*, Susan B. Anthony List v. Driehaus, 573 U.S. 149, 155, 163 (2014); Virginia v. Am. Booksellers Ass'n, 484 U.S. 383, 392–93 (1988); Steffel v. Thompson, 415 U.S. 452, 475 (1974).

[60] Free Enter. Fund v. Pub. Co. Accounting Oversight Bd., 561 U.S. 477, 490–91 (2010) (quoting Thunder Basin Coal Co. v. Reich, 501 U.S. 200, 212 (1994); and then MedImmune, Inc. v. Genentech, Inc., 549 U.S. 118, 129 (2007)).

[61] In explaining why it upheld the injunction against the Attorney General, instead of insisting that the railroad violate the statute in order to challenge its constitutionality, the *Ex parte Young* Court emphasized precisely the deterrent effect of requiring the railroads to bet the farm: "A sale of a single ticket above the price mentioned in that act might subject the ticket agent to a charge of felony, and upon conviction to a fine of five thousand dollars and imprisonment for five years. . . . It would not be wonderful if, under such circumstances, there would not be a crowd of agents offering to disobey the law." *Ex parte* Young, 209 U.S. 123, 164 (1908).

[62] At least that was the plaintiffs' allegation. "'It was by many considered a matter of congratulation that the legislature had enacted rate laws which, whether or not constitutional, the railroad companies could not successfully contest.'" Barry Friedman, *The Story of* Ex parte Young: *Once Controversial, Now Canon*, *in* FEDERAL COURTS STORIES 247, 265 (Judith Resnik and Vicki C. Jackson, eds., 2010) (quoting Brief for Respondent at *51, *Ex parte* Young, 209 U.S. 123, 1907 WL 18907).

Ex parte Young had not quite hit upon the S.B. 8 strategy of allowing only private enforcement, but they did the next best thing, deliberately omitting from the statute any provision authorizing state officials to enforce the regulation. When the railroads' shareholders sued the state Attorney General to enjoin the enforcement of the rates, the Attorney General asserted, as Texas did in *Whole Woman's Health*, that the suit was necessarily a suit against the state, barred by sovereign immunity, because no state official was charged with enforcing the statute.[63] The *Ex parte Young* Court rejected that argument by ruling that, under state law, the state Attorney General had "general powers" to "enforce such or any other law."[64]

The relief that the plaintiffs asked for in *Whole Woman's Health*—an injunction forbidding Texas state court clerks from docketing S.B. 8 actions—does not seem very different from the relief the Court awarded in *Ex parte Young*. It would end the ongoing constitutional violation, by removing the deterrent that, in practical terms, made it impossible for women to obtain pre-viability abortions. The Court's first set of reasons for concluding otherwise—that clerks are in various ways not suitable defendants—was scattershot. The Court's principal concern seemed to be that clerks "perform [only] ministerial tasks . . . related to litigation."[65] It is true that, unlike a prosecutor, for example, a clerk does not make the decision to institute an action. But there is nothing anomalous about courts granting specific relief against government officials who have only ministerial responsibilities. One example is mentioned in *Ex parte Young* itself: a writ of habeas corpus is directed at the official responsible for the custodial institution in which the petitioner is held. In *Osborn v. Bank of the United States*,[66] which *Ex parte Young* relied on, a state official—Ohio's state auditor—had what the Court described as a "ministerial" duty to seize the Bank's property;

[63] *See Ex parte* Young, 209 U.S. at 154; Friedman, *supra* note 62, at 261 ("Savvy about the rule of *Fitts v. McGhee*, [172 U.S. 516 (1899),] the legislation also gave no particular state officer . . . authority to enforce the rates."). *Fitts v. McGhee* had rejected a claim for an injunction against state officials' enforcing an allegedly unconstitutional rate regulation on the ground that the officials did not have any "any special duty in connection with the Act," 172 U.S. at 529, so the suit was effectively against the state and therefore barred by sovereign immunity.

[64] *Ex parte* Young, 209 U.S. at 154 ("There was no special provision in the statute as to rates, making it the duty of the Attorney General to enforce it, but under his general powers he had authority to ask for a mandamus to enforce such or any other law.").

[65] Whole Woman's Health v. Jackson, 142 S. Ct. 522, 533 (2021).

[66] 22 U.S. (9 Wheat.) 738 (1824).

the Court ruled that an injunction was the proper remedy against that unconstitutional act.[67]

The Court in *Whole Woman's Health* said that clerks "serve to file cases as they arrive, not to participate as adversaries in those disputes."[68] Engaging in adversary litigation is, of course, not the job responsibility of a clerk the way it is for a prosecutor. But that is true of many state officials who are subject to federal court orders. The custodian who is a respondent in a habeas corpus proceeding "serve[s] to" operate the prison, not to litigate. State government officials who administer social welfare programs do not have litigation in their job description, but they can be sued by beneficiaries and enjoined to conform to federal law.[69]

The more substantial question, suggested by the Court in *Whole Woman's Health* and raised in the briefs, is whether a clerk is sufficiently adverse to a plaintiff seeking a pre-enforcement injunction. Unlike a custodian or the administrator of a social welfare program, the clerks who docket a pre-enforcement challenge are not actively depriving plaintiffs of their liberty or of benefits to which they are allegedly entitled. But the clerks are nonetheless adverse, for essentially the same reason that a prosecutor is adverse in the typical pre-enforcement challenge. A pre-enforcement challenge is justified because the possibility of after-the-fact litigation deters an individual from exercising a constitutional right. The reason prosecutors are adverse to the plaintiff in the ordinary pre-enforcement challenge, when the law is enforced by the government, is that prosecutors can institute the litigation that has the deterrent effect that effectively denies a constitutional right.

The clerk also institutes litigation that has that effect—just at a later stage of the proceedings, and even when the litigation is brought by a private party. The litigation cannot proceed unless the clerk dockets the case and issues a summons. The clerk does not have discretion to do otherwise, but that is generally true of ministerial officials, like the custodian in habeas corpus proceedings; it does not make the clerk less of an adverse party. Clerks also do not act on their own initiative; a private party must file a complaint first. But prosecutors would not be

---

[67] *Id.* at 840.

[68] *Whole Woman's Health*, 142 S. Ct. at 532.

[69] *See, e.g.*, Edelman v. Jordan, 415 U.S. 651, 657 (1974) (citing Graham v. Richardson, 403 U.S. 365 (1971); and Goldberg v. Kelly, 397 U.S. 254 (1970)).

exempt from an *Ex parte Young* injunction even if they could act only after receiving a complaint from a private party. In *Susan B. Anthony List v. Driehaus*,[70] for example, a challenge to the constitutionality of an Ohio law that regulated speech during political campaigns, the Court unanimously allowed a pre-enforcement suit for an injunction against the members of a state administrative agency that could act only upon a complaint.[71]

Clerks perform other ministerial functions that are closely connected to the work of the court, and when they do, they can be sued for an injunction by a plaintiff alleging a violation of federal law. For example, lower courts have held—and the Supreme Court's decision in *Fuentes* v. *Shevin*[72] at least suggests—that an individual can obtain an injunction by showing that a clerk or similar official threatened to seize property, in violation of the Constitution or federal law, in order to satisfy a judgment or to provide a pre-judgment remedy.[73] Those actions would also be nondiscretionary, and they would be taken at the instance of a private party. Those actions by clerks involve the seizure of property; but *Ex parte Young* itself explicitly equated, for purposes of pre-enforcement relief, the seizure of property and the enforcement of an unconstitutional law.[74] Just as the auditor in *Osborn* could be enjoined from taking the ministerial action of seizing the Bank's property, because such an injunction would protect the Bank from an unconstitutional injury, enjoining clerks from docketing S.B. 8 complaints would remove the threat of liability under S.B. 8 and in that way prevent the effective denial of a right to an abortion.

The *Whole Woman's Health* Court's other suggestions were that clerks would have to retain counsel and that they would have to keep track of the cases they were enjoined from docketing. But a clerk acts on behalf of the state and would be represented by the state—just like

[70] 573 U.S. 149 (2014).

[71] *See* Brief for State Respondents at 8, Susan B. Anthony List v. Driehaus, 573 U.S. 149 (2014) (No. 13-193).

[72] 407 U.S. 67 (1972). In one of the cases decided by *Fuentes*, individuals who were subject to a pre-judgment remedy sought an injunction against not just private defendants but also the "prothonotary"—the chief clerk—of the Philadelphia courts. *See* Epps v. Cortese, 326 F. Supp. 127, 128 (E.D. Pa. 1971) (three-judge court). In *Fuentes*, the Court reversed the district court's denial of relief without suggesting that there was an issue about entering an injunction against a clerk.

[73] *See, e.g.*, Strickland v. Alexander, 772 F.3d 876, 886 (11th Cir. 2014); Finberg v. Sullivan, 634 F.2d 50, 53 (3d Cir. 1980) (en banc).

[74] *See Ex parte* Young, 209 U.S. 123, 167 (1908).

a warden or any other state official, including a prosecutor or the Texas state licensing officials against whom the Court allowed the *Whole Woman's Health* plaintiffs to proceed. None of them has to "hire independent legal counsel." And the clerk need not "assemble a blacklist"; the defendant in an S.B. 8 action would let the clerk know that the case should not be docketed, if that became necessary. The weakness of these arguments advanced by the Court is symptomatic of how far *Whole Woman's Health* is removed from the traditional approach courts have taken when a violation of a right might go unremedied.

B. LIMITING PRINCIPLES

The *Whole Woman's Health* Court was apparently concerned that, if the providers were able to sue clerks of the court, other potential defendants who anticipated raising a federal defense could sue state court clerks to enjoin private common law or statutory lawsuits before they were brought. With respect to this issue, too, the *Whole Woman's Health* Court treated the providers' claim as more novel than it actually was and failed to recognize that it had an established analogue. What the providers sought was, in principle, the same kind of relief that would be available in a declaratory judgment action. The providers' claim would be subject to the same limits as a declaratory judgment action, limits that were not at issue in *Whole Woman's Health*.

A principal purpose of a declaratory judgment action is to enable individuals to avoid being in the position of the providers in *Whole Woman's Health*—having to risk liability in order to assert the lawfulness of something they intend to do.[75] A declaratory judgment was not available to the *Whole Woman's Health* plaintiffs because S.B. 8 allowed anyone to sue an abortion provider for damages, so the providers could not identify a private defendant against whom they could bring a declaratory judgment action. That is why they had to sue the clerks.

But ordinarily common law or statutory actions can be brought only by identifiable individuals. There is some form of standing requirement—which, of course, S.B. 8 deliberately omitted, in order to prevent a pre-enforcement challenge. Those individuals might then be defendants in a declaratory judgment action, if they threaten to sue

---

[75] *See, e.g.*, MedImmune, Inc. v. Genentech, Inc., 549 U.S. 118, 128–34 (2007). John Harrison, *Ex parte* Young, 60 STAN. L. REV. 989, 1000 (2008), points out the "resemblance between anti-suit injunctions and declaratory judgments." *See also id.* at 1014 n.100.

the declaratory judgment plaintiff. Declaratory judgment actions are subject to limits imposed by principles of ripeness and concreteness. The threat of liability has to be more than just speculative, and the dispute has to be well-enough defined for a court to decide whether a constitutional right is at stake.[76] Those are the relevant limiting principles, and they would apply to a suit, like the providers', against state court clerks. There would be no reason for a different outcome in *Whole Woman's Health* to have removed those limits.

IV. Sovereign Immunity and *Ex parte Young*

The Court's treatment of sovereign immunity and *Ex parte Young* is potentially the most far-reaching aspect of the opinion. It raises questions not just about the specific result in *Whole Woman's Health* but about whether the Court will, in the future, make it more difficult for federal courts to prevent states from violating the Constitution.

A. A "Narrow Exception"?

The Court described *Ex parte Young* as "a narrow exception" to sovereign immunity that is "grounded in traditional equity practice" and "allows certain private parties to seek judicial orders in federal court preventing state executive officials from enforcing state laws that are contrary to federal law."[77] This description of *Ex parte Young* is more grudging than what the Court has said—and, more important, done—in the past.[78] For decades, the Court has simply treated it as settled law that, by virtue of *Ex parte Young*, sovereign immunity does not bar suits against state officials to enjoin them from violating the

---

[76] *See, e.g.*, *MedImmune, Inc.*, 549 U.S. at 127; Steffel v. Thompson, 415 U.S. 452, 459 (1974).

[77] *Whole Woman's Health*, 142 S. Ct. at 532.

[78] It is also discordant with what contemporaries thought. When it was decided, *Ex parte Young* was intensely controversial; it is generally thought to have prompted the enactment of the statute, 36 Stat. 557 (1910), that required a three-judge district court when plaintiffs sought a preliminary injunction against a state statute on constitutional grounds (later extended to state administrative orders, and to permanent injunctions, before being mostly repealed in 1976). *See, e.g.*, Steffel v. Thompson, 415 U.S. 452, 465 (1974) ("A 'storm of controversy' raged in the wake of Ex parte Young . . . ." (quoting Henry M. Hart, Jr. & Herbert Wechsler, The Federal Courts and the Federal System 967 (2d ed. 1973))). On the history of the three-judge court statute, *see, e.g.*, Michael E. Solimine & James L. Walker, *The Strange Career of the Three-Judge District Court*, 72 Case W. Res. L. Rev. 909 (2022); Hart & Wechsler, *supra* note 28, at 1089–90.

Constitution.[79] The Court has also explained that *Ex parte Young* is a principal way in which federal courts ensure the supremacy of federal law: it is "necessary to permit the federal courts to vindicate federal rights and hold state officials responsible to 'the supreme authority of the United States.'"[80] So, for example, in contrast to the *Whole Woman's Health* Court's hedged references to "traditional equity practice," "certain private parties," and "state executive officials," Justice Scalia, writing for the Court in *Verizon Maryland Inc. v. Public Service Commission of Maryland*,[81] said that whether *Ex parte Young* overcomes state sovereign immunity depends on a "straightforward inquiry into whether [the] complaint alleges an ongoing violation of federal law and seeks relief properly characterized as prospective."[82]

It is not just a matter of verbal formulations. *Ex parte Young*, understood in this way, has been the basis of many of the most far-reaching things federal courts have done. Federal courts issued injunctions forcing state universities to desegregate.[83] They required states to reapportion their legislatures.[84] They have entered injunctions requiring state prisons to reform their conditions in order to meet constitutional standards.[85] They have required states to modify social welfare programs to comply with federal constitutional and statutory provisions.[86] State sovereign immunity would have barred all of these were it not for *Ex parte Young* or an equivalent principle. Although the Court from time to time has described *Ex parte Young* as a "narrow" principle, these decisions, and others like them, are not minor assertions of federal supremacy. And none has an obvious connection to "traditional equity practice."

---

[79] *See, e.g.*, Griffin v. Cnty. Sch. Bd., 377 U.S. 218, 228 (1964) ("It has been settled law since *Ex parte Young*, 209 U.S. 123 (1908), that suits against state and county officials to enjoin them from invading constitutional rights are not forbidden by the Eleventh Amendment.").

[80] Pennhurst State Sch. & Hosp. v. Halderman, 465 U.S. 89, 105 (1984) (quoting *Ex parte* Young, 209 U.S. 123, 160 (1908)).

[81] 535 U.S. 635 (2002) .

[82] *Verizon Maryland Inc.*, 535 U.S. at 645 (quoting Idaho v. Coeur d'Alene Tribe of Idaho, 521 U.S. 261, 296 (1997) (O'Connor, J., joined by Scalia & Thomas, JJ., concurring in part and concurring in judgment)).

[83] *See, e.g.*, McLaurin v. Okla. State Bd. of Regents, 339 U.S. 637 (1950); Meredith v. Fair, 305 F.2d 343 (5th Cir. 1962).

[84] *See, e.g.*, Reynolds v. Sims, 377 U.S. 533 (1964); *id.* at 537 n.2.

[85] *See, e.g.*, Hutto v. Finney, 437 U.S. 678, 684–85 (1978).

[86] *See, e.g.*, Edelman v. Jordan, 415 U.S. 651, 664 (1974).

In fact, most of the time, in cases seeking injunctive relief against state officials, no one even had to mention *Ex parte Young*. If the plaintiffs sought an injunction against unconstitutional state conduct, it went without saying that sovereign immunity was not an obstacle (as long as the plaintiff took care to name a responsible official as the defendant, in keeping with the fiction, if that is the right term, of the traditional officer suits). In the Supreme Court, at least, an issue about *Ex parte Young* came up only when the injunction sought relief that was either unusual—for example, when a state agency sued another state agency in federal court (the Court held that *Ex parte Young* permitted that suit[87])—or when the plaintiffs sought an injunction that arguably required the equivalent of damages payments out of the state treasury for a completed violation, something that *Ex parte Young* does not authorize. Otherwise, it was just taken for granted that *Ex parte Young* authorized injunctions against state officials. It is like the proverbial dog that did not bark: When a principle is applied over and over without being mentioned, because even parties with the strongest incentives to challenge its application did not do so, that is very good evidence of how well established the principle is.

The abortion providers' claim in *Whole Woman's Health* unquestionably satisfied the "straightforward" inquiry described in Justice Scalia's opinion for the Court in *Verizon Maryland*. There was an "ongoing violation of federal law"—state action prevented women from obtaining pre-viability abortions—and the relief that the providers sought was entirely prospective. The relief that the abortion providers sought in *Whole Woman's Health* would have been far less intrusive into the state's operations than many of the high-profile injunctions that *Ex parte Young* permitted. What the *Whole Woman's Health* plaintiffs sought was routine, the same kind of relief as in *Ex parte Young* itself and in "myriad" other cases[88]—an injunction barring an enforcement action.

B. INJUNCTIONS AGAINST CLERKS AND JUDGES

The Court's principal reason for its conclusion to the contrary was an emphatic statement in the opinion in *Ex parte Young*: the power to enjoin a state official from bringing suit "does not include the power

---

[87] Va. Off. for Prot. & Advoc. v. Stewart, 563 U.S. 247, 262 (2011).

[88] Sup. Ct. of Va. v. Consumers Union of Am., Inc., 446 U.S. 719, 737 (1980).

to restrain a court from acting in any case brought before it" because "an injunction against a state court would be a violation of the whole scheme of our Government."[89] That statement apparently reflected a principle of equity practice: an injunction against a lawsuit must be directed to the party, not to the court.[90] The *Whole Woman's Health* Court said that this ruled out injunctions against state court judges. As for clerks, the *Ex parte Young* Court had added that there was also no "power to prevent any investigation or action by a grand jury," which is "part of the machinery of a criminal court."[91] Clerks, the *Whole Woman's Health* Court said, are "part of the machinery" of an enforcement court, and therefore they also could not be enjoined.[92]

Even taking it as a given that judges cannot be enjoined—it is not clear that should be taken as a given[93]—it does not follow that clerks cannot be enjoined. The metaphor about "machinery" is obviously inadequate. Clerks and judges perform different functions; the principle forbidding injunctions against judges cannot simply be extended to clerks. There seems to be no question, for example, that judges can be sued for injunctions related to non-judicial functions, such as enforcing bar membership requirements.[94] In addressing the analogous question of when a judge is entitled to absolute immunity from damages liability, the Court has described "paradigmatic judicial acts" as "resolving disputes between parties" and said that "[a]dministrative decisions, even though they may be essential to the very functioning of the courts, have not similarly been regarded as judicial acts."[95] The clerk of the court (distinguished, if necessary, from a judge's own law clerks) plays no role in resolving disputes between parties; and the clerk's acts, however essential to the functioning of the court, are entirely administrative.

In Texas, the clerk of the court receives a complaint, dockets the case, and issues a citation. An injunction would be directed to those

---

[89] 209 U.S. 123, 163 (1908).

[90] *See, e.g.*, 2 Joseph Story, Commentaries on Equity Jurisprudence, as Administered in England And America ¶¶ 874–75, at 189–90 (Melville Bigelow ed., 13th ed., 1886); *see also* Pulliam v. Allen, 466 U.S. 522, 529 (1984).

[91] *Ex parte* Young, 209 U.S. at 163.

[92] *Whole Woman's Health*, 142 S. Ct. at 533.

[93] See *infra* text accompanying notes 100–111 (discussing Mitchum v. Foster, 407 U.S. 225 (1972); and Pulliam v. Allen, 466 U.S. 522 (1984)).

[94] *See, e.g.*, Sup. Ct. of Va. v. Consumers Union of Am., Inc., 446 U.S. 719 (1980).

[95] Forrester v. White, 484 U.S. 219, 227–28 (1988).

administrative actions by the clerk. The citation says that if the defendant does not file an answer within a specified time, "a default judgment may be taken against you."[96] Issuing a citation, and warning recipients that a judgment will be entered against them if they ignore it, is what police officers do when they issue citations for traffic violations. Lower courts have held that state court clerks can be enjoined from seizing property to enforce judgments by garnishments that are forbidden by federal law.[97] That action could be described as part of the "machinery" of the court; it is a way of bringing litigation to a conclusion. The role of a clerk who dockets a case and issue a summons is a counterpart to that: it initiates the litigation.

Clerks act at the instance of the private party who filed the complaint, instead of on their own initiative, but it is difficult to see why that should matter;[98] the clerk is still a state official enforcing the law. The clerk of the court may be employed by the judicial branch of the state government (and may be supervised by a judge—in the judge's non-judicial role[99]), but a state's internal organization should not affect whether federal relief is available. Functionally, the clerk's actions toward the plaintiff seeking an injunction are like those of an executive branch enforcement official: both direct the apparatus of the state at an individual, after which the court will adjudicate the merits.

All of this assumes that injunctions against judges are forbidden, because of the "whole scheme of government" dictum in *Ex parte Young* and the equity principle it reflects. But the equitable rule grew out of a system of separate (and to some degree competing) courts of law and equity that was different in many ways from ours, so the implications of that rule for our system are not obvious. In several cases that are much more recent than *Ex parte Young*, the Court has signaled that it is comfortable with federal injunctions against state judges. In *Mitchum v. Foster*,[100] the Court ruled that Section 1983 authorized a federal court to enjoin a state court public nuisance action, notwithstanding the Anti-Injunction Act, which generally forbids federal

[96] Tex. R. Civ. P. 22–27, 99(a)–(c).

[97] *See supra* note 73 and accompanying text.

[98] *See supra* text accompanying notes 70–71.

[99] *See Forrester*, 484 U.S. at 229–30 (holding that a judge is not entitled to absolute immunity from damages for a "personnel decision[]").

[100] 407 U.S. 225 (1972).

courts from enjoining state court proceedings.[101] But *Mitchum* held that Section 1983 is within an exception to the Anti-Injunction Act.[102] The defendants in *Mitchum* included a state court judge, as well as a prosecutor and a sheriff.[103] The Court in *Mitchum* not only did not distinguish among the defendants but said that Section 1983 was enacted in part to protect against state judicial violations of constitutional rights, as well as violations by state legislatures and executive officials.

The *Whole Woman's Health* Court said that *Mitchum* "did not purport to pass judgment on any sovereign immunity defense, let alone suggest any disagreement with" the *Ex parte Young* dictum about judges.[104] The Court also said that *Mitchum* "went out of its way to emphasize that its decision should not be taken as passing on the question whether 'principles of equity, comity, and federalism' might bar the suit."[105] It is true that *Mitchum* did not discuss the *Ex parte Young* dictum or sovereign immunity, but this is another example of the dog not barking: even in a case involving a statute that explicitly forbade injunctions against state court proceedings, no one thought to raise a sovereign immunity defense or to rely on the "whole scheme of our government" dictum as a basis for saying that the *Ex parte Young* exception to sovereign immunity did not apply. As usual, everyone assumed that sovereign immunity was not an issue when a plaintiff sought an injunction against a state official for unconstitutional actions. As for the principles of equity, comity, and federalism, the Court in *Mitchum* actually identified the cases that it had in mind.[106] None of those cases involved sovereign immunity.

The *Whole Woman's Health* Court's discussion of *Pulliam v. Allen*, a case that relied in part on *Mitchum*, also brushed aside a plausible

---

[101] The Anti-Injunction Act, 28 U.S.C. § 2283, provides that "[a] court of the United States may not grant an injunction to stay proceedings in a State court except as expressly authorized by Act of Congress, or where necessary in aid of its jurisdiction, or to protect or effectuate its judgments." *Mitchum* held that Section 1983 was an express authorization.

[102] The Anti-Injunction Act incorporated some of the equity principles governing injunctions against lawsuits; for an argument that it does so very imperfectly, see James E. Pfander & Nassim Nazemi, *The Anti-Injunction Act and the Problem of Federal-State Jurisdiction Overlap*, 92 TEX. L. REV. 1 (2013).

[103] Mitchum v. Foster, 315 F. Supp. 1387, 1388 (N.D. Fla. 1970); *see* Mitchum v. Foster, 407 U.S. 225, 227 (1972) ("state judicial and law enforcement officials").

[104] *Whole Woman's Health*, 142 S. Ct. at 534.

[105] *Id.* (quoting *Mitchum*, 407 U.S. at 243).

[106] *See Mitchum*, 407 U.S. at 243 (citing, among others, Younger v. Harris, 401 U.S. 37 (1971)).

basis for allowing the providers' suit. *Pulliam* held that judicial immunity does not bar a Section 1983 suit seeking an injunction against a state judge.[107] After *Pulliam* was decided, Section 1983—which explicitly authorizes "suit[s] in equity"—was amended in a way that contemplates that judges will be subject to suits for specific relief: Section 1983 now includes a proviso that says that "in any action brought against a judicial officer for an act or omission taken in such officer's judicial capacity, injunctive relief shall not be granted unless a declaratory decree was violated or declaratory relief was unavailable." The Court in *Whole Woman's Health* said *Pulliam* "had nothing to do with state-court clerks, injunctions against them, or the doctrine of sovereign immunity."[108] But if judges can be enjoined, then, according to the *Whole Woman's Health* Court's own reasoning, clerks can also be, a fortiori.

In addition, there was, in fact, a sovereign immunity issue in *Pulliam*, and once again the lesson is that the conventional view of *Ex parte Young*—sovereign immunity does not bar injunctions against state officials alleged to be violating the Constitution—was an unquestioned part of the legal landscape. The question presented in *Pulliam* was whether attorney's fees could be awarded to a plaintiff who obtained an injunction against the state court judge;[109] the Court addressed judicial immunity because the attorney's fees issue would not arise unless the plaintiff was entitled to prevail. The state of Virginia, defending the judge, explicitly argued that the fee award violated state sovereign immunity, because it would require a payment out of the state treasury. Virginia made that argument even though it seemed to be squarely foreclosed by the recent decision in *Hutto v. Finney*;[110] the best Virginia could do was to cite the dissenting opinion in *Hutto*.[111] But, as aggressive as Virginia was in challenging settled law, even it did not make the argument that the Court endorsed in *Whole*

[107] *See also* Mireles v. Waco, 502 U.S. 9, 10 n.1 (1991) ("The Court . . . has recognized that a judge is not absolutely immune from criminal liability or from a suit for prospective injunctive relief." (citing *Ex parte* Virginia, 100 U.S. 339, 348–49 (1880); and Pulliam v. Allen, 466 U.S. 522, 536–43 (1984)).

[108] 142 S. Ct. at 533.

[109] *See Pulliam*, 466 U.S. at 527.

[110] 437 U.S. 678, 689–700 (1978).

[111] *See* Brief on Behalf of the Petitioner at 24–25, Pulliam v. Nicholson, 466 U.S. 522 (1984) (No. 82-1432), 1983 U.S. S. Ct. Briefs LEXIS 162, at *47–49.

Woman's Health—that the injunction against the judge was barred by sovereign immunity.

C. *EX PARTE YOUNG* REVISIONISM

The *Whole Woman's Health* Court's treatment of *Ex parte Young*—the description of the principle, the grudging application, and the failure to recognize the consistent and unquestioning acceptance, over time, of a broad understanding of that principle—is troubling for reasons that go beyond S.B. 8. Some members of the Court have endorsed a view that would limit *Ex parte Young* to, essentially, the circumstances presented in that case: *Ex parte Young*, on that view, allows only a claim that "assert[s] 'in equity . . . a defense that would otherwise have been available in the State's enforcement proceedings at law.' "[112] The *Whole Woman's Health* Court did not go that far, and it is unclear whether the Court intended to redefine the principle of *Ex parte Young* in a way that would have effects beyond the specific issue presented in *Whole Woman's Health*. But the Court's backward-looking association of *Ex parte Young* with "traditional equity practice" risks undermining what *Ex parte Young* has been for more than a century in the United States: a way of "hold[ing] state officials responsible to 'the supreme authority of the United States' "[113] across the full range of official action.

There are many reasons to reject *Ex parte Young* revisionism. As a matter of principle, there is no reason to try to recover "traditional equity practice" as it developed in a system very different from ours—assuming that can even be done—and to prefer it to what *Ex parte Young* has become, over the last century, in our system. That is especially true because *Ex parte Young*, as it has been understood—as authorizing injunctive relief against unconstitutional state actions generally—has not been a peripheral feature of our law. It has not led to discredited results, and it has not been widely questioned. On the contrary: it has been the basis of some of the most celebrated decisions of the last century, and it has, unusual cases aside, been taken for granted as a fixture of our system.

---

[112] 142 S. Ct. at 540 (Thomas, J., concurring in part and dissenting in part) (quoting Va. Off. for Prot. & Advoc. v. Stewart, 563 U.S. 247, 262 (2011) (Kennedy, J., concurring) (ellipsis in original); and citing Harrison, *supra* note 75).

[113] Pennhurst State Sch. & Hosp. v. Halderman, 465 U.S. 89, 105 (1984) (quoting *Ex parte* Young, 209 U.S. 123, 160 (1908)).

In addition, as often happens with long-standing principles, a network of complementary principles has grown up around *Ex parte Young*. The usual argument about unanticipated consequences applies: changing one feature of a complex arrangement, even in a way that is attractive in isolation, risks perverse effects elsewhere in the system. Specifically, the system of remedies for unconstitutional state actions has developed in a way that is arguably comprehensive and coherent;[114] it is an imperfect but plausible approximation to what Blackstone's maxim envisioned. *Ex parte Young* is a critical element of that system.[115]

If injunctions were not available, states' constitutional violations would be prevented only by the deterrent effect of damages awards.[116] Section 1983 does authorize damages awards against officials who violate individuals' constitutional rights, but damages are an imperfect deterrent in the best of circumstances, and the deterrent effects of Section 1983 damages awards are further limited by immunity doctrines: Officials will not be held liable for violations of constitutional rights that are not clearly established, and officials are absolutely immune when they exercise judicial, legislative, or prosecutorial functions.

As long those gaps remain in the availability of damages, the principal way of preventing constitutional violations by state officials—officials acting mistakenly in good faith, or simply taking advantage of the immunities—is injunctive relief. If the Court were to decide that *Ex parte Young* extends, for example, only to injunctions against enforcement actions brought by a state executive official, it would do more than open the door to laws like S.B. 8 that substitute private enforcement for government enforcement. Unless the Court were

---

[114] See, for this important argument, John C. Jeffries, Jr., *In Praise of the Eleventh Amendment and Section 1983*, 84 VA. L. REV. 47 (1988), especially *id.* at 59.

[115] The Court has, in a way, acknowledged that injunctive remedies are crucial to maintaining an acceptably comprehensive remedial system. In *Alden v. Maine*, 527 U.S. 706 (1999), the Court held that, because of state sovereign immunity, a state could not be sued for damages in its own courts for a violation of a federal statute. But the Court explicitly relied on *Ex parte Young* to claim that the overall system of remedies still "strikes the proper balance between the supremacy of federal law and the separate sovereignty of the States" because it "provide[s] ample means to correct ongoing violations of law." *Id.* at 757.

In the analogous context of claims that the federal government violated the Constitution, *Ziglar v. Abbasi*, 137 S. Ct. 1843 (2017), which refused to allow an implied right of action for damages against federal officials for a constitutional violation, took for granted that injunctive relief would be available. The Court said that the question was "whether traditional equitable powers suffice to give necessary constitutional protection—or whether, in addition, a damages remedy is necessary." *Id.* at 1856.

[116] An official who violates an individual's constitutional rights can also be guilty of a federal crime. *See* 18 U.S.C. § 242; United States v. Lanier, 520 U.S. 259 (1997).

also prepared to modify immunity doctrines, the effect might be to permit state governments to violate the Constitution on a much larger scale.

A state government official could freely engage in actions—electronic surveillance of individuals, for example, or a regulation of property, or a punitive practice in a state custodial institution, or a restriction on speech, or any number of other acts—that were unconstitutional but were in the area of uncertainty to which qualified immunity applies. Damages actions might eventually establish clear rules, if courts, dismissing cases on qualified immunity grounds, were willing to do that in dictum[117] (and if plaintiffs had sufficient incentives to sue in those circumstances). Officials with absolute immunity might be limited in other ways, if constitutional objections to their actions could be raised defensively, or by internal government controls or disciplinary rules. Ultimately, something roughly as effective as the current remedial combination of damages and injunctions might emerge. But it is not clear that the Court, in apparently entertaining a radical change to the *Ex parte Young* exception to sovereign immunity, has a grasp of how extensive the effects of such a decision would be.

And, of course, the question is why such a change is warranted. *Ex parte Young*, understood as a general exception to sovereign immunity that applies whenever a state official is sued for violating the Constitution, is an established principle that is integral to maintaining constitutional limits on the government. If the Court is to change it in a substantial way, there should be a better reason than making the law conform to what the Court conceives to be traditional equity practice.

V. Conclusion

Whole Woman's Health presented the question whether a state can enact a law that is plainly inconsistent with the Supreme Court's decisions and then devise an enforcement scheme that serves no purpose other than to make it difficult or impossible, as a practical matter, for anyone to challenge that law. The dissenting opinions explicitly cast

[117] The Court has struggled with the question whether lower courts should first rule on the merits before dismissing a claim on qualified immunity grounds when the law is not clearly established. *See, e.g.*, Camreta v. Greene, 563 U.S. 692 (2011); Pearson v. Callahan, 555 U.S. 223 (2006); John C. Jeffries, Jr., *Reversing the Order of Battle in Constitutional Torts*, 2009 Sup. Ct. Rev. 139.

the case in those terms. But the Court never acknowledged that that was the question.

If the Court agreed that it was consistent with the basic presuppositions of our constitutional system to allow a state to do what Texas did, it could have said so, and explained why. Or perhaps the Court believed that that characterization of the case was mistaken, and that no fundamental issue was presented because Texas had legitimate reasons to limit remedies in the way it did and was not just evading the Court's decisions. If so, the Court could have identified what those reasons were and, in that way, provided some guidance about what the Court thought were acceptable reasons to limit remedies for constitutional violations. Or, finally, if the Court thought that what Texas had done was on some basic level unacceptable but that the state had managed to find a gap in the rules governing constitutional remedies that left the Court with no choice but to accept a problematic law, it could have said that. That would have invited efforts to find ways around the unjustified limits on remedies in the way that courts have in the past. The general point is that if the Court had made its position on these questions explicit, it would have revealed how it conceived of its role in implementing constitutional remedies.

Instead, the Court reduced the case to a series of narrower issues and then reached questionable conclusions about those issues, without ever explicitly addressing the more basic issue that the case raised. It is a familiar point that the text of the Constitution leaves the law of constitutional remedies almost entirely in the hands of legislatures and courts. One historic commitment of the courts is to try to provide a system of remedies that makes constitutional rights effective—that acknowledges the aspiration expressed by Blackstone's maxim. The decision in *Whole Woman's Health* is not alone among recent decisions in raising questions about whether this Court shares that historic commitment.

MICHELE GOODWIN

OPPORTUNISTIC ORIGINALISM: DOBBS v. JACKSON WOMEN'S HEALTH ORGANIZATION

In June 2022, the United States Supreme Court overturned *Roe v. Wade*[1] and *Planned Parenthood v. Casey*[2] in *Dobbs v. Jackson Women's Health Organization*,[3] destabilizing a legacy of cases[4] that endorsed

Michele Goodwin is a Chancellor's Professor at the University of California, Irvine School of Law.

Author's note: I thank Naomi Cahn, June Carbone, Dale Carpenter, Erwin Chemerinsky, David Cruz, Victoria Nourse, Dorothy Roberts, and Reva Siegel. I also thank Priscila Coronado for invaluable research assistance and the research librarians at the Georgetown University Law Center and the University of California, Irvine School of Law.

[1] 410 U.S. 113 (1973), *overruled by* Dobbs v. Jackson Women's Health Org., 142 S. Ct. 2228 (2022).

[2] 505 U.S. 124 (2007), *overruled by Dobbs*, 142 S. Ct. 2228.

[3] 142 S. Ct. 2228.

[4] *See* June Med. Servs. L.L.C. v. Russo, 140 S. Ct. 2103, 2132 (2020) (explaining that in light of "the District Court's significant factual findings—both as to burdens and as to benefits— . . . . Louisiana's [hospital admitting privileges] law poses a 'substantial obstacle' to women seeking an abortion" and that "in light of these findings and our precedents, [the law] violates the Constitution"), *abrogated by Dobbs*, 142 S. Ct. 2228; Whole Woman's Health v. Hellerstedt, 136 S. Ct. 2292, 2299 (2016) (stating that "[u]nnecessary health regulations that have the purpose or effect of presenting a substantial obstacle to a woman seeking an abortion impose an undue burden on the right" and as such are "constitutionally invalid" (citing Planned Parenthood of Se. Pa. v. Casey, 505 U.S. 833, 878 (1992)), *abrogated by Dobbs*, 142 S. Ct. 2228; Doe v. Bolton, 410 U.S. 179, 201 (1973) (holding gatekeeping measures that constrained abortion rights unconstitutional), *abrogated by Dobbs*, 142 S. Ct. 2228; Connecticut v. Menillo, 423 U.S. 9, 11 (1975) (explaining, even while upholding a statute that banned nonphysicians from performing pregnancy terminations, that "*Roe* teaches that a State cannot restrict a decision by a woman,

The Supreme Court Review, 2023.
© 2023 The University of Chicago. All rights reserved. Published by The University of Chicago Press. https://doi.org/10.1086/724759

Roe's principal holding, which decriminalized abortion[5] and established the constitutional right for a woman to terminate a pregnancy.[6] Spanning nearly fifty years of jurisprudence, the "legacy cases" did not equivocate on the constitutional right to terminate a pregnancy across a broad spectrum of laws seeking in various ways to undermine and infringe the right to an abortion. Even as the Court upheld harsh, economically burdensome state and federal laws that denied government funding for abortion, in each instance the Court reaffirmed *Roe* and the fundamental right to terminate a pregnancy.

Rather than repudiating Mississippi's law, which banned abortion after fifteen weeks of pregnancy and made no exceptions for cases of rape or incest[7]—a clear retreat from *Roe's* principal holding[8] adopted as the framework in *Casey*[9]—or scrutinizing the state's implausible claim that the law advanced the state's interest and commitment to

---

with the advice of her physician, to terminate her pregnancy during the first trimester because neither its interest in maternal health nor its interest in the potential life of the fetus is sufficiently great at that stage"); Bigelow v. Virginia, 421 U.S. 809, 812–13, 822 (1975) (striking down a Virginia law that prohibited "any person, by publication, lecture, advertisement, or by the sale or circulation of any publication" that promoted "the procuring of abortion or miscarriage" and noting "the activity advertised pertained to constitutional interests"); La. State Bd. of Med. Examiners v. Rosen, 419 U.S. 1098 (1975) (affirming the lower court's ruling that held unconstitutional a Louisiana law that prohibited doctors from performing abortions); Planned Parenthood of Cent. Mo. v. Danforth, 428 U.S. 53, 53–54 (1976) (striking down a Missouri law requiring husband consent and parental consent, among other antiabortion provisions); Bellotti v. Baird, 443 U.S. 622, 651 (1979) (striking down a Massachusetts statute, finding that the law "falls short . . . in two respects: First, it permits judicial authorization for an abortion to be withheld from a minor who is found by the superior court to be mature and fully competent to make this decision independently. Second, it requires parental consultation or notification in every instance, without affording [her] an opportunity to receive an independent judicial determination that she is mature enough to consent or that an abortion would be in her best interests"); City of Akron v. Akron Ctr. for Reprod. Health, Inc., 462 U.S. 416, 416–18 (1983) (holding an Ohio law that requires all abortions performed in the first trimester to be facilitated in a hospital and that imposes impermissible restrictions on a minor seeking an abortion, among other mandates, unconstitutional), *overruled by Casey*, 505 U.S. 833; Thornburgh v. Am. Coll. of Obstetricians and Gynecologists, 476 U.S. 747, 759 (1986) (holding that a Pennsylvania law that imposed various restrictions "wholly subordinate[d] constitutional privacy interests and concerns with maternal health"), *overruled by Casey*, 505 U.S. 833.

[5] *Roe*, 410 U.S. at 113–15.

[6] *See id.*

[7] *See* Adeel Hassan, *What to Know About the Mississippi Abortion Law Challenging Roe v. Wade*, N.Y. Times (May 6, 2022), https://www.nytimes.com/article/mississippi-abortion-law.html ("The law bans abortions if 'the probable gestational age of the unborn human' is determined to be more than 15 weeks, with narrow exceptions for medical emergencies or 'a severe fetal abnormality.'").

[8] *See Roe*, 410 U.S. at 161 (1973) ("Viability is usually placed at about seven months (28 weeks) but may occur earlier, even at 24 weeks."), *overruled by Dobbs*, 142 S. Ct. 2228.

[9] *See Casey*, 505 U.S. at 846, *overruled by Dobbs*, 142 S. Ct. 2228.

promoting the health and safety of women in Mississippi,[10] the Supreme Court undermined the rule of law in three critical aspects.

First, it eliminated an established right that for nearly fifty years served as a crucial element of women's civil liberties, one no less vital than a man's constitutional right to be free from the state dictating his reproductive destiny.[11] Second, in overturning *Roe*, it erased "a right that has safeguarded women's ability to participate fully and equally in society."[12] Third, as discussed below, the Court "renounce[ed] this fundamental right, which it had repeatedly recognized and reaffirmed."[13] Given its history of repeatedly recognizing the constitutional validity and primacy of *Roe* and later *Casey*, as recently as *June Medical Services L.L.C. v. Russo*,[14] the Court seriously erred, "upend[ing] the doctrine of *stare decisis*,"[15] and calling into question whether the Court acted arbitrarily, failed to apply law fairly, or delivered impartial justice.[16] Commentators might be less suspicious of the Court's motives were it not for the fact that the Court neglected to offer a meaningful analysis related to reliance interests and, ironically, barely acknowledged the serious national and state-level concerns related to pregnancy, rape, maternal mortality, and morbidity[17] in a decision that at the most fundamental level should acknowledge pregnancy and related contemporary concerns.

Yet there are other concerns. Even while the term "originalism" does not appear in the *Dobbs* opinion, it is nonetheless clear that

---

[10] *See Dobbs*, 142 S. Ct. at 2284.

[11] *See* Skinner v. Oklahoma *ex rel.* Williamson, 316 U.S. 535, 541 (1942) (opining that "invidious discriminations" are made when the state enacts a law that interferes with "the basic civil rights of man" to determine his own reproductive and procreative destiny). *But see* Buck v. Bell, 274 U.S. 200–07 (1927) (upholding a Virginia eugenics law that permitted the sterilization of indigent young women to advance the "health of the patient and the welfare of society").

[12] *See* Press Release, U.S. Dep't of Just., Attorney General Merrick B. Garland Statement on Supreme Court Ruling in *Dobbs v. Jackson Women's Health Organization* (June 24, 2022), https://www.justice.gov/opa/pr/attorney-general-merrick-b-garland-statement-supreme-court-ruling-dobbs-v-jackson-women-s (explaining that "few rights are more central to individual freedom than the right to control one's own body").

[13] *Id.*

[14] 140 S. Ct. 2103 (2020), *abrogated by Dobbs*, 142 S. Ct. 2228.

[15] U.S. Dep't of Just., *supra* note 12.

[16] Dahlia Lithwick & Neil S. Siegel, *The Lawlessness of the* Dobbs *Decision*, SLATE (June 27, 2022, 2:58 PM), https://slate.com/news-and-politics/2022/06/dobbs-decision-glucksberg-test-lawlessness.html ("Alito's majority opinion is not in fact constitutional law, and betrays little sense of a felt obligation to attempt to be.").

[17] *See Dobbs*, 142 S. Ct. at 2261–68.

Justices in the majority embrace it as a methodology for interpreting the Constitution and analyzing legal disputes, including on matters of urgent contemporary concern such as abortion and gun control.[18] Or at least the Justices feign filtering their analysis through the sieve of originalism. In *Dobbs*, Justice Alito derides the *Casey* plurality for what he describes as a departure from an "original constitutional proposition" that "courts do not substitute their social and economic beliefs for the judgement of legislative bodies."[19] In *New York State Rifle & Pistol Ass'n v. Bruen*,[20] Justice Thomas included a prologue to underscore the historical relevance of the majority's analysis to current questions related to gun control.[21]

This Article argues that *Dobbs* is rooted in opportunistic originalism, whereby the Court implies its commitment to text, original meaning, and long, unbroken ties to tradition, but seriously veers from that methodology or performs its adjudication in such a neglectful and unsystematic manner that it belies fidelity to the methodology. The Court's articulated willingness to cabin off privacy protections in *Griswold v. Connecticut*,[22] *Lawrence v. Texas*,[23] and *Obergefell v. Hodges*[24] underscores this point. This Article makes the case that the Court exudes selective decision-making in *Dobbs*, taking shelter in the smokescreen of originalism. This should not have been the basis for overruling *Roe*.

For example, to justify its holding in *Dobbs* "that the Constitution does not confer a right to abortion"[25] and that "the authority to regulate abortion must be returned to the people and their elected

---

[18] The majority implies fidelity to the original intent of the Reconstruction ratifiers and yet relies less on their writings than that of British legal scholars whose theories predated ratification of the Constitution, the Civil War, and Reconstruction. The majority implies a faithful interpretation of the Reconstruction amendments and promises a searching investigation of the debates and the social and legal concerns that shaped them. However, the Court leaves serious factual and analytical gaps in its historical analysis. Namely, it decontextualizes the era leading to the Reconstruction—avoiding serious interrogation of American slavery. The majority strips racial discrimination from their reading of what animated the Reconstruction amendments as if operating from a universe entirely apart from the political debates and controversies concerning American slavery.

[19] *Dobbs*, 142 S. Ct. at 2277 (quoting Ferguson v. Skrupa, 372 U.S. 726, 729–30 (1963)).

[20] 142 S. Ct. 2111 (2022).

[21] *Id.* at 2150–53.

[22] 381 U.S. 479 (1965).

[23] 539 U.S. 558 (2003).

[24] 135 S. Ct. 2584 (2015).

[25] *Dobbs*, 142 S. Ct. at 2279.

representatives,"[26] the Court suggests that it is unclear whether "[t]he ability of women to participate equally in the economic and social life of the Nation has been facilitated by their ability to control their reproductive lives."[27] The Court hypothesized that "[w]omen are not without electoral or political power,"[28] referencing demographic population data in Mississippi (that women represent 51 percent of the state's population), while ignoring that men constitute 86 percent of that state's legislature,[29] and within that body, women wield marginal leadership and comparatively little power.[30] Only three of the nearly fifty House committees, which are appointed by the Speaker, "are chaired by women,"[31] which may contribute to Mississippi's relatively neglectful leadership response to alarming rates of infant mortality, maternal mortal, and maternal morbidity.[32]

According to *Mississippi Today*, Mississippi's male-dominated legislature shows little will in addressing distressing concerns related to prenatal care, childbirth, delivery, and postnatal medical services. Whether from a lack of capacity or deliberate indifference, Mississippi's legislature contributes to the state's appalling record on birthing outcomes as "Mississippi babies are more likely to die before their first birthday than infants anywhere else in the country" and "Black babies are twice as likely to die as their white counterparts in Mississippi."[33] Compelling data further underscore the myriad harms very likely to be compounded by *Dobbs*, because "Mississippi has the highest preterm birth rate and the lowest birth weight rate in

---

[26] *Id.*

[27] *Id.* at 2276 (alteration in original); *id.* at 2277 (claiming that a reliance interest is hard to find or assess, including "the effect of the abortion right on society and in particular on the lives of women").

[28] *Id.* at 2277.

[29] *Women in State Legislatures 2023*, CTR. FOR AMERICAN WOMEN & POL., https://cawp.rutgers.edu/facts/levels-office/state-legislature/women-state-legislatures-2023 (last visited Feb. 1, 2023).

[30] *See, e.g.*, Adam Ganucheau, *Mississippi Mothers and Babies Are Dying. One Man and His 87% Male House Are Blocking Help*, MISS. TODAY (Jan. 23, 2023), https://mississippitoday.org/2023/01/23/postpartum-extension-philip-gunn ("Looking around the House chamber, it quickly becomes apparent how Gunn could comfortably justify his decision to kill a bill that most directly affects Mississippi women and why there hasn't been an uprising of lawmakers pleading with him to change his mind. In the chamber Gunn leads, just 15 members out of 120 current members (13%) are women.").

[31] *Id.*

[32] *Id.*

[33] *Id.*

America, and one of every seven babies born here are preterm."[34] As of yet, "lobbying legislators, voting, and running for office"[35] has not rescued women from the harms endured as a result of pregnancy and giving birth in Mississippi, where the legislature and its leadership demonstrate uninterest and a lack of commitment to the health and safety of women, especially Black women.[36] Indeed, the Court's appeal to the Mississippi legislative process to resolve abortion as a constitutional concern rings particularly hollow and naïve given enduring legacies of slavery, Jim Crow, and racial gerrymandering, reflected in the fact that "[n]o African American has won statewide office in Mississippi in 129 years."[37]

To the extent that the Court invokes originalism to answer whether abortion *is or should be* a right protected by the Constitution, *Dobbs* then becomes a case about what ratifiers knew, understood, and intended.[38] Yet, to understand *Dobbs* as a decision framed only by method ignores both the political[39] and judicial maneuvering to reach

---

[34] *Id.* ("These numbers are made worse because of the state's high rates of poverty and uninsured people. Across Mississippi, about 65% of babies are born to mothers on Medicaid. Because of lag times in getting approved for coverage and a 60-day cutoff of postpartum care coverage, mothers often do not receive the prenatal and postpartum care they need—care that could prevent major problems.").

[35] Dobbs v. Jackson Women's Health Org., 142 S. Ct. 2228, 2279 (2022) ("Our decision returns the issue of abortion to those legislative bodies, and it allows women on both sides of the abortion issue to seek to affect the legislative process by influencing public opinion, lobbying legislators, voting, and running for office.").

[36] Ganucheau, *supra* note 30 ("Mothers who give birth in Mississippi are more likely to die here than in 45 other states, with a pregnancy-related mortality rate nearly double the national average. A whopping 86% of pregnancy-related deaths occur postpartum, including 37% after six weeks. Black women are three times likelier than white women to die of a pregnancy-related cause.").

[37] John A. Tures & Seth Golden, *No African American Has Won Statewide Office in Mississippi in 129 Years—Here's Why*, Conversation (June 17, 2019), https://theconversation.com/no-african-american-has-won-statewide-office-in-mississippi-in-129-years-heres-why-118319 ("Mississippi requires winners to receive more than 50% of the votes. When no one receives a majority, the Mississippi legislature, not the voters, chooses the winner. In late May 2019, four African American Mississippians sued in federal court to end this practice, which they say was designed to keep black candidates from winning.").

[38] *Dobbs*, 142 S. Ct. at 2266–68.

[39] *See, e.g.*, Emily Schultheis, *Trump Talks to "60 Minutes" About Same-Sex Marriage, Abortion and the Supreme Court*, CBS News (Nov. 13, 2016), http://www.cbsnews.com/news/trump-promises-pro-life-justices-supreme-court-same-sex-marriage. Both before and after his 2016 election as President of the United States, Donald Trump expressed the view that *Roe v. Wade* should be overruled. "I'm pro-life," Trump explained to crowds and journalists, "The judges will be pro-life." *Id.*

the end of *Roe*, including reinterpreting other areas of law to arrive at the holding in *Dobbs*.[40] That is,

> Rather than overturn the abortion right incrementally, and narrowly, and endeavor to reaffirm the equal citizenship of those whose rights the Court was abrogating, the *Dobbs* Court defined women's liberties in terms of nineteenth-century norms under a new history-and-traditions standard that upended a half-century of abortion law and threatened many other rights, transforming protected liberties in ways that make our constitutional order less democratic.[41]

In other words, it is virtually impossible to reconcile the outcome in *Dobbs* in terms of current constitutional law standards as well as application[42] on matters of childbearing,[43] abortion,[44] contraception,[45]

---

[40] *See* Burwell v. Hobby Lobby Stores, Inc., 134 S. Ct. 2751 (2014) (recognizing closely held corporations, which make up more than 90 percent of US corporations, as having protected religious interests and holding that lawful exercise of those interests extends to denying female employees contraceptives); Nat'l Inst. of Fam. & Life Advocs. v. Becerra, 138 S. Ct. 2361, 2378 (2017) (finding that California's notice requirements that crisis pregnancy centers post whether they are medically licensed and information about state-related services for indigent patients seeking reproductive healthcare violates the First Amendment, constituting viewpoint discrimination); Shelby Cnty. v. Holder, 570 U.S. 529, 560 (2013) (Ginsburg, J., dissenting) ("'[V]oting discrimination still exists; no one doubts that.' But the Court today terminates the remedy that proved to be best suited to block that discrimination. The Voting Rights Act of 1965 (VRA) has worked to combat voting discrimination where other remedies had been tried and failed. Particularly effective is the VRA's requirement of federal preclearance for all changes to voting laws in the regions of the country with the most aggravated records of rank discrimination against minority voting rights.").

[41] *See* Reva Siegel, *Memory Games:* Dobbs*'s Originalism As Anti-Democratic Living Constitutionalism—and Some Pathways for Resistance*, 101 TEX. L. REV. (forthcoming 2023) (manuscript at 5), https://ssrn.com/abstract=4179622 ("On this account, executive branch appointments politics matter critically to originalism's authority, as do originalism's appeals to constitutional memory to legitimate the exercise of public power.").

[42] *Id.*

[43] Skinner v. State of Okla. *ex rel.* Williamson, 316 U.S. 535, 541 (1942) ("[Under laws serving the purpose of denying the liberty to choose one's reproductive future, individuals are] forever deprived of a basic liberty. We mention these matters not to reexamine the scope of the police power of the States. We advert to them merely in emphasis of our view that strict scrutiny of the classification which a State makes in a sterilization law is essential, lest unwittingly, or otherwise, invidious discriminations are made against groups or types of individuals in violation of the constitutional guaranty of just and equal laws. The guaranty of 'equal protection of the laws is a pledge of the protection of equal laws.'" (quoting Yick Wo v. Hopkins, 118 U.S. 356, 369 (1886))).

[44] *See supra* note 4.

[45] Griswold v. Connecticut, 381 U.S. 479 (1965); Eisenstadt v. Baird, 405 U.S. 438, 443 (1972) ("Instead, the court concluded that the statutory goal was to limit contraception in and of itself—a purpose that the court held conflicted 'with fundamental human rights' under *Griswold v. Connecticut*, where this Court struck down Connecticut's prohibition against the use of contraceptives as an unconstitutional infringement of the right of marital privacy." (citation omitted)); *id.* at 453 ("It is true that in *Griswold* the right of privacy in question

interracial marriage,[46] and marriage equality.[47] According to Justice Brennan in *Eisenstadt v. Baird*, "[i]f the right of privacy means anything, it is the right of the individual, married or single, to be free from unwarranted governmental intrusion into matters so fundamentally affecting a person as the decision whether to bear or beget a child."[48]

For all its bold moves and odious claims, *Dobbs* is also a case of omissions and neglect of historical and contemporary facts as well as social and legal concerns. In a cynical and highly injudicious turn, the Court invoked *Plessy v. Ferguson*.[49] The Court resurrected *Plessy* seemingly as an analogue to *Roe*, dangerously tying an odious history of race and sex discrimination and suppression known as Jim and Jane Crow—that included myriad horrific laws[50] and state-sanctioned violence targeted at Black women, men, and their families[51]—to the reproductive liberation of all women, and particularly Black women. However, the majority never acknowledges or even references *Black women*, although they are the population most at risk of death due to pregnancy-related risks in Mississippi and throughout the nation. The majority and concurring Justices obfuscate, speaking around Black women; they are seemingly and simply vessels in service to the "African-American population."[52]

---

inhered in the marital relationship. Yet the marital couple is not an independent entity with a mind and heart of its own, but an association of two individuals each with a separate intellectual and emotional makeup.").

[46] *See, e.g.*, Loving v. Virginia, 388 U.S. 1, 12 (1967) (striking down Virginia's ban on interracial marriage); *id.* ("These statutes also deprive the Lovings of liberty without due process of law in violation of the Due Process Clause of the Fourteenth Amendment. The freedom to marry has long been recognized as one of the vital personal rights essential to the orderly pursuit of happiness by free men.").

[47] Obergefell v. Hodges, 576 U.S. 644, 666 (2015) ("Indeed, the Court has noted it would be contradictory 'to recognize a right of privacy with respect to other matters of family life and not with respect to the decision to enter the relationship that is the foundation of the family in our society.'" (quoting Zablocki v. Redhail, 434 U.S. 374, 386 (1978)).

[48] *Baird*, 405 U.S. at 453.

[49] 163 U.S. 537 (1896).

[50] *See generally* States' Laws on Race and Color (Pauli Murray ed., 2016).

[51] *Id.*

[52] Dobbs v. Jackson Women's Health Org., 142 S. Ct. 2228, 2256 n.41 (2022) (foregrounding disputed arguments made by Justice Thomas that "some such supporters [of abortion rights] have been motivated by a desire to suppress the size of the African-American population").

Cloaked in the unconvincing guise of impartial adjudication, the majority neuters *Dobbs* of any relationship to racial discrimination in pregnancy, gestation, childbirth, or women's lived lives—avoiding any mention of incest, rape, ectopic pregnancy, miscarriage (save in an addendum), gestational diabetes, eclampsia, or bleeding. Confoundingly, the majority cabins discussion of maternal mortality to one paragraph related to 1973 rather than 2022,[53] even though maternal deaths in the United States exceed that of all industrialized nations, earning it the moniker of the deadliest nation in the "developed world" to give birth.[54] Justice Alito offers one, lone passing reference to the "burdens of motherhood," primarily to rebut the dissent,[55] but bypasses the opportunity to engage with complex problems and informed literature reflecting research and scholarship developed by civil society organizations,[56] and in the fields of

---

[53] *Id.* at 2268.

[54] Alexa Lardieri, *U.S. Hospitals Do Little to Protect Mothers During Birth*, U.S. News (July 27, 2018), https://www.usnews.com/news/health-care-news/articles/2018-07-27/report-us-most-dangerous-place-to-give-birth-in-developed-world ("The United States is the most dangerous place in the developed world to give birth, with more than 50,000 mothers suffering severe injuries during or after childbirth and about 700 mothers dying. According to a USA Today investigation, at least half of these deaths could have been prevented and half of the injuries reduced or prevented with better care. And the investigation revealed that hospitals and healthcare workers do not intervene quickly enough when something goes wrong."); Ognjen Š Miljanić, *America Is the Most Dangerous Place to Give Birth in the Developed World—It's Only Getting Worse*, Hill (Sept. 20, 2022, 10:30 AM), https://thehill.com/opinion/healthcare/3650993-america-is-the-most-dangerous-place-to-give-birth-in-the-developed-world-its-only-getting-worse ("A report by the World Health Organization (WHO) found that all world regions experienced significant drops in [maternal mortality rates] between 2000 and 2017: the European rate went down by 53 percent, sub-Saharan Africa improved by 38 percent, Latin America by 23 percent and the global average decreased by 38 percent. The only region that saw a sharp rise is North America, whose maternal mortality increased by 52 percent. Don't blame Canada—their [maternal mortality rates] stayed roughly the same during the past two decades.").

[55] *Dobbs*, 142 S Ct. at 2261.

[56] *See, e.g.*, Sarah Jane Glynn, *Breadwinning Mothers Continue to Be the U.S. Norm*, Ctr. for Am. Progress (May 10, 2019), https://www.americanprogress.org/article/breadwinning-mothers-continue-u-s-norm (acknowledging that even though "[g]ender norms are changing . . . women still carry the brunt of unpaid work within the home and spend more time caring for children and performing household labor than men with similar demographic characteristics and parental status."); *id.* ("Women and the paid labor they perform are important economic drivers in the United States, but current policies do little to reflect this reality. Too many workplaces, and most existing public policies, are based on a faulty underlying premise: that most families have access to an unpaid caregiver and that if someone needs to take time away from work to provide care it will most likely be a wife, mother, or adult daughter—and that doing so will have no negative impact on her family's economic well-being."); *The Motherhood Penalty*, AAUW, https://www.aauw.org/issues/equity/motherhood (last visited Feb. 1, 2023) ("Because of old workplace paradigms, becoming a mother can take a toll on a woman's

economics,[57] public health,[58] family law,[59] race law,[60] psychology,[61]

---

earnings and jeopardize her family's economic security."); *id.* ("Due to the Motherhood Penalty, mothers make 58 cents for every dollar paid to fathers.").

[57] Katrin Elborgh-Woytek et al., Women, Work, and the Economy: Macroeconomic Gains from Gender Equality 13 (Int'l Monetary Fund ed. 2013) (stating that "[t]he average duration of paid parental leave in advanced economies is 26 weeks, and all [Organization for Economic Development and Cooperation] countries except the United States and Korea currently offer paid parental leave"); Christian Neumeier et al., *The Implicit Costs of Motherhood over the Lifecycle: Cross-Cohort Evidence from Administrative Longitudinal Data*, 84 S. Econ. J. 716, 716 (2018) ("The gender wage gap, and its evolution over time, is among the most studied topics at the intersection of economics and demography. One often touted explanation for this phenomenon is the variety of constraints typically placed upon women as a result of motherhood, generating what the literature refers to as the 'family gap', or the earnings gap between mothers and non-mothers. The historical lack of family-friendly labor market policies in the United States, likely widening the family gap, is cited as one reason that the U.S. has a large gender wage gap relative to many other developed countries." (citation omitted)); Wei-hsin Yu & Hara Yuko, *Motherhood Penalties and Fatherhood Premiums: Effects of Parenthood on Earnings Growth Within and Across Firms*, 58 Demography 247, 248 (2021) (finding that as a consequence "of the wage penalty associated with motherhood and the wage premium tied to fatherhood, the gender pay gap widens as men and women move through the life course. Parenthood is therefore a key contributor to gender inequality").

[58] *See* J. Phillip Gingrey, *Maternal Mortality: A US Public Health Crisis*, 110 Am. J. Pub. Health 462, 462 (2020) (arguing, from a physician's perspective, that the United States faces a crisis in maternal health and safety and explaining that "[r]acial and socioeconomic factors are impossible to overlook. African American, Alaska Native, and American Indian women die at a rate almost three times as high as White women"); Daniel B. Nelson et al., *Population-Level Factors Associated with Maternal Mortality in the United States, 1997–2012*, 18 BMC Pub. Health 1007 (2018).

[59] Dorothy Roberts, *Welfare Reform and Economic Freedom: Low-Income Mothers' Decisions About Work at Home and in the Market*, 44 Santa Clara L. Rev. 1029, 1032 (2004) ("A defining aspect of this welfare vision, however, was the social control of poor immigrant families and the neglect of Black women. Worried about urban immigrants' threat to the social order, the elite women reformers treated welfare as a means of supervising and disciplining recipients as much as a means of providing charity. They feared that welfare that was too generous might provide an incentive for state dependency, moral degeneracy, and family breakdown."); June Carbone & Naomi Cahn, Marriage Markets: How Inequality Is Remaking the American Family (2014) (noting generally a pattern of burdened outcomes for families and family planning in red states); Martha Minow, *The Welfare of Single Mothers and Their Children*, 26 Conn. L. Rev. 817, 819 (1994) (arguing "that the welfare reforms" dating back twenty-five years now "are not reforms but instead reiterations of longstanding lines of social cleavage," and urging for the development of a "responsible social policy").

[60] *See* Khiara Bridges, *Racial Disparities in Maternal Mortality*, 95 N.Y.U. L. Rev. 1229, 1293–99 (2020) (discussing racial erasure in discourse about preventing maternal deaths); Dorothy E. Roberts, *Welfare and the Problem of Black Citizenship*, 105 Yale L. J. 1563, 1590 (1996) (addressing how white supremacy undermines Black parenting, because "[m]any white Americans remain uninterested in advancing the welfare of Black Americans; many others see helping everyone as contrary to their self-interest because they perceive Black people's social position in opposition to their own," and under this ideology welfare and other social mobility efforts that benefit Black women will "necessarily be antithetical to white interests because Blacks' social advancement diminishes white superiority").

[61] Jay Belsky, *The Determinants Of Parenting: A Process Model*, 55 Child Dev. 83, 85 (1984) ("Not only is there evidence that such young mothers express less desirable child-rearing attitudes and have less realistic expectations for infant development than do older mothers

sociology,[62] or theology,[63] that expresses how persistent patterns of sex discrimination, unequal pay, single-parenthood, and other barriers to equality exacerbate health, social, and economic burdens on mothers, creating a motherhood penalty.[64]

In their blistering response to *Dobbs*, Supreme Court journalist Dahlia Lithwick and law professor Neil S. Siegel queried, "Why is *Dobbs* not just wrong, but *lawless*?"[65] and answered, "Because it is utterly unprincipled. It articulates a reason for overruling *Roe* out of one side of its mouth, then repeatedly protests that it will not be bound by this reason out of the other side of its mouth."[66] Yet, this was not the first time that a majority opinion written by Justice Alito

---

but, from a more behavioral standpoint, it has been observed that they also tend to be less responsive to their newborns and to engage infants in less verbal interaction.").

[62] A particularly important study by Kristin Turney addresses a largely overlooked aspect of contemporary motherhood, namely incarceration and its impacts on children and mothers. Her research is particularly relevant to childbirth and rearing, given intensified policing of pregnancies in the United States, which lead to incarceration. *See, e.g.*, Kristin Turney, *Stress Proliferation Across Generations? Examining the Relationship Between Parental Incarceration and Childhood Health*, 55 J. HEALTH & SOC. BEHAV. 302 (2014) (finding parental incarceration is associated with physical and psychological health harms in children); *see also* Michelle J. Budig & Melissa J. Hodges, *Statistical Models and Empirical Evidence for Differences in the Motherhood Penalty Across the Earnings Distribution*, 79 AM. SOC. REV. 358 (2014) (investigating "how social processes may affect individuals differently based on their locations in inequality distributions" and finding that the motherhood penalty has a significant impact on the earnings of mothers); Shelley J. Correll et. al., *Getting a Job: Is There a Motherhood Penalty?*, 112 AM. J. SOC. 1297, 1324 (2007) (finding that "regardless of race, mothers experience negative biases in workplace evaluations").

[63] Ariane Kemkes-Grottenthaler, *God, Faith, and Death: The Impact of Biological and Religious Correlates on Mortality*, 75 HUM. BIOLOGY 897, 900 (2003) (studying differentials of mortality, including maternal mortality between and among religious denominations); MARGARET D. KAMITSUKA, ABORTION AND THE CHRISTIAN TRADITION (2019) (offering ethical arguments related to pregnancy and motherhood); *see also* M.T. Dávila, *Hyperfocus on Abortion Issue Obscures Call of Common Good*, NCR (Oct. 13, 2020), https://www.ncronline.org/news/opinion/theology-en-la-plaza/hyper-focus-abortion-issue-obscures-call-common-good (noting that the article "is not an argument for reproductive choice" but explaining that "[f]or the past 45 years, the public voice of the Catholic Church has been progressively limited to the pro-life cause" and that this perspective "has had disastrous consequences for the fullness of the Catholic vision of the common good in the public square," particularly because it "obscures the multiple ways in which we participate in a culture of [maternal] death").

[64] Sarah Kliff, *A Stunning Chart Shows the True Cause of the Gender Wage Gap*, Vox (Feb. 19, 2018, 11:00 AM), https://www.vox.com/2018/2/19/17018380/gender-wage-gap-childcare-penalty (finding that "what we often think of as a gender pay gap is more accurately discussed as a childbearing pay gap or motherhood penalty" because "childless women have earnings that are quite similar to men's salaries, while mothers experience a significant wage gap").

[65] Lithwick & Siegel, *supra* note 16.

[66] *Id.* They conclude that while *Dobbs* "is now the law . . . this is not legal reasoning that can or should be respected." *Id.*

could be accused of the same or exhibiting "startling breadth"[67] in reaching a decision that denies women access to reproductive healthcare.[68]

This Article addresses these concerns and proceeds in four parts. It argues that in *Dobbs*, the Court claims to be guided by looking

---

[67] Burwell v. Hobby Lobby Stores, Inc., 573 U.S. 682, 739–40 (2014) (Ginsburg, J., dissenting) ("In the Court's view, RFRA demands accommodation of a for-profit corporation's religious beliefs no matter the impact that accommodation may have on third parties who do not share the corporation owners' religious faith—in these cases, thousands of women employed by Hobby Lobby and Conestoga or dependents of persons those corporations employ."); *see also* Aaron James Spencer, *Hobby Lobby Is One of the Most Dangerous Supreme Court Decisions Ever, and Here's Why*, WHYY (July 17, 2014), https://whyy.org/articles/hobby-lobby-is-one-of-the-most-dangerous-supreme-court-decisions-ever-and-heres-why ("The ruling misinterprets religious freedom in the United States, restricts the rights of women everywhere, raises religious belief above medical and scientific fact in lawmaking, and is easily expandable to accommodate the religious objections of a host of corporation owners."); Marci A. Hamilton, *What's Really Wrong With the Decisions in* Burwell v. Hobby Lobby *and* Conestoga Wood v. Burwell?, VERDICT (June 30, 2014), https://verdict.justia.com/2014/06/30/whats-really-wrong-decisions-burwell-v-hobby-lobby-conestoga-wood-v-burwell ("Specifically, female employees can be denied coverage for contraception—either some or all (depending on the particular religious beliefs of the employer)—because the employer doesn't believe anyone should use it, and does not want to have to pay into a plan where women might make a choice to use it, say, after they were raped."); Lawrence Sager & Frederick Mark Gedicks, *Why Every American Should Worry About the* Hobby Lobby *Ruling*, UT NEWS (July 11, 2014), https://news.utexas.edu/2014/07/11/why-every-american-should-worry-about-the-hobby-lobby-ruling ("For the first time in U.S. history, the Supreme Court has endorsed the principle that for-profit businesses may be excused from obeying laws and regulations to which they or their owners object on religious grounds."); Noah Marks, *Dangerous Precedents in* Hobby Lobby, HARV. L. & POL'Y REV. BLOG (July 1, 2014), https://harvardlpr.com/2014/07/01/dangerous-precedents-in-hobby-lobby ("[I]t's one thing to say that abortion is against my religion and therefore the Government cannot compel me to have an abortion. It's quite another thing to say that abortion is against my religion and therefore I have the right to prevent others from having the choice to have (what I consider to be) an abortion. The Court accepts at face value the equivalence between them, and because it does not specify a limiting principle, its [sic] hard to imagine any amount of attenuation that would override an objection to any action in the public sphere (except to taxes!).").

[68] In *Burwell*, 573 U.S. 682, a 2014 5-4 split decision along ideological lines, the Court ruled that the Department of Health and Human Services's reproductive health mandate, requiring that companies provide their female employees no-cost contraceptive access, violated the Religious Freedom Restoration Act as applied to for-profit, "closely held" corporations. *Id.* at 688–91. The Court cabined its decision, making clear that it applied only to contraceptive access and does not reach religious objections to other medical concerns. *Id.* at 733. According to Justice Alito, "In any event, our decision in these cases is concerned solely with the contraceptive mandate. Our decision should not be understood to hold that an insurance-coverage mandate must necessarily fall if it conflicts with an employer's religious beliefs. Other coverage requirements, such as immunizations, may be supported by different interests (for example, the need to combat the spread of infectious diseases) and may involve different arguments about the least restrictive means of providing them." *Id.* In other words, based on religious objection, closely held corporations could not assert an interest to deny their employees access to blood transfusions, vaccinations, prescribed narcotics, or other medical services based on a religious objection; they could only deny their employees access to female contraception. Notably, closely held corporations comprise over 90 percent of American companies.

backward. And in various ways, the Court delivers, frequently citing authors of British treatises who died before the United States existed, ratified its Constitution, or developed its current form of government. However, the Court's deference to the past is selectively, if not opportunistically, deployed with outcomes seemingly predetermined. In its opportunistic originalism, the Court wields the methodology as cover to deflect criticism for what on the one hand resembles careless and imprecise historical analysis. On the other hand, the Court's opportunistic originalism masks the Court's political decision-making. Rather than originalism serving its purported goal to lessen judicial discretion and dampen political urges, it has been retrofitted in the Roberts Court with dangerous consequences in the present, particularly for vulnerable communities that were divested from citizenship and meaningful political representation, participation, and power at the nation's Founding. Moreover, the Court's opportunistic originalism is further hampered by inattention to robust historical archives that awaken relevant debates and surface abolitionist ratifiers' intentions. The result is the Court's indifference to and neglect of Reconstruction narratives, and its failure to see that the vestiges and badges of slavery manifest in Mississippi's abortion case. As such, *Dobbs* is difficult to rescue from claims that it represents unprincipled, dishonest, and amoral decision-making—taking the Court at its terms.

In its backward, tilting view, the Court catapults over key arguments, materials, and facts relevant to its inquiry related to Reconstruction, female bodily autonomy, and abortion. The Court fastens to some facts, while relevant, material details and archives are left dusty without evidence of a fingerprint. The Court correctly argues that abortion is not mentioned in the Constitution, but incorrectly implies that matters involving and adjacent to abortion were not acknowledged, debated, reviewed, or countenanced by the Framers and ratifiers. This Article shows the incorrectness of that logic.

Part I argues that the path toward dismantling *Roe* was set in motion by cases largely overlooked in the abortion canon. Part II analyzes the Mississippi law and its path to the Supreme Court, illuming *Casey* as both pivotal to the preservation of *Roe* and also a weak link in preserving abortion rights. Part III addresses originalism and historical memory. It queries and scrutinizes the prioritization of certain social and legal narratives over others. Part III brings to the fore overlooked speeches and debates that better inform the cause of Reconstruction. It revives debates, speeches, and other archives relevant to a more rigorous, informed, inclusive, and honest account of reproductive privacy. Part IV turns to the Court's racial turn in dismantling *Roe*. The Article then concludes.

I. The *Roe* Era

The obvious explanation for the elimination of an abortion right can be pulled from the pages of *Dobbs*. The majority casts nearly fifty years of precedent aside to shape an opinion free from adhering to the dictates of *stare decisis*. However, *Roe* proved vulnerable in the 1970s and 1980s as the Court took up cases denying state and federal assistance to women seeking abortions. In *Maher v. Roe*[69] and *Beal v. Doe*,[70] reviewed as companion cases, the Court ruled that states did not "impinge upon the fundamental right recognized in *Roe*" by issuing regulations that limited Medicaid benefits to "medically necessary" abortions.[71] Instead, the Court emphasized that its "conclusion signal[ed] no retreat from *Roe* or the cases applying it."[72] As the Court explained in *Maher*, "[t]here is a basic difference between direct state interference with a protected activity and state encouragement of an alternative activity consonant with legislative policy."[73]

The Court reached a similar result less than a decade after *Roe* in *Harris v. McRae*,[74] a case that raised constitutional and statutory questions regarding whether the "so-called Hyde Amendment" could lawfully restrict the use of federal funds to reimburse the cost of abortions under the Medicaid program.[75] The Court concluded that the funding restrictions of the Hyde Amendment do not impinge on the liberty interests safeguarded by the Due Process Clause of the Fifth Amendment held in *Roe*.[76] Justice Powell explained, "[t]he constitutional underpinning of *Wade* was a recognition that the 'liberty' protected by the Due Process Clause of the Fourteenth Amendment includes not only the freedoms explicitly mentioned in the Bill of

---

[69] 432 U.S. 464 (1977).

[70] 432 U.S. 438 (1977).

[71] *See Maher*, 432 U.S. at 474.

[72] *Id.* at 475.

[73] *Id.*

[74] 448 U.S. 297 (1980).

[75] *Id.* at 312.

[76] *Id.* at 322.

Rights, but also a freedom of personal choice in certain matters of marriage and family life."[77] He further stated, "[t]his implicit constitutional liberty, the Court in *Wade* held, includes the freedom of a woman to decide whether to terminate a pregnancy."[78]

Even while the government funding abortion cases decisively reaffirmed *Roe*, leaving virtually no doubt as to its legitimacy and future viability, they were nonetheless troubling holdings because "[a]s the Court well [knew], these regulations inevitably . . . have the practical effect of preventing nearly all poor women from obtaining safe and legal abortions."[79] Indeed, the Court showed a "distressing insensitivity"[80] to the often tragic conditions in the lives of indigent, pregnant women whose physical as well as mental health are often taxed by the burdens of unwanted pregnancy.[81] *Maher*, *Beal*, and *Harris* stand among the harshest abortion decisions post *Roe v. Wade*—until now. The cases essentially established a hierarchy in the right to an abortion.

Prior to *Roe*, wealthy women could already travel to New York or London to end an unwanted pregnancy—as burdensome as even that was. As noted in prior work, "between 1968, when Great Britain liberalized its abortion laws, and 1970, when New York repealed its criminal ban, making legal abortions available in the United States,

---

[77] *Id.* at 312.

[78] *Id.*

[79] Beal v. Doe, 432 U.S. 454, 455 (1977) (Marshall, J., dissenting).

[80] *Maher*, 432 U.S. at 483.

[81] *See, e.g.*, Katherine Barton et al., *Unplanned Pregnancy and Subsequent Psychological Distress in Partnered Women: A Cross-Sectional Study of the Role of Relationship Quality and Wider Social Support*, 17 BMC PREGNANCY CHILDBIRTH 1, (2017), https://bmcpregnancychildbirth.biomedcentral.com/articles/10.1186/s12884-017-1223-x (noting that in a study of 12,462 mothers that reported an unplanned pregnancy, less than 25 percent reported feeling "happy" upon discovering their pregnancies and that "[u]nplanned pregnancy was associated with a significantly increased odds of [psychological distress] compared to planned [pregnancies]"); *see also Maher*, 432 U.S. at 483 (Brennan, J., dissenting) ("The stark reality for too many, not just 'some,' indigent pregnant women is that indigency makes access to competent licensed physicians not merely 'difficult,' but 'impossible.' As a practical matter, many indigent women will feel they have no choice but to carry their pregnancies to term because the State will pay for the associated medical services, even though they would have chosen to have abortions if the State had also provided funds for that procedure, or indeed if the State had provided funds for neither procedure."); Roe v. Wade, 410 U.S. 113, 153 (1973) ("Maternity, or additional offspring, may force upon the woman a distressful life and future. Psychological harm may be imminent. Mental and physical health may be taxed by child care. There is also the distress, for all concerned, associated with the unwanted child . . . ."), *overruled by* Dobbs v. Jackson Women's Health Org., 142 S. Ct. 2228 (2022).

it is estimated that 5,000 abortions per year were performed on American women in Great Britain."[82] Twenty times that number of women made their way to New York to obtain safe, legal abortions. According to the Guttmacher Institute, a research and policy non-governmental organization focused on reproductive health and rights, in 1972, the year before *Roe* was decided, "just over 100,000 women left their own state to obtain a legal abortion in New York City."[83] Notably, however, for poor women in New York City, obtaining a safe, legal abortion was too often out of reach. Rachel Benson Gold, Vice President for Public Policy at the Guttmacher Institute underscores this point. She reports that in New York City, during the 1960s, a physician was involved in only two percent of poor women's abortions.[84]

Poor women were trapped literally and figuratively, too often unable to secure funds to obtain safe abortions and thus shackled to pregnancies they did not desire. As Justice Marshall noted in his dissent in *Beal*, "[i]t is all too obvious that the governmental actions in these cases, ostensibly taken to 'encourage' women to carry pregnancies to term, are in reality intended to impose a moral viewpoint."[85] The "real-world" effects of upholding legislation restricting state funding for abortion, even while providing payment for the higher costs associated with pregnancy delivery, prenatal services, and postnatal care, coercively fastened impoverished women to the intergenerational blight of poverty. Or more bluntly, the legislation challenged in *Maher*, *Beal*, and *Harris* would serve to "brutally coerce poor women to bear children whom society will scorn for every day of their lives."[86]

---

[82] Erwin Chemerinsky & Michele Goodwin, *Abortion: A Woman's Private Choice*, 95 Tex. L. Rev. 1189, 1212 (2017) (citing Richard L. Worsnop, *Abortion Law Reform*, *in* 2 Editorial Research Reports 543, 553 (William B. Dickinson et al. eds., 1970)).

[83] Rachel Benson Gold, *Lessons from Before* Roe*: Will Past be Prologue?*, 6 Guttmacher Pol'y Rev. 8, 10 (Mar. 2003), https://www.guttmacher.org/gpr/2003/03/lessons-roe-will-past-be-prologue ("According to an analysis by The Alan Guttmacher Institute, an estimated 50,000 women traveled more than 500 miles to obtain a legal abortion in New York City; nearly 7,000 women traveled more than 1,000 miles, and some 250 traveled more than 2,000 miles, from places as far as Arizona, Idaho and Nevada.").

[84] *Id.* at 8.

[85] Beal v. Doe, 432 U.S. 454, 454 (1977) (Marshall, J., dissenting).

[86] *Id.* at 456; *see also* Naomi Cahn & June Carbone, *The Blue Family Constitution*, 36 J. Am. Acad. Matrim. Laws. (forthcoming 2023) (on file with author).

Justice Marshall's largely overlooked but prescient dissent, a prophecy that closes the door on *Beal*, opens a future portal to *Dobbs*, which overturned *Roe* and *Casey*.[87] Marshall responded to the predictable consequences flowing from the Court's decisions in *Beal* and *Maher*. Namely, the foreseeable currents that trap women of color in poverty and drown them in debt,[88] that shame them for their poverty,[89] and the riptides that sweep their children into various systems of surveillance,[90] which result in "blighted lives," tethered to "foster

---

[87] *See* Dobbs v. Jackson Women's Health Org., 142 S. Ct. 2228 (2022).

[88] Only weeks after *Dobbs*, the National Women's Law Center issued a stunning report on the intersections of race, sex, and poverty. *See Amid Rising Household Costs, Black, Non-Hispanic Women and Latinas Continue to Face Economic Fallout from the Pandemic*, NAT'L WOMEN'S L. CTR. (Aug. 18, 2022), https://nwlc.org/resource/amid-rising-household-costs-black-non-hispanic-women-and-latinas-continue-to-face-economic-fallout-from-the-pandemic ("Black, non-Hispanic and Latinas were more likely than white, non-Hispanic men and women to: [h]have lost employment income; [n]ot have enough food to eat; [n]ot be able to afford enough food for their children to eat if they themselves were also experiencing food insufficiency; [b]e behind on their rent. . . . "); *see also* Robin Bleiweis et al., *The Basic Facts About Women in Poverty*, CTR. FOR AM. PROGRESS (Aug. 3, 2020), https://www.americanprogress.org/article/basic-facts-women-poverty ("In the United States, more women than men live in poverty. According to U.S. Census Bureau data, of the 38.1 million people living in poverty in 2018, 56 percent—or 21.4 million—were women [and disproportionately among them are women of color]."); AAUW, DEEPER IN DEBT 2 (2017), https://www.aauw.org/app/uploads/2020/03/DeeperinDebt-nsa.pdf ("Women—especially women of color—are most likely to experience difficulties, 34 percent of all women and 57 percent of black women who were repaying student loans reporting that they had been unable to meet essential expenses within the past year.").

[89] In the same period that the Court heard *Maher* and *Beal*, Black women came to be associated with the "welfare queen" stereotype, a trope trotted out by politicians who promised to be tough on crime and punish people cheating the government and exploiting the public fisc. *See 'Welfare Queen' Becomes Issue In Reagan Campaign*, N.Y. TIMES, Feb. 15, 1976, at 51 ("'There's a woman in Chicago,' the Republican candidate said recently to an audience in Gilford, N.H., during his freeswinging attack on welfare abuses. 'She has 80 names, 30 addresses, 12 Social Security cards and is collecting veterans' benefits on four nonexisting deceased husbands.' He added: 'And she's collecting Social Security on her cards. She's got Medicaid, getting food stamps and she is collecting welfare under each of her names. Her tax-free cash income alone is over $150,000.'"); *see also* KAARYN S. GUSTAFSON, CHEATING WELFARE: PUBLIC ASSISTANCE AND THE CRIMINALIZATION OF POVERTY (2011); Carly Hayden Foster, *The Welfare Queen: Race, Gender, Class, and Public Opinion*, 15 RACE, GENDER & CLASS J. 162, 163 (2008) ("Invocation of the Welfare Queen by politicians, policymakers, and journalists cues common stories and media images of a sexually promiscuous poor single African American woman who scams taxpayers by having babies then demanding public support.").

[90] *See generally* DOROTHY ROBERTS, TORN APART: HOW THE CHILD WELFARE SYSTEM DESTROYS BLACK FAMILIES—AND HOW ABOLITION CAN BUILD A SAFER WORLD (2022); DOROTHY ROBERTS, SHATTERED BONDS: THE COLOR OF CHILD WELFARE (2002); Saki Knafo, *Here Are All the Countries Where Children Are Sentenced to Die in Prison*, HUFFPOST (Dec. 6, 2017), https://www.huffpost.com/entry/juvenile-life-without-parole_n_3962983 ("That's right: The United States is the only country in the world that condemns people to spend their lives behind bars for crimes they committed before they turned 18."); Michele Goodwin, *How Mass Incarceration Trickles Down to Children*, HUFFPOST (Aug. 19, 2017), https://www.huffpost.com

homes, orphanages, and 'reform' schools."[91] Shortly after *Maher* and *Beal*, Elisabeth Landes and Richard Posner published *The Economics of the Baby Shortage*,[92] an article whose economic theory to address foster care and adoption received blistering criticism,[93] but which presented a profound, though crudely framed truth that "there is a

/entry/when-mass-incarceration-t_b_11554242 ("In the wake of a staggering mass incarceration problem that predicts one in three Black men will be involved in the criminal justice system at some point in his lifetime (in jail, on probation, paroled, or in a halfway house) politicians and civil society organizations frequently overlook what the so called race to incarcerate means for women and children. . . . [C]hildren are caught in the spiral too.").

[91] *See* Beal v. Doe, 432 U.S. 454, 456 (1977) (Marshall, J., dissenting).

[92] Elisabeth M. Landes & Richard A. Posner, *The Economics of the Baby Shortage*, 7 J. LEGAL STUD. 323 (1978).

[93] *See, e.g.*, Robin West, *Submission, Choice, and Ethics: A Rejoinder to Judge Posner*, 99 HARV. L. REV. 1449, 1449 (1986) ("For Posner's ethical world is, I suspect, quite far from that of the *Harvard Law Review* reader. I don't know, but I doubt that the reader shares Posner's belief that we should be allowed to sell our babies for profit."); John J. Donohue III & Ian Ayres, *Posner's Symphony No. 3: Thinking About the Unthinkable*, 39 STAN. L. REV. 791, 793 (1987) (reviewing RICHARD A. POSNER, ECONOMIC ANALYSIS OF LAW (3d ed.) (1986)) ("Posner seems willing to dare opprobrium by questioning the prohibition against selling babies. . . . While economics is a powerful and valuable tool in the analysis of legal and public policy issues, those who use it need an appreciation of its limitations."); Jerome McCristal Culp, Jr., *Posner on Duncan Kennedy and Racial Difference: White Authority in the Legal Academy*, 41 DUKE L.J. 1095, 1113–14 (1992) ("This Judge Posner is not the caricature who wants to sell babies, but the Judge Posner who genuinely cares about people, his wife, his children, and society. . . . To me, this 'real' Judge Posner, however, is no different than the white slaveowners in the antebellum South who were kind to their slaves (some of the males genuinely loved and indirectly married black female slaves) or the white southerner who is particularly kind to 'Uncle Joe' or 'Aunt Susie,' or the white northerner who patronizes her black friends. These people are not intrinsically, eternally, or completely bad people. They demonstrate real kindness to the black people with whom they interact, but they also demand the right to decide when a black person is 'out of place.'" (footnote omitted)); Margaret F. Brinig, *The Effect of Transaction Costs on the Market for Babies*, 18 SETON HALL LEGIS. J. 553, 555–56 (1994) ("Perhaps the greatest problem with Posner's market theory remains largely unexplored" in that it fails to seriously account for "concerns about abusive adoptive parents and a potential oversupply of older or handicapped children" (footnote omitted)); *see also* Jana B. Singer, *The Privatization of Family Law*, 1992 WIS. L. REV. 1443, 1486–88. Scholars and commentators predicted that Posner's market theory on the supply and demand of babies and children available for adoption would upend the likelihood of a nomination to the Supreme Court. *See Selling Babies*, FORBES (Feb. 24, 1997, 12:00 AM), https://www.forbes.com/forbes/1997/0224/5904096a.html?sh=18a4ee082108 ("In the article which may have cost him a Supreme Court bench, Posner argued that the application of free market principles to the adoption process would benefit children. As it is the process is painfully slow and has created a black market, in good part because demand far outstrips the available supply."); Margaret J. Radin, *What, If Anything, Is Wrong with Baby Selling?*, 26 PAC. L.J. 135, 139 (1995) ("I think that his baby selling articles are probably going to keep him from ever being on the Supreme Court even if the Republicans get back in. So he has the courage of his convictions."). *But see* Ronald A. Cass, *Coping With Life, Law, and Markets: A Comment on Posner and the Law-and-Economics Debate*, 67 B.U. L. REV. 73, 73–74 (1987) (noting that there was less objection to Posner's theory a decade after the original publication).

shortage of white babies for adoption [and] . . . there is a glut of black babies . . . for adoption."[94] The authors surmised, "[t]he very high ratio of illegitimate black births to black adoptions suggests why there is no shortage of black babies for adoptions."[95]

Marshall, perhaps better than any other Justice at the time—given his nearly three dozen civil rights cases argued before the Court,[96] especially litigation stemming from the persistent, invidious conditions of Jim Crow and lasting vestiges of slavery[97]—could wisely predict that in constraining poor women's access to abortion, Black women would disparately suffer, that "[m]any children of the poor, sadly, [would] attend second-rate segregated schools," and that opposition to aid for poor families would strengthen, leaving "impoverished mothers and children . . . little chance for the children to grow up in a decent environment."[98] Similar predictions are offered here post-*Dobbs* because the material conditions underlying Marshall's fears have yet to dissipate. Indeed, today's resurgence in antipathy toward the concerns and conditions of poor women and their children, particularly women of color, is packaged in race-neutral terms from the dismantling of freedoms associated with reproductive rights,[99]

---

[94] Landes & Posner, *supra* note 92, at 324–25.

[95] *Id.* at 325.

[96] *See Thurgood Marshall*, NAACP, https://naacp.org/find-resources/history-explained/civil-rights-leaders/thurgood-marshall (last visited Jan. 19, 2023) ("Marshall became one of the nation's leading attorneys. He argued 32 cases before the U.S. Supreme Court, winning 29.").

[97] *See, e.g.*, Brown v. Bd. of Educ., 347 U.S. 483 (1954) (holding that separate but equal educational facilities for racial minorities is inherently unequal, violating the Equal Protection Clause of the Fourteenth Amendment); Sweatt v. Painter, 339 U.S. 629 (1950) (holding that the law school had to admit the black plaintiff, Heman Sweatt, to the state's flagship law school partly because of the intangible and interpersonal elements of a quality education); Shelley v. Kraemer, 334 U.S. 1 (1948) (holding that state enforcement of restrictive covenants violates the Equal Protection Clause of the Fourteenth Amendment); Smith v. Allwright, 321 U.S. 649 (1944) (holding that restricting the primary vote to white people violates the Equal Protection Clause of the Fourteenth Amendment).

[98] *See* Beal v. Doe, 432 U.S. 454, 456 (1977) (Marshall, J., dissenting) (citing Milliken v. Bradley, 418 U.S. 717 (1974)).

[99] *See* Dobbs v. Jackson Women's Health Org., 142 S. Ct. 2228 (2022); *see also State Court Abortion Litigation Tracker*, Brennan Ctr. for Just. (Jan. 6, 2023), https://www.brennancenter.org/our-work/research-reports/state-court-abortion-litigation-tracker ("As of January 6, 2023, a total of 34 cases have been filed challenging abortion bans in 19 states, of which 27 remain pending at either the trial or appellate levels.").

voting rights,[100] religious liberties,[101] to speech,[102] and association.[103] In other words, such consequences, borne out of the antiabortion position, exposed an "ethical bankruptcy" according to Justice Marshall, because it means under social policies present then, and sadly persisting today, "a bare existence in utter misery for so many poor women and their children."[104]

---

[100] *See* Shelby Cnty. v. Holder, 570 U.S. 529 (2013) (gutting the Voting Rights Act's most effective provision, its preclearance regime); Rucho v. Common Cause, 139 S. Ct. 2484 (2019) (holding that partisan gerrymandering claims are nonjusticiable political questions); Brnovich v. Democratic Nat'l Comm., 141 S. Ct. 2321 (2021) (making it harder to block discriminatory voting laws). The Supreme Court is also deciding two cases with major implications for democracy this term. *See* Zach Montellaro, *Supreme Court to Hear 2 Cases with Major Implications for 2024*, POLITICO (Sept. 29, 2022, 4:31 AM), https://www.politico.com/news/2022/09/29/supreme-court-2024-election-cases-00059380 ("One lawsuit out of North Carolina could have broad ramifications, with Republicans asking the Supreme Court to revoke the ability of state courts to review election laws under their states' constitutions. The reading of the Constitution's Elections Clause that underpins the case—called the 'Independent State Legislature' theory—has gotten buy-in from much of the conservative legal world, and four Supreme Court justices have signaled at least some favorability toward it. The decision in the case could upend American elections. And another case out of Alabama that will be heard on Tuesday involves a challenge to the state's congressional map—and whether Black voters' power was illegally diluted. The result could kick back open congressional redistricting in several states two years after the entire nation went through a redraw."); *see also* Michael Wines, *Maps in Four States Were Ruled Illegal Gerrymanders. They're Being Used Anyway.*, N.Y. TIMES (Aug. 8, 2022), https://www.nytimes.com/2022/08/08/us/elections/gerrymandering-maps-elections-republicans.html.

[101] *See* Kiara Alfonseca, *Map: Where Anti-Critical Race Theory Efforts Have Reached*, ABC NEWS (Mar. 24, 2022, 6:03 AM), https://abcnews.go.com/Politics/map-anti-critical-race-theory-efforts-reached/story?id=83619715 ("A total of 35 states so far have signed into law or proposed legislation banning or restricting the teaching of critical race theory, the academic discipline at the center of the debate."); Patricia Mazzei & Anemona Hartocollis, *Florida Rejects A.P. African American Studies Class*, N.Y. TIMES (Jan. 19, 2023), https://www.nytimes.com/2023/01/19/us/desantis-florida-ap-african-american-studies.html ("Florida will not allow a new Advanced Placement course on African American studies to be offered in its high schools, stating that the course is not "historically accurate" and violates state law.").

[102] *See id.*

[103] *See* Summer Ballentine & John Hanna, *Missouri Considers Law to Make Illegal to 'Aid or Abet' Out-of-State Abortion*, PBS (Mar. 16, 2022, 2:45 PM), https://www.pbs.org/newshour/politics/missouri-considers-law-to-make-illegal-to-aid-or-abet-out-of-state ("A proposal that could be debated in the Legislature as soon as next week seeks to make it illegal to "aid or abet" abortions outlawed in Missouri, even if they are performed in other states."); Maggie Astor, *Here's What the Texas Abortion Law Says*, N.Y. TIMES (Sept. 9, 2021), https://www.nytimes.com/article/abortion-law-texas.html ("S.B. 8 invites lawsuits against anyone who performs or 'abets' an abortion. That could mean an array of people and groups, including clinics and their employees, from doctors to receptionists; friends, relatives or strangers who pay for an abortion, including people who donate to or administer abortion funds; insurers that approve a claim; ride-share drivers who drive a patient to a clinic; and anyone who shares information about abortion options.").

[104] *See* Beal v. Doe, 432 U.S. 454, 456–57 (1977) (Marshall, J., dissenting).

In *Dobbs*, a grave insensitivity to the "human dimension"[105] or what Justice Brennan might have described as the "stark real[i]ty" of poor women's lives in Mississippi is inherent in the Court's analysis.[106] The distressing records of maternal mortality[107] and morbidity[108] in the United States gave light to the certain risks American women would face if *Roe* and *Casey* were overturned. The amicus brief jointly submitted by the American College of Obstetricians and Gynecologists, American Medical Association, American Academy of Family Physicians, American Academy of Nursing, American Academy of Pediatrics, and American Association of Public Health Physicians made this point: "maternal mortality rates have increased staggeringly" since 2005.[109] Further, as noted in prior work,[110] Mississippi is a deadly state for pregnant women, especially Black women. For example, "it is nearly 118 times more dangerous for a Black woman to give birth [in Mississippi] than it is for the average American woman to have an abortion."[111]

* * *

The Court's grave insensitivity that the dissenting Justices spoke to in *Beal* resurfaces in *Dobbs*. In 1977, Justice Marshall voiced suspicion and wariness "that the Court's decisions" in *Maher* and *Beal* would invite lawmakers, "already under extraordinary pressure from well-financed and carefully orchestrated lobbying campaigns, to approve

---

[105] *Id.* at 457.

[106] *See* Maher v. Roe, 432 U.S. 464, 483 (1977) (Brennan, J., dissenting).

[107] *See* Julia Belluz, *We Finally Have a New US Maternal Mortality Estimate. It's Still Terrible.*, Vox (Jan. 30, 2020, 10:40 AM), https://www.vox.com/2020/1/30/21113782/pregnancy-deaths-usmaternal-mortality-rate (noting that maternal mortality in the United States ranks fifty-fifth in the world, after Russia, and is climbing).

[108] *See Severe Maternal Morbidity*, Ctrs. for Disease Control & Prevention, https://www.cdc.gov/reproductivehealth/maternalinfanthealth/severematernalmorbidity.html (last visited Jan. 19, 2023) ("Severe maternal morbidity (SMM) includes unexpected outcomes of labor and delivery that result in significant short- or long-term consequences to a woman's health. Using the most recent list of indicators, SMM has been steadily increasing in recent years and affected more than 50,000 women in the United States in 2014.").

[109] Brief for American College of Obstetricians and Gynecologists et al. as Amici Curiae in Support of Respondents at 18, Dobbs v. Jackson Women's Health Org., 142 S. Ct. 2228 (2022) (No. 19-1392).

[110] *See* Michele Goodwin, *Complicit Bias and the Supreme Court*, 136 Harv. L. Rev. F. 119, 155 (2022).

[111] *See id.* at 155.

more [abortion] restrictions."[112] Justice Marshall predicted the potential for poor women to incrementally lose access to abortion, making the right more illusory than real. Likely, he did not countenance a time in which the right itself would fall vulnerable to the Court. However, as *Dobbs* reveals, *Roe* would *also* need to withstand unprincipled judicial decision-making. If *Maher*, *Beal*, and *Harris* revealed the possibilities of what *could* and *would* emerge later in the form of targeted regulations of abortion providers (TRAP laws) and myriad abortion bans, then it must be taken into account that the path to *Dobbs*—and along with it injurious abortion bans—was in some sense presaged and not accounted for in the legal movements that advocate for and defend reproductive rights. Justices Marshall, Blackmun, and Brennan sounded alarms that in hindsight went unheeded, likely because those most negatively affected by states' regulations of and bans on abortion are at the margins of society—those who are poor and of color.

II. Abortion Rights and the Mississippi Gestational Age Act

Part II argues that a deeper, more nuanced reflection on *Dobbs* starts from the premise that Mississippi's imperative—the dismantling of *Roe*—was not a project that sprang from the Mississippi legislature or the Roberts Court alone. Rather, the overturn of *Roe* can be traced to strategic litigation and policies that affected the lives of poor women, much as in *Beal*, *Maher*, and *Harris*, decades ago, with devastating reach and impacts on the lives of overlooked and neglected women of color.

For the individual women concerned in those cases, as well as those who are now most impacted by *Dobbs*, geographically constrained,

[112] Beal v. Doe, 432 U.S. 438, 462 (1977) (Marshall J., dissenting). Notably, in the present, other hard-won constitutional rights may also prove vulnerable to the blueprint articulated by Justice Marshall. That is, unprincipled decision-making at the Court stands to open new paths for the dismantling of not only abortion rights, but also other areas of recently achieved privacy rights, including LGBTQ rights and marriage equality. For this reason, recognizing examples of the battles won become important to archive. *See* Dale Carpenter, Flagrant Conduct: The Story of *Lawrence v. Texas* (2012) (providing an important accounting of the pivotal LGBTQ case that sets in motion the dismantling of harmful anti-gay fornication laws); Dale Carpenter, *Windsor Products: Equal Protection from Animus*, 2013 Sup. Ct. Rev. 183, 191. As with preserving the core holding of *Roe* in *Casey*, Justice Kennedy played a crucial role in the LGBTQ liberation cases. Carpenter recognizes this as "an outpouring of decades of constitutional development whose fountainhead is *Carolene Products* and whose tributaries are the gay-rights and federalism streams." *Id.* His point was that animus, built into the legislative process, found succor at the Court until the animus was dislodged. *Id.*

"indigent and financially helpless," as Justice Blackmun acknowledged in dissent,[113] the results are cruel and punitive. Yet, what was then only implicit—the "condescension that she may go elsewhere for her abortion"[114]—is now made explicit in *Dobbs*. What Justice Blackmun described then as the disingenuous posture of those in the majority in *Maher*, *now* seeks no shelter behind the cloak of *Roe* to claim that a constitutional right to abortion—no matter how impaired—exists in the United States. That is, the Court no longer concedes the existence of a constitutional right to abortion.

This Part prefaces the Mississippi story with a broader view of antiabortion legislative activism. It explains that multiple campaigns and strategies galvanized to secure legislative support in chipping away not only abortion rights, but reproductive rights more broadly, which ultimately delivered a viable case to eliminate the constitutional right to an abortion. In other words, the strategies to undercut abortion rights hid in plain sight, because the tactics deployed by the antiabortion movement, and aided by male-dominated state legislatures, fell hard on low-income women of color—already stereotyped and stigmatized—and politically disenfranchised. These women were not the central focus of mainstream reproductive rights organizations, possibly because of racism and classism within the reproductive rights movement, which prioritized the concerns of wealthier white women. For those who were shut out, they became the canaries in the coal mines.

Additionally, the campaigns of the antiabortion movement were buttressed by federal efforts, such as the Hyde Amendment[115] and Helms Amendment.[116] Later, President Ronald Reagan's Mexico City Policy,[117] otherwise known as the "Global Gag Rule"[118] would further impede abortion access for poor women of color in developing

---

[113] *Beal*, 432 U.S. at 462 (Blackmun, J., dissenting).

[114] *Id.*

[115] Consolidated Appropriations Act, 2014, Pub. L. No. 113-76, §§ 506–07, 128 Stat. 5, 409 (2013).

[116] 22 U.S.C. § 2151b(f)(1) (2012).

[117] Policy Statement of the United States of America at the United Nations International Conference on Population, 2d Sess., Mexico City (Aug. 6–14, 1984), *reprinted in* 10 Pop. & Dev. Rev. 574 (1984) [hereinafter Mexico City Policy].

[118] *See What Is the Global Gag Rule?*, Planned Parenthood, https://www.plannedparenthoodaction.org/communities/planned-parenthood-global/end-global-gag-rule (last visited Jan. 21, 2023).

nations.[119] And certainly, the legacy of *Roe* was further undermined by efforts to dismantle Title X during the Trump Administration.[120]

While the broader process of *Roe*'s demise is not the subject of this contribution, Part II foregrounds key aspects. Section II.A explains that the path to overturning *Roe* and *Casey* included a wider patchwork of antiabortion legislation across the nation, strategic legal activism, successful enactments of "TRAP laws"[121]—and decades-old campaigns to normalize fetal personhood. Section II.B then briefly traces the history of Mississippi's Gestational Age Act,[122] a law whose sponsors made numerous attempts to dismantle *Roe* and constrain abortion rights in that state. Section II.C concludes Part II by addressing the order issued by Judge Carlton Reeves, which blocked Mississippi's Gestational Age Act from going into effect but was overruled by *Dobbs*.

A. THE BROADER PATH TO OVERTURNING *ROE* AND *CASEY*

Data from the Guttmacher Institute suggest that nearly half of the pregnancies in the United States are unintended.[123] Women who experience poverty are at the highest risk for unintended pregnancy. For instance, "[t]he rate of unintended pregnancy among women with incomes less than 100% of the poverty [line] was . . . more than five times the rate among women with incomes of at least 200% of poverty."[124] Yet, this data has not slowed the pace of legislative action intended to limit or ban abortion access. To the contrary, in early

---

[119] *See* Michele Goodwin, *Challenging the Rhetorical Gag and Trap: Reproductive Capacities, Rights, and the Helms Amendment*, 112 Nw. U. L. Rev. 1417, 1444–50 (2018).

[120] Julie Hirschfeld Davis & Michael D. Shear, *Trump Rule Would Bar Some Abortion Advice at Federally Funded Clinics*, N.Y. Times (May 22, 2018), https://www.nytimes.com/2018/05/22/us/politics/trump-abortion-limits.html; *see also* Marie Solis, *Here's What the Trump Administration's Proposed Title X Rule Would Do to Abortion Access in America*, Newsweek (May 2, 2018, 12:22 PM), http://www.newsweek.com/hereswhat-trump-administrations-proposed-title-x-rule-would-do-abortion-908474.

[121] *See Targeted Regulation of Abortion Providers (TRAP) Laws*, Guttmacher Inst. (Jan. 22, 2020), https://www.guttmacher.org/print/evidence-you-can-use/targeted-regulation-abortion providers-trap-laws.

[122] Miss. Code Ann. § 41-41-191 (2022).

[123] *Unintended Pregnancy in the United States*, Guttmacher Inst. (Jan. 2019), https://www.guttmacher.org/fact-sheet/unintended-pregnancy-united-states ("Nearly half (45%, or 2.8 million) of the 6.1 million pregnancies in the United States were unintended. Specifically, 27% of all pregnancies were 'wanted later' and 18% of pregnancies were 'unwanted.'" (citation omitted)).

[124] *Id.*

2021, before the term that delivered *Dobbs* began, 561 abortion restrictions were introduced.[125] By August, nearly one hundred of "those introduced abortion-restrictive statutes" were "enacted into law."[126] Possibly emboldened by a significant gain at the Supreme Court in *Hobby Lobby*, or the commitment and follow-through of President Trump to only appoint judges committed to overturning *Roe*,[127] state legislatures passed laws intended to significantly curtail abortion rights or virtually eliminate the right altogether.

---

[125] Elyssa Spitzer & Nora Ellmann, *State Abortion Legislation in 2021*, CTR. FOR AM. PROGRESS (Sept. 21, 2021), https://www.americanprogress.org/article/state-abortion-legislation-2021.

[126] *Id.*

[127] For example, President Trump expressed a commitment to replace Justice Antonin Scalia's vacated seat on the Supreme Court with a staunch opponent of abortion rights. He promised that "[t]he judges will be pro-life." *See* Brit Mccandless Farmer, *Trump in 2016: The Judges Will be Pro-Life*, CBS NEWS (Sept. 23, 2020, 4:17 PM), https://www.cbsnews.com/news/donald-trump-2016-pro-life-judges-60-minutes-2020-09-23. At the time, one of his stated contenders for nomination to the Supreme Court, Judge William Pryor, referred to *Roe v. Wade* as an "abomination." *See, e.g.*, Ariane de Vogue, *How Trump's Election Reignites the Abortion Wars*, CNN (Dec. 14, 2016, 8:39 AM), http://www.cnn.com/2016/12/14/politics/trump-abortion-supreme-court. When appointing Justices Gorsuch and Kavanaugh, and later Justice Barrett, President Trump fulfilled his campaign promise. Prior to their appointments, each of the Justices articulated hostility to reproductive privacy. Arguably, his first nominee, Justice Gorsuch, set the tone for those that would follow. In an amicus brief written in 1996, prior to his federal judicial appointment, he stated that numerous problems "plagued the Court's abortion jurisprudence." Brief for the American Hospital Association as Amicus Curiae in Support of Petitioners at 7, Washington v. Glucksberg, 521 U.S. 702 (1997) (Nos. 96-110, 96-1858), 1996 WL 656278 ("[T]he plurality's opinion rests at heart upon stare decisis principles, upholding the abortion right largely because of the need to protect and respect prior court decisions in the abortion field."). In a prescient conjecture twenty-five years before *Dobbs*, he conjectured that *Planned Parenthood v. Casey* was a case rooted in stare decisis rather than the Court affirmatively upholding abortion rights. *Id.*

While sitting on the Tenth Circuit Court of Appeals, Justice Gorsuch's statements on privacy rights, judicial record on matters related to defunding Planned Parenthood and birth control access, as well as his views on pregnancy discrimination indicated opposition and enmity to women's reproductive rights. *See* Hobby Lobby Stores, Inc. v. Sebelius, 723 F.3d 1114, 1157 (10th Cir. 2013) (Gorsuch, J., concurring) (referring to the Religious Freedom Restoration Act as "something of a 'super-statute'" that supersedes all other legislation, including federal laws like the Affordable Care Act, which mandates contraceptive health coverage for women and their dependents (citation omitted)), *aff'd sub nom.* Burwell v. Hobby Lobby Stores, Inc., 573 U.S. 682 (2014).

Later, as a judge on the Tenth Circuit Court of Appeals, in a case involving efforts to defund Planned Parenthood, Justice Gorsuch issued a dissenting opinion from the denial of en banc review where the circuit court upheld an injunction against Utah Governor Gary Herbert's plans to defund Planned Parenthood. *See* Planned Parenthood Ass'n of Utah v. Herbert, 839 F.3d 1301, 1307 (10th Cir. 2016) (Gorsuch, J., dissenting)

When questioned by Senator Richard J. Durbin (D-Ill.), during his confirmation hearing whether he asked "students in class . . . to raise their hands if they knew of a woman who had taken maternity benefits from a company and then left the company after having a baby," Justice Gorsuch responded, "No." Sean Sullivan, *Gorsuch Denies Former Student's Allegation on Maternity Benefits Question*, WASH. POST (Mar. 21, 2017, 1:42 PM), https://www.washingtonpost

For example, Kentucky lawmakers enacted a law that granted its attorney general authority to close abortion clinics;[128] South Carolina's legislature enacted a ban as early as six weeks in pregnancy;[129] and Arkansas's former Governor Asa Hutchinson signed into law one of the most restrictive abortion bans in the United States, which made no exception for rape or incest[130]—a bridge thought too far just a few years before that.[131] However, even before *Dobbs*, a handful of states had only one abortion clinic remaining, including in Mississippi.[132] For example, a few years after *Casey* was decided, there were 452 clinics that provided pregnancy termination, but by 2014, "the number was down to 272."[133] One hundred days after the *Dobbs* ruling, an additional

.com/politics/2017/live-updates/trump-white-house/neil-gorsuch-confirmation-hearings-updates-and-analysis-on-the-supreme-court-nominee/gorsuch-denies-former-students-allegation-on-maternity-leave-question. Nevertheless, Justice Gorsuch refrained from clarifying whether he believes women abuse maternity leave policies or whether employers should be entitled to ask family planning questions that currently violate federal law. *See Confirmation Hearing on the Nomination of Hon. Neil M. Gorsuch to Be an Associate Justice of the Supreme Court of the United States: Hearing Before the S. Comm. on the Judiciary*, 115th Cong. 118–19 (Mar. 21, 2017) (statement of Judge Neil Gorsuch). In response to Senator Durbin's question, "whether employees should or should not make inquiries into whether an applicant or employee intends to become pregnant," Justice Gorsuch deflected, responding to Senator Durbin that "it sounds like you are asking me about a case or a controversy" and, "with all respect, when we come to cases and controversies, a good judge will listen." *See id.* at 118. *But see* Arnie Seipel & Nina Totenberg, *Amid Charges by Former Law Student on Gender Equality, Former Clerks Defend Gorsuch*, NPR (Mar. 20, 2017, 12:15 AM), http://www.npr.org/2017/03/20/520743555/former-law-student-gorsuch-told-class-women-manipulate-maternal-leave.

[128] *See* Bruce Schreiner, *Lawmakers Hand AG New Authority to Regulate Abortion Clinics*, AP News (Feb. 2, 2021), https://apnews.com/article/legislature-kentucky-laws-101425ea52dcdfe10fea39024423e7bc.

[129] *See* Jeffrey Collins, *South Carolina House Passes Bill Banning Most Abortions*, AP News (Feb. 17, 2021), https://apnews.com/article/south-carolina-passes-abortion-ban-bill-85449b10f9186589a6656f2245754adb. *But see* James Pollard, *South Carolina Supreme Court Strikes Down State Abortion Ban*, AP News (Jan. 5, 2023), https://apnews.com/article/abortion-politics-health-south-carolina-state-government-6cd1469dbb550c70b64a30f183be203c.

[130] *See Arkansas Passes Near-Total Abortion Ban—and a Possible* 'Roe v. Wade' *Test*, NPR (Mar. 10, 2021, 5:06 AM), https://www.npr.org/2021/03/10/975546070/arkansas-passes-near-total-abortion-ban-as-lawmakers-push-for-supreme-court-case.

[131] *See* Michele Goodwin & Mary Ziegler, *Whatever Happened to the Exceptions for Rape and Incest?*, Atlantic (Nov. 21, 2021), https://www.theatlantic.com/ideas/archive/2021/11/abortion-law-exceptions-rape-and-incest/620812.

[132] Holly Yan, *These 6 States Have Only 1 Abortion Clinic Left. Missouri Could Become the First With Zero*, CNN (June 21, 2019, 12:48 PM), https://www.cnn.com/2019/05/29/health/six-states-with-1-abortion-clinic-map-trnd/index.html ("In just a few days, Missouri could become the first state in the country with zero abortion clinics. But several other states are just one closure away from that fate. Six states are now down to one abortion clinic—Kentucky, Mississippi, Missouri, North Dakota, South Dakota and West Virginia.").

[133] *Id.*

sixty-six clinics that previously provided abortion care were forced to close or stop delivering those services.[134]

While legislative restrictions could not be said to map onto public sentiment,[135] efforts to chip away at abortion access were quite successful and materialized in the successful shuttering of clinics across the country, creating reproductive healthcare deserts. Yet, according to Pew Research polling, the only cohort of Americans that by a large margin supports making abortion illegal is "White evangelical Protestants."[136] Notably and "by contrast, 84% of religiously unaffiliated Americans say abortion should be legal in all or most cases," as do the majority of Catholics.[137]

In considering the judicial path to the present, perhaps the most significant victory in dismantling *Roe* was the case that affirmed it, namely *Casey*.[138] The Court emphasized that *Roe's* statement that states may have "important and legitimate interest in potential life," had been "given too little acknowledgment and implementation by the Court in its subsequent cases."[139] In counterpoise, this retreat was not necessary, nor just. The Court could have defended *Roe*'s progeny as reflecting the protection of a woman's constitutional interest in liberty and that further, a fundamental aspect of a woman's liberty interest *is* the right to bodily integrity as well as the longstanding right articulated by the

---

[134] Marielle Kirstein et al., *100 Days Post-Roe: At Least 66 Clinics Across 15 US States Have Stopped Offering Abortion Care*, Guttmacher Inst. (Oct. 6, 2022), https://www.guttmacher.org/2022/10/100-days-post-roe-least-66-clinics-across-15-us-states-have-stopped-offering-abortion-care.

[135] *See Public Opinion on Abortion*, Pew Rsch. Ctr. (May 17, 2022), https://www.pewresearch.org/religion/fact-sheet/public-opinion-on-abortion ("Currently, 61% say abortion should be legal in all or most cases, while 37% say it should be illegal in all or most cases.").

[136] *Id.* ("About three-quarters of White evangelical Protestants (74%) think abortion should be illegal in all or most cases.").

[137] *Id.*

[138] In *Casey*, while the Court stated that "we are led to conclude this: the essential holding of *Roe v. Wade* should be retained and once again affirmed," Planned Parenthood of Se. Pa. v. Casey, 505 U.S. 833, 846 (1992), the Court also underscored that "the state has legitimate interests from the outset of the pregnancy in protecting the health of the woman and the life of the fetus that may become a child." *Id.* at 834. The Court did not find these principles in conflict and adhered to each. Thus, in *Casey*, the Court overruled parts of *Thornburgh v. American College of Obstetricians and Gynecologists*, 427 U.S. 747 (1986), and *City of Akron v. Akron Center for Reproductive Health, Inc.*, 462 U.S. 416 (1983), that it found "inconsistent" with the principle that "the State has a legitimate interest in promoting the life or potential life of the unborn," even while emphasizing that the "central premise of those cases represents an unbroken commitment by this Court to the essential holding of *Roe*." *Casey*, 505 U.S. at 870.

[139] *Casey*, 505 U.S. at 871 (citation omitted).

Court to "control one's person,"[140] dating back to *Skinner v. Oklahoma ex rel. Williamson*,[141] *Rochin v. California*,[142] and *Stanley v. Georgia*.[143] Certainly if the Court could reach the conclusion even prior to *Roe* that the United States's "whole constitutional heritage rebels at the thought of giving government the power to control men's minds,"[144] it could defend the principle that this value, embedded in law, holds true for women to control their bodies.

Instead, the Court capitulated, surrendering to an implicitly coercive[145] and paternalistic logic (under the guise of protecting women's health and safety) that states could take steps to enact laws to ensure that a woman's decision to terminate a pregnancy "is thoughtful and informed."[146] According to *Casey*, "It follows that States are free to enact laws to provide a reasonable framework for a woman to make a decision that has such profound and lasting meaning."[147] As post-*Casey* abortion legislation demonstrated, such measures proved coercive,

---

[140] *Id.* at 915 (Stevens, J., dissenting) (citations omitted). While it is true these principles had not always been imprinted in American common law, which tolerated marital rape as well as domestic violence, society as well as the Court evolved from indifference to its heritages of coverture and slavery to recognize the civil rights and full citizenship of women. *See* Michelle J. Anderson, *Marital Immunity, Intimate Relationships, and Improper Inferences: A New Law on Sexual Offenses by Intimates*, 54 HASTINGS L.J. 1465, 1477–85 (2003) (describing the history of the marital rape exemption from English common law to the law of the United States in the 1970s); RAQUEL KENNEDY BERGEN, WIFE RAPE: UNDERSTANDING THE RESPONSE OF SURVIVORS AND SERVICE PROVIDERS 4–5 (1996); Diana E.H. Russell & Nancy Howell, *The Prevalence of Rape in the United States Revisited*, 8 SIGNS: J. WOMEN CULTURE & SOC'Y 688, 690 (1983) (explaining that 44 percent of women were victims of either rape or attempted rape in their lifetimes).

[141] 316 U.S. 535 (1942).

[142] 342 U.S. 165 (1952).

[143] 394 U.S. 557 (1969).

[144] *Id.* at 565.

[145] According to the Court, "Even in the earliest stages of pregnancy, the State may enact rules and regulations designed to encourage her to know that there are philosophic and social arguments of great weight that can be brought to bear in favor of continuing the pregnancy to full term and that there are procedures and institutions to allow adoption of unwanted children as well as a certain degree of state assistance if the mother chooses to raise the child herself." Planned Parenthood of Se. Pa. v. Casey, 505 U.S. 833, 872 (1992). The end result was that given the political milieu at the time to cut back state and federal assistance to poor families, persisting social stereotyping of poor mothers, and myriad problems associated with the American foster care system, forced pregnancy or coerced pregnancy could legitimately outweigh a woman's liberty interests. *See generally* GUSTAFSON, *supra* note 89; *see also* KHIARA M. BRIDGES, THE POVERTY OF PRIVACY RIGHTS (2017).

[146] *Casey*, 505 U.S. at 872 ("Though the woman has a right to choose to terminate or continue her pregnancy before viability, it does not at all follow that the State is prohibited from taking steps to ensure that this choice is thoughtful and informed.").

[147] *Id.* at 873 ("This, too, we find consistent with *Roe*'s central premises, and indeed the inevitable consequence of our holding that the State has an interest in protecting the life of the unborn.").

burdensome, and dangerous to reproductive liberty.[148] And the material, harmful effects of *Casey* began to appear at the Court. In *Gonzales v. Carhart*,[149] the Court allowed government to make decisions based on what it regards as best for pregnant women.[150] It allowed government to regard the fetus as an "unborn" child with potentially chilling civil and criminal consequences for pregnant women and medical providers. And Justice Kennedy offered a troubling dictum, implying that abortions may cause mental health harms—an assertion widely touted by antiabortion advocates, which is unsupported by medical and sociological research.[151]

To be fair, Justice O'Connor opined, "the reservations any of us may have in reaffirming the central holding of *Roe* are outweighed by the explication of individual liberty we have given combined with the force of *stare decisis*"[152] and further that "the central premise of those cases represents an unbroken commitment by this Court to the essential holding of *Roe*."[153] These matters are pointed out not to suggest that the decision in *Dobbs* was inevitable or correct. Likely, the Justices that voted to uphold *Roe* could not have foreseen that *Casey* would be weaponized to undue *Roe* by a "thousand cuts,"[154] or that the Court would abandon its commitment to stare decisis to deal a final blow to abortion rights. Even as *Casey* came before the Court, Justices Anthony Kennedy, David Souter, and Sandra Day O'Connor—each appointed by a Republican President—"had been meeting secretly" to preserve *Roe*.[155] Justice Kennedy wrote a note to Justice Blackmun, who at the

---

148 *See generally* Whole Woman's Health v. Hellerstedt, 136 S. Ct. 2292, 2299 (2016), *abrogated by* Dobbs v. Jackson Women's Health Org., 142 S. Ct. 2228 (2022).

149 550 U.S. 124 (2007).

150 *See id.*

151 *See, e.g.*, Diana Greene Foster, The Turnaway Study 21 (2020) (reporting on a ten-year study involving one thousand women that addressing the consequences of having an abortion or being denied an abortion, "[w]e find no evidence that abortion hurts women. For every outcome we analyzed, women who received an abortion were either the same or, more frequently, better off than women who were denied an abortion").

152 *Casey*, 505 U.S. at 853.

153 *Id.* at 870 ("It is that premise which we reaffirm today.").

154 *See* Serena Mayeri, *How Abortion Rights Will Die a Death By 1,000 Cuts*, N.Y. Times (Aug. 30, 2018), https://www.nytimes.com/2018/08/30/opinion/brett-kavanaugh-abortion-rights-roe-casey.html ("And it seems more likely that abortion rights will die a death by a thousand cuts through restrictive state laws allowed under *Casey* than by a reversal of *Roe*." (italics added)).

155 *See e.g.*, Evan Thomas, *How the Supreme Court Justice Sandra Day O'Connor Helped Preserve Abortion Rights*, New Yorker (Mar. 27, 2019), https://www.newyorker.com/news/news-desk/how-the-supreme-court-justice-sandra-day-oconnor-helped-preserve-abortion-rights.

time "was near despair," stating, "I need to see you as soon as you have a few free moments. I want to tell you about some developments in *Planned Parenthood v. Casey*, and at least part of what I say should come as welcome news."[156] By no means uniform in their thinking about the exact extent of a state's legitimate interests in regulating the right to an abortion, the "Troika," as they came to be known, along with Justices Blackmun and Stevens, were firm in the belief of stare decisis and protecting the "health and safety—and autonomy—of women."[157] Today, there is no majority on the Court committed to these two principal views. Likely, the Justices would be distressed by the Court's unprincipled stance in *Dobbs*.[158]

Nevertheless, through *Casey* the path was delineated and set for the promulgation of TRAP laws,[159] which ultimately resulted in the closure of clinics (through hundreds of laws intended to shutter clinics or restrict who could perform abortions and under what circumstances such as admitting privileges laws),[160] interference with medical providers' relationship to their patients (by mandating that physicians and other abortion providers read state-developed abortion counseling materials as a condition of lawfully performing abortions in a given state),[161] imposition of "waiting periods" to terminate pregnancies (under the guise of promoting the health and safety of the pregnant

---

[156] *Id.* (italics added).

[157] *Id.*

[158] *See id.* ("It is almost certain that [Justice O'Connor] never favored outlawing abortion altogether, but it is also likely that she struggled in her own mind to settle on the proper legal limits.").

[159] *See supra* note 121.

[160] *See* Whole Woman's Health v. Hellerstedt, 136 S. Ct. 2292, 2299 (2016) (holding that H.B. 2 did not advance the state's interest in protecting women's health but did place a substantial burden in the path of a woman seeking an abortion by forcing about half of the state's abortion clinics to close), *abrogated by* Dobbs v. Jackson Women's Health Org., 142 S. Ct. 2228 (2022); June Med. Servs. L.L.C. v. Russo, 140 S. Ct. 2103, 2132 (2020) (holding that a Louisiana law that requires abortion providers to hold admitting privileges at local hospitals was unconstitutional), *abrogated by Dobbs*, 142 S. Ct. 2228.

[161] As of 2006, thirty-two states enacted "informed consent" mandates that specifically relate to abortion. Chinué Turner Richardson & Elizabeth Nash, *Misinformed Consent: The Medical Accuracy of State-Developed Abortion Counseling Materials*, 9 GUTTMACHER POL'Y REV. 6–7 (Oct. 23, 2006), https://www.guttmacher.org/gpr/2006/10/misinformed-consent-medical-accuracy-state-developed-abortion-counseling-materials; *see also id.* at 7 ("Nonetheless, the U.S. Supreme Court in 1992 ruled in favor of abortion-specific informed consent mandates in *Planned Parenthood of Southeastern Pennsylvania v. Casey*. In upholding Pennsylvania's law requiring preabortion counseling, the Court said such mandates are permissible as long as the information the law requires to be given to the woman is 'truthful and nonmisleading.'").

patient),[162] and the requirement of medically unnecessary procedures (such as vaginal ultrasounds).[163] As well, "[i]n the *Casey* decision, the Court also affirmed the states' authority to develop written materials detailing the abortion procedure and to require providers to distribute them."[164] Of concern with the latter were the ways in which states sought to interfere with abortion rights by imposing conditions on providers to make available and distribute materials that made medically inaccurate claims, and likely intended to coerce a pregnant person from terminating a pregnancy, including falsities "related to breast cancer, psychological impact, [and] fetal pain" as well as "possible detrimental psychological effects" of abortion.[165] Moreover, TRAP laws have serious consequences.

In addition to TRAP laws, efforts to defund or restrict Title X payments to clinics that provide care to the underserved produced devastating consequences. A study published in the peer-reviewed *New England Journal of Medicine* demonstrated the grave effects of taking away reproductive health care access to poor women.[166] Using Texas as a case study, the study's authors pointed out how this affected a broad range of health services beyond abortion, including reducing access to long-acting contraception.[167] Shutting down Planned Parenthood meant making STD screenings less accessible.[168] Further, it meant that breast cancer examinations, a medical service provided to low-income

---

[162] *See Waiting Periods for Abortion*, GUTTMACHER INST. (Jan. 22, 2020), https://www.guttmacher.org/evidence-you-can-use/waiting-periods-abortion ("Waiting periods serve no medical purpose; these hurdles are intended to make abortion less accessible." (citations omitted)).

[163] *See* David Kroll, *State-Mandated Transvaginal Ultrasounds: What are the Medical Societies?*, FORBES (July 7, 2013, 10:39 AM), https://www.forbes.com/sites/davidkroll/2013/07/07/state-mandated-transvaginal-ultrasounds-where-are-the-medical-societies/?sh=443defe0524b.

[164] *See* Turner Richardson & Nash, *supra* note 161; *see also id.* ("In July 2006, we conducted a 50-state investigation to discover in which states the health department had developed materials; we found that 22 states had done so, all under the direction of their legislatures. (One additional state, Oklahoma, has a law requiring the health department to develop materials, but the department has yet to do so.) In most cases, the topics included in the materials were specifically required by law; in other cases, the topics had been selected at the discretion of the health department.").

[165] *See id.*

[166] *See generally* Amanda J. Stevenson et. al., *Effect of Removal of Planned Parenthood from the Texas Women's Health Program*, 374 NEW ENG. J. MED. 853 (2016).

[167] *Id.*

[168] *Id.*

women who sought care at Planned Parenthood would be more difficult to obtain. In the wake of Texas legislators restricting funding to Planned Parenthood, the rates of contraceptive use dropped and unintended pregnancies increased.[169] According to the researchers, "the exclusion of Planned Parenthood affiliates from a state-funded replacement for a Medicaid fee-for-service program in Texas was associated with adverse changes in the provision of contraception."[170] The researchers noted that among "women using injectable contraceptives, there was a reduction in the rate of contraceptive continuation and an increase in the rate of childbirth covered by Medicaid."[171]

By the time *Dobbs* was decided, a broad array of sophisticated strategies had been deployed to undermine the core holding in *Roe* and *Casey*. Beyond the variety of TRAP laws, and efforts to defund Planned Parenthood, lawmakers instituted "non-surgical abortion restrictions."[172] For example, "some women have been forced to have a surgical procedure when they would have chosen non-surgical abortion instead."[173] Personhood amendments were another strategy in the arsenal of antiabortion lawmakers.[174] For years, such efforts were defeated, including in Mississippi.[175] However, in 2018, Alabama secured a victory in passing "an amendment to the constitution declaring the state's policy to recognize the rights" of fetuses.[176] Antiabortion alliances became more sophisticated in working with state legislators to defeat abortion access in their states and to enact model legislation

[169] *Id.* at 858.

[170] *Id.* at 853.

[171] *Id.* Possibly, by producing more unintended pregnancies, these measures actually increased abortion rates—the opposite of what the legislatures claimed to want.

[172] *See Types of State Attacks on Abortion*, Planned Parenthood, https://www.plannedparenthoodaction.org/issues/abortion/types-attacks#nonsurgicalabortionban (last visited Jan. 22, 2023).

[173] *Id.*

[174] *See* Madeleine Carlisle, *Fetal Personhood Laws Are a New Frontier in the Battle over Reproductive Rights*, Time (June 28, 2022, 4:40 PM), https://time.com/6191886/fetal-personhood-laws-roe-abortion.

[175] *See* Frank James, *Mississippi Voters Reject Personhood Amendment by Wide Margin*, NPR (Nov. 8, 2011, 11:28 PM), https://www.npr.org/sections/itsallpolitics/2011/11/08/142159280/mississippi-voters-reject-personhood-amendment.

[176] *See* Abbey Crain, *Alabama Passes Anti-Abortion Amendment*, AL.Com (Nov. 7, 2018, 12:04 PM), https://www.al.com/news/2018/11/alabama-passes-anti-abortion-amendment.html ("The amendment recognizes the rights of the unborn and ensures state funds will not go to funding abortion care.").

circulated nationwide.[177] Finally, federal and state gestational bans, like Mississippi's Gestational Age Act, were introduced across the country.[178] In fact, by 2015, twelve states had already banned abortions after twenty weeks of pregnancy,[179] violating a key tenet of *Casey*—even while such laws were successfully challenged. Ultimately, this strategy proved to deliver the ultimate defeat of *Roe*.

B. THE GESTATIONAL AGE ACT

Simply put, Mississippi's Gestational Age Act[180] was clearly part of the larger strategy to dismantle *Roe*, but it was not necessarily the plan. Instead, it was among a constellation of "gestational age" bans promulgated in recent years to challenge *Roe*'s viability framework, adopted in *Casey*.[181] A sophisticated, well-funded, strategic campaign over the thirty years[182] since *Casey* coalesced around multiple strategies, but found the most devastating traction with gestational age bans. As Megan K. Donovan, who tracks these laws, explained in 2020, "[g]estational age bans are nothing more than a smokescreen" that had "long been a favored tactic of antiabortion activists and politicians."[183]

The process of enacting the law was relatively straightforward, although nonetheless controversial.[184] Already, Mississippi had only

---

[177] *See id.* (noting that Alabama's personhood amendment was sponsored by the Alliance for a Pro-Life Alabama).

[178] Miriam Berg, *How 20-Week Abortion Bans Are Spreading Across the U.S. (in 1 GIF)*, PLANNED PARENTHOOD (Feb. 13, 2015, 2:09 PM), https://www.plannedparenthoodaction.org/blog/how-20-week-abortion-bans-are-spreading-across-us-one-gif.

[179] *See id.* These states included Alabama, Arkansas, Indiana, Kansas, Louisiana, Mississippi, North Carolina, Nebraska, Oklahoma, Texas, and West Virginia. *Id.* North Dakota enacted a six-week ban, which was blocked from going into effect. *See id.*

[180] MISS. CODE ANN. § 41-41-191 (2022).

[181] *See* Megan K. Donovan, *Gestational Age Bans: Harmful at Any Stage of Pregnancy*, GUTTMACHER POL'Y REV. 1 (Jan. 9, 2020), https://www.guttmacher.org/gpr/2020/01/gestational-age-bans-harmful-any-stage-pregnancy.

[182] *See generally* MARY ZIEGLER, DOLLARS FOR LIFE: THE ANTI-ABORTION MOVEMENT AND THE FALL OF THE REPUBLICAN ESTABLISHMENT (2022) (arguing that over the course of decades, the anti-abortion movement laid the groundwork for an insurgent candidate like Donald Trump).

[183] Donovan, *supra* note 181, at 1.

[184] *See* Sarah Fowler, *Mississippi Banned Most Abortions to Be the 'Safest State' for the Unborn. Meanwhile, One in Three Mississippi Kids Lives in Poverty*, INSIDER (Nov. 26, 2021, 9:23 AM), https://www.businessinsider.com/mississippi-defends-abortion-ban-one-in-three-kids-in-poverty-2021-11 ("In March of 2019, Mississippi drew national headlines when Governor Phil Bryant signed into law one of the most restrictive abortion bans in the country, making Mississippi—as backers of the bill frequently put it—'the safest place in the country for unborn babies.'").

one abortion clinic in the state to serve all the women, girls, and people with the capacity for pregnancy that would need the care they provided.[185] That clinic was owned by Diane Derzis, derisively described by antiabortion activists as the "abortion queen."[186] In November 2021, before oral arguments in *Dobbs*, I interviewed Derzis.[187] As she explained to me,

> For the last, I'd say, 30 years, we've been under assault, and each year it has worsened. We have had absolutely no response in the last five years. I'm going to point Jackson out right now, Mississippi. The police will no longer come. We have sometimes 100 protesters in the street, blocking the driveway. It is absolutely . . . I can't even portray the picture to you of how bad it is for those women who have to enter that every day, but they do so. They do so with tears running down their faces, grimaces, but they come in because that's what women have always done, and they do whatever it takes to obtain an abortion.[188]

According to Derzis, in Mississippi, the patients that came to her clinic along with the providers that helped them, were subjected to

---

[185] Caroline Kitchener & Casey Parks, *How Mississippi Ended Up with One Abortion Clinic and Why It Matters*, WASH. POST (Nov. 30, 2021, 9:00 AM), https://www.washingtonpost.com/dc-md-va/2021/11/30/abortion-mississippi-closed-clinics ("For 25 years, lawmakers and antiabortion activists had worked tirelessly to put [Nancy] Atkins's clinic out of business. When they succeeded in 2004, she knew she'd be leaving the state with just one abortion provider. Thousands of people would have a harder time accessing the procedure, but Atkins felt she'd fought as long as she could. She was tired, and the work had become too difficult.").

[186] *See* Timmerie Millington, *The Abortion Queen in Mississippi*, SECULAR PRO-LIFE (July 12, 2012), https://secularprolife.org/2012/07/abortion-queen-in-mississippi ("The 'Abortion Queen's' clinic has been all over the news today because she has filed a lawsuit seeking to stop the new law that puts women's safety at the front of the issue. She insists the law is unnecessary and could drive her out of business."). Even a decade ago, Derziz was battling to keep Mississippi's one abortion clinic open. *See id.* More recently, Derzis talked about the devasting effects of Mississippi's abortion laws. *See* Natalie Gontcharova, *The US Supreme Court Is Now Posed to Overturn* Roe v. Wade *—Here's What to Know*, REFINERY29 (May 18, 2021, 2:58 PM), https://www.refinery29.com/en-gb/2021/05/10477918/supreme-court-mississippi-abortion-case ("Diane Derzis, owner of Jackson Women's Health Organization, said in a statement provided to Refinery29. 'If this ban were to take effect, we would be forced to turn many of those patients away, and they would lose their right to abortion in this state. Mississippi politicians have created countless barriers for people trying to access abortion, intentionally pushing them later into pregnancy. It's all part of their strategy to eliminate abortion access entirely.' "); *see also Morning Edition: Mississippi's Only Abortion Clinic Has Closed Its Doors for Good*, NPR, at 00:53 (July 7, 2022), https://www.npr.org/2022/07/07/1110222245/mississippis-only-abortion-clinic-has-closed-its-doors-for-good. ("Oh, they're devastated—people sending things, calling. And you would be surprised at how many have no idea that this bill has passed and that abortion is now illegal in Jackson, as well as other states surrounding, so lots of anger.").

[187] On the Issues with Michele Goodwin, *American Terrorism: Abortion Clinics on the Front Lines (with Diane Derzis, Dr. Monica McLemore and Frances Kissling)*, MS. MAG. (Nov. 22, 2021), https://msmagazine.com/podcast/49-american-terrorism-abortion-clinics-on-the-front-lines-with-diane-derzis-dr-monica-mclemore-and-frances-kissling.

[188] *Id.* at 04:05.

violence and "behaviors unaccounted for."[189] Derzis told me that while some of the legislative and even activist strategies were the same as usual—such as protesting her clinic while brandishing images of fetuses—in other ways, the antiabortion strategies had become more violent. When I asked whether these concerns were featured in news broadcasts or made more accessible to the public, she told me, "No. No. It doesn't, and no, this is violence, violence, violence. They have not only dead baby pictures. It's gone way beyond that."[190] She continued by describing that "the men are clearly armed. . . . [P]olice will not come. . . . I mean it just goes beyond anything you would imagine."[191]

Derzis is no stranger to violence targeting her, the staff, or patients.[192] Twenty-five years ago, she owned a clinic in Birmingham, Alabama, which was bombed by an antiabortion activist, Eric Rudolph.[193] Rudolph's terroristic activities were not confined to targeting abortion clinics. He also planted and detonated bombs at the Olympics held in Atlanta, Georgia, injuring over 100 people and killing three.[194] At Diane's clinic, he killed their security officer—Officer Sanderson, an off-duty police officer—who, as Diane told me, was not "pro-choice," but believed that women should be afforded protection and safety when terminating a pregnancy.[195] As well, Emily Lyons, one of her nurses, was severely injured, and though she survived the bombing, she subsequently endured eighty surgeries.[196]

---

[189] *Id.* at 05:05.

[190] *Id.* at 06:19.

[191] *Id.* at 06:55; *see also id.* at 06:28 ("In Jackson, Mississippi, we have a doctor who has 12 children, and they are stairsteps, and they are in the road grabbing hold of cars. I'm surprised one of them has not been killed, but they use these children to stop traffic. They're screaming to the women and to the patients. They're banging on doors, on patient doors. The men are clearly armed. You can see the guns in their pockets, and many of them will stand across the street and use their hands to tell the women that are with them what to do, where to go.").

[192] *See id.* at 21:15.

[193] *Id.*

[194] *Id.*

[195] *Id.* at 22:44 ("I received the call that said, the sirens you hear are coming for us. They've just blown up the clinic. Officer Sanderson, who was absolutely wonderful, was our Birmingham policeman that we paid as security, and he saw the plant and bent over because it wasn't normally there, and Rudolph was right across the street, and he detonated the button because he knew the game was up.").

[196] *Id.* at 23:13 ("Emily Lyons was our RN. She was at the door, opening it, and when that bomb went off, it cut him in half, and it nearly killed Emily. She's had over 80 surgeries since that time because her insides are so messed up. She lost an eye.").

The United States Department of Justice lists federal and civil prosecutions in recent years related to "violence against reproductive health care providers."[197] In just the last year, their list includes old destruction of property tactics to destroy buildings and block entrances; fastening doors and gates to clinic entrances by glue and chains to prevent access; throwing projectiles, including concrete, slingshots, and ball bearings through windows; as well as prosecutions for physical violence.[198] In just the past three years, the list includes prosecutions for physical assaults, threats to kill patients, bomb threats, tossing "Molotov cocktails" at facilities, shooting police officers, the murder of at least one officer and a patient, as well as arson.[199] Not on the list is the deadly attack on a Colorado Springs Planned Parenthood, which resulted in three deaths and nine injuries.[200] Since 1977, "there have been eight murders, 17 attempted murders, 42 bombings, and 186 arsons targeted at abortion clinics and providers across the United States."[201] And while violence has been an open secret of the anti-abortion cause,[202] so too has the movement to reshape polices, build influence through legislative action, cultivate influence within the Republican party, and ultimately move their battleground to the courts.[203] In Mississippi, the tactics and strategies were no different.[204]

Antiabortion strategists were already quite successful in their appeals to the Mississippi legislature, such that at least a dozen TRAP laws preexisted the Gestational Age Act.[205] Mississippi Republicans controlled both chambers of the legislature and the governor's office, ensuring a likelihood of success with enacting the ban. The state's attorney general regularly spoke at antiabortion rallies, claiming to

---

[197] *Recent Cases on Violence Against Reproductive Health Care Providers*, U.S. Dep't of Just. (June 24, 2022), https://www.justice.gov/crt/recent-cases-violence-against-reproductive-health-care-providers.

[198] *See id.*

[199] *Id.*

[200] *See* Sarah Frostenson, *40 Years of Attacks on Abortion Clinics, Mapped*, Vox (Dec. 1, 2015, 12:30 PM), https://www.vox.com/2015/12/1/9827886/abortion-clinic-attacks-mapped.

[201] *See id.*

[202] *See id.* (detailing attacks on abortion providers, clinics, patients, and escorts).

[203] *See* Goodwin & Ziegler, *supra* note 131 ("The anti-abortion movement had a plan: to win over as many Americans as possible, to make moderate Republicans comfortable working with them, and to maximize the chances of success before the Supreme Court.").

[204] *See* Kitchener & Parks, *supra* note 185.

[205] *See State Facts About Abortion: Mississippi*, Guttmacher Inst. (June 2022), https://www.guttmacher.org/fact-sheet/state-facts-about-abortion-mississippi.

value religious principles that prioritize fetal life.[206] In March 2018, Mississippi lawmakers enacted The Gestational Age Act.[207] Notably, the group that sponsored the legislation largely reflected the demographics of the state's legislature—they were nearly all men, except for the bill's sponsor, Becky Currie.[208] In Mississippi, nearly 83 percent of the legislators are men.[209] Among the co-authors of The Gestational Age Act, the majority had previously sponsored antiabortion legislation (or drafted such legislation after H.B. 1510).[210] For example, Representative Dan Eubanks sponsored H.B. 1511, The Abortion Complication Act, as well as H.B. 580. The former was a law that would have mandated reporting every abortion performed in Mississippi to the Mississippi Department of Health.[211] And the latter was legislation that would have criminalized all abortions as felony murder.[212] H.B. 580 died in committee four months before the *Dobbs* ruling.[213] Another bill, H.B. 1368, was sponsored by Representative Lester Carpenter and proposed to amend Mississippi's Code to prohibit abortion except to save the pregnant person's life, and violation of this law would have resulted in a $100,000 fine.[214] This law also failed in committee months before *Dobbs*.[215]

---

[206] *See* Holly Honderich, *US Supreme Court: The Woman Who Helped to End* Roe v. Wade, BBC News (June 21, 2022), https://www.bbc.com/news/world-us-canada-61789443 (noting Attorney General Lynn Fitch's public antiabortion beliefs).

[207] Miss. Code Ann. § 41-41-191 (2022); *see also* Jenny Gathright, *Mississippi Governor Signs Nation's Toughest Abortion Law Into Law*, NPR (Mar. 19, 2018, 6:44 PM), https://www.npr.org/sections/thetwo-way/2018/03/19/595045249/mississippi-governor-signs-nations-toughest-abortion-ban-into-law.

[208] *See* H.B. 1510, 2018 Leg., Reg. Sess. (Miss. 2018) (enacted).

[209] *See* Renuka Rayasam et al., *Why State Legislatures Are Still Very White—and Very Male*, Politico (Feb. 22, 2021, 9:19 PM), https://www.politico.com/interactives/2021/state-legislature-demographics.

[210] The co-authors of the Act included Representatives Dan Eubanks, Andy Gipson, Kevin Ford, Lester Carpenter, Nick Bain, Donnie Scoggin, William Tracy Arnold, Chris Brown, Tom Weathersby, Ray Rogers, Robert Foster, Stacy Hobgood-Wilkes, Steve Hopkins, Carolyn Crawford, and Ashley Henley. *See* H.B. 1510, 2018 Leg., Reg. Sess. (Miss. 2018) (enacted).

[211] *See* H.B. 1511, 2022 Leg., Reg. Sess. (Miss. 2022).

[212] *See* H.B. 580, 2022 Leg., Reg. Sess. (Miss. 2022).

[213] *See Mississippi House Bill 580 (Prior Session Legislation)*, LegiScan, https://legiscan.com/MS/research/HB580/2022 (last visited Jan. 27, 2023).

[214] *See* H.B. 1368, 2022 Leg., Reg. Sess. (Miss. 2022).

[215] *See Mississippi House Bill 1368 (Prior Session Legislation)*, LegiScan, https://legiscan.com/MS/research/HB1368/2021 (last visited Jan. 27, 2023).

Mississippi's legislature, like that of Texas, adopted a more aggressive approach than states with similar gestational age bans, which sought to ban abortion at twenty weeks gestational age, commonly and inaccurately described as "heartbeat bills."[216] The Gestational Age Act restricted abortion to fifteen weeks of gestation.[217] Like similar laws enacted in recent years, the Act established the definition of "human being" as beginning at the point of conception.[218] While the legislature did not provide exceptions for rape or incest, lawmakers included a provision that permitted abortions for medical emergencies or severe fetal abnormality.[219]

The state defined "severe fetal abnormality" as a "life-threatening physical condition that, in reasonable medical judgment, regardless of the provision of life-saving medical treatment, is incompatible with life outside the womb."[220] Further, the Act articulated a medical emergency as

> a condition in which, on the basis of the physician's good faith clinical judgment, an abortion is necessary to preserve the life of a pregnant woman whose life is endangered by a physical disorder, physical illness, or physical injury, including a life-endangering physical condition arising from the pregnancy itself, or when the continuation of the pregnancy will create a serious risk of substantial and irreversible impairment of a major bodily function.[221]

---

[216] States like Texas claim that cardiac activity is present as early as six weeks of pregnancy. *See* Julie Carr Smyth, *Explainer: Abortion Landscape Under State 'Heartbeat' Laws*, AP News (June 29, 2022), https://apnews.com/article/abortion-us-supreme-court-health-ohio-tennessee-0056dcfb4e5fe1590f07b5993c52078a. Their claims are discredited by medical experts. *See* Jay Janner, *'Heartbeat Bills': Is There a Fetal Heartbeat at Six Weeks of Pregnancy?*, NBC News (May 3, 2022, 5:39 PM), https://www.nbcnews.com/health/womens-health/heartbeat-bills-called-fetal-heartbeat-six-weeks-pregnancy-rcna24435 ("[A]ccording to experts, the term 'fetal heartbeat' is misleading and medically inaccurate."). However, the terminology of "heartbeat bills" proved to be a successful "name reference" in the antiabortion movement. Bethany Irvine, *Why "Heartbeat Bill" Is a Misleading Name for Texas' Near-Total Abortion Ban*, Tex. Trib. (Sept. 2, 2021, 4:00 PM), https://www.texastribune.org/2021/09/02/texas-abortion-heartbeat-bill ("[T]he moniker has helped rally supporters around the law in Texas and nationwide where other states are considering similar legislation.").

[217] Miss. Code Ann. § 41-41-191 (2022).

[218] *Id.* § 41-41-191(3)(g). The Act established the definition of gestational age to begin on the first day of the last menstrual period of the woman. *Id.*

[219] Miss. Code Ann. § 41-41-191(4)(a) (2022).

[220] *Id.* § 41-41-191(3)(h).

[221] *Id.* § 41-41-191(3)(j).

The Mississippi House of Representatives initially voted to implement the law on February 2, 2018.[222] Typical of the legislative policy debates undergirding abortion bans, the votes split along party lines. Seventy-nine representatives voted in favor of the law, while thirty-one opposed.[223] Nearly a dozen did not cast a vote.[224] When the measure passed, nearly all those who voted in favor were Mississippi Republican lawmakers.[225] Similarly, in a vote a few weeks later, on March 6, 2018, the Mississippi Senate considered an amended version of the Act and the bill passed along party lines.[226] Less than two weeks later, on March 19, 2018, Governor Bryant signed the Gestational Age Act into law.[227]

C. THE PATH TO THE SUPREME COURT

On March 19, 2018, the same day that Mississippi Governor Phil Bryant signed the Gestational Age Act into law, Diane Derzis prepared for litigation. Her clinic, the Jackson Women's Health Organization (JWHO), along with its clinic medical director, Dr. Sacheen Carr-Ellis, filed a complaint, petitioning for injunctive relief.[228] Dr. Carr-Ellis, a board-certified obstetrician who is licensed to practice in Mississippi and performs pre-viability abortions, filed on behalf of herself and her clients.[229] Their concerns about the Mississippi ban

---

[222] *See Mississippi House Bill 1510 (Prior Session Legislation)*, LEGISCAN, https://legiscan.com/MS/bill/HB1510/2018 (last visited Jan. 27, 2023).

[223] *See Mississippi House Bill 1510 (Prior Session Legislation)*, LEGISCAN, https://legiscan.com/MS/rollcall/HB1510/id/693321 (last visited Jan. 27, 2023).

[224] *Id.*

[225] *Id.* Eight Democratic lawmakers voted in favor of the law. *Id.*

[226] *See Mississippi House Bill 1510 (Prior Session Legislation)*, LEGISCAN, https://legiscan.com/MS/rollcall/HB1510/id/717959 (last visited Jan. 27, 2023). Thirty-five senators voted to support the amended legislation, while fourteen voted against the bill. Several Democrats also joined ranks with Republican lawmakers. *Id.* Two days after the Senate passed the amended bill, the House voted again and on March 8, 2018, seventy-five representatives voted yes, thirty-four representatives voted no, and eleven representatives abstained. *See Mississippi House Bill 1510 (Prior Session Legislation)*, LEGISCAN, https://legiscan.com/MS/rollcall/HB1510/id/719308 (last visited Jan. 27, 2023). Like the votes before, the bill passed along party lines. *Id.* However, this time, only five Democratic representatives voted yes for the bill. *Id.*

[227] *See* Emily Wagster Pettus, *Mississippi Governor Signs 'Heartbeat' Abortion Law*, AP NEWS (Mar. 21, 2019), https://apnews.com/article/fdce54d41bb74d6a974f7b39bcb83cf7.

[228] *See* Complaint, Jackson Women's Health Org. v. Currier, No. 3:18-CV-171-CWR-FKB, 2018 WL 6112313 (S.D. Miss. Mar. 19, 2018).

[229] *See id.* at 3.

could be framed under three key arguments: a) the law is unconstitutional because it bans abortion pre-viability; b) abortions are fundamental to women's liberty and equal participation in society; and c) permitting the law to stand would irreparably harm their patients and the ban's exceptions would not cure the constitutional violations.[230]

Their complaint argued that the law "includes a provision banning abortion after 15 weeks . . . which is months before viability."[231] As such, they contended, "the Act prohibits Plaintiffs from providing pre-viability abortions to their patients."[232] This ban, they asserted, was unconstitutional in that it violated "decades of United States Supreme Court precedent" and regardless of any exceptions, Mississippi is not free to ban "abortion prior to viability."[233] The plaintiffs sought a declaration that the Gestational Age Act was unconstitutional and preliminary and permanent injunctive relief, barring its enforcement "as applied to pre-viability abortions.[234] They noted that other existing Mississippi bans imposed significant burdens, stating that "[b]ecause of patients' work and family commitments combined with the fact that the Clinic does not provide abortions every day of the week, Mississippi's mandatory-delay and 2-trip requirement delays many patients by several days or more in obtaining an abortion."[235] The next day, Judge Carlton Reeves issued a temporary restraining order.[236]

However, it was the plaintiff's amended complaint, filed three weeks later, which exposed the extreme methods Mississippi already deployed to restrain abortion rights even before enactment of the

[230] *See id.* at 5–8. The complaint stated that "over the past forty-five years, safe and legal abortion has been important to facilitating women's equal participation in society, including in the economic and social life of the nation. The availability of abortion enables women to decide whether to forego educational and economic opportunities due to unplanned pregnancy, whether to raise children with an absent or unwilling partner, and whether to accept the risk of carrying medically-compromised pregnancies to term." *Id.* at 7.

[231] *Id.* at 2.

[232] *Id.* They also argued that the Mississippi law imposed significant constraints on the provider and patients. *Id.* at 7. Mississippi law already mandated an unnecessary and burdensome waiting period, as well as "two separate trips to an abortion provider," before a pregnancy could lawfully be terminated pre-viability. *Id.* at 5. "Accordingly, each of the Clinic's patients must make two separate visits to the Clinic at least one full day apart." *Id.*

[233] *Id.* at 2.

[234] *Id.*

[235] *Id.* at 5 ("Thus, even patients who contact the Clinic and are able to schedule their first visit before 15 weeks . . . may be after 15 weeks . . . by the time of their procedure, because of the mandatory delay and 2-trip law.").

[236] *See* Temporary Restraining Order, Jackson Women's Health Org. v. Currier, No. 3:18-CV-171-CWR-FKB, 2018 WL 1567867 (S.D. Miss. Mar. 20, 2018).

fifteeen-week ban.[237] JWHO's amended complaint argued that the Mississippi abortion ban violated the "liberty rights of [the] [p]laintiffs' patients"[238] and asserted multiple claims, including that Mississippi's various existing laws were intended to and had the effect of curtailing abortion rights.[239] They argued that the wait period law (connected to the "two trip" scheme), the TRAP licensing scheme, the "two trip" requirement scheme,[240] biased counseling law, the telemedicine ban,[241] and the physician only requirement[242] cumulatively violated their patient's right to liberty and privacy guaranteed by the Fourteenth Amendment.[243] They argued that imposing pre-viability conditions created undue burdens and substantial obstacles that were not reconcilable with precedent.[244] JWHO also argued that each of the laws individually violated the Fourteenth Amendment right of liberty and privacy by interfering with women's right to choose an abortion.[245]

---

[237] *See* Amended Complaint, Jackson Women's Health Org. v. Currier, No. 3:18-cv-00171-CWR-FKB, 2018 WL 6120525 (S.D. Miss. Apr. 9, 2018).

[238] *Id.* at 9.

[239] *See id.* at 1–9.

[240] Mississippi required that a patient seeking an abortion make two trips to her clinician's office, first to receive the prescribed information "orally and in person" by the physician who was to perform the abortion and, at least twenty-four hours later, to obtain the abortion. *See* MISS. CODE ANN. § 41-41-33 (2022). Mississippi also required that a physician who failed to comply with the mentioned requirement would be subject to criminal penalties of six months imprisonment, a $1,000 fine, or both. *Id.* § 41-41-39. Petitioners argued that the mandate was selectively imposed as no other medical providers must oblige by such a requirement. Amended Complaint, *supra* note 237, at 36.

[241] *See* MISS. CODE ANN. §§ 41-41-33, 41-41-107 (2022). They argued that even while Mississippi authorizes the use of telemedicine generally, it excludes this service for abortion-related care. Amended Complaint, *supra* note 237, at 6. In their petition, the plaintiffs argued that the ban "creates undue burdens for women seeking abortions by: (1) requiring the physician to physically examine the patient prior to administering medication abortion; (2) prohibiting anyone other than a physician from providing abortion-inducing drugs to patients; and (3) requiring that the abortion-inducing drug be administered 'in the same room and in the physical presence' of the physician." *Id.* at 46 (citation omitted).

[242] Mississippi requires that "[a]bortions shall only be performed by physicians licensed to practice in the State of Mississippi." *See* MISS. CODE ANN. § 41-75-1(f) (2022). The law bans midwives, advanced practice clinicians–certified nurse practitioners, certified nurse midwives, and physician assistants–from providing abortion services. *See id.*; *see also* Miss. Code Ann. §§ 41-41-107(1), -111(1); *id.* § 41-75-26 (stating that violation of these laws constitutes a misdemeanor).

[243] Amended Complaint, *supra* note 237, at 5–6.

[244] *See id.* at 3.

[245] *See id.* at 52. Similar to arguments made in *Whole Woman's Health v. Hellerstedt*, 136 S. Ct. 2292, 2299 (2016), *abrogated by* Dobbs v. Jackson Women's Health Org., 142 S. Ct. 2228 (2022), JWHO claimed that Mississippi's laws arbitrarily and irrationally singled out abortion providers, even while the provisions overall did not advance patients' health or provide medical benefits. Amended Complaint, *supra* note 237, at 23, 53. For example, they argued that the laws

Perhaps the most remarkable aspect of the pre-*Dobbs* petitions, orders, and decisions, is the permanent injunction issued by Judge Carlton Reeves, striking down Mississippi's abortion ban.[246] As a technical legal matter, the order reflected a clear application of existing law.[247] Undoubtedly, this is likely why one year later, a three-judge panel on the conservative Fifth Circuit Court of Appeals affirmed his order.[248] Judge James Ho in his concurrence explained, "Like every other court to consider the issue, the majority concludes that *Casey* prohibits any and all bans on pre-viability abortions."[249] Judge Ho continued by stating, "I am aware of no judicial opinions that read *Casey* differently, and Mississippi and its amici provide none."[250] In fact, as Judge Ho pointed out, "Mississippi concedes that HB 1510 would be held unconstitutional in every circuit that has addressed such issues to date."[251] This was not a complicated set of facts or findings on which to analyze and frame a judicial decision. Mississippi's law clearly violated the tenets of *Casey*.

Yet it was Judge Reeves's order that gave voice to the concerns of communities too frequently overlooked and their interests too often neglected. In the years leading to *Dobbs*, antiabortion strategies such as gestational age bans like Mississippi's "were usually cloaked in supposed justifications that obscured the end goal"[252]—much like informed consent strategies coupled with waiting periods, which ultimately imposed burdensome obstacles in the path of obtaining an abortion.[253] Such obstacles became normalized with states claiming a legitimate interest in protecting a pregnant woman's health, while neglecting or leaving unresolved crisis-level maternal mortality and morbidity concerns.[254]

---

impose more burdensome requirements on the plaintiffs than similarly situated providers of medical services with no benefits. *See id.* This imposition, JWHO argued, arbitrarily and irrationally impeded their Fourteenth Amendment right to equal protection. *Id.* at 53.

[246] *See* Jackson Women's Health Org. v. Currier, 349 F. Supp. 3d 536 (S.D. Miss. 2018), *aff'd sub nom.* Jackson Women's Health Org. v. Dobbs, 945 F.3d 265 (5th Cir. 2019), *rev'd and remanded*, 142 S. Ct. 2228 (2022).

[247] *Id.* at 538–39.

[248] *Dobbs*, 945 F.3d 265 (5th Cir. 2019), *rev'd and remanded*, 142 S. Ct. 2228 (2022).

[249] *Id.* at 278 (Ho, J., concurring in the judgment).

[250] *Id.* at 279 (majority opinion).

[251] *Id.*

[252] *See* Donovan, *supra* note 181, at 1.

[253] *See id.* at 3.

[254] *See supra* notes 107–111 and accompanying text.

In Mississippi, as the state's data shows, being pregnant and carrying pregnancies to term has become a deadly proposition, especially for Black women.[255] The data raises serious questions related to legislative attempts to thwart what could best be described as measured, logical, responsible, medical self-defense. In Mississippi, it is nearly 118 times more dangerous for a Black woman to give birth than it is for the average American woman to have an abortion.[256] Mississippi's legislature and governor do not feign ignorance of this crisis, nor have they taken aggressive steps to resolve it. Further contributing to the realization of how deadly pregnancy is for Black women in Mississippi, the Mississippi Department of Health acknowledges that Black women accounted for "nearly 80% of pregnancy-related cardiac deaths" in their state investigation of maternal health and mortality.[257] Nevertheless, Mississippi Attorney General Lynn Fitch stunningly claimed in the period leading to *Dobbs* that "[t]he Mississippi legislature enacted [the abortion ban] . . . to promote women's health."[258]

This was one of the positions articulated by Mississippi along its path to the Supreme Court. Mississippi's Attorney General Lynn Fitch adopted this strategy in appealing not only to the Supreme Court, but also to the public, attending rallies, speaking on talk shows and in other venues to defend the state's fifteen-week abortion ban.[259] According to Fitch, the ban empowered women, while it "promote[d] life."[260] Fitch used her life as an example, giving speeches at antiabortion rallies and in news media, telling the story of her divorce, single parenting, and

---

[255] *See id.*

[256] *See* Goodwin, *supra* note 110, at 155. In Mississippi, between 2013 and 2016, "the pregnancy-related mortality ratio for Black women was 51.9 deaths per 100,000 live births, nearly three times the White ratio of 18.9." MISS. STATE DEP'T OF HEALTH, MISSISSIPPI MATERNAL MORTALITY REPORT 2013–2016, at 5 (2019), https://msdh.ms.gov/msdhsite/_static/resources/8127.pdf. "The national legal induced abortion case-fatality rate for 2013–2017 was 0.44 legal induced abortion-related deaths per 100,000 reported legal abortions." Katherine Kortsmit et al., *Abortion Surveillance—United States, 2018*, MORBIDITY & MORALITY WKLY. REP. 7 (Nov. 27, 2020), https://www.cdc.gov/mmwr/volumes/69/ss/pdfs/ss6907a1-H.pdf.

[257] MISS. STATE DEP'T OF HEALTH, *supra* note 256, at 16.

[258] Lawrence Hurley, *U.S. Supreme Court Takes Up Major Challenge to Abortion Rights*, REUTERS (May 17, 2021, 9:41 AM), https://www.reuters.com/world/us/us-supreme-court-takes-up-case-that-could-limit-abortion-rights-2021-05-17.

[259] *See* Caroline Kitchener, *The Woman Who Could Bring Down* Roe v. Wade, LILY (Dec. 1, 2021), https://www.thelily.com/the-woman-who-could-bring-down-roe-v-wade.

[260] *See id.*

elevation to the role of the Attorney General for the state of Mississippi—a role no woman had previously occupied.[261]

How then, after all, should a district court, responsible for resolving disputes by "determining the facts and applying legal principles to decide who is right,"[262] interpret the known truths about the state of Mississippi, with its catastrophic rates of maternal mortality and devastating health harms associated with maternal morbidity? Arguably, it calls for dispensing with the sophistry of sex and race blindness in the search for truth in analyzing the petitioner's claims and the state's responses. And in a searching review and establishing of facts to arrive at truths related to abortion, bodily autonomy, privacy, and forced pregnancy in Mississippi, may one look squarely at Mississippi's blighted history related to discriminatory treatment of women, its remarkably cruel legacy of racial animus against Black women and their daughters, and its heritage of race and sex violence? Does one ignore and disregard? Or is there a responsibility for truth-telling, because there is the story that Mississippi tells of itself, and there is the other that history records?

Rather than adopting the approach later taken by the Supreme Court in *Dobbs*, which failed to contemplate the consequences of Mississippi's ban on communities that would be most impacted, particularly poor women of color, and especially Black women, Judge Reeves undertook what the Supreme Court described in *Gonzales v. Carhart*[263] as an "independent constitutional duty"[264] to review factual findings and surely this meant interrogating the matters in plain sight.[265] In a compelling passage, found in footnote 22, he wrote that legislation like H.B. 1510 "is closer to the old Mississippi—the Mississippi bent on controlling women and minorities."[266] According to Judge Reeves, it is the "Mississippi that just a few decades ago, barred women from serving on juries 'so they may continue their service as

---

[261] *See id.*

[262] *Court Role and Structure*, U.S. Courts, https://www.uscourts.gov/about-federal-courts/court-role-and-structure (last visited Jan. 29, 2023).

[263] 550 U.S. 124 (2007).

[264] *Id.* at 165.

[265] *See* Jackson Women's Health Org. v. Currier, 349 F. Supp. 3d 536, 540 n.22 (S.D. Miss. 2018) (citation omitted), *aff'd sub nom.* Jackson Women's Health Org. v. Dobbs, 945 F.3d 265 (5th Cir. 2019), *rev'd and remanded*, 142 S. Ct. 2228 (2022).

[266] *Id.*

mothers, wives and homemakers.' "[267] He compared the Mississippi that inflicted harms on Black women through coercive sterilization to that which today would force them to remain pregnant against their will.[268] As he explained, that Mississippi "sterilized six out of ten black women in Sunflower County at the local hospital—against their will.[269] This, he reminded, is the Mississippi that became the last state to ratify the Nineteenth Amendment, "the authority guaranteeing women the right to vote."[270] He concluded:

> Mississippi Legislature's professed interest in "women's health" is pure gaslighting. In its legislative findings justifying the need for this legislation, the Legislature cites *Casey* yet defies *Casey*'s core holding. The State "ranks as the state with the most [medical] challenges for women, infants, and children" but is silent on expanding Medicaid. Its leaders are proud to challenge *Roe* but choose not to lift a finger to address the tragedies lurking on the other side of the delivery room: our alarming infant and maternal mortality rates.[271]

At the crux of the matter, according to Reeves, was that Mississippi brought the matter on itself: "The state chose to pass a law it knew was unconstitutional to endorse a decades-long, campaign, fueled by national interest groups, to ask the Supreme Court to overturn *Roe v. Wade*."[272]

But it was Judge Reeves's order which seemingly struck a nerve, drawing ire from Judge James C. Ho, who, although concurring in the opinion to affirm the district court order, expressed that he was "nevertheless deeply troubled by how the district court handled this case."[273] He claimed that the "district court displays an alarming disrespect for the millions of Americans who believe that babies deserve legal protection during pregnancy as well as after birth, and that abortion is the immoral, tragic, and violent taking of innocent human life."[274] He accused Judge Reeves of failing to show proper

---

[267] *Id.* (citing State v. Hall, 187 So. 2d 861, 863 (Miss. 1966)).

[268] *Id.*

[269] *Id.* (citation omitted).

[270] *Id.* (citation omitted).

[271] *Id.* (citation omitted).

[272] *Id.* at 542.

[273] Jackson Women's Health Org. v. Dobbs, 945 F.3d 265, 278 (5th Cir. 2019) (Ho, J., concurring in the judgment), *rev'd and remanded*, 142 S. Ct. 2228 (2022).

[274] *Id.* at 278.

deference and respect to the views of those who take an antiabortion position, explicating that "instead of respecting all sides, the district court opinion disparages the Mississippi legislation as 'pure gaslighting.'"[275] For Judge Ho, it was a "smear[]" for Judge Reeves, an African American, to link the Mississippi law "to the state's tragic history of race relations," while in his view "ignoring abortion's own checkered racial past"[276]—a rebutted view, which nonetheless Justice Thomas has promoted in concurring opinions.[277]

For the individual women and people with the capacity for pregnancy concerned in Mississippi, as well as all those who are now most impacted by *Dobbs*, geographically constrained, "indigent and financially helpless," as Justice Blackmun might say,[278] the results of *Roe*'s overturn are cruel and punitive. Yet, what was then only implicit in *Maher*, *Beal*, and *Harris*—the "condescension that she may go elsewhere for her abortion"[279]—is now made explicit in *Dobbs*. What Blackmun described then as the disingenuous posture of those in the majority in *Maher* now seeks no shelter behind the cloak of *Roe* to claim that a constitutional right to abortion—no matter how impaired—exists in the United States. That is, the Court no longer concedes the existence of a constitutional right to abortion.

* * *

Today, implicit in the Court's analysis in *Dobbs* is a chilling disregard for women, girls, and people with the capacity for pregnancy, particularly in light of the devastating health consequences, including death, associated with pregnancy in the United States,[280] no less

---

[275] *Id.*

[276] *Id.*; *see also* Melissa Murray, *Race-ing* Roe*: Reproductive Justice, Racial Justice, and the Battle for* Roe v. Wade, 134 Harv. L. Rev. 2025, 2036-37 (2021) (explaining that eugenicists believed that "society should encourage the procreation of those of superior lineage, while discouraging procreation among—and public support for—those of inferior lineage").

[277] *See Dobbs*, 142 S. Ct. at 2256 n.41 (referencing Justice Thomas's concurrence in Box v. Planned Parenthood of Ind. & Ky., Inc., 139 S. Ct. 1780 (2019) (per curiam)); *see also Box*, 139 S. Ct. at 1782–84 (Thomas, J., concurring) ("The use of abortion to achieve eugenic goals is not merely hypothetical.").

[278] Beal v. Doe, 432 U.S. 438, 462 (1977) (Blackmun, J., dissenting).

[279] *Id.*

[280] *See generally* Elizabeth G. Raymond & David A. Grimes, *The Comparative Safety of Legal Induced Abortion and Childbirth in the United States*, 119 Obstetrics & Gynecology 215 (2012) (noting that a woman is fourteen times more likely to die by carrying a pregnancy to term than a legal abortion).

acknowledged by the Court in *Whole Woman's Health v. Hellerstedt*[281] in 2016.[282] Given that a woman is fourteen times more likely to die during childbirth or shortly thereafter than by having an abortion,[283] the Court failed to justify why such a risk could possibly be legitimate or in pursuit of an established constitutional principle or value.[284] Indeed, the opinion shows a particular malevolence against the lives of those directly concerned, but who are too often rendered invisible, namely Black and Brown women.[285] According to the Centers for Disease Control and Prevention, Black women are at least three times more likely to experience maternal mortality than their white counterparts—as a national average, not accounting for local trends.[286]

III. Originalism as Proxy

Dobbs represents opportunistic originalism—a brand that suggests there is no originalism at all. For example, Justice Alito claims that a right should only be protected if it is found in the Constitution's text, comports with its original meaning, or is part of a long, unbroken tradition. Yet, how does the latter point fit within the scope of originalism if it includes post-ratification history? Such contradictions are embedded in *Dobbs*. The Court cautions that its holding will not defeat the promise of *Griswold*, *Lawrence*, and *Obergefell* because they do not involve potential life. In other words, it is appropriate for the Court to protect non-textual, non-originalist rights so long as they do not involve potential life. Yet, the Court's

---

[281] 136 S. Ct. 2292 (2016).

[282] *See generally id.*

[283] Raymond & Grimes, *supra* note 280, at 215.

[284] Indeed, Justice Alito and those in the majority did not even try. The Court barely acknowledged the crisis of maternal mortality in the United States and in the passage where it addressed such concerns, it was not in relation to current national trends or those in Mississippi, but rather to scrutinize *Roe*'s attention to such concerns. *See* Dobbs v. Jackson Women's Health Org., 142 S. Ct. 2228, 2268 (2022).

[285] *See generally id.*

[286] Donna L. Hoyert, *Maternal Mortality Rates in the United States*, Ctrs. for Disease Control & Prevention, https://www.cdc.gov/nchs/data/hestat/maternal-mortality/2020/maternal-mortality-rates-2020.htm (last visited Jan. 29, 2023) ("Rates for non-Hispanic Black women were significantly higher than rates for non-Hispanic White and Hispanic women. The increases from 2019 to 2020 for non-Hispanic Black and Hispanic women were significant."); Cecilia Lenzen, *Facing Higher Teen Pregnancy and Maternal Mortality Rates, Black Women Will Largely Bear the Brunt of Abortion Limits*, Tex. Tribune (June 30, 2022) ("Black women are three to four times more likely to experience a pregnancy-related death than white women, and the risk spans income and education levels.").

clarification suggests that its commitment to originalism is weak and conditional. It is exactly the opportunistic originalism that this Article speaks of. Simply put, the Court's originalism is not about text, original meaning, or history. Rather, it is about the state's interest in protecting what it regards as potential life, and such originalism consists of little more than a conservative political and religious judgment.

Nevertheless, *Dobbs* does parade as originalist. Indeed, some commentators might suggest that the Court ignoring the extreme consequences facing women in Mississippi and across the country as a result of *Dobbs* is consistent with an originalist methodology. They might submit that it is wrong to consider the human consequences, offering up *Bruen* and *Dobbs* as consistent. The former pays little attention to the consequences of gun violence in the United States, and the latter ignores maternal mortality, maternal morbidity, infant mortality, and myriad other pregnancy-related concerns. Originalists would dismiss the relevance of the impact of these decisions.

Despite the Court's embrace of originalism,[287] and claims to search for original meaning in texts, salient criticisms underscore the illogic and "dangers" of its incorporation in judicial decision-making. First, "for most constitutional provisions, there is no 'original meaning' to be discovered."[288] Second, the myriad brands of originalism confuse its meaning and the ability to understand it, such that there is no "clear statement of the doctrine itself."[289] Third, originalism "reduces the complex task of judging to an overly simplistic and faux-historical

---

[287] A number of right-leaning commentators have claimed that *Dobbs* is not originalist, mainly on the grounds that it did not go far enough in completely repudiating substantive due process. Ilan Wurman, *Hard to Square* Dobbs *and* Bruen *with Originalism*, Denver Post (July 13, 2022, 3:47 PM), https://www.denverpost.com/2022/07/12/roe-vs-wade-originalism-dobbs-bruen-abortion-guns. However, the argument put forth in this Article is that the Court is deploying originalism strategically—simply rejecting rights it doesn't like and keeping those it supports.

[288] *See, e.g.*, Erwin Chemerinsky, Worse Than Nothing: The Dangerous Fallacy of Originalism, at xi (2022) ("Instead, there is a range of possibilities that allows for exactly the judicial discretion that originalism seeks to eliminate."); William N. Eskridge, Jr. & Victoria F. Nourse, *Textual Gerrymandering: The Eclipse of Republican Government in an Era of Statutory Populism*, 96 N.Y.U. L. Rev. 1718, 1755 (2021) ("Recent [Supreme Court] cases demonstrate [] that the slicing and dicing techniques generating imaginary plain meanings have now become standard practice within the Supreme Court."); Leah Litman & Katherine Shaw, *Textualism, Judicial Supremacy, and the Independent State Legislature Theory*, 2022 Wisc. L. Rev. 1235, 1247 ("It is impossible to miss that probing beneath textualism's surface claims reveals significant variety, as well as remarkable inconsistency, in actual practice.").

[289] Mitchell Berman, *Originalism Is Bunk*, 84 N.Y.U. L. Rev. 1, 5 (2009).

inquiry that would lead to intolerably retrograde decisions."[290] In a nutshell, the latter reflects the sentiment of many pundits and scholars.[291] After all, none of the Justices are historians. Fourth, the Court may be unprincipled in its application of originalism, preferencing some Founders over others. The Court may extol certain virtues articulated by Founders, while denying the existence of incompatible principles and values expressed by the same or other Founders. Originalists may exalt certain historical debates in their analysis while discounting, ignoring, or even dismissing others that depart from their reasoning. The Court may cherry-pick eras in history that tailor to their analysis and desired outcomes while dismissing or neglecting other periods, historical texts, documents, and truths because they contradict or fail to vindicate the desired outcome. To the extent that originalism may vindicate constitutional values in the present day, underscoring liberty, equality, and autonomy—*if Dobbs is an example*—the Court failed miserably.

Fifth, the Court may be sloppy, inept, and inaccurate in its application of originalism. Sixth, originalism heightens the likelihood that the concerns, arguments, and debates of historically marginalized communities will be written out of the Court's jurisprudence as necessarily insignificant and irrelevant given that the focus centers Framers and Founders—their debates, intentions, meanings, and laws—in periods in which white women were denied suffrage, Black people were legally deemed property, Black women suffered the weight of sexual and physical chattel, and Indigenous people were marched off their lands. Indeed, a fallible and corruptible Court may do this even if the Framers and Founders were keenly interested in the concerns of the vulnerable and marginalized. *Dobbs* presents that example. Seventh, as noted above, originalists may ignore the consequences of upholding harmful state laws.

Lastly, even while originalism professes to simply discover "law" that is fixed at the time of enactment, this too overlooks the fact that it necessarily smuggles in normative principles in a number of ways. One example is by selecting the level of generality at which legal meaning is specified.[292] Adrian Vermeule's devastating critique of

---

[290] Justin Driver, *Ignoble Specificities*, New Republic, (Mar. 15, 2012), https://newrepublic.com/article/101724/driver-scalia-law-brown-education-originalism.

[291] *See, e.g.*, Lithwick & Siegel, *supra* note 16.

[292] Adrian Vermeule, Common Good Constitutionalism 91–116 (2022).

originalism in *Common Good Constitutionalism* makes this point, underscoring it in a chapter titled "Originalism As Illusion."[293] Vermeule's rigorous critique builds on Ronald Dworkin's criticism of originalism as inevitably grounded in normative principles of political morality.[294] The example offered is the meaning of "cruel and unusual punishment," which could be interpreted as only the exact punishments the Framers thought were cruel and unusual. On the other hand, it could signify something much broader and evolving. The critique is that there is no way interior to originalism to make the informed choice between meanings. Similarly, the principle could apply to the Reconstruction Amendments, which could be interpreted only to forbid those laws the Reconstruction Congress deemed impermissible or could be understood in broader and evolving ways. The choice between those two levels of generality is inevitably normative.

Part III turns to these concerns—not as a comprehensive literature review. Rather, Section III.A briefly argues that the Court's 2021 Term foreshadows the centrality of originalism and constitutional memory[295] to future jurisprudence such that, despite the consistent academic inclination to reject the methodology (and the temptation of some scholars to mock it), readers might be wise to revisit it—not necessarily as a project to endorse or legitimize, but as a pragmatic matter. At least five Justices on the Supreme Court articulate some level of fidelity to originalism as a methodology for interpreting constitutional law. Ignoring this will be to the detriment of defending and vindicating civil liberties and civil rights in the present before an ultraconservative Supreme Court bent on looking to the past. Nor might the application or purported application of originalism be limited to the deliberations of the Supreme Court. As the procedural posture of *Dobbs* reveals, originalist instincts infused Judge Ho's Fifth Circuit concurring opinion.[296]

---

[293] *Id.*

[294] *Id.* at 95–97; *see* Ronald Dworkin, Law's Empire 4–11, 228–39 (1986); Ronald Dworkin, Justice in Robes (2006); *see also* Ronald Dworkin, *Bork's Jurisprudence*, 57 U. Chi. L. Rev. 657, 664 (1990); Ronald Dworkin, *The Moral Readings of the Constitution*, N.Y. Rev. Books (Mar. 21, 1996), https://www.nybooks.com/articles/1996/03/21/the-moral-reading-of-the-constitution.

[295] *See generally* Siegel, *supra* note 41.

[296] Jackson Women's Health Org. v. Dobbs, 945 F.3d 265, 277 (5th Cir. 2019) (Ho, J., concurring) ("Nothing in the text or original understanding of the Constitution establishes a right to an abortion.").

Section III.B takes up the majority's invitation to seriously examine history in search of original intent and meaning leading to and at the period of Reconstruction. Indeed, a rich archive exists that illumes debates, speeches, and concerns the Court overlooked related to involuntary reproductive servitude, unwanted and forced pregnancies, coerced childbirth, and motherhood. That is, there are other important reasons for closely studying the Reconstruction Amendments (on which the Court relies in *Dobbs*), including clarifying and reconciling inaccuracies in the historical record, contending with social and legal transgressions embedded in American history and law, and confronting the troubling history of American slavery—the impetus undergirding the Amendments.

In *Bruen*, for example, confronting the disturbing history of American slavery necessitated a prologue to substantiate that Black men were thwarted at every turn from the ability to protect their bodily autonomy prior to and immediately following Reconstruction. By this, Justice Thomas and the majority found it vital to address the consequences of denying Black men access to firearms. By reaching to the past, Justice Thomas suggested that the Constitution's meaning at the time of Reconstruction could be uncovered. Notably, neither Justice Alito nor Thomas engaged in a similar practice to uncover constitutional meaning in relation to the Founders' serious concerns related to Black women, their bodily autonomy, childbearing, and forced reproduction, revealing the non-neutral manner in which this Court practices originalism. Nevertheless, stamping out the forced involuntary sexual and reproductive servitude of Black girls and women served as a powerful motivator for abolitionists in Congress who framed the Thirteenth and Fourteenth Amendments. Section III.B offers a corrective, drawing from the debates, speeches, and archives left untouched by the Court, but that yield meaningful content from which to illustrate the nature of the ratifiers' concerns.

A. ORIGINALISM AND CONSTITUTIONAL MEMORY

In *Memory Games:* Dobbs*'s Originalism as Anti-Democratic Living Constitutionalism—And Some Pathways for Resistance*, Professor Reva Siegel addresses the disjuncture found in the Court's holding in *Dobbs*, a glaring inconsistency. As she points out, although the "Supreme Court has reversed many cases. . . . [it] never has [overturned] a right that the Court itself had justified as important to a group's equal participation 'in the economic and social life of the Nation.'"[297] As recently as 2020, in *June Medical Services L.L.C. v. Russo*, striking down a Louisiana TRAP law, the Court cautioned, "at the risk of repetition, we remind the reader of the standards we described [in *Casey* and *Whole Woman's Health*]"[298]: "a statute which, while furthering [a] valid state interest has the effect of placing a substantial obstacle in the path of a woman's choice cannot be considered a permissible means of serving its legitimate ends."[299]

As Professor Siegel clarifies, to only understand originalism as a method of interpreting the Constitution would obscure originalism as a "politics whose longstanding goal has been reversing *Roe*,"[300] evidenced in no small part through "specialized judicial appointment practices," dating back to President Ronald Reagan, "designed to achieve movement-party goals and through constitutional memory work that can justify a new court's doctrinal innovations as restoring the Framer's Constitution."[301] So too, it seems, originalism can serve as the Trojan Horse—a distraction. This distraction calls attention away from "[a] Supreme Court that the Republican Party composed by a series of norm-busting appointments practices."[302]

Thus, the turn to originalism as the selected method of constitutional interpretation in *Dobbs* necessarily conflicts with contemporary

[297] Siegel, *supra* note 41 (manuscript at 4) (quoting Planned Parenthood of Se. Pa. v. Casey, 505 U.S. 833, 856 (1992)). According to Reva Siegel, "it is now urgent that we attend to the many kinds of arguments from constitutional memory in our constitutional tradition, inside and outside of courts. It is time we focus, carefully, on the ways that arguments from constitutional memory can be deployed in the service of rights deprivation. And it is time we begin to explore some ways that arguments from constitutional memory might be deployed in resistance." Siegel, *supra* note 41 (manuscript at 6) (citation omitted).

[298] June Med. Servs. L.L.C. v. Russo, 140 S. Ct. 2103, 2120 (2020) (plurality opinion) (quoting Whole Woman's Health v. Hellerstedt, 579 U.S. 582 (2016)).

[299] *Id.* (citation omitted). "As we stated in *Casey*, a State's abortion-related law is unconstitutional on its face if 'it will operate as a substantial obstacle to a woman's choice to undergo an abortion' in 'a large fraction of the cases in which [it] is relevant.' In *Whole Woman's Health*, we reaffirmed that standard. We made clear that the phrase refers to a large fraction of 'those women for whom the provision is an actual rather than an irrelevant restriction.' That standard, not an 'every woman' standard, is the standard that must govern in this case." *Id.* at 2132–33 (citations omitted).

[300] Siegel, *supra* note 41 (manuscript at 4) ("This goal was so important that during the Court's deliberations in *Dobbs*, originalists were threatening to break with the movement and embrace a more openly values-based method if the Court did not overturn *Roe*." (citation omitted)).

[301] *Id.* (manuscript at 5) ("On this account, executive branch appointments politics matter critically to originalism's authority, as do originalism's appeals to constitutional memory to legitimate the exercise of public power.").

[302] *Id.*

understandings of constitutional law, which recognizes women's personhood, freedom, and equal citizenship.[303] Simply put, the modern Constitution assumes, among other liberties, women's emancipation from involuntary labor, sex, and motherhood. However, even if one were to reach into the archives of constitutional jurisprudence and congressional debate, dating back more than 150 years, the Constitution more than acknowledges Black women's emancipation from systems of bondage—the Reconstruction Amendments directly address it. Given the implicit and explicit questions regarding race laced and lodged in *Dobbs*, the central questions seem whether conditions imposed upon Black women, prior to the Civil War, including forced sex, involuntary reproductive labor, coerced motherhood, and compelled domestic servitude and care of white families fall within the scope of liberation demanded by the Reconstruction Amendments.[304] As a matter of modern constitutional interpretation, it is beneath the dignity of the Court and Constitution to debate such questions.

Yet, the Court's decisions in *Dobbs* as well as *Bruen* foreshadow its centrality in future jurisprudence. Justice Thomas's prologue in *Bruen* suggests that fidelity to the past in order to "vindicate the sovereignty of the living" should be the standard for the Court, and that the Justices must hold firm against all other constitutional interpretations, including "calls for a 'living Constitution.'"[305] Indeed, his prologue specifically references "[e]vidence from around the adoption of the Fourteenth Amendment" to ascertain whether the respondents' claims can be vindicated—finally concluding that they cannot.[306]

---

[303] *See id.* (suggesting that the "history-and-traditions standard and its application are designed" to clash with an evolved understanding of the constitution, civil liberties, and civil rights).

[304] Peggy Cooper Davis, Neglected Stories: The Constitution and Family Values 180 (1997) ("As abolitionists saw the matter, exploitation of the labor, sexual and physical integrity, and reproductive autonomy of slaves violated the slave's 'inalienable . . . right to his own body.'" (omission in original)); *see also Plessy v. Ferguson*, 163 U.S. 537, 555 (1896) (Harlan, J., dissenting) ("These two amendments, if enforced according to their true intent and meaning, will protect all the civil rights that pertain to freedom and citizenship.").

[305] *See* Joel Alicea, *Originalism and the Rule of the Dead*, 23 Nat'l Affs. 149, 149 ("The new trend in originalism abandons this heritage. It self-consciously rejects the authority of the past and the duties rightfully imposed by our forebears, elevating instead the will of the present and the ideologies of its theorists. Having internalized the basic assumptions of living constitutionalism, it is but one step away from becoming what has always been considered originalism's intellectual adversary.").

[306] *See* N.Y. State Rifle & Pistol Ass'n v. Bruen, 142 S. Ct. 2111, 2150 (2022) ("Evidence from around the adoption of the Fourteenth Amendment also fails to support respondents' position. For the most part, respondents and the United States ignore the "outpouring of

Likely for this reason, Professor Siegel urges that "we renew conversations about what originalism is," and she probes "what it might portend for the nation."[307] Others have answered these questions in recent decades, but even their responses may deserve revisiting.[308] Perhaps rhetorically, William Baude asks whether "originalism is our law" and "[i]f not, what is?"[309] His project suggests that the debates on originalism can be reoriented to allow positivists and those committed to originalism "to move forward" and that "[t]his move is the 'positive turn.'"[310] To be sure, originalism attracts adherents[311] as well as critics and detractors who argue that originalism is a smokescreen or cover for Justices leaning into ideological or even political preferences and judgments.[312] Among the former, Joel Alicea has argued,

---

discussion of the [right to keep and bear arms] in Congress and in public discourse, as people debated whether and how to secure constitutional rights for newly free slaves" after the Civil War.").

[307] *See* Siegel, *supra* note 41 (manuscript at 6).

[308] *See, e.g.*, Daniel A. Farber, *The Originalism Debate: A Guide for the Perplexed*, 49 Ohio St. L.J. 1085 (1989); Jack M. Balkin, Living Originalism (2014) (arguing that originalism based on the expectations of constitutional Framers and Founders fails because it cannot comprehend the evolving nature of constitutional developments); Randy E. Barnett & Evan D. Bernick, The Original Meaning of the Fourteenth Amendment: Its Letter and Spirit, at xvi (2021) (establishing that their theory of original meaning of the Fourteenth Amendment includes among other principles, "The Equal Protection of the Laws Clause is not limited to barring state action but also imposes an affirmative duty on states to act. . . . A denial of 'the equal protection of the laws' extends beyond a state's official failing to protect the privileges or immunities of its citizens; it extends as well to failures of states to enact laws for their protection. . . ."); Aileen Kavanagh, *Original Intention, Enacted Text, and Constitutional Interpretation*, 47 Am. J. Juris. 255 (2002); Earl Maltz, *Foreword: The Appeal of Originalism*, 1987 Utah L. Rev. 773; Jane S. Schacter, *The Confounding Common Law Originalism in Recent Supreme Court Statutory Interpretation: Implications for the Legislative History Debate and Beyond*, 51 Stan. L. Rev. 1 (1998).

[309] William Baude, *Is Originalism Our Law?*, 115 Colum. L. Rev. 2349 (2015).

[310] *Id.* at 2351. Seemingly less a deep inquiry than a conceptual concession, Baude writes, "If originalism is the law, then neither the conceptual nor normative justifications need to bear much weight" and argues "that a version of originalism is indeed our law." *Id.* at 2352.

[311] *See* Ilan Wurman, A Debt Against the Living: An Introduction to Originalism (2017) (defending originalism as a methodological approach to constitutional interpretation and holding forth that interpreting the constitution as it was originally written and understood will allow Americans to discharge the debt owed to its Framers); Alicea, *supra* note 305, at 149.

[312] Chemerinsky, *supra* note 288, at 78 ("It is factually wrong to say that 'the people' consented to the Constitution because less than 5 percent of the population participated in its ratification."); Eric J. Segall, Originalism as Faith 192 (2018) (building the case that originalism serves as a proxy for judges expressing personal values and reaching politically driven outcomes); *id.* ("Originalism fails to limit the effects of personal values on judging. The only effective way to restrain judges from overturning state and federal laws based on ideology rather than law is for them to adopt a clear error rule where they would act only when the evidence of constitutional error is overwhelmingly clear."); Driver, *supra* note 290

"[i]f originalism is to avoid collapsing into that which it has always opposed, legal conservatives must begin by remembering why they believe in originalism. They must look to originalism's past in order to preserve its future."[313] In his calculation, legal scholars Jack Balkin and Randy Barnett concede far too much, "replac[ing] originalism's core philosophical assumptions with those of living constitutionalism."[314] This contribution does not referee that debate.

Legal scholars are not aligned in one view on originalism save that it is perplexing and confusing because even among its adherents, there lacks coherence. Consider *Bostock v. Clayton County*.[315] Justices Gorsuch, Kavanaugh, and Alito—three originalism adherents—articulated vastly different analyses on the question of whether Title VII protects lesbian, gay, and transgender employees against sex discrimination and reached vastly different conclusions.[316] According to Justice Gorsuch, author of the majority opinion,

> [O]nly the words on the page constitute the law adopted by Congress and approved by the President. If judges could add to, remodel, update, or detract from old statutory terms inspired only by extratextual sources and our own imaginations, we would risk amending statutes outside the legislative process reserved for the people's representatives. And we would deny the people the right to continue relying on the original meaning of the law they have counted on to settle their rights and obligations.[317]

As Justice Gorsuch and the majority rationalized it, that's "just a starting point. The question isn't just "what 'sex' meant, but what Title VII says about it."[318]

("If the liberal adoption of originalism's terminology becomes much more widespread, it also could have the subtly corrosive effect of suggesting that originalism provides the only legitimate method of constitutional interpretation.").

[313] *See* Alicea, *supra* note 305, at 149.

[314] *Id.* at 160.

[315] 140 S. Ct. 1731 (2020).

[316] *See id.*

[317] *Id.* at 1738 ("With this in mind, our task is clear. We must determine the ordinary public meaning of Title VII's command that it is 'unlawful . . . for an employer to fail or refuse to hire or to discharge any individual, or otherwise to discriminate against any individual with respect to his compensation, terms, conditions, or privileges of employment, because of such individual's race, color, religion, sex, or national origin.' . . . To do so, we orient ourselves to the time of the statute's adoption, here 1964, and begin by examining the key statutory terms in turn before assessing their impact on the cases at hand and then confirming our work against this Court's precedents." (first omission in original) (citation omitted)).

[318] *Id.* at 1834. For Justice Kavanaugh, in dissent, the key question is one of correct decisions based on "original public meaning." *See id.* at 1835 n.10 (Kavanaugh, J., dissenting).

However, in his dissent, Justice Alito accuses Justice Gorsuch and the majority of attempting a sleight of hand, erecting smokescreens in order to convince readers "that it is merely enforcing the terms of the statute, but that is preposterous."[319] According to Justice Alito, "the concept of discrimination because of 'sex' is different from discrimination because of 'sexual orientation' or 'gender identity,'" and at any rate, he interprets the Court's duty "to interpret statutory terms to 'mean what they conveyed to reasonable people *at the time they were written.*'"[320]

The Court's current composition of at least five Justices frame their judicial methodology and approach as being shaped in whole or part by assessing original meaning.[321] In 2009, Professor Mitchell N. Berman, offered one of the more forceful critiques yet at the time, concluding "originalism is bunk" and in nearly one hundred pages refuted its clarity, coherence, and value.[322] Today, a similar conclusion is reached by Erwin Chemerinsky in *Worse Than Nothing: The Dangerous Fallacy of Originalism*.[323]

B. ORIGINALISM AND AUTHORING, REMEMBERING, AND MISREMEMBERING

According to Annette Gordon Reed, "The notion of who gets to pronounce a historical truth is very, very contested territory. The heresy was believing black people."[324] In *Neglected Stories: The Constitution and Family Values*, Professor Peggy Cooper Davis resurrects

[319] *Id.* at 1754 (Alito, J., dissenting).

[320] *Id.* (quoting Antonin Scalia & Bryan A. Garner, Reading Law: The Interpretation of Legal Texts 16 (2012)).

[321] Randy E. Barnett, *An Originalism for Nonoriginalists*, 45 Loy. L. Rev. 611, 613 (1999) ("[O]riginalism is now the prevailing approach to constitutional interpretation.").

[322] Berman, *supra* note 289, at 4 ("Notwithstanding their enviable energy and commitment to the cause, advocates of originalism have not made good on its core claim. Moreover, dispassionate analysis suggests that the prospects for greater success are slim indeed. Or so I will argue.").

[323] *See, e.g.*, Chemerinsky, *supra* note 288, at 12 (reflecting that "[o]riginalism would have been relegated to dissents, especially by Justice Thomas, who would have been the sole originalist on the Court," had Donald Trump not been elected. However, "[e]verything . . . is different because Donald Trump won the presidency and put Gorsuch, Kavanaugh, and Barrett on the Court").

[324] Anne Gibson, *Renowned Historian Recounts Jefferson-Hemings Controversy for Clark Audience*, ClarkNow, (Dec. 12, 2018), https://clarknow.clarku.edu/2018/12/12/renowned-historian-recounts-jefferson-hemings-controversy-for-clark-audience.

neglected constitutional narratives. One collection of narratives—the doctrinal—lawyers and judges construct. They "explain the Fourteenth Amendment of the United States Constitution" and how it "protects the independence and integrity of families."[325] The other set of narratives—Motivating Stories—are "of enslaved families in the United States—families composed of people who were denied, in law and often in fact the right to create and maintain ties of affection and responsibility to spouses, to children, to parents, or to extended kin."[326] These individuals were denied the right to marry; "they had no legal tie to their parents; and they had no legal tie to the children born of their bodies."[327] Professor Cooper Davis perceptively observed that enslaved Black people "therefore developed a commitment to family freedom and . . . inspired this constitutional commitment."[328]

Thoughtful readers may wonder: *Do such stories tell us anything about the context in which Founders framed debates, speeches, or law related to Reconstruction and family, and more specifically related to reproductive freedom? Were the Founders even aware of the "motivating stories" of enslaved Black women? If Black families' stories—and Black women's in particular—were so important to the passage of Reconstruction Amendments, why are they invisible in legal discourse? Why are they buried and hidden, obscured by other narratives such that they need excavation?*

Sally Hemings comes to mind—described by researchers at Thomas Jefferson's estate, Monticello, as "one of the most famous—and least known—African American women in U.S. history." For more than 230 years, she was "linked to Thomas Jefferson as his 'concubine,' obscuring the facts of her life and her identity."[329] From 1787–89, starting at the age of fourteen, Hemings worked as a maid

---

[325] *See* DAVIS, *supra* note 304, at 4.

[326] *Id.*

[327] *Id.*

[328] In bridging the neglected stories to contemporary constitutional debates, Professor Cooper Davis presented the compelling case for expanding the interpretative lens of constitutional analysis and review and with it, she demonstrated the value of rescuing the obscured, ignored, cast aside Reconstruction narratives. *Id.*

[329] *The Life of Sally Hemings*, MONTICELLO, https://www.monticello.org/sallyhemings (last visited Feb. 2, 2023) ("Hemings bore children fathered by her owner. Female slaves had no legal right to refuse unwanted sexual advances. Sally Hemings was the child of an enslaved woman and her owner, as were five of her siblings. At least two of her sisters bore children fathered by white men. Mixed-race children were present at Monticello, in the surrounding county, across Virginia, and throughout the United States. Regardless of their white paternity, children born to enslaved women inherited their mothers' status as slaves.").

and domestic servant in Thomas Jefferson's Paris household.[330] Meticulously preserved records by researchers and archivists at Monticello reveal that "unlike countless enslaved women, Sally Hemings—sixteen years old at the time was able to negotiate with her owner [Thomas Jefferson]" for "extraordinary privileges," including "freedom for her unborn children."[331] Pregnant by Thomas Jefferson in 1789, Sally Hemings "refused to return to Virginia and be 're-enslaved.'"[332] Seventy-five years prior to Reconstruction, Hemings, not naïve to her own complex, propertied status (and that of two sisters who also "bore children fathered by white men"),[333] could and did calculate the costs of enslavement and the value of freedom and articulated that to Thomas Jefferson, a politician, statesman, principal author of the Declaration of Independence, and future President (1801–1809) of the United States at a level of intimacy perhaps unmatched by anyone else, including Founders.[334]

Nevertheless, as Professor Gordon-Reed explains, "the scholarship written to disprove her connection to Jefferson routinely diminished Hemings's humanity."[335] For example, prominent historians argued "that the story couldn't be truc because Jefferson would never be involved with 'a slave girl' and that such a person was too low to have

[330] Sally Hemings's path to Paris and back to Virginia exposes other confounding truths bound in the journey of American slavery. Sally was a girl—barely a teen—when sent to Paris. She returned at sixteen, pregnant with a child who is now believed to be the offspring of Thomas Jefferson. Commentators generally acknowledge Sally's enslaved status as denying her any legal agency upon her return to the United States, where the age of consent in Virginia was ten years old in the 1780s. Also, Hemings's family lineage unmasks other overlooked truths within the Jefferson household and slavery more generally. She was the half-sister of Jefferson's wife, Martha Wayles Skelton Jefferson. Their father, John Wayles, died the year of Sally Hemings's birth. *See Thomas Jefferson and Sally Hemings: A Brief Account*, MONTICELLO, https://www.monticello.org/thomas-jefferson/jefferson-slavery/thomas-jefferson-and-sally-hemings-a-brief-account (last visited Feb. 2, 2023).

[331] MONTICELLO, *supra* note 329; *see also* Annette Gordon-Reed, *Sally Hemings, Thomas Jefferson and the Ways We Talk About Our Past*, N.Y. TIMES, (Aug. 24, 2017), https://www.nytimes.com/2017/08/24/books/review/sally-hemings-thomas-jefferson-annette-gordon-reed.html.

[332] Gordon-Reed, *supra* note 331 ("This was a bold gambit for a 16-year-old, even at a time, unlike today, when females that age were not thought of as children. The age of consent in Virginia in the 1780s was 10 and was raised to 12 in the 1820s. It is easy to see how her experiences, and the energy and optimism of youth, led her to think she could do this.").

[333] MONTICELLO, *supra* note 329.

[334] *Id.*; *see also Thomas Jefferson: The 3rd President of the United States*, WHITE HOUSE, https://www.whitehouse.gov/about-the-white-house/presidents/thomas-jefferson (last visited Feb. 2, 2023).

[335] Gordon-Reed, *supra* note 331.

influenced Jefferson . . . as if the designation 'slave girl' told readers all they needed to know."[336] Renowned historians disputed Jefferson's relationship to Hemings, including fathering her children, until DNA evidence published in *Nature* and corroborated by the Thomas Jefferson Foundation quieted their skepticism.[337]

Journalist Clint Smith suggests that an ugly political truth resonates throughout history, privileging certain stories while troublingly denying others—or even appropriating them.[338] From a political perspective, this process may be complete or near-complete exclusion from the historical record or may include misremembering harms inflicted on vulnerable communities and groups. Equally, it can signify inaccurately recalling beneficence and magnanimity where none existed. The District Court in *Dobbs* used the term "gaslighting" to describe the phenomenon in Mississippi whereby the state claimed its abortion ban reflected beneficence—in this case protecting the health and safety of pregnant women.[339] Judge Reeves recognized the claim was a mere smokescreen for lawmakers' true objective—abolish abortion rights. The state's engagement in centuries of invidious discrimination

---

[336] *Id.* ("To get Sally Hemings to come back with him, Jefferson promised her a good life at Monticello, and that their children would be freed when they became adults. Madison Hemings said that his mother "implicitly relied" on Jefferson's promises.").

[337] Eugene A. Foster et al., *Jefferson Fathered Slave's Last Child*, 396 NATURE 27 (1998).

[338] Smith offers the recasting of Memorial Day as an example:

> Confederates had converted Charleston's Washington Race Course and Jockey Club into an outdoor prison for captured Union soldiers. The conditions were so terrible that nearly 260 men died and were buried in a mass grave behind the grandstand. After the Confederates retreated, Black men reburied the dead in proper graves and erected an archway bearing the words MARTYRS OF THE RACE COURSE. An enormous parade was held on the track, with 3,000 Black children singing "John Brown's Body," the Union marching song. The first Memorial Day, as Blight describes it, received significant press coverage. But it faded from public consciousness after the defeat of Reconstruction.

Clint Smith, *Why Confederate Lies Live On*, ATLANTIC (May 10, 2021), https://www.theatlantic.com/magazine/archive/2021/06/why-confederate-lies-live-on/618711.

[339] *See* Jackson Women's Health Org. v. Currier, 349 F. Supp. 3d 536, 540 n.22 (S.D. Miss. 2018) ("In that spirit, this Court concludes that the Mississippi Legislature's professed interest in 'women's health' is pure gaslighting. In its legislative findings justifying the need for this legislation, the Legislature cites *Casey* yet defies *Casey*'s core holding. The State 'ranks as the state with the most [medical] challenges for women, infants, and children' but is silent on expanding Medicaid. Its leaders are proud to challenge *Roe* but choose not to lift a finger to address the tragedies lurking on the other side of the delivery room: our alarming infant and maternal mortality rates." (citations omitted)), *aff'd sub nom.* Jackson Women's Health Org. v. Dobbs, 945 F.3d 265 (5th Cir. 2019), *rev'd and remanded*, 142 S. Ct. 2228 (2022).

against women, including coerced sterilizations of Black girls and women, did not reflect a commitment to equality and justice.

Psychologists refer to nonintentional projecting, misremembering, and fabricating of memories as confabulation,[340] when recollections are disrupted, leading to distorted or inaccurate memories. They distinguish the unintentional creation of memories from that which involves purposeful, conscious lying, meant to deceive others. American history seems riddled with both.[341] In *Slavery, Propaganda, and the American Revolution*, Patricia Bradley suggests that the die was cast for the project of politicizing memories during the American Revolution when prominent newspapers such as the *Boston Gazette* misinformed readers about the impeding revolution by misrepresenting abolitionists, refusing to publish writings by abolitionists, and mischaracterizing Black people.[342]

In 2021, Clint Smith visited American plantations, historical landmarks, cemeteries, prisons, museums, and memorials to understand why race-based distortions in American history persist, particularly those that link to slavery. As he tells it, in *Why Confederate Lies Live On*, he was "struck by the many people . . . who believe a version of history that rests on well-documented falsehoods."[343] That is, for a broad swath of Americans "history isn't the story of what actually happened; it is just the story they want to believe,"[344] and this too affects and infects law.[345] Indeed, the *Dobbs* majority begs readers to

---

[340] *See, e.g.*, Armin Schnider, *Spontaneous Confabulation and the Adaptation of Thought to Ongoing Reality*, 4 Nat. Rev. Neuroscience 662 (2003).

[341] *See, e.g.*, Patricia Bradley, Slavery, Propaganda, and the American Revolution (1999).

[342] *See id.*

[343] This is particularly the case involving slavery and its legacy: rape, involuntary pregnancy, forced reproduction, infanticide, the Civil War, and Reconstruction. Smith offers several examples about obscuring historical facts associated with Memorial Day, false claims about Black people fighting on the behalf of the Confederacy, as well as the stunning absence of discourse related to kidnapped, enslaved children crossing the Atlantic Ocean to enter the trail of commerce in the colonies and later the United States. *See* Smith, *supra* note 338.

[344] *Id.*

[345] According to Smith, "in the late 1800s" the "myth of the Lost Cause began to take hold" in an "attempt to recast the Confederacy as something predicated on family and heritage rather than what it was: a traitorous effort to extend the bondage of millions of Black people." *Id. So, whose family narratives prevail or survive given that historical context?* The myth contends that "the Civil War was fought by honorable men protecting communities" and had nothing to do with slavery. *Id.*

suspend curiosity, and perhaps their intelligence too, by its neglect of the record that originalism demands be explored.

1. *Authoring: Ratifiers' Debates.* That *who* is permitted to narrate history, resurrect old facts, add to the canon of American legal thought, and share controversial American truths "is very, very contested territory"[346] should not surprise readers of law, given prohibitions on women serving on juries, thereby forbidding them from hearing hard truths[347] or bans prohibiting Black people from testifying on their own behalf[348]—thus silencing them and hampering their ability to tell their stories. These vestiges and badges of coverture and slavery reveal much about troubling inconsistencies embedded in constitutional interpretation, the politicization of American truths, and democratization of the sources deemed credible as well as those critiqued as lacking merit.[349] Equally perplexing is the authority

---

[346] *See* Gibson, *supra* note 324 (quoting Annette Gordon-Reed).

[347] *See, e.g.*, Taylor v. Louisiana, 419 U.S. 522, 538 (1975) (dismantling Louisiana's centuries-long structural barrier to women's jury service, which required that they file a written declaration in order to participate in jury service, and holding, "Louisiana's special exemption for women operates to exclude them from petit juries, which in our view is contrary to the command of the Sixth and Fourteenth Amendments").

[348] *See, e.g.*, Fugitive Slave Acts of 1850, Act of Sept. 18, 1850, ch. 60, 9 Stat. 462 (repealed 1864) (barring apprehended Black people under the Act from providing testimony, including to defend their freedom from capture and trafficking); *see also* Strauder v. West Virginia, 100 U.S. 303 (1879).

[349] Such matters erupted in the legal academy twenty-five years ago after the publication of a series of articles relating to "race and law" and critical race theory. *See* DAN FARBER & SUSANNA SHERRY, BEYOND ALL REASON: THE RADICAL ASSAULT ON TRUTH IN AMERICAN LAW 4–6 (1997) (describing Derrick Bell by example as a "radical multiculturalist" that has abandoned "moderation" and has "a dearth of common sense," stating in particular that Bell "illustrates the nature of . . . extremism," and asserting that Bell's theories on social constructions of race and racism pose "serious risks to our intellectual and political community"). Farber and Sherry argued that the "radical multiculturalists "includes adherents of a broad assortment of theories, including critical race theory, radical feminism, and legal writing about gays and lesbians, often called 'gaylegal' theory." *Id.* at 6. They argued that the radical multiculturalists caused significant distortions in legal discourse as well as presented "anti-Semitic implications." *Id.*; *see also* Richard A. Posner, *The Skin Trade*, NEW REPUBLIC, Oct. 13, 1997, at 40 (reviewing FARBER & SHERRY, *supra*) ("Every intellectual movement has a lunatic fringe. Radical legal egalitarianism is distinguished by having a rational fringe and a lunatic core. The latter is constituted by the critical race theorists and the other legal academics who have swallowed postmodernism hook, line, and sinker. . . ."); Heather Macdonald, *Rule of Law: Law School Humbug*, WALL ST. J., Nov. 8, 1995, at A21 (arguing that feminist jurisprudence is "antithetical to the very notion of law"). *But see* Peggy C. Davis, *Bah, "Humbug,"* CITY J., Spring 1996 ("'Law School Humbug' is such an extreme example of an absence of balance and fairmindedness that I feel compelled to protest. The article reflected an utter inability (or staunch refusal) to comprehend the rich and complex body of work that it ridiculed. Working from the astounding pretense that feminist and critical race scholarship is now dominant in the legal academy, the article cruelly undercut some of the most humane, most intellectually sophisticated, and most devoted teachers and scholars I have known.").

granted a story depending on who vocalizes it, which in American history has more often favored white storytellers.[350]

In 1896, Justice Harlan wrote in dissent that "[t]he Thirteenth Amendment does not permit the withholding or the deprivation of any right necessarily inhering in freedom."[351] Recounting the intention behind the Reconstruction Amendment, he reminded the majority, "[i]t not only struck down the institution of slavery as previously existing in the United States, but it prevents the imposition of any burdens or disabilities that constitute badges of slavery or servitude."[352] Three questions seem important to answer. First, whether ratifiers decreed or at least perceived universal civil freedom and "the security of personal liberty" to apply to Black women as the Court already acknowledges Black men were included in this promise of freedom and equality.[353] Second, did the ratifiers' explicit language that "no State shall make or enforce any law which shall abridge the privileges or immunities of citizens of the United States; nor shall any State deprive any person of life, liberty or property without due process of law, nor deny to any person within its jurisdiction the equal protection of the laws"[354] apply equally to Black women as it did the Black men that Justice Thomas speaks of in *Bruen*? Even if we assume bodily autonomy was protected under the Thirteenth Amendment, did that extend to sexual autonomy—freedom from sexual abuse and rape? Simply put, were Black women even in the purview of ratifiers?

Was it the intent of the ratifiers to apply the clause "[a]ll persons born or naturalized in the United States . . . are citizens of the United States and the State wherein they reside"[355] to Black women? Was

---

[350] *See, e.g.*, JAMES HASKINS, THE SCOTTSBORO BOYS (1994) (presenting the story of nine Black youths falsely accused of sexually assaulting two white women passengers on a train and the subsequent trial); Richard Pérez-Peña, *Woman Linked to 1955 Emmett Till Murder Tells Historian Her Claims Were False*, N.Y. TIMES (Jan. 27, 2017), https://www.nytimes.com/2017/01/27/us/emmett-till-lynching-carolyn-bryant-donham.html ("For six decades, she has been the silent woman linked to one of the most notorious crimes in the nation's history, the lynching of Emmett Till, a 14-year-old black boy, keeping her thoughts and memories to herself as millions of strangers idealized or vilified her. But all these years later, a historian says that the woman has broken her silence, and acknowledged that the most incendiary parts of the story she and others told about Emmett—claims that seem tame today but were more than enough to get a black person killed in Jim Crow-era Mississippi—were false.").

[351] Plessy v. Ferguson, 163 U.S. 537, 555 (1896) (Harlan, J., dissenting).

[352] *Id.*

[353] *Id.* ("This court has so adjudged [Black men's freedom].").

[354] U.S. CONST. amend. XIV, § 1.

[355] *Id.*; *see Plessy*, 163 U.S. at 555 (Harlan, J., dissenting).

there general acceptance that the Thirteenth and Fourteenth Amendments applied to Black women as well as men? A search of congressional records from 1789 to 1865, as well as archives on American slavery, and other sources, important debates, speeches, and other materials that help to answer these questions materialize.[356] Arguably, the leading and most persuasive arguments wielded by abolitionists in Congress (who became the Reconstructionists) was the depravity of involuntary reproductive servitude and sexual debasement of Black women and girls. In an 1853 speech to Congress, Representative Joshua Giddings of Ohio remarked,

> I hesitate not to say, that, for its barbarity, that law is unequalled in the history of civilized legislation. Is there a reflecting man who will pretend that this barbarous enactment imposes upon those people any moral duty to obey it? Will preachers of righteousness tell them to submit, to let the slave-dealer rivet the chains upon the father, tear the mother form her children, and doom her to a life of wretchedness? Will such preachers advise the daughter *peacefully* to surrender herself into the hands of slave-hunters, and submit to a life of pollution and shame? And will such men be called promoters of *holiness and purity*?[357]

Similarly, in 1851, Representative Horace Mann of Massachusetts, delivered a speech where he admitted to lending assistance to "several young and healthy girls" who had features "southern connoisseurs in sensualism so highly prize."[358] Numerous records point to the broad public discourse available to and generated by abolitionists in Congress that condemned the usurpation of Black women's bodies in service to the Southern agenda of enslavement, coercive pregnancy, and sexual predation.[359] In the same speech, Mann tells the story of "one girl, who after her recapture . . . , was sold six times in seven

---

[356] Ronald G. Walters, *The Erotic South: Civilization and Sexuality in American Abolitionism*, *in* AFFECT AND POWER: ESSAYS ON SEX, SLAVERY, RACE AND RELIGION (David J. Libby, Paul Spickard & Susan Ditto, eds., 2005).

[357] Joshua R. Giddings, Annual Message of the President (Dec. 9, 1850), *in* SPEECHES IN CONGRESS 438 (Boston, J. P. Jewett & Co. 1853).

[358] Horace Mann, Speech on the Fugitive Slave Law, Delivered at Lancaster, Massachusetts (May 19, 1851) *in* SLAVERY: LETTERS AND SPEECHES 510–11 (1851); *see also id.* at 511 ("Elizabeth Russell was one of them. She fell immediately into the slave-traders' fangs, and was doomed for the New Orleans market. The hearts of those who saw her and foresaw her fate were touched with pity. They offered eighteen hundred dollars to redeem her, and some there were who offered to give, who would not have had much left after the gift. But the fiend of a slave trader was inexorable. He knew how he could transmute her charms into gold through the fires of sin.").

[359] *See id.* at 511–12. Further imploring:

weeks, in Maryland and Virginia, for her beauty's sake."[360] He concludes, noting that "her body was found scarred and waled with whip marks, which the villains inflicted upon her because she would not come to their bed."[361]

Speeches more than a decade before ratification were infused with the promise of liberating Black girls and women not only from commonly understood forms of slavery—physical labor in the fields—but specifically from sexual and reproductive abuse. In a speech before Congress on March 16, 1852, Joshua Giddings expressed, "No, Mr. Speaker, I blush for my country, when her representatives take shelter behind unmeaning generalities, and refuse to avow their honest sentiments. If gentlemen intend to support the compromise, they must of course intend to chase down the trembling female, as she flees from the inhumanity of a worse than savage oppressor."[362]

In a series of addresses in Congress and among constituents, Massachusetts Senator Charles Sumner condemned slavery, particularly pointing to the manner in which the system was maintained by

---

> But she proved heroically and sublimely intractable. Like Rebecca, the Jewess, she would have flung herself from the loftiest battlement, rather than yield her person to a villain. Notwithstanding her masters' pretence that they had bought her with their money, and owned her soul, yet she had wealth, which, though all the earth were "one entire and perfect chrysolite," it could not buy. It was not difficult, therefore, to purchase her, and she was redeemed and came to New York; and I have been informed in the most authentic manner from the lady of the very respectable family of which she became an inmate, that, on an examination of her person, [she was brutally abused]."

Id.

[360] *Id.* at 511.

[361] *Id.* at 512.

[362] Joshua R. Giddings, The Compromise Measures (Mar. 16, 1852), *in* Speeches in Congress, *supra* note 356, at 483. In a speech given by Joshua Giddings on June 23, 1852, he extolled:

> You, Sir, lately saw an advertisement in the leading whig paper of this district, in these words: "*For sale, a handsome and accomplished lady's maid, aged just sixteen years.*" Except in this city and New Zealand, I do not think any government within the bounds of civilization, would have permitted; such an outrage upon decency. I speak of New Zealand, without intending any disrespect to the people of that island by comparing their habits with ours. They buy men and women for food only. The object is far more honorable and Christian-like than that for which the young women of this city are advertised and sold. Mr. Chairman, General Scott and General Pierce are both pledged to maintain this traffic in the bodies of women, and the advocates of liberty are asked to aid in electing them.

Joshua R. Giddings, The Baltimore Platforms (June 23, 1852), *in* Speeches in Congress, *supra* note 356, at 495–96.

upholding inhumane wrongs inflicted on Black women and their children. In a speech delivered on August 11, 1855, he admonished the "institution which separates parent and child, which stamps woman as a concubine, which shuts the gates of knowledge, and which snatches from the weak all the hard-earned fruits of incessant" labor.[363] In another address given on June 4, 1860, Sumner described the institution that he would later shut down with other Radical Republicans, as one that trafficked in violence and "stripped [Black women] of every defence," thereby "the chastity of a whole race is exposed to violence, while the result is recorded in the tell-tale faces of children, glowing with a master's blood, but doomed for the mother's skin to Slavery through descending generations."[364]

Other leaders in Congress made similar arguments, including Iowa Senator James Harlan,[365] Ohio Representative Samuel Cox,[366] Iowa Representative Josiah Grinnell,[367] and California Representative Thomas Shannon.[368] Further, the American Freedmen's Inquiry Commission (Department of War) took special notice of the unique bodily harms and involuntary reproduction Black women were subjected to, noting that Black women were not equally protected as white women "against murder and rape" because "the word white is not to be found."[369]

---

[363] Charles Sumner, New Outrage for the Sake of Slavery, Letter to Passmore Williamson, in Moyamensing Prison (Aug. 11, 1855), *in* 5 Charles Sumner: His Complete Works 56 (George Frisbie Hoar ed., Statesman ed. 2015) (1872).

[364] Charles Sumner, The Barbarism of Slavery, Speech in the Senate, on Bill for the Admission of Kansas as a Free State (June 4, 1860), *in* 6 Charles Sumner: His Complete Works, 133 (George Frisbie Hoar ed., Statesman ed. 2015) (1872). Further arguing:

> The Senator from Mississippi [Mr. Brown], galled by the comparison between Slavery and Polygamy, winces. I hail this sensibility as the sign of virtue. Let him reflect, and he will confess that there are many disgusting elements in Slavery, not present in Polygamy, while the single disgusting element of Polygamy is more than present in Slavery. By license of Polygamy, one man may have many wives, all bound to him by marriage-tie, and in other respects protected by law. By license of Slavery, a whole race is delivered over to prostitution and concubinage, without the protection of any law. Surely, Sir, is not Slavery barbarous?

Id.

[365] Cong. Globe, 37th Cong., 2d Sess. app. 321 (1862) (statement of Sen. James Harlan).

[366] Cong. Globe, 38th Cong., 1st Sess. 708–13 (1864) (statement of Rep. Samuel Cox).

[367] *Id.* at 1602 (statement of Rep. Josiah Grinnell).

[368] *Id.* at 2948 (statement of Rep. Thomas Shannon).

[369] Edwin M. Stanton, Report of the Secretary of War, S. Exec. Doc. No. 53, at 65 (1st Sess. 1864) (speaking directly to the changes in law that were necessary to give Black women "self-defence").

2. *Remembering: Abortion Truths.* The Court presents an overly simplistic account of abortion, its regulation, and its history in the United States. Justice Alito observes that abortion is not mentioned in the Constitution. He follows by highlighting states' laws that criminalized abortion prior to the ratification of the Reconstruction Amendments. He dismisses centuries of non-criminalization of abortion by stating that states could have criminalized abortion had they wished to do so. He buttresses these observations with a curious and frequent turn to Blackstone and Hale. However, much is neglected in Justice Alito's historical accounting of abortion in America.

Notably, abortion was not illegal in "early America." In fact, abortions were common in the colonies. Records dating back to the 1600s point to "pregnancies that terminated in elective abortions."[370] Before the early 1800s, Great Britain did not criminalize abortion.[371] Later, in 1803, when Great Britain did criminalize abortion, the new prohibition had no effect on women in the colonies.[372] Even while abortions remained illegal in France and Spain, induced abortions were common in early America and not penalized in their U.S. colonies.[373] In fact, "until about 1880, abortions were allowed under common law and widely practiced. They were illegal only after 'quickening,' the highly subjective term used to describe when pregnant women could feel the fetus moving."[374] A typical advertisement, such as this, spoke directly to women.

> TO MARRIED WOMEN.—Is it not but too well known that the families of the married often increase beyond what the happiness of those who give them birth would dictate? . . . Is it moral for parents to increase their families, regardless of consequences to themselves, or the well being of their offspring, when a simple, easy, healthy, and certain remedy is within our control? The advertiser, feeling the importance of this subject, and estimating the vast benefit resulting to thousands by the adoption of means

[370] Zoila Acevedo, *Abortion in Early America*, 4 Women & Health 159, 159 (1979).

[371] *Id.* at 161.

[372] *Id.*

[373] *Id.*

[374] Jessica Ravits, *The Surprising History of Abortion In The United States*, CNN, (June 27, 2023), https://www.cnn.com/2016/06/23/health/abortion-history-in-united-states/index.html. (referencing historian Leslie Reagan as saying "[a]bortions would become criminalized by 1880, except when necessary to save a woman's life, not at the urging of social or religious conservatives but under pressure from the medical establishment—and the very organization that today speaks out in support of abortion access").

> prescribed by her, has opened an office, where married females can obtain the desired information.[375]

Abortion tonics were advertised in newspapers across the nation. Advertisements responding to "female trouble" and "uterine tonics" were not uncommon marketing terms for abortion.[376] Some abortionists became popularly known, such as Ann Lohman—otherwise known as Madame Restell—whose career performing abortions for an elite clientele spanned four decades.[377] When demand for her services expanded, she sent pills and powder-based birth control in the mail.

In a peer-reviewed article published six years after *Roe*, Dr. Zoila Acevedo describes early American methods to achieve pregnancy terminations as well as the legal landscape during the colonial period.[378] Acevedo explains that elective abortions were performed by colonists as well as Indigenous tribes and Black peoples trafficked to the Americas.[379] She writes, "[j]ust as America was considered to be undiscovered before the white men found it, so the Indian [abortion] drugs were considered" ineffective at achieving an abortion "until white men discovered them."[380] In the 1800s, herbal medications such as red cedar were "accredited with having properties equivalent to those of Sabina or Savin, an abortive agent used by colonial women, among whom elective abortions were not unknown."[381]

Women were not alone in developing strategies and technologies to prevent pregnancy or terminate and abort. As Emily Feng, Megan Lim, and Sarah Handel report, Benjamin Franklin wrote about how to procure a safe abortion[382] and Thomas Jefferson,

---

[375] Karen Abbott, *Madame Restell: The Abortionist of Fifth Avenue*, SMITHSONIAN MAG., (Nov. 27, 2012), https://www.smithsonianmag.com/history/madame-restell-the-abortionist-of-fifth-avenue-145109198 (quoting an abortion advertisement).

[376] *Id.* (noting that contraception was also commonly advertised as "cathartic pills to remove obstructions").

[377] *Id.* ("Around 1836, she met 27-year-old Charles Lohman, a printer at the *New York Herald*. He was well-educated and literate . . . and he began publishing tracts about contraception and population control.").

[378] Acevedo, *supra* note 369.

[379] *Id.* at 159–60 ("Substances used to induce abortion by the Indians of North America included herbs and medicines such as those derived from 'black root' and 'red cedar,' which when finally subjected to testing were found to be effective.").

[380] *Id.* at 160.

[381] *Id.*

[382] Emily Feng et al., *For Ben Franklin, Abortion Was Basic Arithmetic*, NPR, (May 16, 2022), https://www.npr.org/2022/05/16/1099244635/for-ben-franklin-abortion-was-basic-arithmetic.

Patrick Henry, and John Marshall, were of the view that abortion was a private, family matter best addressed privately without state, criminal intervention.[383]

3. *Campaigning: Antebellum Discourse on Black Women.* Numerous sources inform the curious reader that nineteenth-century Americans were sufficiently aware and informed about the sexual exploitation of Black women and girls as a fundamental part of slavery. Discussions can be traced throughout the eighteenth and nineteenth centuries, with particular intensification during the period leading to Reconstruction. In essays spread through abolitionist networks and shared in fiction, visual arts, theater, travelogues, and pamphlets, the story of American slavery's sexual subjugation and appropriation of Black women's bodies is ubiquitous.[384] The key points made in the arguments against the sexual subordination of Black women were that if slavery were abolished, it would spare Black women from such harms. Further, it was perceived as a moral stain on the new nation that Black women were subjected to sexual violence inflicted by white men as part of the economic condition of slavery. Yet, abolitionists were not only concerned about white men directly inflicting sexual harm on Black women and girls. Slavery's predations included forcing Black men to impose themselves on Black women.[385]

For example, John D. Paxton wrote, "What effect is likely to be produced on the morals of the whites, from having about them, and under their absolute authority, female slaves who are deprived of the

---

[383] *See, e.g.*, Sarah Hougen Poggi & Cynthia A. Kiemer, *A 1792 Case Reveals That Key Founders Saw Abortion As a Private Matter*, WASH. POST (July 19, 2022, 6:00 AM), https://www.washingtonpost.com/made-by-history/2022/07/19/1792-case-reveals-that-key-founders-saw-abortion-private-matter.

[384] *See* THOMAS HAMILTON, MEN AND MANNERS IN AMERICA 174 (William Blackwood ed., 1843) ("The moral character of Jefferson was repulsive. Continually puling about liberty, equality, and the degrading curse of slavery, he brough his own children to the hammer, and made money of his debaucheries. Even at his death, he did not manumit his numerous offspring, but left them, soul and body, to degradation, and the cart-whip. A daughter of Jefferson was sold some years ago, by public auction, at New Orleans, and purchased by a society of gentlemen, who wished to testify, by her liberation, their admiration of the statesman, 'Who dreamt of freedom in a slave's embrace.' This single line gives more insight to the character of the man, than whole volumes of panegyric. It will outlive his epitaph, write it who may.").

[385] *Voices from the South, in* THE OASIS 192, 199 (Lydia Marie Child ed., 1834) ("Slavery is a poisonous and deadly vine twining about the sanctuary of domestic life. The unnatural and embarrassing relation that so often exists between [Southern ladies'] slaves and their husbands, their brothers, and their sons, is a sufficient argument against the brutal and degrading system. There is no denying this; the fact is proved on the very *face* of it."). The next section categorizes Black people by degree of white ancestry.

strongest motives to purity, and exposed to peculiar temptations to opposite conduct!"[386] Consider this passage from William Ellery Channing in *Slavery*, published in 1835: "That unprotected females, stripped by their degraded condition of woman's self-respect, should be used to minister to other passions in men than the love of gain, is next to inevitable."[387] In further condemnation of slavery and its odiousness, he wrote, "A slave-country reeks with licentiousness. It is tainted with a deadlier pestilence than the plague. But the worst is not told. As a consequence of criminal connexions, many a master has children born into slavery."[388] He articulated as others did at the time of the need for "protection" for the women and children tethered to slavery, worrying, "These cases must have increased, since the difficulties of emancipation have even multiplied. Still more, it is to be feared, that there are cases, in which the master puts his own children under the whip of the overseer, or else sells them to undergo the miseries of bondage among strangers."[389]

The archive of public narratives is extensive, including memoirs, novels, and essays distributed throughout the United States. In Richard Hildreth's 1836 book, *The Slave: Or Memoirs of Archy Moore*, he writes of the commonality of married planters being unrestrained in their "indulgences" with enslaved, young Black women and the frequency of births. In an interesting passage, perhaps pointing to infanticide, the narrator notes that of the six children born to his mother, only he is the survivor of these complicated births, explaining, "my mother was for several years, distinguished by colonel Moore's very particular regard: and she brought him no less than six children—all of whom, except myself, who was the eldest, were lucky enough to die in infancy."[390]

One of the most striking accounts of slavery was documented by Harriet Martineau, whose treatise, *Society in America*, published in 1837 details the striking and "boundless licentiousness caused by the

---

[386] John D. Paxton, Letters on Slavery: Addressed to the Cumberland Congregation, Virginia 129 (Abraham T. Skillman ed., 1833) ("The condition of female slaves is such, that promises and threatenings and management can hardly fail to conquer them. They are entirely dependent on their master. . . . That the vice prevails to a most shameful extent is proved from the rapid increase of mulattoes.").

[387] William Ellery Channing, Slavery 54 (1835).

[388] *Id.*

[389] *Id.*

[390] 1 Richard Hildreth, The Slave: Or Memoirs of Archy Moore 7 (1836).

practice," including the double horrors of rape from planters and physical violence inflicted by the wives of plantation owners.[391] To underscore the harms she witnessed, Martineau wrote, "There is no occasion to explain the management of the female slaves on estates where the object is to rear as many as possible, like stock, for the southern market: nor to point out the boundless licentiousness caused by the practice: a practice which wrung from the wife of a planter, in the bitterness of her heart, the declaration that a planter's wife was only 'the chief slave of the harem.'"[392]

* * *

Rather than anecdotal, these stories simply marked slavery for what it was—harsh, accepted, and normalized—prior to the ratification of the Thirteenth Amendment. Ratifiers were well aware of the system and the punishments unleashed on Black women who complained. The narratives were told in multiple voices, all reinforcing the other. In this passage from Lewis Clark in *Leaves from a Slave's Journal of Life*, published in 1842, he writes, "I had a pretty sister; she was whiter than I was, for she took more after her father. When she was sixteen years old, her master sent for her. When he sent for her again, she cried, and didn't want to go. She told her mother her troubles. . . . Her master was so mad to think she complained to her mother, that he sold her right off to Louisiana; and we heard afterward that she died there of hard usage."[393]

Notably, the *Dobbs* Court's failure to give a more rigorous accounting of the period leading to Reconstruction and the Reconstruction-era speeches and debates disserves any serious accounting of the Framers' and ratifiers' thinking, objectives, strategies, and plans. This vacuum lodges in American legal history and the

---

[391] 2 Harriet Martineau, Society in America 118 (1837).

[392] *Id.* "A gentleman of the highest character, a southern planter, observed, in conversation with a friend, that little was known, out of bounds, of the reasons of the new laws by which emancipation was made so difficult as it is. He said that the very general connexion of white gentlemen with their female slaves introduced a mulatto race whose numbers would become dangerous, if the affections of their white parents were permitted to render them free. The liberty of emancipating them was therefore abolished, while that of selling them remained." *Id.*

[393] Lydia Maria Francis Child, *Lewis Clarke, Leaves from a Slave's Journal of Life*, Anti-Slavery Standard, Oct. 20, 1842, at 78–79.

Court's jurisprudence. Rather than an outlier, *Dobbs* exemplifies the manner in which the Court strips the impetus for the Reconstruction from the Constitution, while yet claiming to center its Framers' concerns.

IV. Originalism as Proxy for Reinscribing Sex and Race-Sex Harms

In *Dobbs*, the Court selectively yielded to the asserted state interest of Mississippi lawmakers to abolish a fundamental right central to the health, safety, and bodily autonomy concerns of women generally, and Black women particularly, while in the same term (only one day before) in *New York State Rifle & Pistol Ass'n v. Bruen*,[394] issuing an opinion that rejected the constitutionality of New York's centuries-old gun law.[395] That opinion, authored by Justice Thomas, urged that a "prologue is in order"[396] to underscore the importance of centering the troubling history of racial animus in the United States that unjustly and unjustifiably targeted Black men who desired and were denied the opportunity to own guns.[397] He described the various ways in which Black men "were often denied" the ability to defend their bodily autonomy by owning guns "in antebellum America"[398] and later, "[a]fter the Civil War" were "systematically thwarted" in exercising this right.[399] Justice Thomas recounted the myriad circumstances that led Black men to assert "their self-protection, even if not always with success."[400] From Black music professors to a Black man "in Tennessee . . . murdered on his way to get book subscriptions,"[401] Justice Thomas took great pains to humanize the antebellum and contemporary concerns of Black men targeted by racism,[402]

[394] 142 S. Ct. 2111 (2022).

[395] *Id.* at 2122.

[396] *Id.* at 2150.

[397] *See id.* at 2150–53.

[398] *Id.* at 2151.

[399] *Id.*

[400] *Id.*

[401] *Id.*

[402] *See id.* (citing a report from the 40th Congress explaining "how a black man in Tennessee had been murdered . . . with the murderer taking, among other things, the man's pistol" (citation omitted)).

racist laws,[403] and an unsympathetic Supreme Court and Chief Justice.[404]

By stark and stunning contrast, the Court gave *no* consideration or equal weight to the lengthy and enduring invidious discrimination against Black women in the United States.[405] The sexual violence perpetrated and perpetuated against Black women and girls can be traced to their DNA.[406] This legacy stains society as much as it marred Black bodies. A study published in 2020 in the peer-reviewed *American Journal of Human Genetics*[407] explains that "the brutal treatment of enslaved people has shaped the DNA of their descendants"

[403] *See id.* (citing a report from the 39th Congress explaining how "South Carolina's Black Codes . . . prohibited firearm possession by blacks" (citation omitted)).

[404] *Id.* ("Chief Justice Taney offered what he thought was a parade of horribles that would result from recognizing that free blacks were citizens of the United States. If blacks were citizens, Taney fretted, they would be entitled to the privileges and immunities of citizens, including the right 'to keep and carry arms *wherever they went.*'" (citation omitted)).

[405] *See, e.g.*, Harriet A. Jacobs, Incidents in the Life of a Slave Girl 35 (Lydia Maria Child ed., Belknap Press 1987) (1861) "[M]y master was, to my knowledge, the father of eleven slaves. But did the mothers dare to tell who was the father of their children? Did the other slaves dare to allude to it, except in whispers among themselves? No, indeed! They knew too well the terrible consequences."). *See generally* Dorothy Roberts, Killing the Black Body (1997) (documenting the myriad ways in which Black women were subjected to sexual violence and the centrality of such harms to the founding of the United States); Davis, *supra* note 304 (1997) (providing a rigorous and thorough account about how individual liberty related to family life, personal autonomy, and reproductive freedom played a significant role in the debates leading to Reconstruction and were central to the arguments put forth by abolitionists that drafted the Reconstruction Amendments); Rachel A. Feinstein, When Rape Was Legal: The Untold History of Sexual Violence During Slavery (2019) (providing an extensive account of the widespread sexual violence inflicted on Black enslaved women by white men in the United States); Gregory D. Smithers, Slave Breeding: Sex, Violence, and Memory in African American History (2013) (noting that two different threads of history emerge related to slavery—one that Black Americans acknowledge and understand and another that is cabined and confined to the priorities of white male scholars); Melton A. McLaurin, Celia, a Slave (1991) (excavating the story of Celia, convicted of murdering her "owner" who serially raped her starting when she was a child. Celia was hanged shortly after giving birth and denied the opportunity to testify in her defense given that Missouri barred enslaved persons from testifying in court—even in cases of life and death); *see also* State v. Mann, 13 N.C. (2 Dev.) 263 (1829) (holding that held that no legal process could protect an enslaved person from physical violence imposed by her owner); *The Issue in the United States—The North and Slavery*, N.Y. Times, Jan. 18, 1860, at 2 ("The man who holds his fellow-man in slavery, treats him as a chattel, breeds from him with as little regard for marriage ties as if he were an animal, is a moral outlaw; society may find, or fancy it finds, its interest in protecting his life and his 'property' but it does so at its own peril. Before long a certain retribution overtakes it. In all ages of the world men have acknowledged rights which are older than civil society, and immutable. . . .").

[406] *See* Christine Kenneally, *Large DNA Study Traces Violent History of American Slavery*, N.Y. Times (Sept. 20, 2021), https://www.nytimes.com/2020/07/23/science/23andme-african-ancestry.html; Steven J. Micheletti et al., *Genetic Consequences of the Transatlantic Slave Trade in the Americas*, 107 Am. J. Hum. Genetics 265 (2020).

[407] *See* Micheletti et al., *supra* note 405.

such that it further confirms "one of the darkest chapters of world history, in which 12.5 million people were forcibly taken from their homelands in tens of thousands of European ships."[408] The study's authors report that DNA evidence reveals centuries of sexual violence and rape inflicted by white men on Black women.[409] Rapes were so extensive in the United States by white men that they "contributed three times more to the modern-day gene pool of people of African descent than European women did."[410] Importantly, these findings, along with a robust historical archive,[411] raise questions that are not extraneous to law, irrelevant to historical and contemporary debates, or insignificant to questions related to the original meaning embedded in the Thirteenth and Fourteenth Amendments.

As explained in prior work:

> [B]ecause the matter of sexual abuse and violence against Black women and girls was so frequent and common, legislatures sought to resolve parentage and legal status with alacrity. *Was the offspring of a white man and an enslaved Black woman free or enslaved?* Early American law answered this question with painful clarity: offspring would take the status of their enslaved mothers, which further tied capitalism to rape, embedding a horrific practice into the culture of southern economics.[412]

---

[408] *See* Kenneally, *supra* note 405 (summarizing the study published in the American Journal of Human Genetics).

[409] *See id.*

[410] *Id.*

[411] *See Born in Slavery: Slave Narratives Froom the Federal Writers' Project, 1936 to 1938*, Library of Cong., https://www.loc.gov/collections/slave-narratives-from-the-federal-writers-project-1936-to-1938/about-this-collection (last visited Feb. 6, 2023); *The African-American Mosaic: Influence of Prominent Abolitionists*, Library of Cong., https://www.loc.gov/exhibits/african/afam006.html (last visited Feb. 6, 2023); *see also* Anti-Slavery Tracts. First Series, Nos. 1-20 (David M. Hart ed., New York, Am. Anti-Slavery Soc'y 1855–1856); John Hope Franklin & Loren Schweninger, Runaway Slaves: Rebels on the Plantation 64 (1999) (detailing pregnant, enslaved Black women attempting to run away to freedom).

[412] Michele Goodwin, *A Different Type of Property: White Women and the Human Property They Kept*, 119 Mich. L. Rev. 1081, 1090 (2021). In 1662, the Virginia Grand Assembly enacted one of its first "slave laws" to settle this point, expressing:

> Whereas some doubts have arrisen whether children got by any Englishman upon a negro woman should be slave or ffree, *Be it therefore enacted and declared by this present grand assembly*, that all children borne in this country shalbe held bond or free only according to the condition of the mother, *And* that if any christian shall committ ffornication with a negro man or woman, hee or shee soe offending shall pay double the ffines imposed by the former act.

Act XII, *reprinted in* The Statues at Large; Being a Collection of all the Laws of Virginia 170 (William Waller Hening ed., New York, R., W. & G. Bartow 1823); see also *id.* at 260, 266, 270.

The Court shed no ink to illustrate Black women's advocacy and fight for their bodily autonomy dating back to the earliest slave ships docking on these shores, holding captive kidnapped and soon-to-be-trafficked Black girls and women whose forced physical and sexual labor would become central to the burgeoning economies of the colonies and what would later become the United States.[413]

Moreover, it would be a mistake to interpret the foregoing as the misdeeds of outcasts. Rather, as the Pulitzer-Prize-awarded historian Eric Foner writes, "[e]ven the most gentlemanly and prominent owners inflicted brutal, often sadistic punishments."[414] In an undated interview, Walter Calloway, purchased as a child by an Alabama enslaver, recounted this memory of an overseer who "whupped a n– gal 'bout thirteen years old so hard she nearly die."[415] Mary Reynolds, also formerly enslaved, and interviewed at a convalescent home in Dallas, Texas, explained that the terrors of slavery extend beyond anything that words can convey. Instead, she confided, "I got the scars on my old body to show to this day."[416] Ms. Reynolds witnessed cruelties more severe than what she experienced: "I seed them put . . . women in the stock with they hands screwed down through holes in the board and they feets tied together and they naked behinds to the world. Solomon the . . . overseer beat them with a big whip and massa look on. . . . They cut the flesh most to the bones and some . . . never got up again."[417]

Somehow, the Court's originalism finds no room for the unsparing accounts of Black women's suffering and their long appeal for bodily autonomy and reproductive freedom. Sadly, the Court's interpretation of the Fourteenth Amendment neuters its meaning and origins in combatting the vestiges and badges of American slavery, which included matters of family life, bodily autonomy, and reproductive

---

[413] *See generally* THE DEVIL'S LANE: SEX AND RACE IN THE EARLY SOUTH (Catherine Clinton & Michele Gillespie eds., 1997) (addressing the sexual violence inflicted on Black women on plantations); SLAVERY'S CAPITALISM: A NEW HISTORY OF AMERICAN ECONOMIC DEVELOPMENT (Sven Beckert & Seth Rockman eds., 2016); Robert WILLIAM FOGEL & STANLEY L. ENGERMAN, TIME ON THE CROSS: THE ECONOMICS OF AMERICAN NEGRO SLAVERY (1974).

[414] ERIC FONER, FOREVER FREE: THE STORY OF EMANCIPATION AND RECONSTRUCTION 14 (2006).

[415] Interview by W.P. Jordan with Walter Calloway, in Birmingham, Ala. (undated), http://xroads.virginia.edu/~hyper/wpa/callowa1.html.

[416] Interview with Mary Reynolds, in Dallas, Tex. (undated), http://xroads.virginia.edu/~hyper/wpa/reynold1.html.

[417] *Id.*

freedom and privacy.[418] As exemplified by its analysis, the Court misconstrued and misinterpreted race as having no meaning associated with or connected to the Fourteenth Amendment, despite the *undebatable* historical record that proves otherwise.

Dobbs is a case about many things. On its surface, it is a case about abortion; motivating the decision is an appeal to the Court to overturn a permanent injunction imposed by a District Court against Mississippi's abortion ban, which was upheld by the Fifth Circuit. Yet, it is also a case about the constitutional gerrymandering against abortion rights at the Court.[419] To root out *Roe*, the Court must contort the law and overcome its longstanding respect for and commitment to precedent. It must shoehorn *Dobbs* to fit the exception—a case worth overturning. In its survey of potential cases, and with biting cynicism, the Court turns to *Plessy v. Ferguson*,[420] a case rooted in contempt for Black women, men, and children, which shepherds and embeds "separate but equal," an "apparent . . . injustice,"[421] into the legal lexicon, states' laws, and court jurisprudence.[422]

In other words, when the Court places on equal footing a discriminatory race law (such as that at issue in *Plessy*), which reaches Black women, with laws that would seek to liberate them from the enduring vestiges of slavery—involuntary reproductive servitude—it shows a particularly glaring disregard for their personhood, well-being, safety, and dignity. It practices and shows complicity in race-sex harms. Indeed, *Plessy* reserves a special, overlooked indignity toward Black women, virtually uncaptured in the literature or addressed in jurisprudence. The third section of Louisiana's ban on interracial travel in its train cars "provides penalties for the refusal or neglect of the officers, directors, conductors, and employees of railway companies to comply with the act, with a proviso that 'nothing in this act shall be construed as applying to nurses attending children

---

[418] In *Bruen*, Justice Thomas concedes that the Fourteenth Amendment concerned the interests of Black men, yet even his concurring opinion in *Dobbs* does not extend that interpretation to Black women.

[419] *See e.g.*, Erwin Chemerinsky & Michele Goodwin, *Constitutional Gerrymandering Against Abortion Rights:* NIFLA v. Becerra, 94 N.Y.U. L. Rev. 61 (2019).

[420] 163 U.S. 537 (1896).

[421] *Id.* at 553 (Harlan, J., dissenting).

[422] *Id.* at 541 (majority opinion).

of the other race.'"[423] On the one hand, this provision expresses contempt against the civil liberties of Black women of means, banning them from associational dignity—sitting in well-appointed, comfortable train cars alongside white women of equal status.

On the other hand, the law presents something more insidious and pernicious. It reinscribes one of the most enduring vestiges of slavery—Black women's maternal obligations and bondage to white families as protected by state and federal laws. This virtually inescapable bondage entrenches the stereotype of Black women[424] as bound to the expressions and occupation of motherhood with little else to associate with their potential and capacities save housework and hard, field labor. The Louisiana law protects the interests of wealthy white families to tether Black women to tending to white babies and children, which often included nursing them[425]—another component of surrendering their bodies in service to structural inequalities and discrimination, which became a point of trauma for Black women.[426] While the Thirteenth Amendment rescued Black women from the

---

[423] *Id.*

[424] *See, e.g.*, Emily West & R.J. Knight, *Mothers' Milk: Slavery: Wet-Nursing, and Black and White Women in the Antebellum South*, 83 J.S. HIST. 37, 39–40 (2017) ("European colonial travelers to West Africa frequently commented on black women's breasts as large and droopy and compared them to goats' udders. Referring to these black African women's breasts as dugges—an archaic word that meant either a woman's breast or an animal's teat—also 'connoted a brute animality.'").

[425] *Id.* at 39 ("Philip Vickers Fithian, a tutor in Robert Carter's Virginia household, wrote in his 1773 diary, 'I find it is common here for people of Fortune to have their young Children suckled by the Negroes!' Utilizing enslaved women as wet nurses, in the colonial era and thereafter, undoubtedly made sound economic sense for white slaveholders. Paying for the services of a wet nurse was unnecessary when one could be procured for free from one's own chattel."). The iconography associated with the "image of black woman nursing white baby" is embedded in the nation's earliest images, including sketches, paintings, and sepia-hued photographs. A Google search conducted on January 31, 2023 using that phrase yielded 197,000,000 results in 0.47 seconds. Many of these images are now commercialized, adding to an element of *the vestiges of slavery* still untold. In modern times, Black women's breast-feeding has been associated with trauma and "distant echoes of slavery." Rose Hoban, *Distant Echoes of Slavery Affect Breast-Feeding Attitudes of Black Women*, N.C. HEALTH NEWS (Mar. 3, 2016), https://www.northcarolinahealthnews.org/2016/03/03/distant-echoes-of-slavery-affect-breastfeeding-attitudes-in-black-women ("Devane-Johnson explained that often slave women had to be away from their own children, at the cost of their own children's lives. 'A lot of slave babies died during slavery because they weren't breast-fed. They were fed concoctions of dirty water and cows [sic] milk,' she said. Meanwhile, those children's mothers were giving white children their milk." (citation omitted)).

[426] *See, e.g.*, MARCUS WOOD, BLACK MILK: IMAGINING SLAVERY IN THE VISUAL CULTURES OF BRAZIL AND AMERICA 2 (2013) ("Black milk, slave mother's milk, was stolen in vast, unknown, incalculable quantities as generation after generation of white infants 'drank, and drank.'").

formal status of property, Jim Crow and *Plessy* signaled how slavery could be legally preserved through transformation.

This stigmatism in the Court's vision, allows it to see unconstitutional, racial discrimination when limits are placed on Black male gun ownership (see *Bruen*), but the aperture yields no light when states enact abortion bans that deny Black women bodily autonomy.[427] In effect, the Court shows blindness to sex-race discrimination and its injurious effects, such as when states conscribe women's bodies in service of political goals, or when states force women and people with the capacity for childbirth to continue pregnancies that credibly and predictably place their lives at risk, especially in the cases of Black women, who are nearly four times more likely to die due to maternal mortality than their white counterparts. For all that the Court professes to see in the arc from *Plessy* to *Brown*, its sightlessness is apparent, and as such it misses in Justice Harlan's instructive dissent, that the Thirteenth and Fourteenth Amendments, "if enforced according to their true intent and meaning, will protect all the civil rights that pertain to freedom and citizenship."[428]

V. Conclusion

Dobbs represents a deep cynicism and deceit—a smokescreen to cover the Court's politically influenced decision-making. It is a brand of opportunism unconcerned with a moral principle or consequences of injurious, odious legislation. If the Court's true commitment is to originalism, then its execution is neglectful of the history it claims to point to, namely the work of the ratifiers and the history of the Reconstruction Amendments. This Article's criticism of *Dobbs* is based in part on the evidence of the cruel realities that will manifest in the wake of the Court's decision, reflecting existing, alarming rates of maternal mortality, maternal morbidity, and harsh abortion bans that now make no exceptions for rape or incest, and other harms already manifesting. Arguably, under the guise of deconstruction, the Court has reconstructed Reconstruction, stripping it of any relevance to race generally, and particularly the involuntary reproductive servitude that Black women endured, which they sought to stamp out.

---

[427] *See* Bridges, *supra* note 60.

[428] Plessy v. Ferguson, 163 U.S. 537, 555 (1896) (Harlan, J., dissenting).

No longer is it the vision that Justice Marshall projected that with *Roe*, the Court "properly embarked on a course of constitutional adjudication no less controversial than that begun by *Brown v. Board of Education*."[429] Rather, dismantling vestiges of race discrimination in public education in 1954 is used as a foil or smokescreen to undermine women's reproductive freedom in 2022, with a particularly cruel blow to Black and Brown women.

In *Dobbs*, the Court shows great deference to common law theorists such as Sir William Blackstone and Sir Matthew Hale—English jurists that most notably advanced the dangerous theory of coverture, which among other horrors justified marital rape and domestic violence, seemingly as matters of a married woman's unfortunate, but inherent destiny.[430] According to coverture theory, a woman was subsumed under her husband's identity.[431] They possessed no independent identities or personhood for themselves to resist sexual violence or defend themselves from physical chastisement. Sadly, coverture endured in American jurisprudence with marital rape finally giving way in the 1990s.[432]

Yet, as informative as the Court might perceive Blackstone, Hale, Coke, and Bracton to be as guides to contemporary matters of women's reproductive health, no one would agree that these lauded theorists played an active part in the debates involving nineteenth-century American slavery, American Reconstruction, and the ratification of the Reconstruction Amendments. Simply put, Hale and Blackstone were both dead at the time, buried on English soil, unable to speak to the ills of nineteenth-century American slavery, the threats to Black girls' and women's bodily autonomy, and the brutal discipline inflicted on those who resisted. They did not live in a time of perverse maximization of wealth on the forced pregnancies and births of enslaved Black women. Hale died over 250 years before the Framers drafted and ratified the Reconstruction Amendments and Blackstone died roughly 170 years after Hale.

---

[429] Beal v. Doe, 432 U.S. 454, 461 (1977) (Marshall, J., dissenting) (citation omitted).

[430] Hale claimed that a "husband cannot be guilty of rape" because marriage provides unconditional consent, whereby a wife "hath given up herself in this kind unto her husband, which she cannot retract." 1 Matthew Hale, The History of the Pleas of the Crown *629; *see also* 2 William Blackstone, Commentaries *442–45.

[431] Hale, *supra* note 428, at *629.

[432] Briana Bierschbach, *This Woman Fought to End Minnesota's 'Marital Rape' Exception, and Won*, NPR (May 4, 2019), https://www.npr.org/2019/05/04/719635969/this-woman-fought-to-end-minnesotas-marital-rape-exception-and-won.

In other words, the legal authorities cited by Justice Alito and the majority played no role in dismantling slavery. They were not the Framers or ratifiers of the Reconstruction Amendments. That is, the majority opinion references Blackstone over a dozen times, while not once offering mention of abolition or those who inspired that legal movement.

The Court's omissions essentially confuse the key questions and obscure clear evidence from the American constitutional record. This erasure manifests in rendering the concerns of enslaved Black women and the Framers that agitated on their behalf invisible in constitutional jurisprudence. If the Court is committed to addressing contemporary concerns through the filter of the Framers of the Thirteenth and Fourteenth Amendments, then it must begin by understanding them as abolitionists committed to freeing Black women from the conditions of slavery in all its manifestations, which were widely reported in the eighteenth and nineteenth centuries.

Nineteenth-century Americans were keenly aware that bodily autonomy was the matter at hand and sexual exploitation was the key element. Their writings, speeches, debates, and deliberations on slavery reflected this deep social and legal concern. Images and descriptions of enslaved "mulattas" were referenced across genres of writing (fiction, visual arts, theatre, legal treatises, travelogues, pamphlets), in both abolitionist publications and popular cultural production for the general public.

In fiction as well as essays, news articles, and speeches, sexual exploitation of Black women and girls featured prominently in writings by prominent authors such as Lydia Maria Child, William Weld Brown, William Ellery Channing, Frederick Law Olmstead, and Harriet Martineau. Politicians wrote riveting speeches decrying the sexual coercions and punishments inflicted on Black women and girls. In *The Barbarism of Slavery: Senate Speech on the Bill for the Admission of Kansas as a Free State*, Republican Senator Charles Sumner of Massachusetts referred to slavery as a "disgusting" enterprise, because "[b]y license of slavery, a whole race is delivered over to prostitution and concubinage, without the protection of any law."[433] Senator Sumner, a key architect of the Reconstruction Amendment—a Framer and ratifier—rebuked

[433] Charles Sumner, The Barbarism of Slavery. Speech in the Senate, on Bill for the Admission of Kansas as a Free State (June 4, 1860), *in* 6 Charles Sumner: His Complete Works, *supra* note 363, at 133.

slavery, stating, "Slavery paints itself again in its complete *abrogation of marriage*, recognized as a sacrament by the Church, and as a contract by the civil power, wherever civilization prevails. Under the Law of Slavery no such sacrament is respected, and no such contract can exist," because "[t]he ties formed between slaves are all subject to the selfish interests or more selfish lust of the master, whose license knows no check."[434]

[434] *Id.*

JEFFREY S. SUTTON

WHAT SHOULD BE NATIONAL AND WHAT SHOULD BE LOCAL IN AMERICAN JUDICIAL REVIEW

There is much to say about *Dobbs* and the various opinions in it. My jurisprudential sympathies, truth be told, run in favor of the decision. But I plan to say little about it. I would prefer to focus on the source of those sympathies rather than on the decision itself. My concern is that we are asking too much of the U.S. Supreme Court. My claim is that we should decentralize more of our debates about American constitutional law.

One question dominates every other in American history: What should be national and what should be local? Over the last 100 years or so, we have tended to favor national answers over local ones when it comes to American constitutional law. Often with good reasons: dealing with the imperatives of the Great Depression; bringing Jim Crow to heel; addressing policy challenges that have emerged from an increasingly national and global economy. Even as we recall the reasons not to forget these chapters in American history and even as we contend with chapters still unfolding, I wonder whether, halfway through our third century, we should pay more attention to the localism side of federalism and be more patient when it comes to the nationalism side

Jeffrey S. Sutton is the Chief Judge of the U.S. Court of Appeals for the Sixth Circuit, and a Lecturer on Law at Harvard Law School.

The Supreme Court Review, 2023.
© 2023 The University of Chicago. All rights reserved. Published by The University of Chicago Press. https://doi.org/10.1086/724658

of federalism. In debates about American constitutional law, there are many ways in which our fifty state constitutions and fifty state courts have critical roles to play, all true before *Dobbs*, all true after it, all with the potential to alleviate pressure on the U.S. Supreme Court as the perceived sole arbiter of American constitutional values.

One can appreciate why Herbert Wechsler in 1954 would say that "Federalism was the means and price of the formation of the Union."[1] Had I been around, I would have had the same thought, if not his way with words. But today is not 1954. And it's worth revisiting his formulation nearly seventy years later: What is today's price of federalism? Does it offer benefits as well as costs? Are there competing costs of nationalizing so many matters of constitutional law? Can we still afford those costs?

The footprint of the federal courts. Let me describe three features of federal constitutional law that in combination strike me as remarkable—and, I worry, unsustainable in combination over our next century. First of all, it is the rare court system that has embraced judicial review—the power of a court to invalidate democratically enacted laws or executive branch orders—in the way we Americans have come to embrace it for our federal courts. For decades, we have leaned heavily on the federal courts to resolve many of our most intense and intractable policy debates.[2]

Second, this tradition of muscular interpretation of the Federal Constitution concerns a charter that is exceedingly difficult to amend.[3] Article V of the U.S. Constitution requires three-quarters of the states to approve any amendment to the document.[4] When the American

---

[1] Herbert Wechsler, *The Political Safeguards of Federalism: The Role of the States in the Composition and Selection of the National Government*, 54 COLUM. L. REV. 543, 543 (1954).

[2] *See, e.g.*, JEFFREY S. SUTTON, WHO DECIDES? STATES AS LABORATORIES OF CONSTITUTIONAL EXPERIMENTATION 17 (2021) (describing "the growth of federal judicial power over the last seventy-five years"); KEITH E. WHITTINGTON, REPUGNANT LAWS: JUDICIAL REVIEW OF ACTS OF CONGRESS FROM THE FOUNDING TO THE PRESENT Ch. 7 (2019) (similar); ANTONIN SCALIA, A MATTER OF INTERPRETATION 3 (1986) (describing "our national obsession with the law"). Some might say that the American obsession with courts—if not federal courts—has lasted for centuries rather than decades. *See, e.g.*, 1 ALEXIS DE TOCQUEVILLE, DEMOCRACY IN AMERICA 280 (Phillips Bradley ed., 1945) ("Scarcely any political question arises in the United States that is not resolved, sooner or later, into a judicial question.").

[3] *See* SUTTON, *supra* note 2, at 334 ("[N]o state constitution is harder to amend or has been amended less frequently than the U.S. Constitution."); JOHN DINAN, THE AMERICAN STATE CONSTITUTIONAL TRADITION 29 ("Whereas Article V of the U.S. Constitution places substantial barriers in the way of amendment and revision, state procedures are generally more accessible . . . ."); DONALD S. LUTZ, *Toward a Theory of Constitutional Amendment*, 88 AM. POL. SCI. REV. 355, 369 tbl. C-1 (1995).

[4] U.S. CONST. art. V.

people disagree with a decision of our high court, they thus have little recourse, as they cannot realistically respond to mistaken decisions by seeking to overrule them by amendment. Has any court in history united muscular court interpretation with the near impossibility of overturning the decisions? I know of none.[5]

Third, while we may be a government of laws, the men and women who have nearly unreviewable authority over the meaning of the U.S. Constitution have life tenure.[6] Article III does not impose term limits or age limits on federal judges.

All in all, this is a remarkable aggregation of features in a court system, one without parallel within or without the country. One branch of the national government has the power to invalidate democratically enacted state and federal laws based on interpretations of a document that, practically speaking, cannot be amended, and it's a branch whose membership cannot be changed in the short term and is left to caprice in the long term. Ours may be a system that features separation-of-powers checks. But is it one that still combines checks *and* balances? Put another way, when does the check provided by judicial review create a problem of its own—imbalance of power—the key problem the Framers set out to curb in the first place?

The 2016 election illustrates the stakes. By too many accounts to deny, the election of a President of the United States turned on a sufficient number of Americans treating their vote for a president as a vote for a Justice, as a proxy to fill a single vacancy on a nine-member Court. One can illustrate the burgeoning intensity of the judicial selection process in plenty of other ways. Say that Justice Scalia was confirmed by a 98-0 vote and Justice Ginsburg by a 96-3 vote, that both would get few crossover votes today, that neither likely would be confirmed today if different political parties controlled the presidency and the Senate. Say that Senate votes for lower court federal judges are increasingly partisan, often nearly 100 percent partisan. Or say that a divided national government, with a President from one party and a Senate controlled by the other, might fill few federal vacancies at all.

A difficult truth is that the intensity of the judicial selection process is deeply rational if the American people think that federal judges control the disposition of issues they care deeply about. How

---

[5] *See* Lutz, *supra* note 3, at 367 tbl. A-1, 369 tbl. C-1; Richard Albert, *American Exceptionalism in Constitutional Amendment*, 69 ARK. L. REV. 217, 223–24 (2016); AREND LIJPHART, PATTERNS OF DEMOCRACY 229 (1999).

[6] U.S. CONST. art. III, § 1.

else can they respond? They cannot answer decisions they dislike with an amendment. And if the federal courts plant a constitutional flag in an area the people care deeply about, they cannot take a democratic stand on the issue through their elected representatives. Rational though the people's response to the growth in federal judicial power may be, it runs the long-term risk of politicizing the Court and of tarnishing a justifiably esteemed branch of American government.

In thinking about this predicament, it is useful to contrast the just-described trio of features and practices under the National Constitution with their absence in the state courts. It is the rare state court, I submit the non-existent state court, that has exercised judicial review with the same alacrity as the federal courts over such a lengthy period of time. The people in the states indeed might not let them. All state constitutions are easier to amend than the Federal Constitution. Usually much easier, as forty-six states permit amendments by a 51 percent threshold.[7] New Hampshire has the highest threshold, a two-thirds vote requirement,[8] followed by Florida at 60 percent, and Colorado at 55 percent.[9] Delaware does not permit a popular vote but requires the legislature in two separate sessions to approve the amendment by a two-thirds vote. Predictable consequences flow from this juxtaposition between the federal and state amendment requirements. "[E]very state constitution," John Dinan confirms, "is amended more frequently than the U.S. constitution."[10] On top of that, 90 percent of state court judges must face electoral accountability. Every state but Rhode Island has age or term limits.[11] How, it's fair to wonder, could the same people choose such disparate judicial systems for the same country?

The contrast does not end at our shores. No western democracies have constitutions that require a three-quarters vote or more for amendment, and none of them permits their judges to serve for life.[12]

---

[7] 53 Council of State Gov'ts, Book of the States 8 (2021).

[8] *Id.*

[9] *Id.*

[10] John Dinan, State Constitutional Politics: Governing by Amendment in the American States 12 (2018). In particular, as Dinan has noted, no state has anything close to a 75 percent supermajority requirement in any part of its constitutional amendment process. *Id.* at 12–19; *see* Council of State Gov'ts, *supra* note 7, at 8–11 tbl. 1.3 & 1.6.

[11] R.I. Const. art. X, § 4; *see* Sutton, *supra* note 2, at 375–78.

[12] Steven G. Calabresi & James Lindgren, *Term Limits for the Supreme Court: Life Tenure Reconsidered*, 29 Harv. J. L. Pub. Pol'y 769, 819 (2006); Jeffrey Fisher, *The Supreme Court*

The federal courts for some time, it's fair to say, have exercised authority in a way that is unique in American and world history.

Two of these three traits of the federal judicial system, one must acknowledge, cannot be changed without constitutional amendments—lowering the threshold for a constitutional amendment or imposing age or term limits—something the people in the states have not had trouble approving but something the people at the federal level would have a hard time changing given the three-quarters requirement. Both possibilities may deserve consideration at some point.[13] For now, any realistic reform in the near term must come from changes to how we think about federal constitutional law and how we think about the role of the U.S. Supreme Court—the number of areas of law it constitutionalizes under the federal charter and the pace with which it does so. My submission is that not every constitutional right needs to be nationalized and that, when the Court decides to nationalize some rights, it may wish to moderate the pace of change and consider input from the states before doing so.

How can the state courts and state constitutions help? Let me start, and for today's purposes largely end, with one of the most tricky, the most power-enhancing, areas of federal constitutional law: substantive due process.

Substantive due process. Our trio of federal constitutional features and practices—frequent judicial review of a document that the people cannot amend by a Court that the people cannot change—creates more difficulties in some areas than it does in others. For specific constitutional guarantees, say the requirement that the President be thirty-five, the absence of room for interpretation leaves little room for material risks. But the more general a constitutional guarantee, the more room to contest its application. That reality generates vexing constitutional debates about the application of generally worded

---

Reform that Could Actually Win Bipartisan Support, POLITICO (July 21, 2022), https://www.politico.com/news/magazine/2022/07/21/supreme-court-reform-term-limits-00046883 ("[A]most every state in the union imposes term limits on its state supreme court justices, a mandatory retirement age, or both.... The United States ... is the *only* major constitutional democracy in the world to impose neither term nor age limits.").

[13] Less promising is Court expansion. In today's climate of one-upmanship, it's hard to see an end point, whether with 13, 19, 29, or still more justices. *See* Adam Chilton, Daniel Epps, Kyle Rozema & Maya Sen, *The Endgame of Court-Packing* (May 3, 2021) (unpublished manuscript), https://ssrn.com/abstract=3835502 (modeling the end game of court packing, which gets to thirty-nine justices in 100 years and presumably continues from there). Shame indeed might not kick in until we get to 100 justices, say two from each State. At some point, the institution will no longer be a court. And to what end: to create a second Senate?

provisions to unforeseen settings: When is a search "unreasonable"?[14] What is "the freedom of speech"?[15] What process is "due"?[16] When is punishment "cruel and unusual"?[17] As to these difficult questions, as I can attest, the federal model of judicial review has generated plenty of challenges for a democratic republic.

For now, however, let me leave those challenges to the side. A more difficult near-term problem faces us: rights pulled out of thin constitutional air. Here the accountability challenges of federal judicial review become excruciating when disputes arise over potential rights premised on "substantive due process" but unanchored in any enumerated right. This area of the law potentially leaves judges at sea with little, nothing really but personal intuition and instinct, to go on.

Let me illustrate the point by comparing this feature of American constitutional law to a feature of American football. At the beginning of every football game, the referee tosses a coin in the air to decide who gets the ball first. We use coin flips to make the decision because they are neutral and because the arbitrariness of them does little damage. If your team loses the coin toss, you merely have lost control over whether you get the ball first or after halftime, whether you face the wind now or later. We tolerate the whim of a coin flip because it is neutral and the price of losing is nearly non-existent.

Imagine a federal judge, however, who opted to simplify his decision-making process by resorting to coin flips to decide constitutional cases. Heads always means the individual wins. Tails goes to the government. This approach would have the virtue of being neutral and, I suppose, efficient. But any judge who did it would face well-earned criticism in no time. No matter the exact price of defeat—for the individual in losing liberty, property, or even life, or for the people in losing democratic control over the matter—the American people would not permit this approach to last long. If there is one commitment our country has consistently and proudly made, it has been to prohibit arbitrary exercises of power when it comes to the parties to any legal dispute.[18]

---

[14] *See* U.S. CONST. amend. IV.

[15] *See id.* amend. I.

[16] *See id.* amends. V, XIV.

[17] *See id.* amend. VIII.

[18] *See* Zeon Chems., L.P. v. United Food & Comm. Workers, Loc. 72D, 949 F.3d 980, 985–86 (6th Cir. 2020) (Sutton, J.).

Return to our substantive due process decisions. Most of those decisions, happily, were not pulled out of thin air. They used the Fourteenth Amendment to incorporate rights spelled out in the Bill of Rights, such as free speech (the First Amendment), the guarantee against unreasonable searches and seizures (the Fourth Amendment), and bans on cruel and unusual punishment (the Eighth Amendment).[19] It's still a trick to say that the Fourteenth Amendment's Due Process Clause incorporated the first eight Amendments when one of those Amendments, the Fifth Amendment, already has an identically worded Due Process Clause. But at least each segment of the incorporation journey was anchored in guarantees written in the Bill of Rights.

Not so for substantive due process decisions that find/identify/enumerate new rights in an unamendable Constitution. They turn on implications and inferences from silence—silence perhaps amplified by prior Court decisions but decisions that started with silence all the same. If the end of the "telephone game" is hard to predict with an initial sentence, imagine how difficult it would be to predict if it had no beginning.

How do substantive due process decisions that enumerate new rights in the Constitution compare to the coin-flipping judge? Not favorably, I fear. How often in American history have the outcomes of such decisions, all without any written guidance in the Constitution, been consistent with the world views of the judges who issued them? Certainly often, maybe usually. Who can say? But it is fair, maybe even worrisome, to wonder. The more one fears that substantive due process cases have a tendency to constitutionalize the world views of federal judges, the more one should be worried about comparisons to coin flips. Substantive due process decisions that overrule democratic preferences based on nothing more than the policy preferences of a judge are not only unfair. They are less fair than coin flips, as a government by the individual preferences of men and women, not law, is the epitome of arbitrariness. At least coin flips come with a guarantee of neutrality.[20]

---

[19] At this point, by the way, nearly all of them have been incorporated, save for the grand jury, quartering of soldiers, and civil jury rights. *See, e.g.*, F. Andrew Hessick & Elizabeth Fisher, *Structural Rights and Incorporation*, 71 ALA. L. REV. 163, 168 (2019); Timbs v. Indiana, 139 S. Ct. 682, 687 (2019).

[20] *See Zeon*, 949 F.3d at 985–86.

History, tradition, and state constitutions. What of the possibility that the substantive due process test comes with limits—the need to show that the right is "deeply rooted in this Nation's history and tradition" and "implicit in the concept of ordered liberty"?[21] Whatever work that test supplied at the outset, the Court has not consistently stood by it. Compare a few cases to see.

In 1949, in *Wolf v. Colorado*, the U.S. Supreme Court incorporated the Fourth Amendment and made it applicable to the states through the Due Process Clause of the Fourteenth Amendment.[22] The case arose from a seizure of a doctor's records that were later used in a prosecution against the doctor for violating, of all things, a Colorado law restricting abortion.[23] In extending this protection to the doctor, the Court did not start from scratch. Every state constitution at that time had a search and seizure clause or an equivalent guarantee. What did that signal? Not just that each state thought the guarantee was indispensable to a free society; legislation could have conveyed the same sentiment. The key message was that each state thought it essential to remove this aspect of law enforcement from the legislative process, to take it off the plate when it comes to the election of law enforcement leaders, and to prevent individual police officers from taking it upon themselves to create self-identified dispensations from the search-and-seizure guarantee.

Not one of the Justices in *Wolf* seemed troubled by whether to incorporate the Fourth Amendment given its uniform treatment as a constitutional guarantee throughout the states. To nationalize what every state had embraced on its own seemed unobjectionable, and hardly disrespectful to state governments, as it turned on a deeply and widely rooted tradition reflected in *all of the state charters*.[24] And that was so even though this incorporation decision would generate

---

[21] Washington v. Glucksberg, 521 U.S. 702, 721 (1997).

[22] 338 U.S. 25, 28 (1949).

[23] *See* Wolf v. Colorado, 187 P.2d 926, 926–28 (Colo. 1947).

[24] *See id.* at 25 app'x; 2 JENNIFER FRIESEN, STATE CONSTITUTIONAL LAW: LITIGATING INDIVIDUAL RIGHTS, CLAIMS, & DEFENSES § 11.02[1] & n.12 ("Every state constitution contains [a search-and-seizure or related] provision."). Although it might seem otherwise on a first glance, the Arizona and Washington constitutions do not count as exceptions. Both of them say: "No person shall be disturbed in his private affairs, or his home invaded, without authority of law." ARIZ. CONST. art. II, § 8; WASH. CONST. art. I, § 7. The high courts of both States have interpreted the guarantees to protect against unreasonable searches. *See* Arizona v. Peoples, 378 P.3d 421, 424–25 (Ariz. 2016) (construing the provision to protect against "unlawful searches and seizures"); Washington v. Bowman, 498 P.3d 478, 483–85, 489 n.6

many second-order debates about which searches were "unreasonable" and which were not.[25] What divided the Court was the question whether the Court should nationalize the exclusionary rule, the requirement that evidence unreasonably seized or obtained after an unreasonable search should be suppressed in a subsequent criminal trial—a remedy nowhere mentioned in any of America's search and seizure clauses and a remedy that stands in contrast to the remedy for violation provided by the equally prominent guarantee against compelled testimony in the adjacent Fifth Amendment.[26] Even so, the exclusionary rule at that point applied to federal investigations and many state investigations under those States' constitutional guarantees. In deciding not to take the step of extending the remedy of exclusion to all jurisdictions in the country, Justice Frankfurter pointed to the many States that had not adopted such a rule, thirty-one in total, as well as the many democracies that had not done so.[27] Better, he and most of his colleagues thought, to wait for more input from the state courts and more opportunities to establish widespread evidence of this tradition. Justices Murphy and Rutledge by contrast would have embraced the exclusionary rule then and there.[28]

In 1961, twelve years later, the Court nationalized the exclusionary rule in *Mapp v. Ohio*.[29] In adding this protection to the national guarantee, the *Mapp* Court found it sufficient that *half of the States* by that point had adopted an exclusionary rule by judicial interpretation or legislation.[30] In a dozen years, then, the Court went from nationalizing a right that every State had adopted in some form (a guarantee against unreasonable searches and seizures) to nationalizing a right

---

(Wash. 2021) (explaining that the state guarantee protects no less than does the Federal Constitution's Fourth Amendment).

[25] The American people, it's fair to acknowledge, could agree that arbitrary searches and seizures should be barred without agreeing about what makes a search and seizure arbitrary. On the facts of *Wolf* itself, it is not obvious that every state court would have agreed with the U.S. Supreme Court about the application of the search-and-seizure guarantee to that fact pattern. But that shows only that nationalization decisions by the U.S. Supreme Court have many collateral interpretive consequences for federal judges—difficult enough when illuminating the meaning of illegal "searches" and "seizures," excruciating when divining the meaning of unenumerated rights.

[26] U.S. Const. art. V.

[27] *Wolf*, 338 U.S. at 28–30.

[28] *Id.* at 46 (Murphy, J., dissenting).

[29] *See* 367 U.S. 643, 655 (1961).

[30] *See id.* at 651.

that half of the States had adopted (a remedy of excluding any evidence found during an illegal search).[31]

This innovation was not cost free. Few people debate the *Wolf* decision today. Many people debate the *Mapp* decision today. Nor was continued debate about the legitimacy of a decision the only risk created by premature nationalization. Later U.S. Supreme Court decisions cut back deeply on *Mapp*'s breadth (take *Leon*'s good-faith exception to the exclusionary rule) and cut back on the underlying scope of the guarantee against unreasonable searches and seizures (take *Terry*'s stop and frisk exception).[32]

Move forward another twelve years to 1973, the year the Court decided *Roe v. Wade*. Consider the body of state constitutional evidence for incorporating an unenumerated right in *Roe*—and for nationalizing that right. By that point, *no state* had mentioned abortion by name in its constitution when it came to protecting reproductive rights. Not one state in other words had decided that reproductive rights were so fundamental to human liberty that they should be removed from its democratic process. The same was true in 1992 at the time of the *Casey* decision. The same indeed remained true when the Court decided *Dobbs* in 2022.[33] While eleven state constitutions

---

[31] *Mapp*, for what it is worth, counted aggressively. Just one year before *Mapp*, the Court observed that many states had applied the exclusionary rule only in very narrow circumstances—Maryland, for example, applied it in "the trial of most misdemeanors," and Alabama applied it "in the trial of certain alcohol control cases." Elkins v. United States, 364 U.S. 206, 226–28 (1960). Yet *Mapp* treated these states as having embraced the exclusionary rule in full. 367 U.S. at 651 (citing *Elkins*).

[32] *See* United States v. Leon, 468 U.S. 897, 922–24 (1984); Terry v. Ohio, 392 U.S. 1, 30 (1968); Linkletter v. Walker, 381 U.S. 618, 639 (1965) (refusing to apply *Mapp* retroactively); *see also* JEFFREY S. SUTTON, 51 IMPERFECT SOLUTIONS 42–83 (2018) (discussing the exclusionary rule at the state and federal levels). An even more powerful backlash occurred in response to *Furman v. Georgia*, 408 U.S. 238 (1972), when the Court construed the Eighth Amendment to require a halt to all executions in the country. Not only did the Court ultimately alter course in *Gregg v. Georgia*, 408 U.S. 238 (1976), but the number of executions increased dramatically in the years after these decisions. Between 1968 and 1972, no one was executed in the United States. After the *Furman*/*Gregg* decisions, in a type of parabola, the number of executions increased to nearly one hundred per year (in 1999), before leveling off and returning to about twenty executions per year. Tracy L. Snell, *Capital Punishment, 2020—Statistical Tables*, Bureau of Justice Statistics Statistical Tables (Dec. 2021), https://bjs.ojp.gov/content/pub/pdf/cp20st.pdf. Some take from this experience the lesson that ground-up development of social change in the States can be more lasting and effective. *See, e.g.*, MAURICE CHAMMAH, LET THE LORD SORT THEM: THE RISE AND FALL OF THE DEATH PENALTY (2021); Carol S. Steikert & Jordan M. Steikert, *Little* Furmans *Everywhere: State Court Intervention and the Decline of the American Death Penalty*, 107 CORNELL L. REV. 1021 (2022); Anand Giridhardas, *Why the Death Penalty is Dying: A New Book Tells the Surprising Story*, N.Y. TIMES (Jan. 26, 2021).

[33] *See, e.g.*, Becky Sullivan, *With* Roe *Overturned, State Constitutions Are Now at the Center of the Abortion Fight*, NPR (June 29, 2022), https://www.npr.org/2022/06/29/1108251712/roe-v-wade-abortion-ruling-state-constitutions.

by then had included a right to "privacy"[34] and some state courts had construed the guarantee to protect a right to abortion,[35] no state mentioned a constitutional right to an abortion. Since *Dobbs* and in direct response to it, three States—California, Michigan, and Vermont—have proposed and passed constitutional amendments that directly protect reproductive rights.[36] One could imagine other states taking, or at least trying to take, a similar path before long.

In just twenty-four years, the Court revealed three distinct faces of substantive due process. One incorporated a specific guarantee in the Fourth Amendment that the states had already uniformly constitutionalized. Another took a judicial gloss added to the national guarantee and nationalized it based on similar judicial (and roughly five legislative) decisions in half of the states. The third nationalized a right found nowhere in a single state or federal constitution. Whatever one thinks about this development or of *Roe* itself, the progression exemplifies a striking increase in judicial power—not just because there were few objective sources for the decision but also because the people had little recourse to change the decision. In reality, they had just one form of recourse—changes to the composition of the Court—a point not lost on the American people for the next fifty years.

Roe of course was not the first Supreme Court decision to nationalize a right found nowhere in the U.S. Constitution and it may not have been the first to nationalize a right found in none of our state constitutions.[37] It is fair to characterize some other cases that way or at

---

[34] *See* ALASKA CONST. art. I, § 22; ARIZ. CONST. art. II, § 8; CAL. CONST. art. I, § 1; FLA. CONST. art. I, § 23; HAW. CONST. art. I, § 6; ILL. CONST. art. I, § 6; LA. CONST. art. I, § 5; MONT. CONST. art. II, § 10; N.H. CONST. art. 2-b; S.C. CONST. art. I, § 10; WASH. CONST. art. I, § 7.

[35] *See, e.g.*, *In re* T.W., 551 So.2d 1186, 1192 (Fla. 1989) (concluding that the Florida Constitution's privacy provision "is clearly implicated in a woman's decision whether or not to continue her pregnancy"); *see also* Planned Parenthood v. Reynolds *ex rel.* Iowa, 915 N.W.2d 206, 237 (Iowa 2018) (finding a right to abortion in the due process clause of Iowa's Constitution); Hodes & Nauser, MDs, P.A. v. Schmidt, 440 P.3d 461, 645 (Kan. 2019) (finding a right to abortion in the Kansas Constitution's guarantee of natural rights and liberty).

[36] *See* CAL. CONST. art. I, § 1.1; MICH. CONST. art. I, § 28; VT. CONST. art. I, § 22.

[37] Before *Roe*, no state recognized a right or entitlement to abortion. Only four states—Alaska, Hawaii, New York, and Washington—permitted abortion without justification. Approximately fourteen other state legislatures allowed for abortions that protected life or health, and all were construed to protect the life of the mother. *See* Julie Conger, *Abortion: The Five-Year Revolution and Its Impact*, 3 ECOLOGY L.Q. 311, 314–15 (1973).

least that way in part.[38] But this reality only confirms another problem with substantive due process. In the abstract, there might not seem to be anything deeply objectionable about looking the other way when the Court exercises such ungrounded and unreviewable power to extricate us from a seemingly intractable political thicket—so long as it does so with considerable restraint and quite infrequently. Think in other words of unrooted substantive due process decisions as a judicial wand to wave over this or that urgent political problem in the country. But restraint rarely accompanies such power. Who among us would have the self-control to use such power just once a generation or even once every ten years? Such people may well exist. But they tend not to pursue jobs in government, warns Madison.[39] The era from 1949 to 1973 for better and sometimes worse illustrates precisely what can happen if the Court has atextual, unrooted, and unreviewable authority to identify new federal constitutional rights.[40]

One last observation on this score. In the *Ramos* decision in 2020, the Court seemed to return to the original approach reflected in *Wolf*. The Court determined that the Sixth Amendment's unanimity requirement for criminal jury trials applied to the states via incorporation under the Fourteenth Amendment.[41] By the time of the decision, all of the states, save for one outlier, Oregon, had agreed to require unanimity in criminal jury convictions under their state constitutions.[42] From the near consensus in 2020, the Court traced

---

[38] *See, e.g.*, Pierce v. Soc'y of Sisters, 268 U.S. 510, 534–35 (1925) (recognizing a "liberty of parents and guardians to direct the upbringing and education" of their children); Skinner v. Oklahoma, 316 U.S. 535, 541–43 (1942) (recognizing a right of procreation); Griswold v. Connecticut, 281 U.S. 479, 485–86 (1965) (recognizing a right to contraceptives stemming from the confluence of "several fundamental constitutional guarantees"); Moore v. City of E. Cleveland, 431 U.S. 494, 502 (1977) (recognizing a right to define the boundaries of the nuclear family).

[39] *See* THE FEDERALIST No. 51, at 322 (James Madison) (Clinton Rossiter ed., 1961) (recognizing that angels do not sit in our government offices).

[40] If the Privileges and Immunities Clause, as originally understood, was designed to permit federal judges to identify some set of unenumerated rights, similar problems could arise and a similar solution—invocation of state laws and constitutions—would apply. *Compare* KURT T. LASH, THE FOURTEENTH AMENDMENT AND THE PRIVILEGES OF AMERICAN CITIZENSHIP (2014) (arguing that, as a matter of original meaning, the Privileges and Immunities Clause protects only *enumerated* constitutional rights), *with* RANDY E. BARNETT & EVAN D. BERNICK, THE ORIGINAL MEANING OF THE FOURTEENTH AMENDMENT: ITS LETTER AND SPIRIT (2021) (arguing otherwise).

[41] *See* Ramos v. Louisiana, 140 S. Ct. 1390, 1397 (2020).

[42] *See id.*

the unanimity requirement to the first state constitutions, to Blackstone's writings, and ultimately to fourteenth-century England.[43] With such past and present consensus as the standard, many of the most difficult-to-justify features of substantive due process begin to fade.

Much as I remain a skeptic of substantive due process of any kind, *Wolf* and *Ramos* suggest a middle ground, a compromise of sorts. That approach would permit the nationalization of rights, whether in the Bill of Rights or not, so long as a significant majority of the states have recognized them, ideally in state constitutions, sometimes through state legislation as well.

Significant? Where does that come from? How would anyone define it? These questions take me to another feature of *Dobbs*, its stare decisis analysis.

Stare decisis. Anyone worried about the absence of neutral principles underlying some substantive due process decisions should be doubly worried about identifying a principled way to apply stare decisis to them. Here we have two inquiries, not one, grasping for objective limitations on judicial innovation.

Justice Thomas, to his credit, has done his best to identify a neutral principle for resolving which constitutional mistakes to overrule and which ones to leave in place. He takes the position that, if the underlying decision is "demonstrably erroneous" as a matter of original public meaning, it must be overruled, no matter the jolt to the legal system it might cause, no matter the extent to which the people had come to rely on the underlying decision.[44] So far, no other Justice, including Justice Scalia, has adopted this highly principled, yet highly consequential, approach to stare decisis. The price of this approach apparently remains too steep.

Is there any other place to look? Here, again, state constitutions in particular and states in general have something to offer. Recall that stare decisis debates occur only because a majority of the Court thinks a prior decision is wrong. Absent that reality, there is nothing to talk about. Judges do not debate whether to preserve precedents

---

[43] *See id.* at 1396.

[44] *E.g.*, Gamble v. United States, 139 S. Ct. 1960, 1981 (2019) (Thomas, J., concurring) (contending that the Court should overrule a constitutional decision when it is "demonstrably erroneous"). *But see* William Baude, *Precedent and Discretion*, 2019 Sup. Ct. Rev. 313, 319–24 (suggesting that even Justice Thomas's view of stare decisis may leave judges too much discretion to tolerate some erroneous precedents while discarding others).

they think are still right. But once judges determine a precedent is wrong, they need handholds for deciding whether to preserve it based on something other than their own world view about the underlying issue. How intensely an individual judge dislikes an individual decision is not an ideal stare decisis test.

Where to look? Recall what it means to misconstrue the Federal Constitution. It means that the Court has *added* a guarantee to the Constitution that is not there or *subtracted* one that is there.[45] That's problematic because the point of Article V is to set forth a process for adding and subtracting constitutional guarantees.[46] Confirmation of a Justice to the Court does not come, or at least should not come, with a Framer's pencil and eraser. The Framers would not have created an amendment process if they thought the Court could amend it by interpretation or update it by interpretation. A mistaken constitutional ruling thus represents a trespass on Article V—and an un-remediable trespass at that given the difficulty of invoking Article V to check the Court. This dynamic suggests an added reference point for orienting a decision whether to overrule a wayward precedent or leave it on the books.

If amending the Constitution by interpretation circumvents Article V, it's worth asking whether the trespass still deserves correction. The point is not that a mistaken Court decision from, say, 1961 at some point becomes sacrosanct. That is too much to ask, particularly if the 1961 decision disrespected the original 1791 or 1868 meaning of the guarantee. No one credibly thinks we should have fixed meaning of precedents but not fixed meaning of the underlying guarantees. The 1961 decision still counts as an error, and the Court should say as much. The question is whether the error deserves correction. The concept of harmless error is not new. And it's hardly a contrivance. Judges use it all the time to acknowledge mistakes in

---

[45] Judicial errors, it is true, come in all sizes and shapes. A federal court might expand or shrink a guarantee without necessarily adding or subtracting one. Perhaps that was what Madison meant when he said this: "There has been a fallacy in this case, as, indeed, in others, in confounding a question whether precedents could expound a Constitution, with a question whether they could alter a Constitution. . . . None will deny that precedents of a certain description, fix the interpretation of a law. Yet who will pretend that they can repeal or alter a law?" Letter from James Madison to N.P. Trist (Dec. 1831), *reprinted in* 4 Letters and Other Writings of James Madison 204, 211 (1865). Given my focus on substantive due process—a rights-addition setting by my lights and, I might suggest, Madison's—I have kept the focus there. Either way, many of my criticisms and suggestions apply to both settings.

[46] U.S. Const. art. V.

the past in a case, including constitutional mistakes, that do not deserve correction today because the case would have come out the same way.[47]

Why not here? If the people have moved on, as evidenced by the laws and constitutional amendments and other related political activity in their states, it's not obvious that correcting the decision is necessary to correct the Article V breach. Why require a supermajority of the states to approve an amendment if they have already signaled their support for, or at least acquiescence in, the ruling?

Say what you will about *Roe* or *Casey*, it's difficult to maintain that the people had moved on by 2022. Over the last fifty years, no set of decisions has been more central to the politicization of the selection process for federal judges, the one realistic avenue for input the U.S. Constitution gives the people. Whatever metric one wants to use, it's hard to say that the public had come to accept *Roe*'s or *Casey*'s approach to the problem. *Casey*, in retrospect, looks like a second effort to generate public acceptance of judicial dominion over the issue. In backing off *Roe*'s rigid approach to the issue, the Court adopted a flexible "undue burden" approach to the matter. But thirty years after *Casey*, a lack of consensus remained, especially if one focuses on the kind of super-majority consensus needed to satisfy Article V. That absence of consensus continued to affect elections for public office. Think back to the 2016 election. Might the election have come out differently without *Roe* and *Casey*? The question is worth pondering even if there is no conclusive way to answer it. The fifty or so years between *Roe* and *Dobbs* in the end came with two edges. The length of time understandably gave adherents of the decision greater reliance interests. But it also gave opponents of the decision a sound footing for maintaining that the issue had not been, and could not be, finally or for that matter speedily settled by the Court.

Consider by contrast the consensus that emerged over the underlying policy issues in three other cases that the Supreme Court

---

[47] *See, e.g.*, Chapman v. California, 386 U.S. 18, 22–23 (1967) (applying harmless-error rule to violations of Fifth Amendment right against self-incrimination); Harrington v. California, 395 U.S. 250, 254 (1969) (similar); Delaware v. Van Ardsall, 475 U.S. 673, 674 (1986) (applying harmless-error rule to violations of the Sixth Amendment's Confrontation Clause); Rushen v. Spain, 464 U.S. 114, 118 (1983) (per curiam) (right against trials *in absentia*); Chambers v. Maroney, 399 U.S. 42, 52–53 (1970) (Fourth Amendment); *see also* Rose v. Clark, 478 U.S. 570, 576–77 (1986) (listing these examples and more). At one point, the Supreme Court even said that "most constitutional violations" are subject to harmless error review. United States v. Hasting, 461 U.S. 499, 509 (1983).

removed from the democratic process and that arose in other high-profile substantive due process settings: *Griswold* (contraceptives), *Lawrence* (consensual sexual conduct), and *Obergefell* (same-sex marriage). Even if one thinks these decisions were not healthy for the federal judiciary and even if one thinks they were wrong when decided, that does not show that they remain harmful errors from the perspective of Article V.[48] It's doubtful that overruling any of these decisions would lead to the need for a constitutional amendment (because a state at that point would need to re-ban the conduct) or, even if necessary, that it would take long for a state or federal constitutional amendment to be ratified.

There's another reason this inquiry is worth including in the mix of stare decisis considerations. What is it about overruling constitutional decisions, of not standing by them, that bothers us? A central fear is that the meaning of the Constitution turns on the people construing it, not the ratified charter itself. Invocation of state constitutions, laws, and practices as a way of measuring public acceptance separates the subjective opinions of a judge from the objective truth of what is going on in the country. It helps to show either that the decision should be left alone, even if wrong, because the people have accepted it, or that the people have not accepted it and the offense to Article V remains real and deserves correction.[49]

This inquiry casts a reassuring light on *Dobbs*. Let us say for the sake of argument that a majority of the current Court would have rejected the claims in *Griswold*, *Roe*, *Lawrence*, and *Obergefell* in the first instance. And let us say for the sake of argument that a majority of the current Court thinks those decisions are wrong today. The reality that the *Dobbs* majority showed no interest in taking on these other decisions, and Justice Kavanaugh in concurrence accepted two of them by name and the third by implication, offers a useful piece of proof that the composition of the Court does not dictate the meaning of constitutional law and which precedents stay and which do not. One possibility that differentiates the four assumed errors is that

---

[48] What may be a harmless error from "the perspective of Article V," I appreciate, may not be cost free for the individual litigant in the underlying stare decisis case or for that matter the people who enacted the challenged law. But that will usually be true in a constitutional dispute.

[49] *See* Antonin Scalia & Bryan A. Garner, Reading Law: The Interpretation of Legal Texts 413 (2012) (describing one of the stare decisis factors as "whether the decision has been generally accepted by society"); Mitchell v. United States, 526 U.S. 314, 331–32 (1999) (Scalia, J., dissenting) (treating "wide acceptance in the legal culture" as a reason not to overrule a decision).

three are harmless under Article V in America circa 2023 and one is not.

I appreciate that not every one of these inquiries will lend itself to objective benchmarks in state constitutions, state laws, or state court decisions. There's a difference in particular between mistakenly subtracted rights and mistakenly added rights. If a Court decision errs in *subtracting* matters from federal constitutional protection, that leaves more room for the states to fill in the gaps than Court decisions that err in *adding* constitutional protection. The former leaves plenty of room for state activity because it allows the local democratic and judicial process to flourish and to innovate in the area. The latter leaves less room for state input and explicitly bans it in some settings. Even in this second setting, however, the state courts remain free to express disagreement with federal constitutional decisions that mistakenly purport to occupy a field in discussing the meaning of the counterpart guarantees in their own constitutions.[50] Some have done just that.[51] As shown by the distinct post-*Roe* and post-*Griswold* stories, moreover, objective ways remain to determine whether a federal decision that mistakenly adds constitutional protection has gained acceptance and when it has not.

For those still uncomfortable with the judicial discretion that remains in assessing whether an Article V violation has become harmless,[52] I submit that the inquiry nonetheless at least deserves

[50] *See* Idaho v. Guzman, 842 P.2d 660, 671 (Idaho 1992) (rejecting *Leon*'s good faith exception); DeRolph v. Ohio, 677 N.E.2d 733, 737 (Ohio 1997) (recognizing a right to equal public school funding under the Ohio Constitution despite the Supreme Court's refusal to do so under the U.S. Constitution); Tetra Tech EC, Inc. v. Wis. Dep't of Revenue, 914 N.W.2d 21, 63 (Wis. 2018) (rejecting *Chevron* deference under the Wisconsin Constitution); Harper v. Hall, 868 S.E.2d 499, 535 (N.C. 2022) (concluding that political gerrymandering claims are justiciable under the North Carolina Constitution).

[51] *See, e.g.*, Tex. Dep't of State Health Servs. v. Crown Distrib. LLC, 647 S.W.3d 648, 676–78 (Tex. 2022) (Young, J., concurring) (suggesting that the U.S. Supreme Court's reading of the Fourteenth Amendment's Due Process Clause to create substantive due process rights might depart from the meaning of an analogous Texas constitutional limitation); State v. Rowan, 416 P.3d 566, 575 (Utah 2017) (Lee, J., concurring) (suggesting similarly for Utah's Fourth Amendment analogue); MKB Mgmt. Corp. v. Burdick, 855 N.W.2d 31, 41–46 (N.D. 2014) (opinion of Vande Walle, C.J.) (rejecting state constitutional right to an abortion); Mahaffey v. Att'y Gen. of Mich., 564 N.W.2d 104, 109–10 (Mich. Ct. App. 1997) (similar).

[52] One could imagine at least two other concerns. First, the Court's efforts to apply "evolving standards" of decency" have generated considerable debate about how to measure a consensus in the States, much less a super-consensus. *See, e.g.*, Miller v. Alabama, 567 U.S. 460, 481–87 (2012); Roper v. Simmons, 543 U.S. 551, 564–67 (2005). Second, to the extent "tradition" is permitted to evolve after the date a constitutional guarantee is ratified, say 1868, that makes it difficult to see when (if ever) tradition should stop evolving. *See* Sherif Gergis, *Living Traditionalism*, 98 N.Y.U. L. Rev. (forthcoming 2023).

consideration as part of the mix. The inquiry improves on—it does not make worse—the ineffable inquiries already used to determine which substantive due process decisions stand and which do not under the traditional stare decisis inquiry. And the inquiry has the virtue of orienting us around the serious structural problem created by a mistaken constitutional ruling in the first place.

Trial and error, pacing, and variety. Time will tell. But history may show that the two most consequential decisions of the Roberts Court were (1) *Dobbs* and (2) its 2019 *Rucho* decision, which determined that efforts to end extreme partisan gerrymandering present a "political question" that is not amenable to judicial resolution under the U.S. Constitution. Both decisions have some things in common: They concern deeply salient and complex policy matters; they do not end the debate; they remain neutral about the people's answers to these challenges; and they give citizens many local avenues for change, whether by seeking state legislation, state court relief, or state constitutional amendments.

In the short time since the decisions, the intensity of state activity in all directions has been striking. The state responses confirm the array of American views and approaches to the underlying issues and, so far, the elusiveness of a winning insight suitable for export to the whole country. Only after the Court left the stage did the auditioning of these approaches occur in full.[53] It's worth remembering what Justice Brandeis said about federalism *before* mentioning the role of States as "laborator[ies]" of policymaking: "To stay experimentation in things social and economic is a grave responsibility. Denial of the right to experiment may be fraught with serious consequences to the nation."[54] What might those consequences be? Denial of the right to try all kinds of approaches to the issues. Denial of the right to use the gravitational forces of democracy to generate

---

[53] With freedom to experiment comes the chance for productive talk about what it means to be pro-life or pro-choice in full—and where common ground might exist. Consider, for example, two scholars who urge pro-life States to carry their convictions on the sanctity of the unborn child forward to children who have left the womb. *See* Leah A. Plunkett & Michael S. Lewis, *The Wages of Crying Life: What States Must Do to Protect Children After the Fall of* Roe, 2022 PEPP. L. REV. 14, 47 (arguing pro-life States that do not take child abuse and neglect seriously are not true to their ideals).

[54] New State Ice Co. v. Liebmann, 285 U.S. 262, 311 (1932) (Brandeis, J., dissenting). The rest of the quotation is: "It is one of the happy incidents of the federal system that a single courageous State may, if its citizens choose, serve as a laboratory; and try novel social and economic experiments without risk to the rest of the country. This Court has the power to prevent an experiment." *Id.*

compromise. Denial of the right to allow different approaches to flourish in different parts of the country. Not every difficult problem deserves a singular solution.

Because *Dobbs* and *Rucho* have conspicuously unleashed an assortment of trial-and-error approaches to these policy issues, they offer a useful way to think more deeply about how we decide what questions to answer with national solutions, what questions to answer with local solutions, and who should make the choices at either level: legislators, agencies, or judges. Consider the four distinct routes to change at the state level left open by the U.S. Supreme Court in these cases: legislation, direct democratic changes to laws and constitutions, indirect state constitutional amendments, and state courts.

As for legislation, the people's representatives may take matters into their own hands. Consider some of the state activity on abortion policy. Roughly twenty-three states place only modest limits on abortion access.[55] Roughly sixteen additional states permit abortion access up to a certain point in the pregnancy—whether a fetal heartbeat or a time such as fifteen weeks.[56] And roughly eleven remaining states have bans save for the life of the mother, incest, or rape.[57] The legislative activity, one suspects, is just getting started.[58] Legislative efforts to curb gerrymandering have been less apparent, but only by comparison. In total, seven state legislatures have shifted redistricting authority to independent commissions.[59] At least one State—Virginia—made the change post-*Rucho*, and other states may follow over time.

Direct democracy allows the people directly to alter a state statute or constitution. Twenty-four states give the people a way to circumvent an unaccommodating legislature, whether due to political

---

[55] *See State Policies Protecting or Restricting Legal Status of Abortion*, KAISER FAM. FOUND., https://www.kff.org/womens-health-policy/state-indicator/state-policies-protecting-or-restricting-legal-status-of-abortion (last visited Nov. 21, 2022).

[56] *Id.*; *An Overview of State Abortion Laws*, GUTTMACHER INST., https://www.guttmacher.org/state-policy/explore/overview-abortion-laws tbl. 1 (Nov. 1, 2022).

[57] KAISER FAM. FOUND., *supra* note 55.

[58] John Dinan, The Constitutional Politics of Abortion Policy After *Dobbs*: State Courts, Constitutions, and Law Making 49 (unpublished manuscript) (on file with author). State legislatures, due in part to differences in their powers, politics, and rules, are more active on abortion policy than is the U.S. Congress. *See id.* at 41–49.

[59] *Creation of Redistricting Commissions*, NAT'L CONF. OF STATE LEGIS. (Dec. 10, 2021), https://www.ncsl.org/research/redistricting/creation-of-redistricting-commissions.aspx.

headwinds in the legislature or the Governor's office or to gerrymandering itself.[60] In all of these states, the people have the power directly to amend a state statute or constitution. In the 2022 elections, for example, two of the three amendments to the Michigan constitution to protect reproductive rights arose from initiatives proposed by the people directly.[61] Taking a broader lens, five states, starting with Arkansas in 1956, have adopted redistricting reforms by citizen's initiative.[62] All told, eighteen states permit the people directly to amend the state constitution to create new fundamental rights or to curb dysfunctional features of democracy such as extreme partisan gerrymandering. For citizens frustrated that their representatives do not reflect the will of a 51 percent majority in the state, a constitutional initiative or for that matter a statutory referendum offers a way out.[63]

All states permit what I will call indirect constitutional amendments, created through constitutional conventions or through amendments proposed by the legislature and approved by the people, save in Delaware.[64] These paths are particularly relevant for the thirty-two states that do not permit constitutional initiatives proposed directly by the people. The promising news for those in favor of adding fundamental liberty or representation rights to their constitutions is that forty-six states require a mere 51 percent popular vote for approval.[65] Of the thirty-two states that do not offer the option of a direct-democracy initiative, all but New Hampshire permit a constitutional amendment with a 51 percent vote. What a legislature must do to refer an amendment to a popular vote varies widely by the state.[66] The legislatures in ten states, for example, may send an amendment to the

---

[60] *See* Dinan, *supra* note 10, at 16–17 (identifying 18 States that allow citizens to initiate constitutional amendments); Sutton, Who Decides, *supra* note 2, at 342 (identifying another six States that allow citizens to directly amend statutes).

[61] *See Michigan 2022 Ballot Measures*, Ballotpedia, https://ballotpedia.org/Michigan_2022_ballot_measures (last visited Nov. 21, 2022); *see also 2020 Ballot Measures*, Ballotpedia, https://ballotpedia.org/2020_ballot_measures (last visited Nov. 21, 2022) (reporting that 40 of the 120 measures on the ballot in 2020 were citizen initiated).

[62] Nat'l Conf. of State Legis., *supra* note 58.

[63] As one would expect, evidence suggests direct democracy leads to a "congruence between policymaking and public opinion." Dinan, *supra* note 57 (manuscript at 54). Perhaps more surprising, the mere existence of a method for direct participation, aside from its use, leads to more majoritarian outcomes in state legislatures. *See id.*

[64] Sutton, Who Decides, *supra* note 2, at 340–43.

[65] *Id.* at 343.

[66] *See* Dinan, *supra* note 10, at 12–16.

ballot with a simple majority vote.[67] The legislatures in nine more states may do so with a 60 percent vote.[68] Twelve other states require that their legislatures approve the amendments in two successive sessions.[69] Four other states employ a hybrid approach, allowing the legislature to place an amendment on the ballot with a majority vote in two sessions or a supermajority vote in one session.[70] While approval by the legislature of a constitutional amendment may be easier in some states than others, it is *always easier* than the federal constitutional amendment process.

That leaves the fifty state courts, which have the final say over the meaning of their constitutions and laws. There are many reasons why state courts might chart their own courses when it comes to interpreting their state constitutions, whether with respect to abortion, gerrymandering, or any other issue. Here are just a few.[71] Sometimes the text of the state constitution differs from its federal counterpart.[72] Different terms tend to generate different meanings. Sometimes the history behind the guarantee differs. A distinct history might lead one state, say Utah, to treat free exercise of religion distinctly from another state. Sometimes the state court uses a different method of constitutional interpretation from the one the U.S. Supreme Court used in the parallel federal precedent. A formal, fixed-meaning approach to a constitutional question is not apt to lead to imitation by a court that uses an informal, fluid-meaning approach to interpretation. Sometimes the federal doctrine has been shown to be bankrupt or sufficiently filled with difficulties of application to raise eyebrows about unthinking imitation at the state level. If a federal court—if, to personalize the point, Judge Sutton—makes a mistake in construing the federal guarantee, why should a state court judge repeat it?

---

[67] *Legislatively Referred Constitutional Amendment*, BALLOTPEDIA, https://ballotpedia.org/Legislatively_referred_constitutional_amendment (last visited Dec. 19, 2022).

[68] *Id.*

[69] *Id.*

[70] *Id.*

[71] For a much fuller discussion of these and other reasons, *see* SUTTON, WHO DECIDES, *supra* note 2, at 101–43.

[72] With respect to abortion, for example, recent attempts to amend state constitutions tend to substitute general, judge-empowering phrases, think of "privacy" or "due process," for specific, judge-constraining language, think "reproductive autonomy" or "abortion." *See* Dinan, *supra* note 57, at 29–31. Other amendments remove space for judicial gap-filling by clarifying that "[n]othing in this Constitution secures or protects a right to abortion." *Id.* at 33.

All of these considerations, and plenty more, create the possibility for an innovative market of American constitutional interpretation in which States may experiment by offering *more or less* protection under their own constitutions. The use of these trials and occasional errors becomes a promising platform for determining whether to nationalize one approach or another to a question—the more difficult the more valuable the input—or to allow distinct approaches to take root. Just as we do not nationalize all matters of legislative policy, we should not assume that all matters of constitutional interpretation must end up in the same place. The country has not had any problems with fifty-one distinct methods of raising revenue by taxation or fifty-one distinct approaches to spending it. The same may be so, and correctly so, for certain areas of American constitutional law.

The possibility of state court input is not for lack of cases filed in the state courts. The vast majority of legal actions in this country, in inexorable reality, run through the state courts, not the federal courts, meaning the first place most Americans will interact with our legal system will be in the state courts. In the last year for which we have numbers, there were some 40 to 50 million cases filed in the fifty state court systems and 400,000 cases filed in the ninety-four federal district courts.[73] In nearly every one of those state court cases, whether civil or criminal, the state constitution stands as a potential backstop against wrongful state action.

Some state court decisions after *Rucho* illustrate this last point particularly well. After the 2020 census, and *Rucho* notwithstanding, state courts in Maryland, New York, North Carolina, and Ohio set aside alleged partisan gerrymanders under state law.[74] Most of these decisions attracted heated dissents—dissents that voters, at least in North Carolina and Ohio, arguably endorsed in 2022's judicial

[73] *See* Nat'l Ctr. for State Cts., State Court Caseload Digest 7 (2020), https://www.courtstatistics.org/__data/assets/pdf_file/0014/40820/2018-Digest.pdf (reporting more than 83 million State court cases in 2018); *Federal Judicial Caseload Statistics 2018*, Admin. Off. of the U.S. Cts., https://www.uscourts.gov/statistics-reports/federal-judicial-caseload-statistics-2018 (reporting 358,563 cases filed in federal district court between March 2017 and March 2018); *Federal Judicial Caseload Statistics 2022*, Admin. Off. of the U.S. Cts., https://www.uscourts.gov/statistics-reports/federal-judicial-caseload-statistics-2022 (reporting 380,213 cases filed in federal district court between March 2021 and March 2022).

[74] Szeliga v. Lamone, No. C-02-CV-21-001816 (Md. Cir. Ct. Mar. 25, 2022); Harkenrider v. Hochul, 197 N.E.3d 437 (N.Y. 2022); Harper v. Hall, 868 S.E.2d 499 (N.C. 2022); Adams v. DeWine, 195 N.E.3d 74 (Ohio 2022).

elections.[75] *Rucho*, in short, has not prevented state courts from acting as backstops when they think partisan gerrymanders have gone too far. And state *voters* have not shrunk from that task either, including when they think their own state *courts* have gone too far. In the full view of all of these options, above all that forty-six states ultimately permit a 51 percent vote to alter a state's constitution, it's fair to say that any federal constitutional ruling criticized for under-enforcing the U.S. Constitution, as some have criticized *Dobbs* and *Rucho*, does not leave the people empty handed. It's also fair to say that the political accountability risks facing the federal courts when it comes to the development of new rights does not apply in the states. The state constitutions are more readily amendable. And 90 percent of their judges face elections and all of their judges, save Rhode Island's, face age or term limits. These features of the state judiciary assuredly make state judges more accountable than their federal counterparts, may make them more humble given the down-to-earth risks that come with facing the voters on a regular basis, and most definitely make them the best positioned judges for experimenting with innovative constitutional individual and structure guarantees. If we are to have constitutional innovations in this country, they should develop from the ground up, not the top down.

In *A Matter of Interpretation*, Justice Scalia diagnosed our American proclivity to embrace far-reaching federal judicial review as arising in part from the common law tradition we inherited from Great Britain—and the "Mr. Fix-it mentality" that comes with it.[76] Law students trained to understand common law reasoning, as he saw it, eventually became judges who use similar reasoning in construing the unamendable U.S. Constitution.[77] Many scholars agree with this assessment, though many of them applaud the approach to constitutional interpretation rather than deride it.[78] Either way, one

---

[75] *See* Laura Benshoff, *How GOP State Supreme Court Wins Could Change State Policies and Who Runs Congress*, NPR (Nov. 22, 2022), https://www.npr.org/2022/11/22/1138344117/republican-state-supreme-court-abortion-voting-redistricting-ohio-north-carolina (observing that, in North Carolina and Ohio, voters selected more conservative state Supreme Court candidates in 2022).

[76] SCALIA, *supra* note 2, at 13–14.

[77] *See id.* at 37–41.

[78] *See, e.g.*, DAVID A. STRAUSS, THE LIVING CONSTITUTION (2010); GOODWIN LIU ET AL., KEEPING FAITH WITH THE CONSTITUTION (2010).

commendable feature of the common law is that it turns on the reasoning of fifty state court systems, and the lending and borrowing associated with the enterprise, not on the reasoning of one court. Anyone open to common law constitutionalism should be particularly open to enlisting the insights of our fifty high courts before the national high court creates one constitutional rule for the entire country.

Structural headwinds for decreasing the footprint of the federal courts. The increase in the authority of the U.S. Supreme Court has come in fits and starts—and by and large up to now we Americans have seemed to want it. Many of the Court's most important decisions, including some substantive due process decisions, were consistent with the preferences of most Americans and at other times were consistent with large pluralities.

We Americans are a pragmatic lot, however. What seemed like a good idea in the past—delegating power to the Court to identify new substantive due process rights or broad interpretations of text-based rights—may not seem like such a good idea today. But what was relatively easy to increase—federal judicial power—may not be as easy to decrease.

Even as *Dobbs* and *Rucho* amount to two significant decisions to remove federal judicial power over fraught matters of public policy, many structural headwinds face any committed effort to decrease the role of the U.S. Supreme Court in American policy making. Let us suppose today's Court wishes to lessen its role in American government and wishes to obtain more input from state courts before nationalizing rights. That is still easier said than done. Congress and the President may famously have the power of the purse and sword. But they have another power, sometimes a more potent power: The power not to decide, the power not to handle this controversial issue or that one. Not so for the federal courts.[79] If people take their dispute to the courts, the judiciary usually must answer it. That's true whether the matter involves the Federal Constitution or a federal statute. This makes the idea of pacing the development of federal constitutional doctrine difficult and sometimes unrealizable.

[79] *Cf.* Cohens v. Virginia, 19 U.S. (6 Wheat.) 264, 404 (1821) ("The judiciary cannot, as the legislature may, avoid a measure because it approaches the confines of the constitution. We cannot pass it by because it is doubtful. With whatever doubts, with whatever difficulties, a case may be attended, we must decide it, if it be brought before us.").

How can the Court wait for the state courts if one of our 330 million citizens files a lawsuit today to challenge the validity of this or that state law under the Federal Constitution? *Rucho* illustrates one option—decide that some matters do not implicate judicially ascertainable standards. That of course might also be a good answer to substantive due process claims unanchored in the Bill of Rights and unanchored in any super-majority of state constitutional protections. But absent growth in the political-question doctrine, growth in other gate-keeping doctrines to Article III (e.g., standing or ripeness), or delays in resolving circuit splits through the discretion baked into the certiorari process, the Court faces persistent challenges in deciding not to decide.

This problem by the way also confronts the Court in statutory interpretation cases. In one sense, to be sure, statutory cases lower the political and accountability temperature. If the people disagree with the decision, they can plead to Congress that it should amend the statute. Even so, many statutory cases come to the Court precisely because of congressional inaction. It's not easy for the federal courts to determine the meaning of laws that have not been updated in over a half-century. Some of the most difficult statutory cases in recent years—*West Virginia v. EPA*, *Bostock v. Clayton County*, *Massachusetts v. EPA*, *Arizona v. Biden*, *Sackett v. EPA*—became that way because Congress had not revisited the statutes in over a half century to account for new norms, new technology, new understandings of air or water pollution.[80] Some statutes, quite understandably, are drafted "for the moment but not for the duration."[81] That also explains why agencies have been asked to do so much and perhaps why the Court has pushed back on their assertions of power.

There is one other area of conspicuous inaction that has increased pressure on the federal courts and contributed mightily to the difficulty of recalibrating their role in American government. It's the end of our political tradition of amending the U.S. Constitution when it needed to be changed. The last time we did so in a consequential way was more than fifty years ago, when the people decreased the voting

[80] *See, e.g.*, West Virginia v. EPA, 142 S. Ct. 2587 (2022) (Clean Air Act); Bostock v. Clayton Cnty., 140 S. Ct. 1731 (2020) (Title VII); Massachusetts v. EPA, 549 U.S. 497 (2007) (Clean Air Act); *see also* Jody Freeman & David B. Spence, *Old Statutes, New Problems*, 163 U. Pa. L. Rev. 1 (2014).

[81] Glasser v. Hilton Vacations Co., 948 F.3d 1301, 1308 (11th Cir. 2020).

age to eighteen.[82] For much of our history, that was not our way. Whether it was guaranteeing women the right to vote, permitting Congress to impose an income tax, eliminating the indirect election of senators, or starting and stopping Prohibition, we once took the path of Article V in dealing with new social and political imperatives. This tradition did not, I suspect, grow solely out of a finicky desire to adhere to Article V. It also grew out of what must have been perceived as a political advantage of taking a popular amendment to each State in the country to seek ratification. A good issue could be a good way to increase support for a political party. Perhaps those political incentives have given way to the incentives of nationalizing policy preferences in one Court with unreviewable authority to construe the Constitution—what has become its own form of gerrymandering. But it is not obvious that the political incentives of Article V compliance have disappeared for good. We will see.

Early American experiences with judicial review, uses of history, and sources of humility. Many Americans, I appreciate, wish that today's Court acted more like the Warren Court of the 1960s or the Burger Court of the 1970s. Those courts created many national rights, including *Roe*, and did so using uninhibited theories of constitutional interpretation. Whether those same Americans wish that today's Court—with its composition—would use innovative methods of interpretation to generate new rights may be another matter. Not everyone wants to try that hat on for size.

No matter one's perspective, it can be useful in a democratic republic like ours to allow different approaches to constitutional interpretation to have their time in the sun. Give the current Court a chance to show over a period of years that its approach to deciding cases has virtues of its own—that it can be done neutrally, that it will not lead to policy-laden outcomes, that it might decrease the role of the federal courts in American government. Judges, as I can attest, cannot hide for long from being judged themselves over whether their decisions turn on the rule of law or the rule of individual judicial preferences.

The key challenge for the current Court, for any court, for any judge, will be to show that they can apply their approach to constitutional interpretation in fair and even-handed ways. Here, yet again,

---

[82] *See* U.S. CONST. amend. XXVI; *see also* SUTTON, WHO DECIDES, *supra* note 2, at 335–36.

the state courts have lessons to offer about judicial review. In our federal-centric world, we assume that the federal courts started us down the road of judicial review. That's not what happened. Had John Marshall never been born, had there never been an election of 1800, had neither Madison nor Marbury walked the earth, our country would have had judicial review all the same. That's because the state courts had been doing it ever since their formation soon after 1776.

How the state courts engaged in judicial review at the outset offers lessons for our time. In one direction, they had little doubt about their authority to decline to enforce an unconstitutional law. It was a matter of judicial duty, part of the judicial oath, to honor a superior law when it conflicted with an inferior law.[83] The state and federal constitutions counted as superior laws. In the other direction, they looked for ways to avoid the conflict. They did so by construing statutes narrowly (an early form of constitutional avoidance) or by deciding that the constitution, after using all conventional tools of interpretation at their disposal, did not clearly cover the situation.[84] Under this approach, they did not hesitate to hold local laws unconstitutional, but they did so rarely.[85]

Almost 250 years later, federal judges could learn from these examples, examples that might help to readjust the role of the federal courts in American democracy. Yes, history, text, and fixed meaning all should inform when to set aside a democratically enacted law. And yes, courts should not abdicate their duty to set aside such laws when they truly conflict with the Constitution. But the requirement of a clear meaning of the constitutional provision, as the early state courts required, would go a long way to rightsizing the balance of power between the federal courts and the elected branches, between the federal courts and the states. Only historical inquiries that generate clear readings should count, an approach that might help to alleviate, if not always end, the human proclivity to look for allies and ignore foes in any historical record.

Against this backdrop, perhaps labels like judicial restraint or judicial engagement are the wrong way to think about it. Maybe it's brazen humility that's called for, an attitude, not a doctrinal label. It

---

[83] SUTTON, WHO DECIDES, *supra* note 2, at 51.

[84] *See id.* at 52–53, 56.

[85] *See id.* at 41–45, 57.

may be hard to be humble and patient in an age of little humility and considerable impatience. And it may be that one person's humility is another person's abdication. But one way to create judicial humility in constitutional cases might be to focus on two things: the risk of error and the reality that the federal courts are rarely the only available solution to a problem. Article III courts might reflect on the reality that their constitutional decisions cannot be overruled, that mistakenly added constitutional guarantees cannot be corrected by any democratic process or state court, that the judges or Justices who decide them cannot be replaced in the near term and will be replaced in the long term only in unpredictable ways, and that state court judges face few of these risks and might help in the ground-up development of new rights. Judges concerned about improper delegations of power in particular and separation of powers in general, moreover, should be concerned about the remarkable delegation of policy making power that comes with substantive due process and other inventive doctrines of judicial review. All of this might go a long way to capturing the need for the federal courts not just to check the political branches from time to time but also to check themselves from creating the very imbalance in power that separation of powers was designed to avoid.

The signal accomplishment of the Marshall Court was to legitimize federal judicial review. The destiny of the Roberts Court may be the thankless task of saving it.

FREDERICK SCHAUER

DISESTABLISHING THE ESTABLISHMENT CLAUSE

For most of the years of the Warren, Burger, and Rehnquist Courts, controversies over the relationship between religion and the state were decided under the Establishment Clause. Sporadically, the Free Exercise Clause provided the platform for consideration of demands for exemptions from regulatory laws of general application, as with the (successful) claim of the Old Order Amish in *Wisconsin v. Yoder*[1] for exemptions from the requirement that children go to school until the age of sixteen, and the (sometimes successful) claims of some Sabbatarians that they could not be required to be available for Saturday employment in order to be eligible for unemployment and

---

Frederick Schauer is the David and Mary Harrison Distinguished Professor of Law at the University of Virginia School of Law.

AUTHOR'S NOTE: I am grateful for valuable discussions of the issues dealt with here with my colleagues Doug Laycock, Rich Schragger, and Micah Schwartzman.

[1] 406 U.S. 205 (1972). The Old Order Amish did not succeed, however, in claiming that the Free Exercise Clause exempted them from paying social security taxes, even though, as in *Yoder*, they argued that participating in the system was no less prohibited by their faith than was schooling beyond the age of sixteen. United States v. Lee, 455 U.S. 252 (1982). And *Yoder*-based Free Exercise challenges were also unsuccessful in a line of 1980s cases, including *Goldman v. Weinberger*, 475 U.S. 503 (1986) (no exemption for observant Jew from military's uniform headgear requirement), and *Bob Jones University v. United States*, 461 U.S. 574 (1983) (no exemption from IRS refusal to grant tax-exempt status to university that claimed that racial discrimination, in violation of IRS regulations, was required by its religious commitments).

The Supreme Court Review, 2023.
© 2023 The University of Chicago. All rights reserved. Published by The University of Chicago Press. https://doi.org/10.1086/725160

similar benefits.[2] But these cases were the exceptions, and the Supreme Court's religion docket was dominated by questions arising under the Establishment Clause. The questions were most commonly about schools, in particular questions about religious exercises and symbols in the public schools[3] and the extent of permissible state support for sectarian schools.[4] And the Establishment Clause was also the platform for disputes over the constitutionality of various state-supported or state-permitted public religious symbols or displays.[5] For all of these issues, and more, the Establishment Clause was front and center, even when it was not interpreted to command invalidation.[6] But in the analysis, and often in the outcome, the Free Exercise clause during this period played, at best, a supporting role.

Now, however, the predominance of the Establishment Clause has been upset. Issues that were previously decided under the Establishment Clause have been transformed into free exercise questions. Moreover, the Court now shows greater sympathy for religious groups or adherents who advance free exercise arguments (which would preclude the challenged government action) than earlier Courts had shown for similar arguments opposing Establishment Clause claims (even though those arguments sought only to permit the government

[2] Thomas v. Rev. Bd., 450 U.S. 707 (1981) (upholding exemption); Hobbie v. Unemployment Appeals Comm'n, 480 U.S. 136 (1987) (same); Sherbert v. Verner, 374 U.S. 398 (1963) (same).

[3] Edwards v. Aguillard, 482 U.S. 578 (1987) (barring teaching of evolution unless accompanied by teaching of creation science); Wallace v. Jaffree, 472 U.S. 38 (1985) (religiously-motivated moment of silence requirement); Stone v. Graham, 449 U.S. 39 (1980) (posting of Ten Commandments in classrooms); Epperson v. Arkansas, 393 U.S. 97 (1980) (prohibiting teaching of evolution); Sch. Dist. of Abington Twp. v. Schempp, 374 U.S. 203 (1963) (school prayer); Engel v. Vitale, 370 U.S. 421 (1962) (officially composed school prayer).

[4] Wolman v. Walter, 433 U.S. 229 (1977) (state-provided instructional materials); Meek v. Pittinger, 421 U.S. 349 (1975) (same); Comm. for Pub. Educ. v. Nyquist, 413 U.S. 756 (1973) (financial aid to parents of parochial school students); Lemon v. Kurtzman, 403 U.S. 602 (1971) (reimbursement for teacher salaries and instructional materials). Not all of the Establishment Clause challenges were successful. *See* Comm. for Pub. Educ. v. Regan, 444 U.S. 646 (1980) (subsidizing expenses of preparing for state-prepared examinations); Bd. of Educ. v. Allen, 392 U.S. 236 (1968) (state-lent books on secular subjects).

[5] Capitol Square Rev. & Advisory Bd. v. Pinette, 515 U.S. 753 (1995) (Latin cross on state house grounds); Lee v. Weisman, 505 U.S. 577 (1992) (clergy-led invocation at graduation ceremony); Allegheny Cnty. v. Am. C.L. Union, 492 U.S. 573 (1989) (Christmas crèche, Christmas tree, and Hanukkah menorah on state property); Lynch v. Donnelly, 465 U.S. 668 (1984) (Christmas crèche, Santa Claus, and reindeer on city property).

[6] *E.g.*, *Allegheny Cnty.*, 492 U.S. 573; *Lynch*, 465 U.S. 668.

to act as it wished). And even when the Establishment Clause has not been overshadowed by the Free Exercise Clause, it has been interpreted in a far less robust way than it was during the Warren Court and the early years of the Burger Court. In general, the Establishment Clause has receded into the background, and free exercise claims have become far more important as a new Court majority has signaled an increasing receptiveness to such claims and a decreasing openness to the kinds of Establishment Clause claims that dominated a generation or two ago.

A host of recent cases have punctuated this shift in sympathies. One, *Shurtleff v. Boston*[7] is noteworthy here primarily for its unanimity in rejecting an Establishment Clause argument that earlier would at least have been taken seriously. And *American Legion v. American Humanist Association*[8] is similar in its lack of sympathy with an Establishment Clause challenge to a public monument with clearly Christian origins. Even more importantly, three other cases, *Carson v. Makin*,[9] *Kennedy v. Bremerton School District*,[10] and *Espinoza v. Montana Department of Revenue*,[11] exemplify not only the Court's decreased sympathy with Establishment Clause claims but also its commensurate increase in openness to countervailing free exercise and free speech arguments, together signaling the Court's tendency to see constitutional issues about the relationship between religion and the state less in terms of governmental disempowerment and more in terms of the individual liberty of religious believers and the autonomy of religious organizations. The full implications of this transformation remain to be seen, but it holds out the promise or the peril, depending on one's point of view, of a major shift in the Court's view about the role of religion in public life and state decision-making.

I want to explore in some detail each of the five cases just mentioned. Each of them involves a rejection of an Establishment Clause claim, to be sure, but putting it that way underestimates the substantial transformation that has taken place in recent years. The cliché that the whole is greater than the sum of its parts is particularly apt

---

[7] 142 S. Ct. 1583 (2022).

[8] 139 S. Ct. 2067 (2019).

[9] 142 S. Ct. 1987 (2022).

[10] 142 S. Ct. 2407 (2022).

[11] 140 S. Ct. 2246 (2020).

here. Because these five cases each reject Establishment Clause claims in different ways, and against different countervailing considerations, they represent as a group a development apparent over almost the past two decades: concerns about violations of Establishment Clause have virtually disappeared. My goal here is not to offer a new theory of the religion clauses, nor even to offer a large-scale normative prescription for how the Court ought to decide Establishment Clause cases. It is, more modestly but perhaps more usefully, to document, with interstitial criticism, the dramatic transformation that has taken place over the past almost two decades in Establishment Clause doctrine, and particularly the Court's previous three Terms. Especially with the Court as currently constituted, there appear to be few, if any, Establishment Clause arguments likely to find a receptive audience, and thus it is with only minimal hyperbole that we might say that the Establishment Clause has been "disestablished."

I. Five Cases

A. Shurtleff v. City of Boston

Although the two most significant of the recent religion cases—*Kennedy* and *Carson*—were overtly and predominantly about the religion clauses, of almost equal importance is *Shurtleff v. City of Boston*,[12] nominally about freedom of speech but just as importantly about both of the religion clauses. At issue was Boston's practice of allowing private groups to raise and display, for one day, their group's flag on one of the three flagpoles on the plaza in front of the Boston City Hall. In general, the city was indiscriminate about which groups were permitted to take advantage of this opportunity, and a wide variety of civic groups, advocacy groups, non-profit organizations, and for-profit businesses did so. The only exceptions to this all-comers practice were religious groups, which were excluded based on Boston's belief that including religious groups' flags in such a prominent government position in front of a government building and with the city's approval and assistance would violate the Establishment Clause of the First Amendment.[13]

---

[12] 142 S. Ct. 1583 (2022).

[13] On Boston's motivations, see *infra* note 23.

Harold Shurtleff and his small Christian fundamentalist organization—Camp Constitution—challenged the exclusion, and in the process challenged Boston's Establishment Clause justification for its practice. Justice Breyer's opinion for the Court primarily addressed Boston's claim, which the Court unanimously rejected, that the practice was an instance of government speech and thus outside most of the restrictions of the First Amendment's Speech Clause.[14] Because Boston was not engaging in government speech, the flag-raising opportunity provided to all (non-religious) comers created a (limited) public forum. The Establishment Clause could not justify what would otherwise appear to be Boston's unconstitutional viewpoint discrimination in allowing access to this public forum.

The Court's seeming unanimity on this last point—that Boston's Establishment Clause defense was insufficient—broke no new ground. It punctuated a development that arguably began in 1981 in *Widmar v. Vincent*[15] and that provides the foundation for the Court's most recent marginalization of what might otherwise be plausible Establishment Clause concerns. *Widmar* held that a university may not exclude religious groups from using university facilities on the same terms as other groups. Before *Widmar*, it would have been reasonable to believe, in a case like *Shurtleff*, that religion could be subject to distinctive disabilities precisely because of the demands of the Establishment Clause; that conclusion was, one might have thought, the inevitable consequence of the tension between public forum doctrine

---

[14] Under the comparatively new and increasingly important government speech doctrine, speech *by* the government is largely unconstrained by the First Amendment, and the government is free to engage even in viewpoint discrimination in its choice of what to say, how to say it, and what not to say. *See* Walker v. Tex. Div., Sons of Confederate Veterans, Inc., 576 U.S. 215 (2016). The modern precursors are *Pleasant Grove City v. Summum*, 555 U.S. 460 (2009), and, earlier, *Rust v. Sullivan*, 500 U.S. 173 (1991). On the government speech doctrine's broader implications, see Joseph Blocher, *Viewpoint Neutrality and Government Speech*, 52 B.C. L. Rev. 695 (2011); Joseph Blocher, *Government Property and Government Speech*, 52 Wm. & Mary L. Rev. 1413 (2011); Caroline Mala Corbin, *Government Speech and First Amendment Capture*, 107 Va. L. Rev. Online 224 (2021); Caroline Mala Corbin, *Mixed Speech: When Speech is Both Private and Governmental*, 83 N.Y.U. L. Rev. 605 (2008); Helen Norton, *Government Speech, Democratic Harm, and the First Amendment*, 82 Ohio St. L. J. Online 1 (2021); Helen Norton, *The Meaning of Government Speech: Identifying Expression's Source*, 88 B.U. L. Rev. 87 (2008); and Frederick Schauer, *Not Just About License Plates:* Walker v. Sons of Confederate Veterans, *Government Speech, and Doctrinal Overlap in the First Amendment*, 2015 Sup. Ct. Rev. 265.

[15] 454 U.S. 263 (1981). See also, to much the same effect outside of the university context, *Good News Club v. Milford Cent. Sch.*, 533 U.S. 98 (2001); *Lamb's Chapel v. Ctr. Moriches Union Free Sch. Dist.*, 508 U.S. 384 (1993); and *Bd. of Educ. v. Mergens*, 496 U.S. 226 (1990).

and the Establishment Clause,[16] and, perhaps even more, of the more than occasional tension between the Establishment Clause and the Free Speech Clause. After all, as Justice White observed in his dissenting opinion in *Widmar*, even if a university or a teacher might rely on the Free Speech Clause to protect their decisions about what to include within the curriculum, that would not extend so far as to require state universities to include "Sunday Mass" along with its other permitted speech for fear of violating the Free Exercise Clause or the mandate against viewpoint discrimination under the Free Speech Clause.[17] For Justice White, and for Justice Stevens twenty years later in *Good News Club v. Milford Central School*,[18] in the context of primary and secondary public schools, that which is otherwise permitted or even required by the Free Speech and Free Exercise Clauses may wind up being prohibited by the Establishment Clause for fear of preferring religion or a particular religion, or for fear of creating the perception that the schools were endorsing a particular religion or even religion in general. Because so many forms of "establishment" are verbal, understanding the Establishment Clause as a qualification on rights under the Free Speech Clause is not completely off the mark.

Little point would be served here by relitigating *Widmar* or *Good News Club*. The point now is only that Boston's argument—one that attracted no adherents on the Supreme Court in 2022—is one that a generation earlier would not only have been taken seriously but also seemed to influence a shifting collection of Supreme Court Justices.

[16] To be more precise, if the program were to be understood as government speech, then the question is about the extent to which the Establishment Clause limits what the government is permitted to say, which is the question in fact presented in most Establishment Clause cases. And if the program were to be understood as the creation of some sort of forum, then the question is whether and when the Establishment Clause prohibits what the state would otherwise choose to permit.

[17] 454 U.S. at 282 (White, J., dissenting); *see also* Stone v. Graham, 449 U.S. 39 (1980) (invalidating the posting of the Ten Commandments in a public school classroom). The point of Justice White's argument and of his reliance on *Stone* is that posting the Declaration of Independence or the Constitution or Martin Luther King's "I Have a Dream" speech would at least have been within the discretion of the school, and, insofar as the decision was made by a teacher as part of the teacher's decisions about what to teach and how to teach it, arguably protected by the not-quite-yet-definitively established First Amendment right to academic freedom under the Free Speech Clause. *See* Kennedy v. Bremerton Sch. Dist., 142 S. Ct. 2407, 2424 (2022); Garcetti v. Ceballos, 547 U.S. 410, 425 (2006). But, Justice White argued, the Establishment Clause operates as an exception on what would otherwise be a constitutionally-permissible prerogative or even a constitutional right.

[18] 533 U.S. 98, 131 (2001) (Stevens, J., dissenting). To a similar effect is the opinion of Justice Souter, joined by Justice Ginsburg. *See id.* at 135 (Souter, J., dissenting).

And that influence appeared, in turn, as Justice White made clear, to have been at least partially a product of the Court's approach in 1971 in *Lemon v. Kurtzman*.[19] *Lemon*'s three-part test, about which more presently, includes in its second part the requirement that a governmental action, in order to survive Establishment Clause scrutiny, must be one whose "principal or primary effect . . . neither advances nor prohibits religion."[20] The *Lemon* test, unlike for example the Equal Protection test after *Washington v. Davis*,[21] does not treat governmental purpose or intent (or, more accurately, the absence of such) as dispositive, meaning in this context that some number of actions whose goal is to do something other than help, advance, or promote religion might nevertheless run afoul of the Establishment Clause as long as the unintended consequences of such actions is to create the reality or the perception of government support or endorsement of religion.[22]

As long as unintended effects or perceptions might make a difference, one can understand Boston's worry in *Shurtleff* about the Establishment Clause. We do not know the non-doctrinal source of that worry,[23] but the worry is at least fueled by the Establishment Clause. And to the extent that the Court in *Shurtleff* even further extinguished the possibility of Establishment Clause justifications for what might otherwise have been free speech or free exercise violations,

[19] 403 U.S. 602 (1971).

[20] *Id.* at 612.

[21] 426 U.S. 229 (1976).

[22] Interestingly, and perhaps ironically, incidental effects on constitutionally protected interests generate at least some scrutiny in free speech cases, *see* United States v. O'Brien, 391 U.S. 367 (1968), and some in Dormant Commerce Clause cases, *see* Hunt v. Wash. State Apple Advert. Comm'n, 432 U.S. 333 (1977), but none at all in equal protection cases, as per *Washington v. Davis*. *See generally* Michael C. Dorf, *Incidental Burdens on Fundamental Rights*, 109 HARV. L. REV. 1175 (1996).

[23] One possibility is that the source of the worry is litigation avoidance, even the avoidance of challenges highly unlikely to succeed. For reasons that are often difficult to understand, government officials frequently treat the avoidance of litigation as a goal independent of the likelihood that the challenge will succeed and independent of any possible effect on an official's pocketbook or even budget. *See* Frederick Schauer, *The Political Risks (If Any) of Breaking the Law*, 4 J. LEGAL ANALYSIS 83 (2012). Someone in Boston's bureaucracy, as with many other officials, especially lower-level functionaries, may thus have taken "We don't want to get sued" as a reason for a degree of risk-aversion that often verges on the irrational.

An alternative or additional explanation might have been Boston's Constitution-based belief, regardless of the positive caselaw doctrine, that the kinds of anti-establishment or anti-endorsement worries that are the basis of the Establishment Clause were legitimate, the views of the Supreme Court over the previous forty years notwithstanding. In other words, Boston might have been engaging in sincere departmentalist practice, treating *its* view of what the Establishment Clause required as conclusive and leaving it to the courts to step in as appropriate.

it paved the way for the other decisions that are my principal concern here, and which most exemplify the decreasing importance of the Establishment Clause.[24]

B. AMERICAN LEGION V. AMERICAN HUMANIST ASSOCIATION

Shurtleff rejected Boston's Establishment Clause argument largely on the equality basis that finds its home in the Free Speech Clause's aversion to some forms of content discrimination and most forms of viewpoint discrimination.[25] *Shurtleff* imported these concerns into the religion clauses, implicitly in Justice Breyer's opinion for the Court and explicitly in Justice Kavanaugh's concurring opinion,[26] which stressed that discriminating against religion or religious expression offended not only the Free Speech Clause but the Free Exercise Clause as well. *Shurtleff* thus stands for the proposition that behavior that is undeniably religious would be protected against state restriction based on its religious character, Establishment Clause concerns notwithstanding. And importantly, especially so in light of the other recent decisions discussed here, Justice Kavanaugh's identification of a free exercise violation underscores that the constitutional problem would persist even were free speech concerns for some reason not applicable. It is thus clear that, for him and others, singling out religion for a disability from what would otherwise be a generally applicable regime could not only not be justified on Establishment Clause grounds but also would constitute an independent Free Exercise Clause violation.

In the modern history of the Establishment Clause, such equality and free exercise concerns have often been effective in rebutting

---

[24] Thus, the *Shurtleff* Court held that Boston violated Harold Shurtleff's free speech rights by excluding him on viewpoint-based grounds from what otherwise would have been an open forum. Without an Establishment Clause defense, the unconstitutionality of Boston's exclusion was plain under longstanding free speech principles and doctrine. See, for example, Matal v. Tam, 137 S. Ct. 1744 (2017), which invalidated a content- and viewpoint-based Patent and Trademark Office restriction on trademarks of a "disparaging" nature; as in *Shurtleff*, the Court in *Matal* rejected a government speech argument, paving the way to seeing the restriction as an impermissible form of content regulation.

[25] *See* Simon & Schuster, Inc. v. Members of N.Y. State Crime Victims Bd., 502 U.S. 105 (1991); R.A.V. v. City of St. Paul, 505 U.S. 377 (1992); Leslie Kendrick, *Content Discrimination Revisited*, 98 VA. L. REV. 231 (2012); Geoffrey R. Stone, *Content Regulation and the First Amendment*, 25 WM. & MARY L. REV. 189 (1983). I use "finds its home" metaphorically, in contrast to the view that in something closer to the equal protection sense of equality is at the heart of the First Amendment's Speech Clause. *See* Kenneth Karst, *Equality as a Central Principle in the First Amendment*, 42 U. CHI. L. REV. 20 (1975).

[26] 142 S. Ct. 1583, 1594 (2022) (Kavanaugh, J., concurring).

Establishment Clause concerns, which is why I described *Shurtleff* as breaking no new ground, at least in a post-1981 and post-*Widmar* world. But an alternative strategy for countering Establishment Clause arguments, not available in *Shurtleff*,[27] is to deny the religious nature of seemingly religious government practices. This strategy had its origins in the Sunday closing cases,[28] in which the Court secularized the originally religiously-inspired state practice of prohibiting many forms of retail sales on Sunday; the Court asserted that historical practice had converted what was originally a religiously-motivated regulation into a constitutionally non-problematic secular one. To much the same effect were *Lynch v. Donnelly*[29] and *Allegheny County v. ACLU*,[30] which rejected what might otherwise seem to be public celebrations of religious holidays (Christmas, and, to a lesser extent, Hanukkah) by, in part, concluding that Christmas had become a secular holiday whose official acknowledgment and celebration therefore did not offend the Establishment Clause.

This latter approach was at the heart of *American Legion v. American Humanist Ass'n*,[31] decided three Terms prior to *Shurtleff*, *Carson*, and *Kennedy*, but usefully considered as part of the same modern group. At issue in *American Legion* was a public war memorial in Bladensburg, Maryland. The memorial was a thirty-two-foot high Latin cross, designed and constructed to honor the local soldiers who had been killed in World War One. Rejecting an Establishment Clause challenge to this state-constructed and state-maintained religious symbol, the Court, over the dissenting opinion of Justice Ginsburg (joined by Justice Sotomayor), and the more qualified concurrence of Justice Breyer (joined by Justice Kagan), followed the same secularization strategy that had earlier produced the outcomes in *McGowan*, *Lynch*, and *Allegheny County*. In relying on this secularization strategy, Justice Alito's opinion for the Court noted that the meaning of a monument may "evolve," and that in most contexts a cross had become a way of

---

[27] In fact, Boston urged Shurtleff to secularize the flag he wished to fly, or to raise a different flag, but Shurtleff's commitment to his beliefs would not permit him to do so. *See Shurtleff*, 142 S. Ct. at 1588; Shurtleff v. City of Boston, 986 F.3d 78, 84 (1st Cir. 2021).

[28] McGowan v. Maryland, 366 U.S. 420 (1961).

[29] 465 U.S. 668 (1984).

[30] 492 U.S. 573 (1989).

[31] 139 S. Ct. 2067 (2019). For critical commentary, see Richard Schragger & Micah Schwartzman, *Establishment Clause Inversion in the Bladensburg Cross Case*, 3 ACS Sup. Ct. Rev. 21 (2019).

honoring the military dead now substantially shorn not only of its religious origins but also its current religious meaning and significance.[32]

American Legion v. American Humanist Ass'n presented a different issue from *Shurtleff*, but both cases, in different ways, weakened the Establishment Clause. In *Shurtleff*, equality concerns—avoiding the inequality between religious and non-religious flags and their sponsoring organizations—served to forestall Establishment Clause objections to a plainly religious practice.[33] In *American Legion*, as with a generation or more of previous cases, the response to those objections was that there could be no Establishment Clause concerns if the practices at issue were not (now) understood as religious at all. In *American Legion*, the Court addressed, and dismissed as inapplicable (in fact, the Court suggested, obsolete), the test from *Lemon v. Kurtzman*,[34] which highlighted incidental effects (and especially perceptions) as well as governmental intent,[35] and *American Legion* is more significant for its treatment of the *Lemon* test than for its analysis of the history and sociology of crosses as war memorials. But in dismissing *Lemon*, about which more presently, and especially in relying on a somewhat rare secularization strategy to deny the religious significance of a Latin

[32] In a concurring opinion, Justice Breyer, joined by Justice Kagan, cautiously accepted the secularization argument, 139 S. Ct. at 2090, relying in part on the Court's 2005 Ten Commandments decision in *Van Orden v. Perry*, 545 U.S. 677 (2005). Justice Ginsburg, joined by Justice Sotomayor, dissented, relying on her view that nothing in this context could change the fact that a cross was and remains and undeniably Christian symbol. *See also* Richard Schragger & Micah Schwartzman, *Establishment Clause Appeasement*, 2019 Sup. Ct. Rev. 271. At some level the secularization approach is obviously correct in the abstract. As the Ten Commandments illustrate, the criminal sanctions for murder, fraud, and perjury, among others, have early origins in Biblical prohibitions. But all of the action is in the details, and the question, framed in Establishment Clause issues, is when the "taint" of religious origins has been purged. The inquiry is not unrelated to the inquiry about when a history of prior de jure racial discrimination has been sufficiently purged to rebut the inference that current de facto segregation has de jure origins. *See* Parents Involved in Cmty. Schs. v. Seattle Sch. Dist., 551 U.S. 701 (2007), where the issue was highlighted in the context of the Louisville part of that consolidated decision. *See also* W. Kerrel Murray, *Discriminatory Taint*, 135 Harv. L. Rev. 1190 (2022).

[33] For an earlier criticism of this development, see Noah Feldman, *From Liberty to Equality: The Transformation of the Establishment Clause*, 90 Calif. L. Rev. 673, 729 (2002), arguing that "equality [was] being used not to justify the separation of church and state, but to subvert it." In a very different context, the relationship between equality and problematic behavior was highlighted by former Green Bay Packers defensive lineman Henry Jordan. When Jordan, who was African-American, was asked how he was treated by legendarily abusive coach Vince Lombardi, he replied, "He treats us all the same—like dogs." Jerry Kramer, *Winning Wasn't Everything*, N.Y. Times, Jan. 24, 1997, at A31.

[34] 403 U.S. 602 (1971).

[35] On the demise of the *Lemon* test, see also *infra* Part II.

cross, the Court continued the marginalization of the concerns lying behind the Establishment Clause.

C. ESPINOZA V. MONTANA DEPARTMENT OF REVENUE

Establishment Clause litigation has long been dominated by questions about the extent to which the Clause permits state aid to religious schools. In *Espinoza v. Montana Department of Revenue*,[36] the issue was the constitutionality of a Montana program providing a $150 tax credit to those who made donations to organizations awarding scholarships for private school tuition. The program did not distinguish between religious and secular private schools, and the question therefore was, initially, whether the program violated the Establishment Clause by providing these credits for donations to religious private schools.

The Montana tax credit program, by itself, would appear be permissible under existing Establishment Clause doctrine, although perhaps not so in earlier decades. Under the *Lemon* test, a program of tax credits such as Montana's would have been assessed according to its effects as well as to its stated purposes. Accordingly, the Montana program might well have been unconstitutional under the *Lemon* approach to the Establishment Clause if in fact its effect was to advance religion, as for example by providing most or all of its benefits to religious schools. But after *Zelman v. Simmons-Harris*,[37] such an outcome was no longer available.[38] *Zelman* concluded that as long as the amount or percentage of aid to religious schools was a function of private and non-governmental choice, in particular the choices of parents and not the state, the Establishment Clause would not be violated. Insofar as the assistance granted was available "generally without regard to the sectarian-nonsectarian, or public-non-public nature of the institution benefited,"[39] the demands of the Establishment Clause would be satisfied.

---

[36] 140 S. Ct. 2246 (2020).

[37] 536 U.S. 639 (2002), a 5-4 decision in which Chief Justice Rehnquist wrote the opinion of the Court, with Justice Souter, joined by Justices Stevens, Ginsburg, and Breyer, writing a spirited principal dissent.

[38] *Zelman* culminated a line of cases rejecting Establishment Clause challenges to similar forms of aid that disproportionately went to or benefited religious schools. *See* Zobrest v. Catalina Foothills Sch. Dist., 509 U.S. 1 (1993); Witters v. Wash. Dep't of Servs. for the Blind, 474 U.S. 481 (1986); Mueller v. Allen, 463 U.S. 388 (1983).

[39] 536 U.S. at 652.

Zelman appeared to suggest, without more, the permissibility of Montana's scheme, but the Montana Supreme Court nevertheless concluded that any aid to religious schools under the program violated the provision in the Montana Constitution prohibiting "direct or indirect" funding of religious schools. In a 5-4 decision with Chief Justice Roberts writing for the majority, the Court in *Espinoza* held this exclusion impermissible under the Free Exercise Clause because "otherwise eligible recipients [we]re disqualified from a public benefit 'solely because of their religious character.'"[40] *Shurtleff* relied on free speech principles and doctrine to invalidate what seemed to the state to be an impermissible support of religion; in *Espinoza* the Free Exercise Clause played a similar role, applying an anti-discrimination principle to block state efforts to avoid, as Justice Sotomayor put it in her dissent, "pay[ing] for [a] religious practice."[41] *Espinoza* thus used the Free Exercise Clause to append a non-discrimination principle to the principle of Establishment Clause permissibility. Not only would incidental benefits to religious schools not violate the Establishment Clause, but also any attempts by the state to exclude such incidental benefits would violate the Free Exercise Clause. This went beyond saying that states were permitted to include or exclude religious institutions from programs of general application without violating the Establishment Clause; it meant that states *must not exclude* religious institutions from aid programs of general application for fear of violating the Free Exercise Clause. The optional was converted into the mandatory.

D. KENNEDY V. BREMERTON SCHOOL DISTRICT

As in *Shurtleff*, *Kennedy v. Bremerton School District*,[42] from the 2021 Term, also involved the interplay among the Free Speech, Free Exercise and Establishment Clauses. And as in *Shurtleff*, the Establishment Clause finished in third place among these clauses, reinforcing what seems to be its increasing irrelevance.

At the center of *Kennedy* was Joseph Kennedy, one of the coaches of the Bremerton, Washington, high school football team. The controversy arose from Kennedy's practice of kneeling in prayer on the fifty-yard line at the end of the team's games. Kennedy sued when he

---

[40] 140 S. Ct. at 2260.

[41] *Id.* at 2296 (Sotomayor, J., dissenting).

[42] 142 S. Ct. 2407 (2022).

was dismissed from his (part-time) position for engaging in this public prayer on school property shortly after the conclusion of an official school event, but his dismissal followed a succession of events that started with Kennedy's regularly leading organized prayers for and with the team on school property. In several different iterations, starting with Kennedy's organized prayers in conjunction with the team's games, the school sought to impose limits on Kennedy's actions. Kennedy complied with each of these limitations as they were imposed, but even the Court's presumably sympathetic recounting of the history leaves the impression that Kennedy's compliance was grudging, and that at each step of the school's efforts to deal with Kennedy's public and quasi-organized prayers Kennedy sought to comply as narrowly as possible while still remaining faithful to what he sincerely believed were his commitments to his faith. Finally, the school dismissed Kennedy for praying on school grounds in the presence of team members.

The school claimed in dismissing Kennedy that his practice created the impression of school endorsement of a religious exercise on school property in connection with a school event, but Justice Gorsuch's opinion for the 6-3 majority offered a double-barreled rejection of this claim. First, Justice Gorsuch announced that a test focused on whether the government could be reasonably perceived as endorsing religion,[43] if endorsement or a perception of it had ever been the law, had been rejected by the Court in recent years and was no longer to be considered the law.[44] And second, the majority treated

---

[43] The endorsement test's origins are largely the doing of Justice O'Connor. She first offered the idea in her concurring opinion in *Lynch v. Donnelly*, 465 U.S. 668 (1984), initially as a gloss on impermissible purpose under *Lemon*'s first prong. But it subsequently took on a life of its own. *See* McCreary Cnty. v. Am. C.L. Union of Ky., 545 U.S. 844 (2005); Capitol Square Rev. & Advisory Bd. v. Pinette, 515 U.S. 753 (1995); Jesse H. Choper, *The Endorsement Test: Its Status and Desirability*, 18 J. L. & Pol. 499 (2002).

[44] Justice Gorsuch and the majority announced that the test was to be rejected because it was inconsistent with the original meaning of the Establishment Clause, but there is a bit of an apples and oranges aspect to this conclusion. Even under an originalist analysis, original public meaning is rarely (actually, never) sufficient to provide a test useful in reaching conclusions about specific applications and thus about specific cases. Plainly some such operational tests will be more consistent with original public meaning than others, but it is a mistake to confuse the provenance or authority of an interpretive approach with the method used to operationalize it. The endorsement test might or might not be compatible with the original public meaning of the Establishment Clause, but the Court does not tell us why it is incompatible, or what concrete and usable test might be more compatible.

This is not to denigrate the value of the words or phrases commonly employed to make at least slightly more concrete the words and phrases in the constitutional text. Often such words

endorsement of a prayer or other religious practice as constitutionally problematic if and only if it was produced or accompanied by actual coercion of the students. And thus what we might label a "coercion test" or "coercion standard" has in fact replaced the endorsement test, and, *en passant*, the *Lemon* test as well.

In finding no coercion, the Court concluded that there was no evidence of "direct coercion" of the students on the team to join Kennedy for his prayers. But that conclusion depends heavily on what is to count as evidence of coercion in this context. As in school prayer cases going back at least as far as *Engel v. Vitale*[45] in 1962, the Court has treated what we know about behavior in general as evidence of what was likely the case in individual instances.[46] This kind of inference is at the heart of most (perhaps all) evidentiary inferences, and we do not normally require particularized evidence in order to justify, for example, the (admittedly defeasible) conclusion that people who run out of banks carrying bags and wearing ski masks have most likely just robbed the bank.

As Justice Sotomayor stressed in her lengthy dissenting opinion, much of the evidence of coercion comes from what we know about the operation of social norms and group behavior, especially for teenagers. Teenagers (and the rest of us) often wear the same clothes as others and drive the same cars as our neighbors even absent official coercion, for a range of complex psychological and sociological reasons, and the majority's conclusion that there was no coercion is a conclusion inconsistent not only with *Engel* (and presumably with *Abington School District v. Schempp*[47] as well) but also with almost everything we know about the socially conforming tendencies and behavior of people in general and teenagers in particular.[48]

Having treated endorsement as doctrinally obsolete and coercion as absent on these facts, the majority seemingly found it easy to

and phrases—"viewpoint regulation"; "rational basis"; "compelling interest"—can at least help guide the inquiry even if they do not point unequivocally to particular outcomes. But for years the dismissal of the *Lemon* test has been based in large part on its alleged unworkability or indeterminacy, typically without explaining why it is more so than so many of the standards and tests that pervade all of constitutional law.

[45] 370 U.S. 421 (1962).

[46] *Id.* at 431 (referring, without citation, to "indirect coercive pressure").

[47] 374 U.S. 203 (1963).

[48] For an extensive critique of *Kennedy* along these lines, see Justin Driver, *Three Hail Marys:* Carson, Kennedy, *and the Fractured Détente over Religion and Education*, 136 Harv. L. Rev. 208 (2022).

conclude that the Establishment Clause was insufficient to outweigh Kennedy's free speech and free exercise claims. As to the former, the Court found no basis for concluding that there had been actual disruption of school activities,[49] and no basis for concluding that the prohibition on Kennedy's praying was in pursuit of the kind of curricular decision that was at the center of school speech cases such as *Hazelwood School District v. Kuhlmeier*.[50] And because the school unquestionably treated Kennedy's praying as different from other verbal behavior taking place on a coach's unencumbered time, the majority treated singling out religious but not other speech as also constituting a free exercise violation, one not remedied by a misplaced (according to the Court) reliance on the requirements of the Establishment Clause. Again, for the Court the touchstone was alleged neutrality: The Establishment Clause allows benefits to religion or religious institutions that are generally or incidentally granted by laws applicable to religious and non-religious institutions alike. The Free Exercise Clause permits a restriction that is applied to both religious and non-religious conduct. And, conversely, the Free Exercise Clause is violated when a restriction was imposed *because* of the religious nature of the conduct rather than simply in spite of it.

E. CARSON V. MAKIN

The third case in the trilogy of 2021 Term decisions dismissing the importance of Establishment Clause concerns was *Carson v. Makin*,[51] the latest in a long line of Supreme Court cases dealing with state funding and related forms of support for religious schools.

Carson dealt with Maine's approach to managing the intersection between the requirement of public schooling and the state's largely rural and geographically dispersed population. Because of this demographic, the state was faced with the problem of providing public education in areas whose population base was too small and too spread out to support public middle and high schools in an economically and logistically viable way. The state's solution to this problem was to provide financial support to those parents who chose to send their children to private schools. But because most of these private schools

---

[49] *See* Tinker v. Des Moines Indep. Sch. Dist., 393 U.S. 503 (1969).

[50] 484 U.S. 260 (1988).

[51] 142 S. Ct. 1987 (2022).

were religious, the effect of the state's solution was to provide financial assistance for sectarian education. The question, not a new one, was whether this seemingly incidental benefit to religious schools violated the Establishment Clause.

In upholding Maine's program of financial assistance for sectarian education, the Court took the opportunity finally to (sort of) overrule *Lemon v. Kurtzman*. Even in *Carson*, however, the majority never used the word "overrule," although when a majority of the Court, building on a long recent history of statements from the Court undermining a decision, adds that the decision has not been followed—which is what the *Carson* Court did with *Lemon*—the effect was plainly the same. And it is clear that a majority of the current Court believes that *Lemon*'s concern with unintended perceptions (as in *Kennedy* and *Shurtleff*) was misguided, and *Lemon*'s concern with unintended incidental effects even more so.

With *Lemon*'s effects (and, implicitly, entanglement) concerns out of the way, the Court relied again on the apparent neutrality of Maine's scheme, both on its face and seemingly as a matter of legislative intent as well.[52] And thus the incidental benefits to religious schools were insufficient to invalidate Maine's program.

Carson's focus on the neutrality of state benefit programs, and the Court's general focus on neutrality as the driving force behind Establishment Clause doctrine, operates differently from the Court's approach to neutrality in free exercise cases. Incidental benefits to programs of general application, as with the incidental benefits to religious schools in *Carson* as well as the incidental benefits granted to religious organization in *Shurtleff*, are now treated as constitutionally unimportant, and, indeed, much the same applies to *Kennedy v. Bremerton* as well. But although neutrality is now increasingly important in Establishment Clause cases, the same cannot be said about free

[52] Whatever the percentage of primary and secondary schools in Pennsylvania when *Lemon* was decided, it undoubtedly was lower than the percentage of military veterans who were men in 1979, the year of *Personnel Administrator v. Feeney*, 442 U.S. 256 (1979). If the obvious foreseeability that the beneficiaries of the Massachusetts public employment hiring boost for veterans would be men did not permit the inference of intentional discrimination in *Feeney*, then the lower percentage of private schools that were sectarian in 1971 could not without more support the inference that the goal in Pennsylvania in 1971 or the goal in Maine in 2022 was to assist religious schools because they were religious. The point is underscored by *Zelman v. Simmons-Harris*, 536 U.S. 639 (2002), which the Court explicitly followed in *Carson*, and in which 96 percent of the benefited students were enrolled in religiously-affiliated schools.

exercise issues. The *Carson* majority explicitly referred to the Free Exercise Clause, emphasizing not only that the Establishment Clause was not in its view offended by the incidental benefits to religious schools but also that the refusal to allow those incidental benefits created free exercise problems. In other words, Establishment Clause carve-outs, as in *Shurtleff* and *Espinoza*, which were once arguably required, as Boston believed in *Shurtleff*, and then optional but not required, as Montana believed as a matter of federal constitutional law in *Espinoza*, are now understood to violate the Free Exercise Clause and therefore to be constitutionally impermissible.

This free exercise-based movement from optional-but-permissible carve-outs to constitutionally impermissible carve-outs is a significant shift in the doctrine. The Court and commentators have long acknowledged a tension between the Establishment and Free Exercise Clauses: a decision not to confer benefits on a religious practice forestalls Establishment Clause problems but creates free exercise problems by limiting religiously-motivated conduct, even if only incidentally; and a decision to eliminate free exercise problems by exempting religious conduct from otherwise generally applicable requirements creates Establishment Clause problems by granting to believers benefits not available to others.[53] As Justice Breyer emphasized in his dissent in *Carson*, the Court has addressed this tension by allowing states some leeway in managing it, a leeway that he describes as the "play in the joints."[54] Thus, states have been permitted but not required to adopt policies in pursuit of satisfying what they perceive to be non-Establishment themes without violating the Free Exercise Clause.[55]

When *Carson* followed *Trinity Lutheran Church of Columbus, Inc. v. Comer*[56] and then *Espinoza* in holding such carve-outs not merely required by the Establishment Clause but also prohibited by the Free

---

[53] On the traditionally-recognized tension between the two religion clauses, see Walz v. Tax Comm'n 397 U.S. 664, 668–72 (1970); Paul Freund, *Public Aid to Parochial Schools*, 82 Harv. L. Rev. 1680, 1684 (1969); and Louis Henkin, *Infallibility Under Law: Constitutional Balancing*, 78 Colum. L. Rev. 1022, 1031 (1978).

[54] 142 S. Ct. at 2002 (Breyer, J., dissenting). The origin of the phrase in this context is *Walz*, 397 U.S. at 669.

[55] *See* Locke v. Davey, 540 U.S. 712 (2004), and earlier, *Walz*, 397 U.S. 664.

[56] 137 S. Ct. 2012, 2021 (2017) (impermissible under the Free Exercise Clause for the state to "discriminate[] against otherwise eligible recipients").

Exercise Clause, the Court denied that such a tension exists, saying that the two clauses should be understood as serving the same goal of religious liberty. But in fact the majority resolved the long-accepted tension between the two clauses by denigrating Establishment Clause concerns and elevating free exercise ones: in effect, asserting that Establishment Clause concerns are not important enough to create any genuine tension. How this will play out perhaps remains to be seen, but the signals are clearly there.

II. The Past, Present, and Future of the Establishment Clause

As of this writing, it has been eighteen years since the Supreme Court last decided an Establishment Clause claim in favor of the claimant.[57] And although that fact is perhaps a trifle misleading, because it is a function of the Court's control of its own docket as well as the selection effect, the recent and not-so-recent history of seeming hostility to Establishment Clause claims does tell us at least something about the contemporary importance of the Establishment Clause, especially as compared to the 1960s, 1970s, and 1980s. And that something is that the importance of the Establishment Clause in patrolling the relationship between religion and the state has been diminished substantially, as the survey of recent cases just provided illustrates.

This development is punctuated by the views of Justice Thomas, who has argued that the Establishment Clause is a separation of powers provision and consequently not incorporated at all by the Fourteenth Amendment,[58] a position that would, if adopted by a

[57] McCreary v. Am. C.L. Union, 545 U.S. 844 (2005) (impermissible under the Establishment Clause to post copies of the Ten Commandments in courthouses). A possible exception to the generalization in the text is *Hosanna Tabor Evangelical Lutheran Church and School v. EEOC*, 565 U.S. 171 (2012). In upholding the so-called ministerial exception to what might otherwise have been a violation of the Americans with Disabilities Act, Chief Justice Roberts, writing for a unanimous Court, relied both on the Establishment and Free Exercise Clauses. "The Establishment Clause prevents the Government from appointing ministers, and the Free Exercise Clause prevents it from interfering with the freedom of religious groups to select [those who will personify its beliefs]." *Id.* at 184. Because the latter is more applicable to the actual facts of *Hosanna-Tabor* than the former, it is fair to conclude that it was free exercise and not anti-establishment that was actually doing almost all of the work.

[58] *See* Town of Greece v. Galloway, 572 U.S. 565, 604 (2014) (Thomas, J., concurring in part and concurring in the judgment); Cutter v. Wilkinson, 544 U.S. 709, 726 (2005) (Thomas, J., concurring). Most recently, see Justice Thomas's concurring opinion in Espinoza v. Mont. Dep't of Revenue, 140 S. Ct. 2246 (2020), and his opinion concurring in the judgment in Am. Legion v. Am. Humanist Ass'n, 139 S. Ct. 2067 (2019).

majority of the Court,[59] convert into doctrine what is now just an impression. But even if non-incorporation never becomes the law in a formal sense, the effect is very much the same. There is little indication that the current Supreme Court majority is inclined to take Establishment Clause claims very seriously, with the Court's record over the past almost two decades constituting strong confirming evidence.

The Court has left a considerable amount of constitutional wreckage along the way to this state of affairs. One significant element of this wreckage is the demise of both the *Lemon* and endorsement tests, and it is worth saying a bit more about what brought about this state of affairs and what it might mean for the future.

The Court in *Shurtleff* makes it clear that Boston's proffered Establishment Clause justification for excluding religious flags from its otherwise open policy of raising any organization's flag on one of the flagpoles in front of City Hall has no grounding in the requirements of existing constitutional doctrine, and in fact violated the requirements of existing free speech doctrine. And although it would have been surprising for Justice Breyer, who wrote for the unanimous Court, to have taken the occasion to offer his own views about the Establishment Clause except to the extent necessary to support the outcome, Justice Gorsuch, in a concurring opinion, felt no such reluctance. He laid the blame at the feet of the *Lemon* test, and of Boston's unwillingness to recognize that *Lemon* had been "abandoned," "not applied," "exposed as an anomaly and a mistake," and "interred."[60] And he then went on to seek explanations for Boston's seeming neglect of the obvious, and he proceeded to offer two. One was a Legal Realist explanation: to him Boston was offering a doctrinally unsupported view of the requirements of the Establishment Clause in order to support Boston's own policy-based outcome preferences, in particular Boston's policy preference not to fly religious flags. The other was Boston's refusal to recognize the numerous shortcomings of the *Lemon* test and the now-ascendant requirement of interpreting the Establishment Clause according to its original meaning.

But there is a third explanation. It is a true that the Court has long come up with excuses for not applying or adhering to the *Lemon* test.

---

[59] In *Shurtleff*, Justice Gorsuch, concurring, expresses some cautious openness to this view. 142 S. Ct. 1583, 1606 n.6 (2022) (Gorsuch, J., concurring).

[60] *Id.* at 1603.

And it is true that various members of the Court over the years have long made reference to the (recent) historical fact of the test's non-use. But, tellingly, the Court has not once used the word "overrule" in describing the status of the *Lemon* test. Given that the Court knows how to overrule, and knows how to describe what it is doing using exactly that word, someone—Boston, for example—might be forgiven for applying the maxim *expressio unius est exclusio alterius* to reach the conclusion that the *Lemon* test retained at least some vitality. Indeed, given that Justice Sotomayor in *Kennedy v. Bremerton* explicitly describes the Court as having "overruled" *Lemon*, the reluctance of the Court to describe what it has done in those terms is curious and potentially misleading.

Perhaps the Court's reluctance to describe the overruling of *Lemon* (and the endorsement test) in those terms is a product of uncertainty about what might replace it. It is true that the Court's repeated statements about the obsolescence of *Lemon* have typically been accompanied by importunings to interpret the Establishment Clause according to the "original meaning as illuminated by history."[61] But this nice phrase elides the distinction, now ubiquitous in the originalist literature, between original public meaning, and the one hand, and original intentions, on the other, a distinction obscured by the Court's occasional conflation in the Establishment Clause context between linguistic meaning in 1791 (or 1868) and the "understanding of the Founding Fathers."[62] Given this confusion, it is perhaps not surprising that the Court is trying to say as little as possible about the demise of the *Lemon* test. But while it is attempting to under-commit about the details of what might replace the *Lemon* test, the Court might forgive government actors such as Boston in *Shurtleff* for feeling less constrained than the current Court majority wishes they would be.

III. Implications for Free Exercise

Do the recent cases rejecting Establishment Clause worries in various contexts portend anything about the future of free exercise doctrine? Arguably the answer is yes, and the shrinking of the Establishment Clause has been accompanied by the Court's willingness to expand the scope and strength of the Free Exercise Clause.

---

[61] *Am. Legion*, 139 S. Ct. at 2082.

[62] Kennedy v. Bremerton Sch. Dist., 142 S. Ct. 2407, 2428 (2022).

These conjoined developments were most apparent in Justice Gorsuch's opinion for the Court in *Kennedy v. Bremerton*. In upholding Kennedy's claims against his dismissal for public prayer on school time and school property, the Court emphasized, as noted above, that the school's actions were not only not required by the Establishment Clause but were also prohibited by the Free Exercise Clause as a restriction on Kennedy's individual free exercise rights. So too in *Carson*, over the extensive dissent by Justice Breyer on just this point. In *Carson*, Chief Justice Roberts's opinion for the Court emphasized that the Establishment Clause could provide no defense for a violation, in contradiction of what the Court perceived to be the mandates of the Free Exercise Clause, of the principle that it is constitutionally impermissible to "exclude otherwise eligible" beneficiaries of some state program "on the basis of their religious exercise."

This view of the Free Exercise Clause obviously raises questions about what the Court's decisions portend for the central free exercise question: the extent to which the Free Exercise Clause requires exemptions from laws of general application on the basis of those laws' restrictions on religious practice. And as so framed, this question is about the current vitality and future prospects of *Employment Division v. Smith*,[63] long under attack but still breathing, even if barely.

Smith stands for the proposition that the Free Exercise Clause is not violated by the application to religious believers and religious practices of laws of general application applied to religiously-motivated and non-religiously-motivated behavior alike. At first blush there seems no reason to believe that this principle is necessarily connected to questions about the Establishment Clause. A consistent commitment to neutrality, for example, would entail treating religion and non-religion equivalently, thus prohibiting special benefits to religion and allowing religion to be burdened in an incidental and non-selective way. But incidental burdens are still burdens, and the question, at least since *Sherbert v. Verner*,[64] is whether the Free Exercise Clause requires relief from these incidentally applied, and thus "neutral," burdens. Without getting into details, qualifications, and exceptions here, *Smith* rejects the view that the Free Exercise Clause mandates at least some

---

[63] 494 U.S. 872 (1990).

[64] 374 U.S. 398 (1963).

exceptions for religious practices from restrictive laws of genuinely general application.

The question is whether there is anything in the recent cases dealing nominally with the Establishment Clause that would bear on the vitality and the future of *Smith*. And although one answer to this question might be that there is nothing, because two different clauses are involved, the matter is somewhat more complicated. One way of understanding the basic *Smith* principle is that it rejects special privileges[65] for religious believers whose beliefs require them to engage in otherwise prohibited practices[66] or to refrain from otherwise required practices.[67] But *accepting* an argument for special privileges entails granting those privileges, and "privilege" in the ordinary as well as technical Hohfeldian sense, to religious believers but not to others. A question, then, is whether doing so, arguably in the service of free exercise values, would create Establishment Clause problems.

A host of recent controversies illustrates the issue. Does an exemption from the requirements of a civil rights statute grant to some religious believers benefits not available to those whose objections to the constraints of such statutes are pragmatic or economic?[68] Would granting religious exercises exemptions on free exercise grounds from Covid-based restrictions on group gatherings privilege religious meetings in violation of the Establishment Clause?[69]

Although there is nothing in the narrowest readings of the recent Establishment Clause cases discussed here that directly bears on the

---

[65] Not only in the ordinary and casual sense of a privilege, but in the stricter Hohfeldian sense, according to which a privilege is an exemption from an otherwise generally applicable law or other requirement. The privilege against self-incrimination, for example, is an exemption from what would otherwise be the requirement that all people with information relevant to a criminal prosecution can be required to provide that information to a court. And so too with various other evidentiary privileges, such as the spousal and clergy privileges, each of which immunize from compulsion those who are in possession of relevant and thus otherwise usable information.

[66] As in *Smith* itself, where the issue was about the claim of exemption on religious grounds from laws prohibiting the use of drugs that the *Smith* claimants argued was required by their religion as part of their religious observances.

[67] As with the claim in *Wisconsin v. Yoder*, 406 U.S. 205 (1972), for exemption from the otherwise generally applicable requirement of school attendance until the age of sixteen. Another obvious potential application is military conscription, but the Court persistently avoided the constitutional question by focusing on the interpretation of the statutorily-created conscientious objector exemption. *See* Welsh v. United States, 398 U.S. 333 (1970); United States v. Seeger, 380 U.S. 163 (1965).

[68] *Cf.* Masterpiece Cakeshop, Ltd. v. Colo. C.R. Comm'n, 138 S. Ct. 1719 (2018).

[69] *See* South Bay Pentecostal Church v. Newsom, 141 S. Ct. 716 (2021); Tandon v. Newsom, 141 S. Ct. 294 (2021).

issue, the broader implication of the recent cases is that such Establishment Clause concerns would be washed away just as the Establishment Clause has been washed away more generally. If one barrier to the demise of *Smith*, a decision that is increasingly condemned across otherwise salient political divides, is a fear of granting to religious practices and their practitioners privileges not available to others, and if that fear finds its home in the Establishment Clause, then nothing in the Court's recent history with the Establishment Clause would suggest that this concern will be find a congenial welcome at the current Supreme Court.[70]

Conclusion

The Court's now increasingly longstanding shelving of the Establishment Clause is at least partly a function of the Court's rejection of the view that there is a tension or conflict between the Establishment and Free Exercise Clauses. In light of how long this tension has been accepted,[71] the Court's rejection of it is curious. It is true that the two clauses appear in the same amendment, but that can be said as well about the grab bag of rights in the Fifth Amendment, and indeed about the various rights in the First and Sixth. The textual argument for denying the tension seems to be weak, and the textual argument for slighting something appearing so clearly in the text of the First Amendment seems weaker still.

Recognizing that independently valuable goals may at times conflict with each other is rarely easy. And there are various psychological explanations of people's discomfort with such conflict and how they deal with that discomfort.[72] But whatever the cause, denying the conflict between the worthy goals of the Free Exercise Clause and the equally worthy goals of the Establishment Clause

[70] The Court's most recent opportunity to reconsider *Smith* was *Fulton v. City of Philadelphia*, 141 S. Ct. 1868 (2021), in which the free exercise claim arose in the context of an arrangement by which Philadelphia had agreed with various non-profit organizations to manage the process of foster child placement. Catholic Social Services had been excluded because of its opposition to same-sex couples as foster parents, and it claimed the exclusion violated the Free Exercise Clause. Although *Smith* appeared to be relevant, the Court ruled unanimously that the discretionary aspect of Philadelphia's choosing foster placement agencies made *Smith* inapplicable.

[71] *See* Jesse H. Choper, *The Religion Clauses of the First Amendment: Reconciling the Conflict*, 41 U. Pitt. L. Rev. 673 (1980).

[72] Some of these explanations are explored and applied in Frederick Schauer, *Rights, Constitutionalism, and the Perils of Panglossianism*, 38 Oxford J. Legal Stud. 635 (2018).

seems to have produced some arguable increase in clarity at the cost of coming close to discarding the latter. As John Marshall reminded us in *Marbury v. Madison*,[73] an interpretation that renders part of the text superfluous is rarely to be preferred, and that is a lesson worthy of recognition by today's textualists. Reading the Establishment Clause out of the text is not only inconsistent with recognizing an important aspect of managing the relationship between religion and the state; it is inconsistent with an equally important aspect of respecting the text and thus of being governed by a written constitution.

[73] 5 U.S. (1 Cranch) 137 (1803).

MARIANO-FLORENTINO CUÉLLAR AND
AZIZ Z. HUQ

THE HIDDEN JUDICIAL SPRINGS OF U.S. FOREIGN POLICY

Seen from Foggy Bottom, the Washington, D.C., home of the State Department, today's geopolitical landscape is considerably more uncertain and hazardous than at the century's dawn.[1] For one thing, it is no longer unipolar. Twenty years ago, there were no challenges to U.S. hegemony in sight comparable to the present irredentist ambitions of nuclear-armed Russia unconstrained by arms-control agreements, or the remilitarized, economically, and technologically energized Chinese party-state. Rising tensions between the United States and China have shrunk global trade, decoupling technological links between the world's most powerful nation-states.[2] As unipolarity fades, nation-states are no longer converging toward some form of democratic capitalism buttressed by widely-embraced international-law systems for fomenting trade and reducing geopolitical conflict and

---

Mariano-Florentino Cuéllar is President of the Carnegie Endowment for International Peace. Aziz Z. Huq is the Frank and Bernice J. Greenberg Professor of Law at the University of Chicago Law School.

AUTHORS' NOTE: Thanks to Will Baude and Curt Bradley for helpful comments.

[1] For a survey of these challenges, see William J. Burns, *The United States Needs a New Foreign Policy*, CARNEGIE ENDOWMENT FOR INT'L PEACE (July 1, 2020), https://carnegieendowment.org/2020/07/14/united-states-needs-new-foreign-policy-pub-82295.

[2] *See* Jon Bateman, *US-China Technological Decoupling: A Strategy and Policy Framework*, CARNEGIE ENDOWMENT FOR INT'L PEACE (Apr. 25, 2022), https://carnegieendowment.org/2022/04/25/u.s.-china-technological-decoupling-strategy-and-policy-framework-pub-86897.

The Supreme Court Review, 2023.
© 2023 The University of Chicago. All rights reserved. Published by The University of Chicago Press. https://doi.org/10.1086/724920

economic instability. The halcyon days of a post-Cold War "Pax Americana" are done. Simultaneously, shocks to global order are gathering. Their effects are transmitted with increasing speed thanks to successful globalization—an international trading system linked by "just-in-time" inventory management; the transnational flow of culture and political ideas;[3] the physical migration of bodies; a growing merger of debt, equity, and derivative markets; and—of course—the sheer implausibility of confining epidemic sicknesses to national jurisdiction.

So it should be no surprise that ours should be a moment of anxiety and uncertainty for those making foreign-policy decisions in Foggy Bottom. "Foreign policy" has long been an amorphous and capacious term. It picks out all aspects of state functioning that substantially influence, intentionally or not, the geopolitical context in which the United States operates. It has never been possible to draw a crisp line demarcating foreign from domestic policy. But these new developments add new difficulties to the task. In part, this is because diplomacy and armed conflict, the traditional domains of state-to-state contestation, have been supplemented by new kinds of international contestation, such as cyber conflict and platform-based wars of propaganda and disinformation. And in part, it is because when a nation-state such as the United States makes a seemingly domestic policy determination, its choice often has powerful, unavoidable global spillovers.

Here is just one example: The 2018 discovery of significant natural gas reserves in the Permian basin in West Texas might seem a question of domestic energy law and policy.[4] Four years later, however, the U.S. capacity to supply excess national gas to the European market shaped perceptions and strategy respecting the Ukrainian war.[5] This was in stark contrast to the post-world-war period, in which the U.S. offered security guarantees for a Middle Eastern supply to Europe and Japan, while offering its own domestic reserves as a "last resort."[6]

---

[3] Even new populist strains of nationalism are global. The paradoxically global nature of new populist nationalisms is well developed in Benjamin R. Teitelbaum, War for Eternity: The Return of Traditionalism and the Rise of the Populist Right (2020).

[4] Ariel Cohen, *America's Oil And Gas Reserves Double With Massive New Permian Discovery*, Forbes (Dec. 21, 2018), https://www.forbes.com/sites/arielcohen/2018/12/21/americas-oil-and-gas-reserves-double-with-massive-new-permian-discovery/?sh=7d6711072c91.

[5] Anna Shiryaevskaya, *For the First Time, US Is Sending More Gas to Europe Than Russia*, Bloomberg (July 1, 2022), https://www.bloomberg.com/news/articles/2022-07-01/us-lng-supplies-to-europe-overtake-russian-gas-iea-says.

[6] Helen Thompson, Disorder: Hard Times in the Twentieth Century 41 (2022).

Energy policy is not an outlier. When the United States makes a domestic policy decision about its public health policy, its regime for patenting new technologies or licensing pharmaceuticals, its intensive regulation of its domestic financial markets, its monetary policy, or the range of speech available on its social media platforms, there are foreign-policy spillovers. Domestic policy choices hence shape the nation's geopolitical posture. They change the mix of available options for interacting with geopolitical allies and opponents. And they do so with a speed and an uncertainty that would be lacking in a less multipolar, less disordered and fragile, and less tightly concatenated world lacking in the trade, energy, migration, financial, and intellectual flows nurtured by fifty years' globalization. So it is that what first appears to be a transnational public health threat can rapidly become a domestic financial meltdown, which elicits not just financial but monetary policy changes that reshape both national and the international economic environments.[7] This is to say nothing, of course, of our shared exposure to climatic change and environmental catastrophe.[8] Whether or not the contemporary geopolitical dissolves the distinction between foreign and domestic policy entirely is a matter of reasonable debate. What is clear, however, is that the line has become much harder to draw. Its ambiguities have become more salient.

In this fraught context, the political economy of the Supreme Court has undergone a shift too. New appointments have moved the Court's ideal point sharply away from the preferences of the average American, and toward that of the average Republican.[9] Especially when confronting policies associated with the Democratic Party, the Court has evinced markedly less deference and has been manifestly more willing

---

[7] Consider, for example, the way in which the transnational COVID-19 shock in March 2020 created the prospect of a "horrifying" collapse in the market for U.S. Treasuries. ADAM TOOZE, SHUTDOWN: HOW COVID SHOOK THE WORLD'S ECONOMY 119 (2022). The Federal Reserve's response was to offer "limitless cash," and hence to "act not just as a lender of last resort, but a market marker." *Id.*

[8] Even the "adverse impacts of climate change 'on other countries can have spillover effects on the United States, particularly in the areas of national security, international trade, public health, and humanitarian concerns.'" Cass R. Sunstein, *Climate Change Cosmopolitanism*, 39 YALE J. ON REGUL. 1012, 1026 (2022) (citation omitted).

[9] Stephen Jessee, Neil Malhotra & Maya Sen, *A Decade-Long Longitudinal Survey Shows that the Supreme Court Is Now Much More Conservative Than the Public*, 119 PROC. NAT'L ACAD. SCI. (2022).

to break from precedent or majoritarian sentiment.[10] The result is a more muscular form of judicial review less inhibited by concern for destabilizing repercussions. This newly empowered Court is taking up questions ranging from treatment of migrants at the Southern border[11] and "war on terror" detainees[12]; the power of the federal government to address climate change; the existence and operation of executive branch agencies tasked with key elements of financial-sector oversight[13]; and, indeed, even the physical siting of natural gas pipelines.[14] No policy domain seems insulated from its reach. The questions raised in each of these cases (if not their specific resolutions) fall on the newly febrile seam between domestic and foreign policy.

There is, to be sure, a longstanding place for judicial supervision in foreign policy—and even some occasional recognition in business and academic circles that judicial decisions can shape international policy developments.[15] Indeed, lively debates about where to draw the putative boundary between foreign affairs and public law disputes about domestic policy infuse such architectonic decisions as *Youngstown Sheet and Steel.* Justice Black's majority opinion rejected an expansive account of the "theater of war" as license for President Truman's

---

[10] One of us has presented evidence to the effect at length in Aziz Z. Huq, *The Counter-Democratic Difficulty*, 117 Nw. U. L. Rev. 1099 (2023).

[11] Biden v. Texas, 142 S. Ct. 2528, 2535 (2022) (reviewing the legality of the "Remain in Mexico" program).

[12] United States v. Zubaydah, 142 S. Ct. 959, 997 (2022) (Gorsuch, J., dissenting) (discussing treatment of detainee in "black site").

[13] Seila L. LLC v. Consumer Fin. Prot. Bureau, 140 S. Ct. 2183, 2197 (2020) (discussing the constitutionality of the Consumer Finance Protection Bureau).

[14] PennEast Pipeline Co., LLC v. New Jersey, 141 S. Ct. 2244, 2251–52 (2021) (holding that pipeline companies can exercise statutory eminent domain authority to condemn a rights-of-way in which a State has an interest). To be clear, PennEast was not a decision with immediate geopolitical ramifications. But the Supreme Court's interventions on energy regulation do shape grid reliability in potentially dramatic ways. *See, e.g.*, Alexandra Klass, Joshua Macey, Shelley Welton & Hannah Wiseman, *Grid Reliability Through Clean Energy*, 74 Stan. L. Rev. 969, 982 (2022).

[15] Susan Strange, States and Markets 21 (2d ed. 1994) ("[I]t is a recognized fact *in business circles* that decisions taken by the U.S. Supreme Court, and sometimes by lesser courts, . . . may be of critical importance far beyond the border of the country." (emphasis added)). That Professor Strange mentions court decisions in her seminal tome on international political economy is as telling as how she treats the point. To wit: she considers the importance of courts (or, indeed, of domestic administrative agencies whose powers are ultimately interpreted and bounded by court decisions) to be rarely acknowledged among scholars of international relations and international political economy that instead tend to focus excessively on "inter-governmental relations." After noting that the judiciary's role as a key actor in foreign affairs is recognized instead in "business circles," she does not further address the topic in the book.

seizure orders.[16] Chief Justice Vinson's dissent, in contrast, nested the case in the context of a "terrifying global conflict."[17] In effect, majority and dissent disagreed over how to draw the line between foreign and domestic.

Yet recent developments mark a departure from tradition. There is increasingly widespread awareness of the entanglement between the domestic and the international, and also of the growing assertiveness of the U.S. Supreme Court. Under the influence of these factors, the Court's de facto role in American foreign policy is growing. For two reasons, the changes that the Roberts Court is working to the domestic policy-making environment also contrast with earlier phrase shifts in structural constitutional jurisprudence: First, they are likely to be more consequential in foreign-policy terms. Even when a specific line of cases does not dramatically reengineer the scope or location of regulatory authority, it can seed uncertainty. This alone undermines effective policy-making.

Second, these cases are also a league apart from the prevailing ethos of the Rehnquist Court's so-called federalism revolution, which had only modest effects on the national policy-making environment.[18] The shifts underway therefore ought to rekindle longstanding debates about institutional choice, especially respecting who should weigh, and who should ignore, the complex implications traced by the repercussions of putatively domestic legal interpretations and policies in the global sphere. How and when this expected judicial role came about constitute the core descriptive questions we set out to answer in this Article, before exploring what the resulting recalibration of authority portends geopolitically for the United States, and what costs and dilemmas the country could face as elements of foreign affairs become increasingly judicialized.

Although these questions are not entirely unfamiliar, the increasingly urgent project of mapping answers falls within a persistent gap, one located not only between domains of institutional responsibility anchored at the State Department and the U.S. Supreme Court respectively but also between modes of scholarly analysis of foreign

---

[16] Youngstown Sheet & Tube Co. v. Sawyer, 343 U.S. 579, 587 (1952).

[17] *Id.* at 668 (Vinson, C.J., dissenting).

[18] For a nuanced view of this jurisprudence and its political impact, see Timothy J. Conlan & François Vergniolle De Chantal, *The Rehnquist Court and Contemporary American Federalism*, 116 Pol. Sci. Q. 253, 265 (2001).

affairs. When scholars and analysts investigate into how decision-makers forge foreign policy in the United States, they generally use one of two disciplinary lenses. These are foreign relations law (FRL) on the one hand, and what could be termed foreign policy analysis (FPA) on the other. The first is essentially a subfield of public law familiar to a sprawling assemblage of law professors, military lawyers, sanctions attorneys, and judges. The second is more familiar to foreign policy practitioners with a scholarly bent and sounds in the key of international relations and political science. Our threshold ambition is to demonstrate that neither FRL nor FPA—at least as presently practiced—captures the role of the Supreme Court as newly configured under Chief Justice Roberts. The resulting lacuna is likely to become more important as the effect of new appointments by President Donald Trump cascades through constitutional and statutory jurisprudence.

In Part I, we describe how the paradigmatic exponents of each field—legal scholars in FRL and political scientists in FPA—tend to define their respective domains. Despite some redeeming value in each perspective, we contend that their collective contributions offer no more than a partially occluded picture of how foreign policy emerges. Part II draws distinctions between three elements of foreign-policy creation—instruments, channels, and foundational resources. Part III, the core of the piece, substantiates our central claim of a meaningful gap in understandings of such policy's fabrication. We identify four recent lines of recent jurisprudence, all involving meaningful, judicially-driven changes to the law that have substantial foreign affairs implications. These concern legislative delegations; executive-branch centralization; the management of cross-border flows of bodies; and the infrastructure of national democratic self-rule. Across all four lines of cases, recent decisions by the Roberts Court are changing the instruments, channels, or foundational resources of foreign-affairs policymaking. Concluding, Part IV offers brief reflections on the normative implications of recovering this all-too hidden dimension of judicial interventions.

I. Theorizing the Wellsprings of American Foreign Policy

Lawyers and policymakers play their part in setting the country's foreign policy agenda, broadly defined, within a system subjecting them to a variety of interpretive and practical constraints as they pursue personal or institutional interests. But their actions are also guided by clusters of interrelated ideas shaping how these civic actors make sense of those constraints and interests, how they read from their metaphorical scripts, and indeed, when they start searching for entirely new lines to read.[19] Ideas about the sources of American foreign policy are theorized in two different corners of the academy (to a first approximation)—the law school and the political science department. Yet these two approaches—both valuable and important—leave a blind spot. To map the latter, we start by describing the scope and central questions addressed in these two bodies of scholarship. We then point out a striking gap plainly visible on the map between these approaches.

A. THE CANON OF FOREIGN RELATIONS LAW

The conventional domain of FRL is illustrated by the scope of the field's leading casebook.[20] After an introductory chapter on early constitutional history, the leading FRL casebook carves up its topic into three general domains—the role of Congress, the presidency, and the states in relation to other nations;[21] the handling of international law, including treaties, executive agreements, and customary international law;[22] and finally, areas of policy wherein the United States has access to a variety of legal and policy tools—but also faces legal and practical constraints—as it directly confronts violent transnational threats of crime, terrorism, and war.[23] At first blush, this organization of topics seems arbitrary: It tacks between questions of institutional choice, the choice of law, and specific, narrow policy domains. But its seeming haphazard choice has its own inner logic. It reflects a coherent and defensible pedagogical orientation of the law school curriculum toward the likely professional pathways of graduates. That is, FRL as taught from the casebook offers foundational training for lawyers headed to agencies such as the State and Defense Departments, and perhaps the intelligence community, Treasury, and

---

[19] *See* Judith Goldstein & Robert O. Keohane, Ideas and Foreign Policy: Beliefs, Institutions, and Political Change (1993).

[20] Curtis A. Bradley, Ashley S. Deeks & Jack L. Goldsmith, Foreign Relations Law: Cases and Materials (7th ed. 2020).

[21] *Id.* at 137–264.

[22] *Id.* at 265–552.

[23] *Id.* at 553–846.

Commerce. Reasonably, FRL draws attention to issues of professional concern to these lawyers.

The scholarly literature on FRL reflects the same delimited domain. It is organized along two main axes: the division of institutional power between branches (or levels) of government and the effects of various forms of "international law."[24] Central FRL topics include the scope of exclusive Article II authority over foreign affairs;[25] the choice of international law-making form (and especially the choice between treaties and executive agreements[26]), and the availability of individual redress in domestic courts for violations of international norms.[27] FRL, however, is not concerned with how the United States influences that larger environment *generally*. It makes no claim to address all or even most likely pathways by which domestic dynamics influence the geopolitical sphere. Hence, a central FRL question is the scope of deference afforded other branches (or, by extension, states) in their interactions with other states or in matters of war and terrorism in cases that fall within a heartland definition of foreign relations.[28] The narrow boundaries of that heartland in FRL likely are as much a testament to the undeniably disruptive potential of war as they are to the role war plays as a kind of Weberian-ideal type proxy for the fragility lurking in notions of national sovereignty underlying most versions of international and domestic law.

To observe that FRL offers a partial account of foreign policy-making is not, to be clear, a criticism. All scholarly taxonomies cut the

---

[24] David H. Moore, *The Missing D in U.S. Foreign Relations Law*, 109 Geo. L.J. 1139, 1141 (2021) (defining foreign-relations law in terms of "the distribution of foreign affairs powers between the President, Congress, and the courts; the relative foreign relations roles of state and federal governments; and the domestic legal status of international law"); *see also* Curtis A. Bradley, *What Is Foreign Relations Law?*, *in* The Oxford Handbook of Comparative Foreign Relations Law 3, 4 (Curtis A. Bradley ed., 2019). A similar array of topics is addressed in the classic text Louis Henkin, Foreign Affairs and the Constitution (1996).

[25] *See, e.g.*, Curtis A. Bradley & Martin S. Flaherty, *Executive Power Essentialism and Foreign Affairs*, 102 Mich. L. Rev. 545 (2004).

[26] *See, e.g.*, Curtis A. Bradley & Jack L. Goldsmith, *Presidential Control over International Law*, 131 Harv. L. Rev. 1201 (2018).

[27] *See, e.g.*, Christopher Ewell, Oona A. Hathaway & Ellen Nohle, *Has the Alien Tort Statute Made a Difference?: A Historical, Empirical, and Normative Assessment*, 107 Cornell L. Rev. 1205, 1206 (2022).

[28] *See, e.g.*, Eric A. Posner & Cass R. Sunstein, *Chevronizing Foreign Relations Law*, 116 Yale L.J. 1170, 1178 (2007); Daniel Abebe, *Great Power Politics and the Structure of Foreign Relations Law*, 10 Chi. J. Int'l L. 125 (2009) (exploring "the relationship between the strength of external international political constraints on a state and the levels of judicial deference to the executive in that state").

world up based on pedagogical and disciplinary demands. The descriptive observation we want to stress here, however, is simply that FRL as a subfield of law has been understood and instituted to exclude some potentially important pathways.

B. FOREIGN POLICY ANALYSIS AS A SUBFIELD OF INTERNATIONAL RELATIONS

The leading approaches to international relations (IR) tend to "envisage[] states as unitary and purposive actors that consider what other states will or might do when they choose foreign policies."[29] A widely-cited theorist writing in the realist tradition of IR, Kenneth Waltz, thus strips away all attributes of states, instead conceiving of them as unitary actors pursuing policies to ensure their own survival.[30] Obviously, work in this vein has little to say about the institutional sources of a nation's foreign policy. But certain broad approaches to foreign policy eschew black-box approaches to explain "why particular states make particular foreign policy moves at particular times."[31] FPA is one such subfield of IR that, drawing from a range of theoretical approaches and empirical agendas, instead interrogates how foreign policy in fact emerges from within the state.

Like IR more generally, FPA encompasses different perspectives about the drivers of foreign policy decisions. It has neorealist, liberal internationalist, and "bureaucratic politics" strands. We get a general sense of FPA's scope again by looking at leading textbooks. The main actors in the accounts offered in these textbooks are bureaucracies,[32] a diffuse general public,[33] and intermediating bodies such as parties and unions.[34] FPA scholars offer, for example, detailed accounts of how

---

[29] James D. Fearon, *Domestic Politics, Foreign Policy, and Theories of International Relations*, 1 ANN. REV. POL. SCI. 289, 291 (1998).

[30] KENNETH WALTZ, THEORY OF INTERNATIONAL POLITICS 91–92, 99 (1979); *see also* JOHN MEARSHEIMER, THE TRAGEDY OF GREAT POWER POLITICS 30–31 (2001) (offering a similar set of assumptions).

[31] Fearon, *supra* note 29, at 295.

[32] CHRIS ALDEN & AMNON ARAN, FOREIGN POLICY ANALYSIS 31–45 (2012); Valerie M. Hudson, *The History and Evolution of Foreign Policy Analysis*, *in* FOREIGN POLICY: THEORIES, ACTIONS, CASES 19 (Steve Smith et al. eds. 2012); VALERIE M. HUDSON & BENJAMIN S. DAY, FOREIGN POLICY ANALYSIS: CLASSIC AND CONTEMPORARY THEORY 106–16 (3d ed. 2020).

[33] ALDEN & ARAN, *supra* note 32, at 54–60; Piers Robinson, *The Role of Media and Public Opinion*, *in* FOREIGN POLICY: THEORIES ACTIONS, CASES, *supra* note 32, at 168–87; HUDSON & DAY, *supra* note 32, at 145–50.

[34] ALDEN & ARAN, *supra* note 32, at 60–61; HUDSON & DAY, *supra* note 32, at 146.

public opinion generates distinct policy outputs in different countries as a consequence of being filtered through distinct formations of political parties and coalitions.[35]

Our claim is not that scholars agree on every detail or core presumption when they espouse and labor to enhance what we have termed the FPA perspective. Debating the merits of paradigmatic, big-picture perspectives on international politics is table stakes among most international-relations faculties. Generally, too much is made of these abstract debates.[36] But there is no question that classical realists who recognize the relevance of history and internal politics see the world differently from adherents of hyper-rationalism heavily influenced by rational-expectations economics or structural realists who tend to describe global power in the unforgiving terms of a zero-sum game.

When it comes to analyzing applied policy questions or historical episodes, however, scholars interested in understanding the drivers of particular foreign policy decisions—that is, FPA scholars—are often less divided than the vigor of their theoretical debates might imply. This can be seen in three ways. First, whether or not analysts pledge allegiance to particular theoretical camps, they often devote considerable attention to how national governments use tools of diplomacy, economic statecraft, and security to advance a set of national interests. One need not accept that the attempt to advance such interests always results in coherent or normatively defensible positions. It is enough for scholars working on interpreting the behavior of senior officials in the United States and other countries to embrace, as a working presumption, that policymakers and bureaucracies act as though they can formulate reasonable policies to advance shared national concerns.

Second, aside from the strictest adherents to hyper-rationalism and structural realism, scholars of foreign policy are open to the idea that a variety of historical antecedents, societal factors, and economic realities can shape a nation's priorities as well as its leaders' ability to advance those priorities.

Third, even scholars interested in how American foreign policy priorities are affected by sources of domestic bureaucratic, political,

---

[35] *See* Thomas Risse-Kappen, *Public Opinion, Domestic Structure, and Foreign Policy in Liberal Democracies*, 43 World Pol. 479 (1991).

[36] Jonathan Kirshner, An Unwritten Future: Realism and Uncertainty in World Politics 2 (2022) ("The almost ritual rehearsal of clashes between realism and liberalism—the nadir of which was the academic 'paradigm wars of the 1990s' has been as ubiquitous in IR theory as it has been unproductive.").

and economic power spend relatively little attention focused on the judiciary as a distinctive institution: one putatively defined by values antithetical to conventional partisan or interest group politics yet undeniably entangled with the political process.

To wit: while many exemplars in the FPA literature offer detailed accounts of how bureaucracies, the tides of public opinion, and the vicissitudes of interest-group contestation shape foreign policy, they are largely silent as to the courts. Leading FPA casebooks and scholarly contributions to the field respecting U.S. foreign policy-making include few references to the Supreme Court. None of the main textbooks we examined even reference the Court in their indexes.

C. THE GAP IN FOREIGN POLICY ANALYSIS

The FRL and FPA literatures are both concerned mainly with how policy on geopolitical questions arises through the institutions of national government. Both focus on the production and application of what is formally denominated international law (treaties, executive agreements, or customary international law) as well as foreign policy more generally. FRL tends to take a narrow view of courts. It trains on their role reviewing a quite cabined set of decisions denominated as foreign affairs. In contrast, FPA is silent about the judicial role.

What, then, is missing? If a nation state takes decisions that influence its geopolitical situation outside the ordinary bureaucratic channels—the State Department, the Defense Department, and the White House[37]—and without overt recourse to a formal instrument of international law, it will likely fall outside the existing parameters of either perspective. Neither FRL nor FPA, that is, accounts for the possibility that institutions and policies uncontroversially denominated as "domestic" could have a large effect on geopolitics, and that their decisions could be subject to judicial review.

II. Mapping Foreign Affairs Policy-Making: A Stylized Taxonomy

In analyzing this gap, it is helpful to rely on a standard vocabulary. But FRL and FPA focus on different actors by and large and

---

[37] Foreign relations law also pays some attention to the several states' conduct. *See, e.g.*, Daniel Abebe & Aziz Z. Huq, *Foreign Affairs Federalism: A Revisionist Approach*, 66 Vand. L. Rev. 723 (2013).

advance distinct inquiries. FPA scholarship focuses on how conventional domestic politics, interest-group maneuvering, and "organizational behavior" produce foreign policy outcomes.[38] It describes "spheres of competence of the actors involved, communication and information flow, and motivations of the various players."[39] FRL scholarship takes as its basic objects entities envisaged by the Constitution: a legislative, an executive, and the several states.[40] It then aims to deduce legally permissible modes of policy-making by those bodies. That these approaches lack as much as even a common vocabulary is no surprise.

Defining such a shared vocabulary can better help us map and label the space between these literatures, and describe the connections between legal premises and geopolitical effects. This means defining the scope of "foreign affairs" as we use the term, and then picking out three elements involved in its domestic fashioning: instruments, channels, and foundational resources.

We use the term "foreign relations" or "foreign affairs" of the United States to encompass state actions that alter materially the geopolitical environment influencing the United States, without regard to whether such actions are channeled or executed through the State Department, Defense Department, intelligence community, or other agencies.[41] Our definition is not static. When Robert Kaplan predicted in 2002 that growing environmental degradation would ignite new, violent conflicts around the world,[42] his claim was both novel and controversial.[43] Not so much anymore. Even the Director of National Intelligence has recognized time and again in recent years

---

[38] Valerie M. Hudson, *Foreign Policy Analysis: Actor-Specific Theory and the Ground of International Relations*, 1 For. Pol. Analysis 1, 6 (2005).

[39] *Id.* at 6.

[40] Bradley et al., *supra* note 20, at 137–264.

[41] We exclude the influence of private citizens on geopolitical dynamics. *Cf.* Hannah Getahun, *'Elon, The Everywhere'—Elon Musk Is Worrying Government Officials with His Growing Presence in Global Affairs, Report Says*, Bus. Insider (Oct. 23, 2022), https://www.businessinsider.com/elon-musk-foreign-presidents-prime-ministers-washington-report-2022-10. Of course, it is a familiar point that the power of wealthy individuals such as Musk to exercise influence is itself a function of domestic policies of antitrust, taxation, and corporate governance.

[42] Robert D. Kaplan, The Coming Anarchy: Shattering the Dreams of the Post-Cold War (2002).

[43] *See, e.g.*, Marc A. Levy, *Is the Environment a National Security Issue?*, 20 Int'l Sec. 35 (1995).

the national-security implications of climate change.[44] In a similar vein, public health once seemed a parochial matter, an intricate challenge readily addressed through domestic laws and policies. After the COVID-19 pandemic, however, the idea that the determinants of American mortality and morbidity lie solely within national borders seems fanciful.[45] Even domestic democratic choice has an international dimension. Buttressing the success and stability of American democracy was a material element of American Cold War strategy—though often-elided at the time in the public understanding and imagination.[46] More recently, authoritarian regimes in Russia and China have been spending considerable resources shaping that putatively domestic discourse.[47] As a result of these developments, American policymakers are ever more aware that the United States must address "a range of threats, including nonstate actors and nonmilitary and nonhuman threats, such as economic crises, cybersecurity, infectious disease, climate change, transnational crime, and corruption, which are often unmoored from interstate rivalries."[48]

This extended account of foreign policy is controversial. Partisan identity predicts a person's views about the extent and costs of anthropogenic climate change.[49] The consumption of online misinformation is driven disproportionately by users with the strongest partisan identifications.[50] Many of these users would likely resist the idea that they are entangled in a foreign policy matter. A precise definition of "foreign affairs" is hence elusive, or perhaps incorrigibly partisan.

---

[44] Daniel R. Coats, *Statement for the Record: Worldwide Threat Assessment of the U.S. Intelligence Community*, OFF. OF THE DIR. OF NAT'L INTEL. 21–23 (2019), https://perma.cc/7BFG-N3CN.

[45] Predictions of future pandemics will likely rely on computational tools commonly called artificial intelligence, or AI. But AI prediction requires large pools of data from which models can be derived. Pooling information to create the necessary training data for global disease surveillance necessarily depends on international cooperation.

[46] *See, e.g.*, MARY DUDZIAK, COLD WAR CIVIL RIGHTS: RACE AND THE IMAGE OF AMERICAN DEMOCRACY (2000).

[47] DAVID SLOSS, TYRANTS ON TWITTER: PROTECTING DEMOCRACIES FROM INFORMATION WARFARE 5–112 (2022).

[48] J. Benjamin Heath, *The New National Security Challenge to the Economics Order*, 129 YALE L.J. 1020 (2020).

[49] Alec Tyson & Brian Kennedy, *Two-Thirds of Americans Think Government Should Do More on Climate*, PEW RSCH. CTR. (June 23, 2020), https://www.pewresearch.org/science/2020/06/23/two-thirds-of-americans-think-government-should-do-more-on-climate.

[50] Mathias Osmundsen et al., *Partisan Polarization Is the Primary Psychological Motivation Behind Political Fake News Sharing on Twitter*, 115 AM. POL. SCI. REV. 999, 999–1000 (2021).

Even with this caveat in mind, it is useful to distinguish three different elements of the process for making foreign policy: instruments, channels, and foundational resources.

First, state action does not simply spring from the ether fully-formed in the international system. It is undertaken through specific *instruments*. These include federal and state laws, agency regulations and decisions; government procurement and funding decisions; judicial opinions; and presidential executive orders and hortatory statements from the White House. In the United States, the range of available instruments is determined in part by the Supreme Court's gloss on the Constitution. Its 1983 decision invalidating the legislative veto, for instance, eliminated one tool.[51] Its 2008 decision restricting the president's power to direct state courts via a memorandum to comply with a judgment of the International Court of Justice struck out yet another.[52]

Second, the choice of instruments is related to a range of *channels* through which government policy-making can flow. A channel is a *sequence of institutional steps* used to deploy a specific instrument. Change to instrument choice reshapes the channels of foreign policy-making. Invalidation of the legislative veto, for instance, eliminated a potential step in the policy-making process. It thus pushed policy-making into a different sequence of institutional venues. Available channels differ because they are characterized by different veto-gates staffed by distinct players.[53] Each channel is associated with a different "win set," i.e., a different domain of potential policy outcomes agreeable to all veto players. Choice of channel, and concomitant veto-gates, drives the choice of legal instrument. Hence, recent scholarship has identified a "heavy reliance" on "ex ante congressional-executive agreement" that skirt around the Senate's supermajority rule for committees and ex ante public scrutiny of agreed-upon texts.[54] If the Supreme Court were

[51] Immigr. & Naturalization Serv. v. Chadha, 462 U.S. 919 (1983).

[52] Medellin v. Texas, 552 U.S. 491 (2008).

[53] George Tsebelis, *Veto Player Theory and Policy Change: An Introduction*, *in* Reform Processes and Policy Change (Thomas König, Marc Debus & George Tsebelis, eds., 2011) ("[P]olicy stability is the effect of a constellation of "veto players" . . . individual or collective actors whose agreement is necessary for the change of the status quo.").

[54] Oona A. Hathaway et al., *The Failed Transparency Regime for Executive Agreements: An Empirical and Normative Analysis*, 134 Harv. L. Rev. 629, 633 (2020) ("The authorization

to hold that executive agreements were constitutionally impermissible,[55] that would alter not just the instrument but also the larger channel through which foreign policy can be made.

Finally, we use the term *foundational resources* to refer to capabilities of a national polity, whether traits of government or of society more generally, that de facto or de jure determine the credibility and the efficacy of a state's actions in the geopolitical sphere.[56] The quality and existence of a foundational resource turns on both state and nonstate action. Crucially, the absence of a foundational resource can incapacitate the state from advancing its interests internationally.

The most obvious and salient example of a foundational resource in geopolitics is military or coercive capacity. But in the multipolar context of the early twenty-first century, where plural battlefields are in play, there is more than one kind of relevant coercive authority.[57] The U.S. capacity to supply arms to Ukraine and to credibly commit to NATO partners to respond to Russian military incursions shapes outcomes around the Black Sea. In facing China, it is not just the Seventh Fleet's firepower but relative national competencies in artificial intelligence (AI) and quantum computing that will determine the terms of future contestation.[58] Military capacity hence also turns on the pace of domestic scientific development and the creation of new

---

counts as full consent to the agreement prior to the President's negotiation of it. This means that the President can make and conclude the agreement, and thus render it legally binding on the United States, before Congress or the public even knows of its existence.").

[55] Bradford R. Clark, *Domesticating Sole Executive Agreements*, 93 Va. L. Rev. 1573, 1576 (2007) (rejecting "the conclusion that Article II permits the President to alter preexisting rights under state and federal law simply by making a sole executive agreement").

[56] *See, e.g.*, Kirshner, *supra* note 36, at 12. Kirshner emphasizes the potential consequences of lost social cohesion for a variety of material and more inchoate (but no less important) foundational resources:

> [S]ocieties—even apparent great powers—can rot from within, and . . . this, even more than the external threat environment, can determine the prospects for their survival. A fearsome-looking, muscle-bound fighter might prove to have a glass jaw, and fetishizing the physique (apparent power) risks overlooking less visible but ultimately decisive vulnerabilities (social cohesion).

[57] 2022 National Defense Strategy 2 (Oct. 27, 2022), https://www.defense.gov/National-Defense-Strategy (defining Russia and China as threats, but identifying different resources needed to address them).

[58] *Id.* at 6; *see also* Kai-Fu Lee, AI Superpowers: China, Silicon Valley, and the New World Order 228–32 (2018).

"fabs"[59] on U.S. territory, among other things. It is not just a consequence of what is happening in the State or Defense Departments.

But there are other, more subtle resources a nation might draw upon to act credibly in the geopolitical sphere. One is the scope of executive legal authority to act in emergencies. The greater a state actor's discretionary authority as a matter of domestic law, the larger their capacity to threaten external action. The executive's emergency capabilities may not be evenly distributed across policy domains. It might instead be that the government has greater leeway to respond to security threats such as terrorism than in respect to public health threats such as COVID-19. This might follow from the selection of tools in its repertoire, and the range of instruments ranked as *ultra vires*.

Another potentially salient foundational resource is the systemic robustness of democratic institutions. This might take different forms, as a state might have either shallow or deep democratic institutions. These might be able to channel and sustain public support for a costly foreign affairs policy that would otherwise stand on infirm ground.[60] Trust in government of the sort engendered by democratic choice also operates as a policy-relevant resource because it induces greater compliance with individual-level behavioral adaptations elicited by geopolitical crisis.[61]

A final system-level "foundational resource" pertains to the strength and endurance of the kinds of legal constraints on state power that can support the tacit bargains among people, interests, and institutions underlying the capacity and legitimate authority of the state.[62] Following Lon Fuller, we can call this "legality."[63] It is the ability of a legal

---

[59] *See* Ian King & Debbie Wu, *How 'Chip War' Puts Nations in Technology Arms Race*, WASH. POST (Jan. 31, 2023), https://www.washingtonpost.com/business/how-chip-war-puts-nations-in-technology-arms-race/2023/01/31/b0f15b1e-a153-11ed-8b47-9863fda8e494_story.html.

[60] The idea that democratic threats to use military force are more credible than similar threats from nondemocracies, however, has been cast into doubt. Alexander B. Downes & Todd S. Sechser, *The Illusion of Democratic Credibility*, 66 INT'L ORGS. 457 (2012).

[61] This was true during the COVID-19 pandemic. *See* Stefano Pagliaro et al., *Trust Predicts COVID-19 Prescribed and Discretionary Behavioral Intentions in 23 Countries*, 16 PLOS ONE (2021) (finding that "considering psychological differences in terms of trust toward different agents—governments, citizens, and science—provide a more informative picture of individuals' reactions to COVID-19").

[62] *See, e.g.*, Mariano-Florentino Cuéllar, *From Doctrine to Safeguards in American Constitutional Democracy*, 65 UCLA L. REV. 1398 (2018).

[63] LON FULLER, THE MORALITY OF LAW 39 (1964). This is closely related to the Anglo-American idea of the rule of law. *See* AZIZ Z. HUQ, THE RULE OF LAW: A VERY SHORT INTRODUCTION (forthcoming 2023).

system not just to provide reliable guidance to citizens on the likely legal consequences of their actions, but also its capacity to generate a "particular quality of relationship between the lawgiver and the legal subject," characterized by "reciprocity."[64] The salient intuition here is that the existence of that kind of bond between sovereign and citizen allows for a different range of geopolitical actions.

III. Hidden Judicial Springs of Foreign Affairs Policy

The range of available instruments, channels, and foundational resources of U.S. foreign policy turns on constitutional law. And whatever the theorists may say, constitutional law on the ground is in very large part a function of what a particular iteration of the Supreme Court holds. It follows that when the Court changes tack in respect to salient aspects of constitutional law, and its decisions have implications for foreign policy, the instruments, channels, and foundational resources of policy-making will also change.

The last decade has been a period of sharply felt change in constitutional jurisprudence. Recent appointments to the bench have pushed the Court toward the right (in relation to the country as a whole) in dramatic ways.[65] Some of the resulting changes of law have explicitly concerned FRL. In 2015, for example, the Roberts Court spoke directly to the scope of the president's recognition power in respect to foreign states.[66] It has also narrowed substantially the scope of the Alien Tort Statute (ATS) in several cases.[67] Because our aim in this piece is to illuminate "hidden" springs of judicial influence on foreign policy, we do not directly address these lines of cases here: They are well covered in the FRL literature.

Instead, we ask how recent shifts in "non-foreign-affairs-related" constitutional law have reshaped the instruments, channels, and the

[64] Kristen Rundle, Forms Liberate: Reclaiming the Jurisprudence of Lon L Fuller 90–92 (2012).

[65] Jessee et al., *supra* note 9.

[66] Zivotofsky v. Kerry, 576 U.S. 1, 10 (2015) (recognizing that the Reception Clause provides the President with exclusive authority to determine the sovereign status of foreign countries, regardless of congressional intent).

[67] Nestlé USA, Inc. v. Doe, 141 S. Ct. 1931, 1937 (2021) (holding that allegations of "operational decisions" and other "general corporate activity . . . do not draw a sufficient connection between the cause of action respondents seek—aiding and abetting forced labor overseas—and domestic conduct"); Jesner v. Arab Bank, PLC, 138 S. Ct. 1386, 1403 (2018) (ATS actions cannot be brought against foreign corporations); Kiobel v. Royal Dutch Petroleum, 569 U.S. 108, 119 (2013) (finding that ATS did not apply extraterritorially).

foundational resources upon which the United States's capabilities in the geopolitical domain rest. By focusing on recent changes, we do not mean to suggest that the Court's earlier interventions did not influence the path of U.S. foreign relations, or vice versa. One of us has written previously about how the influence of early twentieth-century Supreme Court decisions on executive-branch structure informed "the relationship between bureaus carrying out legal responsibilities and sources of external pressure such as the White House, lawmakers, or civil society organizations."[68] Plainly, the shifting contours of federal regulatory authority in the first decades of the twentieth century shaped American geopolitical power projection both directly and indirectly. Examining a later period of the twentieth century, Mary Dudziak has demonstrated that the Warren Court's desegregation decisions were understood by many as a part of Cold War geopolitical contest with Communism.[69] The project of juris-centric American constitutionalism has long been nested in causally complex geopolitical dynamics.[70]

We identify four recent lines of decisions that alter instruments, channels, and the foundational resources in material ways. For each, we start by identifying the salient judicial opinions. We then plot the ways in which instruments, channels, or foundational resources of foreign policy have been consequently changed by them. For each line of cases, the change is hard to quantify with precision. We nevertheless posit that the foreign-policy effects of each line of cases rises above the de minimus. It is indeed likely substantial. Over time, we will argue, these decisions seed significant effects on the geopolitical context and choices confronting the United States.

The first concerns the national power to regulate, and changes to the instruments (and hence channels) through which the United States exercises global regulatory leadership. The second pertains to the structure of executive-branch decision-making, and the foundational resource of intragovernmental expertise. A third relates to managing the flow of human bodies across borders—in part what is commonly called

---

[68] Mariano-Florentino Cuéllar, Governing Security: The Hidden Origins of American Security Agencies 214 (2013).

[69] Mary L. Dudziak, Brown *as a Cold War Case*, 91 J. Am. Hist. 32, 42 (2004).

[70] *See, e.g.*, Richard Primus, Note, *A Brooding Omnipresence: Totalitarianism in Postwar Constitutional Thought*, 106 Yale L.J. 423, 424 (1996) (examining "the role of antitotalitarianism in postwar judicial doctrine").

immigration law, and treated separately from foreign-relations matters. The fourth and final zone turns on the quality of democratic choice. We conceptualize this as a foundational resource, the corrosion of which is significant for both domestic and external-facing reasons.

A. THE NATIONAL POWER TO REGULATE

1. *The New Major Questions Doctrine.* In three important merits decisions in October Term 2021, the Court invoked the "major questions" doctrine to invalidate regulatory agency actions aimed at addressing novel and arguably profoundly harmful threats to Americans' health and wellbeing. In the first case, the Court invalidated an Environmental Protection Agency (EPA) regulation that had been crafted to mitigate greenhouse gas emissions.[71] In a second case, the Court invoked the doctrine to turn aside an Occupational Safety and Health Agency (OSHA) regulation designed to mitigate COVID-19 risk.[72] In a third case, decided in August 2022, it invalidated a moratorium imposed on evictions by the Centers for Disease Control and Prevention.[73] In the thirty years prior to the October Term 2021, the Court had invoked the doctrine in only five other cases.[74] In the EPA and OSHA cases, however, the Court reformulated the doctrine into a powerful, anti-regulatory clear statement rule. Any "major" agency policy now requires "clear congressional authorization" as a way of enhancing democracy.[75]

Formally, the doctrine attempts to shift the channel of policymaking from federal agencies to Congress by ruling out the instrument of regulation and demanding the use of statutory law. Under present conditions of partisan polarization among legislative caucuses, however, it is immensely difficult to enact new law even under conditions of unified government. The recent failure of the Republican Party to

---

[71] West Virginia v. Env't Prot. Agency, 142 S. Ct. 2587 (2022).

[72] Nat'l Fed'n of Indep. Bus. (NFIB) v. Dep't of Lab., 142 S. Ct. 661 (2022).

[73] Ala. Ass'n of Realtors v. Dep't of Health & Hum. Servs., 141 S. Ct. 2485, 2489 (2021). *But see* Biden v. Missouri, 142 S. Ct. 647, 650 (2022) (per curiam) (upholding a rule to the effect that Medicare and Medicaid funding recipients' staff, unless exempt for medical or religious reasons, had to be vaccinated against COVID-19).

[74] *See* MCI Telecomms. Corp. v. AT&T, 512 U.S. 218 (1994); Food & Drug Admin. v. Brown & Williamson Tobacco Corp., 529 U.S. 120 (2000); Gonzales v. Oregon, 546 U.S. 243 (2006); Util. Air Regul. Grp. v. Env't Prot. Agency, 573 U.S. 302 (2014); King v. Burwell, 576 U.S. 473 (2015).

[75] *West Virginia*, 141 S. Ct. at 2609.

repeal the Affordable Care Act illustrates the (quite general) difficulty of legislative change absent cross-party agreement.[76] But gridlock is not democracy. To the contrary, the quality of a nation's democracy depends on its capacity to identify and remedy the latter's defects and shortfalls.[77] A judicially created rule that induces gridlock, indeed, is more likely to undermine than to improve democracy. Gridlock today cannot be justified as a way of protecting some distribution of pre-political rights (say, to liberty and property). For inaction today benefits those vested interests that benefit from a status quo characterized by previous statutory, regulatory, and judicial action. In effect, it allows present incumbents to shield themselves from challenge. Worse, political scientists since Juan Linz have flagged the peril to democracy implicit in a presidential system mired in gridlock, when a winning popular coalition is "cheated" of its policy preferences.[78] By condemning national policy on certain questions to an oubliette of policy paralysis, the Court elicits national drift and inaction. This perverse result follows from a fallacy of composition—i.e., the flawed assumption that for government to be democratic as a whole, discrete policy decisions must have a specific kind of electoral pedigree.

It is no surprise that this new version of the major questions doctrine has been "widely criticized" for its "indeterminacy" and its "counter-democratic allocation of power from agencies to judges."[79] Rather than adding further to these persuasive criticisms, our goal here is to understand the effect that an outsize major questions doctrine has on the formation of foreign policy, and on the United States' capacity to compete in a global environment

The immediate impact of the opinions issued in the 2021 Term, of course, bore on the U.S. capacity to respond to present and future

[76] Jordan M. Ragusa & Nate Birkhead, Congress in Reverse: Repeals from Reconstruction to the Present (2020).

[77] Mariano-Florentino Cuéllar, *From Doctrine to Safeguards in American Constitutional Democracy*, 65 UCLA L. Rev. 1398, 1424 (2018).

[78] Juan J. Linz, *The Perils of Presidentialism*, 1 J. Democracy 51, 66 (1990).

[79] Nathan Richardson, *Antideference: Covid, Climate, and the Rise of the Major Questions Canon*, 108 Va. L. Rev. Online 174, 176 (2022); *see also* Leah Litman & Dan Deacon, *The New Major Questions Doctrine*, 109 Va. L. Rev. (forthcoming 2023) (manuscript at 32) (arguing that the doctrine will "exacerbate institutional and political pathologies, undermine the ostensible premises of the major questions doctrine, and generate a doctrine that is likely to undermine delegations in the circumstances in which they are more likely to be used and more likely to be effective"); Mila Sohoni, *The Major Questions Quartet*, 136 Harv. L. Rev. 262, 267 (2022) (doubting whether "the Court [c]ould articulate a concrete and specific constitutional value that justifies that rule").

global threats: The climate-change crisis and the COVID-19 pandemic. By reading statutory authorities respecting environmental law and public health narrowly, the Court cast into doubt not just the regulations being challenged, but also any subsequent regulatory effect to address new environmental or public health challenges with a global reach. Hence, among the proposed rules immediately cast into doubt by the EPA ruling was a March 2022 proposal by the Securities and Exchange Commission to impose new disclosure obligations on regulated firms related to climate risk.[80] It is reasonable to conclude, therefore, that the new major questions doctrine's very first applications have the effect of curtailing significantly the nation's capacity to respond to climatic harms and new, as-yet-unknown epidemic or pandemic outbreaks. They have, quite directly, removed a foreign-policy tool from the nation's arsenal. The Court's decisions, of course, do not obviously rest on any information or expertise in the expected magnitude of these risks. Indeed, there is no reason to think that the Court has any expertise at all on the consequences to the nation of preventing regulatory responses to novel climactic or viral harms.

The Court, in any event, had been made aware of the geopolitical ramifications of domestic greenhouse-gas regulation. Consider the arguments made by the George W. Bush Administration for limiting EPA authority in respect to greenhouse cases. In its notice of denial of the petition for rulemaking, the Agency asserted that "[t]he international nature of global climate change also has implications for foreign policy, which the President directs."[81] The EPA argued that the President's policy "supports vital global climate change research and lays the groundwork for future action by investing in science, technology, and institutions" and that it "emphasizes international cooperation and promotes working with other nations to develop an efficient and coordinated response to global climate change."[82] Invalidating this agency determination, the Court rejected wholesale the

---

[80] Christina Thomas, Andrew Olmem & Katelyn Merick, *Supreme Court Decision Casts Doubt on SEC's Climate Proposal and Other Regulatory Initiatives*, HARV. L. SCH. F. CORP. GOV. (July 12, 2022), https://corpgov.law.harvard.edu/2022/07/12/supreme-court-decision-casts-doubt-on-secs-climate-proposal-and-other-regulatory-initiatives.

[81] Control of Emissions from New Highway Vehicles and Engines, 68 Fed. Reg. 52922, 52925 (2003).

[82] *Id.* at 52923.

relevance of foreign policy considerations.[83] We think that this element of the Court's reasoning is flawed.[84]

2. *Foreign Policy-Making After the New Major Questions Doctrine.* What of the doctrine's future application? The three opinions issued in 2022 offered hints on what counted as "major" regulation: The Court looked to whether Congress or the states had been considering a policy question,[85] and the latter's novelty.[86] This doctrinal inflection tracks the doctrine's deployment by the Department of Justice during the Trump Administration, which "distorted the [past] doctrine by raising whichever factors and metrics it found most helpful in any given case, however devoid of established doctrinal support, often contradicting arguments that it made in other cases."[87] As applied now by the Court, this novel version of the new major questions doctrine means that "even broad, otherwise unambiguous delegations of authority are not enough to support 'major' agency policies."[88] It is thus a putatively "targeted" nondelegation doctrine aimed at reinforcing the democratic credentials of agency action,[89] albeit one guided by the Court's sensibilities about what is politically important or novel.

Notwithstanding its formal justification as democracy-enhancing, the likely future effect of the new major questions doctrine will be to disable the federal government from action on many important policy

---

[83] Massachusetts v. Env't Prot. Agency, 549 U.S. 497, 534 (2007) ("[W]hile the President has broad authority in foreign affairs, that authority does not extend to the refusal to execute domestic laws. In the Global Climate Protection Act of 1987, Congress authorized the State Department—not EPA—to formulate United States foreign policy with reference to environmental matters relating to climate.").

[84] Not least, it misses the fact that "so many domains of social and economic regulation now seem populated by numerous agencies, which—to satisfy their missions—must work together cooperatively or live side by side compatibly." Jody Freeman & Jim Rossi, *Agency Coordination in Shared Regulatory Space*, 125 Harv. L. Rev. 1131, 1138 (2012).

[85] *Id.* at 2614; NFIB v. Dep't of Lab., 142 S. Ct. 661, 665 (2022).

[86] West Virginia v. Env't Prot. Agency, 142 S. Ct. 2587 (2022) (invoking novelty of the regulation as an indicia of a major question); *NFIB*, 142 S. Ct. at 666 (same).

[87] Natasha Brunstein & Richard L. Revesz, *Mangling the Major Questions Doctrine*, 74 Admin. L. Rev. 217, 220 (2022).

[88] Litman & Deacon, *supra* note 79 (manuscript at 6).

[89] Cass R. Sunstein, *There Are Two "Major Questions" Doctrines*, 73 Admin. L. Rev. 475, 491 (2021); *see also* Blake Emerson, *Administrative Answers to Major Questions: On the Democratic Legitimacy of Agency Statutory Interpretation*, 102 Minn. L. Rev. 2019, 2048 (2018) ("[I]n its best light, the doctrine aims to protect and to strengthen the connection between the people and governmental action by presuming that a popular and deliberative process settles major questions of policy.").

questions with geopolitical implications.[90] It seems particularly likely that the new major questions doctrine will tend to have a significant effect on foreign-policy matters when the Court's policy preferences diverge from those of the executive branch's. In other words, we anticipate the doctrine will prevent Democratic, but not Republican, administrations from exploiting broadly worded statutory delegations of authority. In October Term 2021, it was applied in challenges to climate and pandemic regulation favored by the Democratic Party.[91] The reach of the new major questions doctrine beyond EPA, OSHA, and like agencies regulating firms remains uncertain. It could also be narrowly construed with carve outs for certain domains of policy-making favored by the Court where business interests do not have the same kind of interests.

To see evidence of this possibility, consider a challenge to a Biden-Administration decision to rescind a Trump-era immigration policy called "Remain in Mexico."[92] The Court's result hinged on the "discretionary" character of the statutory language in question.[93] It set aside contrary arguments that no such discretion existed.[94] That is, the majority construed the statute in a way that obviated the question whether the creative exercise of discretionary over cross-border transfers was itself a "major question."[95] It also highlighted "the foreign

---

[90] Alexander Bolton, *Gridlock, Bureaucratic Control, and Nonstatutory Policymaking in Congress*, 66 AM. J. POL. SCI. 238, 238, 252 (2022) (arguing that "intrabranch supermajoritarian constraints on passing legislation impair the ability of congressional majorities to influence executive policymaking, particularly in the presence of interbranch preference divergence" and noting that "[a]s the goals of legislative actors diverge, mustering the support and votes to constrain executive branch actors becomes more challenging").

[91] *See supra* notes 71–72. The national eviction moratorium invalidated by the Roberts Court was first imposed by the Trump Administration in September 2020. *See* Temporary Halt in Residential Evictions to Prevent the Further Spread of COVID-19, 85 Fed. Reg. 55292 (2020). By July 2021, however, the measure was supported by a supermajority (82%) of Democrats, and opposed by a majority of Republicans (58%). Kathy Frankovic, *Americans Approve of Extending the Eviction Moratorium Through July*, YOUGOVAMERICA (July 6, 2021), https://today.yougov.com/topics/politics/articles-reports/2021/07/06/americans-approve-extending-eviction-moratorium. The Court's invalidation of the measure is therefore in accord with its broader tendency to take positions close to the median Republican voter, but not the median American.

[92] Biden v. Texas, 142 S. Ct. 2528, 2535 (2022) (describing the "Remain in Mexico" program).

[93] *Id.* at 2544.

[94] *Id.* at 2554–55 (Alito, J., dissenting).

[95] The Trump-era Remain in Mexico policy was criticized for cruelly exposing migrants to "robbery, assault and other crimes, including kidnappings." Marisa Limón Garza, *The Ongoing Devastation of Trump's 'Remain-in-Mexico' Policy*, HILL (Feb. 4, 2020), https://thehill.com/opinion/immigration/480477-the-ongoing-devastation-of-trumps-remain-in-mexico-policy. It is possible to imagine a version of the new major questions doctrine that took such

affairs consequences of mandating the exercise of contiguous-territory return" policies.[96] To be sure, these consequences were highlighted in briefing on the "Remain in Mexico" policy, but absent from the Solicitor General's briefing in the EPA case. Nevertheless, given the breadth of the Court's capacity to craft opinions based on non-record facts of its own choosing, that difference cannot be understood as a constraint on the Justices.

The Remain in Mexico decision supports two possible inferences about the scope of the new major questions doctrine. First, it does not apply when the policy objection to a regulation rests on humanitarian concerns, as opposed to the effect a policy will have on regulated firms. We doubt that this would ever be enunciated as a formal part of the doctrine. But a varying sensitivity to different sorts of costs seems inherent to the doctrine. This is one way, indeed, in which the Justices can flex their substantive policy preferences. Second, it may be that the doctrine has no application when there are direct "foreign affairs" consequences.[97] This would be consistent with an earlier suggestion by Justice Gorsuch that the nondelegation doctrine should not apply in respect to "foreign affairs powers . . . constitutionally vested in the president under Article II."[98] But this assumes that the Court has some straightforward way of dividing the foreign from the domestic. It does not. Instead, the Court can discern a "foreign affairs" angle in the sending of a migrant body across the U.S.-Mexico border—but then blink the "foreign affairs" consequences of stymying the nation's effort to address the global problem of climate change.

Nor is Justice Gorsuch's formulation especially illuminating: If a power is "constitutionally vested" in the President, it is not even clear whether and how Congress would be acting.[99] So does Gorsuch mean to describe a null set? More likely, he means to suggest that there is some class of legislation that acts in aid of Article II foreign affairs powers. But how are the latter defined? Article II does not contain some obvious litmus test for what counts as "foreign affairs" related. If

---

human rights violations as a trigger, and asked for specific statutory authorization. The Roberts Court, however, has selected a different set of policy consequences as triggering conditions. Mere human misery does not seem to count.

[96] *Id.* at 2533.

[97] Biden v. Texas, 142 S. Ct. at 2533.

[98] Gundy v. United States, 139 S. Ct. 2116, 2137 (2019) (Gorsuch, J., dissenting).

[99] *Cf.* Zivotofsky v. Kerry, 576 U.S. 1, 10 (2015).

it does, and if Gorsuch believes himself guided by it in the EPA, OSHA, and CDC cases, it has been rendered otiose by the changing geopolitical environment. It is more likely that Article II contains no such theory of the distinction between the foreign and the domestic—and invocation of such a nonexistent theory simply allows judges to inject their own policy preferences into the breach.

Hence, we think that presently available evidence about how the new major questions doctrine is used suggests that it will have a significant "foreign affairs" effect when the (right-of-center) Court objects on policy grounds to a Democratic administration's efforts to grapple with global problems via regulatory initiatives grounded in interpretations of statutes that would be at least facially plausible but for the major questions doctrine (were it not for its bite against otherwise persuasive interpretations, there would be no point to the doctrine). Such constraints will likely stymie the United States from exercising regulatory leadership through the enactment of regulatory coordination initiatives with major geopolitical resonance such as the Indo-Pacific Economic Framework, path-marking regimes for global problems such as climate change, the regulation of social-media platforms, and responses to public health emergencies.

Note that even as the United States is handicapped in exercising one kind of leadership, this doctrine leaves the nation with a free hand to pursue a deregulatory "race to the bottom." The result will be that where a single regulatory regime emerges globally as a matter of de facto conformity, it is more likely to be wrought by other nations or by states. This Roberts Court doctrine hence binds the nation's hands by eliminating one important instrument for shaping its geopolitical environment even beyond the immediate implications of the EPA case.

Our prediction of an asymmetrical effect is worth explicating. It is a truism that in a globalized economy, the regulatory approach of a nation state as large as the United States necessarily has a spillover effect on other jurisdictions.[100] The spillover may be one of two kinds. Only one of these, however, is affected by the new major questions doctrine. First, a nation might compete to offer firms a favorable regulatory climate by weakening mandates or slackening the pace of

---

[100] Indeed, international law is often designed to minimize the cost of such externalities. Andrew T. Guzman, How International Law Works 132–33 (2008).

enforcement.[101] This places pressure on other nations to deregulate in order to remain competitive. Where deregulation has costs to foreign consumers—for instance, in the antitrust context—domestic firms gain an increased ability to impose externalities overseas.[102] The new major questions doctrine does not seriously impede this kind of deregulatory foreign policy. So long as the executive has discretion to temper enforcement—say, in antitrust or securities regulation—then it can push other countries in a deregulatory direction.

But imagine that the United States wants to be a global leader in respect to, say, climate change or platform regulation policy. It might attempt to generate "[u]nilateral regulatory globalization," which "occurs when a single state is able to externalize its laws and regulations outside its borders through market mechanisms, resulting in the globalization of standards."[103] The U.S. is well placed to exercise this kind of influence. But the new major questions doctrine prevents it from doing so: Agencies are legally prohibited from articulating such a policy. But under increasingly likely and increasingly persistent scenarios, Congress is politically precluded from doing so.

The resulting gap can be filled from either above or below. The European Union (EU), on the one hand, has systematically tried to demonstrate "climate leadership capabilities" and has "adapted its exemplary and diplomatic leadership to maximize impact."[104] On the other hand, states such as California have strived to fill the vacuum left by an absence of U.S. leadership in respect to environmental regulation.[105] A similar dynamic can be observed in respect to data privacy. The European Union has promulgated a comprehensive data privacy directive that has extraterritorial effects.[106] The result of internal efforts

---

[101] Joel P. Trachtman, *International Regulatory Competition, Externalization, and Jurisdiction*, 34 Harv. Int'l L.J. 47, 51 (1993).

[102] *Id.* at 54. Recent empirical studies find evidence for this regulatory "chill," but also identify states in which domestic firms can use higher standards to insulate themselves from competition. Emma Aisbett & Magdalene Silberberger, *Tariff Liberalization and Product Standards: Regulatory Chill and Race to the Bottom?*, 15 Reg. & Gov. 987, 987–88 (2021). So in practice, a regulatory race to the bottom induces some variance in standard-setting.

[103] Anu Bradford, *The Brussels Effect*, 107 Nw. U. L. Rev. 1, 3 (2012).

[104] Sebastian Oberthür & Claire Dupont, *The European Union's International Climate Leadership: Towards a Grand Climate Strategy?*, 28 J. Eur. Pub. Pol. 1095, 1095–96 (2021).

[105] Daniel A. Mazmanian, John L. Jurewitz & Hal T. Nelson, *State Leadership in US Climate Change and Energy Policy: The California Experience*, 29 J. Env't Dev. 51 (2020).

[106] Mark Scott & Laurens Cerulus, *Europe's New Data Protection Rules Export Privacy Standards Worldwide*, Politico (Jan. 31, 2018, 12:00 PM), https://www.politico.eu/article

to "harmonize the data processing practices of EU member states" became over time "a global diffusion story."[107] Once again, California has followed suit with its own privacy law.[108] EU nations are poised to enact a sweeping new regulation of AI in late 2023.[109] With Congress unlikely to enact new legislation to comprehensively regulate AI, the new major questions doctrine means that the executive cannot lean on existing privacy and consumer safety laws to step in. As a consequence, the United States is likely to fall behind Europe as a global leader in deciding how AI is used.

A further irony here is that firms and individuals within the United States are not immune from the expansion in non-U.S. regulatory power. Parties in U.S. court "routinely" lean on foreign regulations to shape "thousands of claims every year—even when foreign regulators disagree with U.S. regulators."[110] And this was prior to the recent shift in the doctrine.

3. *The New Major Questions Doctrine and International Negotiation.* In addition, the new major questions doctrine is likely to influence the dynamics of international negotiations. To see this, it is useful to return to the FPA scholarship for a model of how the domestic and international spheres interact.

In an influential contribution to the FPA literature, Robert Putnam has construed the interaction between domestic and international politics as a "two-level game"[111] His model comprises separate domestic and international "table[s]" on which interest groups and national governments respectively pursue their interests.[112] Connecting the two levels are national executives conducting a Janus-faced dual bargaining game. On the one hand, they are constrained by their

---

/europe-data-protection-privacy-standards-gdpr-general-protection-data-regulation; Adam Satariano, *G.D.P.R., a New Privacy Law, Makes Europe World's Leading Tech Watchdog*, N.Y. TIMES (May 24, 2018), https://www.nytimes.com/2018/05/24/technology/europe-gdpr-privacy.html.

[107] Paul M. Schwartz, *Global Data Privacy: The EU Way*, 94 N.Y.U. L. REV. 771, 810 (2019).

[108] Sara Merken, *States Follow EU, California in Push for Consumer Privacy Laws (1)*, BLOOMBERG L. (Feb. 6, 2019, 3:02 PM), https://news.bloomberglaw.com/privacy-and-data-security/states-follow-eu-california-in-push-for-consumer-privacy-laws-1.

[109] *The Artificial Intelligence Act*, https://artificialintelligenceact.eu (last visited Feb. 6, 2023).

[110] Diego A. Zambrano, *How Litigation Imports Foreign Regulation*, 107 VA. L. REV. 1165, 1170 (2021).

[111] Robert D. Putnam, *Diplomacy and Domestic Politics: the Logic of Two-Level Games*, 42 INT'L ORG. 427, 434 (1988).

[112] *Id.*

domestic play of political forces. On the other hand, they are bargaining with other states. Sometimes moves that are rational for the executive on the domestic level are "impolitic" at the international level.[113] Two-level models of foreign policy-making draw on game theory to draw out implications of different levels of domestic flexibility for international bargaining. Among the intriguing implications of these models is the possibility that greater domestic constraints "reduces the scope for international cooperation" but also "increases the bargaining power of a state."[114] Yet another possibility is that the "win set" at the first domestic stage has no overlap with the "win set" at the international level.[115]

The new iteration of the major questions doctrine shrinks the range of positions to which the executive can commit credibly absent congressional scrutiny. This increases the leverage that the executive has in negotiation. The executive can henceforth more plausibly point to a limit to its freestanding discretion, and the need to secure legislative acquiescence. On the one hand, this means that the executive may be able to obtain greater concessions through international negotiations. On the other hand, the difficult of navigating the congressional channel and its veto-gates means that the range of actual policies that can make it across both the domestic and international "table[s]" will be smaller than it otherwise would be.[116] The net result is fewer international agreements. But those agreements that are reached will have better terms for the U.S.

An example of when this effect might arise can be extrapolated from the recent financial crises. During both the 2007–09 financial crisis and the 2020 COVID-19-related crash, the Federal Reserve interpreted its vague and open-textured statutory authorities to purchase assets to engage in extensive lending to non-bank entities, including other sovereign central banks.[117] The statutory provision at issue,

---

[113] *Id.*

[114] *Id.* at 448.

[115] *Id.* at 437–38. In Putnam's model, a "win set" is the range of possible final terms that a country can accept in an international negotiation and still manage to get the treaty ratified. For a useful overview of the literature after Putnam, see Eugénia da Conceição-Heldt & Patrick Mello, *Two-Level Games in Foreign Policy Analysis, inOxford Research Encyclopedia of Politics* (2017), https://oxfordre.com/politics/view/10.1093/acrefore/9780190228637.001.0001/acrefore-9780190228637-e-496.

[116] *Id.*

[117] Lev Menand, The Fed Unbound: Central Banking in a Time of Crisis 18–19, 35–41 (2022).

Section 13(3) of the Federal Reserve, had "rarely" been used before these crises.[118] In 2008, the provision was deployed to "recast" the central bank's power and enabled "the massive buying of bonds that were unprecedented in scope."[119] Subsequent legislative efforts to constrain this power "narrowed" the Fed's freedom of maneuver, but nonetheless do not prohibit extraordinary fiscal and monetary interventions.[120]

In March 2020, the Federal Reserve was hence able to respond to the potential collapse of key financial markets by buying one million dollars of bonds *per second*, hoovering up some five percent of a $20 trillion global market.[121] If there were to be a rerun of these crises, say centered on runs on bank-like institutions such as cryptocurrency exchanges, literal application of the new major questions doctrine would suggest that the central bank could no longer intervene. To be sure, a private suit against the Federal Reserve would face difficult justiciability hurdles. But one can still imagine a future Fed Chair citing the new major questions doctrine in both domestic and global negotiations, and then conjuring the threat of non-intervention, to influence the course of a subsequent financial crisis.[122] In the limit, the new major questions doctrine might end up being the proximate cause of a global depression.

Whatever one's ultimate position on the wisdom of framing the presidency as the country's preeminent foreign affairs institution, few serious observers would deny that a strong historical and jurisprudential tradition bolsters the presidency-foreign affairs nexus.[123] Even from a perspective that primarily privileges historically-rooted American legal traditions rather than primarily consequentialist concerns, the new major questions doctrine may work best if formally limited by "foreign affairs" concerns writ large. But this boundary is at best enormously fraught and difficult to draw coherently—at least as fraught or perhaps more than the boundary between "major" and non-major

---

[118] Lawrence R. Jacobs & Desmond King, Fed Power: How Finance Wins 180 (2d. ed. 2021).

[119] *Id.*

[120] *Id.* at 181–82; Menand, *supra* note 117, at 40–41, 48 (describing the Fed's purchase of "trillions of dollars of financial assets" in 2020).

[121] Tooze, *supra* note 7, at 129.

[122] Menand, *supra* note 117, at 23, chart 1.2 (documenting scale of foreign lending).

[123] *See, e.g.*, Mariano-Florentino Cuéllar, *American Executive Power in Historical Perspective*, 36 Harv. J.L. & Pub. Pol. 53 (2013).

questions. In practice, its channel- and instrument-shifting effects have significant consequences for the geopolitical power of the United States. Along one, deregulatory, margin, that power is largely unaffected. But along another, the Court is indirectly shaping the nation's global footprint based on its own idiosyncratic views of what counts as a "domestic" political question of importance.

B. THE STRUCTURE OF EXECUTIVE POLICY-MAKING

The second major shift in the jurisprudence concerns the structure of the executive branch itself. The Roberts Court has interpreted Article II to preclude legislative protections against the presidential firing of certain officers within the executive branch, and to limit the mechanisms by which certain officers can be appointed. On the one hand, these opinions to date have not destabilized or emasculated the agencies in question—many of which play key roles in the complex federal network of financial regulation. On the other hand, the net substantive result of these changes is uncertain. As we have already seen with respect to the SEC's climate rule, jurisprudence can cast a pall of uncertainty over regulatory prospects. In addition, the structural constitutional jurisprudence discussed below will likely diminish the availability of expertise within the executive branch in ways that erode the quality of foreign policy decisions. As such, it will corrode indirectly the quality of U.S. tools to address international financial crises in particular.

The leading decision on presidential removal authority from the October Term 2021 concerned the head of the Federal Housing Finance Authority (FHFA). The Court held that even "'modest restrictions' on the President's power" to remove the sole director of a federal agency could not be reconciled with the Constitution.[124] It drew on two earlier decisions. The first of these struck down limits on the president's power to remove the head of Consumer Finance Protection Bureau (CFPB) solely "for inefficiency, neglect, or malfeasance."[125] The second invalidated a two-step removal regime for members of the Public Company Accounting Oversight Board.[126] This trio of decisions

[124] Collins v. Yellen, 141 S. Ct. 1761, 1787 (2021).

[125] Seila L. LLC v. Consumer Fin. Prot. Bureau (CFPB), 140 S. Ct. 2183, 2197 (2020)

[126] Free Enter. Fund v. Pub. Co. Acct. Oversight Bd., 561 U.S. 477, 492 (2010) (holding that "the dual for-cause limitations on the removal of Board members contravene the Constitution's separation of powers").

marks a sharp break from the more accommodating, functionalist approach taken to removal questions during the Rehnquist Court.[127] While purportedly justified on originalist grounds,[128] the new removal regime has been sharply criticized as inconsistent with a mounting body of historical evidence.[129] The Court has addressed this accumulating historical evidence by simply ignoring it.

A second line of cases arises under the Appointment Clause of Article II.[130] The Court in recent cases has enlarged the range of federal officials who must be appointed through presidential nomination followed by Senate approval and diminished the range of officials who can be appointed by channels that do not run through Capitol Hill. More technically, the Court in 2021 held that administrative patent judges are "principal officers" who can be appointed only by the President with the advice and consent of the Senate given those judges' ability to issue "final decisions" on patents' validity.[131] Echoing the justifications offered in its removal jurisprudence, the Court has pointed to "legitimacy and accountability" concerns to justify its decisions.[132] At the same time, the Court has resisted broad application of the Appointment Clause when it comes to territorial officials in respect to Puerto Rico.[133] Just as in the regulatory power/delegation context, the Justices appear to check doctrine at the water's edge.

---

[127] *See, e.g.*, Morrison v. Olson, 487 U.S. 654, 686 (1988).

[128] *Free Enter. Fund*, 561 U.S. at 492; *Seila L. LLC*, 140 S. Ct. at 2197.

[129] The recent scholarship roundly ignored by the Court includes Daniel D. Birk, *Interrogating the Historical Basis for a Unitary Executive*, 73 STAN. L. REV. 175 (2021); Jane Manners & Lev Menand, *The Three Permissions: Presidential Removal and the Statutory Limits of Agency*, 121 COLUM. L. REV. 1 (2021); Jed Handelsman Shugerman, *Presidential Removal: The* Marbury *Problem and the Madison Solutions*, 89 FORDHAM L. REV. 2085 (2021); Christine Kexel Chabot, *Is the Federal Reserve Constitutional? An Originalist Argument for Independent Agencies*, 96 NOTRE DAME L. REV. 1 (2021); and Julian Davis Mortenson, *Article II Vests the Executive Power, Not the Royal Prerogative*, 119 COLUM. L. REV. 1169 (2019).

[130] U.S. CONST. art. II, § 2, cl. 2 ("[The President] shall nominate, and by and with the Advice and Consent of the Senate, shall appoint Ambassadors, other public Ministers and Consuls, Judges of the supreme Court, and all other Officers of the United States, whose Appointments are not herein otherwise provided for, and which shall be established by Law: but the Congress may by Law vest the Appointment of such inferior Officers, as they think proper, in the President alone, in the Courts of Law, or in the Heads of Departments.").

[131] United States v. Arthrex, Inc., 141 S. Ct. 1970, 1979 (2021). In an earlier decision, the Court held that administrative law judges at the Securities and Exchange Commission (SEC) are "officers" covered by the Appointment Clause, and not simply employees. Lucia v. Sec. Exch. Comm'n, 138 S. Ct. 2044, 2051–55 (2018).

[132] *Arthrex, Inc.*, 141 S. Ct. at 1979.

[133] In a 2020 case, the Court rejected an Appointment Clause challenge to the Financial Oversight and Management Board created by the Puerto Rico Oversight, Management, and

In both the challenges to the statutory schemes for the FHFA and patent judges, the Court coupled a categorical, formalist rule of decision with a modest remedy. In the first case, the Court declined to rank the agency decision as void, and instead remanded to the lower court to determine whether the constitutional flaw had an implication for how its decision was implemented.[134] In the second case, the Court held that the statutory provision barring agency-head review was unenforceable. It instead bestowed upon the agency head the power to review patent judges' decisions.[135] If we set aside the "managerial checks" that arise from district-court judges' "extraordinary ability to force legal and public accountability onto the Executive in suits challenging enforcement lawmaking,"[136] the immediate effect of these rulings for regulated parties appears to be modest: The specific challenged agency actions proceed as before, and the trajectory of administration persists relatively unimpeded.

The same, however, cannot be said, however, of an October 2022 decision by the Fifth Circuit Court of Appeals to strike down the CFPB's funding mechanism as a violation of the Appropriations Clause.[137] That decision throws into doubt the Bureau's ability to fund its continued operation. It also raises profoundly destabilizing questions about the Federal Reserve's operation—which, as we have already explained, may well introduce new and profound uncertainty into the foreign policy domain.[138] Of course, the Supreme Court may overrule that decision, but it is striking that the consequentialist

Economic Stability Act, on the ground that "primarily local" officers could be appointed without advice and consent. Fin. Oversight & Mgmt. Bd. for Puerto Rico v. Aurelius Inv., LLC, 140 S. Ct. 1649, 1663 (2020). The Court did not evince any sense of the irony of labeling a decision about an offshore territory as "local."

[134] Collins v. Yellen, 141 S. Ct. 1761, 1788 (2021). The challenge to the CFPB's structure also led to a remand, and a circuit-court ruling that the agency director had "expressly ratified the agency's earlier decisions." CFPB v. Seila L. LLC, 997 F.3d 837, 846 (9th Cir. 2021).

[135] *Arthrex, Inc.*, 141 S. Ct. at 1986–88.

[136] Z. Payvand Ahdout, *Enforcement Lawmaking and Judicial Review*, 135 Harv. L. Rev. 937, 942 (2022).

[137] Cmty. Fin. Servs. Ass'n of Am., Ltd. v. CFPB, 51 F.4th 616, 623 (5th Cir. 2022) ("Congress's decision to abdicate its appropriations power under the Constitution, i.e., to cede its power of the purse to the Bureau, violates the Constitution's structural separation of powers.").

[138] For a brief description of the Fifth Circuit's argument, and its potential reach, see Aziz Huq, *Will the Supreme Court Torpedo the Financial System?*, Politico (Jan. 18, 2023), https://www.politico.com/news/magazine/2023/01/18/supreme-court-financial-system-justices-00078124.

reasons for worry triggered by the opinion have played a very small role in recent jurisprudence.

Commentators are predictably calling for more aggressive remedies in these cases that would destabilize regulatory agencies.[139] The possibility of further challenges to executive agencies, especially those with key roles in financial regulation, interacts with uncertainty over remedies to create a friction on agency action. We want to stress a different way in which the Court's interventions are consequential. For our purposes, what is interesting about the Supreme Court's opinions is their dynamic effect on the *channels* through which policy, including in the foreign affairs domain, will be made. The effect of both the removal- and the appointment-related decisions is to funnel more control of the executive branch into the White House, and in particular to subject a greater proportion of officials to partisan rather than professional criteria of evaluation. This increases, among other things, the power of presidents to use "persistent failures to nominate" as an instrument to "prevent an agency from doing the business assigned to it by statute" and so "seem to thwart congressional design."[140] It also enlarges presidential discretion to "undermine the agency expertise that is foundational to the modern regulatory enterprise."[141] To be sure, presidents have other instruments at their disposal to control officials short of appointment and removal.[142] But the recent constitutional removal decisions at minimum increase this discretion at the margin, and prevent the reinstallation of durable expertise-promoting safeguards.

These effects are likely to be pronounced in the foreign affairs domain because of the importance of personnel quality in that domain. Financial agencies such as the SEC already play an important role in promoting "greater cooperation in the cross-border enforcement of

---

[139] William C. Eisenhauer, Note, *A Responsive Remedy for Unconstitutional Removal Restrictions*, 97 NOTRE DAME L. REV. 2195, 2211 (2022) (arguing that the Court should "hold that unlawful removal restrictions per se taint executive action"). *But see* William Baude, *Severability First Principles*, 109 VA. L. REV. 1, 37–41 (2023) (defending the current approach). Another response has been to argue for new rules for institutional standing in these cases. Elizabeth Earle Beske, *Litigating the Separation of Powers*, 73 ALA. L. REV. 823, 828–29 (2022).

[140] Jody Freeman & Sharon Jacobs, *Structural Deregulation*, 135 HARV. L. REV. 585, 606 (2021).

[141] *Id.* at 616. Freeman and Jacobs focus here on exercises of presidential discretion generally, and not the marginal effect of recent decisions. But their terminology is apt.

[142] *See* Aziz Z. Huq, *Removal as a Political Question*, 65 STAN. L. REV. 1, 5–6 (2013) (enumerating those instruments).

securities laws."[143] A constitutional regime in which experts lack discretion and are at the mercy of partisan winds is one that will not recruit or retain the sort of skilled people needed to navigate the complex international sphere (or even address domestic risk effectively). Political appointees are not likely to be even-handedly responsive to the interests of all sides of the national debate. A 2005 empirical study found that "business may exert the most consistent influence" on foreign policy choices, with public opinion having "little or no" palpable effect.[144]

More generally, White House control has underappreciated costs in the foreign policy domain. Derek Leebeart has argued that U.S. foreign policy has already been seriously undermined through its longstanding reliance on a political appointments system for diplomats. On his account, this system "imposes inexperience, compels urgency, courts risk and foments illusions of being able to manage the ethnic, ideological, and political concerns of other nations."[145] Diplomats themselves have lauded mechanisms such as the dissent channel as means to leverage their expertise in the teeth of bad decisions by political appointees.[146] During the Trump presidency, career staff in the military publicly expressed their unwillingness to follow "illegal" orders, in another signal of the variable quality of presidential orders.[147] Political nominees from the career military have also played important roles in soothing roiled diplomatic relations.[148] The Court's structural

[143] Chris Brummer, *Post-American Securities Regulation*, 98 Calif. L. Rev. 327, 331 (2010). On the role of the SEC in global regulation more generally, see Steven L. Schwarz, *Systemic Risk*, 97 Geo. L.J. 193 (2008) (describing the domestic and international nature of systemic risk and the role of securities regulation in addressing it).

[144] Lawrence R. Jacobs & Benjamin I. Page, *Who Influences US Foreign Policy?*, 99 Am. Pol. Sci. Rev. 107, 121 (2005).

[145] Derek Leebeart, *How Foreign Policy Amateurs Endanger the World*, Politico (Oct. 26, 2022), https://www.politico.com/news/magazine/2022/10/26/political-appointees-endanger-foreign-policy-00063259.

[146] John Brady Kiesling, *The Duty of Diplomatic Dissent*, 1 Hague J. Dipl. 295, 296 (2006).

[147] *U.S. Nuclear General Says Would Resist "Illegal" Trump Strike Order*, Reuters (Nov. 18, 2017), https://www.reuters.com/article/us-usa-nuclear-commander/u-s-nuclear-general-says-would-resist-illegal-trump-strike-order-idUSKBN1DI0QV. Another important example is the resistance to the banning of trans-gender service personnel—although the legal status of that action was contested. *See* Ellen Mitchell, *Military Pushes Back on Trump's Transgender Ban*, Hill (Aug. 13, 2017), http://thehill.com/policy/defense/346261-military-pushes-back-on-trumps-transgender-ban.

[148] Idrees Ali, *Pentagon's Mattis Again Seeks to Reassure U.S. Allies*, Reuters (Aug. 19, 2017), https://www.reuters.com/article/us-usa-mattis-trump/pentagons-mattis-again-seeks-to-reassure-u-s-allies-idUSKCN1AZ0D0.

jurisprudence does not directly affect either the Foreign Service or the military. But the costs of politicization observed in those fields suggest that these decisions will have costs elsewhere.

It is worth noting, finally, an internal tension in this jurisprudence. In cases perceived as implicating geopolitical considerations, the Court repeatedly genuflects before the competence and superior knowledge of the executive branch.[149] For the Court to be actively engaged in the deconstruction of the very expertise on which it so ostentatiously relies is no small irony.

C. THE FLOW OF BODIES ACROSS BORDERS

The anti-regulatory disposition of the Roberts Court does not extend across the board. To the contrary, there are important domains in which the Court has substantially loosened the constraints upon governmental action. In the domain of immigration—an issue of growing global importance as migration pressures increase as a result of climate change, and a quintessential exemplar of domestic and foreign policy entanglement—the Court has used tools of statutory interpretation and its broad authority over remedies to enlarge the scope of executive branch discretion. Coupled to its handcuffing of the executive when it comes to regulatory pace-setting, these decisions alter the available range of *instruments* of foreign policy.

The first strand of relevant decisions here is embodied by the Remain in Mexico decision, which ratified presidential discretion to both maintain and to withdraw a minatory regime of temporary exile on the southern border.[150] Similarly, in 2018, the Court upheld the ban imposed by President Trump on entrants from several Muslim-majority countries.[151] It did so in the teeth of substantial record evidence of unconstitutional bias against Muslims in the ban's creation.[152] Deference, though, does not prevail across the board. In a 2020 decision concerning the Obama-era Deferred Action for Childhood Arrivals

---

[149] *See, e.g.*, Jesner v. Arab Bank, PLC, 138 S. Ct. 1386, 1403 (2018) ("The political branches, not the Judiciary, have the responsibility and institutional capacity to weigh foreign-policy concerns."); Holder v. Humanitarian L. Project, 130 S. Ct. 2705, 2711 (2010) (relying on "the courts' lack of expertise with respect to national security and foreign affairs"); *accord* Trump v. Hawaii, 138 S. Ct. 2392, 2418–19 (2018).

[150] Biden v. Texas, 142 S. Ct. 2528, 2533–35 (2022).

[151] *Hawaii*, 138 S. Ct. at 2418–19.

[152] *See* Aziz Z. Huq, *Article II and Antidiscrimination Norms*, 118 MICH. L. REV. 47, 61–68 (2019) (recounting in detail the formation of that policy).

program, for example, the Court held the termination of that program unlawful under the Administrative Procedure Act of 1946 because it was supported by insufficient, incomplete reasoning from the Secretary of Homeland Security.[153]

A second strand of cases concerns remedies for discrete and individual violations of the Constitution or federal statutes though immigration detention. The substance of immigration law defines eligibility for removal in exceptionally sweeping and categorical terms.[154] This uniformity in migrants' legal classification leaves officials with vast discretion. Judicial review of individual cases in consequence has limited scope for disciplining the immigration system. Nevertheless, what remedies obtain have been systematically pared back by the Court. This remedial recession is part of a more general pattern. As one of us has explained, both the Rehnquist and Roberts Courts have rolled back sharply the availability of individual remedies in public law through a wide variety of doctrinal tools.[155] In a series of decisions in October Term 2021, the Court expanded the ability of the executive branch to use detention as an instrument of immigration policymaking by changing the range of detainees' remedies. In a first pair of cases, it eliminated the ability of noncitizens held by the federal government on immigration grounds to seek relief on their confinement in a periodic bond hearing.[156] In a third case, it extinguished their legal right to obtain class-wide relief on any legal ground.[157] These cases all involved legal challenges to the legality of a detention. While the Court has not addressed conditions-focused challenges of late, the dispersion

---

[153] Dep't of Homeland Sec. v. Regents of the Univ. of Cal., 140 S. Ct. 1891, 1912 (2020) (noting that the Secretary's reasoning "treated the Attorney General's conclusion regarding the illegality of benefits as sufficient to rescind both benefits and forbearance, without explanation," and finding this elision arbitrary and capricious).

[154] Shalini Bhargava Ray, *Immigration Law's Arbitrariness Problem*, 121 COLUM. L. REV. 2049 (2021).

[155] *See generally* AZIZ Z. HUQ, THE COLLAPSE OF CONSTITUTIONAL REMEDIES (2021).

[156] Johnson v. Arteaga-Martinez, 142 S. Ct. 1827 (2022) ("[T]here is no plausible construction of the text of § 1231(a)(6) that requires the Government to provide bond hearings before immigration judges after six months of detention, with the Government bearing the burden of proving by clear and convincing evidence that a detained noncitizen poses a flight risk or a danger to the community.").

[157] Garland v. Aleman Gonzalez, 142 S. Ct. 2057 (2022) (holding that 8 U.S.C. § 1252(f)(1), "deprive[s] the District Courts of jurisdiction to entertain [immigration detainees'] requests for class-wide injunctive relief"); *Areteaga-Martinez*, 142 S. Ct. at 1830 (holding, as a statutory interpretation matter, that bond hearings are not required for detainees ordered removed to be detained beyond the 90-day removal period).

of immigration detainees across private, city, and county-run facilities creates technical and jurisdictional barriers to litigation in many cases.[158]

Beyond the detention context, the Court has also lowered the expected cost of state coercion, including deadly force, by eliminating remedies in tort for noncitizens and citizens subject to state violence close to the border. For example, in a 2020 decision, it foreclosed damages based on the fatal shooting of a child by a border agent on the ground that "regulating the conduct of agents at the border unquestionably has national security implications," and the "risk of undermining border security."[159] One commentator has pithily characterized the Roberts Court's approach since Justice Kennedy's retirement as animated by "a basic hostility to noncitizens, especially those with criminal convictions."[160]

Parallel trends can be discerned beyond the strict immigration context. In a 2022 decision in respect to a military detainee who had been subjected to state torture while in U.S. custody, the Court allowed the government to invoke the "state secrets" privilege to prevent two former contractors with the CIA from being compelled to give testimony for a foreign proceedings.[161] Perhaps the most telling sentence in the Court's opinion was its conclusion that the need of a torture victim to establish a public record of his mistreatment was "not great."[162] One does not need to have experienced state violence to have some sense of how the trauma of recognized torture can inflict persisting psychological harms of first-order magnitude. Yet the Court's disregard of the violence inflicted on transient bodies across international borders is in keeping with its approach in immigration cases more generally: The physical harms inflicted on

---

[158] Fatma Marouf, *Immigration Detention and Illusory Alternatives to Habeas*, 12 U.C. IRVINE L. REV. 969, 974–75 (2022).

[159] Hernandez v. Mesa, 140 S. Ct. 735, 747 (2020). This ruling has been extended to cases involving citizen victims of unlawful violence close to a border—again through an invocation of "foreign policy and national security" considerations. Egbert v. Boule, 142 S. Ct. 1793, 1804 (2022).

[160] Nancy Morawetz, *The Perils of Supreme Court Intervention in Previously Technical Immigration Cases*, 64 ARIZ. L. REV. 767, 804 (2022).

[161] United States v. Zubaydah, 142 S. Ct. 959 (2022).

[162] *Id.* at 971 ("Zubaydah suggested that he did not seek confirmation of the detention site's Polish location so much as he sought information about what had happened there."). The Court here purported to rely on a concession of Zubaydeh's counsel. But there was never any doubt that the detainee's legal action was intended to procure evidence of what happened in the Polish black site. This was less a concession than a statement of the prima facie obvious reason for suit.

moving (generally non-U.S.) bodies are not a cognizable cost. Bereft of legal relevance, they vanish from view.

We draw attention to these cases because they have a dynamic effect on the toolkit of foreign policy. Judicial enforcement of constitutional and statutory limits operates ordinarily as a "tax, a mechanism for making some activities more expensive relative to their substitute."[163] The net effect of the decisions canvassed here is to lower that tax in relation to immigration-related coercion and detention. For example, the Court's novel bar on "class wide" relief means that immigration officials can litigate disputes about the terms of detention on a retail rather than a wholesale basis. It seems likely that this will lead to substantially less pressure to release individuals—especially given the absence of free or subsidized legal aid for indigent immigrants. Even presidential administrations ideologically disposed to ease coercive control of immigration flows are likely to lean into the use of these tools.[164] Such administrations not only face pressure from political forces that frame immigration as violent threat.[165] They also confront internal bureaucratic and public pressure to respond to public perceptions that the state does not have effective control over immigration flows.[166] Increased enforcement is often the path of least resistance even if it is not compatible with an administration's stated ideological commitments.

The lowered cost of coercive immigration control must be understood against the context of other dimensions of jurisprudence explored here. Changing migration flows into the United States are linked to ongoing climate change, particularly in the areas such as the Northeast and Midwest of the country.[167] Of course, climate change

---

[163] William J. Stuntz, *The Distribution of Fourth Amendment Privacy*, 67 Geo. Wash. L. Rev. 1265, 1274 (1999).

[164] In January 2023, the Biden Administration turned to a discretionary immigration power called parole to help manage the build-up of migrants at the southern border. Nick Miroff & Maria Sacchetti, *Biden Border Plan Expands Use of 1950s-Era Immigration Parole Powers*, Wash. Post (Jan. 20, 2023), https://www.washingtonpost.com/nation/2023/01/20/border-migrants-parole. This is an example of creative use of regulatory tools, although not of coercive ones (as described in the main text).

[165] Morawetz, *supra* note 160, at 770–71 (noting the presence of "anti-immigrant tropes about the 'criminal alien' that fueled some of the worst excesses in Congress" in the recent jurisprudence of immigration).

[166] *See* Mariano-Florentino Cuéllar, *The Political Economies of Immigration Law*, 2 U.C. Irvine L. Rev. 1, 50–74 (2012).

[167] Qin Fan, Karen Fisher-Vanden & H. Allen Klaiber, *Climate Change, Migration, and Regional Economic Impacts in the United States*, 5 J. Ass'n Env't & Res. Economists 643, 643–44 (2018).

policy and coercive migration control are not anything like perfect substitutes. But the lower the former's cost, the less political pressure there may be to address the latter. There is hence a subtle way in which the Court's interventions, when understood in net, are pushing the United States toward some instruments of foreign policy problems, and away from others. By raising the cost of some policy options and by lowering the price of others, in short, the Court is picking and choosing how the United States is responding to a global crisis such as climate change.

D. DEMOCRACY AS A FOUNDATIONAL RESOURCE IN FOREIGN POLICY

When the United States acts in the international sphere, it leverages several background sets of foundational resources. These, as we've said, include military capacity, as well as natural resources (such as access to natural gas[168]) and human capital (shaped in part by migration policies). Obviously, the availability of most of these resources does not depend on the Court's actions. There is one important foundational resource, however, that can improve or decline in quality as a consequence of the Court's actions. This is democracy.

There are many competing definitions of democracy.[169] We focus here on the simple intuition that elected actors should be reasonably responsive to the preferences of their constituents for the term to fit. Where an election system leads to results that are invariant to the changing views of the electorate, a legitimating link between the public and their putative representatives is lost. Declines in responsiveness plausibly work as an index of faltering democratic quality. With this in mind, it is possible to pick out an unnervingly wide range of Supreme Court caselaw sharply at odds with the project of preserving and protecting democratic rule. Rather than offering a general canvass of this jurisprudence,[170] we pick out four recent decisions or trends with counter-democratic effects. This suffices to show how a Court that in other domains makes a big fuss about its commitment to democracy is

---

[168] *See supra* text accompanying notes 3–6 for more discussion.

[169] *Cf.* TOM GINSBURG & AZIZ Z. HUQ, HOW TO SAVE A CONSTITUTIONAL DEMOCRACY (2018) (offering a minimalist definition of democracy focused on elections, free speech, and the rule of law).

[170] These points are developed at greater length in Aziz Z. Huq, *The Supreme Court and the Dynamics of Democratic Backsliding*, 699 ANNALS AM. ACAD. POL. & SOC. SCI. 50 (2022). The general point that the Constitution is by design counter-democratic is explained by ROBERT A. DAHL, HOW DEMOCRATIC IS THE AMERICAN CONSTITUTION? 41–64 (2001).

systematically dismantling the electoral premises of democratic rule to the benefit of the political faction with which it is aligned.

We have already observed one way in which the Roberts Court is placing new strains on American democracy: the new major questions doctrine, discussed above, makes it more difficult for a winning electoral coalition to make policy change over the objections of an entrenched and connected political minority.[171] In effect, the EPA decision allows a minority employed in legacy coal plants to hold up the majority who experience climate change's immediate costs. The frustration and dissatisfaction this generates feeds into antisystemic forces and figures who seek power by promising to circumvent the orderly, ordinary processes of democratic choice. By increasing national logjams, stoking public discontent with national governance, the Court is kindling (inadvertently or otherwise) an increased risk of political instability.[172] It is creating the conditions under which presidentialism, as Linz predicted, is most likely to generate democratic crisis.[173]

There are three other recent lines of cases that comprise a (nonexclusive) body of counter-democratic effects. First, campaign finance jurisprudence under the First Amendment[174] prevents the state from moderating the effects of wealth on political choices. It hence strains the linkage between representatives and those who elect them. Money flows from a small number of wealthy districts nationally.[175] It tends predictably to increase Republican representation in state legislatures, without respect to changes in public preferences.[176] Especially under

---

[171] *See supra* text accompanying notes 85–110.

[172] This discontent now spills out into violence justified under the banner of constitutional fidelity. *See* Aziz Z. Huq, *On the Origins of Republican Violence*, BRENNAN CTR. FOR JUST. (June 29, 2021), https://www.brennancenter.org/sites/default/files/2021-06/Huq_final.pdf. The Court's complex and indirect relation to these emergent forms of political violence is an important topic, raising issues beyond our scope here.

[173] Linz, *supra* note 78, at 66.

[174] *See, e.g.*, Fed. Election Comm'n v. Cruz, 142 S. Ct. 1638, 1645 (2022) (invalidating a statutory "limit on the use of post-election funds" to repay a candidate's personal loans).

[175] Anne Baker, *The More Outside Money Politicians Take, the Less Well They Represent Their Constituents*, WASH. POST (Aug. 17, 2016), https://www.washingtonpost.com/news/monkey-cage/wp/2016/08/17/members-of-congress-follow-the-money-not-the-voters-heres-the-evidence ("About a quarter of the members of the House of Representatives raise a larger share of their campaign funds from donors outside of their districts.").

[176] Nour Abdul-Razzak, Carlo Prato & Stephane Wolton, *After* Citizens United: *How Outside Spending Shapes American Democracy*, 67 ELECTORAL STUD. 102190, 102910 (2020).

conditions of extreme wealth inequality,[177] this has the effect of selectively enfranchising only a small, wealthy slice of the public. Second, the Court's decision to permit concededly unconstitutional partisan gerrymanders of legislative districts without federal judicial supervision has the effect of enabling partisan lock-ups of state legislatures and congressional delegations—again, largely to the Republican Party's benefit.[178] While the Court is not responsible for the technological advances that enable far more fine-grained and durable lock-ups of political power, its decision to stay its hand contrasts sharply with its aggressive policing of other domains of the electoral process. Third, and relatedly, the Court has interpreted federal laws such as the Voting Rights Act and the National Voter Registration Act, both intended to promote inclusive forms of democracy, so as either to defang them or to transform them into instruments of disenfranchisement.[179] These three lines of decisions limit the responsiveness of elected officials to changing public opinion in increasingly significant ways.

The question here, of course, is how the ensuing loss of democratic responsiveness has affected foreign affairs. In what ways, that is, is domestic democracy a foundational resource in the geopolitical context, and how does its decay shape the nation's external policy options? We see three possible linkages here. We reject the first, most obvious one. Instead, we emphasize two less expected ways in which the judicial diminution of American democracy corrodes the geopolitical standing of the United States.

The first potential linkage to consider draws on Joseph Nye's well-known idea of "soft power." This is the "ability to set the political agenda in a way that shapes the preferences of others."[180] Such power

---

[177] THOMAS PIKETTY, A BRIEF HISTORY OF EQUALITY 150–51 (2022) (documenting growing inequalities in both the United States and Europe).

[178] Rucho v. Common Cause, 139 S. Ct. 2484, 2507 (2019) ("Federal judges have no license to reallocate political power between the two major political parties, with no plausible grant of authority in the Constitution, and no legal standards to limit and direct their decisions."). On the effects of *Rucho*, see Christopher Warshaw, Eric McGhee & Michal Migurski, *Districts for a New Decade—Partisan Outcomes and Racial Representation in the 2021–22 Redistricting Cycle*, 52 PUBLIUS 428 (2022).

[179] *See* Brnovich v. Democratic Nat'l Comm'n, 141 S. Ct. 2321, 2341 (2021) (limiting scope of Section 2 of the Voting Rights Act); Husted v. A. Philip Randolph Inst., 138 S. Ct. 1833, 1843 (2018) (finding no preemption under the Help America Vote Act of an Ohio law that "removes registrants only when they have failed to vote *and* have failed to respond to a change-of-residence notice").

[180] JOSEPH S. NYE JR., THE PARADOX OF AMERICAN POWER: WHY THE WORLD'S ONLY SUPERPOWER CAN'T GO IT ALONE 9 (2002)

depends on "values … expressed in our culture, in the policies we follow inside our country" as well as the "way we handle ourselves internationally."[181] Nye includes democracy as one of the important "values our government champions in its behavior at home."[182]

It would be convenient for our argument if we were able to lean on Nye's account of soft power. Alas, we are skeptical that it can accomplish much load-bearing work for us. For one thing, it is grounded on an account of the accelerating "diffusion of power away from governments to private actors" that no longer seems plausible.[183] The emergence of a powerful centralized party-state capable of mobilizing surveillance and AI-enabled behavior controls in China suggests that Nye was too optimistic.[184] Empirical studies of soft power also find limited evidence for a broad reading of Nye's theory. Rather, the best available empirical evidence suggests that it is the foreign policy views of other countries' electorates that shape their willingness to cooperate with the United States in the global sphere, not their views of our domestic political arrangements.[185]

We see at least two other pathways, however, that might plausibly matter more. First, the U.S. dollar is "far and away the most important currency for invoicing and settling international transactions," without regard to where they occur.[186] This generates "considerable" benefits for the United States.[187] These include the reduction in transaction costs of global trade; the elimination of currency risk; the seigniorage premium (which comes from the lower cost of acquiring dollars); and the lower interest rate on foreign liabilities.[188] This is the so-called exorbitant privilege of printing the world's leading trading and reserve currency. According to the leading academic account, the United States is able to benefit from this so-called exorbitant privilege since it

---

[181] *Id.*

[182] *Id.* at 11.

[183] *Id.* at 51.

[184] *See* Mariano-Florentino Cuéllar & Aziz Z. Huq, *Privacy's Political Economy and the State of Machine Learning: An Essay in Honor of Stephen J. Schulhofer*, 76 N.Y.U. Ann. Surv. Am. L. 317, 346 (2021) (discussing China's domestic surveillance and political control structures); *see also* Xiao Qiang, *The Road to Digital Unfreedom: President Xi's Surveillance State*, 30 J. Dem. 53 (2019).

[185] Benjamin E. Goldsmith & Yusaku Horiuchi, *In Search of Soft Power: Does Foreign Public Opinion Matter for US Foreign Policy?*, 64 World Pol. 555, 556–57 (2012).

[186] Barry Eichengreen, Exorbitant Privilege: The Rise and Fall of the Dollar and the Future of the International Monetary System 2 (2011).

[187] *Id.* at 3.

[188] *Id.* at 3–4.

benefits from the "transparency of [its] public institutions, the right to free expression, and an unfettered media," all of which are necessary for "building confidence."[189] Such a political "system of open and transparent democracy," which has the institutional capacity to effectively implement policy and is capable of identifying and correcting its own errors, "is crucial for explaining the confidence that foreign banks and other investors have" in dollar assets.[190] That democratic system, to be sure, has come under stress from many directions in recent years. But the Court's hostility to democratic responsiveness, and its willingness to tolerate partisan lock-ups, likely adds to the pressure on the structural conditions of the dollar's role as a reserve currency.

The second pathway of interest pertains to the relationship between domestic democracy and international conflict. A substantial empirical literature in international relations substantiates the idea of a "democratic peace." This is the idea of a "substantial, robust, and statistically significant association between joint democracy and the absence of militarized conflict."[191] The democratic peace theory has been subject to a wide range of empirical critiques, and yet remains a powerful and influential finding.[192] We assume its empirical robustness here. On that assumption, one effect of the Court's democratic recession is to make international conflict more likely. This is so even if democracy for the theory's purposes is understood as a binary rather than a scalar quality.[193] For the Court's interventions may be necessary elements of the dynamic that pushes the political system over a brink.[194] And while the

---

[189] ESWAR S. PRASAD, THE DOLLAR TRAP: HOW THE U.S. DOLLAR TIGHTENED ITS GRIP ON GLOBAL FINANCE 301 (2014).

[190] *Id.* at 302. Prasad's view is not universally shared. Eichengreen, in contrast, identifies monetary stability as key. EICHENGREEN, *supra* note 186, at 181.

[191] Allan Dafoe, *Statistical Critiques of the Democratic Peace: Caveat Emptor*, 55 AM. J. POL. SCI. 247, 247 (2011) (also identifying and rejecting critiques of the thesis); *see also* Bruce Russert, *The Fact of Democratic Peace*, *in* DEBATING THE DEMOCRATIC PEACE 58 (Michael E. Brown, Sean M. Lynn-Jones & Steven E. Miller eds., 1996) (discussing the basis of the democratic peace claim).

[192] Håvard Hegre, Michael Bernhard & Jan Teorell, *Civil Society and the Democratic Peace*, 64 J. CONFLICT RESOL. 32, 32–34 (2020).

[193] *See* Reşat Bayer & Michael Bernhard, *The Operationalization of Democracy and the Strength of the Democratic Peace: A Test of the Relative Utility of Scalar and Dichotomous Measures*, 27 CONFLICT MGMT. & PEACE SCI. 85, 85–86 (2010) (proposing that the theory is better operationalized with democracy as a binary).

[194] An obvious way in which this could happen involves a judicial decision negating the popular vote in a presidential election. *Cf.* Republican Party of Pa. v. Boockvar, 141 S. Ct. 1, 2 (2020) (Alito, J., dissenting). The risk is hardly speculative.

mechanism underlying the democratic peace theory remains contested,[195] the insulation of national political leadership from the interests, concerns, and rebukes of a substantial tranche of the population has an obvious effect of making costly international action opposed by that segment of the people more likely.

In sum, the quality of a nation's democracy is plausibly understood as a foundational resource when it acts in the geopolitical sphere. Although the theoretical and empirical arguments supporting the importance of "soft power" effects may be somewhat weaker than some observers have surmised, we see reason to believe that a weakened democratic foundation will undermine U.S. fiscal power and increase its exposure to international conflict. At a moment of growing geopolitical instability, with both Russia and China manifestly discontent with unipolar hegemonic arrangement, this seems a dangerous development.

IV. The Court as a Hidden Source of American Foreign Policy

Nothing compels the court to shape American foreign policy so substantially. The Court picks virtually its entire docket through discretionary choices about which cases to hear, and which to leave unaddressed. The Justices (or at least four of them) make docket decisions on Article III justiciability grounds. They can intervene either early or late in the litigation process, before or after an evidentiary record has been developed. All these choices shape profoundly the scope and extent of its interventions. And even where the Court decides to grant review and decide a case on the merits, it has considerable leeway in determining how broadly or narrowly to craft rulings. Various Justices may understand their role in constrained jurisprudential terms—to interpret the statutory text, for example, or apply the original understanding of constitutional provisions—but these do not have a predictable effect on the footprint of their decisions. As a result, the judicial shaping of foreign policy instruments, channels, and foundational resources is a function of choices the Justices make. It is judicial policy developed at the near-unfettered discretion of the bench.

---

[195] Hegre et al., *supra* note 192, at 34–38 (finding that social mechanisms of constraint—by which "citizen engagement can impose audience costs on leaders *between* elections, either directly or in anticipation of them"—to be the most powerful index of democratic peace).

There is a profound irony at work in the Justices' exercise of this discretion in the cases we have discussed. As we have seen, the Justices commonly invoke their comparative institutional disadvantage on foreign affairs and national security matters in relation to the executive branch.[196] Deference is commonly elaborated in terms of the executive's greater expertise and informational resources in respect to foreign policy.[197] But while the Court has traditionally taken account of at least some prudential foreign policy and national security concerns in order to limit the scope of its foreign-affairs footprint, it has recently exercised its discretion over its own docket to take up cases and questions with profound, and arguably damaging, effects on the geopolitical instruments resources that the United States can marshal at times of international tension. Its recognition of foreign policy implications, indeed, is highly selective in ways that might well be predicted given the partisan affiliation of specific Justices. The Court has also ignored the fact that the argument for humility based on foreign-affairs-related expertise extends far more broadly than is apparent at first blush.[198] At a minimum, this suggests that the present judicially created category of "foreign affairs and national security" matters is underinclusive, malleable, and vulnerable to ideological taint. Its partial nature leads to serious and unnoticed foreign affairs effects of the kind described in Part III. The result is an unprincipled and erratic reshaping of basic geopolitical instruments, channels, and foundational resources.

Resolving what should then be done as a matter of national interest, institutional competence, and the public's well-being is a challenging enterprise. This is in part because the relevant institutional design criteria do not necessarily converge. The fact of foreign-affairs externalities is not necessarily sufficient by itself to justify a radical narrowing of the judicial role. In *Massachusetts v. EPA*,[199] for example, the Court might have more productively distinguished the argument

---

[196] *See* cases cited in *supra* note 149.

[197] *See, e.g.*, Boumediene v. Bush, 553 U.S. 723, 797 (2008) ("[N]either the Members of this Court nor most federal judges begin the day with briefings that may describe new and serious threats to our Nation and its people."); United States v. Curtiss-Wright Exp. Corp., 299 U.S. 304, 320 (1936).

[198] Of course, that asymmetry might be quite general and transubstantive. *Cf.* Cass R. Sunstein, *The Most Knowledgeable Branch*, 164 U. PA. L. REV. 1607, 1648 (2016) (stressing "[t]he immense epistemic advantages of the executive branch [as] a central and insufficiently appreciated aspect of the real world of checks and balances [that] must play a central role").

[199] 549 U.S. 497 (2007); *see also supra* text accompanying notes 83–84.

that foreign policy consequences should lead to a finding that an agency has a purely discretionary right to not even respond to reasonable arguments about what it may be legally compelled to do, and the quite distinct conclusion (illustrated in the new major questions cases) that an agency lacks authority to ever do something. Nor do we think that federal courts should fail to impose sensible limits on executive assertions of power in the name of foreign affairs, as the Supreme Court did in *Youngstown*. A prudent Supreme Court also does not need to kowtow to executive branch arguments grounded in foreign affairs when the weight of legal materials cuts decisively against such a position, as in *Massachusettsv. EPA*. Even the more constrained jurisprudential explanations of the court's role (reasonably interpreted) do not preclude some engagement with the foreign policy implications of decisions.

Rather, the problem of theorizing and critically evaluating the Supreme Court's early twenty-first century in American foreign policy extends well beyond what can be achieved by rote recitation of the formulations in *Curtiss-Wright* or *Youngstown*. Thoughtful observers must take seriously the analytical frameworks drawn from both FRL and FPA, and beyond, to interpret American foreign policy. These frameworks need to be enlarged, though, to account for the newly emboldened Court. Such an exercise can do more than merely help scholars narrate changes in the allocation of control over America's levers of global power. The resulting insights may also help judges, lawyers, and policymakers better understand how to evaluate judicial choices about the role of government power that cast a long shadow on America's role in the world. Bearing these complexities in mind, we offer the following tentative practical implications.

First, domestic and international observers should at a minimum recognize the pivotal and likely growing role of the Court in shaping the American foreign-policy agenda. Whether they are in think tanks, universities, or government agencies, analysts keen to understand American behavior and capacity on foreign affairs will benefit from an analytic framework with a robust role for the judiciary.

Second, key domestic decision-makers who play a role in setting the agenda of the country's highest federal court should be mindful of foreign policy considerations when shaping the Supreme Court's agenda. State attorneys general and federal officials, for example, may find it wise to be wary of triggering Supreme Court review of certain questions affecting democracy or foreign policy, as the country's geopolitical position may benefit from a degree of ambiguity allowing the executive branch and other key actors to continue mapping the space between domestic and foreign policy on an issue-by-issue basis. Indeed, we think that the Supreme Court itself should be wary of granting certiorari unnecessarily on issues that could benefit from further percolation and negotiation between the branches, levels of government in the American federal system, and national governments. We realize, however, that such cautions may have little effect on a bench with scant regard for certain consequentialist considerations.

Third, to the extent the Court continues to leave room for deference to the executive branch on foreign affairs issues, in the tradition of *Curtiss-Wright*, it should be skeptical of under-theorized distinctions between traditional "foreign affairs" and domestic policy domains. There is a threshold question, akin to the "*Chevron* step-zero" inquiry as to whether the classification offered by a specific executive-branch actor is a plausible one.[200] Should the State Department or the Defense Department get more deference than other agencies? We are skeptical that the traditional cut-point for deference is wise. Arguments for deference merit prima facie consideration whether they concern enemy combatants or international public health cooperation.

But we also think deference is indeed sometimes warranted. Admittedly, the terrain here is complex. There is no cause to "level down" comprehensively.[201] An alternative approach might build on the proposal of influential foreign relations law scholars to install a measured judicial deference when it comes to all (and not just some) matters of geopolitical significance.[202] The scope of such deference, however, may well be set much wider than previously has been appreciated beyond the domain of national security and state-to-state diplomacy.[203] Policymakers, scholars, and the public can also help the courts realize

---

[200] Cass R. Sunstein, Chevron *Step Zero*, 92 VA. L. REV. 187, 198–204 (2006). We're grateful to Curt Bradley on this point.

[201] *Cf.* Aziz Z. Huq, *The Discrete Charm of Leveling Down*, 90 GEO. WASH. L. REV. 1487 (2022) (discussing the merits of leveling up and down remedies).

[202] *See, e.g.*, Curtis A. Bradley, Chevron *Deference and Foreign Affairs*, 86 VA. L. REV. 649, 669–70 (2000).

[203] One of us has previously expressed doubt about the resolving power of structural constitutional principles in this field. *See* Aziz Z. Huq, *Structural Constitutionalism as Counterterrorism*, 100 CALIF. L. REV. 887, 890 (2012) (arguing that "abstract structural constitutional theories frequently provide no useful information as to whether a counterterrorism policy is appropriately tailored, justified by a compelling governmental need, or consistent with a statute's command").

a sensible approach to these opaque or downplayed foreign-affairs effects. Scholars and practitioners should also step up their efforts to use amicus briefs and legal arguments that underscore the cross-cutting foreign policy consequences of court decisions. They should urge judges and Justices to recognize the virtues of reasonable restraint in a wider range of cases. They can also acknowledge a reasonable role of doctrines affording executive branch officials a measure of deference to take account of the country's broader foreign policy context, while maintaining core protections for human dignity that have the potential to affect perceptions of American values throughout the world.

Fourth, particularly given that foreign affairs deference may apply in domains that were traditionally viewed as domestic, the Supreme Court should leave ample room for a variety of policy considerations to provide an appropriate measure of nuance in assessing the strength of executive branch arguments. That is, it should evince greater willingness to defer in light of the evinced expertise of the executive branch, rather than offering deference based on hypothecated categories of constitutional structure. Against the backdrop of courts' core focus on the relevant legal materials, some variation in the extent of deference may depend factors such as the extent of executive branch coordination across agencies with relevant jurisdictions,[204] the level of importance an administration assigns to a particular issue across a broader range of other priorities or responsibilities, and the extent to which the relevant statutory and historical context highlights the importance of executive discretion.[205] As with a range of other presumptions, arguments for deference, and rules of construction, such considerations could have a material impact on the interpretation of ambiguous legal materials, or offset the court's willingness to rely on recently-invigorated jurisprudential tools such as the "major questions" doctrine.[206]

---

[204] *Cf.* Amanda Shami, *Three Steps Forward: Shared Regulatory Space, Deference, and the Role of the Court*, 83 Fordham L. Rev. 1577 (2014).

[205] *Cf.* Youngstown Sheet & Tube Co. v. Sawyer, 343 U.S. 579 (1952).

[206] In *West Virginia v. EPA*, 142 S. Ct. 2587, 2608 (2022), the Court wrote that "where there is something extraordinary about the history and breadth of the authority" an agency seeks to deploy or the "economic and political significance" of that assertion, courts should "hesitate before concluding that Congress meant to confer such authority." Even taking the "major questions" doctrine on the terms embraced by the majority in *West Virginia v. EPA*, it follows that a carefully targeted, historically grounded executive branch argument about the relevance of a particular interpretive or regulatory decision to a national foreign affairs function can have a bearing on at least the "political" significance of the executive branch action, or whether there is something "extraordinary" about that action's history or breadth.

V. Conclusion

That courts will continue to play a material role at the fertile intersection between domestic and international issues is all but guaranteed by the American civic tradition. To achieve reasonable outcomes, however, policymakers and the public must begin by understanding the stakes of having a less-constrained high court operating increasingly as a kind of surrogate ministry of foreign affairs. It is better, in short, to grasp the hidden judicial springs of foreign policy than to be perplexed about its weird physics. As the United States is navigating a period of exceptional and fraught global turmoil, judges and the public alike should be wary of a court-imposed framework for foreign affairs where restraint falls by the wayside and policymakers end up with fewer tools, friends, and reservoirs of goodwill to leverage at a time when the world depends on vigorous but nuanced American engagement. The fruits of that engagement depend on blending greater candor about the blurred lines between realms that are foreign and domestic with sensible decision-making from key institutional actors capable of being judicious rather than merely judicial.

GREGORY ABLAVSKY

TOO MUCH HISTORY: CASTRO-HUERTA AND THE PROBLEM OF CHANGE IN INDIAN LAW

> [T]he controlling and conclusive position assumed as the basis of that judgment [*Worcester v. Georgia*], that the Indian Nations were distinct and sovereign political communities, independent of the States, is confidently believed to be incorrect, and that sooner or later it must be abandoned. (*State v. Foreman*, 16 Tenn. 256, 335 (1835).)

> Despite bitter criticism and the defiance of Georgia which refused to obey this Court's mandate in Worcester the broad principles of that decision came to be accepted as law. Over the years this Court has modified these principles in cases where essential tribal relations were not involved and where the rights of Indians would not be jeopardized, but the basic policy of Worcester has remained. (*Williams v. Lee*, 358 U.S. 217, 219 (1959).)

> '[C]riminal jurisdiction in Indian country is an indefensible morass of complex, conflicting, and illogical commands layered in over decades via congressional policies and court decisions and without the consent of tribal nations.'. . . [T]hat leads me to think that one of the things we should be trying to do here is—is to do no harm because there's lots of ripple effects

Gregory Ablavsky is the Marion Rice Kirkwood Professor of Law at Stanford Law School; Professor of History (by Courtesy) at Stanford University.

Author's note: Thanks to Will Baude, Afroditi Giovanopoulou, Liz Reese, and the FSU Law Faculty for feedback. Thanks also to Tanner Allread for exceptional comments and research assistance, and to Rachel Shields, Kai Wiggins, and the staff of the Crown Library for additional research.

The Supreme Court Review, 2023.
© 2023 The University of Chicago. All rights reserved. Published by The University of Chicago Press. https://doi.org/10.1086/724831

from a broad decision. (Justice Kavanaugh, Transcript of Oral Argument at 56–57, *United States v. Cooley*, 141 S. Ct. 1638 (2021) (No. 19-1414).)

In a vivid scene in *The Round House*, Louise Erdrich's powerful novel of criminal justice in Indian country, the main character's father, a tribal court judge, tries to explain federal Indian law to his son by pulling a rancid casserole from the fridge.[1] The judge analogizes each rotten bit of sludge to a different terrible precedent—*Johnson v. M'Intosh*,[2] *Lone Wolf v. Hitchcock*,[3] *Tee-Hit-Ton Indians v. United States*.[4] Only one opinion, symbolized by a stabilizing carving knife, provides a solid footing for the precarious pile of cutlery that represents Indian law. "*Worcester v. Georgia*,"[5] the judge explains. "Now, that would be a better foundation."[6]

Chief Justice Marshall's 1832 opinion, in the midst of Georgia's attempted ethnic cleansing of the Cherokee Nation, invalidated the state's claim to jurisdiction over Cherokee lands.[7] The "treaties and laws of the United States," Marshall ruled, "contemplate the Indian territory as completely separated from that of the states."[8] Scholars have celebrated the decision as the rare moment when the Court resisted the overwhelming tide of nineteenth-century Anglo-American racism.[9] They have also described the ruling—which substantially limited state authority in Indian country and affirmed tribal autonomy—as "[t]he foundational case in federal Indian law."[10]

---

[1] Louise Erdrich, The Round House 227–30 (2012).

[2] 21 U.S. (8 Wheat.) 543 (1823).

[3] 187 U.S. 553 (1903).

[4] 348 U.S. 272 (1955).

[5] 31 U.S. (6 Pet.) 515 (1832).

[6] Erdrich, *supra* note 1, at 229.

[7] *Worcester*, 31 U.S. (6 Pet.) at 561–63.

[8] *Id.* at 557.

[9] Tim Alan Garrison, The Legal Ideology of Removal: The Southern Judiciary and the Sovereignty of Native American Nations 240 (2002) (describing *Worcester* as a "decision that represented judicial courage and clarity"); Gerard N. Magliocca, *The Cherokee Removal and the Fourteenth Amendment*, 53 Duke L.J. 875, 897 (2003) ("*Worcester* is often praised as a model of judicial integrity for reversing the missionaries' conviction in the teeth of intense political pressure.").

[10] Sidney L. Harring, Crow Dog's Case: American Indian Sovereignty, Tribal Law, and United States Law in the Nineteenth Century 25 (1994); *see also* Charles F. Wilkinson, American Indians, Time, and the Law: Native Societies in a Modern Constitutional Democracy 30 (1987) ("*Worcester* is one of the Supreme Court's most lasting statements; modern federal and state courts rely upon it more than upon nearly any other pre-Civil War opinion in any field."); David H. Getches, *Conquering the Cultural*

Yet *Worcester* now wears an ominous red flag in Westlaw: "Abrogation Recognized by Oklahoma v. Castro-Huerta."[11]

The Supreme Court's decision in *Castro-Huerta*[12]—which held that states possess criminal jurisdiction over non-Indian offenses against Indians in Indian country concurrently with the federal government—was at once one of the *least* and *most* mysterious outcomes of the past Term. Least, because, even in a Term rife with decisions that critics derided as nakedly ideological, the realist take on this decision is hard to contest. Two Terms ago, in *McGirt v. Oklahoma*, Justice Gorsuch, joined by the then-four liberal Justices, held that the Muscogee Creek Reservation in eastern Oklahoma had never been disestablished and therefore remained, legally, Indian country.[13] The four other conservative Justices vigorously dissented.[14] So when Justice Barrett replaced Justice Ginsburg, Oklahoma, sensing an opportunity, waged a prolonged campaign to overturn, or at least substantially cabin, *McGirt*.[15]

Castro-Huerta vindicated this strategy: Gorsuch and the other three remaining members of the *McGirt* majority found themselves dissenting against the four *McGirt* dissenters, who, bolstered by Justice Barrett, now constituted the majority. Writing for that majority, Justice Kavanaugh's opinion did not hide its underlying motivation: Its opening section replicated Oklahoma's contested sky-is-falling narrative of post-*McGirt* criminal justice that was, as a formal matter, legally irrelevant.[16] Few other cases this past Term—even,

---

Frontier: The New Subjectivism of the Supreme Court in Indian Law, 84 CALIF. L. REV. 1573, 1582 (1996) ("*Worcester* lays the cornerstone for the legal system's continuing recognition of tribal sovereignty.").

[11] *Worcester*, 31 U.S. (6 Pet.) 515 (1832), 1832 WL 3389.

[12] 142 S. Ct. 2486 (2022).

[13] 140 S. Ct. 2452, 2459 (2020).

[14] *See id.* at 2482–83 (Roberts, C.J., dissenting).

[15] *See, e.g.*, *Attorney General O'Connor Calls on U.S. Supreme Court to Overturn or Limit* McGirt *Decision*, OFF. OF THE OKLA. ATT'Y GEN. (Aug. 2021), https://www.oag.ok.gov/articles/attorney-general-o%E2%80%99connor-calls-us-supreme-court-overturn-or-limit-mcgirt-decision.

[16] *Castro-Huerta*, 142 S. Ct. at 2492–93 ("Castro-Huerta's case exemplifies a now-familiar pattern in Oklahoma in the wake of *McGirt*."). *But see id.* at 2523–24 (Gorsuch, J., dissenting) (critiquing the majority for "[a]dopting the State's representations wholesale"); Rebecca Nagle & Allison Herrera, *Where Is Oklahoma Getting Its Numbers From in Its Supreme Court Case?*, ATLANTIC (Apr. 26, 2022), https://www.theatlantic.com/ideas/archive/2022/04/scotus-oklahoma-castro-huerta-inaccurate-prosecution-data/629674 (questioning Oklahoma's claim

perhaps, *Dobbs* and *Bruen*[17]—demonstrated quite so vividly Justice Kagan's anxiety about the optics of dramatic legal shifts when a single Justice changes.[18]

Yet doctrinally, *Castro-Huerta* is much more mysterious. Because the Court declined to overrule *McGirt*, many commentators portrayed *Castro-Huerta* as a narrow decision.[19] Legally, though, the decision swept much more broadly than a simple reversal of *McGirt* would have. Fundamentally altering long-assumed rules, it judicially expanded the scope of state criminal jurisdiction in Indian country for the first time since 1882. Moreover, the holding applied not just in Oklahoma but in *every* state, even though only a handful of predictable red-state attorneys general—notably *excluding* conservative-leaning states like Utah, Idaho, or Montana that have large Native populations—had filed to support Oklahoma.[20]

In reaching this result, the Court argued that it was construing for the first time the relevant statutory text in the General Crimes Act, which reads: "[T]he general laws of the United States as to the punishment of offenses committed in any place within the sole and

---

that "*McGirt* wreaked havoc" because the number of cases it cited that *McGirt* had supposedly affected "seems to have come out of nowhere").

[17] *Castro-Huerta* was only one of three 5-4 decisions last Term in which only five conservative Justices constituted the majority. Angie Gou et al., SCOTUSblog, STAT PACK for the Supreme Court's 2021–22 Term 11 (2022). In all three cases Justice Gorsuch broke with his conservative colleagues and joined the three liberal justices in dissent. *Id.* The other two cases—*Shoop v. Twyford*, 142 S. Ct. 2037 (2022), and *Patel v. Garland*, 142 S. Ct. 1614 (2022)—were less politically salient and did not involve very recently decided Court decisions.

[18] Nick Ehli & Robert Barnes, *Kagan Says Questions of Legitimacy Risky for Supreme Court*, Wash. Post (July 21, 2022), https://www.washingtonpost.com/politics/2022/07/21/elena-kagan-supreme-court-legitimacy (quoting Justice Kagan as saying, "People are rightly suspicious if one justice leaves the court or dies and another justice takes his or her place and all of sudden the law changes on you.").

[19] *See, e.g.*, Adam Liptak, *Supreme Court Narrows Ruling for Tribes in Oklahoma*, N.Y. Times (June 29, 2022), https://www.nytimes.com/2022/06/29/us/politics/supreme-court-ruling-oklahoma-tribes.html ("The ruling [in *Castro-Huerta*] left in place the basic holding of the 2020 decision, McGirt v. Oklahoma."); James Blenk et al., *What the Supreme Court's Decision in* Castro-Huerta *Means for Tribal Sovereignty*, JD Supra (Aug. 30, 2022), https://www.jdsupra.com/legalnews/what-the-supreme-court-s-decision-in-8538029 ("The Court only accepted review of the narrow issue of whether there is concurrent jurisdiction with the state over crimes committed on Indian country. So, the *McGirt* challenge was, at least for the time being, deflected.").

[20] Brief for the States of Texas et al. as Amici Curiae in Support of Petitioner at 29, Oklahoma v. Castro-Huerta, 142 S. Ct. 2486 (2022) (No. 21-429), 2022 WL 729117, at *29. Notably *Castro-Huerta* will not even affect a majority of these states, since Texas, Nebraska, and Kansas already enjoyed the jurisdiction in dispute under various congressional statutes.

exclusive jurisdiction of the United States . . . shall extend to the Indian country."[21] This statutory language had been unchanged since first drafted in 1834, and derived from still earlier provisions tracing back to 1790.[22] Six prior Supreme Court decisions had concluded that states *lacked* criminal jurisdiction over the category of crime at issue here—non-Indian perpetrators against Indian victims—but the Court distinguished these statements as *dicta* because they had not explicitly relied on the statutory text.[23] As for the relevant background jurisdictional rules, the majority acknowledged that what it dubbed "territorial separation"—that Indian country lay outside state control—was not only the original Founding-era legal understanding but also the holding of the Court's seminal decision in *Worcester v. Georgia*.[24] Yet, despite the absence of any change in either statutory or constitutional text, this principle, the Court argued, "was abandoned" by the Supreme Court over the course of the nineteenth and twentieth centuries.[25]

Like Justice Gorsuch's sharply worded dissent, I disagree with these conclusions.[26] But rather than relitigating the case, I think it is

---

[21] 18 U.S.C. § 1152.

[22] For the original enactment, see Act of June 30, 1834, ch. 161, § 25, 4 Stat. 729, 733; for the 1790 predecessor, see Act of July 22, 1790, ch. 33, § 5, 1 Stat. 137, 138.

[23] United States v. Bryant, 579 U.S. 140, 146 (2016) ("Most States lack jurisdiction over crimes committed in Indian country against Indian victims."); Nevada v. Hicks, 533 U.S. 353, 365 (2001) ("Sections 1152 and 1153 of Title 18 . . . give United States and tribal criminal law generally exclusive application . . . *in Indian country*. . . ."); Washington v. Confederated Bands & Tribes of the Yakima Indian Nation, 439 U.S. 463, 470–71 (1979) ("[C]riminal offenses by or against Indians have been subject only to federal or tribal laws, except where Congress . . . has expressly provided that State laws shall apply." (citation omitted)); Williams v. Lee, 358 U.S. 217, 220 (1959) ("[I]f the crime was by or against an Indian, tribal jurisdiction or that expressly conferred on other courts by Congress has remained exclusive [of state jurisdiction]."); Williams v. United States, 327 U.S. 711, 714 (1946) (concluding that the "law and courts of the United States, rather than those of Arizona, have jurisdiction over offenses committed there . . . by one who is not an Indian against one who is an Indian"); Donnelly v. United States, 228 U.S. 243, 271–72 (1913) (declining to extend *McBratney* and *Draper* to permit state jurisdiction over crimes "committed by white men against the persons or property of the Indian tribes").

[24] *Castro-Huerta*, 142 S. Ct. at 2497.

[25] *Id.* at 2493–94, 2497–98.

[26] I helped co-author a brief supporting Castro-Huerta in the case. Brief of Amici Curiae Federal Indian Law Scholars and Historians in Support of Respondent, *Castro-Huerta*, 142 S. Ct. 2486 (No. 21-429), 2022 WL 1052058; *see also* Gregory Ablavsky & Elizabeth Hidalgo Reese, *The Supreme Court Strikes Again—This Time at Tribal Sovereignty*, WASH. POST (July 1, 2022), https://www.washingtonpost.com/opinions/2022/07/01/castro-huerta-oklahoma-supreme-court-tribal-sovereignty.

more fruitful to place *Castro-Huerta* in broader perspective. *Castro-Huerta* exemplified many of the flaws that scholars have criticized in the Court's Indian law jurisprudence for decades—especially the Justices' reliance on their own subjective assessments to craft law.[27] Building on this work, this Article examines a less explored aspect of the Court's undisciplined, subjective jurisprudence in Indian law that *Castro-Huerta* also epitomized.[28] I call this the problem of too much history.

What I mean by "too much history" is that, in Indian law, judges and litigants must make sense of over two centuries of conflicting debates over jurisdiction in Indian country, recorded largely not in statutes or constitutional provisions but in dozens of shifting Supreme Court decisions. The key question in *Castro-Huerta*, and the core of the dispute between majority and dissent, was the issue of *change*: how the law on state jurisdiction in Indian country had shifted over time. But the sheer mass of history makes it hard for the Justices to identify legitimate legal change in Indian law, partly because of the volume of potential evidence and partly because of the uncertain weight that long-ago precedents deserve in the present.

This conundrum leads, in my view, to two broad types of judicial use of history in Indian law. "Good" history decisions, epitomized by the Court's decision this Term in *Ysleta del Sur Pueblo v. Texas*,[29] employ specific context to examine narrowly defined legal questions. By contrast, "bad" history opinions, exemplified by *Castro-Huerta*, turn to the past as an independent source of law, ask broad, unanswerable

---

[27] For examples of some of the many scholarly criticisms of the Court's Indian-law jurisprudence, see Robert A. Williams, Like a Loaded Weapon: The Rehnquist Court, Indian Rights, and the Legal History of Racism in America (2005); Matthew L.M. Fletcher, *Muskrat Textualism*, 116 Nw. U. L. Rev. 963 (2022); Kathryn E. Fort, *The Vanishing Indian Returns: Tribes, Popular Originalism, and the Supreme Court*, 57 St. Louis U. L.J. 297 (2013); Philip P. Frickey, *(Native) American Exceptionalism in Federal Public Law*, 119 Harv. L. Rev. 431 (2005); Getches, *supra* note 10; and David H. Getches, *Beyond Indian Law: The Rehnquist Court's Pursuit of States' Rights, Color-Blind Justice and Mainstream Values*, 86 Minn. L. Rev. 267 (2001).

[28] Unsurprisingly, given the central role of history in federal Indian law, other scholars have explored different aspects of the role of history in the field. *See* Robert N. Clinton, *The Curse of Relevance: An Essay on the Relationship of Historical Research to Federal Indian Litigation*, 28 Ariz. L. Rev. 29 (1986); Fort, *supra* note 27. An important exploration of the question of change in Indian law—though focused primarily on the question of *factual*, not legal, change—appears in Wilkinson, *supra* note 10, at 32–52.

[29] 142 S. Ct. 1929 (2022).

questions of it, and provide no clear way to assess the inevitable heap of conflicting evidence.

Having laid out this challenge, I reexamine the question of the specific historical change at the core at *Castro-Huerta*. Rather than the majority's narrative of abandonment and the dissent's narrative of continuity, I think a more accurate account of what the Court has done with respect to state jurisdiction in Indian country is *translation*—trying to make sense of older legal principles within a new jurisprudential frame.[30] But this approach makes the Court's decisions in this area especially prone to misreading and selective citation, as *Castro-Huerta* underscored.

History loomed especially large in this past Term. For Indian law, though, this is not new: history has long played a key, even outsized, role in the field. Many commentators have now vociferously contested the claim that the Court's inquiry into history provides an objective, value-neutral basis for legal judgments.[31] But this, too, is not news for Indian country. The field has long offered something of a cautionary tale of what happens when an undisciplined Court is presented with a bonanza of historical evidence and few methodological guidelines or much rigor to constrain its use. In this sense, the entire nation is now confronting what Native peoples have long known.[32]

I. The Problem of Historical Change

Federal Indian law has a problem: too much history and not enough text.

The legal relationship between Native nations and the United States has been one of the central questions of American jurisprudence since, as the cliché goes, the dawn of the republic. Even if we privilege Anglo-American perspectives, as U.S. law has long done, the so-called

---

[30] *Cf.* Lawrence Lessig, Fidelity & Constraint: How the Supreme Court Has Read the American Constitution 49–69 (2019) (arguing, in the context of constitutional law, that the Supreme Court seeks to translate historical meaning based on changes in surrounding context).

[31] For perhaps the most insightful take on ongoing arguments over history in constitutional interpretation, see Jack Balkin, *Lawyers and Historians Argue About the Constitution*, 35 Const. Comment. 345 (2020).

[32] Elizabeth Hidalgo Reese, *Conquest in the Courts*, Nation (July 6, 2022), https://www.thenation.com/article/society/supreme-court-castro-huerta ("[A]fter this Supreme Court term, I suspect a great deal of Americans now understand the feeling [that Native peoples have toward the Supreme Court].").

"Indian problem" spawned centuries' worth of treaties and treaty negotiations, judicial opinions, executive memoranda, administrative regulations, and intense political controversies and debates. When Felix Cohen wrote his canonical treatise first defining federal Indian law in 1940, the volume's introduction, by Solicitor Nathan Margold, counted over four thousand statutes in the area.[33]

And yet this torrent of words and shifting policies produced surprisingly little enduring, positive text clearly defining the structural relationship between Native nations and the United States. The three brief mentions of Indians in the U.S. Constitution, for instance, do very little to demarcate Native nations' legal status; they mostly say what tribes are not.[34] As for the wealth of treaties, legislation, and administrative action, anyone who has spent any time in Indian law knows that these texts created, not a single history of "federal Indian law," but at least 574 distinct histories. Each federally recognized tribe and reservation, in other words, has its own long, complex, and often highly specific legal history. The federal executive did craft "Indian policy" for those it classified, generically, as "Indians," but its constant vacillation makes it difficult to say any of it was especially well-settled or clear. In the trinity of "text, structure, and history,"[35] in other words, there is little of the first two and lots of the third.

This imbalance between history and text has had important consequences for Indian law. It means that, for better or worse (often worse), much of Indian law has been made by the U.S. Supreme Court. Students in a course in, say, federal tax closely study the federal tax code and cases interpreting it. Students in a federal Indian law class, by contrast, seldom encounter statutes. Instead, they almost exclusively read dozens of U.S. Supreme Court cases, spanning two centuries, that crafted structural rules defining tribal, state, and federal authority in Indian country.

Yet, even as it has empowered the Court, the imbalance between history and text has made actually deciding Indian law cases harder. Rarely has the term "opinion" been as appropriate as in the Court's

---

[33] Nathan R. Margold, *Introduction* to FELIX S. COHEN, HANDBOOK OF FEDERAL INDIAN LAW, at viii (1942).

[34] *See* U.S. CONST. art. I, § 2, cl. 3 ("excluding Indians not taxed" from representation); *id.* art. I, § 8, cl. 3 (providing congressional authority to "regulate Commerce . . . with the Indian tribes"); *id.* amend. XIV, § 2 (again "excluding Indians not taxed" from representation).

[35] United States v. Lopez, 514 U.S. 549, 585 (1995) (Thomas, J., concurring).

decisions in the area. In them, the Justices enshrined their wide-ranging and often freeform thoughts on sovereigns' respective powers in Indian country by distilling the mélange that is Indian law's history, often in ways only loosely tied to a specific treaty or law.[36]

The history-text imbalance also exacerbates a jurisprudential problem already especially acute in Indian law given its lengthy pedigree: the challenge of change. Law, like anything else, is not static. But authoritatively adopted texts make identifying legal change much easier. When did the Fair Housing Act become law? 1968.[37] When was it amended to include discrimination based on familial status? 1988.[38] By contrast, extracting legal principles from the muck that constitutes federal Indian law's past, let alone identifying what changed and when, turns out to be much harder.

This task is made all the more difficult by the fact that policymakers' understanding and vision of the place of Native nations has changed dramatically. Much of the legal history of federal Indian law, though still good black-letter law, rests on logic and premises that Congress and the executive long ago rejected. Federal Indian law scholars resolved this tension by distilling certain foundational and enduring principles from the Supreme Court's precedents.[39] But, as last Term vividly reminded us, seemingly even the most durable of precedents can be distinguished away, or even discarded, if a majority can be persuaded that they rested on faulty premises.

Few cases have presented this problem quite so squarely as *Castro-Huerta*. Ostensibly a statutory interpretation case, the decision turned into the latest in the Court's attempts to define broadly and abstractly

---

[36] Other scholars have noted this aspect of federal Indian law. *See* Matthew L. M. Fletcher, *Muskrat Textualism*, 116 Nw. U. L. Rev. 963 (2022); Getches, *supra* note 10. Fletcher critiques this aspect of federal Indian law as what he calls "Canary Textualism." Fletcher, *supra* note 27, at 974–99. He concedes that there was a "great deal of force" to the argument that "this textualism is not textualism at all," but argues that he "takes the judges at their word." *Id.* at 975. However, though Justices like Scalia described themselves as textualists, they rarely characterized their approach to Indian law as textualist. Instead, as discussed below, their approach to questions in Indian law was more akin to structural federalism or separation of powers questions.

[37] Civil Rights Act of 1968, Pub. L. No. 90-284, §§ 800–820, 82 Stat. 73, 81–89.

[38] Fair Housing Amendments Act of 1988, Pub. L. No. 100-430, 102 Stat. 1619.

[39] Margold, *supra* note 33, at ix–xiii (outlining "four leading principles" of federal Indian law); Getches, *supra* note 10, at 1574 (describing how Indian law possessed "entrenched" and "foundation principles"); Philip P. Frickey, *Marshalling Past and Present: Colonialism, Constitutionalism, and Interpretation in Federal Indian Law*, 107 Harv. L. Rev. 381, 381 (1993) (describing aspects of Chief Justice Marshall's "foundational federal Indian law opinions").

the authority of a sovereign ("states," in this case) in Indian country. As with the Court's prior forays in the field, there is surprisingly little text and much more history. And at the case's center is the problem of change. The core of the ruling is the Court's conclusion that the law used to be one way (embracing Indian country's "territorial separation" from states) but is now another (states enjoy jurisdiction in Indian country unless expressly preempted by federal law).

This Part explores this problem of change of Indian law and how the Court grappled with this challenge in *Castro-Huerta*. It first examines Indian law in the context of theories of legal change more generally, examining why the field presents particular jurisprudential difficulties. It then considers two different uses of history in Indian law cases: what I not-so-creatively term "good" and "bad" history. By good history, I don't mean cases where the history is well done; rather, I mean cases where it is *appropriate* to look to history, which can offer a helpful, targeted guide for explicating other sources of law. I use this Term's decision in the Indian law case *Ysleta del Sur* as an exemplar of "good history." "Bad history," by contrast, results when the Court relies on history as a freestanding source of law, asks broad and effectively unanswerable questions of the past, and provides no criteria in assessing the complex mass of historical evidence. *Castro-Huerta*, I argue, could have been a "good history" case, but ended up repeating the flaws of the Court's bad-history decisions.

A. Theories of Legal Change and Indian Law

The problem of change is at the core of much of law, and is perhaps the key challenge of constitutional law. Explaining when and how legitimate legal change occurs has spawned some of the field's most important theories. Answers have encompassed original public meaning originalism's emphasis on text,[40] constitutional moments of extratextual transformation derived from social movements,[41] "original-law originalism" and its stress on positive legal change,[42] and

---

[40] *See* Keith E. Whittington, *The New Originalism*, 2 Geo. J.L. & Pub. Pol'y 599, 609–11 (2004).

[41] *See generally* 1 Bruce Ackerman, We the People: Foundations (1991); Bruce Ackerman, *Constitutional Politics/Constitutional Law*, 99 Yale L.J. 453 (1989).

[42] Stephen E. Sachs, *Originalism as a Theory of Legal Change*, 38 Harv. J.L. & Pub. Pol'y 817, 887–88 (2015).

practice-based accounts that highlight "constitutional liquidation"[43] or "historical gloss."[44]

Indian law presents a challenge for all these theories. Few questions have seen as many dramatic swings in both federal policy and law as the status of Native nations within the United States. Yet none of these shifts produced a change in constitutional text, nor were they subject to the kind of sustained, dramatic national social movements and public debate seemingly required of a "constitutional moment."[45] Liquidation and gloss seem more plausible, but it is hard to see how a regime characterized by constant contestation and inconsistency ever reached the kind of stability and widespread settlement that these theories seem to require.

Consider, for instance, the doctrine of federal plenary power over Native nations, which grants Congress full authority under Article I to control tribes' internal affairs. The best evidence from the Founding era suggests that, although most among the early American political elite thought that the federal government enjoyed exclusive authority as against the states to govern relations *with* Native nations, most thought that Native peoples, subject in some way to U.S. sovereignty, nonetheless lay outside federal statutory jurisdiction.[46] After a century of conquest, though, many Anglo-Americans had changed their views: they now believed that Native peoples were entirely subject to federal authority, and acted accordingly.[47] Neither group of policymakers explicitly codified their views into constitutional or even

---

[43] William Baude, *Constitutional Liquidation*, 71 Stan. L. Rev. 1, 47 (2019).

[44] Curtis A. Bradley & Trevor W. Morrison, *Historical Gloss and the Separation of Powers*, 126 Harv. L. Rev. 411, 424 (2012); Curtis A. Bradley & Neil S. Seigel, *Historical Gloss, Madisonian Liquidation, and the Originalism Debate*, 106 Va. L. Rev. 1, 3 (2020).

[45] *Cf.* Maggie Blackhawk, *On Power & the Law*: McGirt v. Oklahoma, 2020 Sup. Ct. Rev. 367, 402–10 (noting how work on social movements often neglects Native advocacy because Native peoples often did not "focus on changing the public mind").

[46] *See* Robert N. Clinton, *There Is No Federal Supremacy Clause for Indian Tribes*, 34 Ariz. St. L.J. 113, 147 (2002) ("At the time the Constitution of the United States was drafted, the Framers generally accepted the notion that the Indian tribes constituted separate sovereign peoples who were totally self-governing within their territory. . . ."); *see also* Gregory Ablavsky, *Beyond the Indian Commerce Clause*, 124 Yale L.J. 1012, 1083 (2015) ("[T[he Constitution's drafters and early interpreters regarded most tribes as separate sovereigns largely outside U.S. authority. . . .").

[47] *See* Clinton, *supra* note 46, at 164–205; Sarah H. Cleveland, *Powers Inherent in Sovereignty: Indians, Aliens, Territories, and the Nineteenth Century Origins of Plenary Power over Foreign Affairs*, 81 Tex. L. Rev. 1, 42–80 (2002).

statutory text. Perhaps the principle of plenary power was "liquidated"—but, though the principle is now black-letter doctrine, it is hard to see a moment when this claim of federal authority went unchallenged by either Native peoples or non-Native judges and commentators.

Descriptively, the theory of change that best fits Indian law's development is common-law constitutionalism, with its emphasis on how courts, especially the Supreme Court, have construed the Constitution over time.[48] Explaining how plenary power arose under this view is straightforward: despite the doctrine's ambiguous grounding in text and original understandings, the Supreme Court enshrined the principle in *Kagama*[49] and has reiterated it ever since, despite repeated challenges.[50]

Yet common-law constitutionalism presents its own challenges within federal Indian law. One reflects a broader problem with precedent, which is that it often empowers, not constrains, judges.[51] Textualism's promise of certainty and clarity is almost certainly illusory, but text at least forces judges to fight over the same thing—the meaning of a discrete set of words. By contrast, common-law constitutionalism permits judges to select not only the relevant evidence but often the terms of the debate. The past here functions as a buffet—an all-you-can-eat extravaganza in which the Justices can choose whichever precedent they like to reach the outcome they favor.

This problem is especially acute in federal Indian law, because the reality of too much history means the buffet is Las Vegas-on-seafood-night-sized.[52] The Supreme Court has been adjudicating Native status under federal law for over two centuries, often giving wildly different answers to fundamental questions. In my federal Indian law class, I ask students whether, if I paid them a bunch of money, they

---

[48] *See generally* David A. Strauss, *Common Law Constitutional Interpretation*, 63 U. Chi. L. Rev. 877 (1996).

[49] 118 U.S. 375 (1886).

[50] *See, e.g.*, United States v. Lara, 541 U.S. 193, 200 (2004); Washington v. Confederated Bands and Tribes of Yakama Indian Nation, 439 U.S. 463, 470–71 (1979); Lone Wolf v. Hitchcock, 187 U.S. 553, 565 (1903).

[51] *See* Richard M. Re, *Precedent as Permission*, 99 Tex. L. Rev. 907, 909–11 (2021); William Baude, *Precedent and Discretion*, 2019 Sup. Ct. Rev. 313, 314.

[52] Westlaw records 2,748 Supreme Court cases containing the word "Indian." Clearly, this search is wildly overinclusive—it encompasses cases dealing with the nation of India, as well as random uses of the word that have little to do with the field of Indian law—but does give a crude sense of the overabundance of precedent.

could craft a narrative from caselaw that argued that the federal government and states have consistently subordinated tribes and denigrated their authority. They always say yes. I then ask them, if I paid them a bunch of money, they could argue the reverse: that the Supreme Court has upheld and defended tribal sovereignty and autonomy against assaults. They say yes to this, too. The point—as they readily grasp—is that both narratives have substantial evidentiary support and capture considerable truth, and this history is unlikely to offer determinate answers to current legal questions.

But common-law constitutionalism also has a second challenge more distinctive to federal Indian law. In *Castro-Huerta*, the majority skewered the dissent for emphasizing the "history of mistreatment of American Indians," which it argued was irrelevant to the case's "legal questions."[53] This attempt to separate out positive law from morality in Indian law is quite venerable: in *Johnson v. M'Intosh*, Chief Justice Marshall acknowledged the absurdity and dubious morality of the doctrine of discovery but concluded that he was bound to follow its dictates anyway.[54]

But closer examination reveals that this divide between law and the "history of mistreatment" is untenable. Many of the Court's most important Indian law precedents explicitly rested on dubious pseudo-anthropology and white perceptions of Native peoples that we have little trouble recognizing as racist.[55] This was especially true of the Court's key late nineteenth- and early twentieth-century precedents—the historical moment when *Worcester* was supposedly abandoned, according to the *Castro-Huerta* Court—when the Court routinely and explicitly invoked Native peoples' supposed savagery to justify its rulings.[56] Moreover, these decisions were not value-neutral. They believed Indian law served an end—the colonization

---

[53] Oklahoma v. Castro-Huerta, 142 S. Ct. 2486, 2502 (2022).

[54] 21 U.S. (8 Wheat.) 543, 591–92 (1823) (describing the doctrine of discovery as "opposed to natural right, and to the usages of civilized nations").

[55] *See* Adam Crepelle, *Lies, Damn Lies, and Federal Indian Law: The Ethics of Citing Racist Precedent in Contemporary Federal Indian Law*, 44 N.Y.U. Rev. L. & Soc. Change 529, 532 (2021) ("Jurisprudence loaded with grotesque 19th-century racist stereotypes and factual errors about American Indians remains valid precedent.").

[56] *See, e.g.*, United States v. Sandoval, 231 U.S. 28, 47 (1913) (concluding that federal supervision of Pueblo peoples was appropriate given their "isolated and communal life, primitive customs and limited civilization"); *Ex parte* Crow Dog, 109 U.S. 556, 571 (1883) (declining to extend federal criminal jurisdiction over Indian-on-Indian crime because it would "measure[] the red man's revenge by the maxims of the white man's morality").

and paternalist protection of "a weak and defenseless people, who are wards of the nation"[57]—that the Justices deemed normatively good, and so the Court sought to craft law to further that goal. In other words, the nation's history of colonialism, dispossession, and at times genocide against Indigenous communities is not simply context, backdrop, or subtext for the Court's Indian law precedents. They were *text*, an integral part of the opinions' legal reasoning. As Chief Justice Marshall forthrightly acknowledged in *M'Intosh*, the Court itself was an agent of what he called conquest.[58]

Our society has hopefully rejected many of the pernicious stereotypes that undergird many of these decisions, even if elements remain worryingly common. And the federal government has explicitly rejected its past paternalism in favor of Native self-determination. The question for common-law constitutionalism, then, is what weight the Court's prior precedents should now enjoy given that they depend on repudiated foundations.[59] Perhaps these cases' reliance on repugnant principles means they should be discarded as the product of "discriminatory taint"[60] or be deemed "overruled in the court of history."[61] Indeed, this past Term, Justice Gorsuch called for overruling the *Insular Cases* governing the U.S. territories—which themselves heavily cited some of the Court's most dubious Indian law precedents—because they "rest . . . on racial stereotypes."[62]

Yet *rejecting* past Indian law precedents outright is not necessarily a solution for furthering decolonization today. Many of these precedents explicitly subordinated Native peoples, but others relied on now-repudiated racial logics to *affirm* Native sovereignty and autonomy. The clearest example of this is *Worcester* itself, penned (to the befuddlement of commentators past and present) by Marshall less

[57] Choate v. Trapp, 224 U.S. 665, 675 (1912).

[58] 21 U.S. at 588 ("Conquest gives a title which the Courts of the conqueror cannot deny. . . .").

[59] *See, e.g.*, Crepelle, *supra* note 55, at 532 ("Attorneys in the present day United States routinely use cases based on white supremacy to argue against American Indian rights, and judges unblinkingly cite these opinions in federal Indian law cases.").

[60] W. Kerrel Murray, *Discriminatory Taint*, 135 Harv. L. Rev. 1190, 1194 (2022).

[61] Trump v. Hawaii, 138 S. Ct. 2392, 2423 (2018); *see also* Daniel B. Rice, *Repugnant Precedents and the Court of History*, 121 Mich. L. Rev. 577 (2023) (discussing how to identify precedents that should be similarly overruled).

[62] United States v. Vaello Madero, 142 S. Ct. 1539, 1552 (2022) (Gorsuch, J., concurring).

than a decade after his ruling in *M'Intosh*.[63] But there are late nineteenth-century instances, too. In *Crow Dog*, the Court invoked the "strongest prejudices of [Natives'] savage nature" to vindicate Native *independence* from federal law.[64] *Kagama*, decided only a few years later, explicitly and openly predicated expansive federal authority on Native communities' supposed "weakness and helplessness"—but invoked these same principles to reaffirm *Worcester* and Native autonomy from *states*.[65]

Indeed, this past Term presented the Court, in *Denezpi*, with a case that underscored the challenge of making sense of federal Indian law's nakedly colonial past in construing current law.[66] In the late nineteenth century, the federal government established what were known as "courts of Indian offenses."[67] These courts were overtly assimilationist institutions, explicitly intended to extinguish "heathenish" Native customs.[68] In the modern self-determination era, most tribes replaced these courts with tribal courts, but they survived on some reservations in the form of C.F.R. courts.[69] Today, these C.F.R. courts are functionally *tribal* institutions, using tribal personnel to enforce tribal law—but with the benefit of direct federal financial and administrative support.[70]

Whether to classify the C.F.R. courts as federal or tribal institutions posed a difficult question of historical change. Given the C.F.R. courts' unambiguous original purpose to subordinate Native peoples, are the C.F.R. irredeemably colonial institutions, forever tainted by

---

[63] *See* State v. Forman, 16 Tenn. 256, 286 (1835) ("[T]he reasoning in [*Johnson v. M'Intosh*], by the same distinguished jurist, it must be admitted, if not in direct opposition, is greatly in conflict with Worcester's case. . . ."). *Worcester*, too, has been critiqued for its paternalism and denigration of tribal independence. *See* Crepelle, *supra* note 55, at 546.

[64] *Ex parte* Crow Dog, 109 U.S. 556, 571–72 (1883).

[65] United States v. Kagama, 118 U.S. 375, 384 (1886).

[66] Denezpi v. United States, 142 S. Ct. 1838 (2022).

[67] *See* William T. Hagan, Indian Police and Judges; Experiments in Acculturation and Control 107–11 (1966).

[68] *Id.* at 107–10 (citation omitted).

[69] *See CFR Court History*, U.S. Dep't of the Interior: Indian Affs., https://www.bia.gov/CFRCourts/southern-plains-cfr-court/History (last visited Feb. 14, 2022).

[70] Cohen's Handbook of Federal Indian Law § 4.04[3][c][iv][B], at 266 (Nell Jessup Newton ed., 2012) ("CFR courts are now viewed as vehicles for the exercise of tribal jurisdiction.").

their flawed origins, as Denezpi argued?[71] Or can colonized peoples reclaim institutions rooted in colonialism to further their own autonomy, as the tribe argued in its brief?[72] (Justice Gorsuch derided this view as "deeply revisionist.")[73] Ultimately, the Court ducked the question and decided the case on other grounds. But this issue of anticolonial repurposing of colonial institutions and doctrines cropped up elsewhere this Term. In his call to repudiate the *Insular Cases*, Justice Gorsuch noted—and dismissed—a similar academic and judicial effort to reclaim the concept of "unincorporated territories" as a tool to safeguard distinctive territorial practices from a looming assault of colorblind constitutionalism.[74]

This issue in *Denezpi* underscored the fraught relationship between legal and societal change: what weight should the law of the past enjoy today if it rests on views that no longer hold? And how should we assess the implications of those changes for institutions that began with one purpose but have since developed another? These questions sweep much more broadly than common-law constitutionalism alone, but it is especially pressing for this legal theory because it lacks a clear criterion to define change. Administrative and executive policies can be rewritten; legislatures can enact new laws. But—except in the rare instances where they are expressly overruled—precedents live forever, enduring long after the world that produced them has vanished.

---

[71] Brief for Petitioner at 11–12, 17–22, *Denezpi*, 142 S. Ct. 1838 (No. 20-7622), https://www.supremecourt.gov/DocketPDF/20/20-7622/204356/20211207150515943_20-7622.Brief%20for%20Petitioner.pdf.

[72] Brief Amici Curiae Ute Mountain Ute Tribe et al. in Support of the United States at 16–17, *Denezpi*, 142 S. Ct. 1838 (No. 20-7622), https://www.supremecourt.gov/DocketPDF/20/20-7622/210538/20220118132503638_207622UteMountainUTETribeEasternShawneeTribeOfOklahomaAndOtoeMissouriaTribeOfIndians.pdf. This was also the key theme of an amicus brief that I co-authored in the case. Brief of Amici Curiae Federal Indian Law Scholars and Historians in Support of Respondent at 1–4, *Denezpi*, 142 S. Ct. 1838 (No. 20-7622), https://www.supremecourt.gov/DocketPDF/20/20-7622/210553/20220118134834173_20-7622%20bsac%20Federal%20Indian%20Law%20Scholars%20and%20Historians.pdf.

[73] *Denezpi*, 142 S. Ct. at 1855 (Gorsuch, J., dissenting).

[74] United States v. Vaello Madero, 142 S. Ct. 1539, 1557 n.4 (2022) (Gorsuch, J., concurring). For the academic debate, compare *American Samoa and the Citizenship Clause: A Study in* Insular Cases *Revisionism*, 130 HARV. L. REV. 1680 (2017), with Christina Duffy Ponsa-Kraus, *The* Insular Cases *Run Amok: Against Constitutional Exceptionalism in the Territories*, 131 YALE L.J. 2449 (2022); *see also* James T. Campbell, Aurelius's *Article III Revisionism: Reimagining Judicial Engagement with the* Insular Cases *and "The Law of the Territories,"* 131 YALE L.J. 2542 (2022).

This reality creates a double bind in Indian law. For the past forty years, much of the scholarly field has devoted itself to attacking the credibility of the Court's old precedents as undergirding present harms to Native communities.[75] So long as they remain good law, these cases' most pernicious language and holdings remain available to the judges "like a loaded weapon," in Justice Robert Jackson's memorable words.[76] There is, moreover, a dignitary harm that comes from a jurisprudence that continues to memorialize racism.[77]

And yet, there is no guarantee that what would replace these foundational cases would be better for tribes, a truth that *Castro-Huerta*, in discarding centuries-old precedent, demonstrated especially vividly. There is the danger, too, that calls to dismantle the tainted foundations of federal Indian law will come, disingenuously, from critics of tribal sovereignty. This dynamic undergirded *Denezpi*: at oral argument, Justice Kagan astutely noted that the appellant was effectively accusing the Ute Tribe of a kind of "false consciousness" by purportedly embracing colonialism.[78] It has appeared, too, in the context of the current legal challenge to the Indian Child Welfare Act, with well-intentioned commentators advocating, in the name of protecting tribal sovereignty, that the Court invalidate a federal statute that nearly every federally recognized tribe in the country united to defend.[79]

These problems of change have no easy solution. But it seems very unlikely that any good resolution to them will come from within the Court.

---

[75] *See, e.g.*, Robert A. Williams, *The Algebra of Federal Indian Law: The Hard Trail of Decolonizing and Americanizing the White Man's Indian Jurisprudence*, 1986 WIS. L. REV. 219; ROBERT J. MILLER ET AL., DISCOVERING INDIGENOUS LANDS: THE DOCTRINE OF DISCOVERY IN THE ENGLISH COLONIES (2010).

[76] Korematsu v. United States, 323 U.S. 214, 246 (1944) (Jackson, J., dissenting). This phrase has been explicitly invoked in the context of the Court's recent Indian law jurisprudence: WILLIAMS, *supra* note 27, at 136.

[77] *Cf.* Justin Simard, *Citing Slavery*, 72 STAN. L. REV. 79, 109–13 (2020) (recounting the "dignitary harms" stemming from the continued citation of slave cases).

[78] Transcript of Oral Argument at 25–26, *Denezpi*, 142 S. Ct. 1838 (No. 20-7622).

[79] Eugene Volokh, *What's Missing in the* Brackeen *Argument: An Indian Affairs Clause*, VOLOKH CONSPIRACY (Nov. 23, 2022, 2:07 PM), https://reason.com/volokh/2022/11/23/whats-missing-in-the-brackeen-argument-an-indian-affairs-clause. For tribal support of the statute, see Brief of 497 Indian Tribes and 62 Tribal and Indian Organizations as Amici Curiae in Support of Federal and Tribal Defendants, Haaland v. Brackeen, No. 21-376 (U.S. filed Aug. 19, 2022), 2022 WL 3682220.

B. "GOOD" HISTORY IN INDIAN LAW: *YSLETA DEL SUR*

Identifying the problem of "too much history" is not a call to eliminate history from Indian law jurisprudence; that is impossible. *All* law is historical in a broad sense of the term, in that it requires construing rules created in the past, whether two centuries ago or last week.[80] Even if we talk about history the way the term is usually used—to discuss things that happened long ago, flagging the problem that the past was different from the present—Indian law is full of cases requiring courts to go far back in time. We might think of these sorts of cases as mandating a kind of "ordinary" legal history.[81]

A brief survey of the Court's recent Indian law cases makes concrete what I mean by "ordinary" here. The Court's 2021 decision in *Yellen v. Confederated Tribes of the Chehalis Reservation* involved a provision from a very recent statute, the 2020 CARES Act.[82] But other recent cases litigated laws that were over a century old—a common occurrence in Indian law. *McGirt* required deciding whether Congress had disestablished the Muscogee Creek Reservation in a series of late nineteenth- and early twentieth-century statutes transforming the Indian Territory into the state of Oklahoma.[83] *Cougar Den* evaluated whether a 1855 treaty recognizing the Yakama Nation's right "to travel upon all public highways" preempted a state fuel tax.[84] The so-called Culverts case also involved treaty interpretation—in that case, whether heavily litigated mid-nineteenth-century treaty provisions recognizing Native nations' rights to fish required Washington State to replace culverts that impeded salmon spawning.[85] All of these cases interpreted language from the past; all sought to reconstruct how these texts were understood in historical context, especially the provisions from long ago.

---

[80] Will Baude and Steve Sachs have forcefully articulated this point. William Baude & Stephen E. Sachs, *Originalism and the Law of the Past*, 37 Law & Hist. Rev. 809, 809–11 (2019).

[81] *Cf. id.* at 809–10 (describing digging into past law as "ordinary lawyering").

[82] 141 S. Ct. 2434, 2438 (2021).

[83] McGirt v. Oklahoma, 140 S. Ct. 2452, 2461–68 (2020).

[84] Wash. State Dep't of Licensing v. Cougar Den, Inc., 139 S. Ct. 1000, 1006–07 (2019) (citation omitted).

[85] United States v. Washington, 853 F.3d 946, 953–54 (9th Cir. 2017), *aff'd by an equally divided court*, 138 S. Ct. 1832 (2018).

This past Term saw an Indian law case, *Ysleta del Sur Pueblo v. Texas*, that similarly required this kind of "ordinary" legal history.[86] *Ysleta* offers a striking counterpoint to *Castro-Huerta*. Not only did both cases hinge on whether a state could exercise jurisdiction within Indian country, but the vote lineups were nearly identical: Justice Gorsuch and the three liberal Justices against state authority and Chief Justice Roberts and Justices Thomas, Alito, and Kavanaugh in favor. The only difference was Justice Barrett, who, in *Ysleta*, broke with the other conservative Justices to side with Gorsuch's *majority* opinion invalidating state jurisdiction.

Justice Barrett did not write in either case, so we can only speculate why her (outcome-determinative) vote in *Ysleta* differed from her view in *Castro-Huerta*. Still, the *methodological* contrast between the two cases is striking. *Ysleta* involved a provision in the 1987 act restoring the federal trust relationship with the Ysleta del Sur Pueblo outside El Paso, Texas.[87] The statute barred "[a]ll gaming activities which are prohibited by the laws of the State of Texas" on the tribe's lands.[88] Current Texas law permitted some charitable bingo but forbade the kind of bingo that Ysleta del Sur sought to operate. Were such gaming activities "prohibited" by Texan law under the terms of the statute?

In answering no, Justice Gorsuch engaged in close textual analysis, but conceded that "fair questions remain after a look at the ordinary meaning of the statutory terms."[89] He resolved them by looking to "important contextual clues"[90]—specifically, the fact that, only months before the statute's adoption, the Supreme Court had decided *California v. Cabazon* by crafting a somewhat idiosyncratic dichotomy between *prohibitory* and *regulatory* state laws.[91] "For us, that clinches the case," Gorsuch wrote.[92] In dissent, Chief Justice Roberts read the statute in a different context—alongside Public Law 280, which expanded state jurisdiction in Indian country, as well as in light of a tribal

---

[86] 142 S. Ct. 1929 (2022).

[87] Ysleta del Sur and Alabama and Coushatta Indian Tribes of Texas Restoration Act, Pub. L. No. 100-89, 101 Stat. 666 (1987).

[88] *Id.* § 107(a).

[89] 142 S. Ct. at 1937–40.

[90] *Id.* at 1940–41.

[91] 480 U.S. 202 (1987).

[92] *Ysleta*, 142 S. Ct. at 1940.

resolution during the negotiations of the law that embraced barring *all* gaming, including bingo, on the Pueblo's lands.[93]

In my view, this debate in *Ysleta del Sur* represents a "good" use of legal history in Indian law, as is true for the Court's other Indian law decisions discussed above. I mean "good" here in a very specific, thin sense—as a synonym for appropriate. "Good" does not imply that the Court will necessarily get the history *right*: it is easy to imagine situations where it is appropriate to look to legal history but where the Justices nonetheless mangle the past. Similarly, "good history" does not mean good as judged by the professional standards of the historical profession. Greater attention to nuance, complexity, and contingency—some of the virtues praised in strong historical scholarship[94]—would make the Justices better consumers of legal history, but at the end of their day, their job is to decide cases, not win the Bancroft Prize.

Several features of *Ysleta*—shared by other recent decisions like *McGirt* and *Cougar Den*—made the Court's turn to history to help resolve the legal dispute particularly appropriate. First, and perhaps most important, the turn to legal history was essential to the decision. All the Justices acknowledged that not only was the language of the statute (or treaty) in question dispositive of the case, but that its meaning could only be understood within the relevant historical context. Second, the majority and the dissent broadly agreed on what counted as relevant evidence, even as they fought over its interpretation and the weight that it deserved. Third, the dispute was concrete. In *Ysleta*—as in *Cougar Den*, and even to some extent in *McGirt*—the argument over the meaning of a specific statute or treaty focused attention on a specific place, time, and legal question. Judicial opinions can never fully reflect the complexity of the past, but a careful reader of *Ysleta* could glean the relevant evidence from the decision alone and reach their own conclusion on the merits.

It is an interesting question whether "good history" in this thin methodological sense yields outcomes that are *substantively* good for tribal interests. As discussed above, history as imagined by the Court is often biased *against* tribes: most long-ago enactments were crafted by white policymakers who not only anticipated but sometimes

---

[93] *Id.* at 1950–52 (Roberts, C.J., dissenting).

[94] *Cf.* Sam Wineburg, *Historical Thinking and Other Unnatural Acts*, 80 Phi Delta Kappan 488 (1999) (describing the challenges and virtues of emphasizing the strangeness of the past).

sought the elimination of Native peoples.[95] Yet the Court rarely considers this context in deciding what weight to give the past in Indian law. Given this challenge, it is striking that a number of these recent "good history" cases have, in fact, come out in favor of tribes. This outcome partly reflects nose-counting: for a brief moment, Justice Gorsuch plus the then-four liberal Justices constituted a pro-tribal majority. But I also think that "good history" can sometimes serve to protect tribal interests. In the treaty cases, for instance, well-established precedent forces the Court to take seriously the views and expectations of Native peoples in construing law.[96] Good history also helps to the perhaps limited extent that it actually constrains the Court, whose policy preferences have evinced suspicion and hostility toward tribal institutions for decades. Given this barrier, getting the Justices to think not in abstract terms about which sovereign they favor but in specific terms about actual evidence might, perhaps, redound to tribes' benefit.

C. "BAD" HISTORY IN INDIAN LAW: *CASTRO-HUERTA*

In contrast with "good" history, "bad" history frequently results when the Court seeks to plug holes in the broad structures of federal Indian law left by text's absence. Justice Scalia pithily expressed this sense of the field's unmooring from a solid positivist foundation when he quipped, "You know, when it comes to Indian law, most of the time we're just making it up."[97] But of course the Court's opinions don't say it is making things up; they claim to be discerning law. And, in the absence of text, the Court often turns to the past as a free-standing source of this law. Justice Scalia encapsulated this approach, too: the Court's Indian law precedents, he observed in a memo, were based less on "explicit legislation" than on attempts to "discern what

---

[95] *See supra* text accompanying notes 55–79.

[96] *See, e.g.*, Wash. State Dep't of Licensing v. Cougar Den, Inc., 139 S. Ct. 1000, 1012 (2019) ("It is our responsibility to see that the terms of the treaty are carried out, so far as possible, in accordance with the meaning they were understood to have by the tribal representatives at the council") (quoting Tulee v. Washington, 315 U. S. 681, 684 (1942)).

[97] Matthew L. M. Fletcher, *Montana Native Law Student Recalls* Babbitt v. Youpee *and Meeting Justice Scalia*, TURTLE TALK (Feb. 18, 2016) (citation omitted), https://turtletalk.blog/2016/02/18/montana-native-law-student-recalls-babbitt-v-youpee-and-meeting-justice-scalia.

the current state of affairs ought to be by taking into account all legislation, and the congressional 'expectations' that it reflects."[98]

Scalia's memo reflected a venerable tradition in Indian law. Chief Justice Marshall's foundational opinions initiated a sort of grand style in Indian law, with the Justices making broad, sweeping pronouncements based on amateur historicizing. In *Johnson v. M'Intosh*, Marshall transformed what could have been a straightforward decision grounded in British and federal statutes into an extended (and historically dubious) narrative musing on the origins of U.S. sovereignty.[99] And in *Worcester* itself, Marshall penned what one scholar has termed the Supreme Court's "first preemptive opinion" that sought to settle all outstanding issues and entrench Federalist politics.[100]

The most noteworthy modern instantiation of this approach was the Court's 1978 decision in *Oliphant v. Suquamish*.[101] The legal question in the case was whether the Suquamish Tribe—which, in the Court's mind, stood in for all tribes—enjoyed criminal jurisdiction over non-Natives within its territory. The Tribe's treaty was silent on this question, as were, the Justices conceded, most treaties and legislation.[102] The long-established rule in Indian law was that if the federal government had not removed an attribute of tribal sovereignty, it endured.[103] But this approach was not how the Court resolved the issue. It instead wove together a historical narrative from many disparate pieces of evidence to conclude that the United States had implicitly stripped tribal criminal jurisdiction over non-Indians.

Few opinions in Indian law are as reviled as *Oliphant* is among Indian law scholars, who have comprehensively challenged its historical account.[104] But here, I want to focus on the case's *methodological*

---

[98] Getches, *supra* note 10, at 1575 (citation omitted).

[99] 21 U.S. (7 Wheat.) 543 (1823). On Marshall's dubious history, see Lindsay G. Robertson, *John Marshall as Colonial Historian: Reconsidering the Origins of the Discovery Doctrine*, 13 J.L. & Pol. 759, 764 (1997) ("Marshall was not up to the task of shedding much light on the colonial era.").

[100] Gerard N. Magliocca, *The Cherokee Removal and the Fourteenth Amendment*, 53 Duke L.J. 875, 900–03 (2003).

[101] Oliphant v. Suquamish Indian Tribe, 435 U.S. 191 (1978).

[102] *Id.* at 201–08.

[103] *See, e.g.*, Powers of Indian Tribes, 55 Interior Dec. 14, 20 (1934).

[104] *See* Russel Lawrence Barsh & James Youngblood Henderson, *The Betrayal:* Oliphant v. Suquamish Indian Tribe *and the Hunting of the Snark*, 63 Minn. L. Rev. 609 (1978). Critiques of *Oliphant* are too numerous to list. For a recent example, see Fletcher, *supra* note 27, at 986–99. For just one example of the Court's sometimes egregious history in *Oliphant*: the Court

shortcomings. To my mind, the decision epitomized three key features of what I will call "bad history" in Indian law.

First, the Court lacked a clear legal explanation of *why* the history was dispositive. The Court did not say, because it couldn't, that the history it recounted explained or gave meaning to a particular treaty or a statute. Rather, seemingly because the Court did not like the answers that those sources provided, it elevated history itself into a separate and independent source of law.

Second, the Court pitched the historical question at a very broad level of abstraction and generalization: Did the federal government permit tribes to exercise criminal jurisdiction over non-Indians? This framing—*all* tribes at *all* places and times—made the question effectively unanswerable: As anyone who has ever worked in Native history could have readily predicted, the history was all over the place, with contradictory statements and conflicting views from federal policymakers. Such confusion was a naked invitation to cherry-pick—which the Court proceeded to do.

This led to the third, related flaw of *Oliphant*'s "bad history." The Court gave no guidance on how to weigh these many pieces of inconsistent historical evidence against each other. Lower court decisions, committee reports from unenacted bills, attorney general opinions, random unrelated treaties—all got thrown into the pot, at least if they agreed with the Court's conclusion. Contrary evidence got relegated to the footnotes or simply ignored altogether.[105]

All these flaws reappeared in *Castro-Huerta*. Indeed, *Castro-Huerta* is perhaps more surprising than *Oliphant*, because, under my crude typology, the Court transformed what could have been a "good history" case into a "bad history" case.

What I mean by this is that *Castro-Huerta*, like *Ysleta*, required deciding whether a federal statute preempted state law in Indian country. The relevant statute in *Castro-Huerta* was much older than the law at issue in *Ysleta*—from 1834, not 1987—but the legal inquiry was the same: did this language, read in historical context, foreclose the exercise of state jurisdiction? Answering this question seemingly called

cited Justice Johnson's concurrence in *Fletcher v. Peck* as a restriction on *tribal* jurisdiction. 435 U.S. at 209. But in context, it is clear that in the quoted language Johnson was referring to the scope of *federal* authority over tribes. Fletcher v. Peck, 10 U.S. (6 Cranch) 87, 147 (1810) (Johnson, J., concurring).

[105] *See, e.g.*, *Oliphant*, 435 U.S. at 198–99 n.8 (distinguishing Founding-era treaty provisions that authorized tribal punishment of non-Indian invaders in a footnote).

for the same sort of cabined historical inquiry as the Court engaged in in *Ysleta.*

But that was not how the Court framed the legal question. Only a single page of the majority's opinion analyzed the language of the General Crimes Act, and even there, the majority said that what was doing the actual legal work was the "background principle of state jurisdiction over crimes committed within the state, including in Indian country."[106] This, then, was the actual dispositive legal question in *Castro-Huerta*—what *is* the foundational principle governing state criminal jurisdiction in Indian country?

Answering that question goes to the core of federal Indian law, which arbitrates among tribal, federal, and state jurisdiction in Indian country. Given the deep issues involved, one might imagine that the Court would look to text, history, and structure, as well as precedent, to resolve the issue. And the *Castro-Huerta* majority gestured in that direction, announcing an extremely broad extratextual constitutional principle: "[A]s a matter of state sovereignty, a State has jurisdiction over all of its territory, including Indian country. See U.S. Const., Amdt. 10."[107]

But this conclusory codification of what Justice Thomas in a prior opinion called the principle of "States' territorial integrity" is not serious constitutional analysis, as the bare, question-begging citation of the Tenth Amendment suggests.[108] Had the Court actually dug into the conventional sources of constitutional law, it would have found this principle difficult to justify, as I have traced more fully elsewhere.[109] To be sure, states throughout the Founding and, even more, the antebellum eras routinely insisted on their absolute territorial sovereignty, usually as part of an effort to seize lands that belonged to someone else—either Native peoples or the federal government.[110] But the Constitutional Convention explicitly *rejected* a proposal that the Guarantee Clause encompass state territory over fears about

---

[106] Oklahoma v. Castro-Huerta, 142 S. Ct. 2486, 2494–95 (2022).

[107] *Id.* at 2494.

[108] Upstate Citizens for Equality, Inc. v. United States, 140 S. Ct. 2587, 2588 (2017) (Thomas, J., dissenting from the denials of certiorari).

[109] *See* Gregory Ablavsky, *What Justice Thomas Gets Wrong About the Constitutional History of Indian Affairs*, Stan. Legal Aggregate (Nov. 28, 2017), https://law.stanford.edu/2017/11/28/what-justice-thomas-gets-wrong-about-the-constitutional-history-of-indian-affairs.

[110] *See* Gregory Ablavsky, *Empire States: The Coming of Dual Federalism*, 128 Yale L.J. 1792, 1855–62 (2019).

states' sweeping territorial claims.[111] And, after the drafting of the Constitution, Founding-era legal thinkers similarly rebuffed states' aggressive claims for complete territorial jurisdiction over Indian country and over federal lands.[112]

What was remarkable about *Castro-Huerta* was that the majority fully conceded this point. That is, the opinion acknowledged that the understanding at the Founding and for a century afterward was not the Court's newly announced rule but, in fact, its opposite: a "history of territorial separation," under which state jurisdiction did not extend to Indian country at all.[113] The Court likely made this damning concession because it had to: besides all the evidence from the era's statutes and executive practice, territorial separation was, after all, the explicit holding of *Worcester*, which even a very determined Court could not ignore.

The Court thus confronted a substantial problem of legal change. The text of the Constitution concerning Indian affairs had not meaningfully changed since "territorial separation" was the law. Nor, remarkably, had the relevant statutory text. The correct legal conclusion, then, would seem to be that, if not lawfully altered, the earlier law remained binding. Or so the Court's other decisions this Term suggest. The *Dobbs* decision did not ask whether the right to abortion might have become deeply rooted since the Fourteenth Amendment's ratification.[114] The *Bruen* opinion did not assess whether state gun control measures that had endured for nearly a century might have become valid under the Second Amendment.[115] The Court assumed that the meaning of *those* texts was fixed at the moment of adoption.

It was in grappling with this problem of historical change, I argue, that the Court replicated the three flaws of "bad history" in Indian law. First, seemingly because it did not like the answers given by other sources of law, the Court turned to history. The *Castro-Huerta* Court did not seriously examine whether territorial separation was the *right*

---

[111] *See id.* at 1838 (recording opposition to proposed federal guarantee of state "territory" as a "perpetual source of discord" and noting that the Convention subsequently removed the term (citation omitted)).

[112] *See* Ablavsky, *supra* note 46, at 1045–50.

[113] Oklahoma v. Castro-Huerta, 142 S. Ct. 2486, 2497 (2022).

[114] *See* Dobbs v. Jackson Women's Health Org., 142 S. Ct. 2228 (2022).

[115] *See* New York State Rifle & Pistol Ass'n v. Bruen, 142 S. Ct. 2111 (2022).

legal principle, mandated by conventional sources of law. It did not ask, in other words, whether *Worcester* was correctly decided. Instead, the Court asked whether the law on state jurisdiction in Indian country had subsequently *changed*.

This framing of the question epitomized the second flaw of "bad history." With the Court's question pitched at that level of generality—did the law change?—the answer was inevitably yes, almost as a matter of logic. In pretty much any area of law, it is all but a historical certainty that, in the course of two centuries, the law will have grown, developed, shifted—changed.

The more pertinent question was not whether but *how* the law had changed—and whether that change had legitimately altered the law on state criminal jurisdiction in Indian country. This is where the Court repeated *Oliphant*'s third flaw. Unlike the *Oliphant* Court, the *Castro-Huerta* majority at least confined the world of evidence largely to the Court's own decisions. But as discussed above, two centuries of dozens of conflicting Indian law precedents is hardly a cabined source base from which to construct a historical account, especially with the Court's entire corpus at hand for full-text search. And, as in *Oliphant*, the Court gave no clear criterion or rule to decide how to weigh this evidence when it inevitably conflicted. Canonical cases like *Worcester* were dismissed in favor of obscure nineteenth-century deep-cuts like *Dibble*[116]—a case that, in the ensuing 150 years, the Court had cited in only two Indian law cases, mostly to dismiss the *Dibble* language as "dictum."[117]

The predictable result of this lack of discipline was a triumph of subjectivism. On six prior occasions spanning from 1913 to as recently as 2016, the Court had already seemed to decide the issue in *Castro-Huerta*, suggesting that the federal government held exclusive jurisdiction over non-Indian-on-Indian crime in Indian country.[118] But those decisions did not determine the question, the *Castro-Huerta* majority ruled, because they "did not even purport to interpret the text of the General Crimes Act" and so were "tangential dicta."[119] So,

---

[116] New York *ex rel.* Cutler v. Dibble, 62 U.S. (21 How.) 366 (1859).

[117] New York *ex rel.* Ray v. Martin, 326 U.S. 496, 500 (1946); Oneida Indian Nation of N.Y. v. Cnty. of Oneida, 414 U.S. 661, 672 n.7 (1974). The *Oneida* Court observed that, in subsequent cases, it had significantly limited *Dibble*. *Id.*

[118] *See supra* note 23.

[119] Oklahoma v. Castro-Huerta, 142 S. Ct. 2486, 2498–99 (2022).

having distinguished away much contrary precedent, the Court blithely asserted that state jurisdiction here was a question of first impression.

The Court adopted a different approach, however, when it traced the trajectory of *Worcester* and the principle of territorial separation. Now brief remarks in prior decisions discussing *Worcester* were not "tangential dicta," even though these observations were often far removed from the cases' holdings, and none of them seriously engaged with the reasoning of *Worcester*. Instead, the Court interpreted these snippets as proof that the Court had "long ago" abandoned the canonical case.[120]

The following Part takes up this historical narrative much more fully and tries to suggest what a more faithful, thoughtful construction of this historical change might look like. But here, some concluding methodological observations are in order. It is easy, and well-deserved, to skewer the *Castro-Huerta* majority for motivated reasoning. But we also should not miss the broader problems of method that facilitated their unvarnished subjectivism. In some ways, the most egregious part of *Castro-Huerta*—the Court's rather blatant manipulation of precedent and historical evidence—is also the least surprising. I am an expert in this history—or so I'm told—and yet I still struggle to grasp the complex and often strange jurisprudential twists of Indian law's past. Even if we presume good faith and the absence of an agenda (a questionable assumption, as the history of Indian law bears out), it is remarkably hard for judges applying familiar common-law reasoning to make sense of the morass of two centuries of conflicting precedents.

Given this reality, arguably the core problem in *Castro-Huerta* was the Court's decision to turn to history as an independent source of law in the first place. I suspect that the Justices largely assume that history is relevant in most Indian law cases, and don't distinguish between "good" and "bad" history, as I've attempted to do loosely here. But I believe that this distinction can be useful. In the one instance, history can help clarify other sources of law through targeted, cabined inquiries. In the other instance, the Court turns to history to save it from ambiguity (or, more worryingly, to find an answer it prefers), but then asks questions of the past pitched so broadly and

---

[120] *Id.* at 2502.

loosely as to all but guarantee evidentiary mush. The *Castro-Huerta* majority was certainly unprincipled, but even principled jurists can get lost in a swamp.

II. Addressing Change: Translating Past to Present

Castro-Huerta is a legal decision, but the central disagreement between majority and minority reflects a classic historian's conundrum—the balance between continuity and change in relating past and present.[121] Justice Kavanaugh's opinion spun a narrative of rupture: territorial separation was once the law of the land, but courts subsequently abandoned it. Dissenting, Justice Gorsuch challenged this account. "[E]xactly when and how did this change happen?" he queried.[122]

This Part answers this question as a matter of legal history rather than in the heat of litigation. In the process, it tries to address the deeper problem of how to grasp past precedent in light of change. In a historicist sense, as Jack Balkin suggests, we can understand how any prior case was plausibly decided under the law of the time; indeed, those decisions *constituted* the law of that time.[123] The challenge comes as time passes and jurisprudence changes. No decision is freestanding: all court rulings rest, implicitly and explicitly, on all the *other* laws and legal understandings of the era. Thus, even if their black-letter holdings endure, past decisions inevitably become less comprehensible as the jurisprudential world that created them fades.

Worcester and the principle of territorial separation are a striking example of such a transformation. I examined all ninety-six Supreme Court citations of *Worcester*, as well as every nineteenth-century citation of the case. This evidence reveals how the antebellum jurisprudential premises supporting *Worcester* subsequently shifted. In the late nineteenth century, *Worcester* collided with a resurgence of positivism and states' rights arguments that cabined its reach—though these cases rest on principles that have eroded even more sharply than the concept of territorial separation. In the twentieth century, courts

---

[121] *See generally* Alexander Gerschenkron, *On the Concept of Continuity in History*, 106 Proceedings Am. Philosopical Soc'y 195 (1962) (exploring the various meanings of continuity and discontinuity and their application to history).

[122] *Castro-Huerta*, 142 S. Ct. at 2520 (Gorsuch, J., dissenting).

[123] Jack M. Balkin, Constitutional Redemption: Political Faith in an Unjust World 177–86 (2011).

rejected the zero-sum jurisdictional logic of the world of *Worcester* in favor of principles of concurrence and preemption that gave greater space for state authority.

But this problem of jurisprudential change is not unique to *Worcester*. It is, in fact, a routine question for the Court. As other scholars have pointed out, the Court regularly tries to update and reinterpret past precedents for current circumstances.[124] Indeed, that is what the Court previously said it was doing with *Worcester*: revisiting the ruling in light of "changed circumstances," by which it really meant legal change. The Court did *not* say it was overruling or abandoning the decision; it said it had "departed from" its reasoning, and that *Worcester* had "undergone considerable evolution" or "yielded to closer analysis."[125]

This Part seeks to chart a more faithful approach to addressing the problem of change by translating past precedents from the foreign world that created them into the present. It attempts to demonstrate this methodology by delving into the history of *Worcester*'s reception, reconstructing the intricacies of past jurisprudence to better understand what changed and what endured from the past.

Revisiting *Castro-Huerta* and its historical narrative is perhaps quixotic. The decision's questionable claim of *Worcester*'s abandonment will now be written into the U.S. Reports: "an embarrassing new entry into the anticanon of Indian law," according to Justice Gorsuch;[126] a new bit of rancid sludge, in the conceit of *Round House*. Yet *Castro-Huerta* is not the last time that the Court will confront the problem of jurisprudential change in Indian law or elsewhere. Allowing the Court to cast earlier attempts to resolve the gap between past and present as *abandonment* grants the Justices a very powerful tool to disregard precedents they dislike without having to grapple with their merits. Taking precedent seriously requires a more thoughtful approach to assessing historical change.

---

[124] Richard Fallon labels this approach—a bit baroquely—as "Historical Trend Lines Bearing on Intra-Temporal Coherence." Richard H. Fallon, *The Many and Varied Roles of History in Constitutional Adjudication*, 90 Notre Dame L. Rev. 1753, 1795 (2015). Fallon describes the challenge of "disparities between the jurisprudential or methodological assumptions exhibited in past judicial decisions and those that have more recently gained approval," which he illustrates with the example of recent decisions on prudential standing doctrine. *Id.* at 1795–96.

[125] *See infra* notes 241, 245, 251.

[126] *Castro-Huerta*, 142 S. Ct. at 2521 (Gorsuch, J., dissenting).

A. *WORCESTER* IN THE NINETEENTH CENTURY

The *Castro-Huerta* majority somewhat implausibly argued that—even though Indian Removal spawned an intense nationwide political battle and a constitutional crisis over state jurisdiction in Indian country—the question of state jurisdiction "did not previously matter all that much."[127] In fact, courts cited *Worcester* 122 times before the end of the nineteenth century.[128] Though a majority of these citations focused on treaty interpretation and status of Indian tribes more generally, many of them also addressed the scope of state jurisdiction in Indian country.

Grasping *Worcester*'s evolution over the nineteenth century requires understanding two key aspects of the era. First, people in the antebellum United States struggled intensely with the concept of concurrent jurisdiction, as Alison LaCroix's important forthcoming book traces.[129] It was hotly contested whether the federal government had exclusive authority over areas within its enumerated powers, or whether states could share regulatory jurisdiction.[130] This problem proved especially sharp for territorial jurisdiction, which was imagined not as divisible and overlapping but as complete, exclusive, and zero-sum.[131] Within each territory, this logic ran, only a single authority could exist.

The fight over Removal epitomized this either-or framing of state and federal jurisdiction. (As so often in this history, *tribal* jurisdiction got lost in the dominance of this state-federal binary). Georgia and the other states did not argue that they enjoyed *concurrent* jurisdiction with the federal government over Indian country. They asserted full,

---

[127] *Id.* at 2499 (majority opinion). On Removal as an enormous political controversy, see Claudio Saunt, Unworthy Republic: The Dispossession of Native Americans and the Road to Indian Territory, at xvii (2020) ("[S]tate-sponsored expulsion of the 1830s was a turning point for indigenous peoples and for the United States.").

[128] For these results, I searched "Cases" within the "Citing References" for *Worcester* on Westlaw, filtering for the years before January 1, 1900.

[129] *See* Alison L. LaCroix, The Interbellum Constitution: Union, Commerce, and Slavery in the Age of Federalisms (forthcoming 2024) (manuscript at 1–16) (on file with author).

[130] *See id.*

[131] *Cf.* Stephen A. Gardbaum, *Nature of Preemption*, 79 Cornell L. Rev. 767, 785–86 (1994) ("Unlike modem preemption doctrine, which is focused exclusively on the (express or implied) intent of Congress, the earlier [nineteenth-century] doctrine operated automatically whenever Congress entered a field of regulation; thus, any federal regulation of a given area automatically preempted all state regulation in the same area.").

complete, and exclusive jurisdiction over *all* people there, including over Indians and Indian-on-Indian crimes.[132] In their minds, this position necessarily required *rejecting* the existence of any federal jurisdiction within Indian country, on the implicit assumption that federal presence alone divest states of their authority. Chief Justice Marshall's wholesale rejection of Georgia's claim in *Worcester* relied on the same zero-sum logic. He, too, assumed that the existence of federal jurisdiction over Indian country necessarily implied that this jurisdiction was exclusive as against the states for Natives and non-Natives alike.[133]

The second key idea was the trope of the "vanishing Indian"—the idea that Native peoples would eventually either disappear or fully assimilate into Anglo-American society.[134] This view had important legal consequences, since it meant that Native lands under federal jurisdiction would pass to state control—at some point. "[S]elf-government by the Indians, within a state is undoubtedly contemplated to be temporary," Justice McLean wrote in his important but neglected *Worcester* concurrence.[135] But McLean punted on the key questions of when, how, and most significantly *who* decided that this "temporary" period ended. "The point at which this exercise of power by a state [in Indian country] would be proper need not now be considered: if indeed it be a judicial question," he observed.[136]

Taken together, these two ideas established two subsequent jurisprudential tracks governing *Worcester* and state jurisdiction in Indian country—collision or avoidance. Georgia and the other southern states opted for the first. They simply rejected *Worcester*'s authority outright, continuing to assert jurisdiction over Native land notwithstanding Marshall's holding.[137] A despairing Cherokee

---

[132] *See, e.g.*, State v. Forman, 16 Tenn. 256, 312 (1835) ("[I]f our [state] legislative and judicial powers extend to the country; they extend to all its inhabitants; to govern, we must have power over all. This we have, or we have it not.").

[133] Worcester v. Georgia, 31 U.S. (6 Pet.) 515, 560–62.

[134] On this persistent trope in antebellum thought, see JEAN M. O'BRIEN, FIRSTING AND LASTING: WRITING INDIANS PUT OF EXISTENCE IN NEW ENGLAND (2010); and Fort, *supra* note 27, at 308–19.

[135] *Worcester*, 31 U.S. (6 Pet.) at 593 (McLean, J., concurring).

[136] *Id.*

[137] *See, e.g.*, Forman, 16 Tenn. at 304 (noting that the "principle" adopted in *Worcester* "is impossible for the States to abide by" and so "should be most solemnly re-considered by the distinguished tribunal asserting it, before its enforcement is attempted").

minority reluctantly accepted Removal, provoking the Nation's mass deportation.[138]

The more common approach, though, was the course suggested in Justice McLean's concurrence: to acknowledge the principle of territorial separation while arguing that the land in question was, for sundry reasons, no longer under federal jurisdiction. After *Worcester*, a half-dozen or so state, federal, and territorial courts—sympathetic to avaricious white settlers—held that *Worcester* did not apply because the Indians in question were not under federal jurisdiction or because the land in question was not legally "Indian country" (the question settled in *McGirt*).[139]

In reaching these results, courts employed methods of statutory interpretation that, to modern readers, seem wildly purposive and results-oriented. In 1853, for instance, Oregon's Territorial Supreme Court ruled that, even though nearly the entire Oregon Territory remained unceded Native land, the Territory nonetheless contained no "Indian country" as defined in the Trade and Intercourse Act.[140] "Whatever militates against the true interests of a white population," the court's Chief Justice reasoned in interpreting the statute, "is inapplicable."[141] Other courts reached similar results through less explicit reasoning. Many found ambiguities in federal definitions of Indian country that they construed in white settlers' favor; others argued that changed circumstances made applying the statute's literal meaning "absurd."[142]

---

[138] Saunt, *supra* note 127, at 237–38, 275–81; Theda Perdue & Michael D. Green, The Cherokee Nation and the Trail of Tears 91–140 (2007).

[139] United States v. Seveloff, 27 F. Cas. 1021, 1022–24 (D. Or. 1872) (No. 16,252); United States v. Leathers, 26 F. Cas. 897, 897–98 (D. Nev. 1879) (No. 15,581); United States v. Kie, 26 F. Cas. 776, 777–81 (D. Alaska 1885) (No. 15,528A); State v. Doxtater, 2 N.W. 439, 441–50 (Wis. 1879); State v. McKenney, 18 Nev. 182, 188, 202–06 (1883); *In re* Narragansett Indians, 20 R.I. 715, 771–80 (1898); cf. Pickett v. United States, 1 Idaho 523, 530 (Idaho Terr. Sup. Ct. 1874) (finding that federal jurisdiction over Indian country "could not be applicable to the whites, in their intercourse with one another"); Indian Country, 14 Op. Att'y Gen. 290, 292–94 (1873) (describing uncertainty about the meaning of Indian country under federal law).

[140] United States v. Tom, 1 Or. 26, 27 (Or. Terr. Sup. Ct. 1853).

[141] *Id.* at 27. Attorney General Caleb Cushing sharply challenged this reasoning at the time. "'Applicability' in a law has nothing to do with the question whether the persons to whom it is to be applied think it for their interest," he insisted. Indians in Oregon, 7 Op. U.S. Att'y Gen. 293, 294 (1855). "That question Congress determines." *Id.*

[142] *See, e.g.*, *Leathers*, 26 F. Cas. at 898.

The U.S. Supreme Court played an ambiguous role in these efforts. It, too, sometimes manipulated the category of Indian Country on strikingly subjective grounds. In 1876, for instance, it ruled that Pueblo peoples in the New Mexico Territory were too civilized to "be classed with the Indian tribes for whom the intercourse laws were made."[143]

But the Court also rejected state efforts to evade *Worcester*. In two canonical post-Civil War cases, it reaffirmed the principle of territorial separation by rejecting efforts by Kansas and New York to tax Indians within Indian country.[144] The Kansas case was especially significant. Like many other state courts, the Kansas Supreme Court had distinguished away *Worcester* by reasoning that the Shawnees in question were differently situated from the Cherokees.[145] It had then permitted taxes under the same logic as the *Castro-Huerta* majority applied: in the absence of any federal statute or treaty *prohibiting* the exercise of state jurisdiction, state authority was permitted.[146]

The U.S. Supreme Court flatly rejected both sets of arguments. observing, "There can be no question of State sovereignty in the case."[147] Instead, the Court applied *Worcester*'s logic. As long as the "political department" recognized the Shawnees' existence, the Court reasoned, they were "separated from the jurisdiction of Kansas, and to be governed exclusively by the government of the Union."[148] The Court then reiterated the era's zero-sum jurisdictional approach: "If under the control of Congress, from necessity there can be no divided authority."[149]

Then, in 1877, the Court made evading *Worcester* much more difficult. In *Bates v. Clark*,[150] the Court for the first time defined "Indian country" in the Trade and Intercourse Act, holding that it

---

143 United States v. Joseph, 94 U.S. 614, 617 (1876).

144 *In re* New York Indians, 72 U.S. (5 Wall.) 761 (1866); *In re* Kansas Indians, 72 U.S. (5 Wall.) 737 (1866).

145 Blue-Jacket v. Comm'rs of Johnson Cnty., 3 Kan. 299, 357–59 (1865) (distinguishing *Worcester* on the ground that, unlike the Cherokees, "[t]he Shawnees do not hold their lands in common, nor are they contiguously located"—in other words, that their territory was not legally Indian country).

146 *Id.* at 363–64.

147 *Kansas Indians*, 72 U.S. (5 Wall.) at 756.

148 *Id.* at 755.

149 *Id.*

150 95 U.S. 204 (1877).

encompassed land "so long as the Indians retain their original title to the soil, and ceases to be Indian country whenever they lose that title."[151] This definition implicitly overruled the many creative efforts to redefine Indian country that courts had employed.

By the start of the 1880s, then, the logic of *Worcester* endured. Skeptical courts had found ways to dodge its strictures, but, after *Bates*, they were less legally plausible. Then, in the span of four years (1882–86) the Court decided three cases that significantly altered the jurisprudential landscape of state authority in Indian country: *United States v. McBratney*,[152] *Utah & Northern Railway v. Fisher*,[153] and *United States v. Kagama*.[154]

1. *United States v. McBratney* (1882): Transferring Federal Criminal Jurisdiction in Indian Country to New States. *Worcester* had concerned *state* jurisdiction. By contrast, the 1882 case of *United States v. McBratney*—which the *Castro-Huerta* Court called the "leading case in the criminal context" departing from *Worcester*[155]—involved a distinct but equally long-standing question in federal Indian law: what happens to *federal* criminal jurisdiction in Indian country when a territory becomes a state? This question was much-debated but little resolved in the century preceding the decision.

Since 1796—when Tennessee became the first territory admitted to statehood—new states had claimed that federal power over Indian affairs was incompatible with their right to "equal footing" with existing states.[156] This argument only swelled as southern states waged their Removal-era assault on federal jurisdiction.[157] Yet these vociferous claims consistently failed to sway federal courts for the better part of a century.

Other federalism arguments, however, did meet limited success. During Removal, two widely cited circuit decisions by Justice McLean had suggested that federal criminal jurisdiction within

---

[151] *Id.* at 209.

[152] 104 U.S. 621 (1882).

[153] 116 U.S. 28 (1885).

[154] 118 U.S. 375 (1886).

[155] Oklahoma v. Castro-Huerta, 142 S. Ct. 2486, 2494 (2022).

[156] Gregory Ablavsky, Federal Ground: Governing Property and Violence in the First U.S. Territories 203–05 (2021).

[157] *E.g.*, Caldwell v. State, 1 Stew. & P. 327, 351 (Ala. 1832) (Saffold, J.); State v. Forman, 16 Tenn. 256, 303–04 (1835).

Indian country within *states* exceeded Congress's Indian affairs or commerce power, even though this jurisdiction had been enacted by the First Congress.[158] But McLean's idiosyncratic opinions—difficult to reconcile with McLean's own concurrence in *Worcester*[159]—made seemingly little difference to subsequent law or practice.

In 1846, the Supreme Court heard its first significant case, *United States v. Rogers*,[160] construing federal criminal jurisdiction under the Trade and Intercourse Act, the statute at issue in *Castro-Huerta*. Justice Taney's majority opinion focused mostly on whether the defendant and victim—both "white men" who had become tribal citizens—were legally "Indians" under the Trade and Intercourse Act.[161] (Taney said no).[162] But the decision also addressed the source of federal criminal jurisdiction in Indian country. Confusingly, it posited *three* distinct foundations: federal power over the territories (the crime happened in the Indian Territory), treaties entered with Native nations, and federal authority over Indian affairs more generally.[163] But these sources of authority were not interchangeable, and suggested different scopes. *Rogers*'s ambiguity begged the obvious question: could the federal government exercise criminal jurisdiction *within* a state, not a territory, in the absence of a specific treaty provision?

---

[158] In *United States v. Cisna*, Justice McLean echoed his *Worcester* concurrence by determining that the Trade and Intercourse Act's criminal jurisdiction within the Wyandot reservation had "become inoperative by the progress of time and change of circumstances." 25 F. Cas. 422, 424–25 (C.C.D. Ohio 1835) (No. 14,795). More significantly, in *United States v. Bailey*, Justice McLean invalidated federal criminal jurisdiction within a state on the grounds that it exceeded the federal government's power under the Indian Commerce Clause. 24 F. Cas. 937, 940 (C.C.D. Tenn. 1834) (No. 14,495). McLean's decision—parts of which effectively defied *Worcester*—thus invalidated as unconstitutional a provision tracing to the First Congress. The law, however, remained in effect.

[159] *See* Worcester v. Georgia, 31 U.S. (6 Pet.) 515, 591 (1832) (McLean, J., concurring) ("Does not the constitution give to the United States as exclusive jurisdiction in regulating intercourse with the Indians, as has been given to them over any other subjects?").

[160] 45 U.S. (4 How.) 567 (1846).

[161] *Id.* at 572–73.

[162] *Id.*

[163] *See id.* at 572 ("[W]e think it too firmly and clearly established to admit of dispute, that the Indian tribes residing within the territorial limits of the United States are subject to their authority, and where the country occupied by them is not within the limits of one of the states, Congress may by law punish any offense committed there. . . ."); *see also id.* at 573 (discussing federal authority under the Treaty of New Echota).

This question dominated *United States v. Ward*, an 1863 case decided by Justice Miller while riding circuit.[164] Like *Rogers*, the case involved federal jurisdiction over a murder of a non-Indian by a non-Indian within Indian country—in this instance, on the reservation of the Kaw Nation.[165] But unlike *Rogers*, the reservation at issue lay within the newly admitted state of Kansas.

In admitting Kansas to the Union, Congress had made an important legal innovation. Since the early nineteenth century, the federal government had attached conditions to statehood by requiring, for instance, that new states forswear any claim to federal lands.[166] With Kansas's statehood, the federal government added a new requirement: for the first time, the state had to acknowledge that statehood would preserve all existing Indian rights and not affect federal authority over Indian affairs.[167] The enabling act further stipulated that the new state would not "include any territory which, by treaty with such Indian tribe, is not, without the consent of said tribe, to be included within the territorial limits or jurisdiction of any State or Territory; but all such territory shall be excepted out of the boundaries, and constitute no part of the State of Kansas."[168] The purpose of this condition was clear: to preserve existing federal authority and Native treaty rights against state claims. It was, in effect, an attempt to enforce *Worcester* through congressional power over state admission.

But the disclaimer did not succeed in its aim in *Ward*, which invalidated the federal prosecution. Justice Miller's reasoning was striking. He rejected the defendant's argument, ubiquitous at the time, that the "changed condition of affairs" of white settlement had withdrawn federal authority.[169] He instead cited *Worcester*, but in a way that captured a core ambiguity in the decision. In affirming territorial separation, *Worcester* had pointed to *both* structural principles rooted in the Constitution *and* to federal statutes and treaties concerning Indian country.[170] Justice Miller, though, read *Worcester* as solely

[164] 28 F. Cas. 397 (C.C.D. Kansas 1863) (No. 16,639).

[165] *Id.* at 397. The Court described the Kaw Nation as the "Kansas Indians."

[166] *See generally* Eric Biber, *The Price of Admission: Causes, Effects, and Patterns of Conditions Imposed on States Entering the Union*, 46 Am. J. Legal Hist. 119 (2004).

[167] *Id.* at 130–31, 204.

[168] Act of Jan. 29, 1861, ch. 20, § 1, 12 Stat. 126, 127.

[169] *Ward*, 28 F. Cas. at 398.

[170] On this complexity in *Worcester*, see William Walters, *Review Essay: Preemption, Tribal Sovereignty, and* Worcester v. Georgia, 62 Or. L. Rev. 127, 132–41 (1983).

about statutory and treaty law.[171] The question, for him, thus became what the Trade and Intercourse Act required.

Here, Justice Miller made a peculiar statutory reading. The Act did not establish federal jurisdiction on *all* Indian reservations, he concluded. Federal courts "have jurisdiction only when such reservation is 'within the sole and exclusive jurisdiction of the United States,'" he reasoned, thereby converting the statute's language clarifying which laws applied into a *requirement*.[172]

But, Miller continued, the requirement he had just concocted could not be satisfied because of Kansas's newfound statehood. "[P]ower to enforce her criminal laws throughout her boundaries, which is unqualified and exclusive, and is to be possessed, enjoyed, and exercised by the new state alone, and not concurrently with the federal government," Justice Miller reasoned, "is a necessary incident to her equality with the original states."[173] The equal footing doctrine, in words, invalidated federal jurisdiction here.

Ward represented seemingly the first time that a federal court had accepted an equal footing argument in Indian law, after three-quarters of a century of failure. Why the sudden shift? Part of the answer likely lay, paradoxically, with Congress's innovation of the Kansas jurisdictional disclaimer. The Kaw Nation, unlike other Kansas tribes, lacked a treaty explicitly exempting it from state jurisdiction. Justice Miller thus drew a negative inference from Congress's action: because Congress had *excluded* tribes like the Kaw Nation from the jurisdictional disclaimer, he reasoned, it had implicitly endorsed state jurisdiction over them.[174] In other words, by attempting to codify *Worcester*, Congress may have unwittingly undermined the principle it sought to protect.

This question of the persistence of federal criminal jurisdiction after statehood recurred when the Colorado Territory became a state in 1876. In *United States v. Berry*, the circuit court upheld federal jurisdiction over non-Indian-on-non-Indian crime in Indian country

[171] *Ward*, 28 F. Cas. at 398.

[172] *Id.*

[173] *Id.* at 398–99.

[174] *See id.* at 399 (reading the silence in the statutory language as evidence that Congress intended that "all territory which was not covered by such treaties was included within the state, within its jurisdiction and within its territory; and this irrevocably, unqualifiedly, and exclusively").

under the provision of the Trade and Intercourse Act.[175] The court reasoned that Colorado's statehood did *not* eliminate federal jurisdiction. The Ute Tribe had a treaty with the federal government that promised that the federal government would punish offenses against the tribe, (though it said nothing about inclusion within a state). Fulfilling this treaty, the court reasoned, required that the "reservation shall remain under the sole and exclusive jurisdiction of the United States."[176] The court specifically rejected the argument that state jurisdiction over non-Indian crime would not impair the tribe's treaty rights. "If the state has jurisdiction over any part of the territory and people of the reservation," the court reasoned, "it has jurisdiction over the whole of the territory, and over all the people, whether whites, blacks, or Indians."[177] But this, the court noted, was precisely what Georgia had attempted to do, and what Chief Justice Marshall had invalidated in *Worcester*.[178]

The U.S. Supreme Court reached the opposite conclusion, however, three years later in *McBratney*, which presented the same legal question, based on nearly identical facts, as *Berry*.[179] The appeal resulted when Judge Hallett—one of the judges who decided *Berry*—disagreed with Justice Miller, the author of *Ward*, who was riding circuit.[180]

Perhaps unsurprisingly, the Justices sided with their colleague: federal jurisdiction over the crime, they held, no longer existed. Unlike *Berry*, the *McBratney* Court nowhere discussed or cited *Worcester*. Its decision instead closely mirrored the logic of *Ward*, which it cited repeatedly. When Congress wished to preserve federal jurisdiction, the Court reasoned, it did so through the "express words" of the jurisdictional disclaimers.[181] Without such words, the Court reasoned, the equal footing doctrine meant that the Ute "reservation is no longer within the sole and exclusive jurisdiction of the United States."[182] In reaching this conclusion, the *McBratney* opinion nowhere discussed

---

[175] 4 F. 779, 782–83, 792 (D. Colo. 1880).

[176] *Id.* at 787.

[177] *Id.* at 789.

[178] *Id.*

[179] United States v. McBratney, 104 U.S. 621, 624 (1882).

[180] *Id.* at 621.

[181] *Id.* at 623–24.

[182] *Id.* at 624.

the fact that, only seven years earlier (and after *Ward*), the Court, in its first ruling on the topic, had flatly *rejected* equal footing's application to an Indian law case involving the Trade and Intercourse Act.[183]

McBratney's principal holding was that statehood had implicitly repealed the Trade and Intercourse Act and so *federal* jurisdiction no longer existed. Consistent with the era's zero-sum logic of exclusivity, that meant, according to the Court, that jurisdiction fell to the state. In theory, this logic arguably justified state jurisdiction over *all people* within Indian country, as the *Berry* court had suggested. But the Court limited its holding solely to encompass only "offenses committed by white men against white men."[184] The treaty with the Utes establishing federal jurisdiction, the Court reasoned, suggested a different outcome for crimes committed by or against Indians.[185] (Once again, the Court assumed that the mere existence of federal jurisdiction barred state jurisdiction; it did not contemplate concurrence.)

The *Castro-Huerta* Court suggested that the congressional response to *McBratney* was silence.[186] In fact, after *McBratney*, the enabling acts of nearly every subsequent state admission included the "express words" that the decision had demanded to foreclose all state jurisdiction over Indian country.[187] These new disclaimers made no reference to treaties: "Indian lands," they provided, "shall remain under the absolute jurisdiction and control of the Congress of the United States."[188]

---

[183] *See* United States v. Forty-Three Gallons of Whiskey, 93 U.S. 188, 197 (1876) (upholding the criminal restrictions on the sale of liquor in Indian country because the statute "so far from making a distinction between the States, appl[ies] to them all alike").

[184] *McBratney*, 104 U.S. at 624.

[185] *Id.*

[186] Oklahoma v. Castro-Huerta, 142 S. Ct. 2486, 2497 (2022) ("Since 1882, therefore, Congress has been specifically aware that state criminal laws apply to some extent in Indian country. Yet since then, Congress has never enacted new legislation that would render federal jurisdiction exclusive or preempt state jurisdiction over crimes committed by non-Indians in Indian country.").

[187] David E. Wilkins, *Tribal-State Affairs: American States as 'Disclaiming' Sovereigns*, 28 PUBLIUS 55, 68–71 (1998). The acts of admission for Idaho and Wyoming did not contain such limitations, likely because their constitutions already contained such disclaimers. *Id.* at 68.

[188] *E.g.*, Act of Feb. 22, 1889, ch. 180, § 4, 25 Stat. 676, 677 (Enabling Act covering North and South Dakota, Montana, and Washington); Act of July 16, 1894, ch. 138, § 3, 28 Stat. 107, 108 (Utah Enabling Act).

Multiple scholars have argued that these disclaimers were a congressional attempt to overrule *McBratney*.[189]

Yet these provisions also succumbed to the equal footing doctrine. Montana, for instance, was one of the states that had explicitly disclaimed jurisdiction within Indian country upon admission.[190] But in the 1896 decision *Draper v. United States*, the Court held that Montana had to enjoy jurisdiction over non-Indian-on-non-Indian crime, because it would otherwise be on unequal footing with non-disclaiming states like Colorado.[191] In other words, having previously held in *McBratney* that Colorado enjoyed jurisdiction because of the absence of "express words" limiting its authority, the Court now held that Montana enjoyed jurisdiction notwithstanding those "express words"—because of the Court's ruling in *McBratney*.[192] The Court had created a legal Catch-22.

1896 was, in fact, the apex of the equal footing doctrine's brief ascendence in Indian law. That same year, the Court held in *Race Horse* that a treaty preserving the hunting rights of the Bannock Nation was also impliedly repealed by Wyoming's statehood.[193] The Court again relied on the equal footing doctrine, as well as the absence of a jurisdictional disclaimer in Wyoming's admission (even though Wyoming's Constitution included such a disclaimer).[194] But, as we shall see, the Court quickly retreated from the jurisprudence that undergirded these conclusions.

2. *Utah & Northern Railway v. Fisher* (1885): Expanding State Civil Jurisdiction in Indian Country. Compared with the Court's robust caselaw on criminal jurisdiction, cases involving state *civil* jurisdiction in Indian country were rare. Among the few Court decisions on the question were the 1866 decisions, noted above, in which the Court invalidated state taxes within Indian country on the principles of *Worcester*.[195]

---

189 *See* Wilkins, *supra* note 187, at 69 & n.66–67 (citing multiple scholars' arguments that the disclaimers were a response to *McBratney*).

190 *Id.* at 68 tbl.2.

191 164 U.S. 240, 247 (1896).

192 *Id.* at 247.

193 Ward v. Race Horse, 163 U.S. 504, 507, 514 (1896).

194 *Id.* at 514–16.

195 *See supra* text accompanying notes 144–149.

There were, however, a handful of Court cases involving the loosely analogous question of the jurisdiction of U.S. *territories* in Indian country. Like state enabling acts, the organic acts of the territories routinely included language prohibiting the exercise of territorial jurisdiction in Indian country. In 1878, using *Worcester*-like logic, the Court in *Harkness v. Hyde* invalidated an attempt by the Territory of Idaho to exercise jurisdiction within the Fort Hall Reservation.[196] The lands there were "as much beyond the jurisdiction, legislative or judicial, of the government of Idaho, as if it had been set apart within the limits of another country."[197]

Yet the Court soon began to pare back the scope of its holding. As in *Ward*, the Court grappled with an ambiguity about territorial separation: did it reflect principles of structure, or did it rest principally on statutory and treaty law? Two years after *Harkness*, the Court embraced the statutory approach.[198] Citing *Ward*, the Court reversed course and held that state process *could* run onto the Nez Perce Reservation in a suit between two non-Indians.[199] *Harkness* was inapplicable, it concluded, because it had erred—no treaty language exempted the reservation from territorial jurisdiction.[200]

Five years later, the Court decided *Fisher*, in which the Court permitted the Idaho Territory to tax a non-Indian railway corporation's right of way through the Fort Hall Reservation.[201] The Court's six-page decision nowhere addressed *Worcester* or its progeny.[202] It instead rested solely on treaty law: a federal treaty protected the Shoshone-Bannock right to the reservation, but the Territory's jurisdiction "may rightfully extend to all matters not interfering with that protection."[203] The tax at issue, it reasoned, would not impair "any just rights of the Indians," and so was permissible.[204]

Fisher gained little of the attention that *McBratney* received: prior to *Williams* (discussed below), the Court only cited it three times in

---

[196] 98 U.S. 476, 478 (1878).

[197] *Id.*

[198] Langford v. Monteith, 102 U.S. 145, 147 (1880).

[199] *Id.* at 146–47.

[200] *Id.* at 147.

[201] Utah & N. Ry. v. Fisher, 116 U.S. 28, 32 (1885).

[202] *See id.* at 28–33.

[203] *Id.* at 30–31.

[204] *Id.* at 31–32.

minor cases dealing with taxation of railroads in Indian country or federal enclaves.[205] Few at the time seemed to have thought it was a seminal case on state jurisdiction—likely because it did not involve a state. But *Fisher* did echo *McBratney* in seeming to convert the structural principle of territorial separation into a question of treaty interpretation and statutory intent. Only in hindsight, though, did this brief, unheralded case come to be seen as a key turning point in the Court's jurisprudence on state jurisdiction in Indian country.

3. *United States v. Kagama* (1886): Clarifying Federal Authority over Indian Affairs. Finally, in 1886, the Court decided *United States v. Kagama*, which upheld the Major Crimes Act establishing federal criminal jurisdiction over Indian-on-Indian crime.[206] *Kagama* is now rightly seen as a watershed in federal Indian law, though not necessarily for the reasons that it was thought significant at the time. Today, *Kagama* is most famous—or infamous—for enunciating the doctrine of federal plenary power *over* Native nations and their internal affairs.[207]

The Court conceded that this doctrine found little support in the Indian Commerce Clause, but nonetheless reasoned that Native peoples' "weakness and helplessness" meant that the federal government had a "duty of protection, and with it the power."[208]

In the context of the late nineteenth-century United States, however, it was legally unsurprising that the Court rejected Native sovereignty as a meaningful check on federal power. The entire decision, after all, was predicated on the era's either-or jurisdictional logic: "The soil and the people within these limits [of the United States] are under the political control of the government of the United States," it reasoned, "or of the States of the Union."[209] In the Court's telling, the most significant objection to the Major Crimes Act was not Native autonomy but that the law encompassed "reservations lying within a

[205] Maricopa & P.R. Co. v. Arizona, 156 U.S. 347, 351 (1895); Thomas v. Gay, 169 U.S. 264, 273 (1898); United States v. Unzeuta, 281 U.S. 138, 145 (1930).

[206] 118 U.S. 375, 385 (1886).

[207] *See, e.g.*, United States v. Bryant, 579 U.S. 140, 160 (2016) (Thomas, J., concurring) ("*Kagama* endures as the foundation of this doctrine [of federal plenary power].").

[208] *Kagama*, 118 U.S. at 384.

[209] *Id.* at 379.

state."[210] But the Court rejected this argument. Indians already lay outside state jurisdiction, it observed, describing state residents as Natives' "deadliest enemies."[211] In support of this proposition, the Court cited *Worcester* and its progeny—though it rewrote *Worcester* to be about state jurisdiction over *tribes* rather than over Indian country more generally.[212] What was most legally significant about the case at the time, then, was that it endorsed *federal* over state jurisdiction.

What made this holding so important was that it clarified the question, lurking since *Rogers* and reiterated in *Ward* and *McBratney*, about the *source* of federal criminal jurisdiction. The decision made clear that federal power over Indian country stemmed from the federal power over Indian affairs, not its lingering power over the territories, resolving the issue of whether federal criminal jurisdiction could exist within *states*. It similarly found federal authority absent any specific treaty provision. This forceful defense of federal power was significant because it grounded the principle of territorial separation in *structure*, not just in specific treaty provisions or jurisdictional disclaimers.

Kagama did not immediately foreclose the equal footing doctrine—which, as we have seen, briefly flourished even after the ruling[213]—but it dramatically undercut its jurisprudential foundation. *McBratney*'s logic could easily have expanded state jurisdiction to encompass *all* crime in Indian country if a treaty provision did not forbid it, but *Kagama* rendered the case's rationale a dead letter. In 1913, for instance, in *Donnelly v. United States*, the Court explicitly rejected an attempt to extend *McBratney*'s reasoning.[214] Citing *Kagama*, the Court said, "[W]e are satisfied that offenses committed by or against Indians are not within the principle of the *McBratney* or *Draper* cases."[215] *Donnelly* was only one of a number of early twentieth-century cases in which the Court, newly emboldened in its defense of federal authority in Indian affairs, repeatedly and forcefully rejected a slew of equal

---

[210] *Id.* at 383.

[211] *Id.* at 384.

[212] *See id.* (reading Worcester as holding that "the State could not, while [the Indians] remained on those lands, extend its laws, criminal and civil, over the tribes").

[213] *See supra* text accompanying notes 191–194.

[214] Donnelly v. United States, 228 U.S. 243, 271 (1913).

[215] *Id.* at 271–72.

footing arguments.[216] It did so on the strength of *Kagama*, which it cited for the proposition that federal power over Indian affairs persisted *within* states.

4. *Assessing the Nineteenth Century.* What, then, to make of the Court's brief and confusing late nineteenth-century flurry of caselaw authorizing some state jurisdiction in Indian country? On the one hand, the cases demonstrate that there was widespread resistance to *Worcester*'s principle of territorial separation, much of it in state and territorial courts. This skepticism stemmed from a widespread hostility to federal power in Indian affairs because federal jurisdiction impeded the "interests of the white population" championed by state and local governments. But *Worcester*'s reach was also limited by ambiguity about the source and nature of federal authority in Indian affairs, especially the source of federal criminal jurisdiction in Indian country. *McBratney* was a product of this uncertainty. The decision's embrace of the equal footing doctrine and its treatment of Congress's conditions on state admission only made sense if federal criminal jurisdiction in Indian country stemmed from federal power over the territories.

Yet *McBratney* is a relic and a jurisprudential dead-end. The case's black-letter holding remains good law, if only because the Court invalidated Congress's efforts to cabin the decision's scope.[217] But *Kagama*, whatever its faults, unambiguously established that federal criminal jurisdiction within Indian country was separate from federal power over the territories. This conclusion also destroyed whatever vitality the equal footing doctrine had briefly held in Indian law. The doctrine had never made all that much sense applied to Indian affairs.[218] After all, *Worcester* had held that Georgia, one of the original states, lacked jurisdiction in Indian country. It was thus highly implausible to argue that new states were not on equal footing when, under *Worcester*'s logic, *no* state enjoyed this jurisdiction.[219]

---

[216] *See, e.g.*, Winters v. United States, 207 US. 564, 577–78 (1908); Dick v. United States, 208 U.S. 340, 359 (1908); *Ex parte* Webb, 225 U.S. 663, 690–91 (1912); United States v. Sandoval, 231 U.S. 28, 49 (1913); Johnson v. Gearlds, 234 U.S. 422, 439–40 (1914).

[217] *See* Oklahoma v. Castro-Huerta, 142 S. Ct. 2486, 2494 (2022).

[218] *See* Wilkins, *supra* note 187, at 59.

[219] *See* United States v. Forty-Three Gallons of Whiskey, 93 U.S. 188, 197 (1876) ("It is not easy to see how [a federal restriction on liquor sales] infringes upon the position of equality which Minnesota holds with the other States. . . . Based as it is exclusively on the

Nearly a century later, the Court made explicit the equal footing doctrine's underlying weakness. In 1999, in *Mille Lacs*, litigants invoked equal footing after years of doctrinal desuetude to challenge the ongoing validity of an Indian treaty. The Supreme Court rejected the effort: the earlier equal footing cases, it opined, "rested on a false premise."[220] Then, just two Terms ago, in *Herrera*, the Court thoroughly and completely buried the equal-footing argument in Indian law, which—despite other disagreements—all the Justices, both majority and dissenters, acknowledged had been "repudiated" in *Mille Lacs*.[221]

Mille Lacs and *Herrera* both involved treaty interpretation, overruling *Race Horse*. But it is difficult to argue that this repudiation did not encompass *McBratney*'s logic as well. If, in the absence of explicit congressional language, statehood alone does not implicitly extinguish treaty rights, then it is difficult to claim that statehood implicitly repeals a valid and applicable federal statute, as *McBratney* held. In short, *McBratney*'s underlying equal footing argument, too, "must be regarded as retaining no vitality" in light of the Court's recent jurisprudence.[222]

This history reiterates the problem of how to interpret precedents in light of change. The *Castro-Huerta* majority was right: the Court limited *Worcester* and the principle of territorial separation in the late nineteenth century, even if did so silently, ambiguously, and on

---

Federal authority over the subject-matter, there is no disturbance of the principle of State equality.")

[220] Minnesota v. Mille Lacs Band of Chippewa Indians, 526 U.S. 172, 204 (1999). In reaching this conclusion, the Court pointed to the fact that it had upheld limited state authority to conserve natural resources. *Id.* at 204–05. In part, this transformation reflected a significant shift, discussed below, in the understanding of concurrent jurisdiction. *See infra* text accompanying notes 227–228. But it would be an error to conclude that *Mille Lacs* invalidated equal footing only because the Court had subsequently acknowledged (limited) state jurisdiction over treaty rights. At core, equal footing ensures that federal authority within former territories derives from the federal government's general enumerated powers rather than reflecting the persistence of federal territorial authority. *See* Coyle v. Smith, 221 U.S. 559, 574 (1911) (noting that restrictions on state power can be upheld when "the power of Congress extended to the subject"). Thus, the most fundamental misunderstanding in *Race Horse*, the *Mille Lacs* Court itself suggested, was the decision's failure to recognize the federal government's general power to authorize tribal hunting through treaties, including the power to preempt state law. *Id.* at 204 ("Although States have important interests in regulating wildlife and natural resources within their borders, this authority is shared with the Federal Government when the Federal Government exercises one of its enumerated constitutional powers, such as treaty making. U.S. Const., Art. VI, cl. 2.").

[221] Herrera v. Wyoming, 139 S. Ct. 1686, 1695–96 (2019); *id.* at 1708 (Alito, J., dissenting).

[222] *Id.* at 1697 (citation omitted).

grounds different from *Castro-Huerta*. But it also did so based on doctrines—equal footing and debates over the source of federal criminal jurisdiction in Indian country—that *Kagama* and subsequent caselaw themselves rejected. The problem, though the Court failed to grapple with it in *Castro-Huerta*, is that the late nineteenth century was also a long time ago, and jurisprudential change did not stop with *McBratney*.

B. *WORCESTER* IN THE TWENTIETH CENTURY

Much changed in Indian law between the late nineteenth and the twentieth centuries. Politically, it became apparent that Native peoples had no intention of disappearing, and so the tension between Native and state jurisdiction was an enduring, not a temporary, challenge. Over the course of the mid-twentieth century, Congress attempted to resolve this issue by statutorily affirming the jurisdiction of selected states in Indian country.[223] This policy culminated in the 1953 statute known as Public Law 280, which mandated state jurisdiction over Indian country in six states and allowed other states to opt in.[224] But the federal government soon reversed course, affirming Native self-determination and autonomy from state control. Public Law 280 was amended to require tribal consent to further extensions of state jurisdiction—which no tribe has ever given.[225] Many jurisdictions are now in the process of retroceding the authority granted them under the statute.[226]

The law of concurrent jurisdiction changed, too. In the twentieth century, the zero-sum reasoning that had carved territories into exclusively federal or state jurisdictions gave way to an embrace of concurrent, overlapping authority.[227] By mid-century, this shift had led to the development of the modern law of preemption, which, instead of presuming that the existence of federal legislation alone

---

[223] *See, e.g.*, Act of July 2, 1948, ch. 809, 62 Stat. 122 (conferring Indian country criminal jurisdiction on New York); Act of June 8, 1940, ch. 276, 54 Stat. 249 (conferring concurrent Indian country criminal jurisdiction on Kansas).

[224] Act of Aug. 15, 1953, Pub. L. No. 83-280, 67 Stat. 588.

[225] *See* 25 U.S.C. § 1323.

[226] *See* Carole Goldberg, *Unraveling Public Law 280: Better Late than Never*, 43 HM. RTS., no. 1, 2007, at 11, https://www.americanbar.org/groups/crsj/publications/human_rights_magazine_home/vol—43/vol—43—no—1/unraveling-public-law-280—better-late-than-never.

[227] Gardbaum, *supra* note 131, at 805–07.

displaced state authority, focused on statutory language and congressional intent.[228]

The Supreme Court acknowledged this jurisprudential transformation in the canonical case of *Williams v. Lee* (1959),[229] which, in the words of one scholar, "opened the modern era of federal Indian law."[230] The Court had to decide whether Arizona courts could exercise civil jurisdiction over a suit against a Navajo defendant for on-reservation conduct. In other words, *Williams* returned, after a long hiatus in the Court's caselaw, to the question of state jurisdiction in Indian country.[231]

The *Williams* Court opened its discussion with what remained the key precedent governing state jurisdiction in Indian country: *Worcester*. The case had been bitterly opposed and defied, the Court observed, but the "broad principles of that decision came to be accepted as law."[232] Nonetheless, the Court acknowledged, the law had since changed. The Court's next sentence began the key work of translating territorial separation into the present. "Over the years, this Court has modified these principles [of *Worcester*] in cases where essential tribal relations were not involved and where the rights of Indians would not be jeopardized, but the basic policy of *Worcester* has remained."[233] The Court then surveyed the caselaw, including the *McBratney* principle that states had jurisdiction over non-Indian-on-non-Indian crime (as well as observing that federal jurisdiction over crimes against Indians "remained exclusive").[234] Ultimately, the Court crafted a test: "Essentially, absent governing Acts of Congress, the question has always

---

[228] *Id.* at 806 (observing that, in the New Deal era, a "new constitutional strategy replaced a strict division of powers version of federalism with a new version embodying the presumption that state powers, though no longer constitutionally guaranteed, survive unless clearly ended by Congress."); *see also* Mary J. Davis, *Unmasking the Presumption in Favor of Preemption*, 53 S.C. L. Rev. 967, 978 (2002) ("The Court . . . began in earnest in the 1940s to refocus preemption analysis on the discernment of congressional intent.").

[229] 358 U.S. 217 (1959). For additional background on *Williams*, see Bethany R. Berger, Williams v. Lee *and the Debate Over Indian Equality*, 109 Mich. L. Rev. 1463 (2011).

[230] Wilkinson, *supra* note 10, at 1.

[231] This long gap reflected the fact-specific, narrow nature of much of the Court's Indian-law docket for the first half of the twentieth century. *See id.* at 1–2 (observing that most early-twentieth century Indian-law decisions "lacked the public law overtones of either the modern era or the nineteenth century").

[232] *Williams*, 358 U.S. at 219.

[233] *Id.*

[234] *Id.* at 220.

been whether the state action infringed on the right of reservation Indians to make their own laws and be ruled by them"—citing *Fisher* for support.[235]

What was striking about this new test was how it recast the prior precedents. *McBratney* and its progeny, for instance, had focused entirely on federalism and said nothing about Native autonomy: they had declined to affirm state jurisdiction over *all* Indian country crime only because of a treaty.[236] *Fisher* similarly had turned on the Court's conclusion that a federal treaty did not bar the exercise of territorial jurisdiction.[237] But, disregarding the earlier focuses on the specific treaties, the *Williams* Court constructed a new structural test that read into these precedents a broader animating *purpose*, including a concern for Native autonomy largely absent from the late nineteenth-century caselaw.

Williams anticipated what can only be described as a plethora of Court cases on state civil jurisdiction in Indian country throughout the '60s, '70s, and '80s, many of which continued to try to reconcile *Worcester* with the newly enunciated *Williams* test. Only three years after *Williams*, for instance, the Court decided *Kake*, a case involving Alaska state fishing regulations applied to Alaska Natives.[238] The case had no direct bearing on inherent state jurisdiction over Indian country, since the fishing at issue occurred off-reservation and Congress had already delegated criminal and civil authority in Indian country to Alaska through Public Law 280.[239] Yet the decision nonetheless contained another survey of the history of the "relation between the Indians and the States."[240] Justice Frankfurter wrote:

> The general notion drawn from Chief Justice Marshall's opinion in Worcester v. Georgia, 6 Pet. 515, 561; The Kansas Indians, 5 Wall. 737, 755–757; and The New York Indians, 5 Wall. 761, that an Indian reservation is a distinct nation within whose boundaries state law cannot

[235] *Id.*

[236] *See supra* text accompanying notes 179–185.

[237] *See supra* text accompanying notes 201–204.

[238] Organized Vill. of Kake v. Egan, 369 U.S. 60 (1962).

[239] *See* Dylan Hedden-Nicely, *The Reports of My Death Are Greatly Exaggerated: The Continued Vitality of* Worcester v. Georgia, 51 Sw. L. Rev. (forthcoming 2023) (manuscript at 4–7), https://papers.ssrn.com/abstract=4242770 (delving much more deeply into *Kake* and suggesting that *Castro-Huerta* misread the precedent).

[240] *Kake*, 369 U.S. at 71–75.

> penetrate, has yielded to closer analysis when confronted, in the course of subsequent developments, with diverse concrete situations. By 1880 the Court no longer viewed reservations as distinct nations. On the contrary, it was said that a reservation was in many cases a part of the surrounding State or Territory, and subject to its jurisdiction except as forbidden by federal law.[241]

The *Castro-Huerta* majority placed considerable emphasis on this quotation, but the opinion arguably just restated what *Williams* had already said, albeit in sharper terms: the law of territorial separation had changed as jurisprudence had changed. Frankfurter even cited the same case—*Fisher*—to illustrate his point.[242] The key question was not whether *Worcester* had changed but what Frankfurter meant by "forbidden by federal law." *Williams v. Lee* strongly suggested that this did not require a direct federal bar: after all, the *Williams* Court had refused to acknowledge state jurisdiction despite the absence of an express federal law in conflict, holding instead that federal law preempted state jurisdiction whenever "essential tribal relations" were implicated.

The Supreme Court itself later clarified what "forbidden by federal law" meant in two subsequent decisions addressing the scope of state civil jurisdiction in Indian country—though both decisions are absent from the *Castro-Huerta* majority.[243] In *McClanahan* (1973), Justice Thurgood Marshall invalidated the application of Arizona's income tax to a member of the Navajo Nation for income earned on the reservation.[244] In so doing, he once again felt the need to explicate the doctrine's evolution. Marshall observed that the "Indian sovereignty doctrine, with its concomitant jurisdictional limit on the reach of state law" had "undergone considerable evolution in response to changed circumstances" in the "141 years since *Worcester* was decided."[245] In particular, Marshall observed that the Court had turned away "from the idea of inherent Indian sovereignty as a bar to

---

[241] Id. at 72.

[242] *Id.* (citing Utah & N. Ry. v. Fisher, 116 U.S. 28, 31 (1885)).

[243] A much more thorough investigation of the Court's jurisprudence in this area appears in Dylan Hedden-Nicely, *The Terms of their Deal: Revitalizing the Treaty Right to Limit State Jurisdiction in Indian Country*, 27 Lewis & Clark L. Rev. (forthcoming 2023), *available at* https://papers.ssrn.com/sol3/papers.cfm?abstract_id=4192916.

[244] McClanahan v. Ariz. State Tax Comm'n, 411 U.S. 164, 165 (1973).

[245] *Id.* at 171.

state jurisdiction"—what Marshall described as "platonic notions of Indian sovereignty"—and turned "toward reliance on federal preemption," citing *Williams* for support, looking to "applicable treaties and statutes."[246] Yet again, in other words, Marshall was narrating the legal change from the nineteenth to the twentieth centuries.

But the rest of Marshall's opinion established what he meant by "federal preemption." The Court conceded that no treaty or statute *explicitly* barred the exercise of state jurisdiction in question.[247] But it nonetheless read an 1868 Navajo treaty, the Arizona Enabling Act, and the Buck Act—as well as the fact that, under Public Law 280, Arizona could gain jurisdiction over the reservation if the tribe consented—as evidence of an implicit congressional aim to exempt the tribe from state tax.[248]

Finally, in 1980, the Court decided *Bracker*, the case that gave its name to the test that the Court has established to determine the scope of state civil jurisdiction.[249] By now, the caselaw on state jurisdiction in Indian country had grown so tangled that it took the Court five pages to summarize.[250] But the summary was nonetheless revealing. The decision started with the observation that, "Long ago, the Court departed from Mr. Chief Justice's view [in *Worcester*] that 'the laws of [a State] can have no force within reservation boundaries'"—language that would seem to confirm the *Castro-Huerta* majority's view.[251]

But *Castro-Huerta* majority did not cite *Bracker*'s discussion of *Worcester*, likely because what followed was decidedly *unhelpful* to its project. The *Bracker* Court continued by underscoring that the "standards of pre-emption" in Indian law were unique and could not be resolved through "mechanical or absolute conceptions of state or tribal sovereignty" but rather required "a particularized inquiry into the nature of the state, federal, and tribal interests at stake."[252] The Court noted, "We have thus rejected the proposition that, in order to

---

[246] *Id.* at 172.

[247] *Id.* at 173–79.

[248] *Id.* Subsequent caselaw held that states did not in fact gain civil legislative jurisdiction under Public Law 280. *See* Bryan v. Itasca Cnty., 426 U.S. 373, 390 (1976).

[249] White Mountain Apache Tribe v. Bracker, 448 U.S. 136 (1980).

[250] *Id.* at 141–45.

[251] *Id.* at 141 (citation omitted).

[252] *Id.* at 143–45.

find a particular state law to have been pre-empted by operation of federal law, an express congressional statement to that effect is required."[253] The Court then applied this standard by concluding that the federal regulation of on-reservation timber harvesting was so comprehensive as to preclude the application of state taxes even on non-Indian businesses.[254]

Taken together, this caselaw reveals a Court that did exactly what it *said* it was doing: adapting *Worcester*'s principles for a modern era that had different jurisprudential assumptions. *Worcester* relied on an older understanding predicated on exclusivity and zero-sum jurisdiction—what the second Justice Marshall derided as "platonic ideals." From *Williams* through *Bracker*, the Court sought to translate the "basic policy" of *Worcester* and territorial separation into the modern language of preemption, congressional intent, and countervailing tribal, state, and federal interests.

It is an open question how well this translation succeeded in creating workable law. *Bracker* balancing can be a notoriously unpredictable and subjective assessment given its call for judges to focus on "the specific context." And the Court confused matters by calling its new test for state jurisdiction in Indian country "preemption." Ordinary preemption law focuses on congressional intent rather than balancing interests; it often *does* require an "express congressional statement." By contrast, under *Bracker*, the existence of a comprehensive and long-standing federal scheme of governance has almost always been sufficient to defeat state claims of jurisdiction within Indian country, even over non-Indians.[255] In that sense, the reformulated *Bracker* test, though it rested on a distinct jurisprudential foundation, maintained some of the older zero-sum framework of the "territorial separation" era.

---

[253] *Id.* at 144.

[254] *Id.* at 145–51.

[255] *See, e.g.*, Mescalero Apache Tribe v. Jones, 462 U.S. 324, 338 (1983) (noting that state jurisdiction would disrupt a "comprehensive scheme of federal and tribal management [of game] established pursuant to federal law"); California v. Cabazon Band of Mission Indians, 480 U.S. 202, 217–28 (1987) (rejecting a state effort to prohibit non-Indians playing tribal bingo because of the "important federal interests" embodied in federal regulations and support for tribal gaming).

C. *WORCESTER* IN THE TWENTY-FIRST CENTURY

The Supreme Court's discussion of *Worcester* and state jurisdiction in Indian country has dropped off considerably since its late twentieth-century heyday. Prior to *Castro-Huerta* and *Ysleta del Sur*, no twenty-first-century opinion had squarely presented a similar question concerning the scope of state jurisdiction *within* Indian country.[256] Only one prior twenty-first-century case had seriously discussed *Worcester*: *Nevada v. Hicks* (2001).[257]

Hicks concerned the scope of *tribal* court jurisdiction. However, because the defendants were state officials, Justice Scalia offered a disquisition on state sovereignty in Indian country. "State sovereignty does not end at a reservation's border," he observed, quoting *Bracker*'s language about having "long ago" departed from *Worcester*.[258] In a footnote, Scalia observed that *Worcester* had to be interpreted in light of an 1828 Cherokee treaty that had guaranteed that the signatories' lands "would never be subjected to the jurisdiction of any State or Territory."[259]

This influential assertion[260] rested on astoundingly shoddy research. This 1828 treaty nowhere appeared in *Worcester*, and for good reason: it was a treaty with an entirely *different* group of Cherokees–those who had already removed westward from their homelands, and so was legally irrelevant to *Worcester*.[261] Justice Scalia's citation was a remarkably transparent example of uncritically grasping at historical evidence to rewrite a precedent he disfavored.

Which brings us to *Castro-Huerta*. In a sense, *Castro-Huerta* just continued the process of translating the law on state jurisdiction in Indian country, as it suggested with its extensive reliance on the twentieth-century precedents. After all, the decision ostensibly brought the law of preemption in Indian law closer to the ordinary

---

[256] In Wagnon v. Prairie Band Potawotami Nation, 546 U.S. 95 (2005), the Court permitted a Kansas fuel tax because its incidence fell outside Indian country. *Id.* at 99. Similarly, in City of Sherrill v. Oneida Indian Nation of N. Y., 544 U.S. 197 (2005), the Court permitted a municipal land tax on the ground that tribal sovereignty had been dormant for too long to be revived. *Id.* at 202–03.

[257] 533 U.S. 353 (2001).

[258] *Id.* at 361 (citation omitted).

[259] *Id.* at 361 n.4 (citation omitted).

[260] The Court cited this footnote in United States v. Lara, 541 U.S. 193, 206 (2004).

[261] *See* Articles of a Convention, U.S.-Western Cherokee, May 6, 1828, 7 Stat. 311, 311 (making clear the treaty was signed with Cherokee leaders "West of the Mississippi").

law of preemption, as well as reconciling it to the more expansive readings of state sovereignty favored by this Court. This underscores the broader challenge of common law reasoning: precedent is clay that judges can mold into whatever shape they want.

Castro-Huerta, though, felt more like throwing the clay on the ground and stomping on it. Or, to use a different metaphor, at some point the Court ceases to translate precedent and starts to simply rewrite it. The Court could just overrule *Worcester*, but it is telling that it chose not to. Common-law legitimacy comes from the claim of continuity, which is what supposedly gives the common law coherence and stability. There are different ways to interpret prior cases, and judges can argue over which are better or worse, but they are united by the effort to faithfully understand the past decisions. What feels most galling about *Castro-Huerta*, at least to me, is that, compared to its predecessors, the Court rather obviously lacked that basic fidelity; it did not seriously attempt to get the law right.[262]

There is, however, an ironic sense in which *Castro-Huerta*, even while asserting *Worcester*'s abandonment, was quite faithful to the case—indeed, revived its jurisprudential approach. Much of the *Castro-Huerta* opinion fairly reeked of "mechanical" and "platonic" ideas of sovereignty—precisely the aspects of *Worcester* that the twentieth-century Court had supposedly repudiated. Ultimately, the key difference between *Castro-Huerta* and *Worcester* is less methodological than normative. Unlike *Worcester*, the current Court's "absolute" jurisprudence is in defense of states. It is a sad commentary that this persistent but legally and morally dubious claim could not prevail in the U.S. Supreme Court in 1832 during the height of Removal—but could in 2022.

III. Postscript: Normativity and History

In one footnote in *Castro-Huerta*, Justice Kavanaugh responded to Justice Gorsuch's pointed historical arguments by questioning "why the pre-Civil War history of tribal discord with States—unconnected from any statutory text—should disable States from exercising jurisdiction in 2022."[263] This was a tendentious statement even for an argumentative opinion: it strangely cast aside the Trade

---

[262] *See* Lessig, *supra* note 30, at 9–18.

[263] Oklahoma v. Castro-Huerta, 142 S. Ct. 2486, 2502 n.7 (2022).

and Intercourse Act, the very statute at issue in the case; it ignored the fact that this "pre-Civil War history" had precipitated *Worcester*, the precedent they were arguing over; and it euphemized as "tribal discord with States" what were actually state invasions of Native peoples' homelands guaranteed under federal treaties.

But, while Justice Kavanaugh likely sought only to score cheap rhetorical points, he was asking a significant question. What is history good for? Things change, after all. Few would argue that we should look to the views of a handful of white judges in 1833—or 1883—to decide the optimal jurisdictional arrangements in present-day eastern Oklahoma. In fact, given the fraught history of Indian law,[264] there are good reasons why we wouldn't want to do so.

Arguably, the best way to resolve present-day jurisdictional controversies, especially given the underlying ambiguity, would be negotiation: the parties could express their views, bargain in good faith, and come to a mutually agreed arrangement.[265] Indeed, many tribal-state jurisdictional disputes end in such negotiated settlements.[266] Here, the state, tribes, and federal government all share an interest in ensuring public safety. The *McGirt* decision itself emphasized the many preexisting intergovernmental agreements between Oklahoma and the tribes.[267] Before Justice Ginsburg's death allowed Oklahoma to make a successful end-run, it seemed as though some negotiated arrangement would, in fact, be the ultimate outcome of any post-*McGirt* uncertainty.[268]

But when negotiation fails, a neutral arbiter can intervene, assess the competing claims, and impose what they believe is the optimal outcome. As I have traced more fully elsewhere, many of the Founders envisioned federal power in Indian affairs working just this way: an unbiased federal government would fairly adjudicate disputes between

---

[264] *See supra* text accompanying notes 55–79.

[265] Scholars of federalism have observed a similar dynamic in resolving disputes over state-federal jurisdictional boundaries. *See* Erin Ryan, *Negotiating Federalism*, 52 B.C. L. Rev. 1 (2011); Aziz Z. Huq, *The Negotiated Structural Constitution*, 114 Colum. L. Rev. 1595 (2014); Bridget A. Fahey, *Federalism by Contract*, 129 Yale L.J. 2326 (2019).

[266] *See* Matthew L.M. Fletcher, *Retiring the "Deadliest Enemies" Model of Tribal-State Relations*, 43 Tulsa L. Rev. 73, 81–83 (2007).

[267] McGirt v. Oklahoma, 140 S. Ct. 2452, 2481–82 (2020).

[268] *See, e.g.*, *Cole Introduces Legislation in Response to McGirt v. Oklahoma*, Congressman Tom Cole (May 11, 2021), https://cole.house.gov/media-center/press-releases/cole-introduces-legislation-in-response-to-mcgirt-vs-oklahoma (announcing introduction of a bill that would permit state-tribal law enforcement compacting in Oklahoma).

tribes and states just as it resolves controversies between states.[269] Under this model of the judicial role, all the invocations of history distract from what the Court *should* be doing: weighing the parties' competing interests in the present, as *Bracker* balancing does.[270] Indeed, there is a compelling argument that this was what the Court actually *did* in *Castro-Huerta*; all its discussions of history and law concealed what was, at core, a policy judgment.[271]

A moment's reflection, though, will suggest how deeply flawed this vision of a federal arbiter in Indian affairs proved: it anticipated that the federal government would decide impartially between the nation's own citizens (and voters) and people that federal officials deemed savages. For the Supreme Court specifically, a forthright acknowledgment of the Court's role in deciding optimal jurisdictional arrangements in Indian country would also beg the question why it is best equipped to make this assessment.[272] The Court is hardly expert on the subject, routinely demonstrating ignorance of even the most basic Indian country realities. Moreover, as the Court has grown increasingly solicitous of state sovereignty, it has become the federal institution that many regard as the *most* systematically biased against Native interests.[273] As a result, much of the scholarly field of federal Indian law has been arguing to push as much power *away* from this Court as possible.[274]

Castro-Huerta underscored just how ill-equipped this Court is to fair-mindedly assess the balance of the interests in Indian country. The Court audaciously proclaimed that tilting the jurisdictional balance in favor of state authority would leave tribal autonomy and sovereignty unimpaired in spite of staunch tribal opposition to state

---

[269] Ablavsky, *supra* note 156, at 109–11.

[270] White Mountain Apache v. Bracker, 448 U.S. 136, 145 (1980) (mandating a "particularized inquiry into the nature of the state, federal, and tribal interests at stake").

[271] *See* Oklahoma v. Castro-Huerta, 142 S. Ct. 2486, 2521 (2022) (Gorsuch, J., dissenting) ("[T]he Court concludes that Oklahoma's interests in this case outweigh those of the Cherokee.").

[272] *Id.* at 2525 ("This Court has no business usurping congressional decisions about the appropriate balance between federal, tribal, and state interests.").

[273] *See, e.g.*, Getches, *supra* note 27, at 280–87, 320–21.

[274] *See* Matthew L.M. Fletcher, *Politics, Indian Law, and the Constitution*, 108 Calif. L. Rev. 495, 520–46 (2020); Michalyn Steele, *Comparative Institutional Competency and Sovereignty in Indian Affairs*, 85 U. Colo. L. Rev. 759, 779–815 (2014); *see also* Maggie Blackhawk, *Legislative Constitutionalism & Federal Indian Law*, 132 Yale L.J. (forthcoming 2023) (manuscript at 68–97), https://papers.ssrn.com/abstract=4342652.

jurisdiction, which it deemed a mere "policy preference."[275] Predictably, the tribes understood the situation better. Though the *Castro-Huerta* Court left the question of concurrent state jurisdiction over *Indian* defendants for a future case, lower courts have not been similarly restrained. A concurrence in an Oklahoma Supreme Court decision upholding state jurisdiction to enforce civil protection orders between Indians in Indian country relied extensively on *Castro-Huerta*,[276] as did an Oklahoma Tax Commission decision rejecting a tribal member's challenge to a state income tax.[277] Similarly, in a pending Tenth Circuit case challenging municipal jurisdiction over Indians in Indian country, Oklahoma and the city of Tulsa have relied on *Castro-Huerta* to defend municipal authority over tribal members.[278]

Castro-Huerta's assessment of tribal interests is also predicated on the laughable assumption that concurrent state jurisdiction over non-Indians will have no meaningful effects on tribes and their authority. This is demonstrably false. Layering on an additional jurisdiction will only exacerbate, rather than clarify, the jurisdictional maze in Indian country. In one particularly bitter irony, Texas argued in an amicus brief that state criminal jurisdiction was necessary to stem the epidemic of violence against Native women by citing the experience of Alaska—apparently unaware that Alaska state authorities, who *already* enjoy Indian country criminal jurisdiction under Public Law 280, are reportedly a major cause of the underlying problem.[279] In Montana, the Blackfeet Nation and the state are at odds over a state law that prohibits businesses and governments from enforcing a vaccine

[275] *Castro-Huerta*, 142 S. Ct. at 2501 n.6.

[276] Milne v. Hudson, 519 P.3d 511, 518–20 (Okla. 2022) (Darby, C.J., concurring in result).

[277] Okla. Tax Comm'n, Order No. 2022-10-04-14, at 14–16 (Oct. 4, 2022), https://oklahoma.gov/content/dam/ok/en/tax/documents/resources/rules-and-policies/commission-decisions/income/precedential/2022-10-04-14.pdf.

[278] Brief of Amicus Curiae State of Oklahoma in Support of Appellee City of Tulsa and Affirmance at 5 –7, Hooper v. City of Tulsa, No. 22-5034 (10th Cir. filed Aug. 17, 2022), 2022 WL 3694341, at *5–7; Response Brief on Behalf of Appellee City of Tulsa at 21–22, 36–37, Hooper v. City of Tulsa, No. 22-5034 (10th Cir. filed Aug. 11, 2022), 2022 WL 13530643, at *21–22, *36–37.

[279] Brief for the States of Texas et al. as Amici Curiae in Support of Petitioner at 8–10, Oklahoma v. Castro Huerta, 142 S. Ct. 2486 (2022) (No. 21-429), 2021 WL 4976328; *see also* Amnesty Int'l, Maze of Injustice: The Failure to Protect Indigenous Women from Sexual Violence in the USA 36–37 (2007) (noting the state of Alaska's failure to address violence against Native women due to the lack of state-level law enforcement resources as well as state hostility to tribal institutions and jurisdiction).

mandate, including, the state argues, on the reservation.[280] This controversy highlights a broader reality: because concurrent authority is often unworkable, state regulation and taxation routinely precludes tribes from enforcing their own laws and taxes against non-Indians within Indian country.[281] In all of these instances, *Castro-Huerta* placed a giant thumb on the scale in favor of state interests—despite the substantial consequences that expanded state authority will have for affected Native nations.

Returning to the footnote's question about history's relevance, Justice Kavanaugh's disagreement with Justice Gorsuch reflected a deeper question about when, and how, the past should count in making these normative assessments today. Kavanaugh implicitly assumed that the past's principal legal value is as a source of law. Under this view, history can provide neutral, objective answers to what would otherwise be messy, subjective questions. As we have seen, this perspective fails on both counts: history often fails to provide those answers *and* it is ineluctably normative.

But this normativity makes the past more, not less, valuable for current law. History helps us weigh what should matter as we address thorny legal questions in the present. Intuitively, we know this. Few question whether debates in the *Federalist Papers* or long-ago fights over executive authority matter now. Lawyers, scholars, and the Court all turn to this past not just to reconstruct textual meaning but to highlight foundational values and purposes. In this reading, these sources demonstrate long-standing principles like the diffusion of power or the role of popular sovereignty that address the enduring, structural issues and purposes of our nation.[282]

The legal history of Native peoples in the United States is seldom placed in this frame, but it should be. Many of the Constitution's drafters distrusted state authority over Indian affairs based on hard-won experience. Removal vindicated their fears. The legal campaign

---

[280] *See* Katheryn Houghton, *Blackfeet Nation Challenges Montana's Ban on Vaccine Mandates as Infringement on Sovereignty*, Bozeman Daily Chron. (Nov. 18, 2022), https://www.bozemandailychronicle.com/news/health/blackfeet-nation-challenges-montanas-ban-on-vaccine-mandates-as-infringement-on-sovereignty/article_a9da2cd8-66bc-11ed-ab4e-4bdc22156134.html.

[281] *Cf.* Michigan v. Bay Mills Indian Cmty., 572 U.S. 782, 810 (2014) (Sotomayor, J., concurring) ("States have the power to tax certain individuals and companies based on Indian reservations, making it difficult for Tribes to raise revenue from those sources.").

[282] Jack Balkin refers to this use of history as "the model of admirable ancestors." Balkin, *supra* note 31, at 395–96.

that precipitated *Worcester* was not a dispassionate, disinterested debate over the respective merits of state, federal, and tribal jurisdiction in Indian country. It was a sustained, deliberate attempt to use state authority to crush Native sovereignty and take Native land, and it relied on legal reasoning that one historian has decried "as duplicitous and fallacious as any in American legal history."[283] These principles constituted, in his words, a "legal ideology of removal."[284] Other scholars have offered stronger, less euphemistic terms to describe what this legal rationale justified—"deportation,"[285] "mass expulsion,"[286] "ethnic cleansing,"[287] "genocide."[288]

What the *Castro-Huerta* Court did, then, was take sides in one of the great normative legal contests in U.S. history. Repudiating *Worcester*, the Court revived the logic of the southern states' exterminatory campaign and fulfilled, two centuries later, the dark and dubious visions of southern states' supreme courts.

Much has changed in the ensuing two centuries, but *Castro-Huerta* suggested at least one important continuity. State efforts to claim jurisdiction in Indian country have always been about power—aggrandizing their own and diminishing tribes'. The facts surrounding *Castro-Huerta*, no less than the battles around *Worcester*, suggest less concern for actual governance than the anger of state officials who see in tribes "rivals of sovereignty," in the words of the Removal era.[289] Oklahoma, like Georgia before it, has narrowly succeeded in provoking a partisan battle to get the federal government to endorse its pretensions. Perhaps the states will use their newfound jurisdiction responsibly and not simply seek their own authority at the expense of tribes'. But history gives good reason to doubt.

---

[283] Garrison, *supra* note 9, at 11.

[284] *Id.*

[285] Saunt, *supra* note 127, at xiii–xv.

[286] *Id.* at xiv–xv.

[287] Gary Clayton Anderson, Ethnic Cleansing and the Indian: The Crime That Should Haunt America 6 (2014).

[288] Jeffrey Ostler, Surviving Genocide: Native Nations and the United States from the American Revolution to Bleeding Kansas 363–68 (2019); Benjamin Madley, *Reexamining the American Genocide Debate: Meaning, Historiography, and New Methods*, 120 Am. Hist. Rev. 98, 132–32 (2015).

[289] John Quincy Adams, Fourth Annual Message (Dec. 2, 1828), in 3 A Compilation of the Messages and Papers of the Presidents, at 973, 982 (New York, Bureau of National Literature, James D. Richardson ed., 1897).

LAURA K. DONOHUE

SURVEILLANCE, STATE SECRETS, AND THE FUTURE OF CONSTITUTIONAL RIGHTS

I. Introduction

In 2006, the Federal Bureau of Investigation (FBI) hired Craig Monteilh to become an informant for Operation Flex.[1] The FBI directed him to attend the Islamic Center of Irvine (ICOI), to represent that he wanted to convert to Islam and to record his interactions using audio and video devices they provided.[2] His handlers repeatedly made it clear that they were solely interested in information

---

Laura K. Donohue is the Scott K. Ginsburg Professor of Law and National Security at the Georgetown University Law Center.

Author's note: Special thanks to Omar Haddad and Wilson Holzhaeuser for their help obtaining many of the materials addressed in this Article. It further benefited from discussions with Parker Rider-Longmaid and Brian Levy, with whom I worked on amicus briefs in *Federal Bureau of Investigation v. Fazaga* and *Wikimedia Foundation v. National Security Agency*, respectively. Geoffrey Stone, Kenneth Karas, Alex Abdo, and Ahilan Arulanantham provided additional, helpful comments on the text, which are much appreciated.

[1] Fazaga v. FBI, 965 F.3d 1015, 1026 (9th Cir. 2020). Monteilh appears to have originally been hired by the FBI in 2003 to become an informant for white supremacist investigations after he connected with the Aryan Brotherhood while serving time in prison; Teresa Watanabe & Scott Glover, *Man Says He Was FBI Informant*, L.A. Times (Feb. 26, 2009), https://www.latimes.com/archives/la-xpm-2009-feb-26-me-informant26-story.html.

[2] *Fazaga*, 965 F.3d at 1026.

The Supreme Court Review, 2023.
© 2023 The University of Chicago. All rights reserved. Published by The University of Chicago Press. https://doi.org/10.1086/724432

about Muslims.[3] All "leaders in the Muslim community" constituted "potential threats."[4]

Posing as Farouk al-Aziz, Monteilh subsequently attended classes, collected information about community members, went to daily prayers, memorized verses from the Quran, and tried to obtain compromising information to pressure members to become informants.[5] The FBI directed him to pay particular attention to anyone who was particularly religious or who criticized U.S. foreign policy.[6] He recorded face-to-face meetings with devices secreted in the buttons of his shirt.[7] He left his key fob and mobile phone in prayer rooms, offices, restaurants, cafés, and other areas, allowing him to record conversations when he was not present.[8] Monteilh went through drawers.[9] When directed to date Muslim women, he asked about how he should handle intimacy and was told to "just have sex" with them to obtain more information—which he subsequently did.[10] In addition to going daily to ICOI, he spent a significant amount of time weekly at nine other area mosques, attending up to four mosques a day.[11]

As a result of Monteilh's actions, the Bureau obtained "hundreds of phone numbers; thousands of email addresses; background information on hundreds of individuals; hundreds of hours of video recordings of the interiors of mosques, homes, businesses, and associations; and thousands of hours of audio recordings of conversations,

---

[3] *Id.*; Declaration of Craig F. Monteilh, ¶ 16, FBI v. Fazaga, 142 S. Ct. 1051 (2022) (20-828), https://www.supremecourt.gov/DocketPDF/20/20-828/185460/20210730194627462_20-828ja.pdf (informant directed "to meet and get contact information for a certain number of Muslims per day"), *id.* ¶ 17 (noting instructions "to gather as much information on as many people in the Muslim community as possible); *id.* ¶ 18 (directing him to target "anyone who studied *fiqh*, who openly criticized U.S. foreign policy, including the U.S. military's presence in Muslim countries . . . who was an imam or sheikh; who went on *Hajj*; who played a leadership role" etc. and stating that instead of specific targets, his handlers has tasked him with immersing himself in the Muslim community to gather "as much information on as many people and institutions as possible."); *id.* ¶ 19 (reporting FBI Agent as stating, "We want to get as many files on this community as possible.").

[4] *Id.* ¶ 51.

[5] *Fazaga*, 965 F.3d at 1026–27; Declaration of Craig F. Monteilh, *supra* note 3, ¶ 1.

[6] Declaration of Craig F. Monteilh, *supra* note 3, ¶ 18.

[7] *Id.* ¶ 5.

[8] *Id.* ¶¶ 17, 65–67.

[9] *Id.* ¶ 29.

[10] *Id.* ¶ 52.

[11] *Fazaga*, 965 F.3d. at 1026–27; Declaration of Craig F. Monteilh, *supra* note 3, ¶ 26.

public discussion groups, classes, and lectures."[12] Apparently, Monteilh was not the only collection source: the FBI informed him that all of the mosques he visited had extensive surveillance equipment already installed.[13] He further learned that the Orange County/Los Angeles Muslim community had been "saturated" by other informants, at a level commensurate with East Germany during the Cold War, and Cuba.[14] Similar operations were underway in New York and Dearborn, Michigan.[15]

This information is public. It derives from statements by FBI agents in other cases, Monteilh's sworn declarations, Ninth Circuit adjudication, and media reports, and it raises significant concern about whether and how the FBI is using the Foreign Intelligence Surveillance Act (FISA). It indicates that the Bureau may have acted outside either the ordinary criminal Title III warrant procedure or the statutory limitations in FISA to place entire religious communities under electronic surveillance.[16] By the FBI handlers' admission, there was no pre-existing criminal warrant—according to Monteilh, they indicated that they "could get in a lot of trouble if people found out what surveillance they had in the mosques."[17] One handler suggested that national security investigations do not require any warrant: all its absence means is that the FBI cannot use the information in court. But they can still collect it.[18]

Taking the facts at face value, it would be hard to imagine a more troubling disregard for statutory provisions, as well as constitutional rights enshrined in the First and Fourth Amendments of the Constitution and secured by the Fifth Amendment Due Process Clause. Yet the government's position in the case is that it cannot defend itself without recourse to state secrets and, in a radical departure from how the privilege has been understood for centuries, the Court should dismiss the case at the pleadings stage. Its argument turns what, since the Founding of the United States, has been a common law evidentiary rule into a justiciability standard.

---

[12] *Fazaga*, 965 F.3d at 1027; *see also* Declaration of Craig F. Monteilh, *supra* note 3, ¶ 71.

[13] *Id.* ¶ 28.

[14] *Id.* ¶ 58.

[15] *Id.* ¶ 19.

[16] *See* discussion *infra* Part V.

[17] Declaration of Craig F. Monteilh, *supra* note 3, ¶¶ 17, 28.

[18] *Id.* ¶ 67.

Decided on alternate grounds and sent back to the Ninth Circuit for further consideration, the Supreme Court's 2022 decision in *Federal Bureau of Investigation v. Fazaga* heralds a worrying trend.[19] Over the past fifteen years, as more information about how the government wields its foreign intelligence collection authorities on U.S. soil has become available, it has become clear that the government has repeatedly acted outside its constitutional and statutory limits and at times in flagrant disregard for judicial orders.[20] As a result, dozens of cases challenging surveillance have been making their way through the courts.[21] Unlike in prior eras, in certain cases it has become easier in light of the information available and the programmatic nature of collection for litigants to establish an injury-in-fact. In response, the government has crafted a new state secrets analysis, raised the privilege early in litigation to have suits dismissed, broadened its assertion to encompass entire categories of information, and claimed what for centuries has been a common law rule as an Article II constitutional power.[22] Because of the government's shift, what is now at stake is the

---

[19] *See* FBI v. Fazaga, 142 S. Ct. 1051 (2022).

[20] *See, e.g.*, Memorandum Opinion and Order at 11–13, *[REDACTED]*, No. [REDACTED], GID.C.00050 (FISA Ct. 2009) (Hogan, J.), https://www.dni.gov/files/documents/icotr/702/EFF%2016-CV-02041(HSG)%20Doc%2005%2006.13.17%20-%20REDACTED.PDF (NSA analysts improperly acquiring U.S. persons' communication; CIA exhibiting a "profound misunderstanding of minimization procedures" and inappropriately disseminating reports containing USP information to NSA, FBI, and DOJ; and FBI failing to report compliance incidents to the Court in violation of Rule 10(c) of the Foreign Intelligence Surveillance Court Rules of Procedure); Opinion and Order, *[REDACTED]*, No. [REDACTED], GID.C.00138 (FISA Ct.) (Hogan, J.), https://www.dni.gov/files/documents/icotr/08202018/0820218_Document-27.pdf (government conducted unauthorized surveillance); Opinion and Order, *[REDACTED]*, No. [REDACTED], GID.C.00254 (FISA Ct.) (Hogan, J.), https://assets.documentcloud.org/documents/4780432/EFF-Document-2.pdf (holding that NSA's acquisition of [Redacted] constitutes unauthorized electronic surveillance because it failed to comply with 50 U.S.C. § 1804(a)(2) and (a)(3)(B)); Memorandum Opinion and Order at 41, 43, 49–50, *[REDACTED]*, No. [REDACTED] (FISA Ct. Nov. 18, 2020) (Boasberg, J.), https://www.intel.gov/assets/documents/702%20Documents/declassified/20/2020_FISC%20Cert%20Opinion_10.19.2020.pdf (stating that the court "continues to be concerned about FBI querying practices involving U.S.-person query terms, including (1) application of the substantive standard for conducting queries; (2) queries that are designed to retrieve evidence of crime that is not foreign-intelligence information; and (3) recordkeeping and documentation requirements" and noting that it would "continue to closely monitor the government's reporting in order to evaluate whether the querying procedures are being implemented in a manner consistent with the statute and the Fourth Amendment").

[21] *See, e.g.*, Wikimedia Found. v. Nat'l Sec. Agency (NSA)/Cent. Sec. Serv., 14 F.4th 276 (4th Cir. 2021); *In re* Terrorist Attacks on September 11, 2001, 523 F. Supp. 3d 478 (S.D.N.Y. 2021); Page v. Comey, No. 20-CV-3460 (DLF), 2022 WL 3981135 (D.D.C. Sept. 1, 2022); Trump v. Clinton, No. 22-CV-14102, 2022 WL 4119433 (S.D. Fla. Sept. 8, 2022).

[22] For a detailed discussion of how state secrets evolved from 2001 to 2010, see Laura K. Donohue, *The Shadow of State Secrets*, 159 U. PA. L. REV. 77 (2010).

possibility of any litigant to ever challenge illegal and unconstitutional surveillance. *Fazaga* represents the tip of the iceberg in terms of the risks to individual rights that would follow, should the government ultimately prevail.

This Article begins by according three developments a central role in elevating the evidentiary base on which litigants can rely. First, following the release of classified materials by the media and the government, more information about how the government has been interpreting its legal authorities and what programs are underway is now available. It shows that the *way* in which the intelligence community collects communications has shifted, making it easier to demonstrate an injury-in-fact. Second, parallel efforts to obtain information via the Freedom of Information Act (FOIA) about domestic foreign intelligence collection has forced thousands of more documents into the public domain, providing a factual basis for suits to progress. Third, this information indicates that the government routinely exceeds its constitutional and statutory limits, prompting judicial challenge. With standing met, the government is falling back on a re-interpretation of state secrets to avoid Article III adjudication.

The Article next turns to *Fazaga*, detailing the question addressed by the Supreme Court: whether certain provisions in FISA displaced the state secrets privilege. Siding with the government, the Supreme Court determined that it did not and remanded the case for determination of the state secrets assertion. In doing so, the Court sidestepped the most important question: whether the government can avoid accountability when it acts in apparent disregard of acts of Congress as well as Americans' First, Fourth, and Fifth Amendment rights. This question will not go away: it already is up in a parallel case on a petition of certiorari.[23]

The Article highlights the novel theory of state secrets being advanced in *Fazaga* and dozens of parallel cases, noting that it departs in four critical ways from how, for centuries, state secrets has operated. First, the government has collapsed the *Reynolds* evidentiary rule and *Totten* contractual bar by re-casting the latter in terms of a "very subject matter" analysis.[24] This change discards the entire point of

---

[23] *See* Petition for Writ of Certiorari, Wikimedia Found v. NSA/Cent. Sec. Serv., No. 22-190 (U.S. Aug. 26, 2022).

[24] *See* United States v. Reynolds, 345 U.S. 1 (1953); Totten v. United States, 92 U.S. 105 (1875).

Totten: that parties entering into a secret contractual relationship with the government have *ex ante* notice that future disputes are unlikely to be addressed in open court. In other words, *the contract itself*, contains the understanding that confidentiality will be maintained. This rule does not apply to situations where the government acts unilaterally: neither have both parties agreed, nor has notice been served in regard to the unavailability of judicial redress should a pre-existing contract be breached. Second, the government is asserting the privilege early in suits to request dismissal, instead of employing it as part of discovery or at the merits stage to exclude evidence. Third, it has adopted an overbroad approach, asserting state secrets not over particular documents, but instead over broad categories of information. Such assertions sweep in a significant amount of information already in the public domain and often do not withstand judicial scrutiny. Fourth, the government is attempting to transform a common law rule into a constitutional power, increasingly claiming that it derives from the executive's Article II authorities.

The Article concludes by emphasizing the implications of the government's re-casting of state secrets. What is at stake in *Fazaga* extends well beyond the case to protecting the rights of the People against government malfeasance as well as the ability of the democratic structures to hold the executive branch to account.

II. Programmatic Surveillance and Standing

Programmatic surveillance has long been a feature of U.S. intelligence collection.[25] The stakes are high: on the one hand,

---

[25] *See, e.g.*, S. Select Comm. to Study Governmental Operations with Respect to Intel. Activities 1975–76 (Church Committee) Final Report, S. Rep. No. 94-755, bk. II, Intelligence Activities and the Rights of Americans, at 21 (1976) (recounting CIA, NSA, and FBI warrantless collection of domestic and international communications 1936–1976 and noting that "[s]ince the re-establishment of federal domestic intelligence programs in 1936, there has been a steady increase in the government's capability and willingness to pry into . . . the political activities and personal lives of the people. The last forty years have witnessed a relentless expansion of domestic intelligence activity"); *see also* Commission on CIA Activities within the United States 1975 (Rockefeller Commission), at 101–15 (detailing numerous CIA mail intercept programs dating back to the 1950s); *id.* at 116–29 (discussing CIA collection of information on anti-war activists and dissidents in the 1960s and 1970s); *id.* at 130–50 (examining CIA and FBI participation in Operation CHAOS 1967–1974); Interception of International Telecommunications by the National Security Agency, Report by the Committee on Government Operations (Pike Committee), at 5–11 (discussing operation of the Black Chamber 1919–1929, which analyzed cables obtained from Western Union Telegraph Company and the Postal Telegraph Company); *id.* at 11–25 (detailing the collection of international telecommunications traffic 1945–1975 by the Army Signal Security

information gleaned may help to detect and mitigate national security threats. In an age of weapons of mass destruction, the costs of failing to do so may be catastrophic. On the other hand, directed against U.S. citizens, such programs may threaten foundational constitutional rights by chilling freedom of speech, religion, and association, violating the right to privacy, and running roughshod over the right against self-incrimination—to say nothing of due process concerns. Accordingly, the revelation that the government is collecting information on citizens is often accompanied by efforts to hold the government to account.

Following disclosures by President Ford's Commission on CIA Activities Within the United States, for instance, twenty-one individuals and five anti-war organizations brought suit against officials in the CIA, FBI, Department of Defense, and Secret Service asserting violation of their First, Fourth, Fifth, and Ninth Amendment rights.[26] As soon as the *New York Times* revealed that in the aftermath of the September 11, 2001, attacks President George W. Bush had authorized warrantless surveillance, lawsuits alleging a range of constitutional violations mounted.[27] Just over a month after the attack, the Center for Constitutional Rights filed the first action.[28] By February, two more cases had been filed.[29] In May, another twenty-seven cases were filed, with six more in June.[30] As the filings picked up steam, by August 2006, a multidistrict panel had ordered the consolidation of more than fifty cases and their transfer to the Northern District of California.[31] The release of the Snowden documents and the revelation that the government was conducting bulk collection under Section 215 of the USA PATRIOT Act proved no different.[32]

Agency and later the National Security Agency); *id.* at 975–1018 (addressing the Drug Enforcement Administration's domestic and international intelligence programs); *id.* at 1019–86 (highlighting the FBI Domestic Intelligence programs).

[26] Halkin v. Helms, 690 F.2d 977, 981 (D.C. Cir. 1982).

[27] *See* James Risen & Eric Lichtblau, *Bush Lets U.S. Spy on Callers Without Courts*, N.Y. Times (Dec. 16, 2005), https://www.nytimes.com/2005/12/16/politics/bush-lets-us-spy-on-callers-without-courts.html (stating that Bush had signed a presidential order in 2002 giving NSA the authority to monitor international telephone calls and international email messages "of hundreds, perhaps thousands, of people inside the United States without warrants over the past three years in an effort to track possible 'dirty numbers' linked to Al Qaeda").

[28] Complaint at 1, Ctr. for Const. Rights v. Bush, No. 06-00313 (N.D. Cal. Jan. 17, 2006).

[29] *See* Hepting v. AT&T, 439 F. Supp. 2d 974 (N.D. Cal. 2006); Complaint, Al-Haramain Islamic Found., Inc. v. Bush, 451 F. Supp. 2d 1215 (D. Or. 2006) (No. 06-0274).

[30] *See generally* Donohue, *supra* note 22, at 148.

[31] *Id.* at 149.

[32] *See* 50 U.S.C. § 1861 and discussion, *infra*.

Historically, though, the case or controversy requirement often prevented suits from moving forward: it could be devilishly difficult to establish an injury-in-fact when the underlying documents were classified. Following the release of government documents by the media in June 2013, subsequent declassification, and successful FOIA litigation, however, the context changed. The documents revealed the breadth of surveillance underway as well as a government acting outside statutory and constitutional limits. With litigants increasingly able to demonstrate standing, the government turned to state secrets as a way to head off litigation.

A. THE SHOALS OF STANDING

In 1992 in *Lujan v. Defenders of Wildlife*, the Supreme Court articulated the three-part test for standing:

> First, the plaintiff must have suffered an "injury in fact" — an invasion of a legally protected interest which is (a) concrete and particularized, [] and (b) "actual or imminent, not 'conjectural' or 'hypothetical,'" [] Second, there must be a causal connection between the injury and the conduct complained of—the injury has to be "fairly ... trace[able] to the challenged action of the defendant, and not ... th[e] result [of] the independent action of some third party not before the court." [] Third, it must be "likely," as opposed to merely "speculative," that the injury will be "redressed by a favorable decision."[33]

To meet the first criteria, plaintiffs in surveillance cases have to be able to demonstrate that *their* communications in particular are being monitored and/or collected. The mere existence of a program, in the past, has generally proven insufficient. Lacking access to classified information, it could be extremely difficult to make the case for standing.

In the 1982 case *Halkin v. Helms*, for example, the D.C. Circuit noted that just because individuals found themselves on a watch list (in this case, Operation MINARET), it did not necessarily mean that their communications had been collected.[34] Without being able to demonstrate actual interception, the program's constitutionality could not be challenged.[35] Similarly, following revelation of STELLARWIND,

---

[33] Lujan v. Defs. of Wildlife, 504 U.S. 555, 560–61 (1992) (internal citations omitted).

[34] Halkin v. Helms, 690 F.2d 977, 998 (D.C. Cir. 1982).

[35] *Id.*

residential telephone customers brought a class action against government agencies and officers, raising First and Fourth Amendment claims, and asserting, *inter alia*, violation of separation of powers.[36] Chief Judge Vaughn R. Walker of the Northern District of California dismissed the suit on the grounds that neither the plaintiffs nor their purported class representatives had alleged a sufficiently particular injury.[37] Generalized harm would not suffice. In a parallel case, plaintiffs in regular communication with individuals overseas filed a suit in Michigan, asserting a First and Fourth Amendment challenge, as well as violation of separation of powers. The Sixth Circuit dismissed the suit on grounds that the plaintiffs lacked standing.[38]

Even where statutory language has been introduced to govern programmatic collection, the standing bar has proven high. In 2008, Congress added FISA Section 702.[39] The provision empowered the Attorney General in conjunction with the Director of National Intelligence (DNI) to place non-U.S. persons reasonably believed to be outside the United States under surveillance.[40] Human rights, labor, legal, and media organizations in contact with individuals they believed to be likely targets brought suit on Fourth Amendment grounds.

Justice Alito, writing for the Supreme Court in 2013 in *Clapper v. Amnesty International*, dismissed the case for lack of standing: "[R]espondents' theory of *future* injury is too speculative to satisfy the well-established requirement that the threatened injury must be 'certainly impending.'"[41] Even if such certainty could be established, it could not be demonstrated that such injury was "fairly traceable" to the

---

[36] Complaint, Jewel v. NSA, MDL Docket No. C 06-1791 VRW, 2010 WL 235075 (N.D. Cal. Jan. 21, 2010).

[37] *Id.* at *1; *cf.* Shubert v. Obama, C 07-0693 Doc. #38 (MDL Doc. #680). *But see* Jewel v. NSA, 673 F.3d 902 (9th Cir. 2011) (finding standing and remanding for consideration of state secrets assertion).

[38] ACLU v. NSA, 493 F.3d 644 (6th Cir. 2007). *cf. In re* Telecomms. Recs. Litig., 522 F. App'x 383 (9th Cir. 2013) (applying *Clapper* to dismiss plaintiff's claims on the grounds that they lacked actual knowledge of the Government's targeting practices). *see also* Al-Haramain Islamic Found., Inc. v. Bush, 507 F.3d 1190 (1205) (9th Cir. 2007) (suit dismissed on grounds that standing not met).

[39] FISA Amendments Act of 2008, Pub. L. No. 110-261, 122 Stat. 2436.

[40] *See* 50 U.S.C. § 1881a.

[41] Clapper v. Amnesty Int'l, 568 U.S. 398, 401 (2013) (Alito, J.) (emphasis in original).

statutory authority.[42] Nor could respondents avail themselves of the argument that they had already had to adopt "costly and burdensome measures to protect the confidentiality of their international communications.[43] The respondents could not "manufacture standing by choosing to make expenditures based on hypothetical future harm that is not certainly impending."[44] Should the government elect to use FISA-derived information in a criminal trial, it would have to provide advance notice, giving the affected person the opportunity to challenge the lawfulness of the acquisition.[45]

In making this last assertion, the Court relied on statutory provisions (which require the government to notify anyone against whom any information obtained from electronic surveillance has been used), as well as repeated representations to the Court by the U.S. Solicitor General that the Department of Justice (DOJ) would so notify individuals.[46] This representation turned out to be false.[47] Not only was

---

[42] *Id.* at 402.

[43] *Id.*

[44] *Id.*

[45] *Id.* at 421 ("[I]f the Government intends to use or disclose information obtained or derived from a § 1881a acquisition in judicial or administrative proceedings, it must provide advance notice of its intent, and the affected person may challenge the lawfulness of the acquisition.").

[46] *See* 50 U.S.C. § 1806(c). During oral argument, Justice Sotomayor asked, "General, is there anybody who has standing?" She continued, "As I read your brief, standing would only arise at the moment the government decided to use the information against someone in a pending case. To me, that [] would seem to say that the Act . . . if there was a constitutional violation in the interception, that no one could ever stop it until they were charged with a crime, essentially." Transcript of Oral Argument at 3–4, Clapper v. Amnesty Int'l, 568 U.S. 398 (2012) (No. 11-1025), https://www.supremecourt.gov/oral_arguments/argument_transcripts/2012/11-1025.pdf. Don Verrilli, the Solicitor General, replied with a "clear example" of a situation in which litigants would have standing: namely, where "an aggrieved person, someone who is a party to a communication, gets notice that the government intends to introduce information in a proceeding against them. They have standing. That standing could include a facial challenge like the one here." *Id.* at 4. Similarly, in his brief to the court, the Solicitor General argued in response to Amnesty International's argument that failure to find standing would result in immunizing section 702 from constitutional challenge: "That contention is misplaced. Others may be able to establish standing even if respondents cannot. As respondents recognize, the government must provide advance notice of its intent to use information obtained or derived from [702] against a person in judicial or administrative proceedings and that person may challenge the underlying surveillance." Reply Brief for the Petitioners at 15, *Amnesty Int'l*, 568 U.S. 398 (No. 11-1025).

[47] *See* Adam Liptak, *A Secret Surveillance Program Proves Challengeable in Theory Only*, N. Y. Times (July 15, 2013). At the time the Solicitor General made his representation, no criminal defendant had ever received notice of Section 702 surveillance. After the *New York Times* article was published, the government issued five notices from October 2013 to April 2014—including for cases in which individuals had already been tried and convicted. After that, however, the notices stopped for nearly two years. *See* Patrick Toomey, *Why Aren't Criminal Defendants Getting Notice of Section 702 Surveillance – Again?*, Just Sec. (Dec. 11, 2015).

it not the practice of the DOJ to inform defendants that FISA had played a role in their case, but law enforcement agencies had been trained to use "parallel construction" to shape evidentiary chains to ensure that neither the prosecution nor the defense would know that the information leading to a criminal case derived from FISA-related surveillance.[48] With such policies in place, standing, and the ability of citizens to pursue their constitutional claims, proved elusive.

B. PRIMARY SOLIDIFICATION OF THE INJURY-IN-FACT REQUIREMENT

Less than four months after *Clapper*, *The Guardian* and other newspapers began publishing information obtained by Edward Snowden, a former National Security Agency (NSA) contractor.[49] Drawing further on investigative reporting, these articles revealed the surprising scope of surveillance then underway.[50] The public outcry that followed stemmed from deep concern about the impact of the programs on Americans' constitutional rights. The Privacy and Civil Liberties Oversight Board (PCLOB), previously floundering, suddenly took form, held hearings, and issued a scathing report on the government's use

---

[48] *See generally* Responsive Documents, Drug Enf't Admin., MuckRock, http://www.documentcloud.org/documents/1011382-responsivedocuments.html.

[49] *See, e.g.*, Glenn Greenwald, *NSA Collecting Phone Records of Millions of Verizon Customers Daily*, Guardian (June 6, 2013), https://www.theguardian.com/world/2013/jun/06/nsa-phone-records-verizon-court-order; Charlie Savage, *NSA Said to Search Contents of Messages To and From America*, N.Y. Times (Aug. 8, 2013), https://www.nytimes.com/2013/08/08/us/broader-sifting-of-data-abroad-is-seen-by-nsa.html; Barton Gellman & Ashkan Soltani, *NSA Collects Millions of Email Address Books Globally*, Wash. Post (Oct. 14, 2013), https://www.washingtonpost.com/world/national-security/nsa-collects-millions-of-e-mail-address-books-globally/2013/10/14/8e58b5be-34f9-11e3-80c6-7e6dd8d22d8f_story.html.

[50] *See, e.g.*, Barton Gellman & Ashkan Soltani, *NSA Infiltrates Links to Yahoo, Google Data Centers Worldwide, Snowden Documents Say*, Wash. Post (Oct. 30, 2013), https://www.washingtonpost.com/world/national-security/nsa-infiltrates-links-to-yahoo-google-data-centers-worldwide-snowden-documents-say/2013/10/30/e51d661e-4166-11e3-8b74-d89d714ca4dd_story.html; Barton Gellman & Ashkan Soltani, *NSA Tracking Cellphone Location Worldwide, Snowden Documents Show*, Wash. Post (Dec. 4, 2013), https://www.washingtonpost.com/world/national-security/nsa-tracking-cellphone-locations-worldwide-snowden-documents-show/2013/12/04/5492873a-5cf2-11e3-bc56-c6ca94801fac_story.html; Nicole Perlroth, Jeff Larson & Scott Shane, *NSA Able to Foil Basic Safeguards of Privacy on the Web*, N.Y. Times (Sept. 5, 2013) https://www.nytimes.com/2013/09/06/us/nsa-foils-much-internet-encryption.html; Glenn Greenwald & Ryan Grim, *Top-Secret Document Reveals NSA Spied on Porn Habits As Part of Plan to Discredit "Radicalizers,"* HuffPost (Nov. 26, 2013), https://www.huffpost.com/entry/nsa-porn-muslims_n_4346128; Siobhan Gorman, *NSA Officers Spy on Love Interests*, Wall St. J. (Aug. 23, 2013), https://www.wsj.com/articles/BL-WB-40005.

of FISA.[51] Congress held dozens more hearings, calling government officials to testify to ensure that legislators had full information about the extent of the programs that had been secretly operating for years, collecting Americans' telephony and internet metadata and content without their knowledge. From three bills on FISA the prior year (when three of the statute's provisions had been due to sunset), over the twelve months following Snowden's disclosures, Congress considered forty-two different bills, with proposed amendments to foreign intelligence collection ranging from radical restructuring of the Foreign Intelligence Surveillance Court (FISC) to defunding and withdrawing surveillance authority.[52]

Such was the tenor of the public outrage that the government immediately had to respond. The administration first tried to correct the record by issuing statements and documents to offset the information being leaked and the accompanying news articles.[53] President

[51] *See* Priv. & C.L. Oversight Bd., *Report on the Telephone Records Program Conducted under Section 215 of the USA PATRIOT Act and on the Operations of the Foreign Intelligence Surveillance Court* (Jan. 23, 2014), https://documents.pclob.gov/prod/Documents/OversightReport/ec542143-1079-424a-84b3-acc354698560/215-Report_on_the_Telephone_Records_Program.pdf.

[52] *See, e.g.*, 159 Cong. Rec. 12237 (2013) (debating whether to prohibit certain kinds of collection under FISA and to defund section 215 collection).

[53] *See, e.g.*, Press Release, Joint Statement: NSA and Office of the Director of National Intelligence (Aug. 22, 2013), https://www.dni.gov/index.php/newsroom/press-releases/press-releases-2013/item/917-joint-statement-nsa-and-office-of-the-director-of-national-intelligence (stating that an article published in the *Wall Street Journal* mischaracterized the NSA's Section 702 collection activities); Press Release, DNI Statement on the Collection of Intelligence Pursuant to Section 702 of the Foreign Intelligence Surveillance Act (June 8, 2013), https://www.dni.gov/index.php/newsroom/press-releases/press-releases-2013/item/872-dni-statement-on-the-collection-of-intelligence-pursuant-to-section-702-of-the-foreign-intelligence-surveillance-act (criticizing the *Guardian* and the *Washington Post* for not providing the "full context" of the programs underway resulting in "significant misimpressions" and "inaccuracies"); Press Release, DNI Statement on Activities Authorized Under Section 702 of FISA (June 6, 2013), https://www.dni.gov/index.php/newsroom/press-releases/press-releases-2013/item/869-dni-statement-on-activities-authorized-under-section-702-of-fisa (stating that the *Guardian* and *Washington Post* articles "contain numerous inaccuracies"); Press Release, DNI Statement on Recent Unauthorized Disclosures of Classified Information (June 6, 2013), https://www.dni.gov/index.php/newsroom/press-releases/press-releases-2013/item/868-dni-statement-on-recent-unauthorized-disclosures-of-classified-information (stating that the *Guardian* "article omits key information regarding how a classified intelligence collection program is used to prevent terrorist attacks and the numerous safeguards that protect privacy and civil liberties"); *see also* Press Release, Facts on the Collection of Intelligence Pursuant to Section 702 of the Foreign Intelligence Surveillance Act (June 8, 2013), https://www.dni.gov/index.php/newsroom/press-releases/press-releases-2013/item/871-facts-on-the-collection-of-intelligence-pursuant-to-section-702-of-the-foreign-intelligence-surveillance-act; *Administration White Paper: Bulk Collection of Telephony Metadata Under Section 215 of the USA PATRIOT Act* (Aug. 9, 2013), https://irp.fas.org/nsa/bulk-215.pdf.

Barack Obama next directed DNI James Clapper to declassify and make certain information about FISA surveillance programs public—which he did.[54] Obama then constituted a Review Board to examine foreign intelligence collection and to consider the best course of action to respond to the concerns raised.[55]

The cumulative result of these actions was the infusion of a tremendous amount of previously classified material into the public discourse, providing details about the breadth of the surveillance programs underway.[56] Lawsuits, representing a range of political views, demographics, and professional interests, quickly followed. Less than a week after the first Snowden release and the Obama Administration's response, for instance, the ACLU filed suit in the Southern District of

---

[54] *See, e.g.*, Press Release, DNI Announces the Declassification of the Existence of Collection Activities Authorized by President George W. Bush Shortly After the Attacks of September 11, 2001, (Dec. 21, 2013), https://www.dni.gov/index.php/newsroom/press-releases/press-releases-2013/item/991-dni-announces-the-declassification-of-the-existence-of-collection-activities-authorized-by-president-george-w-bush-shortly-after-the-attacks-of-september-11-2001; Press Release, DNI Clapper Declassifies Additional Intelligence Community Documents Regarding Collection Under Section 501 of the Foreign Intelligence Surveillance Act (Nov. 18, 2013), https://www.dni.gov/index.php/newsroom/press-releases/press-releases-2013/item/964-dni-clapper-declassifies-additional-intelligence-community-documents-regarding-collection-under-section-501-of-the-foreign-intelligence-surveillance-act; Press Release, DNI Clapper Declassifies Additional Intelligence Community Documents Regarding Collection Under Section 501 of the Foreign Intelligence Surveillance Act (Oct. 28, 2013), https://www.dni.gov/index.php/newsroom/press-releases/press-releases-2013/item/954-dni-clapper-declassifies-additional-intelligence-community-documents-regarding-collection-under-section-501-of-the-foreign-intelligence-surveillance-act; Press Release, DNI Clapper Declassifies Intelligence Community Documents Regarding Collection Under Section 5 of the Foreign Intelligence Surveillance Act (FISA) (Sept. 10, 2013), https://www.dni.gov/index.php/newsroom/press-releases/press-releases-2013/item/927-dni-clapper-declassifies-intelligence-community-documents-regarding-collection-under-section-501-of-the-foreign-intelligence-surveillance-act-fisa; Press Release, DNI Clapper Directs Annual Release of Information Related to Orders Issued Under National Security Authorities (Aug. 30, 2013), https://www.dni.gov/index.php/newsroom/press-releases/press-releases-2013/item/922-dni-clapper-directs-annual-release-of-information-related-to-orders-issued-under-national-security-authorities; Press Release, DNI Clapper Declassifies and Releases Telephone Metadata Collection Documents (July 31, 2013), https://www.dni.gov/index.php/newsroom/press-releases/press-releases-2013/item/908-dni-clapper-declassifies-and-releases-telephone-metadata-collection-documents.

Starting in 2014, the DNI began annually issuing a Statistical Transparency Report Regarding the government's use of national security authorities. *See Statistical Transparency Report Regarding Use of National Security Authorities: Annual Statistics for Calendar Year 2013* (2014).

[55] *See Liberty and Security in a Changing World: Report and Recommendations of The President's Review Group on Intelligence and Communications Technologies* (Dec. 12, 2013), https://www.justsecurity.org/wp-content/uploads/2013/12/2013-12-12_rg_final_report.pdf.

[56] For a compilation of the NSA documents released in 2013 alone, see *NSA Documents Released to the Public Since June 2013*, Am. C.L. Union, https://www.aclu.org/nsa-documents-released-public-june-2013.

New York.[57] The following day, Anna Smith, a nurse and mother of two children, filed suit in Idaho,[58] as did conservative former DOJ prosecutor Larry Klayman in the D.C. District Court.[59] Within a month, the Electronic Privacy Information Center had filed a writ of mandamus before the Supreme Court, seeking an end to bulk collection.[60] Churches, telephone service providers, and even members of Congress stepped up, seeking to stop what they alleged was a gross violation of their constitutional rights.[61]

In contrast to prior eras, standing no longer proved insurmountable. What had changed was not the mere existence of programmatic (or bulk) collection, but the information available. What it revealed about the processes mattered. In an era of global communications, the *way* in which surveillance was being given effect—such as sitting on the backbone of the internet and scanning all traffic as it crossed certain points (Upstream), or by collecting information in bulk about internet service providers customers (PRISM), made it easier to demonstrate that it was highly likely that plaintiffs' communications were being monitored by the government. In case after case, the courts began to find that plaintiffs had standing.[62]

---

[57] Complaint, Am. C.L. Union v. Clapper, No. 1:13-cv-03994 (S.D.N.Y. June 11, 2013).

[58] Complaint, Smith v. Obama, No. 2:13-cv-00257 (D. Idaho June 12, 2013).

[59] Complaint, Klayman v. Obama, No. 1:13-cv-00881 (D.D.C. June 11, 2013); *see also* Jerry Markon, *Northern Idaho Mom Sues President Over Government Surveillance Program*, WASH. POST (July 25, 2013), https://www.washingtonpost.com/politics/northern-idaho-mom-sues-president-over-government-surveillance-program/2013/07/25/4994d1d4-f092-11e2-bed3-b9b6fe264871_story.html.

[60] Petition for a Writ of Mandamus and/or Prohibition, *In Re* Elec. Priv. Info. Ctr., No. 13-58 (U.S. July 8, 2013).

[61] *See, e.g.*, First Unitarian Church of L.A. v. NSA, No. 4:13-cv-03287 (N.D. Cal. July 16, 2013); Opinion and Order, *In re* Application of the Fed. Bureau of Investigation for an Order Requiring the Prod. of Tangible Things, No. BR 14-01, FISC, (Mar. 20, 2014); Ellen Nakashima, *Surveillance Court Rejected Verizon Challenge to NSA Calls Program*, WASH. POST (Apr. 25, 2014), https://www.washingtonpost.com/world/national-security/surveillance-court-rejected-verizon-challenge-to-nsa-calls-program/2014/04/25/78d430c2-ccc2-11e3-93eb-6c0037dde2ad_story.html; Complaint, Paul v. Obama, No. 1:14-cv-00262 (D.D.C. Feb. 18, 2014).

[62] *See, e.g.*, Klayman v. Obama, 957 F. Supp. 2d 1 (D.D.C. 2013) (finding standing), *vacated*, Obama v. Klayman, 800 F.3d 559 (D.C. Cir. 2015); Smith v. Obama, 24 F. Supp. 3d 1005 (D. Idaho 2014) (finding standing); Smith v. Obama, 816 F.3d 1239 (9th Cir. 2016) (finding standing); ACLU v. Clapper, 959 F. Supp. 2d 724 (S.D.N.Y. 2013) (finding standing); ACLU v. Clapper, 785 F.3d 787 (2d Cir. 2015) (finding standing); Schuchardt v. President of the United States, 839 F.3d 336 (3d Cir. 2016) (finding standing); Wikimedia v. NSA/CSS, 14 F.4th 276 (4th Cir. 2021) (finding standing).

C. SECONDARY STREAM: FREEDOM OF INFORMATION ACT LITIGATION

While the release of documents directly linked to the media coverage that started in June 2013 provided the primary grounds for the shift in standing, parallel developments in the FOIA realm provided a second, powerful force and demonstrated a judiciary increasingly reluctant to give the executive branch a free pass. The government was far from cooperative. It initially ignored formal requests for more details about these programs. When brought into court, the government fought vigorously, asserting FOIA exemptions 1(A), (3), (5), (6), and 7(A), (C), (D), and (E).[63] The government even went so far as to argue to the FISC that it should not allow a suit to proceed because the litigants could use FOIA to seek access to FISC opinions in a district court, only to then show up in a district court to argue that FOIA could *not* be used to obtain FISC opinions.[64] But non-specialized Article III courts became less and less enamored of the government's overbroad

---

[63] The most common FOIA exemptions asserted in the national security context include 5 U.S.C. § 552(b)(1) (materials authorized by Executive Order to be kept secret in the interest of national defense or foreign policy and properly classified under such order); 5 U.S.C. § 552(b)(3) (exempting material specified by statute, including, in the national security context, material specified under the National Security Act of 1947 as amended); 5 U.S.C. § 552(b)(5) (inter-agency or intra-agency memoranda or letters); 5 U.S.C. § 552(b)(6) (personnel files); and 5 U.S.C. § 552(b)(7) (records or information compiled for law enforcement purposes).

[64] *Compare* Government's Opposition Brief at 5, *In re* Motion for Release of Court Records, No. Misc. 07-01 (FISC) ("The ACLU can use FOIA [] to seek access to FISC orders and Government briefs in the Executive's possession. The FOIA process ... is the proper means for the ACLU to seek records of [the FISC's] proceedings form the Executive Branch. Moreover, FOIA's judicial remedies must be sought only in district court, not in [the FISC]."), *with* Declaration of Mark A. Bradley ¶ 7, Elec. Frontier Found. v. DOJ, 57 F. Supp. 3d 54 (D.D.C. Apr. 1, 2013) (No. 12-1441-ABJ), ECF No. 11-3, *and* Department of Justice's Statement of Material Facts Not in Genuine Dispute at 3, Elec. Frontier Found. v. DOJ, 57 F. Supp. 3d 54 (D.D.C. Apr. 1, 2013) (No. 12-1441-ABJ), ECF No. 11-2 (arguing that FISC opinions were not available via FOIA because FISC had not ordered their publication). The government's position is that because the FISC's rulings contain classified information, only the government can decide when to make them public. *See, e.g.*, U.S. Reply Brief at 6–7, *In re* Certification of Questions of L. to the FISCR, 18-01 (FISA Ct. Rev. Mar. 5, 2018) ("Movants would place in the FISC the power to make independent national security judgments and to order the release of information that the Executive Branch has properly classified pursuant to its constitutional power.... [T]his claim of unilateral FISC power to override the Executive's classification decisions is completely devoid of merit."); *id.* at 21 ("[T]he powers 'to classify and control access to information bearing on national security' are constitutionally committed to the Executive Branch, necessarily granting the Executive 'broad discretion to determine who may have access' to national security information."). The government's argument relies on a mischaracterization of *Department of Navy v. Egan*, a case in which the Court held that a particular statute did not give the Merit Systems Protection Board (an Article II tribunal) control over security clearance determinations. *See* 484 U.S. 518, 530–32 (1988).

assertions. As a result, significantly more documents ended up being released.

In 2011, for instance, when the ACLU submitted a FOIA request for records concerning the government's interpretation or use of Section 215, the government produced just three documents in response.[65] Once the ACLU filed suit, the government produced some more documents.[66] Following the *Guardian*'s publication of the FISC order directing Verizon to provide telephony metadata on a daily basis, a federal district court ordered the government to see if there was anything else that could be released.[67] This time, the government released *over 1,000 pages* of material.[68] In 2014, the ACLU narrowed its request to any fully withheld FISC opinions related to bulk collection. The government responded by providing a *Vaughn* index with eight more entries, and an unspecified number of documents.[69]

The federal district court was irate: "[B]y advancing incorrect and inconsistent arguments, the Government acted without the candor this Court expects from it."[70] It noted, "The Government's argument that it believed until June 2013 that FISC orders could not be produced in response to FOIA requests strains credulity."[71] The government's "assertion on the initial summary judgment motion . . . was incorrect. And it then failed to produce or list on the Vaughn index three documents which the Government had disclosed elsewhere."[72] The court continued,

> These inconsistencies shake this Court's confidence in the Government's submissions. The deference the Government ordinarily receives in FOIA cases is rooted largely in the courts' trust that the Government will comply with its statutory obligations. That compliance is not apparent here.[73]

Following the Snowden leaks, the government continued to stonewall. In January 2015, for example, the *New York Times* submitted a FOIA request to the NSA to obtain information about the agency's

[65] ACLU v. FBI, 59 F. Supp. 3d 584, 586 (S.D.N.Y. 2014).

[66] *Id.* at 587.

[67] *Id.*

[68] *Id.* at 588.

[69] *Id.*

[70] *Id.* at 591.

[71] *Id.*

[72] *Id.* at 591–92.

[73] *Id.* at 592.

collection activities under FISA Amendments Act Section 702 and its predecessor, the Protect America Act; bulk phone records collection activities under Section 215 of the USA PATRIOT Act; and bulk Internet metadata collected under the FISA pen register/trap and trace provisions. The NSA did not respond. On March 31, 2015, the newspaper filed suit. Over the course of litigation, despite repeated invocation of national security exceptions, the government was ultimately forced to make hundreds of pages of information public, including, *inter alia*, reports on overcollection, purging of files, and assessing how effectively management controlled collection under the programs.[74] Similar efforts by the Electronic Frontier Foundation (EFF) eventually led to the declassification and public release of *more than 1,000 pages* responsive to the plaintiff's FOIA request—but only after the court stood up to the government's overbroad assertions.[75]

Many documents over which the government asserted national security exceptions related to routine matters of law. In one case, the government had withheld Westlaw printouts, summaries of FISC legal opinions, descriptions of the scope of the FISC's jurisdiction, and discussions of FISA process improvements.[76] The federal district court noted that DOJ had failed to address why such materials could not be released and required further declarations tailored to the withholdings in question. The government, accordingly, narrowed its claims—a pattern repeated throughout the case as challenges to withheld material arose.[77]

In addition to releasing more documents, FOIA suits also resulted in narrowing redactions. In 2012, for example, EFF submitted a FOIA request to DOJ's National Security Division (NSD).[78] EFF requested any written opinion or order where the FISC had held that collection was unreasonable under the Fourth Amendment or implementation had circumvented the spirit of the law, and any briefing to the Senate Select Committee on Intelligence (SSCI) and the House Permanent

---

[74] N.Y. Times v. NSA, 205 F. Supp. 3d 374 (S.D.N.Y. 2016).

[75] Elec. Frontier Found. v. DOJ, No.: 4:11-cv-05221-YGR, 2014 WL 3945646 (N.D. Cal. Aug. 11, 2014). Numerous cases followed suit. *See, e.g.*, Am. C.L. Union v. U.S. Dep't of Just., 210 F. Supp. 3d 467, 471 (S.D.N.Y. 2016). After the DOJ conducted a new FOIA search following court orders in *American Civil Liberties Union v. United States Department of Justice*, 90 F. Supp. 3d 201 (2015), approximately eighty responsive documents were located.

[76] *See* Elec. Priv. Info. Ctr. v. Dep't of Just., 296 F. Supp. 3d 109, 115 (D.D.C. Nov. 7, 2017).

[77] *Id.* at 116.

[78] Elec. Frontier Found. v. Dep't of Just., 57 F. Supp. 3d 54, 56 (D.D.C. 2014).

Select Committee on Intelligence. The government identified five responsive documents, withholding two in full and redacting the remaining three.[79] Following the 2013 public outcry, the government declassified and redacted all of them.[80] The federal district court, conducting *in camera* review, challenged a dozen redactions, questioning whether they were justified on national security grounds.[81] The government subsequently decided to lift many of them—apparently without any risk to national security.[82]

Fazaga itself arose in part because of FOIA litigation.[83] In early 2006, organizations and individuals in the Southern California Muslim community made a FOIA request to the FBI, seeking information related to the surveillance operation. Almost a year later, the FBI notified some of the plaintiffs that it did not have any responsive documents and provided to others a total of four heavily redacted pages.[84] Plaintiffs responded by filing suit in the district court to challenge the adequacy of the search. The government subsequently produced *over one hundred* pages of heavily redacted documents.[85] It followed this with a motion for summary judgment, attesting that its invocation of the national security exemptions was both necessary and proper. Plaintiffs objected and requested that the district court examine the redacted documents. Upon receiving an order from the court directing it to provide any documents redacted or withheld, the FBI acknowledged to the court that there were *additional* responsive documents, which it has not disclosed either to the court or to the plaintiffs.[86] The district court, and the Ninth Circuit on appeal, expressed significant concern that the government had misled both the plaintiffs and the court.[87] The complaint in *Fazaga* went on to cite to some of the materials obtained over the course of the FOIA litigation.

---

[79] *Id.* at 57.

[80] *Id.* at 57–58.

[81] *Id.* at 58.

[82] *Id.* at 59.

[83] *See* Islamic Shura Council of S. Cal. v. Fed. Bureau of Investigation, 635 F.3d 1160 (9th Cir. 2011).

[84] *Id.* at 1162.

[85] *Id.* at 1163.

[86] *Id.* at 1164.

[87] *Id.* The district court sanctioned the government under Rule 11(c) for deceiving the court. *See* Islamic Shura Council of S. Cal. v. Fed. Bureau of Investigation, 278 F.R.D. 538, 539 (C.D. Cal. 2011). The Ninth Circuit subsequently reversed the ruling on technical grounds. See Islamic Shura Council of S. Cal. v. Fed. Bureau of Investigation, No. 12-55305, 2013 WL 3992123 (9th Cir. 2013) (per curiam).

D. RULE OF LAW CONSIDERATIONS

The material that entered the public domain from each of these streams produced important information about the scope of the surveillance programs underway and how the law was being interpreted.[88] It also revealed that the government had repeatedly acted beyond its statutory authority and in contravention to judicial direction, raising concerns about the impact on citizens' constitutional rights.[89]

From the beginning of the bulk telephony program under the FISA's business records provision (Section 215), for example, FISC opinions showed that the NSA routinely ran queries on U.S. persons (USPs) using terms that did not meet the judicially-required standard of reasonable, articulable suspicion.[90] The FISC in 2009 concluded that it had been "so frequently and systematically violated that it can fairly be said that this critical element of the overall ... regime has never functioned effectively."[91] That same year, the FISC discovered that the government had been picking and choosing which of its misdoings it reported, omitting, for example, failures to de-task accounts even after the NSA knew that the targets were on U.S. soil (thereby continuing to collect their communications in violation of the statute).[92]

In another case, the FISC called attention to the government's "chronic tendency to mis-describe the actual scope of NSA [Title I] acquisitions in its submissions to this Court," noting,

> These inaccuracies have previously contributed to unauthorized electronic surveillance and other forms of statutory and constitutional deficiency. It is

---

[88] *See, e.g.*, Opinion and Order, *[REDACTED]*, No. [REDACTED], GID.C.00004 (FISA Ct. [REDACTED]) (Baker, J.), https://www.dni.gov/files/documents/icotr/EFF-FOIA-Jan-31-Doc-2.pdf. [Local] (evaluating Fourth Amendment implications and holding that the FBI marking procedures violated the statutory minimization requirements).

[89] *See, e.g.*, Opinion and Order Regarding Fruits of Unauthorized Electronic Surveillance, *[REDACTED]*, Nos. [REDACTED], GID.C.00059 (FISA Ct. Dec. 10, 2010) (Scullin, Jr., J.), GID.C.00059 https://www.dni.gov/files/documents/icotr/EFF-FOIA-Jan-31-Doc-11.pdf. [Local]; Opinion and Order Requiring Destruction of Information Obtained by Unauthorized Electronic Surveillance, *[REDACTED]*, Nos. [REDACTED], GID.C.00067 (FISA Ct. May 13, 2011) (Scullin Jr., J.), https://www.dni.gov/files/documents/icotr/EFF-FOIA-Jan-31-Doc-10.pdf. [Local]

[90] *In re* Prod. of Tangible Things from [REDACTED], GID.C.00036 at 11, 2009 WL 9150913, at *5.

[91] *Id.*

[92] *Id.*

> evident that the government needs every incentive to provide accurate and complete information to the Court about NSA operations, whenever such information is material to the case.[93]

The executive branch made similar, inaccurate representations about the post-tasking review process.[94]

In 2011, the FISC discovered that the NSA had misled it about Section 702 Upstream collection, acquiring (in violation of the statute) tens of thousands of domestic USP communications.[95] In response, the court forbade the NSA from using USP identifiers to query upstream data. Six years later, the FISC discovered that, nevertheless, "NSA analysts had been conducting such queries in violation of that prohibition [and] with much greater frequency than had previously been disclosed."[96] The court underscored its concern about the government's failure to disclose noncompliance.[97] The institutional "lack of candor" presented "a very serious Fourth Amendment issue."[98]

In December 2019, the FISC similarly noted widespread violations of the FBI's querying standards, including agents undertaking queries

---

[93] *See* Opinion and Order at 10–11, [REDACTED], No. [REDACTED], GID.C.00254, 13014 (FISA Ct. [REDACTED]) (Hogan, J.) (NSA's acquisition of [REDACTED] constituted unauthorized electronic surveillance because it failed to comply with 50 U.S.C.A. § 1804(a)(2), (a)(3)(B) (West)). Citations have been redacted.

[94] *See* Memorandum Opinion and Order at 20–21, *[REDACTED]*, No. [REDACTED], GID.C.00062 (FISA Ct. 2010) (McLaughlin, J.), https://repository.library.georgetown.edu/handle/10822/1052735.

[95] *See* Memorandum Opinion and Order, *[REDACTED]*, No. [REDACTED], GID.C.00073, 2011 WL 10945618 (FISA Ct. Oct. 3, 2011) (Bates, J.), https://www.dni.gov/files/documents/0716/October-2011-Bates-Opinion-and Order-20140716.pdf. *see also* Memorandum Opinion, *[REDACTED]*, No. PR/TT [REDACTED], GID.C.00092 (FISA Ct. [REDACTED]) (Bates, J.), https://www.dni.gov/files/documents/1118/CLEANEDPRTT 2.pdf (stating "NSA exceeded the scope of authorized acquisition continuously during the more than [REDACTED] years of acquisition under these orders").

[96] Memorandum Opinion and Order at 19, *[REDACTED]*, No. [REDACTED], GID.C.00130 (FISA Ct. Apr. 26, 2017) (Collyer, J.), https://repository.library.georgetown.edu/bitstream/handle/10822/1052702/gid_c_00130.pdf?sequence=1&isAllowed=y; *see also* Exhibit A: Procedures Used by the National Security Agency for Targeting Non-United States Persons Reasonably Believed to be Located Outside the United States to Acquire Foreign Intelligence Information Pursuant to Section 702 of the Foreign Intelligence Surveillance Act of 1978, as Amended at 2, *[REDACTED]* (FISA Ct. Mar. 30, 2017) (No. [REDACTED]), https://repository.library.georgetown.edu/handle/10822/1053027; Exhibit B: Minimization Procedures Used by the National Security Agency in Connection with Acquisitions of Foreign Intelligence Information Pursuant to Section 702 of the Foreign Intelligence Surveillance Act of 1978, as Amended, § 3(b)(4)b, at 4, *[REDACTED]* (FISA Ct. Mar 30, 2017) (No. [REDACTED]), https://repository.library.georgetown.edu/handle/10822/1053259.

[97] Memorandum Opinion and Order, *supra* note 96, at 4.

[98] *Id.* at 19 (quotations omitted).

of unminimized Section 702 information to vet sources and candidates; investigate college students participating in a "Collegiate Academy"; conduct background checks on individuals who had visited an FBI office; and dig up information on nearly 16,000 persons, of which NSD later assessed only seven persons satisfied the querying standard.[99] The FBI, moreover, had *never* applied for an order under FISA Section 702(f)(2), which is *required* before querying unminimized contents using a USP query term unrelated to national security or the effort to find and extract foreign intelligence.[100] Numerous other problems came to light.[101]

The government's failure to follow the letter of the law extended well beyond those portions added by the 2008 FISA Amendments Act. Applications under Titles I and III (germane to *Fazaga*) also have been called into question. The DOJ Inspector General's (IG) investigation of Crossfire Hurricane, an operation initiated in 2016 to ascertain whether individuals involved with Donald Trump's presidential campaign coordinated with Russian efforts to interfere in the election, proves illustrative.[102] Following a twelve-month inquiry, IG Michael Horowitz issued an exhaustive, nearly 500-page report based on upwards of 170 interviews with more than 100 witnesses, as well as more than one million documents held by the DOJ and FBI. He found significant discrepancies between law, policy, and FBI practice.[103] FISA applications for electronic surveillance left out information that cut against the FBI or was inconsistent with what they were

---

[99] Memorandum Opinion and Order at 65–67, *[REDACTED]*, No. [REDACTED], GID.C.00282 (FISA Ct. Dec. 6, 2019) (Boasberg, J.), https://www.intelligence.gov/assets/documents/702%20Documents/declassified/2019_702_Cert_FISC_Opinion_06Dec19_OCR.pdf.

[100] *Id.* at 69–71.

[101] The NSA Office of the Inspector General, for instance, has repeatedly noted that the NSA is unable to ensure that data is queried in compliance with Section 702 targeting and minimization procedures. The NSA was supposed to address the concern by December 2017. As of March 2022, however, it still had not been done. *See* Nat'l Sec. Agency, Off. Inspector Gen., Semiannual Report to Congress, 1 October 2021 – 31 March 2022 34 (2022), https://oig.nsa.gov/Portals/71/Images/NSAOIGMAR2022.pdf?ver=jbq2rCrJ00HJ9qDXGHqHLw%3d%3d&timestamp=1657810395907.

[102] Off. of the Inspector Gen., Dep't of Just., Review of Four FISA Applications and Other Aspects of the FBI's Crossfire Hurricane Investigation i (revised Dec. 20, 2019) [hereinafter Crossfire Hurricane Report], https://repository.library.georgetown.edu/handle/10822/1058716. During the campaign, a foreign government informed the Obama Administration that Russia had reached out to the Trump team to offer to release information that would be damaging to the Democratic Party candidate. *Id.* at ii.

[103] *See id.* at i, ii–xviii.

telling the court—information directly relevant to the probable cause determination.[104] They contained information that had not been corroborated, misstatements, inaccurate data, and other errors, and they omitted important information.[105] Horowitz expressed his deep concern,

> That so many basic and fundamental errors were made by three separate, hand-picked teams on one of the most sensitive FBI investigations that was briefed to the highest levels within the FBI, and that FBI officials expected would eventually be subjected to close scrutiny, raised significant questions regarding the FBI chain of command's management and supervision of the FISA process.[106]

The IG further discovered that an FBI lawyer had falsified an email to tip the scales in favor of the FISC granting the application.[107]

Horowitz was sufficiently disquieted that he launched a second audit focused on FBI compliance with the Woods procedures—guidelines the FBI provided to the FISC and indicated that it would follow to ensure that FISA applications contain reliable information.[108] He randomly selected 29 applications targeting U.S. persons

---

[104] *Id.*

[105] *Id.* at viii–ix, xi–xii.

[106] *Id.* at xiv.

[107] *Id.* at ix; *see also* Matt Zapotosky, *Ex-FBI Lawyer Avoids Prison After Admitting He Doctored Email in Investigation of Trump's 2016 Campaign*, WASH. POST (Jan. 21, 2021), https://www.washingtonpost.com/national-security/kevin-clinesmith-fbi-john-durham/2021/01/28/b06e061c-618e-11eb-afbe-9a11a127d146_story.html.

[108] In March 2000, the FISC discovered that in four or five separate cases, DOJ had been disseminating FISA information to the FBI and U.S. Attorney's Office without the required authorizations of the Court. *In re* All Matters Submitted to Foreign Intel. Surveillance Ct., 218 F. Supp. 2d 611, 620, GID.C.00002 (FISA Ct. 2002) (Lamberth, J.), *rev'd*, *In re* Sealed Case, 310 F.3d 717, GID.CA.00001 (FISA Ct. Rev. 2002) (per curiam). A few months later, the government confessed to errors in seventy-five separate FISA applications relating to major terrorist attacks, including:

a. an erroneous statement in the FBI Director's FISA certification that the target of the FISA was not under criminal investigation;
b. erroneous statements in the FISA affidavits of FBI agents concerning the separation of the overlapping intelligence and criminal investigations, and the unauthorized sharing of FISA information with FBI criminal investigators and assistant U.S. attorneys;
c. omissions of material facts from FBI FISA affidavits relating to a prior relationship between the FBI and a FISA target, and the interview of a FISA target by an assistant U.S. attorney.

Id. at 620. These actions prompted the adoption of the Woods procedures, to ensure that the Court could rely on the information in future applications.

from 2014 to 2019.[109] He discovered that *every application* contained errors.[110] Although DOJ had indicated to the FISC that it was following the Woods procedures, it was not in fact doing so. In many cases, there were no files at all.[111] In cases where there were files, facts lacked support or corroboration, or were inconsistent with claims being made to the FISC. On average, each application had approximately 20 errors, with up to 65 in just one.[112] Horowitz continued his examination, in September 2021 issuing the final results in which he identified 400 instances of non-compliance in the 29 applications, as well as an additional 179 applications for which the original Woods File was missing, destroyed, or incomplete.[113]

This is the context within which *Fazaga* arose: information in the public domain indicates that the government has repeatedly acted outside its constitutional and statutory limits—including in regard to the authorities at issue in the case. Despite repeated efforts by the FISC to hold the government to account, it has proven difficult to do so. With standing met, litigants like Fazaga are coming forward to assert their constitutional rights.

III. The *Fazaga* Claims

In 2011, Yassir Fazaga, an imam at the Orange County Islamic Foundation (OCIF), and two ICOI community members, Ali Malik, and Yasser Abdelrahim, brought a class action suit against the FBI, asserting eleven causes of action related to unconstitutional searches as well as unlawful discrimination on the basis of, or burdens on, or

---

[109] Management Advisory Memorandum from Michael E. Horowitz, Inspector Gen., to Christopher Wray, Dir., FBI 2 (Mar. 30, 2020), https://repository.library.georgetown.edu/handle/10822/1058475 (discussing the Audit of the Federal Bureau of Investigation's Execution of its Woods Procedures for Application Filed with the Foreign Intelligence Surveillance Court Relating to U.S. Persons).

[110] *See id.* at 7–8.

[111] *Id.* at 7.

[112] *Id.*

[113] U.S. Dep't. of Justice Office of the Inspector General, Audit of the Federal Bureau of Investigation's Execution of Its Woods Procedures for Applications Filed with the Foreign Intelligence Surveillance Court Relating to U.S. Persons 7 (Sept. 2021), https://oig.justice.gov/sites/default/files/reports/21-129.pdf; *see also* U.S. Dep't. of Just. Off. of the Inspector Gen. A Message from the Inspector General: Audit of the FBI's Execution of its Woods Procedures for Applications Filed with the Foreign Intelligence Surveillance Court Relating to U.S. Persons (Sept. 2021), https://oig.justice.gov/sites/default/files/2021-09/09-30-2021.pdf.

abridgement of the right to religion—implicating the First Amendment, the equal protection guarantee of the Due Process Clause of the Fifth Amendment, the Privacy Act, the Religious Freedom Restoration Act, FISA, and the Federal Tort Claims Act.[114] Their concerns centered on Operation Flex, the surveillance program apparently launched in 2006 to collect information about the Orange County Muslim community. As aforementioned, the FBI had hired Monteilh as an informant. It later directed him to begin asking pointed questions in the Islamic community about jihad.[115] He represented to members of ICOI that he felt it his duty as a Muslim to take violent action.[116] Several members became alarmed and reported him to the mosque's leadership. One called the FBI directly and told others to call the Irvine Police Department, which they did, obtaining a restraining order against Monteilh.[117] In 2009, the government brought a criminal prosecution for naturalization fraud against one of the members who had called the police for the restraining order.[118] An FBI Special Agent's testimony during the bail hearing revealed an informant at ICOI, whom the plaintiffs identified as Monteilh.

As soon as the suit commenced, the Government asserted state secrets privilege in regard to three categories of information and moved to dismiss the discrimination claims on state secrets grounds.[119] The district court dismissed all but one of the plaintiffs' claims—including their Fourth Amendment assertion—based on state secrets.[120] On appeal, the Ninth Circuit reversed on the grounds that "the district court should have reviewed any state secrets evidence necessary for a determination of whether the alleged surveillance was unlawful

[114] First Amended Complaint, Fazaga v. FBI, No. 8:11-cv-00301-CJC-VBK (C.D. Cal. Sept. 1, 2011), https://www.aclu.org/legal-document/fazaga-v-fbi-first-amended-complaint.

[115] Declaration of Craig F. Monteilh, *supra* note 3, ¶ 73. Monteilh claims that the FBI provided him with a letter granting him immunity to allow him to engage in jihadist rhetoric and criminal activity. *Id.* ¶ 72; *see also* Fazaga v. FBI, 965 F.3d 1015, 1027 (9th Cir. 2020).

[116] Declaration of Craig F. Monteilh, *supra* note 3, ¶ 73; *see also Fazaga*, 965 F.3d at 1027.

[117] Declaration of Craig F. Monteilh, *supra* note 3, ¶ 73; *see also Fazaga*, 965 F.3d at 1028.

[118] Niazi, who had lived in the United States since 1998 and obtained citizenship five years previously, was related by marriage to Amin al-Haq, a member of a designated terrorist organization. He failed to disclose the relationship during his application, giving rise to charges of perjury, naturalization fraud, misuse of a passport obtained by fraud, and making false statements to a federal agency. Teresa Watanabe & Scott Glover, *Man Says He Was FBI Informant*, L.A. Times (Feb. 26, 2009), https://www.latimes.com/archives/la-xpm-2009-feb-26-me-informant26-story.html.

[119] *Fazaga*, 965 F.3d at 1024–25.

[120] *Id.* at 1025.

following the secrecy-protective procedure set forth in FISA."[121] For the court, FISA had supplanted state secrets.

The statute creates a private right of action for an individual subject to electronic surveillance in violation of FISA's procedures. For civil liability,

> An aggrieved person, other than a foreign power or agent of a foreign power . . . who has been subjected to an electronic surveillance or about whom information obtained by electronic surveillance of such person has been disclosed or used in violation of section 1809 of this title shall have a cause of action against any person who committed such violation.[122]

In the FISA world, "foreign power" (FP) and "agent of a foreign power" (AFP) are terms of art: the statute defines them to ensure that targets of surveillance are affiliated with foreign governments or are members of international terrorist organizations.[123] The government must demonstrate probable cause that the target is an FP/AFP *and* that the target is using the facilities to be placed under surveillance.[124] The aim is to ensure that individuals who do not meet the probable cause standard—for instance, the entire Orange County Islamic community—are not put under surveillance. The statute *prohibits* electronic surveillance outside of FISA or the ordinary Title III warrant procedure.[125]

Monteilh's recording of conversations to which he was not party fell within FISA's fourth definition of electronic surveillance.[126] The plaintiffs, though, had both an objective and subjective reasonable expectation of privacy in the inner sanctums of the mosque. The Ninth Circuit explained,

> [T]he prayer hall "is [a] sacred space where particular rules and expectations apply. Shoes are prohibited, one must be in a state of ablution, discussing worldly matters is discouraged, and the moral standards and

---

[121] *Id.*

[122] 50 U.S.C. § 1810.

[123] *See id.* § 1801(a), (b).

[124] *Id.* § 1805(2)(A), (B).

[125] *Id.* § 1809. The Ninth Circuit held that the plaintiffs constituted aggrieved persons within the meaning of the statute. *Fazaga*, 965 F.3d at 1053.

[126] *See* 50 U.S.C. § 1801(f) (defining electronic surveillance as "the installation or use of an electronic, mechanical, or other surveillance device in the United States for monitoring to acquire information, other than from a wire or radio communication, under circumstances in which a person has a reasonable expectation of privacy *and* a warrant would be required for law enforcement purposes").

> codes of conduct are at their strongest." Notably, "[g]ossiping, eavesdropping, or talebearing (*namima*—revealing anything where disclosure is resented) is forbidden." And ICOI, which Malik and AbdelRahim attended, specifically prohibited audio and video recording in the mosque without permission.[127]

Fazaga's office and AbdelRahim's apartment and car similarly raised free-standing Fourth Amendment concerns. The case therefore fell within FISA's remit.

The Ninth Circuit looked to two district courts which had previously held that the statute had displaced common law rules like state secrets on matters within FISA's purview.[128] Under FISA Section 1806(f),

> *Whenever* a court or other authority is notified [that the Government intends to enter into evidence or otherwise use or disclose in any trial, hearing, or other proceeding in or before any court, department, officer, agency, regulatory body, or other authority of the United States, against an aggrieved person, any information obtained or derived from an electronic surveillance] . . . or *whenever* any motion or request is made by an aggrieved person pursuant to any other statute or rule of the United States . . . before any court . . . to discover or obtain applications or orders or other materials relating to electronic surveillance or to discover, obtain, or suppress evidence or information obtained or derived from electronic surveillance under this Act, the United States district court . . . *shall, notwithstanding any other law* . . . , review in camera and ex parte the application, order, and such other materials relating to the surveillance as may be necessary to determine whether the surveillance of the aggrieved person was lawfully authorized and conducted.[129]

The language ("notwithstanding any other law," several uses of "whenever," and the command that courts "shall" use the procedures) confirmed "Congress's intent to make the *in camera* and *ex parte* procedure the exclusive procedure for evaluating evidence that threatens national security in the context of electronic surveillance."[130] Those clauses overrode "on the one hand, the usual procedural rules precluding such severe compromises of the adversary

---

[127] *Fazaga*, 965 F.3d at 1034.

[128] *See also* Jewel v. NSA, 965 F. Supp. 2d 1090, 1105–06 (N.D. Cal. 2013); *In re* NSA Telecomms. Recs. Litig., 564 F. Supp. 2d 1109, 1117–24 (N.D. Cal. 2008).

[129] 50 U.S.C. § 1806(f) (emphasis added).

[130] *Fazaga*, 965 F.3d at 1045.

process and, on the other, the state secrets evidentiary dismissal option."[131]

The government disagreed. Nothing in the text of FISA expressly mentions "state secrets."[132] Nor does anything in FISA's legislative history.[133] To overturn such a settled practice, the government argued, Congress needed to give a clear statement.[134] Beyond this, it suggested, Section 1806(f) procedures only apply where the *government* initiates the legal action—not in the case of affirmative challenge to the legality of electronic surveillance.[135]

The Ninth Circuit saw it rather differently, zeroing in on two circumstances in which FISA contemplates *in camera, ex parte* procedures: (1) when the government gives notice of its intent to "use or disclose ... any information obtained or derived from an electronic surveillance"; and (2) where "any motion or request is made by an aggrieved person ... to discover, obtain, or suppress evidence or information obtained or derived from electronic surveillance."[136] In regard to the former, the complaint alleged that information was

---

[131] *Id.* The court further cited to legislative history, noting, "It is to be emphasized that, although a number of different procedures might be used to attack the legality of the surveillance, it is the procedures set out in subsections (f) and (g) 'notwithstanding any other law' that must be used to resolve the question." *Id.* at 1045–46 (quoting H.R. Rep. No. 95-1283, pt. 1, at 91 (1978)). The court made a parallel argument from practice, noting that the same concerns underlay both the FISA procedures and the state secrets privilege—as recognized by multiple district courts in support of the proposition that FISA supplanted state secrets as to ELSUR. *Id.* (citing Jewel v. NSA, 965 F. Supp. 2d at 1106; and quoting *In re* NSA Telecomms. Recs. Litig., 564 F. Supp. 2d at 1119). It also advanced two historical arguments: first, that Congress intended FISA to serve as a comprehensive approach to regulating domestic foreign intelligence collection, by establishing a balancing role for all three branches, repleted with oversight mechanisms, rules, a special court, and a civil enforcement mechanism. Second, the Court suggested that the legislative history confirmed "Congress's intent to displace the remedy of dismissal for the common law state secrets privilege." *Id.* at 1046–47. FISA embodied Congress's response to the revelations of the Church Committee, which had called for "fundamental reform" to "[m]ake clear to the Executive branch that [Congress] will not condone, and does not accept, any theory of inherent or implied authority to violate the Constitution." *Id.* at 1047. FISA was Congress's primary instrument for doing so.

[132] *See* Brief for the Petitioners at 19, Fed. Bureau of Investigation v. Fazaga, 142 S. Ct. 1051 (2022) (No. 20-828).

[133] *Id.*

[134] *Id.* at 20.

[135] See *id.* at 18 ("The government invokes the state-secrets privilege for the same reason that any party asserts any evidentiary privilege: to prevent the introduction or disclosure of the privileged information, not to facilitate its use. Excluding evidence—not using that evidence—is how a litigant claiming any privilege vindicates the interest protected by that privilege.").

[136] 50 U.S.C. §§ 1806(c), (f).

being collected pursuant to FISA Titles I/ III. Although the government refused to confirm or deny it, the fact that the Attorney General asserted state secrets in relation to "any information obtained during the course of" Operation Flex, the "results of the investigation," and "any results derived from" the "sources and methods" used in Operation Flex conveyed notice. The government objected, arguing, "By the panel's reasoning, a litigant who asserts the attorney-client privilege signals her intent to use or disclose private communications with counsel. . . . But, of course, [he does] nothing of the sort."[137] Seeking dismissal did not amount to "a declaration of the govenrment's 'inten[t] to enter into evidence or otherwise use or disclose' any privileged information 'in any trial, hearing, or other proceeding.'"[138] In regard to the latter, the court reasoned that the plaintiffs' effort to "obtain" information gathered or derived from electronic surveillance so that it could be either destroyed or returned, amounted to a "motion or request" to obtain it.[139]

The Supreme Court, unconvinced by the Ninth Circuit's approach, held that in civil litigation, Section 1806(f) does not displace the state secrets privilege.[140] Writing for a unanimous Court, Justice Alito stated that the provision governs circumstances in which the *government* seeks to employ information obtained from electronic surveillance.[141] In reaching this conclusion, the Court relied on two arguments. First, the statute does not refer to "state secrets privilege," which suggests that Congress did not seek to cabin the privilege.[142] Such a limitation, regardless of whether state secrets represented a common law privilege or a constitutional power assigned to the executive branch, could only

[137] Brief for the Petitioners, *supra* note 132, at 25.

[138] *Id.* at 26.

[139] According to the government, as aforementioned, a substantive request for relief on the merits was not the kind of procedural motion that would trigger FISA's procedures. It drew a distinction between a prayer for relief and a "motion," "[a]nd although it might be colloquially described as a 'request,' it is nothing like a motion to suppress or comparable procedural motion at which § 1806(f) is aimed." *Id.* at 18. Instead, the statutory language should be read in conjunction with provisions related to the government giving notice of the intent to use or disclose material, or when the aggrieved person attempted to suppress such material—nor could any litigant try to use a new statute to avoid the statutory language. *Id.* at 29–31. Fazaga responded to this argument by noting that there is nothing in the text that expressly limits motions or requests to "procedural motions"—nor is it limited to adjudication on the merits. *See* Brief for the Respondents at 22, *Fazaga*, 142 S. Ct. 1051 (No. 20-828).

[140] *Fazaga*, 142 S. Ct. 1051.

[141] *Id.* at 1060.

[142] *Id.* at 1060–61.

follow from clear statutory language.[143] While the Court was right that the statute nowhere explicitly replaced "state secrets," the argument sidestepped the fact that at the time of FISA's passage, the term was neither the only nor the most common one employed to describe the evidentiary rule.[144]

Second, the Court determined that the language of Section 1806(f) is not incompatible with the state secrets privilege.[145] They operate in different ways and apply to different circumstances. Justice Alito explained:

> Section 1806(f) is most likely to come into play when the Government seeks to use FISA evidence in a judicial or administrative proceeding, and the Government will obviously not invoke the state secrets privilege to block disclosure of information that it wishes to use. Section 1806(f) is much more likely to be invoked in cases of this sort than in cases in which an aggrieved person takes the lead and seeks to obtain or disclose FISA information for a simple reason: individuals affected by FISA surveillance are very often unaware of the surveillance unless it is revealed by the Government.[146]

Thus, even where an aggrieved party brings suit, the statute and the privilege entail different inquiries, provide for different forms of relief, and adopt different procedures.[147]

The Court did not weigh in on the substantive state secrets claims advanced by the government, remanding the case for further adjudication. The questions presented are critical and have already returned to the Court in the latest state secrets petition for certiorari—the ninth such petition in the past eighteen months.[148] The problem

---

[143] *Id.*

[144] *See, e.g.*, United States v. Nixon, 418 U.S. 683, 701, 711 (1974) (making no reference to "state secrets" and instead describing *Reynolds* as protecting "national security secrets" and "military matters which, in the interest of national security, should not be divulged").

[145] *Fazaga*, 142 S. Ct. at 1061.

[146] *Id.*

[147] *Id.* Under FISA, the question is whether the information was lawfully obtained—an inquiry irrelevant under state secrets, which instead focuses on whether, if made public, the information will damage national security. For FISA, if the court determines that the information was lawfully obtained, it does not have the ability to award relief. In contrast, under state secrets, if a court determines disclosure would not affect national security, it can be disclosed. For Section 1806(f), the court has to award relief if collection violated the law. For state secrets, lawfulness has nothing to do with the relief. Finally, under *Totten*, a case can be dismissed under state secrets—which is not an option under Section 1806(f). *Id.* at 1061–62.

[148] In *Wikimedia Foundation v. National Security Agency*, the operator of one of the busiest websites (more than one trillion international Internet communications annually), is challenging the constitutionality of the government's use of Section 702. Petition for a Writ of Certiorari,

is not going to go away for the simple reason that the manner in which the government has been wielding the privilege in *Fazaga*, and dozens of parallel suits, marks a sharp departure from how it has historically operated and raises serious questions about the future of certain constitutional rights.

IV. Reinventing State Secrets

The government's current interpretation of the state secrets privilege, which is rooted in the early twenty-first century torture and rendition cases, departs in four important respects from how the privilege, for centuries, has operated. First, it collapses the distinction between *Reynolds* and *Totten* to transform an evidentiary rule into a justiciability standard. In *Fazaga* and numerous other cases, the government has adopted a "very subject matter" analysis, sidelining the *ex ante* contractual argument undergirding the *Totten* bar. Second, the government now asserts the privilege early in suits (at the pleadings stage) to request dismissal, instead of employing it to ensure that

supra note 23, at 8. Like *Fazaga*, the suit is based on information in the public domain, much of which the government itself has provided. The DNI reported in April 2022 that the government targeted 232,432 individuals and groups outside the United States over the previous year. Off. of the Dir. of Nat'l Intel., *Annual Statistical Transparency Report* 17 (Apr. 2022). ODNI, PCLOB, the FISC/R, government officials, and FOIA lawsuits have provided details about how Upstream collection occurs on these targets. The NSA routinely scans international Internet communications as it traverses high-speed circuits operated by major communication service providers. *See* Priv. & C.L. Oversight Bd., *Report on the Surveillance Program Operated Pursuant to Section 702 of the Foreign Intelligence Surveillance Act* 36 (2014). Unable to ascertain in advance which communications are unique to the targets, the NSA scans and copies all packets crossing certain international sites, subsequently reassembling them so that they can be examined to determine if they contain certain selectors. Data subsequently can be retained, queried, and disseminated. Because "Wikimedia's trillions of communications cross every international Internet link carrying public Internet traffic into and out of the United States," and "NSA conducts Upstream surveillance on at least one 'international Internet link,'" at least some of Wikimedia's communications are likely being monitored. Petition for a Writ of Certiorari, *supra* note 23, at 11–12. Although the district court initially dismissed the case on the grounds that Wikimedia's allegations were too speculative to establish Article III standing, the court of appeals concluded that the facts put forward were "sufficient to make plausible the conclusion that the NSA is intercepting, copying, and reviewing at least some of Wikimedia's communications." Wikimedia Found. v. NSA, 857 F.3d 193, 211 (4th Cir. 2017); *see also* Wikimedia Found. v. NSA, 14 F.4th 276, 279, 292–94 (4th Cir. 2021) (again finding standing). On remand, the government indicated that it would continue to challenge Wikimedia's standing and argued that discovery should be bifurcated to allow the standing question first to be resolved. The court, however, directed the parties to undertake a limited five-month period of jurisdictional discovery prior to ruling on the merits. Order, Wikimedia Found. v. NSA, 1:15-cv-662 (D. Md. Oct. 3, 2017). The government responded to several requests from the plaintiffs by arguing state secrets privilege. Wikimedia Found. v. NSA, 335 F. Supp. 3d 772 (2018). As in *Fazaga*, the government is requesting that the suit be dismissed altogether on the grounds that the surveillance programs in question are classified and that releasing information covering a range of topics would undermine U.S. national security.

particular evidence is excluded during discovery or trial. Third, it adopts a topical approach, asserting state secrets over broad categories which include information that poses no risk to U.S. national security as well as information already in the public domain. Fourth, the government is beginning to claim as a constitutional power a privilege that has always been a common law rule. All four elements, at issue in *Fazaga*, mark a host of similarly-situated cases making their way through the courts.

A. CREATION OF THE "VERY SUBJECT MATTER" ANALYSIS

State secrets did not, in 1953, spring fully-armed from Zeus's forehead. From the Founding, British and American common law recognized the privilege as a common law evidentiary rule.[149] Where exclusion of evidence prevented a plaintiff from making a prima facie case, the court could order dismissal. The sole exception, established in *Totten*, related to secret contractual relationships between the government and private actors, where both parties had ex ante notice that litigation in open court likely would not be available should a dispute arise.

Post-9/11, with standing met in regard to the Terrorism Surveillance Program (TSP), the government immediately fell back on state secrets to avoid litigation, collapsing the *Reynolds* and *Totten* distinction by re-casting classified contract cases in terms of a "very subject matter" approach. The Ninth Circuit, in a case dealing with the TSP, balked at the assertion—not least because the government itself had already admitted to conducting warrantless surveillance outside of FISA. Undeterred, the government raised the same argument in the context of torture, coercive interrogation, and rendition. As in the surveillance cases, there was enough information in the public domain for standing to be met, prompting the government to re-cast state secrets as a justiciability standard. In *El-Masri* and then in *Jeppesen*, the argument gained purchase. Precedent set, the re-crafted understanding of state secrets worked its way back into surveillance cases like *Fazaga*, heralding potentially devastating effects for the future of citizens' First, Fourth, and Fifth Amendment rights and the ability of the People to hold the government to account.

---

[149] *See, e.g.*, Gen. Dynamics Corp. v. United States, 563 U.S. 478, 485 (2011) ("Reynolds . . . decided a purely evidentiary dispute by applying evidentiary rules."); El-Masri v. United States, 479 F.3d 296, 303–04 (4th Cir. 2007) (considering SSP an evidentiary rule "bas[ed] in the common law of evidence").

1. *State Secrets as an Evidentiary Rule.* English law has long treated state secrets as a common law evidentiary rule.[150] On this side of the Atlantic, starting with Aaron Burr's trial in 1807 and continuing through *Reynolds*, courts routinely approached it in a similar manner.[151] Thus, in the 1912 case *Firth Sterling Steel Co. v. Bethlehem*

---

[150] In *Rex v. Watson*, the court considered whether the criminal defendant could elicit testimony as to the accuracy of a map of the Tower of London. Rex v. Watson, 2 Stark. 116, 148–49 (1817). In the wrong hands, such information could prove problematic. The judge, consistent with the privilege, did not halt the proceedings. Instead, he directed that testimony be limited to establishing whether the plan reflected the interior of the Tower and whether similar documents could be purchased. *Id.* In *Wyatt v. Gore*, the court excluded certain communications between the Lieutenant Governor of Upper Canada and attorney-general, as such communications should be treated as "confidential: no office of this kind could be executed with safety . . . were suffered to be disclosed." Wyatt v. Gore, Holt 299 (N.P. 1816). The case, nevertheless, proceeded, with the plaintiff prevailing. The following year, in *Cooke v. Maxwell*, the court excluded on similar grounds instructions from the governor of Sierra Leone to a military officer. Cooke v. Maxwell, 2 Stark. 183, 183, 185–86 (1817). While the plaintiff could not provide the contents of the communications, he could demonstrate "that what was done was done by the order of the defendant." *Id.* at 186. He prevailed. *Id.* at 187. Similarly, in *Home v. Bentink*, the court withheld minutes taken at a military court of enquiry on the grounds that such "matters, secret in their natures, and involving delicate enquiry and the names of persons, stand protected." Home v. Bentinck, 2 Brod. & B. 130 (1820). Numerous cases from that time through the mid-twentieth century followed suit—each one treating the privilege as a common law evidentiary rule. *See, e.g.*, Regina v. Russell, 7 Dowl Pr. 693 (1839) (excluding the papers of the former Secretary of State); H.M.S. Bellerophon, 44 L.J. Adm. 5, 6–7 (1875) (excluding the log books of a navy ship that collided with another ship as well as government communications about the incident); Mercer v. Denne [1904] 2 Ch. 535, 544 (maps of seashore boundaries prepared for the war office excluded on grounds that it would be too dangerous to admit them); Mercer v. Denne, [1905] 2 Ch. 538, 561 (C.A. 1905) (affirming Mercer v. Denne on appeal); Duncan v. Cammell, Laird & Co., [1941] 1 K.B. 640 644 (C.A.) (excluding documents relating to the design of a submarine which sank in 1939 on the grounds based on the Lord of the Admiralty's claim in an affidavit that it "would be injurious to the public interest"), *aff'd*, Duncan v. Cammell, Laird & Co., [1942] A.C. 624 (H.L.).

[151] In *United States v. Burr*, the court made reference to the evidentiary rule, but did not actually invoke it, explaining its limits in capital cases: "That there may be matter, the production of which the court would not require, is certain; but, in a capital case, that the accused ought, in some form, to have the benefit of it, if it were really essential to his defence, is a position which the court would very reluctantly deny. It ought not to be believed that the department which superintends prosecutions in criminal cases, would be inclined to withhold it." United States v. Burr, 25 F. Cas. 30, 37 (C.C. Va. 1807) (No. 14,692(d)). The Court continued, "What ought to be done under such circumstances present a delicate question, the discussion of which, it is hoped, will never be rendered necessary in this country. At present it need only be said that the question does not occur at this time. There is certainly nothing before the court which shows that the letter in question contains any matter the disclosure of which would endanger the public safety. If it does contain such matter, the fact may appear before the disclosure is made. If it does contain any matter which it would be imprudent to disclose, which it is not the wish of the executive to disclose, such matter, if it be not immediately and essentially applicable to the point, will, of course, be suppressed." *Id.* Chief Justice Marshall, riding circuit, allowed the defense to subpoena President Jefferson to obtain a letter which an alleged co-conspirator, General James Wilkinson, had sent to him. *See* Robert M. Chesney, *State Secrets and the Limits of National Security Litigation*, 75 Geo. Wash. L. Rev. 1249, 1272 (2007).

Steel Co., the federal district court directed that drawings related to the manufacture of armor-piercing projectiles be removed from the record.[152] In *Pollen v. Ford Instrument Co.*, the court ordered that documents detailing sighting mechanics for guns be expunged on the grounds "that any disclosure of the structures used by the Navy or others authorized by it would be detrimental to the national defense and the public interests."[153]

Courts tempered government assertions of the privilege. In *United States v. Haugen*, for instance, the federal district court excluded a government contract but sanctioned proof of a portion of it to be submitted.[154] In *Cresmer v. United States*, the plaintiff sought the report of the Navy Board of Investigation into an airplane that crashed, killing the plaintiff's intestate.[155] Although the document initially had been withheld, the judge requested a copy of it to ensure "that the report in question contained no military or service secrets which would be detrimental to the interests of the armed forces of the United States or to National security."[156] Seeing "nothing in it which would in any way reveal a military secret or subject the United States and its armed forces to any peril by reason of complete revelation," he ordered that it be produced.[157] He explained, "In the absence of a showing of a war secret, or secret in respect to munitions of war, or any secret appliance used by the armed forces, or any threat to the National security, it would appear to be unseemly for the Government to thwart the efforts of a plaintiff in a case such as this to learn as much as possible concerning the cause of the disaster."[158]

The determination of whether to allow the evidence to enter into proceedings resided with the judge.[159] Accordingly, in *O'Neill v.*

---

[152] Firth Sterling Steel Co. v. Bethlehem Steel Co., 199 F. 353 353–56 (E.D. Pa. 1912).

[153] Pollen v. Ford Instrument Co., 26 F. Supp. 583, 584 (E.D.N.Y. 1939).

[154] United States v. Haugen, 58 F. Supp. 436 (E.D. Wash. 1944).

[155] Cresmer v. United States, 9 F.R.D. 203 (E.D.N.Y. 1949).

[156] *Id.* at 204.

[157] *Id.*

[158] *Id.*

[159] Efforts by the government to assert this privilege in realm of criminal prosecution often failed—precisely because such documents could be material to the defense. *See, e.g.*, King v. United States, 112 F. 988, 996 (5th Cir. 1902); Schneiderman v. United States, 320 U.S. 118 (1943); United States v. Coplon, 185 F.2d 629 (2d Cir. 1950); Zimmeman v. Poindexter, 74 F. Supp. 933 (D. Haw. 1947); Haugen v. United States, 153 F.2d 850 (9th Cir. 1946); United States v. Clegg, 846 F.2d 1221 (9th Cir. 1988).

United States, the court took judicial notice of "the general policy of the common law, prohibiting disclosure of state secrets the publication of which might seriously embarrass or harm the government in its diplomatic relations, military operations, or measures for national security."[160] It was for the court to ascertain if and when state secrets applied. As the Second Circuit Court of Appeals noted in the 1947 case *Bank Line v. United States*, "no general principle of refusing discovery on a general statement of prejudice to its best interests can or should be applied to any branch of the government, including the armed forces."[161]

Like the English, American treatises considered state secrets to be a common law evidentiary rule.[162] Simon Greenleaf, author of the first American treatise on evidence that served as the undisputed authority for nearly half a century, underscored the "class of cases, in which evidence is excluded from motives of public policy, namely, secrets of state, or things, the disclosure of which would be prejudicial to the public interest."[163] In the interests of public safety, "the rule of exclusion is applied no further than the attainment of that object."[164] Greenleaf provided as examples "communications between a provincial governor and his attorney-general on the state of the colony, or the conduct of its officers; or between such governor and a military officer under his authority; the report of a military commission of inquiry, made to the commander-in-chief; and the correspondence between an agent of the government and a Secretary of State."[165] All related to military and foreign affairs.

In 1940, John H. Wigmore similarly identified "secrets of state" as one of three types of evidentiary privileges relating to official

---

[160] O'Neill v. United States, 79 F. Supp. 827, 829 (E.D. Pa. 1948).

[161] Bank Line v. United States, 163 F.2d 133, 139 (2d Cir. 1947) (Clark, J., concurring). He did not believe that the information being sought would "aid and comfort some unknown potential enemy if the Navy" stated precisely why "concealment of specific information is material to national defense." *Id.*

[162] *See, e.g.*, 1 Thomas Starkie, A Practical Treatise on The Law of Evidence § 80, at 106 (1826) (drawing attention to the exclusion of documents which, "on grounds of state policy" could be excluded because they were "prejudicial to the community"); John Pitt Taylor, George Pitt-Lewis & 1 Charles Frederic Chamberlayne, A Treatise on the Law of Evidence as Administered in England and Ireland §§ 589, 909 (1895).

[163] 1 Simon Greenleaf, A Treatise on the Law of Evidence § 250, at 323–24 (6th ed. 1852).

[164] *Id.*

[165] *Id.* § 251.

information.[166] The testimonial privilege was limited to international relations, military affairs, and public security.[167] Other scholarly works followed course. In 1949, the *Yale Law Journal* explained that "originally the common law privilege [to withhold government information] protected only the identity of informers and secrets affecting the national security."[168] In the *Vanderbilt Law Review*, William Sanford similarly analyzed the contemporary "State of the Law as to Executive Privileges," considering as the very first "Privilege[] *Established by the Courts*," "Data Affecting National Security (Military and Diplomatic Secrets)."[169] In each case, it was for the court to ascertain whether certain evidence would be allowed as part of the judicial record.

2. Reynolds *and Its Progeny*. By the time of *Reynolds*, the sole exception to this unbroken practice of treating state secrets as an evidentiary rule was the contractual bar established in *Totten*.[170] A Court of Claims action seeking recovery for an espionage contract forged during the Revolutionary War, *Totten* provided for dismissal of the suit on two grounds: first, the contract itself stipulated "a secret service"—"[b]oth employer and agent must have understood that the lips of the other were to be for ever sealed respecting the relation of

[166] 8 WIGMORE, EVIDENCE § 2378, at 792 (3d ed. 1940).

[167] *Id.* at 794.

[168] *Government Privilege Against Disclosure of Official Documents*, 58 YALE L.J. 993, 993 (1949); *see also Evidence-Privileged Communications-State Secrets*, 47 W. VA. L. REV. 338, 340 (1941) (noting that where state secrets are asserted, where it is in the public interest, "the judge may order that the records be opened to the petitioner").

[169] William V. Sanford, *Evidentiary Privileges Against the Production of Data Within the Control of Executive Departments*, 3 VAND. L. REV. 73, 74 (1949) (emphasis added). Sanford wrote, "In the contemporary state of international affairs, where there is always a real danger of a serious international dispute, the security of the state requires efficient armed forces and diplomatic services." *Id.* at 75. While it was virtually impossible to draw a hard line, documents protected by the privilege related to U.S. military and foreign affairs. *Id.*; *see also* Kessler v. Best, 121 F. 439 (C.C.S.D.N.Y. 1903) (excluding documents in the archives of a foreign consulate); Crosby v. Pacific S.S. Lines, 133 F.2d 470, 475 (9th Cir. 1943) ("The special treatment of consuls is caused probably in part by treaty, in part by the fact that they are under the jurisdiction of the diplomatic department of government, and in party by the character of their business and permanent of existence and locality." And noting that in such circumstances (in relation to foreign governments), "the rule to be applied is the one we would apply to a similar department of government here."), *cert. denied*, 319 U.S. 752 (1943); Viereck v. United States, 130 F.2d 945, 961 (D.C. Cir. 1942) (excluding testimony of an employee of the British Censor's Office in Bermuda).

[170] Totten v. United States, 92 U.S. 105, 107 (1875) ("It may be stated as a general principle, that public policy forbids the maintenance of any suit in a court of justice, the trial of which would inevitably lead to disclosure of matters which the law itself regards as confidential, and respecting which it will not allow the confidence to be violated.").

either to the matter."[171] The central concern was not just the "publicity produced" by such an act, but that requiring such information "would itself be a breach of a contract."[172] Second, the Court articulated "a general principle, that public policy forbids the maintenance of any suit in a court of justice, the trial of which would inevitably lead to the disclosure of matter which the law itself regards as confidential."[173]

As the Supreme Court later recognized in *General Dynamics v. United States*, *Totten* stands for an extremely narrow line of precedent resting on the judiciary's "common-law authority to fashion contractual remedies in Government-contracting disputes."[174] The decision captured "what the *ex ante* expectations of the parties were or reasonably ought to have been"—i.e., "that state secrets would prevent courts from resolving many possible disputes under the ... agreement."[175]

Accordingly, in *Tucker v. United States*, a contractor brought a claim for expenses and compensation for secret services he allegedly performed for the Psychological Warfare Branch, Military Intelligence Department, and other agencies of the federal government.[176] The federal district court held in 1954 that secrecy regarding such services was necessary and that publicity would be a breach of contract, thus precluding recovery on the claim.[177] Similarly, in 1988, in *Guong v. United States*, a Vietnamese citizen alleged that he was recruited by the CIA to conduct covert military operations in North Vietnam.[178] The plaintiff claimed that the CIA agreed to pay him and, if captured, rescue him. In the event that the rescue failed, the plaintiff's wife was to receive his pay. After he was captured by the North Vietnamese in 1964, though, he was not rescued and the U.S. government stopped paying his wife. He escaped from prison and in 1986 filed an action to recover damages for the breach of contract. As in

[171] *Id.* at 106.

[172] *Id.* at 107.

[173] *Id.*

[174] Gen. Dynamics Corp. v. United States, 563 U.S. 478, 485 (2011); *see also* Tenet v. Doe, 544 U.S. 1, 3 (2005) (noting "the longstanding rule, announced more than a century ago in *Totten*, prohibiting suits against the Government based on covert espionage agreements").

[175] *Gen. Dynamics*, 563 U.S. at 490 (citing *Totten*, 92 U.S. at 106).

[176] Tucker v. United States, 118 F. Supp. 371 (Ct. Cl. 1954).

[177] *Id.*

[178] Guong v. United States, 860 F.2d 1063 (Fed. Cir. 1998).

Totten, the federal court dismissed the claim on the grounds that the contract was secret at the time of its creation, giving Guong *ex ante* notice.[179]

The mere fact of a government contract was not sufficient; it had to be part of a secret agreement—this is what ensured *ex ante* notice. In other words, *embedded in the contract itself* was the understanding that confidentiality would be maintained, making public access to it a breach.

Outside of this context, the evidentiary rule, formalized in *Reynolds* but previously employed in cases on both sides of the Atlantic, applied.[180] In *Reynolds*, the Air Force refused to release documents related to electronic equipment on board a B-29 bomber, which had crashed and killed three civilian passengers. According to the government, the plane had been on "a highly secret mission," and disclosure of the specific documents in question would "seriously hamper[] national security, flying safety, and the development of highly technical and secret military equipment."[181] Looking to Burr's trial as well as Greenleaf and Wigmore's *Treatise on the Law of Evidence*, the Court noted that state secrets privilege was "well established in the law of evidence."[182]

Over the ensuing decades, outside of the contractual context, *Reynolds* controlled. Thus, in *Greene v. McElroy*, the D.C. Circuit upheld state secrets as an evidentiary rule in regard to a security clearance application.[183] In *Pan Am. v. Aetna*, an insurance dispute arising out of

---

[179] Vu Duc Guong v. United States, No. 21-86C, slip op. at 3 (Ct. Cl. Sept. 30, 1987), *quoted in* Guong v. United States, 860 F.2d at 1064–65 (affirming lower court decisions).

[180] *See, e.g.*, Bernstein v. United States, 256 F.2d 697 (10th Cir. 1958) (ruling, in an action for fraud brought by the government against private parties, that if a report made by witness for government during course of his investigation of sale of war surplus property was relevant and not privileged for security reasons or as an attorney's mental impression, conclusions, opinions or legal theories, it was proper subject for production and inspection).

[181] United States v. Reynolds, 345 U.S. 1, 4–5 (1953).

[182] *Id.* at 6–7; *see also* 1 SIMON GREENLEAF A TREATISE ON THE LAW OF EVIDENCE, § 251 n. 5 (John Henry Wigmore ed., 16th ed. 1899); United States v. Burr, 25 F. Cas. 30, 37 (C.C. Va. 1807) (No. 14,692(d)).

[183] Greene v. McElroy, 254 F.2d 944 (D.C. Cir. 1958). On appeal, the Supreme Court reflected, "Certain principles have remained relatively immutable in our jurisprudence. One of these is that, where governmental action seriously injures an individual . . . the evidence used to prove the Government's case must be disclosed to the individual so that he has an opportunity to show that it is untrue." Greene v. McElroy, 360 U.S. 474, 496 (1959). Embodied in the Sixth Amendment, cross examination, as recognized by Wigmore in his treatise on evidence, had become a cornerstone of American jurisprudence. *Id.* at 497. In this case, neither the President nor Congress had "delegated to the Department of Defense the authority to bypass these traditional and well recognized safeguards." *Id.* at 500.

the 1970 hijacking and subsequent destruction of Pan Am Flight 083 by the Popular Front for the Liberation of Palestine, the Second Circuit upheld the CIA's state secrets claim over certain files, even as the case continued.[184]

Courts did not always acquiesce to government claims. In *Halpern v. United States*, for example, the court rejected the Secretary of the Navy's claim of state secrets over the plaintiff's patent application, and all documents, statements, and testimony related to their technical subject.[185] Remanding the case, the court noted that the district court could adopt procedures to ensure that there would be no danger of public disclosure.[186] In *Heine v. Raus*, a slander action, the Fourth Circuit in 1968 tempered the CIA's invocation of state secrets by issuing rulings on each question calling for information, requiring a witness "to answer those which the Court thought would not impair the privilege while foreclosing answers to those questions which apparently would."[187] In *Committee for Nuclear Responsibility v. Seaborg*, the D.C. Circuit in 1971 considered an environmental challenge to the government's proposed underground nuclear test (Cannikin), to be carried out in Alaska.[188] The court isolated and removed military and diplomatic secrets from documents addressing potential environmental hazards, allowing the case to proceed. While the government's interest in the former was "plain," plaintiffs—in a manner similar to *Fazaga*—stated that they did not seek any secrets.[189]

Cases involving surveillance proceeded in similar fashion. Thus, in *Jabara v. Kelly*, a U.S. citizen sued the FBI director for alleged illegal surveillance, harassment and intimidation. The government asserted state secrets in response to several interrogatories, which the federal district court in 1973 largely upheld even as it required the release of some information (such as the name of the agency intercepting the calls) to the plaintiff.[190] Similarly in *Spock v. United States*, an action was

[184] Pan Am. World Airways v. Aetna Cas. & Sur. Co., 368 F. Supp. 1098, 1140 (S.D.N.Y. 1973); *see also* United States v. Bass, 472 F.2d 207 (8th Cir. 1973) (Court of Appeals confirming that the trial judge used due care to ensure that the defendants in a military contractor fraud/conspiracy case, were afforded maximum but reasonable discovery, and that the government provided enough information to the judge during the *in camera* review.)

[185] Halpern v. United States, 258 F.2d 36, 44 (2d Cir. 1958).

[186] *Id.* at 43–44.

[187] Heine v. Raus, 399 F.2d 785, 788 (4th Cir. 1968).

[188] Comm. For Nuclear Resp. v. Seaborg, 463 F.2d 788 (D.C. Cir. 1971).

[189] *Id.*

[190] Jabara v. Kelly, 476 F. Supp. 561 (E.D. Mich. 1973).

brought against the United States, the Director of the NSA, and several unknown agents for alleged unlawful interception of plaintiff's oral, wire, telephone, and telegraph communications. The federal district court found in 1978 that the state secrets privilege was validly established by a public and a sealed affidavit by the Secretary of Defense, but that it did not prevent the plaintiff from accessing the courts or require dismissal of the complaint.[191] In *Ellsberg v. Mitchell*, plaintiffs sought compensation for warrantless electronic surveillance.[192] The D.C. Court of Appeals ruled in 1983 that the state secrets privilege had been drawn too broadly: the government had failed to disclose the identities of the Attorneys General who had authorized the wiretaps, despite the government conceding, during oral argument, that there was no reason such information would undermine national security.[193]

Under the *Reynolds* standard, if a plaintiff could not make out a prima facie case or carry the burden of persuasion without the privileged information, then the court dismissed the case. Discharging the suit thus acted as a logical consequence of the evidentiary rule. In *Zuckerbraun v. General Dynamics Corp.*, for instance, the Second Circuit in 1991 determined that because information relating to the weapons system's "design, manufacture, performance, functional characteristics, and testing" could not be made public, the appellant could not establish a prima facie case.[194] The following year, *Bareford v. General Dynamics Corp.* was resolved on similar grounds. There, the plaintiffs alleged that a Navy frigate's defense system had failed to counter Iraqi missiles.[195] The Fifth Circuit observed in 1992 that while the plaintiffs had accumulated "considerable evidence," it fell short of establishing that the Phalanx system had failed its basic function.[196]

[191] Spock v. United States, 464 F. Supp. 510 (S.D.N.Y. 1978).

[192] Ellsberg v. Mitchell, 709 F.2d 51 (D.C. Cir. 1983).

[193] *Id.* at 52.

[194] Zuckerbraun v. Gen. Dynamics Corp., 935 F.2d 544, 547–48 (2d Cir. 1991); *see also* Nejad v. United States, 724 F. Supp. 753, 754, 756 (C.D. Cal. 1989) (noting that absent "disclosure of the AEGIS system technology [and rules of engagement], the case could not be tried).

[195] Bareford v. Gen. Dynamics Corp. 973 F.2d 1138, 1140 (5th Cir. 1992).

[196] *Id.* at 1142; *see also* Black v. United States, 62 F.3d 1115, 1118 (8th Cir. 1995) (dismissing the suit on the grounds that absent the prohibited information the plaintiff could not "establish[] a prima facie *Bivens* claim"); Frost v. Perry, 919 F. Supp. 1459, 1468 (D. Nev. 1996) ("Plaintiffs cannot provide the essential evidence to establish its [sic] *prima facie* case for any of its eleven claims due to the Defendants' assertion of the military and state secrets privilege."); Bentzlin v. Hughes Aircraft Co., 833 F. Supp. 1486, 1496 (C.D. Cal. 1993) (plaintiff could not make out a prima facie case without the information excluded due to state secrets).

On the other side of a suit, in the event that a defendant claimed an inability to properly mount a defense without the documents subject to state secrets, following *in camera* review of the information in question and its subsequent exclusion, courts post-*Reynolds*, on occasion, ordered dismissal. As the Fourth Circuit sitting en banc explained in 1980 in *Farnsworth Cannon, Inc. v. Grimes*, a case that dealt with allegedly tortious interference with future contract rights between Farnsworth Cannon and the U.S. Navy,

> The unavailability of the evidence is a neutral consideration, and, whenever it falls upon a party, that party must accept the unhappy consequences. If the assertion of the privilege leaves plaintiff without sufficient evidence to satisfy a burden of persuasion, plaintiff will lose. If plaintiff's case might be established without the privileged information, dismissal is not appropriate. The same standards apply to defendants.[197]

In either case, the action was a logical outcome of the evidentiary rule.

One of the leading cases on the defense side is *Molerio v. FBI*, where then-Judge Antonin Scalia, writing for the D.C. Circuit, dismissed an FBI applicant's First Amendment claim.[198] Only after determining that the privilege had been properly asserted did the court turn to "the difficult issue of the effect" of the privilege. While Molerio could still make out a *prima facie* "circumstantial case permitting the inference that his father's [First Amendment-protected] political activities" contributed to the FBI's refusal to hire him," the court, having looked at the underlying evidence, determined that the decision had been based on entirely different information.[199]

Under the *Molerio* line of cases, where the district court "determine[s] that the defendant will be deprived of a valid defense based on the privileged materials, it may properly dismiss the complaint."[200] Courts understand a "valid defense" to mean a defense which "is meritorious and not merely plausible and *would require judgment for the defendant*."[201] Such a situation would entail a miscarriage of justice. A

[197] Farnsworth Cannon, Inc. v. Grimes, 635 F.2d 268, 270–72 (4th Cir. 1980) (en banc) (per curiam) (citation omitted) (citing various cases).

[198] Molerio v. FBI, 749 F.2d 815 (D.C. Cir. 1984).

[199] *Id.* at 825.

[200] *In re* Sealed Case, 494 F.3d 139, 149 (D.C. Cir. 2007).

[201] *Id.* (citing Kasza v. Browner, 133 F.3d 1159, 1166 (9th Cir. 1998); and Tenenbaum v. Simonini, 372 F.3d 776, 777–78 (6th Cir. 2004) as adopting the "valid defense" standard).

critical first step is judicial scrutiny of the evidence in question. Without that, the D.C. Circuit warned, "virtually every case in which the United States successfully invokes the state secrets privilege would need to be dismissed," replacing "the practice of deciding cases on the basis of evidence" with "a system of conjecture."[202]

From the time of *Reynolds* until the early twenty-first century, *Totten* was cabined to cases involving a secret agreement on the grounds that a classified contract conveyed an *ex ante* understanding that should a dispute arise, litigation in open court would not be available as an option.[203] There was a world of difference between that context and one in which litigants who had never contracted with the government sought relief for violations of their rights. Accordingly, in 1979, in *Clift v. United States*, the Second Circuit reversed the district court's dismissal of a complaint when state secrets prevented government documents relating to a patent application from being produced. According to the court of appeals, *Totten* "afford[ed] no support" as the inventor had not entered into any contract. To the contrary, it was *the government's action* (i.e., placing a secrecy order on his patent application) that was at issue.[204] As Justice Gorsuch later explained, "the general rule" in state secrets cases is that "the privilege protects only against the production of certain evidence—not the inconvenience of lawsuits. If a way exists for a court to discharge its statutory duty to entertain a case without the government's privileged proof, that way must be found."[205]

3. *Creation of the "Very Subject Matter" Analysis.* Post-9/11, as more information about the warrantless electronic surveillance conducted as part of TSP became available and with standing met, the government fell back on state secrets to try to get the suits dismissed. Lacking precedent, it re-crafted the *Totten* bar to read as a "very subject matter" analysis. Initially, the courts rejected this approach: it was hard to uphold a claim that the subject matter was a state secret when the President was giving press conferences about it. The government, nevertheless, persisted. In the torture and rendition cases, with standing met, the government once more collapsed the *Reynolds*

---

[202] *In re Sealed Case*, 494 F.3d at 150.

[203] *See, e.g.*, Clift v. United States, 597 F.2d 826 (1979); Tenet v. Doe, 544 U.S. 1 (2005); Gen. Dynamics Corp. v. United States, 563 U.S. 478 (2011).

[204] *Clift*, 597 F.2d 830 (citing and quoting Totten v. United States, 92 U.S. 105, 106 (1875)).

[205] United States v. Zubaydah, 142 S. Ct. 959, 996 (Gorsuch, J., dissenting).

evidentiary rule and the *Totten* bar. Those cases now serve as precedent for arguments again being made in the surveillance realm. *Fazaga* serves as an example *par excellence*.

The first time the government appears to have made the "very subject matter" argument—in relation to surveillance—the courts did not go along with it. In *Al-Haramain Islamic Foundation v. Bush*, the FBI accidentally provided a Top Secret document containing sensitive compartmented information (TS/SCI) to an individual who had been determined to be a Specially Designated Global Terrorist.[206] Based on the information provided, Al-Haramain and two U.S. citizens who interacted with the organization brought suit for warrantless electronic surveillance conducted outside of FISA.[207] With standing met, the government turned to state secrets and an broad reading of *Reynolds* to argue for dismissal:

> Here, the "very subject matter" of this lawsuit is a state secret. Plaintiffs allege that they have been the subjects of warrantless surveillance under a classified foreign intelligence program. National security matters are not peripheral to this case; the very goal of this lawsuit is to obtain a determination as to whether NSA has undertaken any warrantless surveillance of Plaintiffs and, if so, whether that action was lawful—including whether the President had authority to establish the program and whether its alleged application to Plaintiffs violated their Constitutional rights.[208]

In light of the fact that virtually all foreign intelligence collection programs are classified, the implication of the government's argument was that *no suit* could ever be brought challenging surveillance. This is a remarkable proposition, not least because FISA itself establishes civil liability and contains criminal penalties to address violations.[209] In 2007, the Ninth Circuit rejected the government's assertion on

---

[206] Memorandum in Opposition to Plaintiffs' Opposition to Defendants' Lodging of Material Ex Parte and In Camera at 2, 7, Al-Haramain Islamic Found., Inc. v. Bush, 451 F. Supp. 2d 1215 (D. Or. 2006) (No. 3:06-cv-00274-KI), ECF No. 32. It took two months for the FBI to request its return. *Id.* at 12; *see also* Al-Haramain Islamic Found. v. Bush, 507 F.3d 1190, 1193 (9th Cir. 2007).

[207] Memorandum in Support of the United States' Assertion of the Military and State Secrets Privilege and Defendants' Motion to Dismiss or, in the Alternative, for Summary Judgment at 3, *Al-Haramain*, 451 F. Supp. 2d 1215 (No. 3:06-cv-00274), ECF No. 59.

[208] *Id.* at 5 (quoting United States v. Reynolds, 345 U.S. 1 (1953)).

[209] 50 U.S.C. § 1809 (imposing a fine of not more than $10,000 or imprisonment of up to five years for either engaging in electronic surveillance outside of either FISA or Title III or disclosing or using such information—both of which appear to be at issue in the case of Al-Haramain).

different grounds, noting that the President had already admitted to having undertaken TSP outside FISA's strictures, and remanded the case to the district court to determine whether FISA preempted the state secrets privilege.[210]

As allegations of torture emerged, with standing apparently met, the government once again turned to state secrets and a re-crafting of the traditional test to try to halt litigation. In *El-Masri v. United States*, a German citizen brought suit against the former director of the Central Intelligence Agency and others, alleging that he was captured in Macedonia in December 2003 and subsequently handed over to the CIA, which rendered him to a black site in Afghanistan and proceeded to beat, drug, bind, blindfold, and interrogate him.[211] The complaint asserted three causes of action: a *Bivens* claim regarding El-Masri's Fifth Amendment right to due process and two violations of the Alien Tort Statute (contravention of the international legal norms against arbitrary detention and cruel, inhuman, or degrading treatment).[212]

In its brief to the Fourth Circuit, the government collapsed *Reynolds* and *Totten*, arguing that state secrets were so central to the suit that it could not proceed without disclosing them. In support of this proposition, it cited to the Fourth Circuit's 1985 decision in *Fitzgerald v. Penthouse International*, stating that it stood for the general proposition that where "the unavailability of the information protected by the privilege precludes either the plaintiff or defendant from establishing their respective legal positions on the issues in the case, then the case must be dismissed."[213]

In *Fitzgerald*, however, the central question turned on a contractual claim: whether a scientist could testify about whether he was undertaking marine animal research under a classified contract with the CIA.[214] Similarly, in *Sterling v. Tenet*, a covert agent in the CIA brought a Title VII racial discrimination claim against his employer.[215] The Fourth Circuit in *Sterling* concluded that a trial "would require

---

[210] *Al-Haramain Islamic Found.*, 507 F.3d at 1193.

[211] El-Masri v. United States, 479 F.3d 296, 300 (4th Cir. 2007).

[212] Complaint, El-Masri v. Tenet, No. 1:05-cv-01417 (E.D. Va. Dec. 6, 2005); *see also El-Masri*, 479 F.3d at 300-301.

[213] Brief of the Appellee at 35, *El-Masri*, 479 F.3d 296 (No. 84-1035).

[214] *See* Fitzgerald v. Penthouse Int'l, Ltd., 776 F.2d 1236 (4th Cir. 1985). The government also made brief reference to *Farnsworth Cannon, Inc. v. Grimes*. *See* 635 F.2d 268 (4th Cir. 1980).

[215] Sterling v. Tenet, 416 F.3d 338 (4th Cir. 2005).

disclosure of highly classified information concerning the identity, location, and assignments of CIA operatives."[216] Both cases, like *Totten*, turned on voluntary entrance into a (secret) contractual relationship with the government. In such circumstances, both parties had *ex ante* knowledge that a future suit could risk sensitive national security concerns, making legal action unavailable. In *Totten*, Justice Stephen Field had explained,

> The service stipulated by the contract was a secret service; the information sought was to be obtained clandestinely, and was to be communicated privately; the employment and the service wee to be equally concealed. Both employer and agent must have understood that the lips of the other were to be for ever sealed respecting the relation of either to the matter. This condition of the engagement was implied from the nature of the employment.[217]

The complaint in *El-Masri* was about something entirely different: whether *the government's unilateral actions* had violated El-Masri's rights by subjecting him to extraordinary rendition. Under *Reynolds*, the appropriate response would have been to exclude whichever documents met the standard, and then to proceed. If the litigant could not make out a *prima facie* case, the suit should be dismissed.[218] Alternatively, if the government could not present a meritorious defense without certain information, then the court, after reviewing the material, could dismiss the suit as a logical outcome of applying the evidentiary rule. In *El-Masri*, the Fourth Circuit in 2007 did neither, instead accepting the government's newly-minted "very subject matter" approach.

In a similar action the following year, *Mohamed v. Jeppesen*, five foreign nationals brought a civil suit against a defense contractor, alleging its role in their forced disappearance, torture, and inhumane treatment as part of the CIA's extraordinary rendition program.[219] Once again, the United States intervened and sought dismissal. As in *El-Masri*, the government argued that "the very subject matter of plaintiffs' claims is a state secret."[220] Embedded in this assertion,

---

[216] *Id.* at 341.

[217] Totten v. United States, 92 U.S. 105, 106 (1875).

[218] This was why it mattered that there was already information in the public domain, such as the Council of Europe's 2006 draft report on rendition and detention, and the government's own release of information about the CIA program. *See El-Masri*, 479 F.3d at 302.

[219] Mohamed v. Jeppesen Dataplan, Inc., 539 F. Supp. 2d 1128 (N.D. Cal. 2008).

[220] Redacted, Unclassified Brief for Intervenor-Appellee the United States, Mohamed v. Jeppesen Dataplan, Inc., 539 F. Supp. 2d 1128 (N.D. Cal. 2008) (No. 08-15693), 2008 WL 4973859.

however, was one that had more purchase: the plaintiff's argument turned on a contractual relationship between *Jeppesen* and the CIA. Had the suit been a matter of dispute between Jeppesen and the agency, then *Totten* would at least be available. But it was not. To the contrary, it was a civil action against a private company, making the *Reynolds* evidentiary rule more appropriate. Assumedly, Jeppesen would have sought information about the contractual relationship to mount a defense—which may (or may not) have led to the same conclusion (dismissal of the suit), but this would have been a logical outcome of the evidentiary rule. Instead, the government asserted the much broader claim.

The problem with the "very subject matter" approach is that it puts the government in a position of insulating itself from accountability. *As long as a program is classified*, then *any* constitutional challenge to it can be defeated. This is precisely what has happened as similar arguments have begun to infuse the surveillance realm.

In its brief in *Fazaga*, the government cited the federal district court's 2008 decision in *Jeppesen thirteen times* to support its "very subject matter" claim.[221] Because the issue involved a classified matter, "[l]itigating the case to a judgment on the merits would present an unacceptable risk of disclosing state secrets."[222] The government, quoting a case quoting *Totten*, explained, "Where, as here, the privileged evidence goes to the heart of a claim, that claim cannot be litigated, and dismissal is compelled."[223] But this is not at all what *Totten* held. *Totten* held that individuals entering into a classified contract with the government had *ex ante* notification that future disputes were unlikely to reach the light of day in a courtroom. That was part of the deal. It was a different context than situations challenging unilateral government actions that allegedly violate statutory law and the constitution.[224]

---

[221] Brief for the Federal Appellees at 2, 11, 23, 41, 42, 47, 51, 52, 53, 61, 63, Fazaga v. FBI, 916 F.3d 1202 (9th Cir. 2019) (No. 12-56874).

[222] *Id.* at 2 (internal quotations and citations omitted); *see also id.* at 47, 51, 52.

[223] *Id.* at 53.

[224] In *Wikimedia*, the Fourth Circuit, buying into the government's argument erred. It determined that state secrets could be invoked under three conditions:

> (1) 'the plaintiff cannot prove the prima facie elements of his or her claim without privileged evidence'; (2) 'even if the plaintiff can prove a prima facie case without resort to privileged information, ... the defendants could not properly defend

In tandem with the "very subject matter" analysis is a second argument advanced by the government to support dismissal: namely, "when litigation would risk or require the disclosure of information protected by the state secrets privilege."[225] This assertion appeared in abbreviated form in conjunction with the first re-crafting of state secrets in 2006 in *Al-Haramain*.[226] It rests on the *hypothetical* possibility that privileged information might, in the future, be released which may then impact U.S. national security. This argument evokes the spectre of Justice Scalia's concern in *Molerio*, when he warned against entering the realm of conjecture as grounds on which to deny rights.[227]

The government also argues in conjunction with its "very subject matter" claim that once privileged information is removed, "there is no basis for a court to evaluate that privileged evidence to determine whether a party's contentions are correct or whether a party's claims or defenses would prevail if the state secrets had not been excluded. Indeed, it would be improper for a court to seek to assess the merits of the parties' claims or defenses if doing so would require consideration of privileged evidence."[228] Yet this is precisely how, for centuries, the privilege worked, particularly in light of the potential miscarriage of justice should the defense be deprived of critical evidence.

> themselves without using privileged evidence'; and (3) further litigation would present an unjustifiable risk of disclosure.'

Wikimedia Found. v. NSA/Cent. Sec. Serv., 14 F.4th 276, 303 (4th Cir. 2021) (quoting Abilt v. CIA, 848 F.3d 305, 313–14 (4th Cir. 2017) (footnote omitted)). The first two conditions are entirely consistent with how state secrets have consistently operated. The last condition, however, applies only to the *Totten* line of cases—indeed, the court cited to *Abilt v. CIA* in support of its proposition, a case in which a covert CIA employee sued the agency for discriminating against him for having narcolepsy and taking periodic naps while working. 848 F.3d 305, 309 (4th Cir. 2017).

[225] Brief for Intervenor-Appellee United States at 2–3, Sakab Saudi Holding Co. v. Aljabri, No. 22-01052 (1st Cir. June 17, 2022), 2022 WL 2275195, BL-50.

[226] Al-Haramain Islamic Found., Inc. v. Bush, 451 F. Supp. 2d 1215, 1225 ("Litigating this matter will necessarily require, and risk, the disclosure of state secrets.").

[227] *See, e.g.*, Brief for Intervenor-Appellee United States, *supra* note 225, at 3 ("When full litigation of claims and defenses would require a court to evaluate privileged evidence on which a party relies, thereby risking or requiring the disclosure of privileged information that is excluded from the suit, the court should dismiss the suit.").

[228] *Id.*

B. DISMISSAL AT THE PLEADINGS STAGE

A second novel move made in the post-9/11 environment is the assertion of state secrets early in suits to argue for dismissal. This approach departs from the traditional use of state secrets during the discovery process. Courts historically have jealously guarded this line.

In the 1957 case of *Halpern v. United States*, for instance, the government moved to dismiss without prejudice an action brought by the plaintiff under the Inventive Secrecy Act of 1951 seeking compensation for damage caused by a secrecy order entered by the government against the plaintiff's patent.[229] The federal district court noted that the state secrets privilege provides an exemption from the testimonial duty; as a result, with no testimonial duty present the privilege assertion was premature.[230] Two years later, in *United States v. 62.50 Acres of Land More or Less Situated in Lake County Ohio*, the government sought to condemn a parcel of land and to use it as a ballistic missile launcher site.[231] The government objected to the admission of interrogatories related to an explosion at the Middletown, New Jersey, Nike site and moved that they be stricken.[232] Citing *Reynolds* and *Halpern*, the court ruled that the claim of privilege was premature.[233]

Prior to 9/11, courts took a similar view of efforts to assert state secrets early in suits alleging unconstitutional surveillance. In *In re United States of America*, for instance, the plaintiff brought a claim under the Federal Tort Claims Act, alleging that injuries to her and her deceased husband from COINTELPRO, conducted by the FBI 1950 to 1964.[234] The district court rejected the government's assertion and directed the FBI to answer the complaint.

Post-9/11, the government changed course. In *El-Masri*, it intervened at the pleading stage and asserted state secrets.[235] It argued, "[T]here is no categorical rule that the state secrets privilege can be

---

[229] Halpern v. United States, 151 F. Supp. 183 (S.D.N.Y. 1957).

[230] *Id.*

[231] United States v. 62.50 Acres of Land More or Less Situated in Lake Cnty. Ohio, 23 F.R.D. 287, 288 (N.D. Ohio 1959).

[232] *Id.*

[233] *Id.*

[234] *In re* United States, 872 F.2d 472 (D.C. Cir. 1989).

[235] Brief of the Appellee at 12, El-Masri v. United States, 479 F.3d 296 (4th Cir. 2007) (No. 06-1667), 2006 WL 2726281.

asserted only in response to discrete discovery requests."[236] It then cited three cases, none of which applied. The first, *Sterling v. Tenet*, dealt with an employment contract at the CIA—a case clearly in the *Totten*, not the *Reynolds*, line of cases. The second, *Salisbury v. United States*, was a FOIA case involving a national security exception—not a state secrets case.[237] The third, *DTM Research, L.L.C. v. AT&T Corp*, dealt with trade secrets—not state secrets.[238] The reason there was no rule against it was because state secrets always had operated during discovery or in the course of the trial as an evidentiary rule. The government made a similar move in *Jeppesen*: after the plaintiffs filed the First Amended Complaint, but prior to the defendant's response or any undertaking of discovery, the United States intervened, asserted state secrets, and requested dismissal of the suit.[239]

The government's recourse to state secrets early in litigation carried over to other suits—which then cited to the rendition cases in support. Thus, in *Husayn v. Mitchell*, the government relied heavily on *Jeppesen*. The same occurred in *Sakab Saudi Holding Co. v. Aljbri* and *Kareem v. Haspel*.[240] In *Fazaga*, the government again cited *Jeppesen* in support of asserting its claim at the pleadings stage.[241] According to the government, plaintiffs' claims as to religious discrimination in *Fazaga* could not even move into the discovery stage without jeopardizing national security.[242]

C. FROM PARTICULARITY TO GENERALITY

For centuries, the state secrets privilege applied to *particular* documents or testimonial information that could not enter litigation.[243] In 1910, in *In Re Grove*, for instance, the Secretary of the Navy initially asserted it over "the plans, drawings, specifications, and other illustrative and descriptive papers . . . relating to the construction and operation of the steam turbines" embedded in torpedo

[236] *Id.*; *see* Sterling v. Tenet, 416 F.3d 338, 341–42 (4th Cir. 2005).

[237] *See* Salisbury v. United States, 690 F.2d 966 (D.C. Cir. 1982).

[238] *See* DTM Rsch. LLC. v. AT&T Corp., 235 F.3d 327, 334 (4th Cir. 2001).

[239] Redacted, Unclassified Brief for Intervenor-Appellee the United States, No. 08-cv-15693 (9th Cir. Aug. 27, 2008), 2008 WL 4973859.

[240] *See* Sakab Saudi Holding Co. v. Aljabri, 578 F. Supp. 3d 140 (D. Mass. 2021).

[241] Brief for the Federal Appellees, *supra* note 221, at 3.

[242] *Id.*

[243] *See supra* note 150; Duncan v. Cammell, Laird & Co. [1942] AC 624.

boat destroyers.[244] He later capitulated, allowing the documents to be used in litigation. In *The Wright and the Papoose*, a case in admiralty, the plaintiff sought discovery of the records of the Naval Court of Inquiry on May 31, 1931, regarding the collision.[245] Since no secrecy had been preserved during the hearing, the court considered the argument as to the danger of granting the motion to withhold the information "not very impressive."[246]

Courts repeatedly have focused on which documents, in particular, should be excluded. In *Republic of China v. National Union Fire Insurance*, the federal judge in 1956 excluded an order in council from the British government, two letters from the Secretary of State and replies by Britain, and one statement by the British Foreign Office.[247] In *Zuckerbraun v. General Dynamics Corp.*, the Secretary of the Navy in 1990 exercised the privilege over the rules of engagement authorized for, and military orders applicable to, the USS Stark at the time of the incident being litigated, as well as information regarding the design, performance, and functional characteristics of the combatant ships and the weapons and defense systems installed on them.[248] Case after case followed course.[249]

Over the past fifteen years, however, the government has increasingly begun to assert state secrets over entire categories of information.[250]

---

[244] *In re* Grove, 180 F. 62, 65–66 (3d Cir. 1910).

[245] The Wright, 2 F. Supp. 43 (E.D.N.Y. 1932); *see also* Anglo-Saxon Petroleum Co. v. United States, 78 F. Supp. 62 (D. Mass. 1948) (seeking an order directing the United States to produce the transcript from a particular Naval Board of Inquiry).

[246] *The Wright*, 2 F. Supp. 43.

[247] Rep. of China v. Nat'l Union Fire Ins., 142 F. Supp. 551 (D. Md. 1956).

[248] Zuckerbraun v. Gen. Dynamics Corp., 755 F. Supp. 1134, 1139 (D. Conn. 1990).

[249] *See, e.g.*, Pollen v. Ford Instrument Co., 26 F. Supp. 583, 583–84 (E.D.N.Y. 1939) (noting where the U.S. Navy exerted the privilege over "drawings showing the construction of range keepers or other apparatus for determining sighting data for guns"); Firth Sterling Steel Co. v. Bethlehem Steel Co., 199 F. 353 353–56 (E.D. Pa. 1912) (excluding and expunging certain drawings relating to armor-piercing projectiles); United States v. Haugen, 58 F. Supp. 436 (E.D. Wash. 1944) (detailing a contract relating to construction of a nuclear power plant).

[250] *See, e.g.*, *In re* Terrorist Attacks on Sept. 11, 2001, 523 F. Supp. 3d 478, 499 (S.D.N.Y. 2021) (government asserting state secrets privilege over subject information, reasons for investigation and results, sources and methods, and foreign government information and information sharing with foreign partners); Defendants' Request that the Court Discharge the Order to Show Cause and Deny Plaintiff's Request for Access to the Classified Steinbach Declaration, or, in the Alternative, Motion to Dismiss in Light of the Attorney General's Assertion of the State Secrets Privilege at 10, Twitter v. Barr, No. 4:14-cv-04480-YGR (N.D. Cal. Mar. 15, 2019) (government asserting the state secrets privilege over four categories,

The trend began as early as 2006, with *Al-Haramain*, in which the government asserted the privilege over "(i) information regarding the al Qaeda threat; (ii) information regarding the Terrorist Surveillance Program; and (iii) information that would confirm or deny whether Plaintiffs have been targeted for surveillance under the TSP or any other program."[251] Although there was information in the public domain about the first two categories—including from the President himself—the government cast the net widely to prevent further information from becoming public.

The torture and rendition cases again proved instrumental. In *Mohamed v. Jeppesen*, General Michael Hayden, the Director of the CIA, submitted a classified declaration detailing "the scope of information" subject to state secrets and harms that might follow, should such matters be publicly released.[252] Hayden identified "several categories of information implicated by" the case: any information that would tend to confirm or deny whether any private entity or foreign government assisted the CIA on clandestine intelligence activities; any information about the scope or operation of the CIA's detention and interrogation program—including site locations, methods of interrogation, and identity of prisoners; and any additional information regarding CIA activities, sources, or methods.[253]

This approach continues to mark cases relating to the interrogation and rendition programs. In 2019, for instance, responding to a request from the Polish government that two former CIA contractors testify about the detention and transfer of prisoners for coercive interrogation, the government intervened and asserted state secrets over seven topics:

---

including "Information Regarding How Adversaries May Seek to Exploit Information Reflecting the Government's Use of National Security Legal Process"); Mitchell v. United States, No. 16-MC-0036-JLQ (E.D. Wash. May 31, 2017), ECF No. 91 (government asserting privilege over seven categories, ranging from "information identifying individuals involved with the" CIA interrogation program to "information concerning the CIA's internal structure and administration"); Ibrahim v. Dep't of Homeland Sec., No. 06-00545 (WHA), 2013 WL 4549941, at *4–5 (N.D. Cal. Aug. 23, 2013) (government asserting state secrets privilege over information related to subject identification, the reasons for investigations and their results, and sources and methods employed in counterterrorism operations).

[251] Memorandum of Points and Authorities in Support of the United States' Assertion of the Military and State Secrets Privilege and Defendants' Motion to Dismiss or, in the Alternative, for Summary Judgment at 4, Al-Haramain Islamic Found., Inc. v Bush, 451 F.3d 1190 (9th Cir. 2007) (No. 3:06-cv-00274).

[252] Redacted, Unclassified Brief for Intervenor-Appellee the United States, *supra* note 220. *See generally* Mohamed v. Jeppesen Dataplan, Inc., 539 F. Supp. 2d 1128 (N.D. Cal. 2008).

[253] Redacted, Unclassified Brief for Intervenor-Appellee the United States, *supra* note 220.

> (1) information that could identify individuals involved in the CIA detention and interrogation program; (2) information regarding foreign government cooperation with the CIA; (3) information pertaining to the operation or location of any clandestine overseas CIA station, base, or detention facility; (4) information regarding the capture and/or transfer of detainees; (5) intelligence information about detainees and terrorist organizations, including intelligence obtained or discussed in debriefing or interrogation sessions; (6) information concerning CIA intelligence sources and methods, as well as specific intelligence operations; and, (7) information concerning the CIA's internal structure and administration.[254]

These categories encompassed a report issued by the SSCI on the CIA's use of enhanced interrogation techniques; a ruling issued by the European Court of Human Rights regarding treatment of a prisoner held in Guantánamo Bay; testimony provided in court by former CIA contractors who designed and carried out the post-9/11 interrogation program; and a memoir published by one of the contractors.[255] That such information was public did serve as formal acknowledgement of the location of the black sites, but the CIA's rationale for withholding *all* such information, on the grounds that something contained in these categories might mention "Poland" or Polish officials (particularly in light of the fact that it was polish officials requesting the information), stretches credulity. Under the government's approach, even the CIA's copy of SSCI's Report would fall subject to state secrets. These categories, moreover, incorporate a significant amount of utterly unrelated material. The seventh category, for instance, ("information concerning the CIA's internal structure and administration"), appears to include everything from the name of the current Director of the CIA to the Agency's (unclassified) organization chart—currently available on Wikimedia.[256]

The categorical approach has spread to other kinds of national security-related suits, in some cases verging on the absurd. In *Twitter*

---

[254] Husayn v. Mitchell, 938 F.3d 1123, 1132 (9th Cir. 2019).

[255] *See* Report of the Senate Select Committee on Intelligence, Committee Study of the Central Intelligence Agency's Detention and Interrogation Program, S. Rep. 113-288, (Dec. 9, 2014), https://www.intelligence.senate.gov/sites/default/files/publications/CRPT-113srpt288.pdf; Case of Abu Zubaydah v. Lithuania, Judgment, Strasbourg, Eur. Ct. H.R., May 31, 2018; J. Mitchell & B. Harlow, Enhanced Interrogation: Inside the Minds and Motives of the Islamic Terrorists Trying to Destroy America (2016); United States v. Zubaydah, 142 S. Ct. 959 (2022).

[256] *See CIA Org Chart* (Illustration) (May 14, 2009), https://commons.wikimedia.org/wiki/File:Cia_org_chart_2009_may_14.jpg.

v. Barr, for instance, the executive in 2019 asserted state privilege over four categories, including, inter alia, "Information Regarding How Adversaries May Seek to Exploit Information Reflecting the Government's Use of National Security Legal Process."[257] Assumedly, this would include every official report required by FISA, as well as a significant amount of scholarship.

In *Wikimedia*, a case challenging programmatic collection under Section 702, the government in 2021 asserted the privilege over seven topic areas:

(1) "Entities subject to Upstream surveillance activities";
(2) "Operational details of the Upstream collection process";
(3) "Location(s) on the Internet backbone at which Upstream surveillance is conducted";
(4) "Categories of Internet-based communications subject to Upstream surveillance activities";
(5) "the scope and scale on which Upstream surveillance is or has been conducted";
(6) "NSA decryption capabilities"; and
(7) "Additional categories of classified information contained in opinions and orders issued by, and in submissions made to, the FISC."[258]

These categories included information not just in the public domain but which *the government itself* had already released. The Fourth Circuit did not seem at all bothered by this, reading the lower court's decision as merely limiting "*additional* information related to those categories."[259]

That approach, however, made two false assumptions: first, that all material not yet released threatens national security; and second, that all material *in these categories* threatens national security. The first is demonstrably false: all sorts of information not previously released by the government does not rise to the level of potentially causing

[257] Defendants' Request that the Court Discharge the Order to Show Cause and Deny Plaintiff's Request for Access to the Classified Steinbach Declaration, or, in the Alternative, Motion to Dismiss in Light of the Attorney General's Assertion of the State Secrets Privilege, *supra* note 250, at 10.

[258] Defendants' Memorandum at 7–8, No. 1:15-cv-00662-TSE (D. Md. Apr. 28, 2018), ECF No. 138.

[259] Wikimedia Found. v. NSA/Cent. Sec. Serv., 14 F.4th 276, 302 (4th Cir. 2021); *see also* Wikimedia Found. v. NSA, 427 F. Supp. 3d 582, 611 (D. Md. 2019).

grave harm if released. Such documents are constantly being sought across the government through FOIA requests and, when ignored or blocked, litigation. Simply because information hasn't been released has no bearing on its national security qualities. As for the latter claim, the fact that information in each of these categories is already public—at no apparent risk to national security—defies the assumption.

In *Fazaga*, the government again adopted a categorical approach. The Attorney General asserted state secrets over three general areas:

(1) *Subject Identification*: Information that could tend to confirm or deny whether a particular individual was or was not the subject of an FBI counterterrorism investigation, including in Operation Flex.
(2) *Reasons for Counterterrorism Investigations and Results*: Information that could tend to reveal the initial reasons (i.e., predicate) for an FBI counterterrorism investigation of a particular person (including in Operation Flex), any information obtained during the course of such an investigation, and the status and results of the investigation. This category includes information obtained from the U.S. Intelligence Community related to the reasons for an investigation.
(3) *Sources and Methods*: Information that could tend to reveal whether particular sources and methods were used in a counterterrorism investigation of a particular subject, including in Operation Flex. This category includes previously undisclosed information related to whether court-ordered searches or surveillance, confidential human sources, and other investigative sources and methods, were used in a counterterrorism investigation of a particular person, the reasons such methods were used, the status of the use of such sources and methods, and any results derived from such methods.[260]

The FBI added a catch-all: "This description of the broad categories of information subject to the Attorney General's claim of privilege is not meant to foreclose the possibility that other information related

[260] Public Declaration of Mark F. Giuliano, Fed. Bureau of Investigation, at 52–53, FBI v. Fazaga, 142 S. Ct. 1051 (2022) (No. 20-828) (2022), https://www.supremecourt.gov/DocketPDF/20/20-828/185460/20210730194627462_20-828ja.pdf.

to FBI counterterrorism investigations including Operation Flex may be identified in later proceedings as subject to privilege."[261]

Once again, these categories incorporate information already in the public domain—much of it put there by the government. They envelop law enforcement data that the public has a right to know and which is necessary to ensure that the FBI acts within its lawful limits. Within these categories, moreover, the possibility that the government may overclaim the national security risks is significant, not least because it has a long history of doing precisely that.

In 2018, District Court Judge Anthony Trenga published an article in the *Harvard National Security Journal* analyzing how state secrets privilege plays out in the courts.[262] Addressing the increased frequency and scope of government assertions, he underscored the importance of the judicial role in challenging "overstated or, in fact, baseless" government claims.[263] Interviewing thirty-one of his colleagues on the federal bench, Judge Trenga noted "[t]he willingness of more experienced judges to probe deeper," speculating that it related in part "to their commonly held belief that, with effort, most issues of disclosure can be resolved in a way that allows the litigation to proceed."[264] Several of the judges he interviewed "talked about how the scope of a privilege claim narrows substantially once a judge 'pushes back.'"[265] One put the number as high as 50 percent of the time.[266] Judges could often find a way to allow the suit to move forward "by working through the parties' specific needs of particular pieces of information."[267] It was a matter of being willing to get into the details of a case. Trenga wrote,

> Several other judges, particularly those with substantial state secrets experience . . . , some with a background in law enforcement, saw . . . the initial assertion of privilege claims broader than the government can ultimately defend and attributed this conduct, in various articulations, to an attempt, for the most part, to avoid "the hard analysis' and the sometimes

---

[261] *Id.*

[262] Anthony John Trenga, *What Judges Say and Do in Deciding National Security Cases: The Example of the State Secrets Privilege*, 9 HARV. NAT'L SEC. J. 1 (2018).

[263] *Id.* at 2, 49–51.

[264] *Id.* at 49.

[265] *Id.*

[266] *Id.*

[267] *Id.* at 50.

> tedious and difficult task of separating protected information from non-protected information until a judge reacts adversely.[268]

Some of the judges further highlighted "concern[] about a 'bureaucratic habit' to assert the privilege in 'too rote a fashion.'"[269]

Judge Trenga's findings align with numerous cases in which the government has over-stated the national security risks at stake in releasing information publicly.

The No Fly List cases provide a good example. In the 2014 case *Ibrahim v. Department of Homeland Security*, a visa holder argued that her inclusion on the federal No Fly List violated her constitutional rights.[270] The government initially invoked an evidentiary state secrets privilege but subsequently reversed its position, arguing for summary judgment on state secrets grounds.[271] When the case went to trial, however, the government conceded that the plaintiff had never actually presented a threat to the United States. An FBI agent had merely misunderstood the form that he had filled out which resulted in her being placed on the list.[272] The following year, in *Mohamed v. Holder*, the government once more moved for summary judgment on state secrets grounds.[273] Following *in camera* review, the federal district court "conclude[d] that there is no information protected from disclosure under the state secrets privilege that is necessary" for the litigation to proceed.[274]

Overbroad claims mark government representations in parallel national security realms. The FOIA context has already been addressed.[275] The same could be said of representations to specialized Article III courts about what materials can be made public. In 2013, for example, the ACLU, along with numerous media organizations and legislators from both sides of the aisle, filed a motion requesting that the court release its opinions interpreting Section 215.[276] The

---

[268] *Id.* at 51.

[269] *Id.* (internal citation omitted).

[270] Ibrahim v. Dep't of Homeland Sec., 62 F. Supp. 3d 909 (N.D. Cal. 2014).

[271] *Id.* at 914.

[272] *Id.* at 915–16.

[273] Mohamed v. Holder, 2015 U.S. Dist. LEXIS 92997, at *4–5 (E.D. Va. July 16, 2015).

[274] *Id.*

[275] *See* discussion *supra* Section II.C.

[276] Opinion and Order, *In re* Orders of this Ct. Interpreting Section 215 of the Patriot Act, No. Misc. 13-02, GID.C.00085, (FISA Ct. Sept. 13, 2013) (Saylor IV, J.), 2013 WL 5460064.

government identified one opinion and, without any explanation, stated that it would be withheld in full. The FISC expressed its concern that the government had not provided any rationale as to why the information could not be released.[277] The government shifted course and agreed to publish parts of the opinion.[278] After more pressure from the court and its staff, the government agreed that "certain additional information" could be released—apparently with no risk to national security.[279] These cases underscore the dangers inherent in the categorical approach increasingly adopted by the government: it sweeps in significant amounts of information already in the public domain while preventing other, important information, which does not present any sort of risk to national security, from reaching the light of day.

D. JUDICIAL AUTHORITY VERSUS EXECUTIVE BRANCH CONSTITUTIONAL POWER

In yet another departure from how the state secrets privilege has historically been treated, the government has begun to argue that the state secrets privilege is a constitutionally-derived Article II authority.[280] In its brief to the Supreme Court in *Fazaga*, the government stated, "[a]ny ambiguity in Section 1806(f) must be resolved in favor of retaining the *constitutionally based* state secrets privilege."[281] For the government, the President's commander-in-chief powers, as well as his "authority to 'make Treaties,' to 'appoint Ambassadors,' and to otherwise conduct the Nation's foreign affairs," accords the executive the primary responsibility for the conduct of foreign relations.[282]

Conspicuously missing from the brief was any reference to the foreign affairs and military powers provided to Congress in Article I, Section 8.[283] Absent, too, was any reference to the Advice and Consent

[277] *Id.* at 9.

[278] *Id.*

[279] *Id.*

[280] Brief for the Petitioners, *supra* note 132, at 42 ("The state-secrets privilege is firmly rooted in the Constitution as well as the common law."); *see also id.* at 43 (stating, "The privilege . . . is . . . firmly rooted in the Constitution.")

[281] *Id.* at 42 (emphasis added).

[282] *Id.* at 42–43.

[283] *See, e.g.*, U.S. CONST., art. I, § 8, cl. 10 ("To define and punish Piracies and Felonies committed on the high Seas, and Offences against the Law of Nations"); *id.* cl. 11 ("To

Clause and the role of the Senate in making treaties or appointing ambassadors or other public ministers and consuls.[284]

The government also failed to acknowledge the myriad ways in which Congress gives effect to its foreign affairs and war powers *by regulating sensitive national security information*. The legislature has gone to great length in FISA to ensure that information obtained from electronic surveillance, physical search, pen registers and trap and trace equipment, and access to business records remains secret.[285] The Atomic Energy Act of 1954 requires protection against unauthorized disclosure of any restricted data, as defined in the statute.[286] The Classified Information Procedures Act provides, in turn, for the admission of classified information at trial, establishing rules for, inter alia, pretrial conference, protective orders, discovery, notice, and the sealing of records in in camera hearings.[287] Four different statutes encapsulating five different requirements regulate national security letters.[288] Congress has insisted on gag orders, provided for delayed-notice search warrants, and taken numerous other steps to regulate access to national security information.[289]

In its brief to the Court, however, the government sidelined all of this, instead underscoring the executive's "enormous power in the two related areas of national defense and international relations,"

---

declare War, grant Letters of Marque and Reprisal, and make Rules concerning Captures on Land and Water"); *id.* cl. 12 ("To raise and support Armies"); *id.* cl. 13 ("To provide and maintain a Navy"); *id.* cl. 14 ("To make Rules for the Government and Regulation of the land and naval Forces"); *id.* cl. 15 ("To provide for calling forth the Militia to execute the Laws of the Union, suppress Insurrections and repel Invasions"); *id.* cl. 16 ("To provide for organizing, arming, and disciplining, the Militia"); *id.* cl. 18 ("To make all Laws which shall be necessary and proper for carrying into Execution the Foregoing Powers, and all other Powers vested by this Constitution in the Government of the United States, or in any Department or Officer thereof.").

284 *Id.* art. II, cl. 2.

285 *See* 50 U.S.C. §§ 1801(f).

286 *See* 42 U.S.C. § 2014(y).

287 *See* Classified Information Procedures Act, Pub. L. 96-456; 94 Stat. 2025 (Oct. 15, 1980), as amended through Pub. L. 111-16 (May 7, 2009).

288 *See* Right to Financial Privacy Act of 1978, 12 U.S.C. §§ 3401-22; Electronic Communication Privacy Act of 1986, 18 U.S.C. § 2709; Fair Credit Reporting Act of 1970, 15 U.S.C. §§ 1681u, 1681v; National Security Act of 1947, 50 U.S.C. § 3162.

289 *See, e.g.*, USA PATRIOT Act § 213, Pub. L. 107-56, 115 Stat. 272 (authorizing sneak and peek search warrants); *id.* § 505 (forbidding recipients of NSLs from sharing the information with anyone else). In 2008, the Second Circuit found that the gag orders and lack of judicial review constituted a violation of the First Amendment. Doe v. Mukasey, 549 F.3d 861 (2d Cir. 2008).

suggesting that because state secrets privilege "relate[s] to the effective discharge" of that power, it is a constitutional power.[290]

In addition to overlooking Congress's substantial foreign affairs and war powers and numerous legislative acts regulating sensitive national security information, the government failed to consider several flaws in its argument. First, and most obviously, as an historical matter it is simply false that state secrets has always operated as an Article II constitutional power. As the earlier discussion demonstrates, state secrets has *always* been a common law rule. For the *Reynolds* and *Totten* lines of cases, the privilege derives from the exercise of Article III power, not from an authority assigned to Article II.

Like the collapse of the *Reynolds* evidentiary rule and *Totten* bar, the slippage appears to stem from the post-9/11 torture and interrogation cases and application to the surveillance realm. In its brief in *El-Masri*, the government opined on "The Nature of the State Secrets Privilege," suggesting that it is not merely a privilege, but rather, "the means by which the Executive Branch exercises its critical constitutional responsibility to protect secrets of state in the national interest."[291] Citing *dicta* in *United States v. Nixon* (which focused on executive privilege in relation to presidential recordings), the government rather generously attributed to the Supreme Court the position "that state secrets privilege is rooted in, and is an aspect of, the powers granted to the President by Article II of the Constitution."[292]

In actual fact, all that the Court did in *Nixon* was to mention in passing the President's "Art[icle] II duties" in regard to "military or

---

[290] *Id.* at 44–45 (internal citations and quotations omitted); *see also* Brief for the Federal Appellees, *supra* note 221, at 54 (citing *El-Masri* in support of its reference to "the Executive Branch's constitutional authority to exercise control over information whose disclosure would jeopardize national security"). In its brief to the Fourth Circuit in *Wikimedia*, the government argued against interpreting Section 1806(f) as "Congress's intent . . . to displace the longstanding *and constitutionally grounded* privilege." *See* Appellee's Brief at 11, Wikimedia Found. v. NSA/CSS, 14 F.4th 276 (4th Cir. 2021) (emphasis added); *id.* at 32 ("This longstanding feature of our legal system enables the Executive Branch to fulfill its constitutional duties. As the Supreme Court has made clear, '[t]he authority to protect [national-security] information falls on the President as head of the Executive Branch and as Commander in Chief.' And executive privileges that 'relate[] to the effective discharge of a President's powers' are 'constitutionally based.' This Court has thus correctly recognized that the state-secrets privilege has 'a firm foundation in the Constitution' and 'performs a function of constitutional significance' because 'it allows the executive branch to protect information whose secrecy is necessary to its military and foreign-affairs responsibilities.'" (internal citations omitted)).

[291] Brief of the Appellee, *supra* note 235, at 8.

[292] *Id.*

diplomatic secrets."[293] The government's brief in *El-Masri* went on to cite to *United States v. Curtiss-Wright Export Corp.*, again referencing the President as "the sole organ of the federal government in the field of international relations," to conclude in a non-sequitur that the state secrets privilege has "a substantial constitutional basis."[294]

According to the government, "The privilege is not merely a manifestation of a decision to classify national security information. Rather, it is a separate, carefully delineated exercise of Executive authority pursuant to the Constitution's allocation of responsibility to protect the Nation's Security."[295] The Fourth Circuit in *El-Masri* gave a nod to the government's claim, stating, "Although the state secrets privilege was developed at common law, it performs a function of constitutional significance, because it allows the executive branch to protect information whose secrecy is necessary to its military and foreign-affairs responsibilities."[296] The court stopped short of grounding it in an Article II power, merely stating that state secrets operates in effect as judicial acknowledgment of the role of the executive.[297]

Second, relatedly, there is a distinction to be drawn between an evidentiary rule that takes notice of constitutional responsibilities and one that *derives* from constitutional power afforded the executive. There is a logical disconnect between the courts acting on their inherent power and, from this, the executive deriving an independent constitutional authority to prevent information from being presented in a judicial context. State secrets is a common law, judicially-created rule. Indeed, it could be argued that efforts by the executive branch to usurp the privilege by asserting it as an Article I power amounts to an unconstitutional exercise of Article III authorities. Allowing or excluding evidence goes directly to the Case or Controversy Clause and

---

[293] United States v. Nixon, 418 U.S. 683, 710 (1974). *Pari passu*, the case cited within *Nixon* (Chicago & S. Air Lines, Inc. v. Waterman S.S. Corp., 333 U.S. 103 (1948)) focused on transportation routes between the United States and foreign countries. State secrets was nowhere to be found. Instead, the Court asserted that on such matters of foreign affairs, it should be deferential to the executive branch. *Id.* at 111 ("Such decisions are wholly confided by our Constitution to the political departments of the government, Executive and Legislative. They are delicate, complex, and involve large elements of prophecy.").

[294] Brief of the Appellee, *supra* note 235, at 8.

[295] *Id.* at 45.

[296] El-Masri v. United States, 497 F.3d 296, 303 (2007).

[297] The court picked up on dicta in *Reynolds*, namely that the judiciary did not have to demand that the executive reveal certain information. United States v. Reynolds, 345 U.S. 1, 6 (1953).

the judiciary's obligation to ensure that justice ensues. This is why courts at times dismiss suits based on state secrets either because a plaintiff cannot make out a *prima facie* case without the evidence, or because defendants would not be able to respond sufficiently to prevent juries from being misled in a manner that results in a miscarriage of justice.

Third, the suggestion that the executive branch gets to determine the extent of state secrets evokes the legal principle *nemo iudex in causa sua*. The point of separating and offsetting functional powers is to ensure that each branch not set the bounds of *its own authority*. This reading would contravene the constitutional design, undermining the structural checks and balances. As with the government's other novel interpretations of state secrets, the constitutional claim falls woefully short as a matter of history, logic, and constitutional analysis.

V. Concluding Remarks

In 2009, as Congress became increasingly concerned about the government's use of the state secrets privilege and began drafting legislation to restrict it, the Obama Administration issued guidance, stating it would "invoke the privilege in court only when genuine and significant harm to national defense or foreign relations is at stake and only to the extent necessary to safeguard those interests."[298] The government's move took the political winds from the sails of legislative reform, but it failed to stem the tide driving an ever more expansive interpretation of the common law evidentiary rule. Hearkening back to the torture and rendition cases in the early twenty-first century, the government has in the interim re-crafted the common law rule in four critical ways: (1) collapsing the *Reynolds* evidentiary rule and the *Totten* contractual bar by discarding the *ex ante* argument and adopting a "very subject matter" analysis; (2) asserting state secrets at the pleadings stage to demand dismissal; (3) applying it to broad categories, instead of to specific information; and (4) claiming it as an Article II constitutional power.

[298] Memorandum from the Attorney General to Heads of Executive Departments and Agencies, Heads of Department Components, Policies and Procedures Governing Invocation of the State Secrets Privilege 1 (Sept. 23, 2009), https://www.justice.gov/archive/opa/documents/state-secret-privileges.pdf.

On September 30, 2022, in the midst of the Supreme Court sending *Fazaga* back to the Ninth Circuit, the Attorney General re-issued the 2009 memorandum, again assuring the public that DOJ "is committed to ensuring that the United States invokes the state secrets privilege only when genuine and significant harm to national defense or foreign relations is at stake and only to the extent necessary to safeguard those interests."[299] Based on the cases making their way through the courts, however, that does not appear to be accurate. In the most recent filings in *Fazaga*, now back on remand to the district court, the government casually asserted (without any formal legal declaration) that someone in DOJ had looked at the case and determined that the state secrets assertion remains valid—a remarkable claim seventeen years after the actions in question and despite the considerable amount of information already in the public domain about Operation Flex and Monteilh's actions. It is possible, of course, that the program did not end in 2007 or that a similar initiative continues in a different guise. If so, however, then particularly in light of the significant statutory and constitutional questions at issue there is all the more reason to allow the case to proceed, excluding whatever specific information that may create a genuine risk to U.S. national security as part of the discovery process.

The issues presented in *Fazaga* are not going to go away. In case after case, the privilege is being used in new and more expansive ways. Employed in the context of suits challenging the warrantless collection of information on U.S. citizens, question exists as to whether *any* claim will be able to survive—even where as in *Fazaga*, a significant amount of information in the public domain suggests that the government may be acting outside statutory and constitutional constraints. Whether the government will be held to account remains to be seen. At stake is the future of foundational First, Fourth, and Fifth Amerndment constitutional rights.

[299] Memorandum from the Attorney General to Heads of Executive Departments and Agencies, Heads of Department Components, Supplement to Policies and Procedures Governing Invocation of the State Secrets Privilege 1 (Sept. 30, 2022), https://www.justice.gov/d9/pages/attachments/2022/09/30/supplement_to_policies_and_procedures_governing_invocation_of_the_state_secrets_privilege.pdf.

DAVID A. SKLANSKY

THE NEGLECTED ORIGINS OF THE HEARSAY RULE IN AMERICAN SLAVERY: RECOVERING QUEEN v. HEPBURN

Queen v. Hepburn is the most important hearsay case in American history, but law students almost never learn about it. Decided by the Supreme Court in 1813, with a majority opinion by Chief Justice John Marshall,[1] *Queen v. Hepburn* is the ruling that, more than any other, marked the transformation of the hearsay rule in the United States from a loose, haphazardly applied principle of preference for live testimony, into more or less what it is today: a strict rule of evidentiary exclusion, subject only to a bounded series of exceptions, and applicable even when better proof is unavailable. *Queen v. Hepburn* also had sweeping, and brutal, practical consequences. It buttressed the institution of slavery by closing off one of the few legal avenues through which

David A. Sklansky is the Stanley Morrison Professor of Law at Stanford Law School.

Author's Note: For encouragement, guidance, and helpful criticism, I am indebted to Rick Banks, Jeff Bellin, Rabia Belt, Bennett Capers, Jack Chin, Doug Egerton, Andrew Fede, Jeff Fisher, Henry Louis Gates Jr., Ariela Gross, Jill Lepore, Anna Lvovsky, Doug Melamed, Jisha Menon, Bernadette Meyler, Andrea Roth, Rebecca Scott, Shirin Sinnar, Jeff Sklansky, Norman Spaulding, William Thomas, Laurence Tribe, Langston Ward, and Diego Zambrano. I received superb research assistance from Jacqueline Maldonado, Daniel Slate, and the staff and fellows of the Robert Crown Law Library at Stanford Law School, especially Grace Lo, Aneliese Castro, Jacob Maddox, Yasmin Moreno, Lillian Siegel, and Andrés Felipe Suárez. Wise and careful editing by Justin Driver significantly improved this article.

[1] 11 U.S. (7 Cranch) 290 (1813).

The Supreme Court Review, 2023.
© 2023 The University of Chicago. All rights reserved. Published by The University of Chicago Press. https://doi.org/10.1086/724278

people in bondage could seek their freedom. Seldom has a judicial decision about the laws of proof been so consequential.

Nonetheless, *Queen v. Hepburn* does not appear in any of the widely used evidence casebooks, and most evidence instructors do not know about it. I regret to report that it does not appear in my own casebook, although my co-author and I will remedy that error in the next edition. In my defense, I never encountered *Queen v. Hepburn* when I was a law student, and I taught evidence law for about a decade and a half before stumbling across a reference to the case in the course of research for an earlier article that appeared in this journal, on the background and future prospects of the hearsay rule.[2] Even after reading the decision, though, I did not think it was sufficiently important to assign to my students. It seemed a historical curiosity, not a case that demonstrated anything significant about evidence law. Worse, I feared that the facts of *Queen v. Hepburn* would distract students from the doctrinal questions at the center of the evidence course. The hearsay rule can be notoriously difficult for students to master, or even to believe, and for that reason I introduce the topic early in the course, giving students ample time to work through the rule's convoluted logic and labyrinthine exceptions. I thought that bringing up the enormity of slavery just as we were getting started would throw us off track. I worried, as well, that it might seem exploitative or insensitive, particularly to Black students and other students of color.

---

[2] David Alan Sklansky, *Hearsay's Last Hurrah*, 2009 SUP. CT. REV. 1, 21–23. For earlier commentary on the decision, see LEONARD BAKER, JOHN MARSHALL: A LIFE IN THE LAW 726–27 (1974); DUNCAN J. MACLEOD, SLAVERY, RACE AND THE AMERICAN REVOLUTION 117–18 (1974); CHARLES ALAN WRIGHT & KENNETH W. GRAHAM, 30 FEDERAL PRACTICE AND PROCEDURE § 6321, at 18 (1997); IRVING YOUNGER, HEARSAY: A PRACTICAL GUIDE THROUGH THE THICKET 4–5 (1988); Robert M. Cover, *For James Wm. Moore: Some Reflections on a Reading of the Rules*, 84 YALE L.J. 718, 725–26 (1975); Jason M. Gillmer, *Suing for Freedom: Interracial Sex, Slave Law, and Racial Identity in the Post-Revolutionary and Antebellum South*, 82 N.C. L. REV. 535, 584–85 (2004); and Donald M. Roper, *In Quest of Judicial Objectivity: The Marshall Court and the Legitimation of Slavery*, 21 STAN. L. REV. 532, 533 (1969). For a recent account of the case—far richer and more extensive than anything previous—see WILLIAM G. THOMAS III, A QUESTION OF FREEDOM: THE FAMILIES WHO CHALLENGED SLAVERY FROM THE NATION'S FOUNDING TO THE CIVIL WAR (2020). For other helpful treatments, see PAUL FINKELMAN, SUPREME INJUSTICE: SLAVERY IN THE NATION'S HIGHEST COURT 62–68 (2018); ANDREW FEDE, ROADBLOCKS TO FREEDOM: SLAVERY AND MANUMISSION IN THE UNITED STATES SOUTH 343–45 (2011); Andrew T. Fede, *Not the Most Insignificant Justice: Reconsidering Justice Gabriel Duvall's Slavery Law Opinions Favoring Liberty*, 42 J. SUP. CT. HIST. 7 (2017); and Liza I. Karsai, *The Horse-Stealer's Trial Returns: How* Crawford*'s Testimonial-Nontestimonial Dichotomy Harms the Right to Confront Witnesses, the Presumption of Innocence, and the Beyond a Reasonable Doubt Standard*, 62 DRAKE L. REV. 129, 151–53 (2013).

Those rationales all now strike me as deeply misguided. I can no longer see any justification for teaching evidence law without discussing Mina Queen's lawsuit against her enslaver, and the role the hearsay rule played in keeping her in bondage. It is not just that *Queen v. Hepburn* illustrates, with particular clarity, certain crucial features of the hearsay rule, including the fact that hearsay generally is inadmissible even when live testimony cannot be obtained. Nor is it just that students should see that the hearsay rule, while sometimes serving as a shield against falsehood, can at other times be an instrument of injustice. *Queen v. Hepburn* also offers an object lesson in the law's complicity in the institution of slavery. It serves as a reminder that the history of that complicity lies embedded in virtually every field of American legal doctrine, including the rules of evidence.

I understand why most evidence instructors do not assign *Queen v. Hepburn*, even if they have heard of it, because I made the same choice myself. But that decision now seems to me mistaken, and one purpose of this article is to explain why. I will start by tracing the background of the case and analyzing how the Supreme Court ultimately resolved it. Then I will assess how *Queen v. Hepburn* fits into the history of American slavery and the history of the hearsay rule. I hope that this discussion will interest not just law students and teachers of evidence law, but also legal historians and scholars of race and law. There are important lessons here about the ways in which American law developed hand in hand with slavery—even in areas, like hearsay doctrine, that have not generally been viewed as sites of racial contestation, and even in the decisions of a jurist as celebrated as John Marshall.[3] The article will end with some reflections about the

[3] This article adds to a growing body of scholarship exploring the connections between evidence law and race. For important examples, see Montré D. Carodine, *Race Is Evidence: (Mis)Characterizing Blackness in the American Civil Rights Story*, *in* CIVIL RIGHTS IN AMERICAN LAW, HISTORY, AND POLITICS 64 (Austin Sarat ed., 2014); Bennett Capers, *Evidence Without Rules*, 94 NOTRE DAME L. REV. 867, 885–93 (2017); George Fisher, *The Jury's Rise as Lie Detector*, 107 YALE L.J. 575, 671–97 (1997); Jasmine B. Gonzales Rose, *Toward a Critical Race Theory of Evidence*, 101 MINN. L. REV. 2243 (2017); Chris Chambers Goodman, *The Color of Our Character: Confronting the Racial Character of Rule 404(b) Evidence*, 25 LAW & INEQ. 1 (2007); Thomas D. Morris, *Slaves and the Rules of Evidence in Criminal Trials*, 68 CHI.-KENT L. REV. 1209 (1992); Maria L. Ontiveros, *Adoptive Admissions and the Meaning of Silence: Continuing the Inquiry Into Evidence Law and Issues of Race, Class, Gender, and Ethnicity*, 28 SW. U. L. REV. 337 (1999); and Anna Roberts, *Reclaiming the Importance of the Defendant's Testimony: Prior Conviction Impeachment and the Fight Against Implicit Stereotyping*, 83 U. CHI. L. REV. 835 (2016). The article also builds on the wide-ranging assessment of John Marshall's relationship with slavery in FINKELMAN, *supra* note 2.

continuing significance of *Queen v. Hepburn*, and about how the case should be approached in law school classrooms.

I.

In January 1810 a woman named Mina Queen filed a petition in the Circuit Court of the District of Columbia charging that she and her daughter, Louisa Queen, were being held in slavery illegally by a man named John Hepburn.[4] The basis for the petition was Mina Queen's claim that her great grandmother, Mary Queen, had come to the American colonies as a free woman and an indentured servant, not a slave.[5] Hepburn was summoned into court, and a jury trial was held five months later. The Queens were represented by Francis Scott Key, the young lawyer and amateur poet who four years later would write "The Star-Spangled Banner."[6]

We know much less about Mina and Louisa Queen than about their lawyer, but recent historical work has filled in some of the blanks. Between the 1780s and the outbreak of the Civil War, thousands of enslaved people sued for their freedom in American courts, and a good number of these suits were successful.[7] Prominent

[4] Regarding the background of *Queen v. Hepburn*, see THOMAS, *supra* note 2; and O SAY CAN YOU SEE: EARLY WASHINGTON, D.C., LAW & FAMILY, an extensive online collection of documents assembled by Professor Thomas in collaboration with Jennifer Guiliano and Trevor Muñoz. Mina Queen's first name appears as "Mima" in William Cranch's case reports for the Supreme Court and for the District of Columbia Circuit Court, but that appears to have been an error. *See Mima Queen & Louisa Queen v. John Hepburn*, *in* O SAY CAN YOU SEE, *supra*, https://earlywashingtondc.org/cases/oscys.caseid.0011.

[5] Many scholars today avoid the term "slave" in favor of "enslaved person," because "slave" can suggest that slavery was a matter of a someone's essence, rather than a condition imposed by others. There are also scholars, though, who believe that "slave" more accurately captures the ruthless manner in which "American law and society reduced persons of African descent to items of property." Darren Lenard Hutchinson, *"With All the Majesty of the Law": Systemic Racism, Punitive Sentiment, and Equal Protection*, 110 CALIF. L. REV. 371, 380 n.45 (2022). I have alternated usages in this article, in an effort to keep in mind both the personhood of the enslaved and the brutally dehumanizing nature of the institution of slavery.

[6] THOMAS, *supra* note 2, at 160; MARK CLAGUE, O SAY CAN YOU HEAR?: A CULTURAL HISTORY OF "THE STAR-SPANGLED BANNER" (2022).

[7] A large and illuminating literature on freedom suits has emerged in recent decades. See, for example, FEDE, *supra* note 2; ALEJANDRO DE LA FUERTE & ARIELA GROSS, BECOMING FREE, BECOMING BLACK: RACE, FREEDOM, AND LAW IN CUBA, VIRGINIA, AND LOUISIANA 84–100, 114–31 (2020); KELLY M. KENNINGTON, IN THE SHADOW OF *Dred Scott*: ST. LOUIS FREEDOM SUITS AND THE LEGAL CULTURE OF SLAVERY IN ANTEBELLUM AMERICA (2017); MACLEOD, *supra* note 2, at 109–26; LOREN SCHWENINGER, APPEALING FOR LIBERTY: FREEDOM SUITS IN THE SOUTH (2018); THOMAS, *supra* note 2; ANNE TWITTY, BEFORE *Dred Scott*: SLAVERY AND LEGAL CULTURE IN THE AMERICAN CONFLUENCE, 1787–1857 (2016); LEA VANDERVELDE, REDEMPTION SONGS: SUING FOR FREEDOM BEFORE *Dred Scott* (2014); KIMBERLY M. WELCH,

lawyers like Key often represented these litigants, while in many instances continuing to own slaves themselves.[8] Key, for example, argued more than a hundred freedom petitions between 1807 and 1840, but he was himself a slaveholder and far from an abolitionist.[9] Key was active in the American Colonization Society, which called for expelling freed Black people from the United States and forcibly transporting them to Africa, but not for ending the institution of bondage.[10] On several occasions, as United States Attorney for the District of Columbia in the 1830s, he sought to use the power of his office to suppress abolitionism.[11]

The principal motivation for most of the lawyers who represented enslaved people seeking their freedom does not appear to have been pecuniary. Slaveholders who lost these suits sometimes were ordered to pay the fees of the petitioners' lawyers, and in other cases antislavery societies paid lawyers to bring freedom suits. But by and large these cases were not money makers; the lawyers seem to have acted out of some mixture of careerism and a sense of professional obligation.[12]

BLACK LITIGANTS IN THE ANTEBELLUM SOUTH (2018); and Gregory Ablavsky, Comment, *Making Indians "White": The Judicial Abolition of Native Slavery in Revolutionary Virginia and Its Racial Legacy*, 159 U. PA. L. REV. 1457, 1477–94 (2011).

[8] KENNINGTON, *supra* note 7, at 90; SCHWENINGER, *supra* note 7, at 241–45; TWITTY, *supra* note 7, at 97–101; WELCH, *supra* note 7, at 88–92.

[9] SCHWENINGER, *supra* note 7, at 243. Regarding Key's attitudes toward slavery and recent controversies about the lyrics of *The Star-Spangled Banner*, see CLAGUE, *supra* note 6. Professor Clague concludes that the anthem's most notorious line—"No refuge could save the hireling and slave/From the terror of flight or the gloom of the grave"—probably was not meant, nor originally understood, to refer to Black people in bondage, but instead alluded, metaphorically, to white people in the service of the English Crown, or perhaps to a particular English officer. But this itself reflected deep-seated racism: in the context of the song, Black people held in actual slavery were beneath notice. *Id.* at 189–96.

[10] CLAGUE, *supra* note 6, at 198, 204, 207–09; Timothy S. Huebner, *Roger Taney and the Slavery Issue: Looking Beyond—and Before—*Dred Scott, 97 J. AM. HIST. 17, 19–21 (2010). Regarding the American Colonization Society, which came to be adamantly opposed by the vast majority of abolitionists and freed African Americans, see OUSMANE K. POWER-GREENE, AGAINST WIND AND TIDE: THE AFRICAN-AMERICAN STRUGGLE AGAINST THE COLONIZATION MOVEMENT (2014); THOMAS, *supra* note 2; TWITTY, *supra* note 7, at 112–13; Douglas R. Egerton, *"Its Origin is Not a Little Curious": A New Look at the American Colonization Society*, 5 J. EARLY REPUBLIC 463 (1985); and Nicholas Guyatt, *"The Outskirts of Our Happiness": Race and the Lure of Colonization in the Early Republic*, 95 J. AM. HIST. 986 (2009).

[11] CLAGUE, *supra* note 6, at 206; JEFFERSON MORLEY, SNOW-STORM IN AUGUST: WASHINGTON CITY, FRANCIS SCOTT KEY, AND THE FORGOTTEN RACE RIOT OF 1835, at 108–10, 139, 211–20 (2012).

[12] KENNINGTON, *supra* note 7, at 89–96; SCHWENINGER, *supra* note 7, at 241–42, 260; THOMAS, *supra* note 2, at 63; TWITTY, *supra* note 7, at 96–119; VANDERVELDE, *supra* note 7, at 8–9, 201; WELCH, *supra* note 7, at 92–99.

Historians have been divided over the importance of these cases. Some have argued that they served mainly to buttress the institution of slavery by giving it a veneer of principled legality, while others, especially in recent years, have suggested that freedom cases were opportunities for people in bondage to speak their own truths, to confront their enslavers, to reclaim their own agency—and, in a significant number of cases, to win their liberty.[13]

Many of the freedom suits, like Mina Queen's, were rooted in claims of genealogy, with the petitioners arguing that they were descended from free women, or from women who had lived free until they were themselves illegally enslaved.[14] The specific nature of the claims varied: in Virginia, for example, the plaintiffs usually argued that they were descended from Native Americans; in Maryland, in contrast, the typical freedom plaintiff argued that he or she was descended from a white woman, or from a woman who had come to the United States unenslaved.[15] Years before Mina Queen filed her petition, some two dozen of her relatives had won their freedom in Maryland courts, based on the same claim later advanced by Mina Queen: that their common ancestor, Mary Queen, had come to America in indentured servitude, not in bondage. The Maryland plaintiffs, enslaved on a plantation owned by the Jesuits who founded Georgetown College in 1789, had proven their cases with depositions taken from people who reported hearing that Mary Queen had come from South America by way of England, not from Africa, that she had dressed like a free woman, and that by rights she should have been free.[16] Her geographic origins were important in part because

[13] *E.g.*, Kennington, *supra* note 7, at 8, 10, 192–93; Thomas, *supra* note 2, at 8–9; Vandervelde, *supra* note 7, at 1; Ablavsky, *supra* note 7, at 1460–61, 1477; Michael L. Nicholls, *"The Squint of Freedom": African-American Freedom Suits in Post-Revolutionary Virginia*, 20 Slavery & Abolition 47, 51 (1999).

[14] American law generally specified that slave status passed from mother to child, *see, e.g.*, Darrel A.H. Miller, *The Thirteenth Amendment and the Regulation of Custom*, 112 Colum. L. Rev. 1811, 1826 (2012), and a majority of early freedom suits, concentrated along the eastern seaboard, were based on claims that a matrilineal ancestor had been free, *see* Schweninger, *supra* note 7, at 118–19; Fede, *supra* note 2, at 8; Ariela J. Gross, *Litigating Whiteness: Trials of Racial Determination in the Nineteenth-Century South*, 108 Yale L.J. 109, 141 & n.117 (1998); Ablavsky, *supra* note 7, at 1478. The situation differed further west, and it changed everywhere after the Supreme Court's decision in *Queen v. Hepburn*. *See infra* note 86 and accompanying text.

[15] *See, e.g.*, MacLeod, *supra* note 2, at 110; Thomas, *supra* note 2; Ablavsky, *supra* note 7, at 1459–60.

[16] Thomas, *supra* note 2, at 16–42, 47–85.

Lord Mansfield's 1772 decision in *Somerset v. Stewart* had provided grounds for arguing that no one who had passed through England could lawfully be enslaved.[17]

When Mina Queen's freedom petition came for trial, however, the Circuit Court of the District of Columbia excluded crucial portions of the evidence that her relatives in Maryland had introduced. Some deposition testimony was kept out of evidence because the witnesses failed to specify the sources of their information. The critical parts of other depositions were excluded because they recounted what today would be called "double hearsay": for example, what a witness's mother had heard from the witness's grandfather.[18] (The deposition itself was not considered an additional layer of hearsay: then, as today, it was common for sworn depositions to be introduced in civil cases in lieu of live, in-court testimony.)

Key also tried to introduce new evidence, not available when the Maryland cases had been tried: a deposition from a man named Fredus Ryland, who recalled having spoken directly with Mary Queen. According to Ryland, Mary Queen said that she had been born free in Guayaquil, a port city in what is now Ecuador, and that she had afterward lived in London.[19] (Guayaquil and the surrounding region had a significant population of free Black people and people of mixed ancestry in the eighteenth century, and probably also the 1690s, when Mary Queen was likely born.[20] Some of the hearsay evidence excluded

---

[17] 98 Eng. Rep. 499 (K.B. 1772); THOMAS, *supra* note 2, at 32–34. The meaning of *Somerset* was a matter of controversy at the time it was decided, and continues to be so today. *See, e.g.*, Derek A. Webb, *The* Somerset *Effect: Parsing Lord Mansfield's Words on Slavery in Nineteenth Century America*, 32 LAW & HIST. REV. 455 (2014). Some lawyers and judges understood the decision to mean that an enslaved person became permanently free upon setting foot in England, whereas others suggested that a subsequent departure from England would restore the condition of bondage. *Compare* Mahoney v. Ashton, 4 H. & McH. 210, 211–12 (Md. Gen. Ct. 1798) (noting argument that "as any person sets his foot on British ground he is free"), *and* Mahoney v. Ashton, 4 H. & McH. 295, 304 (Md. Gen. Ct. 1799) (Chase, C.J.) (reading *Somerset* to hold that "a slave was free by coming to England"), *with* Mahoney v. Ashton, 4 H. & McH. 295, 320 (Md. App. 1799) (noting argument that "the moment the master and slave left England, the master's right revived"). *Somerset* was also the basis for the doctrine, applied in several slave states, that residing in a free state put a permanent end to a person's bondage: "once free, always free." TWITTY, *supra* note 7, at 12–13; *see also* KENNINGTON, *supra* note 7, at 85, 218 n.37.

[18] Queen v. Hepburn, 11 U.S. (7 Cranch) 290, 294–95 (1813).

[19] THOMAS, *supra* note 2, at 169–71.

[20] *Id.* at 175; KRIS E. LANE, QUITO 1599: CITY AND COLONY IN TRANSITION 61 (2002); TAMARA J. WALKER, EXQUISITE SLAVES: RACE, CLOTHING, AND STATUS IN COLONIAL LIMA 3 (2017); Michael L. Conniff, *Guayaquil Through Independence: Urban Development in a Colonial System*, 33 AMERICAS 385, 402–04 (1977).

in Mina Queen's lawsuit referred to Mary Queen's complexion as "yellow," a description often applied in eighteenth-century America to people of Spanish descent, but also applied to many people of mixed or African ancestry.[21]) Ryland's deposition was ruled inadmissible as well, apparently on the ground that since an enslaved person could not testify against a white litigant, her hearsay statements were also inadmissible.[22]

The Ryland deposition had also been excluded several days earlier in the trial of a separate, contemporaneous freedom petition that Key had filed in the District of Columbia on behalf of Priscilla Queen, who may have been Mina's first cousin, against Priscilla's enslaver Francis Neale, a Jesuit priest who was later president of Georgetown College.[23] In that case, the judges explained that "[t]he declarations of an ancestor, while held as a slave, cannot be given in evidence."[24] They nonetheless allowed Key to introduce the deposition of a separate witness, John Jiams, who recalled hearing other people, now deceased, describe Mary Queen as a free woman.[25] In Mina Queen's case, though, the judges instructed the jury to disregard the Jiams deposition unless they concluded he had been speaking from his own firsthand knowledge, rather than repeating information he had obtained from unnamed others.[26]

The Maryland and District of Columbia courts thus reached different decisions about the admissibility of the depositions offered to prove that Mary Queen had been free. Nonetheless the courts appeared to share an important, underlying assumption: that there was no across-the-board prohibition of hearsay evidence, either in general or with regard to the freedom of a litigant's ancestor. That was partly, as we will see, because both in Maryland and in the District of

---

[21] THOMAS, *supra* note 2, at 66, 75; SHARON BLOCK, COLONIAL COMPLEXIONS: RACE AND BODIES IN EIGHTEENTH-CENTURY AMERICA 67–68 (2018). Justice Gabriel Duvall, who dissented from the Supreme Court's affirmance of the judgment against Mina Queen, later described Mary Queen as "a yellow woman, a native of South America." *Davis v. Wood*, 14 U.S. (1 Wheat) 6, 7–8 (1816).

[22] 11 U.S. at 294. Even emancipated African Americans were generally barred from testifying against white litigants in slave states, and in some free states. *See* Gonzales Rose, *supra* note 3, at 2247–48; Morris, *supra* note 3, at 1209–10.

[23] THOMAS, *supra* note 2, at 160.

[24] Queen v. Neale, 2 D.C. (2 Cranch) 3 (1810).

[25] *Id.* For reasons that are unclear, Priscilla Queen's case does not appear to have proceeded to a verdict.

[26] 11 U.S. at 294–95.

Columbia the courts were operating within a legal universe in which the rule against hearsay, as a rigid, categorical ban, was still in the process of solidifying.[27]

With little admissible evidence to support Mina Queen's claims about her great-grandmother, the jury rejected her petition for freedom.[28] Key appealed the case to the Supreme Court, where he argued it two years later, assisted by James Morsell, who subsequently served as judge on the District of Columbia Circuit Court.

At the time that *Queen v. Hepburn* was litigated, Maryland law governed in the portion of the District of Columbia north of the Potomac River.[29] But the lawyers who argued the case before the Supreme Court, and the Justices themselves, treated the hearsay question as a matter of "general law": a legal principle that crossed jurisdictional lines.[30] Key told the Justices that the issue in the case was relatively narrow: whether the fact that Mary Queen had been a free woman could be proved by "hearsay of hearsay." If it could not, he argued, "it will be impossible to prove any ancient fact."[31] Hepburn's lawyers argued for affirmance on even narrower grounds: that there was no proof that statements related by the deponents had been made by people who were now dead, and that the initial declarants had not even been identified. But then they broadened their argument, urging that the use of hearsay evidence should be limited to proving "prescription, custom, and pedigree." And in this case, they contended, the hearsay in question had not been offered to prove ancestry, custom, or common usage, but "to prove that a certain ancestor came from England."[32]

In essence, Hepburn's lawyers were arguing that the ban on hearsay should be understood as a rigid, categorical rule, not just a general principle to be balanced against other equities, and not a rule

---

[27] *See, e.g.*, Thomas, *supra* note 2, at 171.

[28] *Id.* at 180.

[29] Thomas, *supra* note 2, at 166; Kent Newmyer, *On Assessing the Court in History: Some Comments on the Roper and Burke Articles*, 21 Stan. L. Rev. 540, 542–43 (1969).

[30] Regarding the concept of "general law" and its centrality to nineteenth-century jurisprudence, see Caleb Nelson, *The Persistence of General Law*, 106 Colum. L. Rev. 503, 505–06 (2006); and Caleb Nelson, *A Critical Guide to* Erie Railroad v. Thompkins, 54 Wm. & Mary L. Rev. 921, 925–49 (2013).

[31] 11 U.S. at 291–92. Cranch's report separates out the arguments made by each of the four lawyers who argued the case in the Supreme Court; the version incorporated into the U.S. Reports does not.

[32] *Id.* at 292.

with a fluid, evolving set of exceptions. If they were right, and if hearsay was flatly inadmissible to prove that an enslaved plaintiff's ancestor had been free, it would mean not only that the Maryland courts had been wrong when they freed Mina Queen's relatives, but that the Circuit Court of the District of Columbia had also been wrong in Priscilla Queen's case, when they ruled the deposition by John Jiams fully admissible. In his reply for the appellants, James Morsell argued against the idea that hearsay was automatically inadmissible. Instead, he said, "[t]he general rule of evidence" was that the best available proof should be admitted: "if the evidence offered be the best which the nature of the case admits, and leaves no presumption that there is better behind, it is admissible." And with regard to whether Mary Queen had been free or enslaved, "[a]fter a lapse of 100 years better evidence than this cannot be expected."[33]

The Supreme Court, though, fully embraced the broad ground for affirmance that had been urged by Hepburn's lawyers. Hearsay, Chief Justice Marshall explained for the Court, "is in its own nature inadmissible," even when testimony from a live witness was impossible. There were narrow exceptions to the hearsay ban, for "cases of *pedigree*, of *prescription*, of *custom*, and in some cases of *boundary*." In addition, Marshall said, "[t]here are also matters of general and public history which may be received without that full proof which is necessary for the establishment of a private fact." Beyond that, though, "[t]he danger of admitting hearsay evidence is sufficient to admonish Courts of justice against lightly yielding to the introduction of fresh exceptions to an old and well established rule." Indeed, given the great importance of the hearsay rule, the Supreme Court was "not inclined to extend the exceptions further than they have already been carried."[34]

The reason for excluding hearsay, Marshall explained, is not just that "some better testimony . . . might be adduced." Even when no better evidence is available, the "intrinsic weakness" of hearsay, "its incompetency to satisfy the mind of the existence of the fact, and the frauds which might be practiced under its cover, combine to support the rule that hearsay evidence is totally inadmissible." Marshall therefore rejected the notion that hearsay should ever be admitted "to

[33] *Id.* at 293.

[34] *Id.* at 295–97.

prove a *specific fact*, because the eye witnesses to that fact are dead." If, on the contrary, "the circumstance that the eye witnesses of any fact be dead should justify the introduction of testimony to establish that fact from hearsay," then "no man could feel safe in any property, a claim to which might be supported by proof so easily obtained."[35]

The fact that the "property" in this case consisted of human beings did not, in Marshall's view, alter the analysis. "However the feelings of the individual may be interested on the part of a person claiming freedom," he explained, "the Court cannot perceive any legal distinction between the assertion of this and of any other right."[36]

The sole dissenter in the case was Justice Gabriel Duvall. Duvall served on the Supreme Court for more than two decades, from 1811 to 1835, and *Queen v. Hepburn* was the only case in which he ever wrote a dissenting opinion.[37] He had personal connections to the underlying litigation: he had represented Mina Queen's relatives when they successfully petitioned for freedom in Maryland, he had collected some of the deposition testimony that Key had tried unsuccessfully to introduce on behalf of Mina Queen, including the deposition of Fredus Ryland, and he had testified as a witness at the trial of Mina Queen's petition.[38] Today, connections of this kind would almost certainly prevent a Justice from participating in the Supreme Court's consideration of a case, but standards for recusal were looser in the early nineteenth century.[39]

Duvall noted in his dissent that Maryland courts had allowed hearsay evidence to prove the freedom of an ancestor who had been dead for so long that live testimony could not be obtained. That was the only sensible result, Duvall argued. Hearsay should be admissible in such cases for the same reason that was admissible to prove custom, pedigree, or the boundaries of land: it was necessary. "From the

---

[35] *Id.* at 295–96.

[36] *Id.* at 295.

[37] Cover, *supra* note 2, at 726; Roper, *supra* note 2, at 533. Duvall also dissented, but without a written opinion, in *Trustees of Dartmouth College v. Woodward*, 17 (4 Wheat.) U.S. 518 (1819). *See Symposium on Hearsay Reform*, 84 FORDHAM L. REV. 1323, 1366–67 (2016) (remarks of Stephen A. Salzburg).

[38] *See, e.g.*, THOMAS, *supra* note 2, at 42, 62–63, 78, 180. Regarding Duvall's representation of freedom plaintiffs in Maryland, see also Fede, *supra* note 2, at 8–11.

[39] THOMAS, *supra* note 2, at 366 n.37. Duvall did recuse himself, when he was a state court judge in Maryland, from participating in a freedom case in which he had represented the petitioner. Mahoney v. Ashton, 4 H. & McH. 210, 215 (Md. Gen. Ct. 1798).

antiquity of the transactions to which these subjects may have reference" he explained, "it is impossible to produce living testimony"; therefore "to exclude hearsay in such cases, would leave the party interested without remedy."[40]

Like Key, Duvall was himself a slaveowner[41]—as was John Marshall, on a considerably larger scale.[42] But it was obvious to Duvall that the arguments for admitting hearsay in support of a freedom petition were far weightier than in cases about land boundaries. "It will be universally admitted," he wrote, "that the right to freedom is more important than the right of property. And people of color from their helpless condition under the uncontrolled authority of a master, are entitled to all reasonable protection."[43] Duvall warned that the decision in *Queen v. Hepburn* would slam the door on any freedom petition based on claims of an ancestor's freedom. "A decision that hearsay evidence in such cases shall not be admitted," he said, "cuts up by the roots all claims of the kind, and puts a final end to them, unless the claim should arise from a fact of recent date, and such a case will seldom, perhaps never, occur."[44]

Duvall's admonition proved prophetic. As we will see, the Supreme Court's decision in *Queen v. Hepburn* helped to close off a key legal channel through which enslaved people in the late eighteenth and early nineteenth century sought their freedom. It also did more than any other single case to turn the hearsay ban into what it is today: a categorical, often unforgiving rule of exclusion, not a flexible, intermittently applied principle of preference. *Queen v. Hepburn* did not by itself create the modern hearsay rule, but it was a landmark in the rule's development.

II.

Nearly two decades before he wrote the Supreme Court's decision in *Queen v. Hepburn*, John Marshall argued, unsuccessfully, for the admissibility of hearsay in a case before Virginia's highest court, then called the Supreme Court of Appeals. The case was

---

[40] 11 U.S. at 298.

[41] Thomas, *supra* note 2, at 183, 219–21.

[42] Finkelman, *supra* note 2, at 36–50.

[43] 11 U.S. at 298–99.

[44] *Id.* at 299.

Claiborne v. Parrish,[45] and it offers a helpful window into how the hearsay rule was generally understood in post-Revolutionary America. The plaintiff sued to recover a large parcel of land, arguing that it had been fraudulently surveyed. In support of this claim the plaintiff called a witness who testified, over objection, that he had heard one of the surveyor's assistants—the "chain carriers"—admit to having measured the land unfairly.[46] The jury ruled for the plaintiff, but the Supreme Court of Appeals reversed the judgment. In so doing, the high court made clear that hearsay was generally inadmissible, but that exceptions were allowed when no better evidence was available.

The lead opinion was by the court's most influential member, Spencer Roane.[47] Roane started by acknowledging that, "by the general rules of law," hearsay was inadmissible—a doctrine that he said was "so well founded and so generally known that it is unnecessary to state it." But there were also exceptions to the rule, "warranted by reasons as sound as those which induced the rule itself." These included cases involving "pedigrees" or "other ancient transactions, in which it is the best evidence which the nature of such cases admit of." These exceptions, in turn, were limited by the requirement that "there is no better evidence behind, and in the power to produce"; otherwise, there was "a presumption against the admission of inferior testimony." Thus, in Roane's view, the basic idea of the hearsay rule was to exclude secondhand testimony when better, firsthand testimony could be introduced instead. The problem in *Claiborne v. Parrish* was that Marshall's client had not provided grounds to believe that any exception to the hearsay rule applied—i.e., "that the transaction was ancient, his witnesses dead, and that their attendance could not be procured, or the like."[48] Even disregarding those last three words—"or the like"—Roane's argument hardly suggested that

---

[45] 2 Va. (2 Wash.) 146 (1795).

[46] *Id.* at 146.

[47] Roane started his career as an Anti-Federalist politician and in later years, when he and John Marshall were both on the bench, he arguably became Marshall's most important adversary. *See* David Johnson, Irreconcilable Founders: Spencer Roane, John Marshall, and the Nature of America's Constitutional Republic (2021); F. Thornton Miller, *John Marshall in Spencer Roane's Virginia: The Southern Constitutional Opposition to the Marshall Court*, 33 J. Marshall L. Rev. 1131 (2000); Note, *Judge Spencer Roane of Virginia: Champion of States' Rights—Foe of John Marshall*, 66 Harv. L. Rev. 1242 (1953).

[48] 2 Va. at 147–48.

hearsay could be admitted only if it fell within the narrow, traditional categories of land boundaries and family lineage; it suggested, on the contrary, that hearsay might be introduced to prove any very old "transaction" when live testimony was unattainable.

The president of the Supreme Court of Appeals, Judge Edmund Pendleton, filed an opinion that was even clearer on this point. Hearsay evidence "may be proper," Pendleton wrote, "where from the nature of the case it is not expected that better testimony could be procured." This would be true, for example, "where the transaction was ancient." In this case, though, there was no claim that the chain carrier in fact "was dead, or that his attendance could not be procured"—nor anything else to justify "a departure from the general rule."[49]

This view of the hearsay rule—that it was an application of the principle that better evidence should be preferred to inferior evidence—was widespread in America in the first decades of independence. For example, Chancellor James Kent of New York wrote in 1812 that the key problem with hearsay was that it was "resorting to an inferior or secondary species of proof, without necessity." In fact, even when a statement made outside of court was "not, technically, hearsay evidence"—because it was introduced for some purpose other than proving the truth of what it asserted—Kent thought that secondhand evidence of the statement should not be introduced, if it was possible to bring the speaker himself into court to testify about what was said. "It is against sound principle, and would at once awaken distrust," he explained, "for a party to resort to a secondary species of evidence, so long as the original and primary evidence exists and can be produced."[50]

Similarly, Chief Justice Zephaniah Swift of Connecticut explained in his 1810 treatise on evidence and commercial law that although

[49] *Id.* at 148. Both Roane and Pendleton also thought that the chain carrier should not have been allowed to testify even if he had appeared in court, because his own statement showed him guilty of fraud, which made him incompetent as a witness. *Id.* at 148–49.

[50] Coleman v. Southwick, 9 Johns. 45, 49–50. Kent's opinion thus was notable not just for treating the hearsay rule as a principle of preference, but also for how strongly and widely it suggested the preference should apply. *See* Thomas, *supra* note 2, at 367 n.44. His proposal that firsthand testimony, if available, should be required even outside the scope of the hearsay rule—that is to say, even with regard to statements not introduced to prove the truth of what they asserted—was not taken up. *See* Esek Cowen & Nicholas Hill, Notes to Phillips' Treatise on the Law of Evidence 569–71 n.434 (1839) (criticizing Kent's suggestion as contrary to the weight of authority). I thank William Thomas for pointing to me this source and highlighting its significance.

"[h]earsay evidence, generally speaking, is not admissible," the rule did not apply "where it is impossible, in the nature of things," to obtain live testimony, and where the hearsay evidence "is sufficient to establish the matter in question." This would typically happen, he noted, "in matters of long standing, where the witnesses, who were knowing to them are not in being."[51]

Like Roane, Kent and Swift were unusually influential judges, even beyond the confines of their respective states.[52] Still, it is important not to exaggerate the clarity or consistency of the rules regarding hearsay in the years leading up to *Queen v. Hepburn*. There were treatises and decisions that suggested the hearsay ban operated more rigidly, or might do so. This was notably true in English sources widely consulted by American judges and lawyers. For example, Blackstone's *Commentaries on the Laws of England*, published in the 1760s, treated the hearsay rule as application of the preference for the best evidence available—just as Roane, Kent, and Swift would do four decades later. "The one general rule that runs through all the doctrines of trials," Blackstone explained, was "that the best evidence the nature of the case will admit shall always be required, if possible to be had; but, if not possible, then the best evidence that can be had shall be allowed." Nonetheless Blackstone suggested that exceptions to hearsay rule were not completely open-ended, but instead were limited to questions of reputation and general opinion: "So, no evidence of a discourse with another will be admitted, but the man himself must be produced; yet in some cases (as in proof of any general customs, or matters of common traditions or repute) the courts admit of *hearsay* evidence, or an account of what persons deceased have declared in their life-time: but such evidence will not be received of any particular facts."[53]

Blackstone's *Commentaries* had an exceptionally broad readership in America,[54] so his brief discussion of hearsay is worth pausing over. Three features of that discussion deserve particular attention for what

---

[51] Zephaniah Swift, A Digest of the Law of Evidence in Civil and Criminal Cases and a Treatise on Bills of Exchange and Promissory Notes 121 (1810).

[52] *See, e.g.*, Farah Peterson, *Expounding the Constitution*, 130 Yale L.J. 2, 73 (2020).

[53] 3 William Blackstone, Commentaries on the Laws of England *368 (1768).

[54] *See, e.g.*, Albert W. Altschuler, *Rediscovering Blackstone*, 145 U. Pa. L. Rev. 1, 1–19 (1996); Julian S. Waterman, *Thomas Jefferson and Blackstone's Commentaries*, 27 U. Ill. L. Rev. 629, 630–34 (1933).

they say about how the ban on hearsay was understood in the late 1700s and early 1800s.

First, Blackstone was hardly alone in thinking that the cornerstone of evidence law was a preference for the best available proof. This was the dominant understanding both in England and in America. It persisted until the end of the nineteenth century, when the modern view—replacing the best evidence principle with the less demanding requirement of mere relevance—was successfully championed in the United States by James Bradley Thayer and then, in the early twentieth century, by Thayer's student, John Henry Wigmore.[55]

Second, the distinction Blackstone drew between proof of "particular facts" and proof of "customs," "traditions," or "repute," was also widespread in the late eighteenth and early nineteenth centuries. As we have seen, Chief Justice Marshall's opinion for the Supreme Court in *Queen v. Hepburn* declared hearsay inadmissible "to prove a *specific fact*," while acknowledging its admissibility in "cases of *pedigree*, of *prescription*, of *custom*, and in some cases of *boundary*," and suggesting it might also be allowed as to "matters of general and public history."[56] The rough idea was that hearsay should be allowed to prove what people generally thought and said, but not what a particular person, who could not be examined in court, reported having observed.

Thomas Peake's treatise on evidence law, published in England and widely consulted in the United States, echoed Blackstone in stressing that "a distinction should be attended to between hearsay evidence of mere *facts*, and of *general reputation*,"[57] although Peake suggested the distinction might not matter with regard to questions of family history. Peake explained that "in cases of *pedigree*, declarations of deceased members of the family as to birth, marriage, or

---

[55] *See* Dale A. Nance, *The Best Evidence Principle*, 73 Iowa L. Rev. 227, 230–56 (1988). Nance notes that arguments similar to those advanced by Thayer and Wigmore also surfaced in England in the late nineteenth and early twentieth century. *See id.* at 248 n.103 (citing Sidney L. Phipson, The Law of Evidence 12–13 (1892); and J.R. Gulson, The Philosophy of Proof in Its Relation to the English Law of Judicial Evidence 448–54 (1905)).

[56] Queen v. Hepburn, 11 U.S. (7 Cranch) 290, 296 (1813).

[57] Thomas Peake, A Compendium of the Law of Evidence 13 (1804). Peake's treatise went through four revisions, the last published in 1822, with American editions following quickly on the heels of the London editions. A companion to the treatise, collecting American cases, appeared in 1810. Samuel Bayard, A Digest of American Cases on the Law of Evidence Intended as Notes to Peake's Compendium of the Law of Evidence (1810). James Morsell cited Peake's treatise in his reply argument to the Supreme Court on behalf of Mina Queen. 11 U.S. at 293.

death of any member of it, are admitted; but in cases of *custom* and *prescription*, which can only be supported by a variety of facts, and by long and uniform usage, the general reputation only can be proved by hearsay evidence." This meant that a "witness may be permitted to state what he has heard from dead persons respecting the reputation of the right; but not to state facts of the exercise of it which the dead persons said they had seen." Peake gave an example:

> Thus in a case tried before Mr. Justice Chambre, at Shrewsbury, where the question on the record was, whether a turnpike was erected within or out of the limits of the town of Wen; that learned judge permitted the plaintiff, who contended it was within the town, to give evidence of general reputation that the town extended to a close called the Townend Piece, and that old people, deceased, had said that such was the boundary of the town; but he would not suffer it to be proved that deceased persons had said that there were formerly houses where none then stood, observing that this was evidence of a particular fact and not of general reputation.[58]

The distinction between reputation and "facts" can sound strange today, but it made more sense in the late eighteenth and early nineteenth century, when physical records were fragmentary and often impossible to access, and "gossip and hearsay formed the backdrop of all legal deliberations at the local level."[59] Then as now, some legal claims and defenses explicitly depended on proving the understanding passed down within a community: this was true of ancient boundaries, and of property rights obtained by prescription—i.e., by longtime, well known use. More generally, the knowledge that witnesses could convey in court, especially in agrarian societies like the American South, typically came from conversations with neighbors and acquaintances.[60] For distant events, secondhand accounts originating from someone who could not now be questioned in court were commonly thought less reliable than the distilled understanding of the community, handed down over the years. This was true even for the questions of family history as to which Peake said hearsay evidence of specific "facts" *was* allowed, although in this case what mattered was the knowledge passed down within a family, not within a community.

---

[58] PEAKE, *supra* note 57, at 13–14.

[59] LAURA F. EDWARDS, THE PEOPLE AND THEIR PEACE: LEGAL CULTURE AND THE TRANSFORMATION OF INEQUALITY IN THE POST-REVOLUTIONARY SOUTH 118 (2009).

[60] *See id.*

Accordingly, to prove matters of family history, such as "who a man married, that A. died abroad, &c.," Peake said that "declarations of deceased persons, who from their situation were likely to know, *and the general belief of the family*, are sufficient."[61]

The third thing worth noting about Blackstone's discussion of hearsay is how rough and undeveloped it was. There was no suggestion that the hearsay ban was limited to statements introduced to prove the truth of what they assert—a critical component of the modern rule. Instead, Blackstone asserted that the rule banned any "evidence of a discourse with another." He also suggested that hearsay was admissible as long as the declarant appeared in court, which is not how the modern rule operates. The only other exception he discussed was limited to statements by people who were no longer living, introduced to prove customs, traditions, or reputation. He gave no hint of what are today the major exceptions to the hearsay rule: dying declarations, party admissions, business records, excited utterances, and so on.

Blackstone suggested that readers interested in a more detailed discussion of the rules governing proof at trial should turn to Geoffrey Gilbert's treatise, *The Law of Evidence*, posthumously published in 1765. If anything, Gilbert's discussion of hearsay is even further than Blackstone's from modern understandings of the rule. Gilbert declares that "a mere Hearsay is no Evidence," because it is unsworn. But this statement comes in the midst of a discussion of the factors that should influence the weight given to oral testimony. Furthermore, Gilbert says that testifying witnesses are free to describe their own, consistent statements, given on earlier occasions. The bulk of Gilbert's treatise is devoted to the admissibility of written documents, which he says is governed by the overarching principle that the best available evidence should be introduced; he does not discuss any limitations on written hearsay.[62]

---

[61] PEAKE, *supra* note 57, at 12 (emphasis added). A version of the preference for proof of "general reputation" over proof of "particular facts" survives today in American evidence rules that allow peoples' characters to be proved with evidence of their reputations, but not—except when character is integral to a legal claim or defense—with evidence of specific things they have done. *See, e.g.*, FED. R. EVID. 405; Michelson v. United States, 335 U.S. 469, 477 (1948).

[62] GEOFFREY GILBERT, THE LAW OF EVIDENCE 152–53 (1765). For helpful discussions of Gilbert's treatise, see William Twining, *The Rationalist Tradition of Evidence Scholarship*, *in* RETHINKING EVIDENCE: EXPLORATORY ESSAYS 32, 35–38 (1990); Richard D. Friedman, *The Mold That Shapes Hearsay Law*, 66 FLA. L. REV. 433, 437–38 (2014); T.P. Gallanis, *The Rise of*

Relying in part on Gilbert's treatise and in part on a review of English trial reports, the legal historian Thomas Gallanis has argued persuasively that the hearsay rule had not yet emerged in its modern form in the 1760s. It is not just that the rule had not yet been authoritatively formulated in its now familiar form. It also was applied inconsistently and at best sporadically; most of the time that hearsay was offered, it came in with little if any discussion.[63] The key period for the development of the hearsay rule, Gallanis concludes, was the closing decades of the eighteenth century, when lawyers began to appear more frequently to represent defendants in criminal cases, and—because they were not yet allowed to address the jury—focused their attention on aggressive cross-examination and evidentiary objections. Lawyers who developed these tactics in criminal cases then began to apply them in civil cases as well.[64]

By the beginning of the nineteenth century, Gallanis says, "the contours of the modern rule against hearsay were largely in place": the rule was applied much more consistently, and most of the major exceptions to the rule were regularly recognized.[65] But even at that point, Gallanis notes that English courts continued to admit hearsay in cases where it was the only available evidence on a particular point, because the person or persons with direct knowledge were no longer alive.[66] They agreed, that is to say, with the position taken in the United States by influential judges like Roane, Kent, and Swift: when hearsay was the best evidence available, it made no sense to exclude it.

The rule may have been stricter with regard to evidence introduced against a criminal defendant, because the use of hearsay in criminal cases implicated the confrontation right codified in the Sixth Amendment and in some state constitutions. Nearly two decades ago, in *Crawford v. Washington*, the Supreme Court concluded that these

---

Modern Evidence Law, 84 IOWA L. REV. 499, 504–09 (1999); and John H. Langbein, *Historical Foundations of the Law of Evidence: A View from the Ryder Sources*, 96 COLUM. L. REV. 1168, 1172–76 (1996).

[63] Gallanis, *supra* note 62, at 512–15. Wigmore thought the hearsay rule took shape in the 1600s and early 1700s, in response to the move away from self-informing juries. But a growing body of scholarship supports Gallanis's account. *See* Sklansky, *supra* note 2, at 26–27 & nn.114–115 (citing sources).

[64] Gallanis, *supra* note 62, at 545–50.

[65] *Id.* at 534–35; *but cf.* Friedman, *supra* note 62, at 473 n.18 (noting that "I know of no articulation of anything like the modern definition of hearsay before 1800").

[66] Gallanis, *supra* note 62, at 536 & n.274 (citing decisions from 1806 and 1814).

provisions codified a common law doctrine that had protected criminal defendants against "testimonial" hearsay.[67] The actual practices with regard to confrontation in colonial trials are contested. There is good evidence that, at least in some states, statements taken from witnesses outside the presence of the accused, and without cross-examination, were regularly introduced against the defendant at trial.[68] In any event, as Richard Friedman has stressed, the confrontation right was not "a rule about the admissibility of hearsay in general."[69] Whatever the practice in criminal cases, in civil cases the weight of authority before *Queen v. Hepburn*, both in England and in the United States, treated the hearsay rule as a flexible principle of preference, requiring sworn, in-court testimony when possible but allowing secondhand evidence when more reliable proof was unavailable. In particular, community and family lore was widely admitted to prove the kinds of things for which live witnesses were likely to be unavailable: customs, traditional boundaries, and family history.

Whether or not a person who lived long ago had been enslaved seems to fit comfortably within that category of admissible hearsay. It is therefore unsurprising that courts before *Queen v. Hepburn* often allowed hearsay evidence to prove whether an ancestor of a plaintiff suing for liberty had been free. As Justice Duvall noted in his dissent, this had long been the practice in Maryland. That is how Mina Queen's relatives had won their freedom. And Maryland law does not appear to have been unusual in this regard. Elsewhere in the South as well, it seems to have been common for petitioners in freedom cases to rely on hearsay evidence about their ancestors. Typically, it was the only evidence available.[70]

---

[67] 541 U.S. 36 (2004).

[68] *See* Bernadette Meyler, *Common Law Confrontations*, 37 Law & Hist. Rev. 763 (2019).

[69] Friedman, *supra* note 62, at 434. In deciding *Crawford v. Washington*, the Supreme Court itself reasoned that a 1790 English decision allowing hearsay—*King v. Eriswell*, 100 Eng. Rep. 815 (K.B. 1790)— was largely irrelevant to the meaning of the Confrontation Clause because it came in a civil rather than a criminal proceeding. 541 U.S. at 54 n.5. The Justices also noted that *Eriswell* was an affirmance by an equally divided court, and that the King's Bench itself rejected the reasoning of *Eriswell* eleven years later. *See id.* (citing King v. Ferry Frystone, 102 Eng. Rep. 289 (K.B. 1801)). As it happens, *Eriswell* was the only decision that Chief Justice Marshall cited in his opinion for the Supreme Court in *Queen v. Hepburn*; there, too, the Court noted that the King's Bench was divided in *Eriswell* and later rejected the result in had reached in that case. *See infra* notes 102–104 and accompanying text.

[70] *See, e.g.*, Fede, *supra* note 2, at 340–43; MacLeod, *supra* note 2, at 110, 112–13; Schwenginger, *supra* note 7, at 10–24, 53, 124–25; Fede, *supra* note 2, at 8; Nicholls, *supra* note 13, at 48–51.

Writing half a century ago, Duncan MacLeod maintained that Southern courts in the late 1700s and early 1800s were departing from "well established" law in allowing hearsay evidence not just to prove the identity of a slave's ancestry, but also to establish the ancestor's race, and whether the ancestor had been free or enslaved.[71] Subsequent scholarship has made clear, however, that the hearsay rule in this period was less settled and more flexible than MacLeod thought. Andrew Fede has suggested more recently that hearsay proof of an ancestor's freedom fell within the "recognized common law exception" for proof of "family history, reputation or pedigree."[72] Yet this, too, overstates the clarity and consistency of evidence law in the decades following the American Revolution, and it slights the distinction that courts and commentators often drew between proof of genealogy—the identity of one's ancestors—and characteristics of those ancestors, such as their race, their appearance, or their status as free or enslaved.[73] As Fede himself has stressed, American slavery developed into an institution that in many ways was unprecedented, raising novel legal questions.[74] English courts rarely if ever had to decide whether to allow hearsay proof of an ancestor's race or status, because slavery in Britain never became an all-encompassing system of racial hierarchy, as it did the American South. That system hardened in the United States just as lawyers were beginning to appear more frequently in civil and criminal trials both in England and in America, creating the conditions for the development of the modern hearsay rule.

In the two decades before *Queen v. Hepburn*, Southern legislatures began to impose new procedural limitations on freedom suits. In 1793, for example, Maryland took jurisdiction over these cases away from the General Court of the Western Shore and redirected them to lower district courts, where slaveholders could be expected to have more influence over jurors and witnesses.[75] Two years later, Virginia

---

[71] MacLeod, *supra* note 2, at 110, 113.

[72] Fede, *supra* note 2, at 340.

[73] This is a distinction that MacLeod rightly emphasized. *See* MacLeod, *supra* note 2, at 113. The same distinction could be drawn under modern exceptions to the hearsay rule applicable to statements about "facts of personal or family history," such as "birth, adoption, legitimacy, ancestry, marriage, divorce, relationship by blood, adoption, or marriage." Fed. R. Evid. 804(b)(4)(A); *see also* Fed. R. Evid. 803(19). Consequently it is unclear whether, as Fede suggests, the hearsay offered in *Queen v. Hepburn* would today be admissible under those exceptions. *See* Fede, *supra* note 2, at 16.

[74] *E.g.*, Fede, *supra* note 2, at 8.

[75] *See* Thomas, *supra* note 4, at 68.

eliminated the ability of freedom plaintiffs to choose their own lawyers, required courts to appoint counsel who would serve without compensation, barred abolitionists from serving on juries in freedom cases, and imposed a fine on anyone who assisted a petition that ultimately proved unsuccessful.[76] During this same period, judges in the South grew more wary of hearsay evidence in freedom cases—especially, although not only, when it was offered by the plaintiff.[77] They began to find ways to exclude secondhand evidence of an ancestor's race or status, for example by distinguishing between simple hearsay and "hearsay of hearsay,"[78] or by allowing only evidence of reputation among free white people no longer living.[79]

Into the second decade of the nineteenth century, however, enslaved people in the United States largely retained the ability to sue for their freedom by introducing secondhand evidence that their ancestors had been free. *Queen v. Hepburn* did more than any other decision to foreclose that opportunity, by making most of the secondhand accounts relied upon by freedom plaintiffs flatly inadmissible under the hearsay rule.

III.

Although *Queen v. Hepburn* was tried under Maryland law, which governed at the time in the portion of the District of Columbia north of the Potomac River,[80] the Supreme Court treated the hearsay question as a matter of "general law," to be decided according to principles not limited to any particular jurisdiction.[81] Either way, however, judges outside Maryland would follow the holding only if they believed that Chief Justice Marshall had set forth principles that should govern the common law of every American jurisdiction. Moreover, as Farah Peterson has pointed out, most lawyers and judges in the early nineteenth century did not view the United States Supreme Court as the nation's preeminent expositor of the law: they were more likely to look for guidance to eminent state-court jurists—including

[76] *See id.* at 80.

[77] Honor Sachs, *"Freedom by a Judgment": The Legal History of an Afro-Indian Family*, 30 Law & Hist. Rev. 173, 199–200 (2012).

[78] Thomas, *supra* note 2, at 168–69.

[79] Fede, *supra* note 2, at 11; *supra* text accompanying note 24.

[80] Thomas, *supra* note 2, at 166; Newmyer, *supra* note 29, at 542–43.

[81] *See supra* note 30 and accompanying text.

Spencer Roane, James Kent, and Zephaniah Swift—than to John Marshall.[82] Even Maryland judges might not apply the reasoning of *Queen v. Hepburn* in later cases if they found it unconvincing.

Nonetheless, the decision proved enormously influential. The Supreme Court itself reaffirmed the decision three years later, in *Davis v. Wood*, ruling that a freedom plaintiff could not introduce hearsay evidence, from people who were now dead, that the plaintiff's ancestor was a white woman who had been born in England. Chief Justice Marshall again wrote for the Court, and he rejected out of hand the suggestion that *Queen v. Hepburn* should be revisited.[83] *Queen v. Hepburn* and *Davis v. Wood* all but eliminated the practice of enslaved people suing for liberty based on the free status of their maternal ancestors. Without the possibility of relying on hearsay, such suits were hopeless.[84] And while not every state court followed the *Queen* and *Davis* decisions in rejecting hearsay proof an ancestor's race or status, most did, including the high courts of Maryland and Virginia, jurisdictions that produced a disproportionate share of freedom cases.[85] Freedom cases did not disappear, but, as Jason Gillmer has noted, their focus shifted away from claims of ancestry and toward other claims, including, in many cases, that the plaintiff was white.[86] *Queen v. Hepburn* thus contributed to a process already underway when it was decided: the repositioning of slavery within an increasingly rigid, increasingly explicit, and increasingly binary system of racial caste.[87]

The Supreme Court's decision in *Queen v. Hepburn* also played an important role in transforming the hearsay rule from a loose preference

---

[82] Peterson, *supra* note 52, at 75–76; *cf.*, *e.g.*, Gregory v. Baugh, 2 Va. (2 Leigh) 665, 695 (1831) (Brooke, J., dissenting) (citing *Queen v. Hepburn* but conceding that it was "not authority here").

[83] 14 U.S. (1 Wheat) 6, 7–8 (1816); *see* Gillmer, *supra* note 2, at 582–85.

[84] MacLeod, *supra* note 2, at 110–12; Schweninger, *supra* note 7, at 119–21.

[85] *See* Gillmer, *supra* note 2, at 585–87. Some states, including Missouri and Tennessee, continued to allow reputation evidence to prove that a slave or his or her ancestor had lived free. *See* Vaughn v. Phebe, 8 Tenn. 5 (1827); Kennington, *supra* note 7, at 137–38.

[86] Gillmer, *supra* note 2, at 587–88; *see also* Schweninger, *supra* note 7, at 121 (observing that in the years following the Supreme Court's decision in *Queen v. Hepburn*, "the number of plaintiffs claiming white material ancestry declined"). In Missouri, the majority of freedom suits were based on claims that the plaintiff had become emancipated by residing in free territory. *See* Kennington, *supra* note 7, at 4, 10, 12.

[87] Gillmer, *supra* note 2, at 601–05; *see also*, *e.g.*, Ariela J. Gross, What Blood Won't Tell: A History of Race on Trial in America 4, 16–47 (2008); Angela Onwuachi-Willig, *Multiracialism and the Social Construction of Race: The Story of* Hudgins v. Wrights, *in* Race Law Stories 147 (Rachel F. Moran & Devon W. Carbado eds., 2008); Ablavsky, *supra* note 7, at 1460, 1477.

for in-court testimony, with a flexible and open-ended series of exceptions, to a flat prohibition, with only a fixed and limited set of carveouts. John Marshall himself, sitting as Circuit Justice, cited *Queen v. Hepburn* and *Davis v. Wood* in 1825 for the proposition that exceptions to the hearsay rule would not be extended "further than they have been already carried."[88] Writing for the Supreme Court in 1836, Justice Joseph Story took *Queen v. Hepburn* to establish that hearsay was admissible only to prove "pedigree"—i.e., ancestry—or, in the form of reputation evidence, to prove land boundaries and other "claims of a public nature" obtained by long usage.[89] In particular, Story read *Queen v. Hepburn*, along with *Davis v. Wood*, to make clear that "hearsay evidence was not admissible to prove a specific fact, although the witnesses to the fact were dead."[90] Echoing Marshall's reasoning in *Queen v. Hepburn*, Story wrote that hearsay was excluded not just because "it often supposes better evidence behind," but because "it is not upon oath," because there was "no opportunity of cross-examination," because "it is peculiarly liable to be obtained by fraudulent contrivances," and "above all" because "it is exceedingly infirm, unsatisfactory and intrinsically weak in its very nature and character." Consequently, Story explained, the admission of hearsay evidence in any situation was "somewhat of an anomaly."[91]

---

[88] Stegall v. Stegall, 22 F. Cas. 1226, 1228 (C.C.D. Va.).

[89] Ellicott v. Pearl, 35 U.S. (10 Pet.) 412, 434–38. Writing for the Court five years earlier, Chief Justice Marshall cited and quoted *Queen v. Hepburn* for the rule that "hearsay evidence is incompetent to establish any specific fact, which fact is in its nature susceptible of being proved by witnesses who speak from their own knowledge." Scott's Lessee v. Ratliffe, 30 U.S. (5 Pet.) 81, 86 (1831) (quoting Queen v. Hepburn, 11 U.S. (7 Cranch) 290, 295 (1813)).

In her excellent book, *What Blood Won't Tell*, Ariela Gross suggests that nineteenth-century judges used the term "pedigree" to refer to race. Gross, *supra* note 87, at 9, 24–25. I read the cases differently. "Pedigree" meant lineage: the identity of a person's ancestors, and possibly the date of their births, deaths, and marriages. *See, e.g.*, Black's Law Dictionary 882 (1st ed. 1891) (defining "pedigree" as "Lineage; line of ancestors from which a person descends; genealogy. An account or register of a line of ancestors. Family relationship"). Sometimes, though—especially before *Queen v. Hepburn*—judges were loose with the application of the pedigree exception to the hearsay rule, and they applied the exception to key facts about the ancestors, like their race or nationality, and whether they had been free or enslaved. Thus, for example, Francis Brooke, then president of Virginia's Supreme Court of Appeals, acknowledged in an 1831 decision that "hearsay evidence has been silently permitted to go to juries, to prove the condition of a person, from whom pedigree has been deduced," although he disapproved of that practice and thought it had been properly rejected in *Queen v. Hepburn* and *Davis v. Wood*. Gregory v. Baugh, 29 Va. (2 Leigh) 665, 695 (1827).

[90] *Ellicott*, 35 U.S. at 438.

[91] *Id.* at 434.

This view of hearsay had become conventional in the United States by 1836, in large part because of *Queen v. Hepburn*. That same year, Connecticut's highest court declared it "a general principle of the law of evidence" that "mere declarations of third persons as to any fact, or what is commonly called *hearsay* evidence, is inadmissible to establish that fact," because it was "too vague and unsubstantial to afford any reasonable test of its truth." The "few exceptions" to this rule, the court explained, should be "strictly construed," and neither "favored nor extended." The authority for this sweeping rule was *Queen v. Hepburn*.[92]

Simon Greenleaf's *Treatise on the Law of Evidence*, which was published in 1842 and remained for decades the foremost exposition of American evidence law, fully embraced the best evidence principle but treated the hearsay rule as an independent restriction on proof.[93] To explain the basis for banning hearsay, Greenleaf quoted *Queen v. Hepburn* and made its reasoning his own:

> That this species of testimony supposes something better, which might be adduced in the particular case, is not the sole ground of its exclusion. Its intrinsic weakness, its incompetence to satisfy the mind as to the existence of the fact, and the frauds, which may be practiced under its cover, combine to the support the rule, that hearsay evidence is totally inadmissible.[94]

In the early twentieth century, Greenleaf's treatise was supplanted in influence by John Henry Wigmore's monumental *Treatise on the System of Evidence in Trials at Common Law*. Wigmore called hearsay the "most characteristic rule of the Anglo-American law of evidence," and "a rule which may be esteemed, next to jury-trial, the greatest contribution of that eminently practical legal system to the world's jurisprudence of procedure."[95] But he found the argument advanced by Marshall and Story—that hearsay was excluded because of its "intrinsic weakness"—utterly unconvincing.[96] In a supplement to his

---

[92] Brown v. Crandall, 11 Conn. 92, 94–95 (1835).

[93] 1 SIMON GREENLEAF, A TREATISE ON THE LAW OF EVIDENCE § 82, at 93; *id.* § 98, at 111–12.

[94] *Id.* at 112 (quoting Queen v. Hepburn, 11 U.S. (7 Cranch) 290, 295–96 (1813)). On Greenleaf and his influence, see Twining, *supra* note 62, at 49–50.

[95] 2 JOHN HENRY WIGMORE, A TREATISE ON THE SYSTEM OF EVIDENCE IN TRIALS AT COMMON LAW § 1364, at 1695 (1904). Regarding Wigmore's treatise and its importance, see, for example, Twining, *supra* note 62, at 59–60.

[96] *See* 2 WIGMORE, *supra* note 95, § 1363, at 1679.

treatise, Wigmore stressed that the "spirit" of the hearsay rule was simply "to insist on testing all statements by cross-examination, *if they can be*," and that when a witness had "passed beyond the power of the law to procure him," the rule should be disregarded. Applying the rule more rigidly struck him as "pedantic."[97]

By then, though, the ship had sailed: hearsay was a per se rule of exclusion, not a rule of preference.[98] Well into the twentieth century, moreover, *Queen v. Hepburn* remained "the leading American case" on hearsay.[99] It continues to be cited today, although far more sporadically, for the proposition that the hearsay rule applies even when live testimony is unavailable.[100]

Queen v. Hepburn did not, of course, singlehandedly create the modern rule against hearsay. There were decisions afterward that continued to treat the hearsay ban as more flexible and the exceptions as more open-ended. Until the end of the nineteenth century, as Paul Kirgis has noted, American courts often "eschewed formalistic definitions in favor of a loose, conceptual approach to hearsay," making "essentially ad hoc judgments based on a range of credibility concerns."[101] Moreover, there were judicial pronouncements before *Queen v. Hepburn* that sounded very much like the rule we know today, especially in England. In *King v. Eriswell*, the only prior decision that Chief Justice Marshall cited in ruling against Mina Queen, Lord Kenyon reasoned that the sworn declaration of a witness was inadmissible against a party who had not had the opportunity to cross-examine the witness, and that the rule should be applied strictly, even

---

[97] John Henry Wigmore, A Supplement to a Treatise on the System of Evidence in Trials at Common Law xxix (2d ed. 1915). Wigmore subsequently incorporated these observations into the second edition of his treatise. 1 John Henry Wigmore, A Treatise on the Anglo-American System of Evidence in Trials at Common Law § 8, at 138 (2d ed. 1923).

[98] *See, e.g.*, Sklansky, *supra* note 2, at 20–21.

[99] Comment, *Ancient Documents as an Exception to the Hearsay Rule*, 33 Yale L.J. 412, 412 n.2 (1924).

[100] *See, e.g.*, Cooper v. Harris, 137 S. Ct. 1455, 1499 n.18 (2017) (Alito, J., concurring in part and dissenting in part); Valman v. State, 5 So. 3d 1079, 1087 (Miss. 2009) (Kitchens, J., dissenting); Garza v. Delta Tau Delta Fraternity Nat'l, 984 So. 2d 84, 91 (La. 2006); State v. Lewis, 619 S.E. 2d 830 (N.C. 2005), *vacated*, 584 U.S. 924 (2006). These cases form part of a larger pattern in which courts continue to cite slavery cases for general legal propositions. *See* Justin Simard, *Citing Slavery*, 72 Stan. L. Rev. 79 (2020). Professor Simard has created a valuable online database of slave cases and modern decisions treating them as precedent: Citing Slavery Project, https://www.citingslavery.org.

[101] Paul F. Kirgis, *Meaning, Intention, and the Hearsay Rule*, 43 Wm. & Mary L. Rev. 275, 279–80 (2001).

when the witness was now unavailable to be examined in court.[102] Lord Kenyon was on the losing side in *Eriswell*: the King's Bench was evenly divided and therefore the decision below, admitting the hearsay evidence, was allowed to stand. But Marshall noted, correctly, that the case had not remained good authority in England[103]: in a subsequent case with very similar facts, Lord Kenyon had written again for the King's Bench, explicitly rejecting the result in *Eriswell*.[104]

In the decades following *Queen v. Hepburn*, English courts and commentators continued to take the lead in expanding the scope of the hearsay prohibition. *Queen v. Hepburn* was not unknown in England,[105] but the decision does not appear to have had a significant influence on courts or commentators there. In the nineteenth century, as in the eighteenth, the hearsay rule developed in England without appreciable reliance on American authorities; the influence tended to run in the other direction. Thomas Starkie's *Practical Treatise on the Law of Evidence*, published in London in 1824 and widely consulted in the United States, has an extensive discussion of the hearsay rule, explaining that "the more declarations of third persons as to a fact, or what is commonly called hearsay-evidence, is not admissible," and that this rule must be followed even when no better evidence is available, because "the law must proceed by general and certain rules," and because statements made without an oath and without the opportunity for cross-examination are simply too unreliable.[106]

The outermost expansion of the hearsay rule came not in the United States but in England, in the case of *Wright v. Tatham*, decided by the Exchequer Chamber in 1837 and by the House of Lords

---

[102] 100 Eng. Rep. 815 (K.B. 1790) (quoted and discussed by Queen v. Hepburn, 11 U.S. (7 Cranch) 290, 295–96 (1813)).

[103] 11 U.S. at 296.

[104] King v. Ferry Frystone, 102 Eng. Rep. 289 (KB 1801). *Eriswell* and *Ferry Frystone* were both disputes about which parish had the responsibility to take care of a pauper and his family under England's settlement laws, and in each case the question was whether a sworn declaration by the pauper should be admitted into evidence. In each case, moreover, the pauper was unavailable to testify in court: in *Eriswell* because of mental illness, and in *Ferry Frystone* because of his death. Regarding the settlement laws and their administration, see, for example, Norma Laundau, *Who Was Subjected to the Laws of Settlement? Procedure Under the Settlement Laws in Eighteenth-Century England*, 43 AGRIC. HIST. REV. 139 (1995).

[105] *See, e.g.*, 1 JOHN PITT TAYLOR, A TREATISE ON THE LAW OF EVIDENCE, AS ADMINISTERED IN ENGLAND AND IRELAND § 385, at 366 (1848) (quoting *Queen v. Hepburn*).

[106] THOMAS STARKIE, A PRACTICAL TREATISE ON THE LAW OF EVIDENCE §§ 23–26 (1824). On Starkie's treatise and its influence, see Twining, *supra* note 62, at 46–47. Regarding Starkie's treatment of hearsay, see Gallanis, *supra* note 62, at 516–23.

in 1838.[107] Widely cited both in England and in the United States, *Wright v. Tatham* held that out-of-court statements were inadmissible not just to prove the truth of what they explicitly asserted, but also to prove the truth of any implicit assumptions the declarants appeared to have made—and that this rule applied even when the declarants could not be brought into court to testify, and even if no better evidence were available.[108] None of the many opinions in the case cited *Queen v. Hepburn*, and it is difficult to trace any influence of the American decision on the resolution of the English dispute.

Nonetheless, and despite the influence that English decisions and English treatises had in the United States throughout the eighteenth and nineteenth centuries, American evidence law did not develop in lockstep with its English counterpart. The hearsay rule, in particular, appears to have been applied for much of this period with greater flexibility in the United States. For American courts, *Queen v. Hepburn* was a critical step in the transition of the hearsay rule from a loose preference for sworn, in-court testimony, to a categorical ban on out-of-court statements offered to prove the truth of what they asserted, even when no better evidence was available.

The Supreme Court might conceivably have reached the same decision in *Queen v. Hepburn* even if the case had not concerned an enslaved plaintiff's petition for freedom. The principal force driving the development of the hearsay rule, in the United States as in Britain, was not slavery but the rise of the legal profession, the expanding role that lawyers played in civil and criminal trials, and the increasing control that lawyers, along with judges, asserted over the factfinding process.[109] John Marshall himself, when sitting as a trial judge in the 1807 prosecution of Aaron Burr, had called the hearsay rule "essential to the correct administration of justice," and had described the rule as "exclud[ing] from trials of a criminal or civil nature

---

[107] 112 Eng. Rep. 488 (Ex. Ch. 1837); 5 Cl. & Fin. 670, 47 Rev. Rep. 136 (H.L. 1838); *see also* Richard D. Friedman, *The Confrontation Right Across the Systemic Divide*, *in* CRIME, PROCEDURE AND EVIDENCE IN A COMPARATIVE AND INTERNATIONAL CONTEXT: ESSAYS IN HONOR OF PROFESSOR MIRJAN DAMAŠKA 261, 263 (John Jackson, Máximo Langer & Peter Tillers eds., 2008) (describing *Wright v. Tatham* as the hearsay ban's "high water mark").

[108] *See* Friedman, *supra* note 107, at 263; Kirgis, *supra* note 101, at 277–80; John M. Maguire, *The Hearsay System: Around and Through the Thicket*, 14 VAND. L. REV. 741, 749–60 (1961); Sklansky, *supra* note 2, at 24 & n.101.

[109] *See* EDWARDS, *supra* note 59, at 40–53; THOMAS, *supra* note 2, at 171–72; Gallanis, *supra* note 62, at 545–50; Sklansky, *supra* note 2, at 27–28.

the declarations of any other individual than of him against whom the proceedings are instituted." Courts, he said, should be "watchful of every inroad on a principle so truly important."[110] These remarks came in the course of proceedings in which lawyers played what was, for the time, an unusually dominant role—indeed, proceedings that arguably signaled "a transitional moment in the emergence of a distinctly American adversarial tradition."[111] (Even so, Marshall did not suggest in the Burr case that exceptions to the hearsay rule were limited to those already established, or that the exceptions should be applied rigidly instead of with an eye to equity. He excluded the out-of-court statements offered against Burr because he concluded they fell outside not just the letter but also the rationale of the exception for statements of coconspirators.[112])

Strictly speaking, moreover, *Queen v. Hepburn* was not a case about slavery; it was a case about procedure. Chief Justice Marshall's opinion for the Supreme Court in *Queen v. Hepburn* stressed that the ban on hearsay evidence was a "general rule," which operated the same regardless of the nature of the substantive right at issue.[113] This was in keeping with the usual practice of nineteenth-century courts to treat enslaved people as just another form of property, to be governed by the same rules.[114] Furthermore, slaveholders sometimes introduced hearsay evidence of their own to disprove a plaintiff's claim that an ancestor had been free, or illegally enslaved.[115] This happened at the trial in *Queen v. Hepburn* itself: John Hepburn's lawyers introduced a deposition by Gabriel Duvall's great-uncle, Benjamin Duvall, relaying hearsay about the origins and legal status of Mina Queen's ancestor, Mary Queen.[116] Therefore, *Queen v. Hepburn* may

---

[110] United States v. Burr, 25 F. Cas. 187, 193 (C.C.D. Va. 1807).

[111] R. Kent Newmyer, The Treason Trial of Aaron Burr: Law, Politics, and the Character Wars of the New Nation 6 (2012).

[112] 25 F. Cas. at 193–95. Nor did Marshall make clear in 1807 whether the application of the hearsay rule applied even when better evidence was unavailable. The out-of-court statements offered against Burr had been made by his alleged coconspirator, Harman Blennerhassett. Whether Blennerhassett could have been made to testify is unclear: he was attending the proceedings but also awaiting trial himself on related charges. Newmyer, *supra* note 111; Harman Blennerhassett, Breaking With Burr: Harman Blennerhassett's Journal, 1807, at 82–89 (Raymond E. Fitch ed., 1988).

[113] 11 U.S. (7 Cranch) 290, 295–96 (1813).

[114] Simard, *supra* note 100, at 88–89.

[115] Fede, *supra* note 2, at 340, 347–49.

[116] Thomas, *supra* note 2, at 75–76, 171–73.

have helped some freedom plaintiffs by barring evidence that otherwise might have been introduced against them.

There is no doubt, though, that the decision hurt freedom plaintiffs much more than it helped them.[117] Slaveholders might or might not have access to documentary proof, but enslaved plaintiffs almost never could introduce any evidence other than hearsay regarding the race or status of their ancestors. This is a general feature of the hearsay rule: it tends to pose a bigger problem for disempowered litigants, who have less access to admissible written records and are less able to track down live witnesses and bring them to court.[118] Furthermore, in the absence of other proof, courts in slave states often applied a presumption that individuals who appeared to be of African ancestry were lawfully in bondage.[119] Justice Duvall understood that the ruling in *Queen v. Hepburn* would all but shut off an important legal avenue through which enslaved people had sued for freedom; that was why he dissented so strongly. And notwithstanding Chief Justice Marshall's protestations, the slavery context of *Queen v. Hepburn* may have mattered to him as well. Marshall was a major, not a minor, slaveholder.[120] As Paul Finkelman has noted, "[s]laves were a constant factor in his personal life, his economic success, and his children's future."[121] Marshall's lifelong membership in the American Colonization Society reflected, in part, his deep and unrelenting hostility to idea of Black people living free in the United States.[122] He wrote the opinion in seven cases involving enslaved people seeking their freedom; each time, the plaintiff lost.[123] The role that a stronger hearsay rule would play in shutting down freedom suits—and in

---

[117] *Cf.* Gonzales Rose, *supra* note 3, at 2252 ("Accordingly, in examining the law of evidence, we must ask: Who benefits? Even if a rule of evidence appears to apply uniformly to all persons, does a racial group disproportionately gain an evidentiary advantage under it?").

[118] Evidence of Native land claims, for example, often takes the form of oral histories. *See* Hope M. Babcock, *"[T]his I Know from My Grandfather: The Battle for Admissibility of Indigenous Oral History as Proof of Tribal Land Claims,"* 37 AM. INDIAN L. REV. 19 (2013). I am indebted to Andrea Roth for drawing this connection for me.

[119] *See, e.g.*, Gobu v. Gobu, 1 N.C. 188, 188 (N.C. Super. L. & Eq. 1802); Hudgins v. Wrights, 11 Va. (1 Hen. & M.) 134, 139, 141(1806); Davis v. Curry, 5 Ky. 238, 240 (1810); Gary v. Stevenson, 19 Ark. 580, 586 (1858); Carodine, *supra* note 3, at 68–69. Jack Chin stressed the importance of this point to me.

[120] FINKELMAN, *supra* note 2, at 36–50.

[121] *Id.* at 48.

[122] *Id.* at 34, 50–52, 74, 110–11, 221–22.

[123] *Id.* at 30–31, 56–75.

blocking in Mina Queen's suit in particular—may not have been incidental to Marshall.

IV.

The hearsay rule was not invented in *Queen v. Hepburn*, and courts did not start banning hearsay just to keep African Americans in bondage. Many other legal tools, alas, existed to accomplish that odious end. But *Queen v. Hepburn* did mark a significant point of transition: more than any other single decision, at least in the United States, it turned hearsay from a loose principle of preference for in-court testimony into a rigid rule of evidentiary exclusion. The consequence of applying this new, stricter version of the hearsay doctrine in *Queen v. Hepburn* was to deny Mina Queen the opportunity to challenge her enslavement, and the enslavement of her daughter, by introducing the only evidence available to them of their ancestor's freedom. And the precedent set by *Queen v. Hepburn* had the same consequence for many other freedom plaintiffs and would-be plaintiffs. By strengthening the hearsay rule in *Queen v. Hepburn*, the Supreme Court helped to shut down one of the few legal channels through which enslaved people could seek their own liberty. That may not have been an unintended result; on the contrary, it may have helped to motivate the decision, at least for its author. The hearsay rule, as a genuine rule as opposed to a flexible principle, has its origins, in significant part, in American slavery.

William Thomas, the foremost authority on Mina Queen's lawsuit, notes the irony that "the hearsay rule's primary definition came in a case where its application was most specious, a case that would become a cornerstone in the perpetuation of slavery in the United States." He stresses that the rule "privileged the kinds of evidence slaveholders could control (documentary records and direct testimony) at the expense of the kinds of evidence the enslaved persons could produce (old depositions and family lore)," and that it "kept the family histories and genealogies of the enslaved people in the shadows for decades, even centuries, afterward."[124]

Law students should learn this history. They should know the full background of the rule that, more than any other, separates the law of

---

[124] THOMAS, *supra* note 2, at 5, 185.

evidence in the United States from the procedures followed in almost every other liberal democracy. The hearsay rule and its series of exceptions are confusing even for seasoned lawyers and judges; for nonlawyers, they are impenetrable. That means that the future of the hearsay rule—whether it will be retained in its current form, strengthened, weakened, or reoriented—will almost certainly be decided by lawyers. No one else understands it. Lawyers need to have an informed comprehension of how and why we came to have it.

The hearsay rule plainly has protected, and continues to protect, against many injustices. The treason trial of Sir Walter Raleigh—often said to have inspired the hearsay rule as well as the confrontation right of criminal defendants—is properly infamous. It was manifestly unfair to convict Raleigh with an out-of-court confession from a witness he never had the opportunity to cross-examine.[125] But the hearsay rule can itself create injustices, and it was not developed solely to protect persecuted scapegoats like Raleigh. *Queen v. Hepburn* is Exhibit A.

Obviously, the connections between the hearsay rule and American slavery—the role that the rule played in keeping Mina Queen, Louisa Queen, and many others enslaved—do not necessarily mean that the rule should be discarded. A rule that served bad purposes, even a rule that was intended, in whole or in part, to serve bad purposes, may today do more good than harm. But tainted history does provide reason to treat an inherited rule with at least a little more skepticism.[126] There is a natural but unfortunate tendency for people to become invested in rules or procedures they have worked hard to master. This can easily happen with the hearsay rule, especially when students begin their study of the rule, as they often do, with a case like Raleigh's, which casts the hearsay rule in a distinctly heroic light. Learning about *Queen v. Hepburn* may help to inoculate them against an overly complacent, or even celebratory, understanding of the hearsay rule and its origins.

All the more so because the features of the hearsay rule solidified in *Queen v. Hepburn*—the rigid, categorical nature of the evidentiary exclusion, and the fact that it operates even when live testimony

[125] Sklansky, *supra* note 2, at 38–39.

[126] W. Kerrel Murray, *Discriminatory Taint*, 135 Harv. L. Rev. 1190, 1202 (2022); *see also*, *e.g.*, Charles L. Barzun, *The Genetic Fallacy and a Living Constitution*, 34 Const. Comment. 429 (2019); Simard, *supra* note 100, at 108–09, 120.

cannot be obtained—are precisely the features that now distinguish American evidence law most sharply from the rules followed in other modern legal systems. Virtually every modern legal system prioritizes in-court testimony. Rules requiring witnesses to testify live in court are found around the world, in common-law and civil-law jurisdictions alike. Outside of the United States, though, these rules generally give way when the witness has died or cannot come to court, and the out-of-court statement is the best evidence available. This has long been true in England, which abolished the hearsay rule entirely in civil cases in 1997 and, for criminal cases, created a broad exception in 2003 applicable to firsthand hearsay from declarants who cannot come to court to testify. The law is largely similar in Scotland, and most other common law countries, including Canada, Australia, and New Zealand, allow the introduction of any "reliable" firsthand hearsay from unavailable declarants.[127] Meanwhile civil law countries, prodded in many cases by the European Court of Human Rights, have adopted stronger requirements of in-court testimony over the past several decades, but these rules operate as principles of preference, or occasionally as rules of evidentiary sufficiency. They are not categorical bans: they do not bar evidence of out-of-court statements from witnesses who are unavailable to be questioned under oath.[128] By contrast, modern American rules of evidence generally prohibit hearsay in both criminal and civil cases even when live testimony cannot be obtained. And although there are catchall exceptions that

[127] Sklansky, *supra* note 2, at 28–30; *see also* Matthew Caton, *Abolish the Hearsay Rule: The Truth of the Matter Asserted at Last*, 26 MAINE B.J. 126, 129–30 (2011); Deborah Paruch, *Testimonial Statements, Reliability, and the Sole or Decisive Evidence Rule: A Comparative Look at the Right of Confrontation in the United States, Canada, and Europe*, 67 CATH. U. L. REV. 105, 125–37 (2018). In Ireland, on the other hand, hearsay remains categorically excluded, subject to a gradually widening set of relatively fixed exceptions. *See* CAROLINE FENNELL, THE LAW OF EVIDENCE IN IRELAND 339–50 (4th ed. 2020); LAW REFORM COMMISSION (IRELAND), CONSOLIDATION AND REFORM OF ASPECTS OF THE LAW OF EVIDENCE 118 (2016); Liz Heffernen, *Hearsay in Criminal Trials: The Strasbourg Perspective*, 49 IRISH JURIST (N.S.) 132, 135 (2013); Eoin O'Connor, *The Right to Confrontation, Hearsay, and the European Court of Human Rights*, 14 U.C. DUBLIN L. REV. 78, 99 (2014).

[128] Lorena Bachmaier, *Rights and Methods to Challenge Evidence and Witnesses in Civil Law Jurisdictions*, *in* THE OXFORD HANDBOOK OF CRIMINAL PROCESS 841 (Darryl K. Brown, Jenia Iontcheva Turner & Bettina Weisser eds., 2019); Caton, *supra* note 127, at 129–30; Paruch, *supra* note 127, at 137–55 (2018); Sklansky, *supra* note 2, at 30–36. South Africa's legal system, which mixes common law and civil law traditions, currently takes a similarly "flexible" approach to hearsay, allowing its introduction "in the interests of justice." P.J. Maartens & P.J. Schwikkard, *A Juryless Jurisdiction and the Epistemic Rules of Evidence*, 128 S. AFRICAN L.J. 513, 527 (2011).

allow hearsay to be introduced when it appears particularly trustworthy,[129] they are "extremely narrow" and apply "only in the most exceptional circumstances."[130]

Not only does the majority opinion in *Queen v. Hepburn* illustrate and defend the central, defining features of the modern American rule against hearsay, it does so plainly and concisely, accompanied by a sharp, even shorter dissent. The case would therefore have considerable pedagogic value even apart from its facts and its historical context. But those facts and that context—Mina Queen's story, and the role of secondhand testimony when enslaved people sued for freedom—are what make me regret having taught evidence law for years without mentioning *Queen v. Hepburn*. It is not just that lawyers should know the background of the hearsay rule and the ways in which it can frustrate rather than promote justice. It is that all Americans, and lawyers in particular, should understand how thoroughly slavery is woven into the history of all of our institutions, including virtually every field of law.[131] In a striking metaphor, Isabel Wilkerson compares the United States to an ancient house built on a faulty foundation; its current residents, she says, need to understand what is lurking underneath, because ignoring the cracks and flaws is a recipe for recurrent disaster.[132] American legal history is part of that foundation. As Justin Simard has urged, "[a]t a time when other groups and institutions, from universities to businesses, are confronting their links to slavery, lawyers must do the same."[133]

There's one final reason to teach *Queen v. Hepburn*. It is a modest form of remembrance. The institution of slavery was constructed in part by erasing the past: obliterating the ancestry and cultural heritage of African Americans and defining them solely through their place in a strict racial hierarchy, unmoored from history. *Queen v. Hepburn* enlisted the hearsay rule in that enterprise of forgetting. Mina Queen's family background was made legally invisible—and, as

[129] *E.g.*, Fed. R. Evid. 807.

[130] *E.g.*, United States v. Mason, 951 F.3d 567, 574 (D.C. Cir. 2020). Many other decisions employ the same or similar language.

[131] Regarding, for example, the significance of slavery in the development of property law and contract law in the United States, see, respectively, K-Sue Park, *The History Wars and Property Law: Conquest and Slavery as Foundational to the Field*, 131 Yale L.J. 1062 (2022); and Dylan C. Penningroth, *Race in Contract Law*, 170 U. Pa. L. Rev. 1199 (2022).

[132] Isabel Wilkerson, Caste: The Origins of Our Discontents 15–20 (2020).

[133] Simard, *supra* note 100, at 84.

a consequence, all but lost to history. The official reports of the case compounded this erasure by getting Mina Queen's name wrong and omitting the name of her daughter, Louisa Queen.[134] Even William Thomas, a dogged and perceptive historian who spent years researching the freedom suits brought by Mina Queen and her relatives, could find no trace of her life after the Supreme Court ruled against her. He surmises that John Hepburn sold Mina Queen within a few years, but that he may have kept Louisa Queen as his slave, separating mother and daughter.[135] Teaching *Queen v. Hepburn* will do nothing to right any of these wrongs. It is nonetheless a small but important act of historical recovery—an exercise of recognition and acknowledgment.

Evidence instructors who decide to teach *Queen v. Hepburn*—and I hope many will make that choice—should take steps to ensure that class discussions maximize the pedagogic value of the case while minimizing its potential to make the classroom less welcoming and inclusive. The racism in *Queen v. Hepburn* is not as flagrant as in, say, *Dred Scott v. Sanford*;[136] the white triumphalism is not as overt. And the fact that slavery was defended not just in racist screeds like *Dred Scott* but also in more genteel texts like *Queen v. Hepburn* is itself an important lesson. Still, encountering Chief Justice Marshall's grotesque solicitude for John Hepburn's rights to his "property" can be dispiriting and disturbing, particularly, but not only, for Black students and other students of color. The discomfort should not be gratuitous.

The key to using *Queen v. Hepburn* effectively in the classroom is to be clear, both before and during lecture, about the reasons the case has been assigned. Students should know that they are not being asked to read the decision just because it is old or because it was written by a famous jurist. They should know that including a freedom lawsuit in

---

[134] *See supra* note 4.

[135] THOMAS, *supra* note 2, at 189. Thomas notes that when Hepburn paid his District of Columbia property taxes five years later, he listed slaves that may have included Louisa Queen, but not her mother. *Id.*

[136] 60 U.S. 393 (1857). Chief Justice Roger Taney, the author of the Supreme Court's decision in *Dred Scott*, was Francis Scott Key's brother-in-law, and early in his career he was active, with Key, in the American Colonization Society. By the 1850s, however, Taney had become a stalwart defender of slavery and of white supremacy. *See, e.g.*, FINKELMAN, *supra* note 2, at 172–218; Huebner, *supra* note 10; Gabriel J. Chin, Dred Scott *and Asian Americans*, 24 U. PA. J. CONST. L. 633 (2022). Because the racism in *Dred Scott* is so raw and unapologetic, the pedagogic issues it raises have become especially fraught. *See* Jeannie Suk Gersen, *The Importance of Teaching Dred Scott*, NEW YORKER, June 8, 2021, https://www.newyorker.com/news/our-columnists/the-importance-of-teaching-dred-scott.

the syllabus is not accidental, and that slavery is part of the course because it is part of the history of evidence law. They should be made aware, both during lecture and in the materials they are assigned to read along with *Queen v. Hepburn*, of the role the case played both in the development of the hearsay rule and in the buttressing of slavery. They should know that Justice Duvall was right to say that the hearsay rule had previously been applied more flexibly, and that his fears about the consequences of the decision proved justified.

There was a time when *Queen v. Hepburn* was widely cited, unabashedly, as the principal authority for the inadmissibility of hearsay in the United States. It counts as progress, however slow and partial, that the case eventually became something of an embarrassment and that citations to it grew sparser. Ignoring *Queen v. Hepburn* was better than celebrating it, but it would be better still to remember the decision, to hold it up to the light, and to try to learn from it.[137] The American hearsay rule took its modern form in a decision that kept a woman and her daughter enslaved by suppressing what was known about one of their ancestors. The implications of this fact for how the hearsay rule should be understood and applied today are debatable, but there is little to be said for avoiding the debate—for matching suppression of family history with suppression of legal history. There is little to recommend keeping law students ignorant about the role the hearsay rule played in supporting American slavery and, conversely, about the role American slavery played in the development of the hearsay rule.

[137] For thoughtful suggestions about the citation and use of slavery cases, see Simard, *supra* note 100, at 119–22. On the costs of erasing slavery from legal pedagogy, see, for example Park, *supra* note 131, regarding property law; and Penningroth, *supra* note 131, regarding the law of contracts.